ANNUAL REVIEW OF ASTRONOMY AND ASTROPHYSICS

EDITORIAL COMMITTEE

ANNUAL REVIEW OF ASTRONOMY AND ASTROPHYSICS

LEO GOLDBERG, *Editor*
Harvard College Observatory

DAVID LAYZER, *Associate Editor*
Harvard College Observatory

JOHN G. PHILLIPS, *Associate Editor*
University of California. Berkeley

VOLUME 5

1967

ANNUAL REVIEWS, INC.
4139 EL CAMINO WAY
PALO ALTO, CALIFORNIA, U.S.A.

ANNUAL REVIEWS, INC.
PALO ALTO, CALIFORNIA, U.S.A.

Library of Congress Catalogue Number: 63-8846

FOREIGN AGENCY
Maruzen Company, Limited
6 Tori-Nichome Nihonbashi
Tokyo

PRINTED AND BOUND IN THE UNITED STATES OF AMERICA BY
GEORGE BANTA COMPANY, INC.

PREFACE

Volume 5 of the *Annual Review of Astronomy and Astrophysics* was organized by the Editorial Committee in Pasadena, California on May 8, 1965 with the valuable assistance of Professor Jesse L. Greenstein, Dr. Beverly Oke, and Professor Harold Zirin. The larger than average size of the volume is a measure both of the growth of astronomy and of the unusually high proportion of authors who succeeded in meeting the prescribed deadline for the submission of manuscripts. We are grateful to the authors for the excellence of their articles and for their cooperation in bringing about their timely publication.

The work of editing the manuscripts has been carried out by Drs. David Layzer and John G. Phillips, Associate Editors. We also acknowledge with thanks the valuable assistance of Joann Huddleston of the staff of Annual Reviews, Inc.

THE EDITORIAL COMMITTEE

ERRATUM

Volume 4 (1966)—"The Structure of Radio Galaxies," by Allan T. Moffet:

Two figures were incorrect in the original article, and a corrected version of one appears below.

In Figure 1 (page 149) the scale of the radio contours of Cygnus A did not match that of the optical field, the contours being too large by a factor of about 1.3. A corrected version is given here, with the original caption.

The source 3C 338, associated with NGC 6166, is not circular in shape, as was shown in Figure 8 (page 162) of the original article, but is a bar, or perhaps a barely separated double, elongated in the east–west direction and $\leq 20''$ in extent north–south (from unpublished measurements by E. B. Fomalont and the author).

I am indebted to C. M. Wade for calling my attention to the error in Figure 1. J. R. Shakeshaft also mentioned to me the error in the structure of 3C 338.

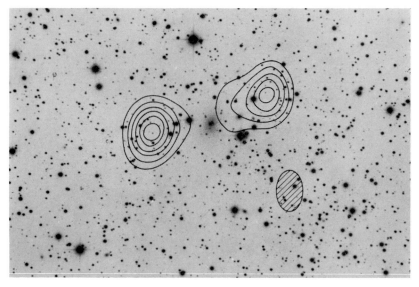

FIG. 1. Cygnus A, radio contours as observed with a $23'' \times 35''$ beam at 21 cm (32) superposed on a photograph taken by Baade with the Hale reflector (1). The shape of the radio beam is shown by the ellipse at lower right. The field is $4' \times 6'$.

CONTENTS

MAGNETIC FIELD OF THE SUN (OBSERVATIONAL)[1]

By Robert Howard

Mount Wilson and Palomar Observatories, Pasadena, California

Introduction

Although nearly sixty years have elapsed since magnetic fields were first measured in sunspots (Hale 1908), it is only within the last fifteen years that measurements of weak magnetic fields outside sunspots have become practical. The solar magnetograph introduced an improvement of about two orders of magnitude in the sensitivity of magnetic-field measurements over the existing visual or photographic techniques. Since the development of the magnetograph, a large amount of observational data has been collected with instruments at many observatories. It seems clear that magnetic fields on both a large and a small scale play an important role in the many puzzling phenomena that constitute solar activity. For this reason there is scarcely a large solar observatory in the world that does not measure solar magnetic fields in some manner now or have plans to begin such measurements in the near future. In recent years several review articles on solar magnetic fields have appeared. Among them are those by Babcock (1963) and Severny (1964).

In quiet areas of the Sun, magnetic fields are found on the boundaries of supergranule cells. Magnetic fields on the solar surface are closely associated with all phases of the development of an active region from the first sudden appearance of a tiny bright plage to the final slow dissolution of the old fragmented plage. The polar field of the Sun may be considered to arise from the poleward drift of following portions of old-plage magnetic fields out of the rather regular pattern of the background fields, and thus to reverse its polarity during each cycle. Evidently the whole range of solar activity is reflected in the study of the magnetic fields on the solar surface. It seemed logical in a review of this sort to divide the subject to some extent by angular resolution. The discussion of fine-scale fields will include the small-scale structure of the solar atmosphere, the development of active regions, and the process of the decay of active regions. The large-scale discussion will include the background-field pattern and the large-scale distribution of solar activity, and will lead to the polar fields and a brief consideration of the Sun as a magnetic star.

Although sunspots form an important part of what is usually regarded as solar activity and provide about one-half the magnetic flux in an active region, I have specifically omitted a discussion of sunspot magnetic fields in this review, partly because this is a very large and specialized field with its own techniques and problems, and partly because a recent monograph has thoroughly reviewed the field (Bray & Loughhead 1964).

[1] The survey of literature for this review was concluded in October 1966.

1

Measurement of Magnetic Fields of the Sun

The solar magnetograph was developed by H. W. Babcock (1953). The principle of the magnetograph and more recent refinements have been discussed in the reviews mentioned above and will not be repeated in detail here. Basically a magnetograph is a double-slit photoelectric differencing photometer with a modulated electro-optic analyzing device. Used with a high-dispersion spectrograph, it can be a very sensitive tool for measuring the longitudinal Zeeman effect in an absorption or emission line formed in the solar atmosphere. The magnetograph is simple in principle and easy to construct, but the peripheral requirements add enormous complications. Among these important auxiliary functions may be mentioned keeping the spectrum line centered on the exit slits, scanning the solar image, and recording the data in a usable form.

There are about twelve photoelectric magnetographs currently in operation. No two of them are identical. Several instruments have been designed to measure both the longitudinal and transverse Zeeman effects, and thus to obtain the magnetic vector in the solar atmosphere (except for a 180° ambiguity in the direction of the transverse field) (e.g. Stepanov & Severny 1962). The sensitivity for the transverse effect is one or two orders of magnitude less than for the longitudinal effect because the differencing technique cannot be used. Additional difficulties arise in measuring the transverse fields because the polarization due to the Zeeman effect is a small part of that caused by the instrument, so the latter must be eliminated. One magnetograph was specifically designed to measure the Zeeman effect in emission lines in solar prominences (Lee, Rust & Zirin 1965).

A very important consideration in the study of the results of magnetograph observations is that of angular resolution. The magnetograph measures the magnetic field averaged over the aperture or slit which is used. It is not likely that anyone has ever measured a magnetic field with a magnetograph over an area on the Sun smaller than about 4×10^6 km^2 (taking into account the smearing effects of astronomical seeing). Since it is not at all inconceivable that separate magnetic features may have areas as small as 10^5 km^2, it is clear that we are very far from resolving the most interesting features in magnetograph measurements. Therefore, when one measures a field of 5 G on the solar surface, it must be remembered that this refers to an average over the aperture, and even for the smallest aperture used it should not be implied that there is a uniform field of 5 G across the aperture. This is a point which has been emphasized consistently by those who observed, since the first days of the magnetograph.

That this resolution problem represents a fundamental limitation of the application of magnetograph results to the understanding of the physics of the solar atmosphere was pointed out at an early stage by Alfvén (1952). Whether or not, as he suggested, the scale and complexity of the magnetic field structures on the solar surface and the correlation of magnetic-field strength with brightness are such as to render magnetograph observations of little value has yet to be demonstrated. It seems unlikely, considering the regularity of the pattern of background fields and the apparently reasonable behavior of the weak magnetic fields on a large scale in the solar atmosphere. For example, weak-field structures do not form on the solar surface except by the expansion and weakening of the strong fields of active regions. If the resolution problem were playing serious tricks on the appearance of the large-scale fields, one might expect to observe very puzzling behavior. The only possible hint of any effect of this sort concerns the decay of the polar field and will be discussed below in the section on polar fields.

Probably the two-dimensional solar magnetic observations with the best angular resolution are those of Leighton and his group (Leighton 1959). This spectroheliograph technique has advantages in speed and angular resolution over the point-by-point magnetograph method, although at the sacrifice of an order of magnitude in sensitivity. In practice, the photographic cancellation technique is tedious and the method is not often used. Recently Title (1966) has devised an automatic means of cancellation which speeds up the reduction process. A technique similar in principle, but employing a Fabry-Perot interferometer and a vidicon system for recording and subtraction, is now under development by Giovanelli and his group (Ramsay 1966). This system will provide rapid magnetic pictures of active regions with good angular resolution.

SMALL-SCALE MAGNETIC FIELDS

It is best first to define what is meant by small-scale magnetic fields. Clearly, if a space-probe could carry a magnetometer into the solar photosphere, it is conceivable that interesting magnetic structures connected with magnetohydrodynamic wave motion might be found on a scale of a few tens of kilometers, or even smaller. Solar physicists are not accustomed to considering such small-scale features. For the purposes of this article we shall consider that small-scale features have dimensions of at least a few hundred kilometers, which is about the size of the smallest structures seen on the best white-light photographs of the solar surface. The upper limit of what we shall call "small scale" is roughly the size of an active region.

Because of the limitations of angular resolution mentioned above, the magnetograph is not the best instrument for observing small-scale magnetic fields. Recently the most interesting results about very small-scale structures have come from magnetic spectroheliograms (Sheeley 1966a) and high-dispersion spectra taken at times of good seeing (Sheeley 1966b, Beckers & Schröter 1966). On magnetic spectroheliograms taken under good observing conditions at Mount Wilson, Sheeley found small points less than 3000 km in diameter where magnetic fields were measured to be of the order of several hundred gauss. These points were in general identical with enhanced network fragments and well outside active regions. Later, with high-dispersion spectra taken at Kitt Peak Observatory, Sheeley, while searching for a granular magnetic field, found instead occasional "gaps" in certain spectrum lines over a length as small as 500 km in which some lines (e.g., λ 5250) were very much weakened, and where the magnetic field was of the order of several hundred gauss. Figure 1 shows an example of such spectra. These gaps were found in regions far removed from sunspot activity. The strong small-scale fields often corresponded to local darkenings in the continuum, which suggests that the fields often appear in the dark lanes between granules. These tiny areas of strong magnetic field correspond to regions of bright Ca II K_{232} emission.

Independently, Beckers & Schröter (1966) discovered the gaps in solar-spectrum lines and concluded after a study of them that they represented regions about 700 km in diameter which contained a vertical magnetic field of about 1100 G. These points appear at the location of slight darkenings in the continuum—in a dark lane between granules, but not in pores—where the continuum darkening is about 12 per cent. As yet a number of interesting questions about these gaps remain unanswered. How much of the magnetic flux outside spots is in the gaps? Because the fields within the gaps are so high, it is likely that a magnetograph measurement will not give a correct magnetic field integrated over the measuring aperture. How serious is this effect and what effect does the weakening of the line profile have on the result? Is a plage magnetic field made up entirely of such small magnetic features? It is clear that this is an exciting new subject for observational and theoretical investigation, and we may anticipate interesting results in the next few years.

Hale (1922a, b) discovered small regions on the Sun where the magnetic fields were measured to be several hundred gauss. He named these "invisible sunspots." Many of these features later developed into sunspots or were the remains of sunspots that had recently disappeared, but some were apparently not associated with sunspots although they were within active regions. It is difficult to speculate now about what Hale's spatial resolution was at that time and whether he was seeing strong magnetic fields inside plages or individual features that were identical to the newly discovered gaps. It is interesting to note that the method used by Hale to search for invisible sunspots was very similar to that used by Sheeley (1966b). The improved spectrographs and telescope optics in use today make a great difference in the angular resolution that is obtainable.

Stenflo (1966) has measured magnetic fields with a magnetograph near the poles of the Sun. Using apertures of different sizes, he finds a greater average net magnetic flux when using small apertures than when using large apertures. The instrument he was using (the magnetograph of the Crimean Astrophysical Observatory) provides compensation for the brightness of the source. The pronounced fine structure of the polar fields is well known and Stenflo's results imply a brightness-field correlation which renders the net flux dependent on aperture size. This interpretation has been criticized, however (Howard 1966), on the grounds that intensity fluctuations of the amount necessary to explain the results simply do not exist on a scale in the

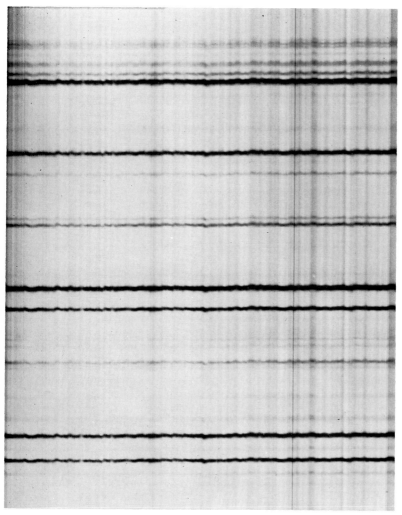

FIG. 1. A portion of the solar spectrum near λ 5250. Line shifts due to the motion of granules and the material between granules are evident. Note the "gaps" in some of the spectrum lines. The total length of the slit corresponds to about 115,000 km. Photograph courtesy of Dr. N. R. Sheeley, Kitt Peak National Observatory.

solar atmosphere large enough to be observed with even the smallest of Stenflo's apertures.

Magnetograph observations of the small-scale structure of the solar magnetic fields are plentiful, although somewhat limited by the resolution problem, as mentioned above. Babcock & Babcock (1955) established the existence of magnetic fields in active regions outside sunspots. Among the first magnetic observations made with relatively high resolution were those of Leighton (1959), using the spectroheliograph technique, and this author (Howard 1959). They both showed the very close connection between the shapes of the plages and the shapes of the magnetic-field structures. Leighton's pictures showed the connection between the small-scale magnetic fields around a plage and the enhanced calcium network. Stepanov (1958) discussed the filamentary structure around an active region which he found in general to lie at right angles to isogauss lines. The close association of the plage with the magnetic fields and other small-scale correlations led this author (Howard 1959) to postulate that there is a one-to-one relation between brightness in the K line and magnetic fields, and that carrying this to the smallest observable features implies that the magnetic field is confined to very small areas on the solar surface, and within these areas the field must be very strong.

Observations using a magnetograph aperture nearly comparable with the size of large granules (Howard 1962) gave a root-mean-square magnetic field of 8.2 G in quiet regions. This number must be qualified by the consideration of the size of the magnetic features which may have been involved. Such a number could easily have come from a distribution of features with dimensions of a few hundred kilometers and magnetic-field strengths of many hundreds of gauss.

Bumba & Howard (1965a), from a study of fine-scan magnetograms made at Mount Wilson, have found that when small apertures are used one can frequently find a mixture of polarities where only one polarity was seen with poorer resolution. The very close association between the magnetic fields and the calcium network was also confirmed.

Severny (1965a), in a study of the polar field, has confirmed its fine structure, and has found both polarities present at each pole. In another paper (Severny 1966), he studied the relation of the magnetic field measured in the photosphere and the chromosphere. Using Hα and λ 5250, he finds that in about 70 per cent of the cases, features in the photospheric level are repeated in the chromosphere. In general, the chromosphere shows less fine structure than the photosphere.

The close relationship of the magnetic fields outside active regions to the chromospheric emission patches (plagettes) on the borders of the supergranular network (Leighton, Noyes & Simon 1962; Simon & Leighton 1964) provides an important clue to the relation of the fine-scale structure of solar magnetic fields to solar activity as a whole. It has been proposed (Parker 1963) that the horizontal flow involved in the supergranular network will sweep aside a weak ambient magnetic field and gather the lines of force at the network boundaries. The chromosphere is heated at the location of the fields by a mechanism such as that of Piddington (1956). Thus the chromospheric network is seen as the consequence of the supergranular network, which itself is seen as a slow convective mode with dimensions about that of the depth of the hydrogen convection zone. The gaps may be presumed to be especially enhanced field structures on the boundaries of the network. The network certainly extends to the polar regions, and the motions of magnetic flux described below concentrate a great deal of this flux in the polar regions; this concentration explains the polar faculae as enhanced network plagettes with enhanced magnetic fields.

Numerous observations of the structure of magnetic fields within active regions have been made. Stepanov (1960) discussed the vertical gradient of the magnetic field in the solar photosphere determined from measures made at different positions on the sodium D lines, and found it to be from 0.026 to 0.035 G/km. Later work with the vector magnetograph of the Crimean Astrophysical Observatory (Stepanov & Gopasyuk 1962; Tsap 1964, 1965a) confirmed earlier work with the longitudinal effect alone related to magnetic fields in an active region. Howard & Harvey (1964) studied magnetograms and chromospheric features and concluded that the filamentary structure around active regions could be divided sharply into two sorts. The fibrils were perpendicular to isogauss lines and showed increased contrast when seen on the blue wing of Hα. The filaments lay in general parallel to isogauss lines separating areas of opposite polarity, similar to the relationship of filaments and fields outside active regions, and, seen on the blue wing of Hα, the filaments show decreased contrast. The filaments are clearly much higher than the fibrils and are, of course, identified with prominences seen at the limb. The fibrils are associated with spicules. Tsap (1965b) has studied the relation between magnetic fields and plage brightness in some detail. He finds that the brightness increases with total field strength until the field strength reaches about 300 G; then it levels off somewhat for further increases in brightness.

One of the aspects of the study of magnetic fields on the solar surface that has practical importance is their relation to solar flares. In the last few years the paucity of flares has somewhat slowed this field from the observational standpoint. Results from the preceding maximum (Severny 1958, 1960; Howard & Babcock 1960; Howard & Severny 1963; Howard 1963) tended to indicate that, although there were variations in the sunspot magnetic fields about the time of major flares, there seemed to be no changes in the weaker magnetic fields of the active-region plage during the course of a major (or minor) flare. The variations in the spot fields could not be directly linked to the hours during which the flare was in progress, and one may consider that during a phase of the spot-group lifetime when the areas of the spots are changing, there is more chance of seeing a flare. Also, when the spot-group magnetic polarities are distributed in a complex manner, the probability of occurrence of a flare is increased. This can be measured by the gradient at a neutral point in the field pattern or by a qualitative appraisal of the complexity and proximity of the spot magnetic fields. For example, a situation where two magnetic polarities exist within the same penumbra is one which has a very high probability of producing flares. While numerous attempts have been made to give a theoretical interpretation of the flare phenomenon, none has succeeded to the point where it can provide a basis for flare prediction. A number of attempts have been made to predict flares, but on the whole it is not likely that any of them perform any better than an intelligent guess by an experienced observer. It is hoped that during the next maximum a great deal more observational evidence will be gathered concerning the connection of magnetic fields and flares, and perhaps this can provide a good basis for a reliable prediction scheme. The field has been reviewed recently by this author (Howard 1964).

The relation of prominences to magnetic fields deserves special mention. Babcock & Babcock (1955), using a low resolution, established that filaments divide areas of opposite magnetic polarity outside active regions or lie on the poleward side of active-region magnetic fields. Later observations with higher angular resolution indicated that filaments outside active regions always lie between areas of opposite magnetic fields, and the cases where the Babcocks' filaments did not, must have been due to their missing weak magnetic features because of their low resolution. A model for the magnetic support of a filament has been proposed (Kippenhahn & Schlüter 1957). These authors postulate that prominence material can collect in depressions in the magnetic lines of force at the peaks of these lines of force above the chromosphere. This would demand that the lines of force within the prominence should be perpendicular to the axis of the prominence. Rust (1966) has observed magnetic fields within prominences with the High Altitude Observatory magnetograph. He observed nearly 100 prominences, most of them quiescent, using the Hα line. He found that the intensity of the field in the prominence depends on the intensity of the surrounding photospheric magnetic fields. The average horizontal field for prominences in young bipolar regions is about 15 G. The polar crown prominences have fields of about 5 G, and the strongest fields, about 50 G, occurred in sunspot prominences. Using observations of the underlying magnetic fields, Rust calculated the field strengths and configurations to be expected at the prominence heights from potential theory, and his prominence magnetic-field observations agreed well with these observations, as far as the directions were concerned. He concluded that the prominence material is probably supported in depressions at the tops of the field loops. The observed magnetic-field strengths in prominences were about ten times the values derived from potential theory. The field intensity in prominences observed at the limb does not appear to depend upon the angle between the axis of the prominence at the line of sight. This suggests that the situation is more complicated than the simple picture of Kippenhahn & Schlüter would imply. Rust concluded that the magnetic fields responsible for the support of a prominence are small-scale fields at the boundaries of supergranules. Small bundles of lines of force near a large-scale polarity boundary contain depressions in which material collects to form a prominence. He did not, however, find it necessary to conclude that the small-scale structure of the field extended to prominence heights. In fact, as Rust points out, there are good theoretical reasons that this may not be the case.

Bumba & Howard (1965b) have studied the early stages of the development of active regions. The plage and the magnetic fields appear at the same time in the first hours of the birth of the region, and at all times the behavior of the chromospheric active region parallels that of the magnetic fields. The position of the actual first appearance of the region is a network cell boundary. The first development of the new region follows the shape of the preexisting supergranular network. The young bright plage extends rapidly around the old cells and fills them in with emission. Generally the leading portions of growing regions appear first and develop more rapidly than the following portions. In some cases there was an apparent imbalance of magnetic flux during the first day or two in the life of the region. A rapid ordering of the filamentary structure surrounding the active region takes place as the young region grows. The boundary of the ordered structure can be seen to expand with a speed of about 200 m/sec. When this boundary reaches a preexisting filament, the filament disappears. It is interesting that the disappearance of the filament occurs when the expanding filamentary structure reaches the immediate neighborhood of the filament, not when it simply enters the large magnetic feature on one side of the filament. This tends to support the view that the prominence is a consequence of the relatively local distribution of surrounding magnetic fields, depending for its support on the photospheric field lines immediately adjacent to the boundary of the magnetic polarities.

Leighton (1964) has discussed the decay of magnetic regions as the eroding effect of the supergranular motions. The magnetic fields of a plage are slowly moved in small pieces from the plage into the network pattern. By a process analogous to a random walk, the field becomes distributed over a large area around the plage and, as we shall discuss later, is stretched by differential rotation to form a part of the background-field pattern. An important quantity in a consideration of this process is the lifetime of the supergranular or network cell. Macris (1962) has established a lifetime of about 24 hours for a typical network cell. This is a very difficult quantity to determine because it is difficult to define the point in time when a cell no longer exists. Slight changes in the boundaries of cells could alter the effects of a correlation study without involving the destruction of any cells. It is clear that enhanced cells around active regions last for at least several days. For example, see Figure 2.

The notion of the erosion of an active region by supergranular motions certainly has some merit from the observational standpoint. It is well known

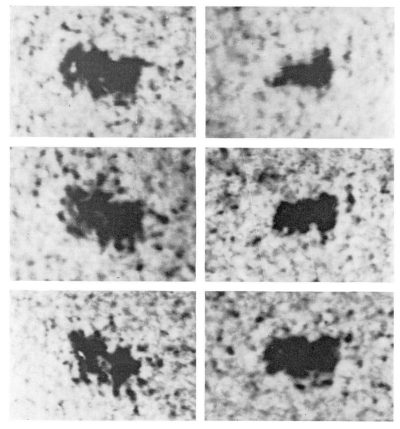

FIG. 2. Portions of calcium K spectroheliograms for successive days. This is an active region in the first week of its growth. The first day is in the upper right (August 15, 1963). The fourth day is in the upper left (August 18, 1963).

that a young plage is bright with a rounded, smooth boundary. As the plage ages, the boundary becomes irregular and rather scalloped, and the surrounding network becomes enhanced (see Figure 2). In its final stages, the plage resembles a highly enhanced network, and gradually the emission at the boundaries of the network cells spreads over the surrounding area, such that all that remains is a slightly enhanced network emission pattern and, eventually, even that merges with the background. This process requires many weeks, in the case of a large active region.

So the picture of the small-scale fields which is taking shape, although it is still incomplete, is the following:

At the photospheric level, at least a large part of the magnetic field is in the form of small bundles of lines of force where the field strength is of the order of several hundred gauss. Higher up in the atmosphere these bundles of lines of force are the loci of the bright points in K-line spectroheliograms. The photospheric fields show more fine structure than do the chromospheric magnetic fields. This may be due to the enlargement of the diameters of the magnetic bundles at higher elevations. These magnetic-field points (or gaps) are distributed around the boundaries of the supergranulation cells. The supergranular flow probably sweeps the lines of force to the boundaries; it at least will ensure that they remain there. Still higher, where the prominences are found, the field no longer exists in discrete bundles. The prominences are found at the boundaries of magnetic polarities—but only at large-scale boundaries, since over a large part of the Sun, magnetic fields on a very small scale include a mixture of both polarities. The calcium or supergranular network is seen to play a fundamental role in the distribution of small-scale magnetic fields. Fields first appear on the Sun at the birth of a new active region, or at the time when an older region becomes enhanced. The new fields appear first at the edges of supergranular cells and the growth of the region follows the pre-existing network pattern. Despite the dramatic appearance of the magnetic fields in the chromosphere, the field distribution is practically invisible in the photosphere. The small bundles of force lines may exist in the dark lanes between some granules. Clearly this is a field for fruitful observational investigation.

Large-Scale Magnetic Fields

Babcock & Babcock (1955) gave the first overall observational picture of the large-scale distribution of magnetic fields on the solar surface. In their discussion of more than two years of data, they established the existence of magnetic regions, which they called bipolar magnetic regions (BMRs), that had a much longer lifetime than the spots sometimes associated with them in their early stages of development. With the spatial resolution used for these observations (52,000 km), the fields measured in active regions were generally lower than those measured in later years with improved resolution. The plages were found where the magnetic fields measured more than about 2 G. In addition, the Babcocks defined multipolar magnetic regions as regions where the distribution of magnetic polarities was somewhat complicated, analogous to the definition of a magnetically complex spot group. Unipolar magnetic regions (UMRs) were recognized. Some of these regions had lifetimes of many months, and there was a good correlation between one of these regions and a 27-day recurrent magnetic storm. Also, the Babcocks' paper gave the first conclusive evidence for the existence of a general magnetic field at the higher latitudes.

Bumba & Howard (1965c) studied four and one-half years of synoptic full-disk magnetic material. These magnetograms were made with considerably better angular resolution (17,000 km) than was used for the earlier data. This increase in resolution proved to be important in the detection of weak magnetic-field features covering large areas of the solar surface. This background-field pattern, which has a lifetime of many months, consists of the weak expanded magnetic fields of old active regions. The pattern is approximately that of alternating polarities from east to west across the solar surface in each hemisphere. The shapes of the weak features that make up the pattern are determined, at least to a large extent, by the shearing effects of the differential rotation of the Sun. Thus, in general, the poleward portions of these features are stretched eastward (see Figure 3). The background-field pattern is slowly undergoing changes all the time. As new active regions form within the pattern, they add their magnetic fluxes, which spread out and merge with the pattern. The appearance of the pattern varies with the solar cycle in such a way that, when the number of active regions is large, the background-field pattern is compact, the fields are relatively strong (about 6 G), and there are many (perhaps 15) alternations of polarity in 360° of longitude (see Figure 3). Later in the cycle, when the level of activity was lower, the field strength of the background fields was lower by perhaps a factor 2. In places there was no pattern, or it was below the limit of detectability and the number of alternations of polarity around the Sun was small.

What Bumba & Howard (1965c) defined as the unipolar magnetic regions were large features, extending more than 100° in longitude, made up of portions of the following polarity for that hemisphere which had moved poleward from the background-field pattern. These UMRs evidently represented following flux polarity moving to the poles. The UMRs lasted for many months and were the most easily visible part of a very large-scale pattern. Ahead of the UMR could be seen in general a very weak and spread-out feature of leading polarity. Also consisting of fields from the background-field pattern, this "ghost" UMR was very much extended and contained a magnetic flux estimated to be about one-fifth that of the UMR itself. The preceding portions of the UMR and the ghost were the strongest and extended to the lowest latitudes. Another part of the very large-scale pattern that was observed only on the magnetograms of finest quality was a very faint reversed-polarity copy of the UMR and its ghost in the opposite hemisphere.

The increase in angular resolution which occurred between the studies of Babcock & Babcock (1955) and Bumba & Howard (1965c) made an enormous difference in the magnetic appearance of the solar disk. It is not possible to say now what the earlier UMRs were in context of the background-field pattern. It may be that they were not the same as the features later defined to be UMRs because their latitudes were, in general, lower than those of the later UMRs. It may be that they were large enhanced portions of what was later called the background-field pattern. Or, perhaps they were the leading low-latitude portions of the large high-latitude UMRs. This subject will arise again in the discussion of solar-geophysical relations.

The poleward migration of following magnetic flux was predicted by Babcock (1961) as a part of his model of solar activity. Leighton (1964) suggested a mechanism for this transport of flux as discussed above. The supergranulation motions provide a persistent eroding action which works to

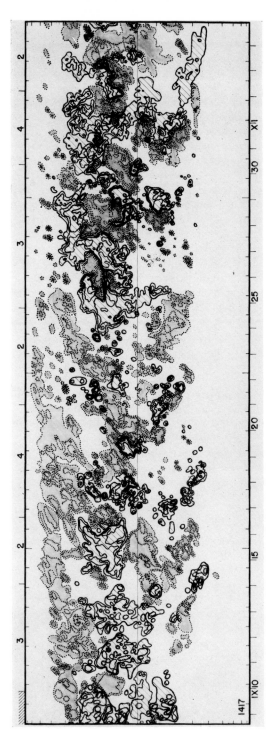

Fig. 3. Synoptic chart of solar magnetic fields for rotation number 1417 (August 1959). Solid lines and hatching represent positive polarity, and dotted lines and shading represent negative polarity. Isogauss lines are for 2, 6, 10, 15, 25 G. Dates are given below with marks representing 10° intervals of longitude. The equator is drawn, and every 10° in latitude is marked at the sides. The numbers at the top give an indication of the quality of the magnetograms from which the synoptic chart was drawn, with 4 the best. The hatching represents an area which had to be drawn more than 40° from the central meridian of the magnetogram.

transport the compact magnetic fields of a plage to far-distant parts of the solar surface. First, flux lines near the boundary of the plage are broken away in little bits; then later, as the plage grows old, the effect of supergranular motion is to break it apart. Later the spreading of the magnetic flux that results from the random-walk effect of the constantly changing supergranular pattern will be modified by differential rotation to form the characteristic pattern of the background fields. The reason for the definite preference of the following magnetic polarities for the poleward migration is not entirely clear, but may be explained as a result of the preferential equatorward tilt of the leading portions of active regions.

The model for the 11-year polar activity cycle that was proposed by Babcock seems to answer, at least qualitatively, many of the puzzling questions in connection with solar activity. Briefly, the model starts with a poloidal field that emerges through the surface in the polar regions. In lower latitudes, the field lines connecting the polar fields are submerged below the surface at a depth of about 0.1 R. The differential rotation after many months will draw out the field lines, with those at the equator leading those at higher latitudes. After about three years the field lines are wrapped around the Sun several times and greatly amplified. At latitudes of about $\pm 30°$, instabilities begin to appear which will bring the flux rope to the surface, forming a bipolar magnetic region. The latitude of this instability moves equatorward as the flux at the higher latitudes rises to the surface in the form of active regions. These regions break apart with age and the following polarities migrate to the polar regions, forming the polar fields. The leading polarity fields of the active regions preferentially move toward the equator, thus canceling with leading polarity fields from the other hemisphere that have also moved toward the equator. Thus the stage is set for the next cycle with the accumulation of new polar fields whose field lines, by a process of severing and reconnection, join under the surface again. Babcock's model of the activity cycle gives a reasonable qualitative explanation of many features of the cycle. The cyclic reversal of the polar fields, Hale's law of sunspot polarity, the dominance of preceding over following spots, the latitude drift of spots during the cycle, the cyclic variation of the appearance of the corona, the chromospheric "whirls" around many spots that are independent of the spot magnetic polarity, and the equatorward tilt of the leading portions of spot groups are all observed quantities explained by Babcock's model.

A preliminary study of the association of green-line coronal features with photospheric magnetic fields (Bumba, Howard & Kleczek 1965) suggests a very definite relation between the age of the photospheric region and the appearance of the corona above the region. For the interval studied, every coronal feature could be matched in position with a magnetic feature below, and every magnetic feature, except some which were extremely old and weak, could be matched with a coronal feature. Coronal loops or arches appear to connect regions of opposite polarity—but not necessarily sunspots. The youngest active regions seem to be associated with amorphous structures in the corona. The older field regions tend to have their leading and following portions widely separated and to show faint arches in the corona. One example of ray structure was seen. This lay above a large stable sunspot. Thus it appears that the structure of the corona is dependent upon the structure of the magnetic field below in the photosphere.

Ness & Wilcox (1966) have found a good correlation between the direction at any time (toward or away from the Sun) of the magnetic field in interplanetary space measured with the earth satellite IMP-1 and the direction (into or out of the Sun) of the magnetic-field lines near the central meridian of the Sun. The latitude extent of the weak solar magnetic features which went into the correlation was so great that it was not possible to determine precisely the solar latitude of the origin of the fields measured near the Earth, but Ness & Wilcox used the rotation period of the magnetic pattern near the Earth to establish that the solar origin must be about 10 or 15 degrees from the solar equator. The time lag which fits the data best was about four and one-half days; that is, the best correlation with interplanetary magnetic fields near the Earth was with solar magnetic fields which had passed the central meridian four and one-half days previously. The magnetic fields on the Sun that appeared to be correlated with the interplanetary fields were not, in general, the strong magnetic fields of young active regions but the weaker large-scale pattern of the background fields. At the time of the IMP-1 observations, the interplanetary magnetic fields were observed to have a large-scale sector pattern (Wilcox & Ness 1965). It appears that this large-scale pattern, which consisted of four sectors within each of which the magnetic field had consistently the same polarity (the polarities of the sectors alternated from one sector to the next), was an extension of a very weak and spread-out background-field pattern across the solar surface. This period was very close to the minimum of solar activity, and the background fields were extremely weak and difficult to observe. It seems likely that close to maximum there would be many sectors in the interplanetary field.

The opportunity to study weak magnetic fields on the solar surface has made vivid a very large-scale distribution of activity. Bumba & Howard (1965c) described "complexes of activity" as active longitude zones within which could be followed a buildup over several rotations of active regions in both hemispheres. This was a rediscovery of an effect which Brunner-Hagger (1944) first noticed. As the density of active regions grows over some longitude zone, the strength of the background-field pattern increases and finally a UMR develops. During 1962 five clearly defined complexes of activity were observed. The two largest of these produced UMRs (see Figure 4). At times of higher activity, it may be presumed that complexes of activity overlap to such an extent that they cannot be followed as individual entities.

The possibility of a very large-scale cellular pattern on the solar surface has been pointed out by Bumba, Howard & Smith (1964). The appearance of the background-field pattern suggests a very large cellular pattern with

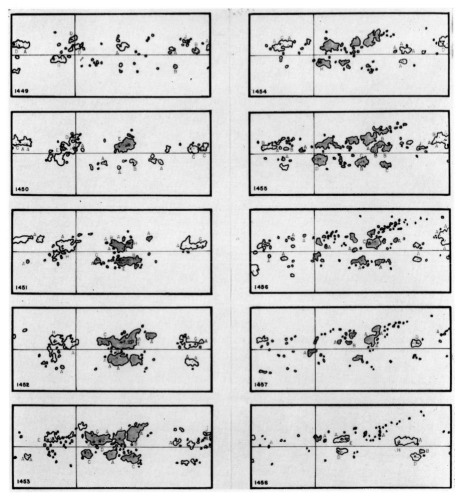

FIG. 4. Portions of Zürich *Heliographische Karten der Photosphäre* for the rotations indicated. Shadings represent one complex of activity. No magnetic polarities are indicated. The thin horizontal line represents the equator and the vertical line represents 0° longitude. Note the UMR visible in the later rotations in the Northern Hemisphere.

dimensions the order of a solar radius, and fields and calcium emission in general concentrated at its boundaries. Whether this apparent cellular pattern is physically real or is the result of the distribution of fields spreading out from a random distribution of active regions has yet to be determined. The existence of large-scale patterns of activity mentioned above and the possibility of very large-scale regularity in some features of activity (Dodson 1966) tend to give some weight to the possibility of the physical reality of a large-scale cellular pattern. The large scale and regularity of the interplanetary sector pattern (Wilcox 1966) is another indication of such a cellular pattern on the Sun.

The well-known 27-day recurrent magnetic storms provide another example of a very large-scale ordering of some sort of feature on the solar disk. A number of attempts have been made to associate solar features of one sort or another with the hypothetical "M" regions (e.g. Mustel 1944, Shapley & Roberts 1946, Bednarova-Novakova 1960). A persistent difficulty has been that none of the features usually identified on photographs of the Sun has a lifetime anywhere near the lifetimes of the "M" regions. As a result, it has been necessary to assign several solar features in succession to the same "M" region. This does not seem to be a very satisfactory solution. Babcock & Babcock (1955) pointed out that the weak UMRs live for many months, and, on the basis of one good example, proposed that the UMRs were identical with the hypothetical "M" regions. This example was a long-lived and strong UMR that reached the central meridian of the Sun several days before the commencement of a long-lived 27-day recurrent storm. Later work (Simpson, Babock & Babock 1955) confirmed the identification. Bumba & Howard (1965d) also discussed the identification of "M" regions and concluded that, although there was a definite connection between the large-scale pattern of magnetic fields on the solar surface and the sources of the recurrent magnetic storms as defined by some index such as that of Saito (1964), it was not possible to say for certain at this time exactly what the source of the storms was (see Figure 5). The features that Bumba & Howard defined as UMRs came to the central meridian somewhat after the commencement of the magnetic storm at the Earth (see Figure 5). Therefore, it seemed likely that the source of the particles responsible for the storms was located ahead of the UMR. Moreover, the rotation period of the UMR was longer than 27 days; this indicated that the "M" region was located at a lower latitude than the UMR—perhaps in the background-field pattern preceding the UMR. (See Wilcox & Ness 1965, Figure 15.) Thus it seems that what Babcock & Babcock described as a UMR was, in the later picture of the field distribution, an enhanced portion of the background-field pattern in the sunspot latitudes. The later, increased resolution had the advantage of showing up the background-field pattern, but it may have had the disadvantage of masking subtle differences in magnetic-field strength which distinguish ordinary background fields from true "M" regions.

Our present view of the large-scale distribution of magnetic fields on the solar surface may briefly be summarized in the following way:

There is no observational evidence that magnetic flux appears on the solar surface except at the time of the birth of an active region or the resurgence of activity in an active region. The active region breaks up during a period of days or weeks, and the magnetic fields begin to be spread about by the effects of the supergranular motions. When the level of activity is high, a strong background-field pattern builds up with many alterna-

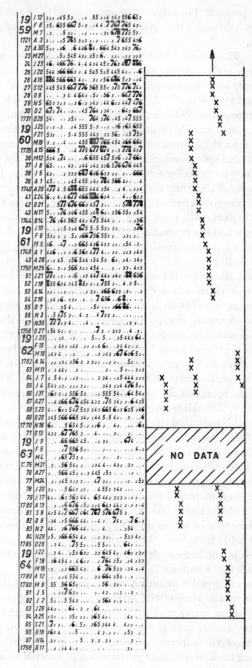

FIG. 5. *Bottom*: the daily geomagnetic character figure C9 as published by the Geophysical Institute, Göttingen. *Top*: to the same scale, diagram showing an estimate of the central meridian passage of the preceding portions of unipolar magnetic regions.

tions of polarity across the surface. The activity may come as a part of a large-scale "complex of activity." In such a case, a large unipolar magnetic region will build up predominantly poleward of the background-field pattern. This UMR will have the polarity of the following sunspots for that hemisphere. The magnetic flux of the UMRs will migrate poleward and build up a general polar magnetic field. Presumably this sets the stage for the next solar cycle, according to the model of H. W. Babcock. The large-scale pattern of the background fields can be seen in the interplanetary magnetic field as far out as the orbit of the Earth. The large-scale pattern is responsible in some way for the particle streams which cause the 27-day recurrent magnetic storms and other related geophysical effects.

POLAR MAGNETIC FIELDS

The first sensitive photoelectric observations which conclusively demonstrated the existence of magnetic fields at the poles of the Sun were those of Babcock & Babcock (1955). They detected a weak field of about 1 G average field at latitudes above about 55°. The field seemed to be centered on the axis of rotation and it showed some fine structure. H. D. Babcock observed the polarity of the polar fields to reverse near the time of the solar maximum. First the south polar field reversed. After the reversal, the polar field of the Sun was parallel to the Earth's magnetic field. Later von Klüber & Beggs (1960, 1962) confirmed the polarity and strength of the field as it was established at Mount Wilson. Waldmeier (1960) pointed out that the polar field in each hemisphere reversed about the time of maximum activity in that hemisphere. About the beginning of 1961 the polar field in the south dropped below the limit of detectability of magnetographs (Howard 1965). The low value of the south polar field was confirmed by von Klüber (1965), and Severny (1965a). It is not easy to understand the disappearance of the magnetic fields in the south. Bumba & Howard (1965c) examined the poleward-moving magnetic flux in the years 1959–1962 and concluded that there was much more flux moving to the North Pole than to the South Pole during these years, but a mechanism for the disappearance of the field has not been suggested. Similarly, Bumba & Howard pointed out that at the North Pole, because of the flux observed to drift to the polar region, we should observe a larger field than we do. It may be that the fine-scale structure of the solar magnetic fields provides the magnetic flux to cancel the fields observed with the magnetograph. Or, it may be that there is a poleward flow of magnetic flux that is too weak to be detected with the sensitivity and angular resolution of the magnetograph.

The fine-scale structure of the polar magnetic fields and their association with the polar faculae was observed by this author (Howard 1959). Later discussions of this include those of Severny (1965a,b) and Howard (1965). Sheeley (1964), on the assumption that the strength and number of the polar faculae are indications of the strength of the polar field of the Sun, has been able to trace the alternations of polar-field strength (presumably representing alternations of magnetic polarity) back to observations of 1906 from Mount Wilson material. The strengths of the faculae are about 180° out of phase with the sunspot number, as one would expect for a polar field that weakens and gradually reverses at each maximum of solar activity.

Earlier measurements of the polar magnetic field of the Sun, which were visual estimates of line shifts made on spectrum plates (Hale et al. 1918), showed a dipole magnetic field of about 20 G. Such fields are near the limit of detectability of the visual method. Certainly there is no such dipole field on the Sun now, and Sheeley's faculae counts indicate that the polar fields at that time were probably not much stronger than the fields measured now. There are probably several reasons for the higher fields measured then. They were looking for a dipole field, so they looked for a maximum field around 50° latitude. In fact, the Sun's polar field is not a dipole and is found for the most part above about 55° latitude. It seems likely that, at latitudes below about 60°, Hale and his co-workers measured portions of the background-field pattern. It does not seem reasonable to include such determinations along with modern measurements of the polar fields in a comparison of either field strength or polarity. The polar fields of the Sun were not measured until modern photoelectric methods were used.

THE SUN AS A MAGNETIC STAR

Bumba, Howard & Smith (1966) have investigated the integrated magnetic field of the Sun from Mount Wilson magnetograph measurements. The object was to reproduce the measurements made of magnetic stars (Ledoux & Renson 1966). The results are shown in Figure 6. The integrated magnetic field of the Sun depends to a large extent on the composition of the background-field pattern. During the period of these observations, the background-field pattern was of predominantly positive magnetic polarity. The reason for this was that the south polar field was very weak. The north polar field showed a negative polarity, and the preceding magnetic fields from the Northern Hemisphere regions collected at the lower latitudes and had a positive polarity. The polar fields contribute very little to the integrated fields, because of foreshortening. Basically the whole imbalance was due to a predominance of active regions in the Northern Hemisphere in the last cycle, which seems to be continuing into this next cycle. The irregular variations in the integrated magnetic fields are large, and they are due primarily to changes in the background-field pattern from the introduction of the magnetic flux of new active regions and the disappearance of old magnetic flux by cancellation at low latitudes. It is very difficult to see the 27-day rotation effect because of the many changes taking place in the pattern. Clearly the appearance of the Sun in light integrated across the disk as a magnetic star would depend upon whether it was viewed from near the plane of its equator or approximately pole-on. In the latter case, the Sun would appear to have nearly a 22-year period.

The introduction of the sensitive photoelectric method of measuring

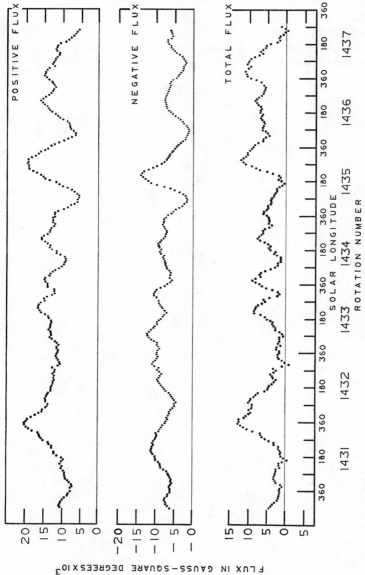

FIG. 6. Magnetic fields integrated over the solar disk as a function of time. *Top:* positive flux. *Middle:* negative flux. *Bottom:* net flux. The units of the ordinate are also 1.5×10^{21} maxwells.

magnetic fields on the solar surface has produced observations which have greatly affected our view of the Sun and solar activity. On the small scale of granules and the large scale of the whole solar surface, our notion of what is happening on the Sun has altered markedly in the last dozen years or so. It seems likely at this point that some very fruitful areas of research have been opened, and that we may expect exciting new results about solar activity in the next decade.

LITERATURE CITED

Alfvén, H. 1952, *Ark. Fys.*, **4**, No. 24

Babcock, H. W. 1953, *Ap. J.*, **118**, 387

Babcock, H. W. 1961, *ibid.*, **133**, 572

Babcock, H. W. 1963, *Ann. Rev. Astron. Ap.*, **1**, 41–58

Babcock, H. W., Babcock, H. D. 1955, *Ap. J.*, **121**, 349

Beckers, J. M., Schröter, E. H. 1966 (Paper presented at A.A.S. Meeting, Boulder, Colo.)

Bednarova-Novakova, B. 1960, *Studia Geophys. Geod.*, **4**, 54

Bray, R. J., Loughhead, R. E. 1964, *Sunspots* (Chapman and Hall Ltd., London)

Brunner-Hagger, W. 1944, *Publ. Eidgen. Sternwarte Zürich*, **8**, 31

Bumba, V., Howard, R. 1965a, *3rd Consultation on Solar Physics and Hydrodynamics, Tatranska Lomnica*, 24–26 (Publ. House Czechoslovak Acad. Sci., Prague)

Bumba, V., Howard, R. 1965b, *Ap. J.*, **141**, 1492

Bumba, V., Howard, R. 1965c, *ibid.*, 1502

Bumba, V., Howard, R. 1965d, *ibid.*, **143**, 592

Bumba, V., Howard, R., Kleczek, J. 1965, *Publ. Astron. Soc. Pacific*, **77**, 55

Bumba, V., Howard, R., Smith, S. F. 1964, Annual Report of the Director, Mount Wilson and Palomar Observatories, *Carnegie Institution of Washington Year Book 63*

Bumba, V., Howard, R., Smith, S. F. 1966, *Magnetic and Related Stars* (Cameron, R. C., Ed., Mono Book Corp., Baltimore, Md.)

Dodson, H. W. 1966 (Private communication)

Hale, G. E. 1908, *Ap. J.*, **28**, 315

Hale, G. E. 1922a, *Publ. Astron. Soc. Pacific*, **34**, 59

Hale, G. E. 1922b, *Monthly Notices Roy. Astron. Soc.*, **82**, 168

Hale, G. E., Seares, F. H., Maanen, A. van, Ellerman, F. 1918, *Ap. J.*, **47**, 206

Howard, R. 1959, *Ap. J.*, **130**, 193

Howard, R. 1962, *ibid.*, **136**, 211

Howard, R. 1963, *ibid.*, **138**, 1312

Howard R. 1964, *AAS-NASA Symposium on Solar Flares, NASA SP-50*, 89–93 (Hess, W. N., Ed., Sci. and Tech. Inform. Div., Washington, D.C.)

Howard, R. 1965, *I. A. U. Symposium No. 22, Stellar and Solar Magnetic Fields*, 129–36 (Lüst, R., Ed., North-Holland, Amsterdam)

Howard, R. 1966, *Observatory*, **86**, 160

Howard, R., Babcock, H. W. 1960, *Ap. J.*, **132**, 218

Howard, R., Harvey, J. W. 1964, *Ap. J.*, **139**, 1328

Howard, R., Severny, A. B. 1963, *Ap. J.*, **137**, 1242

Kippenhahn, R., Schlüter, A. 1957, *Z. Ap.*, **43**, 36

Klüber, H. von. 1965, *I. A. U. Symposium No. 22, Stellar and Solar Magnetic Fields*, 144 (Lust, R., Ed., North-Holland, Amsterdam)

Klüber, H. von, Beggs, D. W. 1960, *Quart. J. Roy. Astron. Soc.*, **1**, 92

Klüber, H. von, Beggs, D. W. 1962, *ibid.*, **3**, 116

Ledoux, P., Renson, P. 1966, *Ann. Rev. Astron. Ap.*, **4**, 293–352

Lee, R. H., Rust, D. M., Zirin, H. 1965, *Appl. Opt.*, **4**, 1081

Leighton, R. B. 1959, *Ap. J.*, **130**, 366

Leighton, R. B. 1964, *ibid.*, **140**, 1547

Leighton, R. B., Noyes, R., Simon, G. 1962, *Ap. J.*, **135**, 474

Macris, C. J. 1962, *Mem. Soc. Astron. Ital.*, **33**, 3

Mustel, E. R. 1944, *Dokl. Akad. Nauk SSSR*, **42**, 117

Ness, N. F., Wilcox, J. M. 1966, *Ap. J.*, **143**, 23

Parker, E. 1963, *Ap. J.*, **138**, 226

Piddington, J. H. 1956, *Monthly Notices Roy. Astron. Soc.*, **116**, 314

Ramsay, J. V. 1966, *Capri Symposium on Solar Fine Structure* (In press)

Rust, D. M. 1966, *Measurements of the Magnetic Fields in Quiescent Solar Prominences* (Thesis, Univ. of Colorado, Boulder, Colo.

Saito, T. 1964, *Rept. Ionosphere Space Res. Japan*, **18**, 260

Severny, A. B. 1958, *Izv. Crimean Astrofiz. Obs.*, **20**, 22

Severny, A. B. 1960, *ibid.*, **22**, 12

Severny, A. B. 1964, *Space Sci. Rev.*, **3**, 451

Severny, A. B. 1965a, *I.A.U. Symposium No. 22, Stellar and Solar Magnetic Fields*, 139 (Lüst, R., Ed., North-Holland, Amsterdam)

Severny, A. B. 1965b, *Astron. Zh. USSR*, **42**, 217

Severny, A. B. 1966, *ibid.*, **43**, 466

Shapley, A. H., Roberts, W. O. 1946, *Ap. J.*, **103**, 257

Sheeley, N. R., Jr. 1964, *Ap. J.*, **140**, 731

Sheeley, N. R., Jr. 1966a, *ibid.*, **144**, 723

Sheeley, N. R., Jr. 1966b (Paper presented at A.A.S. Meeting, Boulder, Colo.)

Simon, G. W., Leighton, R. B. 1964, *Ap. J.*, **140**, 1120

Simpson, J. A., Babcock, H. W., Babcock, H. D. 1955, *Phys. Rev.*, **98**, 1402

Stenflo, J. O. 1966, *Observatory*, **86**, 73

Stepanov, V. E. 1958, *Izv. Crimean Astrofiz. Obs.*, **20**, 52

Stepanov, V. E. 1960, *ibid*, **22**, 42

Stepanov, V. E., Gopasyuk, S. I. 1962, *Izv. Crimean Astrofiz. Obs.*, **28**, 194

Stepanov, V. E., Severny, A. B. 1962, *Izv. Crimean Astrofiz. Obs.*, **28**, 166

Title, A. M. 1966, *A Study of Velocity in the Hα Chromosphere by Means of Time-Lapse Doppler Movies* (Thesis, California Inst. of Technol., Pasadena, Calif.)

Tsap, T. T. 1964, *Izv. Crimean Astrofiz. Obs.*, **31**, 200

Tsap, T. T. 1965a, *ibid.*, **33**, 92

Tsap, T. T. 1965b, *ibid.*, **34**, 296

Waldmeier, M. 1960, *Z. Ap.*, **49**, 176

Wilcox, J. M. 1966, *Science*, **152**, 161

Wilcox, J. M., Ness, N. F. 1965, *J. Geophys. Res.*, **70**, 5793

ON THE INTERPRETATION OF STATISTICS OF DOUBLE STARS[1,2]

By Alan H. Batten

Dominion Astrophysical Observatory, Victoria, British Columbia

INTRODUCTION

The present time is opportune for reviewing the statistics of double stars. Only a few years ago, Worley (1963) published a catalogue of the orbital elements of visual binary systems, containing elements for over 500 systems. At present, the writer is working on a catalogue of the orbital elements of spectroscopic binary systems which, he hopes, will appear about the same time as this review. It will contain elements for nearly 750 systems—an increase of 50 per cent over the last published catalogue (Moore & Neubauer 1948). Fewer than 50 systems are common to Worley's and the writer's catalogues. It would thus seem that there is ample material for the statistical analysis of both visual and spectroscopic binaries. Unfortunately, this is very far from being the case. The quality of published orbits varies widely. Some investigators seem to believe that any determination of orbital elements is good enough for statistical analysis of the properties of both visual and spectroscopic binaries. This notion cannot be too strongly contested. Badly determined orbits are useless, even for statistical studies. In both Worley's catalogue and the writer's, estimates are given of the quality of each set of orbital elements, and about half of the orbits given in each catalogue should be discarded before any statistical investigation is begun. For any investigation of some specialized group of binaries, the well-determined orbits are all too few. As has already been implied, there are fewer than 50 visual binaries with well-determined orbits for which there are also radial-velocity observations; and there are only very few double-lined spectroscopic binaries, which are also eclipsing binaries, for which measurements both of spectra and of the light-curve are all reliable. Thus the determination of masses in close binary systems, particularly at the upper end of the main sequence, is based on very few systems.

Not only are the reliable data relatively few; they are also heavily affected by observational selection. This is clearly indicated by a comparison of the number of binary stars known with the numbers just quoted for the orbit catalogues. The Index Catalogue of Double Stars (Jeffers, van den Bos & Greeby 1963) lists over 64,000 pairs, the vast majority of which must be physical pairs, although admittedly too widely separated for their orbital motions to be readily observed. The General Catalogue of Variable Stars

---

[1] The survey of literature for this review was concluded in October 1966.

[2] *Contributions from the Dominion Astrophysical Observatory No. 114:* published by permission of the Deputy Minister, Department of Energy, Mines and Resources, Ottawa, Canada.

(Kukarkin, Parenago, Efremov & Kholopov 1958) lists over 2500 eclipsing binaries, of which about 300 appear in the new catalogue of spectroscopic binaries. It is evident that it will be impossible to study all these stars individually, and many of them, indeed, present, to those observers who would study them in detail, difficulties that are insuperable with present-day observational equipment. This great difference between the number of systems known to exist and the number that have been even partially studied should serve as a warning against any complacency about the amount of data now available. It also serves to demonstrate how very selected is the sample of binary systems with well-determined orbits. Even statistical conclusions drawn from such a sample may be vitiated by effects of observational selection, which are often difficult to foresee, let alone to eliminate. This warning against observational selection runs like a refrain through the pages that follow.

No attempt has been made in this review to cover the whole subject of the statistics of double stars. Instead, a few topics that appear to represent growing points in the study of double stars, and to which statistical data are relevant, have been selected for detailed examination. Throughout the review the term "double star" is interpreted quite generally to include all binary systems, whether "visual," "spectroscopic," or "eclipsing." These terms, which reflect only the method of observation, never represented a real classification of binary systems, although the possibility that wide and close pairs did have a different origin is discussed in another section. The emphasis of the discussions that follow is, however, on problems of those close binary systems for which spectroscopic data are available.

It remains only to explain one obvious omission from the review—the determination of stellar masses. This is clearly a topic of the utmost importance, which is dependent on the statistics of double stars. Its importance has been recently underlined by the work of Petrie, Andrews & Scarfe (1966) who have shown that estimates of masses made from double-lined spectroscopic binaries may be appreciably affected by systematic errors of measurement. This topic is discussed by Popper in another chapter of this volume (p. 85) and so it has not been included here.

CLASSIFICATION OF BINARY STARS

In an observational science such as astronomy, the classification of any species of object into various subgroups is often the first step to a deeper insight into the nature of the species being studied. A useful classification scheme for binary stars ought to be applicable to all systems, and exhibit clearly both the distinctions and the relationships among the various classes. No such system has yet won general acceptance. This no doubt reflects partly the large range in the mutual separations of the components of binary systems (a range over a factor of at least 10^6) and partly the number of parameters needed to define a binary system completely. Most attention has been concentrated on the close systems, and in particular on eclipsing binary systems. This is natural, because it is among close binary systems that distinct families are most easily recognized, and it is from eclipsing binary systems that have also been observed spectroscopically that the greatest amount of information on which to base a classification can be obtained. It should always be remembered, however, that any proposed classification, to be useful, should be capable of being extended to all types of binary system.

The old system of classifying eclipsing systems by the superficial characteristics of their light-curves into Algol, β Lyrae, and W Ursae Majoris systems is clearly inadequate, although the old terms die hard. Other schemes have been proposed by Krat (1944), Struve (1951), Kopal (1955), and Sahade (1960). The schemes put forward by Krat, Struve, and Sahade are all related to the positions of the component stars in the Hertzsprung-Russell diagram. Kopal's classification depends on three parameters, the radius of each star relative to the separation of the two stars, and the mass ratio of the system. The relative merits of these classifications have been discussed by Plavec (1964) who makes a strong case for a two-dimensional classification of binary systems, and suggests a combination of Kopal's and Sahade's schemes. Thus Plavec would describe a system by a number representing its position in the Hertzsprung-Russell diagram, and a letter representing its position in Kopal's scheme. Even a two-dimensional classification of binary systems has its limitations, since seven parameters (or seven functions of them) are needed to specify a binary system. These parameters are the mass, radius, and luminosity of each component, and the distance between the two components. (For a limited range of separations, the orbital eccentricity may be important too, for in an elliptic orbit there may be some surface interaction between the components near periastron.) The number of parameters could be reduced by taking the ratios of the masses, etc., but this would ignore the information contained in the system's position on the Hertzsprung-Russell diagram. Plavec's scheme of classification does, in fact, implicitly use all seven parameters, since four are required in Sahade's scheme to specify a system's position on the Hertzsprung-Russell diagram, and three are required in Kopal's. Plavec may therefore justly claim to have suggested the most comprehensive scheme yet put forward. A disadvantage, however, is that neither of the criteria used by Plavec can be represented by a continuously varying numerical quantity.

Thus, it is impossible to draw a meaningful diagram, analogous to the Hertzsprung-Russell diagram, in which the different families of binary systems, and their interrelationships, will be clearly shown. To do this successfully, it will probably be necessary to choose functions of a few of the seven parameters, and ignore the information contained in the remainder. For single stars, there are three possible pairs among the quantities mass, radius, and luminosity, and yet the mass-luminosity relation, and the Hertzsprung-Russell diagram (which can be regarded as a radius-luminosity relation) have proved to be much more generally useful than the mass-radius relation. Similarly, out of the 21 possible relations between pairs of seven parameters, some are likely to be much more significant than others. The failure to find a generally acceptable scheme of classification is the failure, so far, to pick out the right pairs of parameters. It may be surmised that one parameter in such a classification scheme will be some function of the radii relative to the separation, since this is a crude measure of the degree to which the two stars are likely to interact. It may be, however, that no single two-dimensional classification will be entirely satisfactory for all purposes.

The lack of an agreed classification is, at present, nowhere felt so keenly as with the group of stars known as W Ursae Majoris or "contact" systems. This is usually considered to be a compact group of systems with periods of less than a day, and whose components are clustered around the main sequence from the late-A spectral types to the early-K types. This definition may, however, be too restrictive. Eggen (1961) has considered as contact systems all eclipsing systems with periods less than 1.5 days whose light-curves show continuous variation, without restriction as to spectral type. He finds for these systems a relation between period and colour, which are functions of all seven of the parameters listed above. A theoretical relation between these quantities can be derived, although its precise form for such highly distorted systems is difficult to obtain. Eggen finds a nearly linear relation between the logarithm of the period and the $(B-V)$ colour, for stars with periods between 0.25 days and 0.65 days. For longer periods there appears to be a discontinuity in the period-colour relation, and all longer-period systems appear to have colours that cluster loosely around $B-V=+0.1$.

Binnendijk (1965) also distinguishes two types of contact systems which he calls A and W. In the A-type systems, the component of earlier spectral type is also the brighter component. It has a spectral type of late A or early F, and apparently lies on the main sequence. In the W-type systems, the component of earlier spectral type is the fainter of the two. Its spectral type is F5 or later, and it usually appears to be farther from the main sequence than does its companion. In Binnendijk's sample of 13 systems with completely determined orbital elements, there are three systems of type A. The question naturally arises whether these two divisions of the contact systems are equivalent, since such a difference in relative colours and luminosities of the two components of a system might well have some effect on the observed total colour. They are not equivalent, for two of Binnendijk's A-type systems and eight of his W-type systems were used by Eggen to determine the linear, short-period portion of his log P-$(B-V)$ relation. It is worth noting that Eggen's division in the period-colour relation depends very heavily on 16 systems for which he has applied reddening corrections to the observed values of $B-V$. Since there is evidence that some of the shorter-period systems have anomalous $(U-B)$ colours, and since the only two long-period systems for which reddening corrections have not been made (μ^1 Scorpii and

IM Monocerotis) fall close to the log P-$(B-V)$ relation defined by the short-period systems, the reality of the division found by Eggen in the period-colour relation seems open to doubt.

Eggen's selection of 1.5 days as the limiting period of a contact system enables him to consider as contact systems many that are earlier in spectral type than those normally included in the term "W Ursae Majoris system." Thus the recent discovery that HD 205372 is a contact binary (Bartolini, Mammano, Mannino & Margoni 1965) consisting of two equal stars of type A2 poses no problem for him, although this binary is distinctly earlier in type than are previously known W Ursae Majoris systems. Indeed the new system fits well the period-colour relation defined by the systems with periods less than 0.65 days, although its own period is about 0.94 days. Thus it seems to be basically similar to the W Ursae Majoris systems. Both components seem to be of the same spectral type, and to lie somewhat above the main sequence, so the system does not fit clearly into either of Binnendijk's subdivisions.

Another interesting system is AO Cassiopeiae. It was explicitly not considered by Binnendijk, and its period of 3.5 days ruled it out of consideration by Eggen. Yet it consists of two O-type stars, separated by about 30 $r_\odot$, and must certainly be nearly a contact system. It does not obey the period-colour relation, if its colour is that of an ordinary O9 star, but then its period is nearly six times the maximum period for which that relation was established by Eggen. At the other end of the main sequence is the system of YY Geminorum, consisting of two similar stars of spectral type dM1e, with an orbital period of 0.81 days. The stars are nearly in contact, and must have a common envelope, but the system was considered by Kopal to be detached. Consistent with this, it flagrantly disobeys the period-colour relation.

It seems, then, that classification as a contact system should be made dependent on the adherence of the system to the empirical period-colour relation derived by Eggen. In contradiction to Eggen, it is supposed here that one relation of approximately the form

$$\log P = -0.7 (B - V) - 0.03$$

holds for all periods less than 1.5 days. More work is needed to decide whether or not the division of contact systems into A-type and W-type corresponds to a real physical distinction. Contact systems seem to extend all the way up the main sequence, at least to the B types, and not to be confined to the spectral types associated with the W Ursae Majoris systems. This was confirmed in a recent paper by Kurochkin & Kukarkin (1966) where a plot of period against spectrum shows a single relation for all contact systems, although these authors distinguish between W Ursae Majoris and β Lyrae systems. It is not clear if there are contact systems with spectral types of late K or M classes; any such systems would be intrinsically very faint, and observational selection would militate strongly against their discovery.

RELATIVE BINARY FREQUENCIES: ROTATION AND BINARY FREQUENCY

The absolute frequency of binary stars in the Galaxy is very difficult to estimate. Petrie's estimate (1962) that 50 per cent of all stars are binaries applies only to spectroscopic binaries, and is therefore probably only a lower limit. Blaauw & van Albada (1966) estimate a binary frequency of 60 per cent in certain groups of early-type stars. The problems of observational selection and of detection of marginal cases are such that no final estimate of this frequency is foreseeably within our grasp. It is possible, however, to discuss the relative frequency of binary systems in different groups of stars, whether these are spatial groupings, or groups of restricted classes of stars. Even in such a discussion, observational selection presents difficulties. The outstanding example of work along these lines, in recent years, is the demonstration by Kraft and his associates that the binary frequency in the class of old novae and eruptive variables appears to be very high, and may be 100 per cent. The evidence for this has recently been surveyed by Kraft (1963) and no further review will be made here. Similarly, Sahade (1958) has suggested that all Wolf-Rayet stars are members of binary systems, although here the evidence is not conclusive. The importance of these suggestions to our understanding of the nova and Wolf-Rayet phenomena is obvious. Other investigations of binary frequency, depending strongly on published catalogues of radial velocities, have been made by Jaschek & Jaschek, who have investigated the relative frequency in different spectral classes and in some clusters (1957, 1959).

One of the most interesting investigations of relative binary frequency so far is that for the A stars undertaken by Abt (1961, 1965). He has considered the relative frequency among the metallic-line A stars (Am stars), and the normal A stars of the same spectral range (about A4 IV and V to F2 IV and V). Abt confirms the high frequency of binary systems amongst Am stars which was suspected by earlier investigators (Roman, Morgan & Eggen 1948), and believes it may be 100 per cent. On the other hand, he finds a relatively low frequency of binary systems among the normal A stars. In addition, he finds that the orbital periods of systems containing Am stars are usually less than 100 days, while the periods of systems containing normal A stars, in his sample, are without exception greater than 100 days. Abt suggests that the abnormality of Am spectra is a result of the low rotational speeds found for Am stars, and that these low rotational speeds, in turn, are the result of tidal forces in close binary systems. Not all Am stars are necessarily members of close binaries, because low speeds of rotation may sometimes be the result of other causes, but Abt's argument leads him to predict that Am stars will be found preferentially in close binary systems (periods less than 100 days), while normal A stars in the same spectral range will be found, preferentially, to be single, or to be members of wide systems. These are interesting ideas which may have far-reaching effects on our understanding of the various types of peculiar spectra found in this region of the spectral sequence. Just because they are interesting, it is necessary to subject the observational data on which they are based to careful scrutiny.

Abt's evidence for the Am stars was presented in his 1961 paper: 25 stars were selected from lists of known Am stars, and their radial velocities were investigated. Abt found evidence that 22 are binaries (many already had known orbits) and he considered that the other 3 represented no more than the expected proportion of undetectable binaries. It is possible to question, however, whether Abt has established the binary nature of all 22 of his stars. The amplitudes of several of the suspected velocity variations are low, and the evidence for any variation at all sometimes rests on a few observations of low weight which happen to differ from the mean velocity of the star. Thus the stars 15 Vulpeculae, μ Aquarii, ξ Cephei A, HR 906, and ω Tauri are all systems which should be further observed in order to establish the variation of their radial velocities. (Observations of ξ Cephei A were undertaken at Victoria by Petrie and are being continued by Scarfe. While occasional line doubling is suspected in its spectrum, no convincing demonstration of its binary nature, or even of its velocity variation, has been obtained.) If these stars are omitted from consideration, only 17 of Abt's sample of 25 Am stars are known to be binary systems. This would reduce the binary frequency to 68 per cent. Such a frequency is higher than the estimated average for all stars, but not by more than might be expected in a small sample.

Probably some of the stars named above are, indeed, binary systems, but the true frequency in the sample may still be less than 100 per cent. Abt's paper stimulated observers at Victoria to determine more precisely the orbital elements of some of the new binary systems he had discovered. As a result, Gutmann (1965a,b) confirmed the binary nature of ζ Ursae Majoris B and v Ophiuchi, although he found the orbital periods to differ from the values found for them by Abt. Similarly, Fletcher (1964) found a value for the period of 47 Andromedae different from that found by Jose (1951) and quoted by Abt. These changes do not alter Abt's conclusion that the majority of binary systems containing Am stars have periods of less than 100 days, but they do emphasize the importance of obtaining well-determined orbits even for statistical discussions of the elements.

The normal A stars were discussed by Abt in his 1965 paper. He investigated a random sample of 55 stars with spectral types between A4 IV, V and F2 IV, V chosen from those stars for which Slettebak & Howard have determined rotational velocities (1955). Amongst these he finds 17 binaries or suspected binaries, that is, about 33 per cent of the sample, at most, are binary systems. There is no doubt that this frequency is lower than that found for the Am stars, and it is probably lower than the average binary frequency in all stars. It is hard to estimate whether or not the difference is greater than would be expected from statistical fluctuations in the small samples. The surprising result is that Abt found no orbital periods, in this sample, shorter than 100 days. There are plenty of systems outside the A4-F2 spectral range with short orbital periods, and a number of W Ursae Majoris systems are known with spectral types as early as A8.

In order to test whether or not Abt's sample of 55 A stars is representative of the whole class of normal A stars, a search was made of the orbital data compiled by the writer for the Sixth Catalogue of Orbital Elements of Spectroscopic Binary Systems. These data were not available in a conveniently collected form when Abt's paper first appeared. It was found that there are 73 systems with known orbits, of which at least one component has a spectral type in the range A4-F2, and whose periods are all less than 50 days. (Most of the periods, indeed, are less than 10 days.) None of these 73 systems had been recognized by the original investigators as having Am spectra. Since many of the original investigations date back before the full recognition of the Am stars as a special group, this list of 73 systems was checked against Bertaud's lists of all known Am stars (1959, 1960, 1965). As a result of this, only 3 systems were found to contain an Am star. Thus there are 70 short-period binary systems with known orbits which contain apparently normal A-type stars: at least 3 of these are W Ursae Majoris systems, and 2 contain stars with spectra classified as "n." It is understood that Abt is investigating whether or not any of these stars should be reclassified as Am stars. Many of the stars are faint, and it will be difficult to obtain good spectrograms that will settle the question. In a private communication, Abt has emphasized that it is important that the classification of the spectra be done by an experienced person on a consistent scheme (the MK system). He states that there is no known short-period system (other than W Ursae Majoris systems) that has been placed in the MK system after careful examination, between the limits A4 IV, V to F2 IV, V. He claims that this fact, coupled with his failure to detect any short-period systems in his random sample of 55 stars, is sufficient evidence of the rarity of such systems. These points have validity, but meanwhile it seems legitimate to suspend judgement on the degree of the deficiency of short-period binaries among the normal A stars.

In a related series of investigations, Abt and his collaborators have studied the relative frequency of binary stars in certain clusters, and correlated it with the mean rotational velocities of stars in these clusters (Abt, Barnes, Biggs & Osmer 1965; Abt & Hunter 1962; Abt & Snowden 1964; Chaffee & Abt 1966). They find that in clusters whose members have high rotational velocities the frequency of binary systems is low, and vice versa. This apparent correlation is confirmed by Heard & Petrie (1966) in their work on the α Persei cluster. The rotational velocities of the stars in this cluster seem to be very high, and among 64 certain members, they find only 2 stars that probably vary in velocity, and 11 that possibly vary. Abt again infers from this correlation that the principal mechanism acting to slow down the rotation of stars is tidal friction in close binaries. Clusters with many close binaries among their members will, therefore, generally contain fairly slowly rotating stars, unless they are very young clusters.

Although the latest work on the α Persei cluster supports Abt's thesis, the history of studies of this cluster illustrates the importance of basing generalizations of this sort on the fullest possible observational data. Jaschek & Jaschek (1959) estimated the binary frequency in this cluster to be 27 per cent. Their estimate was based on measures of the velocities of 15 stars in the cluster made by Roman & Morgan (1950). Later, Abt & Hunter (1962) found that out of 7 stars studied, 3 showed evidence of variable velocity, and on this basis they included it among the clusters having a high binary frequency. Now the detailed investigation by Heard & Petrie, just cited, indicates that the binary frequency is at most 20 per cent. Even now, there are available no orbital elements, and no individual star in the cluster is known for certain to be a binary system. The most that can be said about this cluster is that its binary frequency is apparently low, and its rotational velocities are high. It is of interest that Heard & Petrie found one star, which they believed to be a member of the cluster, that is apparently an Am star.

It is clear that estimates of binary frequency, which depend on estimates of the percentage of stars in a group that show variable velocity, must be affected by the errors of measurement of the spectrograms obtained. These errors are, in turn, determined by the quality and nature of the individual lines in the stellar spectra. The broad lines in the spectra of rapidly rotating stars will be much more difficult to measure accurately than the sharp lines in spectra of the slowly rotating stars of the same spectral type. There is clearly a selection effect making it more difficult to detect variations in the velocities of rapidly rotating stars. Can this effect account for the observed differences of binary frequency? As Abt himself has pointed out, detectability of binary systems is a very difficult thing to define, and it may well be influenced by the way in which the plates are measured. The development of oscilloscopic measuring devices may offer the promise of greater detectability, both by increasing the accuracy of measurement and by providing an easier means of recognizing the effect of a faint secondary spectrum on the line profile.

Experience at Victoria, however, has suggested that no great increase in accuracy will result from the use of an oscilloscopic device for the measurement of rotationally broadened line profiles. A test was made on 49 spectrograms of AR Cassiopeiae, obtained at a dispersion of 15 Å/mm at Hγ. The star has a B3 spectrum with fairly broad lines. The rotational velocity is estimated to be nearly 200 km/sec (Slettebak & Howard 1955, Batten 1961). It is an eclipsing binary, and the spectrographic orbit at several epochs has been derived by Petrie (unpublished). The spectrum of the primary component only is visible. Petrie derived orbital elements for the system from his measures of the 49 plates made with the aid of a projection microscope. On the average, six lines were measured on each plate. The mean error of a single measurement, obtained from the residuals about the velocity-curve, was found to be 7.85 km/sec. Later, the same plates were measured by the writer, on an oscilloscopic device, and a new orbit was derived. The average number of lines measured on each plate was also six. There were no significant differences between the two sets of orbital elements, and the mean error of a single plate was reduced only to 7.15 km/sec.[3] The spectrum is difficult to measure, but probably not as difficult as many of late-B or early-A spectral types.

Thus, if one is to attempt to measure a spectrogram of a rapidly rotating star of spectral type about A0, obtained at a dispersion of 50 or 60 Å/mm at Hγ such as is used by Abt in his survey work, it is reasonable to estimate a mean error per measurement of radial velocity of the order of 10 km/sec, whatever method of measurement is used. [Petrie (1962) estimated a mean error per spectrogram of 5.1 km/sec for A-type spectra, and 6.4 km/sec for B-type spectra, at a dispersion of 50 Å/mm at Hγ. His estimates were concerned with the average mean error for each spectral type. The estimate of 10 km/sec refers to rotationally broadened spectra, and so *should* be larger than Petrie's.]

Thus it seems reasonable to require that a rapidly rotating star should show a range in velocity of at least 30 km/sec before it is accepted that its velocity is variable. A star of similar spectral type whose spectrum shows little or no rotational broadening will yield a mean error per plate of perhaps 3 or 4 km/sec, and so a range in velocity of about 10 km/sec would be sufficient evidence that the star is variable in velocity. The probability that two successive observations of a single-spectrum binary star will yield velocities differing by more than an assigned quantity Δ can be computed rather easily. The probability turns out, of course, to be dependent upon the part of the velocity-curve at which the first observation is made. In most cases, the maximum probability is obtained if the first observation happens to be made at one of the nodes of the curve, although for certain values of Δ/K (where K is the semiamplitude of the velocity-curve) there is a slightly greater chance of the required difference being observed if the first observation is made at the points where the velocity of the star is equal to the systemic velocity. For circular orbits, the probability p that any observation will differ by an amount Δ from an observation obtained at one node of the velocity-curve is given by

$$p = \frac{1}{2} + \frac{1}{\pi} \sin^{-1}\left(1 - \frac{\Delta}{K}\right)$$

This, of course, is a very different thing from the probability of detecting

[3] Dr. Abt states that experience at Kitt Peak National Observatory leads to a more optimistic estimate of the capabilities of an oscilloscopic measuring machine. It is, perhaps, natural that opinions should differ as to the accuracy obtainable with these machines, until many more years' experience are available. It should be emphasized that the question here is not whether an operator can set on a given line more consistently with an oscilloscopic device, but whether he can measure the radial velocity of a star more accurately. If some of the "errors of measurement" result from real effects on the photographic plate or in the star, the potential increase in the accuracy of velocity determination is limited.

a binary system. To calculate that, it is necessary to consider the probability of obtaining an observation at the node, and this will depend on the orbital eccentricity. It is also necessary to take account of the effect of a secondary spectrum, which will depend on the resolution of the photographic plate and on other things. The period of the system is also important, although it does not enter into the above formula explicitly. Long-period systems are difficult to detect, and so are those with periods that are nearly multiples of a day, and the binary nature of these kinds of system would be harder to detect than would be expected on the basis of this formula. If, however, attention is restricted to single-spectrum systems with circular orbits, the formula can serve as a crude estimate of the relative probability of detecting binary systems with given values of K. In Table I are listed the values of p given by the formula for three values of K combined with two of Δ. These figures are based on the assumption that only two spectrograms have been obtained for each star. Of course the probabilities of detection are increased if more spectrograms are taken, but the relative probabilities, for different values of K and Δ, are not changed unless K is so nearly equal to Δ that the velocity variation cannot be detected at all. Nearly nine out of ten binary systems with K greater than 100 km/sec will be detected in a cluster of slowly rotating stars, and three quarters of the similar systems in a cluster of rapidly rotating stars will be found. The difference in the proportion of binary systems detected is not great, if K is large. For $K = 30$ km/sec, however, it is found that about half as many systems again would be detected in the cluster of slowly rotating stars, assuming that the two clusters contain the same numbers of binary stars with each given value of K, and differ only in the mean rotational speeds of their members, by an amount corresponding to these two values of Δ. If $K = 15$ km/sec, *no* binary stars will be detected in the cluster of rapidly rotating stars. Thus there is clearly a selection effect which reduces the probability of detecting low-amplitude binary systems. Increasing the number of spectrograms obtained for each star will increase the disparity between the two probabilities, since p will remain zero for the case $\Delta = 30$ km/sec.

These figures cannot be used to estimate the relative fractions of binary systems that would be detected in the two imaginary clusters, without making some estimate of the distribution function of the K's. This is unknown,

TABLE I

PROBABILITY THAT TWO VELOCITY OBSERVATIONS OF A
BINARY SYSTEM DIFFER BY A GIVEN AMOUNT Δ

K	$\Delta = 10$ km/sec	$\Delta = 30$ km/sec
km/sec		
100	0.86	0.75
30	0.74	0.50
15	0.61	0.00

precisely because of the selection effect being discussed. It would seem likely that small values of K will predominate in any cluster, since they may arise either from long-period or low-mass systems with any orbital inclination, or from other systems with low inclinations, while high values of K will be found only for high-mass, short-period systems with high inclinations. If this is so, most of the binary systems in a cluster of rapidly rotating stars will not be detected, and so it seems likely that there is a fairly substantial selection effect in the discovery of binary systems in star clusters, which must act to produce a correlation, in the same sense as that observed by Abt and his colleagues, between binary frequency and mean rotational velocity. It is not clear whether or not this selection can account for the whole of the observed correlation: its effects might be either mitigated or exaggerated if some clusters should have a preferred direction for the orbital inclinations of their binary systems. One possible approach to the problem would be to study in detail a cluster in which rotational velocities are low, and to determine good orbits for all the detectable binaries. It would then be possible to estimate the distribution function of the K's. The expected number of detectable systems in any other cluster could then be computed on the assumption that the distribution function is the same for all clusters, and the results could be tested against the observed binary frequencies. How far this assumption can be justified is at present uncertain.

In conclusion, it may be of interest to note two situations in which there does not appear to be any correlation between frequency of close binary systems and speed of rotation, although they are not directly related to Abt's discussions. The first is among the group of so-called "Algol" or "semide-tached" systems in which the early-type component is usually found to rotate appreciably faster than would be expected for orbital synchronism. Indeed, the puzzle in these systems has been to find the source of the extra rotational angular momentum. A typical system is U Cephei, with a primary component of type B8 or B9 rotating with an equatorial speed in excess of 200 km/sec, and an orbital period of about 2.5 days. There must be many systems like this, which are unrecognized because they do not happen to show eclipses. Their velocity variations would be of low amplitude and very difficult to detect. The excess rotation is perhaps the result of mass transfer from the secondary component. Their existence emphasizes a point already made by Abt, namely, that rapid rotation can be associated with membership in a close binary system. The second point is that preliminary estimates of the frequency of spectroscopic binary systems along the main sequence seem to show that it is constant for all spectral classes (Jaschek & Jaschek 1957, Petrie 1960). The rotational angular momentum of main-sequence stars, on the other hand, is known to decrease quite sharply at about spectral class F5, except for members of W Ursae Majoris systems. These estimates are, no doubt, subject to confirmation, especially since recent results obtained by Wilson (1966) seem to contradict them; but the two estimates were reached quite independently and by quite different means.

Origin of Binary Stars: Fission Hypothesis

A renewed interest in the problem of the origin of binary stars is apparent. It is becoming obvious that binary systems are so numerous that any acceptable theory of star formation must be able to account for them. Indeed, it is now obvious that a complete theory of star formation must also account for the observed proportions of multiple stars. Recent estimates (Petrie & Batten 1966, Batten 1966, Worley 1966) all agree that at least 30 per cent of binary systems are in fact multiple.

Four plausible hypotheses of the origin of binary systems have been put forward:

First, there is the capture hypothesis. This divorces the origin of the system from those of its component stars, and explains the binary system as the result of a chance encounter. A capture (of this kind) requires a close encounter of three stars, or at least two stars and an interstellar cloud, and such triple encounters are too improbable to account for the observed proportion of binary stars, at least where the mean density of stars is equal to that found in the solar neighbourhood. The capture mechanism may still play a role in star clusters, however, or during the early life of expanding associations. It may also contribute to the formation of triple systems from already existing binary systems.

Second, there is the hypothesis of condensation of an interstellar cloud about two nearby nuclei. Wood (1964) has suggested that this hypothesis can account qualitatively for many of the observed properties of close binary systems. The writer has shown (1966) that the observed frequency of multiple systems is not in conflict with the hypothesis, although, of course, this frequency is subject to observational selection. It is, on the other hand, difficult to envisage simultaneous condensations about very close nuclei.

Third, there is the fission hypothesis, according to which one rotationally unstable protostar breaks into two stars during the pre-main-sequence contraction stage of the star's life. The classical treatment of the fission theory by Jeans (1929) has been criticized by several investigators, because it applies only to incompressible fluids. The derivation of a general theory applying to real stars presents formidable difficulties. Kurochkin & Kukarkin (1966) have argued for fission as the origin of contact systems.

Fourth, there is a hypothesis recently put forward by van Albada (1966) which postulates that close binary systems are the endproduct of a process of escape of stars from an unstable association. Each escape takes energy from the remaining stars, leaving behind first a multiple system, and finally a binary system. If more reliable data on the proportion of multiple systems were available, it would be possible to test this theory.

It may be, of course, that each of these possible origins is a source of some of the observed binary systems. Blaauw & van Albada (1966) have recently suggested that there are at least two classes of binary system, each with its own origin. Their suggestion is based on their work on the associations I Orionis, II Perseus, and I Lacerta, and on the Cassiopeia-Taurus group.

They have studied the frequency of binary systems amongst the early-B stars of these groups as a function of the separation of the two components of each system. They find that the close binary systems define a relation between frequency and separation which, if extrapolated to the separations found within the visual systems, would indicate that considerably fewer visual systems should be found than are actually observed. The observed relation between number of systems and separation has a broad minimum at separations between 1 and 100 a.u. Blaauw & van Albada suggest that there may be two different mechanisms operating in these groups to produce binary systems, each effective in its own range of separations. It is difficult to evaluate the effects of observational selection on these results; the observed minimum in frequency is just in the range where detection of binary systems, either by spectroscopic or visual techniques, is likely to be most difficult. It would be instructive to look for stars with composite spectra in these groups, since these might in fact be binary systems in the range of separations where the observed numbers of systems are low.

When it is recalled that the range of separations of the components of binary stars is from a few solar radii, for contact systems, to tens or even hundreds of thousands of astronomical units for wide common-proper-motion pairs (a range of at least a factor of 10^6), it does not seem unlikely that different mechanisms are needed to produce the objects at each end of the range. The idea is not new, and was put forward over 30 years ago by Bleksley (1934). At that time, most astronomers were disposed to reject the idea, chiefly on the evidence of the period-eccentricity relation for binary orbits. Now, that relation has been called in question (Finsen 1962) and the idea of separate origins should be re-examined. It is relevant to this, to consider whether or not close systems can evolve into wide systems or vice versa. This has been the subject of well-known investigations by Ambartsumian (1937) and Chandrasekhar (1944), and has recently been considered again by Yabushita (1966). Yabushita considers systems of total mass equal to 1 $m_\odot$, consisting of two equal components, and subjected to close encounters with other stars. The surrounding star density is supposed equal to that found in the solar neighbourhood. From the probability that any one encounter will disrupt the system, he finds the expected lifetime of such a system is 2.9×10^{12}, 2.4×10^{11}, and 3×10^{10} years for initial separations of 10^2, 10^3, and 10^4 a.u. respectively. Yabushita has ignored the possibility that successive encounters, each of which was insufficient to disrupt the system, might together have an important cumulative effect on the separation. This may partly account for the fact that the lifetimes he estimates are somewhat larger than those computed by Chandrasekhar and Ambartsumian. Nevertheless, all three authors agree that the lifetime of a binary system of even 100 a.u. separation is several times greater than the estimated age of the Galaxy. It would thus appear that, whatever their origin, close systems were formed as close systems, and wide systems as wide systems. In a recent paper by Huang (1966), however, the possibility is discussed that wide pairs can evolve into close pairs by ejection of mass particles along magnetic lines of force. This paper is germane to many of the discussions in this review, but it was not received in Victoria until after the writer's preliminary draft was completed.

If it is granted that it is at least possible that close and wide systems have different origins, then the most plausible hypothesis to consider for the close systems, despite the limitations of Jeans' argument and the difficulty of formulating a general theory, is fission of a rotationally unstable protostar. This theory has recently been reconsidered by Roxburgh in three interesting papers (1965; 1966a,b). For convenience, the 1966a paper will be referred to as paper I, the 1965 paper as paper II, and the 1966b paper as paper III. This appears to be the order in which they were written. Roxburgh suggests that during the stage of a protostar's life when a radiative core begins to form and grow, the central regions of the star become uncoupled from the outer envelope and continue to contract, conserving their angular momentum, while the surface is losing mass through rotational breakup. The contracting core is virtually incompressible and so fission can occur there. In paper I, Roxburgh compares this theory with observations of W Ursae Majoris systems, which he regards as the products of fission. He gives two observational supports for his theory: 1) although fission resulting in a contact system should take place only if the original protostar has a mass in the range 0.8–4.0 $m_\odot$, he finds that the total masses of the 15 W Ursae Majoris systems listed by Kopal & Shapley (1956) are all between 0.74 and 3.8 $m_\odot$; 2) using the same source, he finds that there is a relation between total mass and total orbital angular momentum of W Ursae Majoris systems. In paper II, Roxburgh finds a similar relation between mass and orbital angular momentum, for all close binaries for which spectroscopic data are given in the catalogue compiled by Kopal & Shapley. He claims that this relation is evidence for the fission theory of their origin.

Because this is an important theoretical development, it is necessary to examine its observational support closely. It is, of course, notoriously difficult to determine the masses of W Ursae Majoris systems, both because of the intrinsic width of the spectral lines, and because of the shortness of the orbital periods, compared with the necessary exposure times. The work of Petrie, Andrews & Scarfe (1966) on the possible systematic errors resulting from the blending of broad double lines has already been mentioned. Morris (1960) has estimated that the error from this source may have resulted in the estimated total mass of the W Ursae Majoris system VW Cephei being too small by nearly 50 per cent. Thus the limiting masses of 0.74 and 3.8 $m_\odot$ found for the systems in the catalogue of Kopal & Shapley must be regarded with reserve. Roxburgh could not have known of the system HD 205372, which has also been discussed elsewhere in this review, since its orbital elements were published about the same time as his own papers. The total mass of this system is estimated to be 5.2 $m_\odot$. Too much weight should not be placed on one system with a total mass in excess of his upper limit, since Roxburgh himself has emphasized the approximate nature of his figures. Moreover, the estimated mass of HD 205372 must itself be subject to the same sort of uncertainty as that of W Ursae Majoris systems. If, however, μ^1 Scorpii is regarded as a contact system, as it was by Eggen, its total mass of at least 15 $m_\odot$ (Struve 1940) is considerably in excess of the upper limit for contact systems derived by Roxburgh.

Consider now Roxburgh's second comparison with observation, the relation between total mass and total orbital angular momentum of a binary system. This is presented for contact systems in paper I, and for all close systems in paper II. As Roxburgh himself shows, the total orbital angular momentum of a binary system about its centre of mass, denoted by H, is given by the expression

$$H = \tfrac{1}{4} G^{1/2} M^{3/2} d^{1/2}$$

where M is the total mass of the system, d is the separation of the components, and G is the constant of gravitation. The formula is only strictly correct for circular orbits, and for mass ratios of unity. The corrections needed for departures from these conditions are not important, however, in the ranges of mass ratio and orbital eccentricity commonly encountered in close systems. Apart from these restrictions, the only assumptions made in deriving the above expression are the definition of angular momentum, and Kepler's third law. Now d enters the formula only as its square root, and since most of the eclipsing systems in the catalogue compiled by Kopal & Shapley have separations that fall in a fairly restricted range, it is obvious that, to a first approximation,

$$H \propto M^{3/}$$

so a plot of log H against log M should be a straight line with slope 3/2, and the points in the plot should be scattered about this line, mainly because of the variations in d from system to system, perhaps partly because of deviations from Keplerian motion in some systems, and very slightly because most systems have mass ratios different from unity, or small orbital eccentricities.

Upon examination of the plot of log H against log M in Roxburgh's paper II, it is found that he has drawn a line through the points of slope 5/2. The extra power of M seems to have come from a substitution for the radius in an expression for the angular momentum of the unsplit protostar. The points plotted do not agree well with the line, lying systematically above it in the middle of the range, and below it at high masses. It is, however, possible to satisfy nearly all the points on this plot by two parallel lines of approximate slope 1.43, which is very close to the value of 1.5 predicted by Kepler's laws. The two lines are displaced by about 0.34 measured as the difference of their intercepts on the log H axis. Now the systems given in Kopal & Shapley's catalogue, that were used by Roxburgh, have a range of separation from about 2 to 40 or 50 $r_\odot$. All the contact systems have separations between 2.1 and 3.5 $r_\odot$, and most of the others have separations between 10 and 20 $r_\odot$, only one or two systems having separations outside this range. Thus the mean separation of the components in contact systems is about 3 $r_\odot$, and for components of other systems the mean separation is about 15 $r_\odot$. The ratio of these two numbers is 5, and the logarithm of the square root of 5 is 0.35. Thus one might very well expect that the plot of log H against log M for all these systems would show a tendency to cluster about two parallel lines of slope 1.5, and displaced along the log H axis by about 0.35. It is difficult to tell from Roxburgh's plot which symbols represent which class of system, but since the contact systems in the diagram are, for the most part, the low-mass systems, it may be fairly safely concluded that the two lines about which his points cluster represent the mean relations between orbital angular momentum and mass for contact and noncontact systems. These relations contain no information that is not already in the laws of two-body motion, and, unfortunately, cannot be used as support for the fission hypothesis.

It seems, then, that the orbital angular momentum is not a quantity that will provide much insight into the origin of binary systems, since it is fixed by the properties of motion under the law of gravitation. It must be required of any theory of the origin of binary systems that it should be able to predict more detailed properties of the component stars. For instance, such a theory might be required to predict the empirical period-colour relation found for contact systems by Eggen (1961) and already discussed in this review. It should also satisfy the condition pointed out by Binnendijk (1965) that within a given contact system the mass of each component is proportional to its luminosity. Such predictions can be made only after detailed computations of the structure of the two stars resulting from the fission have been carried out. It is understood that Roxburgh is making such computations. In paper III, Roxburgh does in fact attempt a more detailed comparison of his theory with observation, with particular reference to the eclipsing system KO Aquilae. He discusses the possibility that the two components of this system are still contracting to the main sequence after a comparatively recent fission. The system shows a single spectrum, the secondary being, in Kopal's terminology, an undersize subgiant. This, unfortunately, means that the masses and radii of the components can only be determined by making some assumption. Roxburgh assumed that the primary component is still contracting to the main sequence, and derived exactly the same values for the masses and radii as Kopal obtained with the assumption that the primary component is a main-sequence star (1958). Roxburgh points out that Kopal's assumption leads to too large a radius for the primary component, if it is in fact a main-sequence star. This can also mean that the observations are not good enough for a choice to be made between the two hypotheses. Thus the impasse reached in the discussion of the origin of binary systems presents as much of a challenge to observers as it does to theoreticians.

Concluding Remarks

It is possible to distinguish two phases in the history of the study of double stars. For a long time after Herschel's discovery of physically related double stars, the only sources of quantitative insight into stellar structure were the few parallaxes that indirectly give the absolute luminosities of the stars, and the few measures of stellar masses that could then be made from visual double stars. The discovery of spectroscopic binary stars and the development of astronomical spectroscopy greatly increased the possibilities of mass determination, and in the early years of this century, almost every major observatory engaged in one form or another of double-star observation. It could almost have been said that astrophysics was double-star astronomy. Then, as new instruments and new theoretical insights became available, the astrophysical study of many more diverse topics was possible, and the proportion of effort put into the study of double stars decreased. At the same time, the work of Struve and his school showed that a double-star system was not the simple two-body system it had been assumed to be, and led to the questioning, which is still evident today, of the validity of many of the mass data. The growth of accurate photoelectric photometry also confirmed the complex nature of the close binary systems. These developments tended to make double-star astronomy turn back upon itself, and to seem, perhaps, like a minor portion of the whole astronomical effort, not closely related to the topics of major importance.

Today, we seem to stand at the threshold of a new phase. It is increasingly recognized that double-star systems are a major component of the Galaxy, and worthy of study, not only for their own sakes, but also for the purpose of achieving a greater understanding of the Galaxy as a whole. This attitude can be clearly seen in the work of Abt on the relative frequency of binary systems in different stellar groups. Failure to agree with all his conclusions does not signify a failure to recognize the importance of the problems to which he has addressed himself. Similarly, the renewed interest in the origin of binary systems, on the part of both theoretical and observational workers, indicates the great possibilities of increased understanding of all stars that are inherent in the application of the powerful computational techniques now available to the complex problems of close double stars. It is fortunate that this upsurge of interest in double stars should coincide with the construction of several large telescopes which will be able to reach a much larger sample of close binary systems than has yet been studied. Double-star astronomy has of necessity received only a fairly small share of time on the existing large telescopes, and it is to be hoped that in a few years' time some attempt can be made to decrease the large gap between the number of systems known and the number studied that was bemoaned at the beginning of this review. The resulting new data will help to eliminate, or at least to estimate, the undesirable effects of observational selection. It is also to be hoped that increasingly accurate orbital elements will be obtained, so that the data can be applied with greater confidence.

I am grateful to several colleagues who have read the manuscript of this review, and made helpful criticisms. I am especially grateful to Dr. C. D. Scarfe for many useful discussions.

LITERATURE CITED

Abt, H. A. 1961, *Ap. J. Suppl. 6*, 37
Abt, H. A. 1965, *Ap. J. Suppl. 11*, 429
Abt, H. A., Barnes, R. C., Biggs, E. S., Osmer, P. S. 1965, *Ap. J.*, **142**, 1604
Abt, H. A., Hunter, J. H. 1962, *Ap. J.*, **136**, 381
Abt, H. A., Snowden, M. S. 1964, *Ap. J.*, **139**, 1139
van Albada, T. S. 1966, *Commun. Roy. Obs. Belgium* (In press)
Ambartsumian, V. A. 1937, *Russian Astron. J.*, **14**, 207
Bartolini, C., Mammano, A., Mannino, G., Margoni, R. 1965, *Asiago Contrib. No. 168*
Batten, A. H. 1961, *J. Roy. Astron. Soc. Can.*, **55**, 120
Batten, A. H. 1966, *Commun. Roy. Obs. Belgium* (In press)
Bertaud, Ch. 1959, *J. Obs.*, **42**, 45
Bertaud, Ch. 1960, *ibid.*, **43**, 129
Bertaud, Ch. 1965, *ibid.*, **48**, 211
Binnendijk, L. 1965, *Kl. Veroeffentl. Remeis Sternwarte, Bamberg*, **4**, 36
Blaauw, A., van Albada, T. S. 1966, *Proc. I.A.U. Symp. No. 30* (In press)
Bleksley, A. E. H. 1934, *Nature*, **133**, 613
Chaffee, F. R., Abt, H. A. 1966 (In preparation)
Chandrasekhar, S. 1944, *Ap. J.*, **99**, 54
Eggen, O. J. 1961, *Roy. Obs. Bull. No. 31*
Finsen, W. S. 1962, quoted in *Publ. Astron. Sci. Pacific*, **74**, 5
Fletcher, J. M. 1964 (M.A. thesis, Univ. of Toronto, unpublished)
Gutmann, F. 1965a, *Publ. Dom. Ap. Obs.*, **12**, 361
Gutmann, F. 1965b, *ibid.*, 391
Heard, J. F., Petrie, R. M. 1966, *Proc. I.A.U. Symp. No. 30* (In press)
Huang, S.-S. 1966, *Ann. Ap.*, **29**, 331
Jaschek, C., Jaschek, M. 1957, *Publ. Astron. Soc. Pacific*, **69**, 546
Jaschek, C., Jaschek, M. 1959, *Z. Ap.*, **48**, 263
Jeans, J. H. 1929, *Astronomy and Cosmogony*, 261–80 (1st ed., Cambridge Univ. Press)
Jeffers, H. M., van den Bos, W. H., Greeby, F. M. 1963, *Publ. Lick Obs.*, 21
Jose, P. D. 1951, *Ap. J.*, **114**, 370
Kopal, Z. 1955, *Ann. Ap.*, **18**, 379
Kopal, Z. 1958, *Close Binary Systems*, 497 (Chapman and Hall, London)

Kopal, Z., Shapley, M. B. 1956, *Jodrell Bank Ann.*, **1**, 141
Kraft, R. P. 1963, *Advan. Astron. Ap.*, **2**, 43
Krat, V. 1944, *Russian Astron. J.*, **21**, 20
Kukarkin, B. V., Parenago, P. P., Efremov, Yu. I., Kholopov, B. N. 1958, *Obschij Katalog Peremmenich Zvyozd* (2nd ed., Moscow)
Kurotchkin, N. E., Kukarkin, B. V. 1966, *Soviet Astron.—AJ*, **10**, 64
Moore, J. H., Neubauer, F. J. 1948, *Lick Obs. Bull.*, **20**, 1
Morris, S. C. 1960 (M.A. thesis, Univ. of Toronto, unpublished)
Petrie, R. M. 1960, *Ann. Ap.*, **23**, 744; *Contrib. Dom. Ap. Obs. No. 67*
Petrie, R. M. 1962, in *Astronomical Techniques*, 78 (Hiltner, W. A., Ed., Univ. of Chicago Press); *Contrib. Dom. Ap. Obs. No. 73*
Petrie, R. M., Andrews, D. H., Scarfe, C. D. 1966, *Proc. I.A.U. Symp. No. 30* (In press)
Petrie, R. M., Batten, A. H. 1966, *Trans. I.A.U.*, **12B**, 476; *Contrib. Dom. Ap. Obs. No. 94*
Plavec, M. 1964, *Bull. Astron. Czechoslovakia*, **15**, 156
Roman, N. G., Morgan, W. W. 1950, *Ap. J.*, **111**, 426
Roman, N. G., Morgan, W. W., Eggen, O. J. 1948, *Ap. J.*, **107**, 107
Roxburgh, I. W. 1965, *Nature*, **208**, 65
Roxburgh, I. W. 1966a, *Ap. J.*, **143**, 111
Roxburgh, I. W. 1966b, *Astron. J.*, **71**, 133
Sahade, J. 1958, *Etoiles à Raies d'Emission, Mem. Soc. Roy. Sci. Liège*, **20**, 46
Sahade, J. 1960, in *Stellar Atmospheres*, 466 (Greenstein, J. L., Ed., Univ. of Chicago Press)
Slettebak, A., Howard, R. F. 1955, *Ap. J.*, **121**, 102
Struve, O. 1940, *Festschrift für Elis Strömgren*, 258 (Lundmark, K., Ed., Munksgaard, Copenhagen)
Struve, O. 1951, *Observatory*, **71**, 197
Wilson, O. C. 1966, *Ap. J.*, **144**, 695
Wood, F. B. 1964, *Vistas Astron.*, **5**, 119
Worley, C. E. 1963, *Publ. U. S. Naval Obs.*, **18**, Part 3
Worley, C. E. 1966, *Comm. Roy. Obs. Belgium* (In press)
Yabushita, S. 1966, *Monthly Notices Roy. Astron. Soc.*, **133**, 133

ASTRONOMICAL OPTICS[1]

By I. S. Bowen

Mount Wilson and Palomar Observatories, Carnegie Institution of Washington
California Institute of Technology, Pasadena, California

Geometrical optics is one of the oldest of the sciences, all of its basic laws having been known since the seventeenth century. While spectacular new discoveries cannot be expected, many advances in astronomical optics have been made in the development of new designs to meet the needs of changing programs and to take advantage of new light sensors.

Up to the end of the last century nearly all astronomical observations were made visually, the retina of the eye being the light sensor. The area of the retina possessing high resolution is very small, having a diameter of less than 1 mm. Because of this, a long range of effective focal lengths of the telescope-eye system could be provided by combining an objective of a convenient focal ratio with a battery of small, inexpensive eyepieces. Furthermore, the wavelength range of appreciable sensitivity of the eye is limited to 1000 Å to 1500 Å centered at about 5500 Å. For this narrow wavelength range, satisfactory achromatization can be achieved with lens optics even of moderately large apertures.

Photographic plates with sensitivity sufficient for many astronomical observations became available late in the last century, and because of their obvious advantages over the eye soon replaced it as the receiver. This, however, immediately required a major revision in telescope design. Thus, while the wavelength range of sensitivity of the early plates was not much greater than that of the eye, it was centered in the blue-violet rather than in the yellow-green, where visual telescopes were achromatized. Later, as panchromatic plates were developed, this sensitivity was extended to a range of 3000 Å to 4000 Å, over which no achromatization is feasible with lens optics of large aperture. Furthermore, in order to record faint stars, galaxies, and nebulae in a practical one-night exposure of 4 to 6 hr, it was necessary that the optical system yield a certain minimum intensity of illumination on the plate. For the early plates this required a focal ratio of $F = 5$ or less.

This change of the receiver from the eye to the photographic plate forced an immediate shift from the refractor to the paraboloid reflector, which could provide both the complete achromatism and the necessary focal ratio. The other new factor introduced by the photographic plate was the large field which it could record, since this is limited only by the size of plate that it is practical to handle and process.

The paraboloid reflector of low focal ratio, however, falls far short of providing an adequate field of good definition to take advantage of this

[1] The survey of literature for this review was concluded September 1, 1966.

possibility. Thus the distance L, off axis of a paraboloid at which the comatic image reaches a size C, is given by Equation 1, in which F is the focal ratio

$$L = \frac{16F^2C}{3} \qquad \qquad 1.$$

At a focal ratio of $F = 5$, $L = 133C$, or the diameter of the field over which the comatic images have a size less than 1 sec is less than $4\frac{1}{2}$ min of arc.

OPTICAL DESIGNS FOR WIDE FIELDS

Most of the advances in the optical designs for telescopes in the past few decades have been concerned with the problem of widening the field of good definition. The first attempts, early in the century, were directed toward the design and construction of large-size anastigmats of the same general type as those then being developed for terrestrial photography. While success was achieved in very substantially widening the field of good definition compared to the older doublets, the basic limitation, set by the lack of achromatism of any lens system, still remained. The exact amount of residual chromatic aberration depends on the glasses used in the design. In general, however, the residual chromatic aberration in a lens designed for the photographic range 3900 Å to 4900 Å yields an image diameter of about $60/F$ sec of arc and for the visual range 5200 Å to 6200 Å about $20/F$ sec, in which F is the focal ratio of the system. Since most photographic observations require a focal ratio of $F10$ or less, it is evident that the definition is not adequate for more than very short-focus patrol instruments. Furthermore, a different lens system is required for each wavelength range.

A major advance in enlarging the field of view was made with the development of the Schmidt (1932) camera. This development started from the principle that a sphere with an aperture stop at the center of curvature has no axis, and consequently gives equally good images at all points in the field. However, a sphere does have spherical aberration. While for a given focal ratio this is smaller than the spherical aberration of any of the simpler anastigmats, such as the Cooke triplet and Tessar types, the aberration becomes objectionably large for apertures and focal ratios needed for many astronomical applications. Schmidt's great contribution was the introduction of a thin glass plate figured aspherically to correct all orders of this spherical aberration at the stop at the center of curvature of the mirror. The theory and residual aberrations of the Schmidt camera have been treated extensively by Strömgren (1935), Caratheodory (1940), Bouwers (1946), Linfoot (1949, 1951), and Bowen (1960). Near the axis, the chief aberration arises from the chromatic aberration of the corrector plate, which is normally made of one material—either crown glass or fused silica. Thus, if n is the refractive index of the corrector plate at the wavelength for which it is designed and figured, and n' is its index at some other wavelength, then the diameter W_1 of the image formed by light of the second wavelength is given by Equation 2, f and F being the focal length and focal ratio of the camera:

$$W_1 = \frac{f(n' - n)}{128F^3(n - 1)} \qquad \text{2.}$$

If W_1 is the maximum permissible image size, the lowest value of the focal ratio F that can be used is given by Equation 3,

$$F = \left[\frac{f(n' - n)}{128W_1(n - 1)} \right]^{1/3} \qquad \text{3.}$$

Or expressing the image diameter in seconds of arc s, one may write $W_1 = sf/206265$ and

$$F = \left[\frac{1611(n' - n)}{s(n - 1)} \right]^{1/3} \qquad \text{4.}$$

If, as is customary, the corrector plate is figured to give zero aberration at $\lambda = 4300$ Å, the maximum absolute values of $(n'-1)/(n-1)$ reached for various wavelength ranges are as listed in Table I.

TABLE I

WAVELENGTH RANGES CORRESPONDING TO VALUES OF $(n'-n)/(n-1)$

$\pm \dfrac{n'-n}{n-1}$	Fused silica	Crown glass
0.01	3850–4960 Å	3920–4860 Å
0.02	3520–6040 Å	3610–5740 Å
0.03	3260–8000 Å	3390–7320 Å
0.04	3030–11600 Å	—

While the mirror has no axis, the corrector plate does, and small aberrations are introduced when the light passes the corrector plate at an angle φ with its normal. The maximum value of φ before the diameter of the image caused by this aberration exceeds a permissible value W_2 is given by Equation 5,

$$\sin \phi = \left[\frac{24F^3W_2}{f} \right]^{1/2} \qquad \text{5.}$$

or, if W_2 is set equal to the on-axis aberration W_1 given by Equation 2,

$$\sin \phi = \left[\frac{3(n' - n)}{16(n - 1)} \right]^{1/2} \qquad \text{6.}$$

One of the chief disadvantages of the Schmidt camera, especially for long focal lengths, is that the distance between the mirror and the corrector plate is equal to twice the focal length and thus a tube length equal to slightly more than twice the focal length is required. In order to avoid this very long tube length, a number of modifications of the Schmidt camera have been investigated by Baker (1940b), Burch (1942), Slevogt (1942), Linfoot (1944, 1955), and Wayman (1950). Most of these are of the two-mirror Cassegrain type. In nearly all cases the tube length is little, if any, larger than the focal length of the system. On the other hand, the aberrations are, in general, substantially larger than those of the standard Schmidt of the same focal length and focal ratio. For example, the concentric-mirror Cassegrain Schmidt requires a focal ratio nearly twice that of a standard Schmidt to satisfy the same criteria as to the maximum permissible image size, and, therefore, requires about the same tube length as a standard Schmidt of the same aperture.

Another modification of the Schmidt camera was introduced independently by Maksutov (1944) and by Bouwers (1946). In this modification the aspheric corrector plate of the Schmidt is replaced by a strongly curved meniscus. This has the advantage that all surfaces are spherical, thus the difficulties of figuring an aspheric surface are avoided. Chromatic aberrations are about the same as in a standard Schmidt, but with the spherical surfaces full correction can be attained only for third-order aberrations of the mirror. Furthermore, the meniscus becomes objectionably thick for the larger apertures. Because of the great advantages of spherical surfaces for mass production methods, the Maksutov design has found its widest use in small, commercially produced instruments.

Another approach to the problem of enlarging the field of a reflector was made by Schwarzschild (1905) and by Chrétien (1922). Both used a two-mirror system—Schwarzschild investigating a Gregorian system and Chrétien a Cassegrainian. Both dropped the condition that the primary alone should form an image without spherical aberration and, with the added degree of freedom thus obtained, explored variations in the shape of the surfaces to achieve optimum correction of the aberrations at the focus of the two-mirror system. In particular, they were able to eliminate coma, which is the major aberration present in other large reflecting systems. From a practical standpoint, the Cassegrain system investigated by Chrétien is the better solution, since for a given focal length the tube length is shorter and the image can be formed at a convenient position behind the primary rather than midway between the two mirrors, as is necessary with the Schwarzschild solution. The following discussion will therefore be limited to the second system, which is now usually designated as the Ritchey-Chrétien system since Ritchey was the first one to carry out the difficult figuring of the aspheric surfaces demanded by the theory. The theory has also been discussed by Theissing & Zinke (1948) and by Malacara (1965).

The coma of a two-mirror Cassegrain system may be expressed as follows:

Assume that the primary is a surface of revolution, the radius of curvature of whose intersection with a plane through optic axis is given by Equation 7 for a point at a distance y from the axis,

$$R = 2f \left(1 + \frac{\alpha y^2}{f^2}\right) \qquad\qquad 7.$$

The focal length of the primary is f, and α is a constant. The secondary is given the figure that eliminates spherical aberration at the Cassegrain focus. The maximum dimension C of the comatic image at a distance L from the axis of such a Cassegrain system is given to a first approximation by Equation 8,

$$\frac{C}{L} = \frac{3}{16m^2F^2} + \left(\frac{3}{8} - \alpha\right)\left(\frac{m+1-K}{K}\right)\frac{1}{4F^2} \qquad 8.$$

The focal ratio of the primary is F and of the Cassegrain, mF. The distance between the focus of the primary and of the Cassegrain divided by the focal length of the primary is K. The value of K is normally slightly larger than 1, except in the case of the coudé focus where it may be as large as 2 or 3.

For a spherical primary, α is obviously 0. For a paraboloid primary, α is $\frac{3}{8}$ and the last term of Equation 8 vanishes. In this case, Equation 8 yields the usual relationship that the coma of a Cassegrain with a paraboloid primary is the same as that of a paraboloid alone with the same focal ratio.

The condition for a Ritchey-Chrétien system with zero coma is then obtained by setting C equal to 0 in Equation 8 and solving for α. This yields the value for α in Equation 9,

$$\alpha = \frac{3}{8}\left[1 + \frac{2K}{m^2(m+1-K)}\right] \qquad 9.$$

Since in most Cassegrain telescopes m is between 2 and 5 and K is between 1 and 1.2, it is evident that α differs by only a few per cent from its value for a parabola, and, likewise, the figure of the primary departs from a sphere by only a few per cent more than the figure of a parabola does.

Like the standard Cassegrain, the Ritchey-Chrétien has field curvature and astigmatism which, since spherical aberration and coma have been eliminated, now become the dominant aberrations. The radii of curvature of the two focal surfaces R_{tang} and R_{sag} for the Ritchey-Chrétien are given by Equations 10 and 11:

$$\frac{1}{R_{\text{tang}}} = \left[\left(\frac{m^2-1}{mK} - 1\right) + \frac{3}{2}\left(\frac{2m^2+m+K-1}{m^2K}\right)\right]\frac{1}{f} \qquad 10.$$

$$\frac{1}{R_{\text{sag}}} = \left[\left(\frac{m^2-1}{mK} - 1\right) + \frac{1}{2}\left(\frac{2m^2+m+K-1}{m^2K}\right)\right]\frac{1}{f} \qquad 11.$$

Table II lists the field diameters out to the point where the image reaches a maximum dimension of $\frac{1}{2}$ sec because of these aberrations for the case of a flat plate and for the case of a plate bent to a surface midway between the two astigmatic surfaces. Since these fields are achieved with mirrors alone, there is no limitation on the wavelength range over which the system may be used.

To increase the size of the field, Schulte (1966) has designed a corrector which consists of two thin fused-silica lenses, one of which has an aspheric surface, and which is placed a short distance in front of the focus. Starting with a mirror system with characteristics similar to the lowest case in Table II, this yields images of $\frac{1}{2}$ sec or less over a field 90' in diameter on a flat plate. Similarly, Wynne (1965) has a design that yields images of less than 0.2" over a 30' flat field for a Ritchey-Chrétien with specification similar to the middle case in Table II. Other designs have been developed by Köhler (1966) and Rosin (1966).

One of the disadvantages of the Ritchey-Chrétien design is that it usually results in an increase in the coma at focal ratios other than the one for which C is set equal to zero. Thus, if m_0 and K_0 are the constants of the focus that

TABLE II

DIAMETER OF FIELD OF RITCHEY-CHRÉTIEN OVER WHICH IMAGE SIZE $\leq \frac{1}{2}$ SECOND

	Flat plate	Curved plate
$K=1.1$ $F=2$ $m=4.5$	12.5 min	21.3 min
$F=3$ $m=3$	17.3 min	25.3 min
$F=4.5$ $m=2$	23.3 min	30.0 min

satisfies the Ritchey-Chrétien condition, at other values of m and K, the coma is given by Equation 12,

$$\frac{C}{L} = \left\{ \frac{1}{m^2} - \frac{(m+1-K)K_0}{m_0^2(m_0+1-K_0)K} \right\} \frac{3}{16F^2} \qquad 12.$$

Another method for increasing the field of a reflector was introduced by Ross (1935). This corrector consisted of a pair of lenses designed to correct for the coma of a paraboloid when placed a short distance in front of the plate at the prime or Newtonian focus. Used first with the 100-inch telescope at a focal ratio of $F5$, the Ross corrector gave a good correction for coma and very substantially increased the field of good definition. However, when a similar system was constructed for the 200-inch, which has a focal ratio of $F3.3$, the results were only partially satisfactory until two additional lenses were introduced. Even with this four-element lens the field of good definition is limited to a diameter of about 15'.

The basic difficulty is that coma and spherical aberration are so closely related that it is almost impossible to correct one without disturbing the correction of the other. To avoid this, Baker (1953), in his design of a coma-corrected system, introduced a full-aperture aspherical plate in front of the paraboloid. The spherical aberration introduced by this plate and the coma were then corrected together by means of a small two-element lens placed a short distance in front of the focus. Similarly, Rosin (1961) designed a large-field reflector in which spherical aberration was introduced by a hyperboloid primary and then corrected along with the coma by a simple lens system near the focus.

Since the primary of a Ritchey-Chrétien departs from a paraboloid in order to eliminate coma at the Cassegrain focus, it also simplifies the design of a field corrector for the prime focus. A number of designs for such a prime-focus corrector for use with a Ritchey-Chrétien system have been developed by Köhler (1966) following an earlier design by Meinel (1953), and by Baranne (1966), and Wynne (1965). In general, these designs yield well-corrected fields 30' to 60' in diameter.

TELESCOPE DESIGN

In this section consideration will be given to the application of the optical systems discussed in the preceding section to the design of telescopes for the solution of present-day observational programs.

Wide-field telescopes for survey or patrol programs.—For very wide fields of over 10°, lens optics in general provide the best solution. Because of the limitations imposed by chromatic aberration, critical definition cannot be obtained with apertures over 10 cm for blue-corrected lenses, or 20 to 30 cm for those visually corrected. A number of lenses have recently been developed for aerial photography in the yellow-red with apertures from 5 to 15 cm that yield resolutions of 35 to 50 lines per millimeter over a 30° field. Baker has designed both blue- and red-corrected lenses for the Harvard patrol cameras that yield 20- to 30-μ images over a 40°×32° field (Ingrao 1964).

For moderately wide fields of up to 10° in diameter and medium focal lengths of 1 to 5 m, the Schmidt camera most nearly satisfies requirements. From Equations 3 to 6 of the preceding section it is evident that the lower limit on the focal ratio F and the limit on the field diameter (2φ) are fixed by the wavelength range to be covered, which determines $(n'-n)/(n-1)$, and the permissible size W of the aberration image. In this focal-length range the diameter of a star image on the plate depends on the combined effect of the plate resolution p, the linear diameter of the seeing image, and W. For a fast plate, p lies between 0.015 to 0.020 mm, and at the better locations "good average seeing" may be taken as 1.0 to 1.5 sec of arc, or the linear image diameter as $6\times10^{-6}f$, in which f is the focal length. Since the apparent light distributions caused by plate resolution and by seeing approximate an error function, the diameter of their combined effect may be written as $q=(p^2+3.6\times10^{-11}f^2)^{1/2}$. Both the chromatic and the off-axis aberrations of a Schmidt camera result in a light distribution that is substantially more concentrated toward the center than an error function. Thus, for the off-axis aberration, 80 per cent of the light is concentrated in a circle whose diameter is $W_2/4$. However, if these aberrations are not to increase the size of the final image by more than 20 per cent, W should not be larger than q. On the other hand, if high speed and a large field are essential for certain programs, W_1 can be permitted to reach $1.5q$, and W_2 as much as $2q$ without increasing image sizes by more than 30 to 50 per cent of q.

For example, the 48-inch telescope on Palomar Mountain was designed for extreme speed and a large field to permit the rapid survey of large areas. Indeed, if either the speed or field had been appreciably reduced, it would have been impossible to carry through such programs as the National Geographic Society-Palomar Observatory Sky Survey in a practical number of years. In order to hold $(n'-n)/(n-1)$ to a value of 0.02, the corrector plate was made of a glass that is nearly opaque below 3600 Å. For a focal length of 3000 mm the diameter of a perfectly imaged star under average seeing conditions is $q=[(0.018)^2+3.6\times10^{11}\times9\times10^6]^{1/2}=0.025$ mm. On the other hand, for the focal ratio of $F2.5$ and the above index range, W_1 equals 0.030 mm. Similarly the W_2 caused by off-axis aberrations at the corner of the plate is 0.049 mm. Under these conditions the image sizes of the fainter stars at all points of the plate are observed to be 0.030 to 0.035 mm, which for most purposes is satisfactorily close to the ideal value of $q=0.025$ mm.

Too large a focal ratio may also impose a limit on the size of the field of a Schmidt. Thus, if the diameter of the field in degrees exceeds about $30/F$, the obscuration of the beam by the plate holder becomes excessive.

The focal surface of a Schmidt is curved with a radius equal to the focal length. Experience has shown that it is feasible to bend thin glass plates covering an area of over $6°\times6°$ to radii of 2 m or more. For shorter focal lengths a simple field flattener may be used.

In cases in which the long tube length of the standard Schmidt is objectionable, one of the Cassegrain Schmidts may be used. However, because of the larger aberrations it is necessary to use larger focal ratios and smaller fields than with a regular Schmidt. Because of these limitations, no modified Schmidt has been constructed of aperture over 85 cm.

Large telescopes of over 2 m of aperture.—The high cost of these large instruments ordinarily limits the number available at any one observatory to one or two. It is therefore very desirable that such an instrument be designed to handle all types of programs that require a large telescope, including those that can be carried out in moonlight as well as those for which a dark sky is necessary. Furthermore, for most types of observation, a doubling of the image size by aberrations or other causes reduces the effectiveness of the instrument to that of one of half the aperture. In order that the telescope be able to take full advantage of occasional periods of exceptionally fine seeing, it is very desirable that aberrations be held to so low a value that little, if any, degradation of the image occurs even under these conditions. It is therefore customary to specify that the optics of the telescope should concentrate nearly all of the light in a circle not over $\frac{1}{2}$ sec in diameter.

Two general types of observations are carried out by a large telescope: direct photography of appreciable areas of the sky and detailed investigations of individual objects. The optical requirements for these are as follows:

(a) Direct photography: In order to obtain the most effective record of faint stars, nebulae, galaxies, or other objects, it is necessary to give the plate an exposure that brings the density of the background up to a value between 0.5 and 1.0. Furthermore, the limiting magnitude of stars recorded increases by 1 mag for each increase in this exposure time by a factor of 6.3. If any telescope is to be able to reach the ultimate limiting magnitude for which its aperture is capable, it should be provided with a focal ratio that will permit it to just attain this sky background density in an exposure of 4 to 6 hr, this being the longest that can be given in one night without going to excessive zenith distances. With modern fast plates such as Eastman 103aO, this condition requires a focal ratio of $F8$ to $F10$. If substantially slower plates are developed with the quantum efficiency equal to that of 103aO plates, the same limiting magnitude can also be reached at other lower focal ratios.

Other programs make it very desirable to have another focus with as low a focal ratio as can be achieved with a wide field of good definition. These include survey programs in which it is necessary to cover a large area of the sky rapidly, even at the expense of losing some limiting magnitude, and programs requiring the use of interference or other very dense filters or of infrared or other slow plates. For both foci it is obviously of the highest importance to have the largest possible field of good definition.

(b) Observations of individual objects: A major part of the work of any large telescope consists of the detailed observation of individual objects with photometers, spectrographs of a wide range of dispersions, spectrum scanners, and infrared receivers. Since some of these measurements may be at wavelengths to which glass and quartz are opaque, it is important that no transmitting material be required in the light path to the focus at which these observations are made.

As the receiver and, therefore, the image of the object under study are placed on the optic axis of the telescope, the field of good definition required for the actual observation is very small. However, present-day photometers, low-dispersion spectrographs, and spectrum scanners are so sensitive that many of the most important observations are made of objects too faint to be seen visually with the telescope used. Every such observation must, therefore, be preceded by a photograph of the area on which the exact distance of the object from two or more visible stars is measured. The receiver is then carefully offset from these stars, one of which is used as a guide star throughout the observation. Both the visibility of these stars and the accuracy with which the offsets can be measured are greatly reduced if the stars are enlarged by aberrations. Since in regions away from the Milky Way the nearest stars of sufficient brightness (mag 14 or 15) for guiding are at an average distance of 5 to 10 min of arc, a field of good definition at least 20′ in diameter is essential. Furthermore, the offsetting operation is greatly simplified if the plates used for measuring the offset distances are taken with the identical optical system that will be used for the final observation, since this eliminates many uncertainties as to the plate scale and the distortions of the field.

For observations of individual objects, a moderately large focal ratio of $F8$, or larger, is preferred by the majority of observers, as this permits the use of simple lenses or mirrors of longer focal length for reimaging the light in the auxiliary equipment. Furthermore, at the larger focal ratios the scale of the star field around the object is greater, and less linear precision is required in making the offsets.

Finally, for some observations, such as high-dispersion spectroscopy, the auxiliary equipment is too large to be mounted on the telescope, thus requiring a coudé focus located at a fixed point on the polar axis. As the length of the optical path to this focus is substantially greater than to the Cassegrain focus, a larger focal ratio of about $F30$ is normally adopted to avoid an objectionably large secondary mirror. Objects observed at this focus are bright enough for direct identification and guiding on the object; the field of good definition may, therefore, be small.

To summarize, these programs seem to require foci having focal ratios of approximately $F3$, $F9$, and $F30$. The first two should have critical definition over the largest possible field. Since observations of faint individual objects will be made at the intermediate focal ratio, at least a moderately large field should be available without the use of any correcting lens or other transmitting material.

Table III indicates how the various optical designs discussed in the first section fulfill these requirements. The optical systems listed are 1) a telescope with a paraboloid primary, 2) a telescope designed to satisfy the Ritchey-Chrétien condition at the $F9$ Cassegrain focus, 3) a Schmidt camera to which hyperboloids are added to give the $F9$ and $F30$ focal ratios, and 4) a Schmidt camera from which the corrector plate is removed and aspheric secondaries are added to the spherical primary to eliminate spherical aberration at the longer focal ratios. The upper unbracketed figure is the diameter of the field in minutes of arc over which coma and astigmatism enlarge the image by less than $\frac{1}{2}$ sec when the mirror system alone or, as in case 3), when the mirror and corrector plate are used. The lower bracketed figure gives the corresponding field diameter as enlarged by one of the corrector lenses discussed in the preceding section.

From the Table it is evident that the requirements for a large telescope summarized above are most satisfactorily fulfilled by the Ritchey-Chrétien system. Thus it gives a larger field for direct photography at the prime focus than does the paraboloid. At the Cassegrain focus the field is larger than that of the paraboloid both with and without a corrector lens. Furthermore, outside the fields of critical definition listed, the aberrations of the Ritchey-Chrétien are of a symmetrical type, which is much less objectionable than the coma of the paraboloid. The coudé is the only focus at which the paraboloid gives the larger field, but the field of the Ritchey-Chrétien is more than adequate for the observations carried out there.

The last two columns indicate the great difficulties encountered in attempting to adapt a Schmidt camera to types of observations other than prime-focus photography. Furthermore, while the Schmidt provides by far the largest field at the prime focus, it should be noted that in order to achieve half-second images at any point in the field with an $F3$ Schmidt it is necessary to limit the range of $(n'-n)/(n-1)$ to ± 0.0085, or the wavelength range to from 3920 Å to 4850 Å, with a fused-silica corrector plate. If this wavelength

TABLE III

DIAMETER OF FIELD WITH IMAGE DIAMETERS $\leq \frac{1}{2}$ SECOND

Focus	(1) Paraboloid	(2) Ritchey-Chrétien	(3) Schmidt plus hyperboloid	(4) Sphere plus aspheric
Prime $F=3$	0.8 min (30 min)	none (60 min)	275 min*	none
Cassegrain $F=3$ $m=3$ $K=1.1$	7.2 min	25 min (60–90 min)	0.9 min	0.56 min
Coudé $F=3$ $m=10$ $K=2$	80 min	4.5 min	0.8 min	0.35 min

* For $(n'-n)/(n-1)=0.0085$.

range is extended to the limits that are very desirable in a large telescope, that is, 3000 Å to 11,000 Å, it is necessary to increase the focal ratio to about $F5$ to attain half-second images. This requires a tube length at least ten times the aperture, and makes the cost of a dome to house a Schmidt telescope prohibitive for apertures substantially larger than the Tautenberg 53-inch or the Palomar 48-inch telescope.

SPECIAL-PURPOSE TELESCOPES

ASTROMETRIC INSTRUMENTS

Because of their great focal length, uniform definition over a moderate field, and stability, large refractors have been the chief instruments used for past studies of parallax and precise proper motions. With the increasing interest in the distances and motions of white dwarfs and other intrinsically very faint stars, it has become necessary to use larger apertures. Because of the chromatic limitations of the refractor, this has forced a shift to reflection optics.

A 60-inch reflector for astrometric observations has recently been constructed at Flagstaff, Arizona, by the Naval Observatory (Strand 1962, 1963). To avoid excessive coma, the primary is given a focal ratio of $F10$. The tube length is reduced to about one-half the focal length by a secondary mirror that reflects the image back through a hole in the primary. The secondary is a flat mirror, thus any variable distortions of the field that might be introduced by the changes in collimation of the mirrors of a conventional Cassegrain are eliminated.

Solar Telescopes

Because of the great surface brightness of the Sun, telescopes for its study are normally designed with very large focal ratios in the $F40$ to $F150$ range. Aberrations are therefore negligible, and a spherical mirror or a single achromat in its range of achromatization gives diffraction or seeing-limited images over as wide a field as the diameter of the Sun. Because of the very great lengths of these instruments, it is not practical to point the telescope at the Sun. The instrument is therefore fixed and one or two moving mirrors are used to reflect the Sun's light into the telescope.

The most serious limitation on the definition achieved by a solar telescope is the disturbance to the "seeing" caused by the heating of all parts of the telescope and its supports by the sunlight falling on them. Two procedures are being tried in an effort to reduce the disturbances to the seeing both in and near the telescope.

At the McMath Solar Telescope of the Kitt Peak Observatory, the tower which supports and houses the telescope is sheathed in copper in which tubes carrying liquid coolant are imbedded. All surfaces of the tower are thereby maintained at a temperature slightly below ambient air temperature. This insures that air drainage is down and away from the incoming light beam (Pierce 1964).

Plans for the solar telescope of the Sacramento Solar Observatory contemplate the further step of evacuating the air from the internal light path, starting at a point just before the beam strikes the first coelostat mirror (Dunn 1964).

Infrared Telescopes

Because of the relatively large size of most infrared receivers, the tolerance on the definition of a telescope for infrared observations can, in some cases, be substantially relaxed. Since all objects at ambient temperature radiate strongly in the 10-μ-wavelength region, the signal-to-noise ratio can be reduced by the use of a very low focal-ratio radiation collector. In view of these special requirements, a number of telescopes designed especially for infrared studies have been constructed (Zirkind 1965; Kuiper 1964; Neugebauer, Martz & Leighton 1965).

In the Neugebauer, Martz & Leighton instrument, the surface of the mirror with a focal ratio of $F1$ was formed by spinning, at a very uniform rate, a shell containing a plastic during its setting period. After aluminization, the paraboloid thus formed concentrates the light of a star in an area about 2 min in diameter.

OBJECTIVE-PRISM TELESCOPE FOR RADIAL-VELOCITY MEASUREMENTS

In order to expedite the process of measuring the radial velocities of large numbers of stars, Fehrenbach (1947) has developed an objective-prism telescope without the field distortions of ordinary instruments of this type. The original instrument located at the Haute-Provence Observatory is a refractor of 39-cm aperture and 399.4-cm focal length and covers a field of $2° \times 2°$. To avoid the distortions introduced by the conventional objective prism, Fehrenbach uses a double prism consisting of two opposed prisms of exactly the same angle. The prisms are made of two kinds of glass with substantially different dispersive powers but with refractive indices that are identical at $\lambda = 4210$ Å. Two or more exposures are taken of each field, the prism being rotated by 180° between exposures.

ASTRONOMICAL SPECTROGRAPHS

In addition to the telescope, other astronomical instruments, and in particular the spectrographs, have been greatly improved by recent advances in optical design.

In spectroscopic observations the light of a star is focused by the telescope at the slit of the spectrograph. After passing the slit, the light is rendered parallel by a collimator of the same focal ratio as the telescope. The light is then dispersed by a prism or grating and brought to a focus by the camera optics. It can be shown that, except for small second-order effects, the telescope-spectrograph system has an effective focal length equal to the telescope aperture D, times the focal ratio F, of the spectrograph camera. Consequently, if the slit is removed, a monochromatic star having an average seeing disk with a diameter of 1–1.5″ will form an image at the plate with a linear diameter $(1-1.5) \ FD/206265 \approx 6 \times 10^{-6} FD$. If the star image is larger than the plate resolution ($p = 0.018$ for modern fast plates), loss of resolving power will result unless a slit is introduced to narrow the image to this value. For a spectrograph to avoid loss of light at the slit, it is therefore necessary that $6 \times 10^{-6} FD$ should not exceed $p = 0.018$ mm, or that

$$F \leq 3000/D \qquad 13.$$

in which D is in millimeters (Bowen 1952, 1962). Thus for a telescope of 1-m aperture, a spectrograph-camera focal ratio of $F3$ will permit operation with no loss of light at the slit under the conditions assumed. On the other hand.

the focal ratio of a spectrograph camera for use with a 200-inch or 5-m telescope must not exceed $F0.6$ if full efficiency is to be achieved.

The steady growth of telescope apertures during the present century has required the development of faster and faster cameras in order to maintain the full efficiency of the larger instruments. Fortunately, the Schmidt camera has made possible great advances in camera speed while retaining critical definition. For example, Equation 3 shows that, for a focal length of 80 mm, critical definition is possible at $F1$ over a range of $(n'-n)/(n-1)$ of ±0.03. Furthermore, the corrector plate can be figured to give optimum definition over the range from 3000 Å to 5000 Å. The added correction required for good definition in the red can be supplied by the introduction of a plane-parallel plate of the appropriate thickness in the converging beam in front of the photographic plate (Bowen 1962). This plate may also serve as the filter required to eliminate high orders of shorter wavelength.

Even lower focal ratios in the range from $F=0.47$ to $F1$ have been achieved with the solid-block or thick-mirror Schmidt cameras (Hendrix 1939, Hendrix & Christie 1939, Baker 1940a) and the aplanatic-sphere Schmidt (Bowen 1952, 1960). With the aid of these modified Schmidts, full efficiency of the 200-inch has been obtained for dispersions from 39 Å/mm to 800 Å/mm.

The last few years have seen the development of the image-intensifier tubes (see references at end of Literature Cited). Present tubes have a quantum effciency of 10 to 20 per cent, compared to about 1 per cent for the fastest photographic plates in the blue and much less than this at longer wavelengths. The tubes have a resolving power comparable to that of photographic plates over an area up to 3 or 4 cm in diameter.

In all types of tubes a spectrum, or other image, is focused on a cathode which is deposited on the inside of a window at one end of the tube. The electrons ejected are speeded up by a potential difference of a few tens of thousands of volts and are focused by suitable electric or magnetic fields. In the Lallamand type of tube, the electrons with their greatly increased energy are recorded on a photographic plate placed in the vacuum at their focus. In the Magee type, the electrons come out of the vacuum through a very thin mica window and are recorded on a photographic emulsion in contact with the outside of the window. The electrons in the Carnegie RCA tube impinge on a sensitive phosphor which glows with a much-intensified image. Additional intensification can be obtained with this tube by the introduction of intermediate membranes having a phosphor on one side and a cathode on the other: the electrons are speeded up and refocused in each stage. With the RCA tube the final image on the phosphor must be reimaged optically on a photographic plate.

In general, these image tubes have a diameter two or three times that of their field of good definition. Moreover, several types must be surrounded by magnets or solenoids to provide the focusing fields, and by magnetic shields or other devices to eliminate stray magnetic fields from the surroundings. These tubes are, therefore, much too large to be placed on the axis of a Schmidt camera in the center of the beam between the corrector plate and the mirror. New fast optical systems are required to replace the Schmidt camera of the spectrograph, and also in the case of the RCA tube to reimage the phosphor on a photographic plate.

Camera Optics for Imaging Spectra on the Cathode of Image-Intensifier Tubes

A concentric mirror system has been developed for microscope objectives by a number of investigators and has been investigated in detail by Burch (1947). Figure 1 shows the adaptation of this design as a spectrograph camera.

If r is the radius of the concave mirror and r/a that of the convex mirror, the focal length f is $r/2 \, (a-1)$, which is also the distance of the focal surface from the common center of curvature. A simple analysis shows that the third-order spherical aberration passes through zero at $a = (3 + \sqrt{5})/2 = 2.618$. By making the ratio a slightly less than this, a small amount of negative

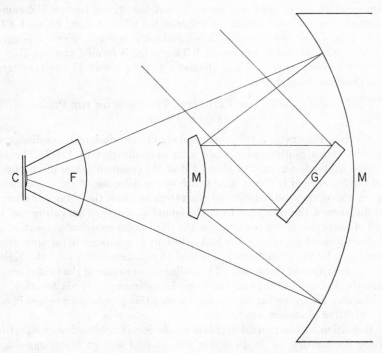

Fig. 1. Fast spectrograph camera for use with image-intensifier tube. G = grating, M = mirrors, F = field flattener, C = cathode of tube.

third-order spherical aberration can be introduced to compensate in part for the fifth- and higher-order aberrations. For example, at a focal ratio of $F1.1$, optimum compensation is obtained if $a = 2.56$. In this case all of the light falls in a circle whose diameter is $7 \times 10^{-5} f$, f being the focal length.

Before reaching the cathode the light must pass through the window, which, since it has to withstand atmospheric pressure without appreciable flexure, is ≥ 2 mm thick. This introduces additional spherical aberration which may be compensated by reducing a by an additional amount of 1.3 T/f, in which T is the thickness of the window.

Since the two mirrors are concentric, the system has no axis, but the focal surface is curved with a radius equal to the focal length. For fields of a very few degrees, flattening can be achieved with a thin plano-concave lens in contact with the window. For larger fields up to 12° to 15° in diameter, the field may be flattened with a thick fused-silica lens of the type shown in the Figure. This yields critical definition for the monochromatic images of a spectrograph and reduces the focal ratio of the system to about 0.8 of that of the mirrors alone.

The chief disadvantage of this two-mirror system is the size of the concave mirror, which must have a diameter about four times that of the camera aperture. For use with smaller telescopes for which a focal ratio of $F2$ to $F2.5$ gives acceptable efficiency, a solid-block concentric-mirror Cassegrain Schmidt of fused quartz can be used. The image is formed at some distance behind the main mirror, whose diameter is only about 25 per cent larger than the camera aperture.

Optical System for Reimaging Phosphor on the Photographic Plate

Figure 2 illustrates a concentric-mirror system designed to reimage the phosphor on a photographic plate at a magnification of 1:1.2. The field covered subtends an angle of about 15° at the common center of curvature, and the linear field is about one twelfth of the diameter of the large mirror. By an appropriate adjustment of the ratios between the three parameters, the distance of the phosphor from the common center of curvature and the radii of curvature of the two mirrors, the third-order spherical aberration can be given a small negative value just sufficient to compensate for aberrations introduced by the phosphor window and to reduce substantially the higher-order aberrations of the system. The radius of curvature of the field is several times the distance of the phosphor from the common center of curvature, and satisfactory flattening can be obtained with a thin plano-concave lens in contact with the phosphor window.

Definition at least equal to plate resolution is achieved with apertures having focal ratios on the phosphor side of $F1.1$ and on the image side of $F1.32$, or, allowing for the rather large obscuration by the convex mirror, the effective focal ratio on the phosphor side is $F1.4$. A number of reimaging

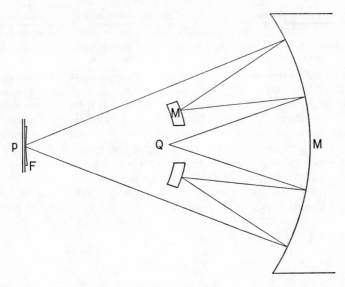

FIG. 2. Camera for reimaging phosphor of image-intensifier tube. $P =$ phosphor,
$F =$ field flattener, $M =$ mirrors, $Q =$ focus at photographic plate.

lenses have also been developed commercially. However, most of these do not
yield the required definition at sufficiently low focal ratios.

SPECTRUM SCANNERS

The solution of many spectroscopic problems has required greater accuracy in intensity measurements than can be achieved with a photographic
plate and has led to the use of photomultiplier tubes for the measurement of
the energy in selected sections of a spectrum. In one technique, a photomultiplier tube with a slit in front of it to isolate a definite wavelength range
is slowly moved along the spectrum and the current recorded as a function of
wavelength of the center of the range. In another procedure, the slit is allowed to remain centered at certain discrete wavelengths while the intensity
is measured, often by photon-counting techniques. Such procedures have
been used both with low-dispersion spectrographs to outline the energy distribution of an object over a broad wavelength range and with spectrographs
of the highest dispersion to study the contour of an individual line (Code &
Liller 1962).

Unlike the photographic plate, the efficiency of the photocathode does not
depend on having a certain minimum intensity of illumination on its surface.
The optics of a spectrum scanner can therefore be designed with any convenient focal ratio, usually in the $F5$ to $F16$ range. Furthermore, the use of
an image slicer (Bowen 1938) to make all of the light of the object pass a

narrow slit does not introduce the difficulties encountered in its use on photographic spectrographs. Its use has the further advantage that, since it causes all of the light of the object to pass the slit, fluctuations induced by changes in seeing are reduced, as compared to those encountered with a slit.

In order to eliminate the effects caused by residual variations in seeing or atmospheric transparency, it is customary to place one or more photomultiplier tubes behind nearby fixed sections of the spectrum and to record the ratio of the currents from the moving to the fixed phototubes. For very faint objects it is also necessary to correct for the intensity of the sky background by running parallel measurements on a nearby region of the sky.

Finally, to shorten the time required for a given investigation, 30 or 40 photomultipliers are placed behind separate sections of the spectrum to permit simultaneous observations of all sections. Because of the high rate of accumulation of data, it is necessary to provide for the automatic recording of the data accumulated on punched cards that can then be given directly to a large computer for processing.

NEW OPTICAL MATERIALS

Large Mirrors

The development of methods for the manufacture of large blanks from new materials of superior properties has led to a great improvement in the performance of large mirrors. Table IV lists the pertinent properties of some of these new materials in comparison with those of materials formerly used. One of the most serious problems encountered with large mirrors is that of maintaining a good figure during and after large changes in the ambient temperature. For example, the 100-inch mirror on Mount Wilson is made of plate glass and has a thickness of 13″. Experience shows that its figure is so disturbed as to cause a noticeable deterioration of the image for two to three nights after the passage of a cold front. Since the time to reach equilibrium varies as the square of the thickness, it is evident that a solid 150-inch or 200-inch mirror with a proportionally greater thickness would rarely reach equilibrium between successive passages of cold fronts.

Two properties of the material are important in determining the effect of the temperature changes. First and most important is the coefficient of expansion, since the disturbance to the figure caused by thermal gradients in the mirror is directly proportional to this. Second is the ratio of the thermal conductivity to the heat capacity per unit volume (specific heat times density), since this is a measure of the rate at which thermal equilibrium is reached after a change in ambient temperature. From the table it is evident that fused silica and the new product, Cervit, produced by Illinois Owens Glass Company, and a similar product, Pyrocyram, made by Corning Glass Company, are very superior to the plate glass of the older telescopes, and even to Pyrex. Cervit and Pyrocyram have been developed so recently that the question of their homogeneity and stability over a period of years has not yet been adequately tested in very large blanks.

While the metals have a much higher coefficient of expansion, they also come to equilibrium much more rapidly. This high thermal conductivity is of special importance for solar mirrors. Thus the solar radiation absorbed by the aluminum coat in the course of a day's observations causes a glass or quartz mirror a few inches thick to build up a temperature difference of

TABLE IV

PROPERTIES OF MIRROR MATERIALS

	T Thermal expansion $\Delta l/l$ per °C $\times 10^7$	c Thermal conductivity $\dfrac{cal}{cm, sec, °C}$	s Specific heat $\dfrac{cal}{g, °C}$	ρ Density $\dfrac{g}{cm^3}$	$\dfrac{c}{s \times \rho}$	E Young's modulus $\dfrac{dyne \times 10^{-11}}{cm^2 \times \Delta l/l}$	ρ/E $\times 10^{12}$
Plate glass	80–90	0.0022	0.21	2.5	0.0042	7.0	3.6
Pyrex	32	.0027	.18	2.23	.0067	6.3	3.5
Fused silica	5.6	.0032	.17	2.20	.0086	7.3	3.0
Cervit	2.5	—	.196	2.51	—	9.1	2.8
Metals							
Aluminum	232	.56	.22	2.70	.94	6.9	3.9
Iron	117	.19	.11	7.86	.22	20.5	3.8
Beryllium	115	0.53	0.44	1.85	0.65	29	0.64

several degrees between the two faces. This not only distorts the mirror, but introduces turbulence in the nearby air, with the resultant disturbance to seeing.

Another significant property of mirror materials is the ratio of density to the elastic constant (Young's modulus), since the amount of flexure of a mirror under its own weight while resting on a given support system is directly proportional to this ratio. Table IV shows that there is little difference between the glasses, fused quartz, aluminum, and steel, but that beryllium is superior by a factor of 5.

The use of metal mirrors has become practical because of the development of Kanogen coating (Gutzeit & Mapp 1956). This is a nickel and phosphorus layer which can be applied to any metal surface in thicknesses up to several thousandths of an inch. The coating is very hard and can be ground and figured easily to give an optical surface that is then coated with aluminum with the normal evaporation techniques.

Materials for Transmission Optics

The high transparency of fused silica between 2000 Å and 20,000 Å and its low dispersion, combined with the great homogeneity attained by new manufacturing processes, makes it an ideal material for transmission optics where long pathlengths are required for such applications as thick-mirror Schmidts, image rotators, and right-angle prisms.

Other materials with unusual properties, such as a very high index of refraction, have recently become available and are valuable for certain design problems, such as field flatteners in thick-mirror Schmidt cameras. These include the rare-earth glasses, which in general have lower dispersion and better ultraviolet transmission than the older glasses of similar index. Unfortunately most of these glasses with indices above 1.7 contain thorium oxide, whose radioactivity precludes their use near a photographic plate.

Another material is synthetic sapphire, which has recently been made in homogeneous disks up to several inches in dameter. Its index of 1.764, dispersion of 70.5, and transparency from 1500 Å to 5.5 μ make it a useful material for a number of design problems, although it is slightly double-refracting.

Finally, for designs in which a still higher index is required, diamond may be used. Its index is 2.42, its dispersion is 42, and it is transparent over most of the range from 2500 Å to short-wave radio wavelengths. In sizes up to 10 or 12 mm the cost of the material is less than that of the labor to shape and figure it.

GRATINGS

Another development of great importance to astronomical spectroscopy has been the ruling of large blazed gratings that concentrate 60 to 70 per cent of the incident light in one order. In large apertures these gratings are more efficient than prisms. This, combined with their much more uniform dispersion over a wide wavelength range and their lower sensitivity to temperature change, has resulted in the replacement of prisms by gratings in most spectrographs.

For all dispersions higher than 25 to 50 Å per mm, the collimator aperture and the camera focal ratio are set by the size of available gratings rather than by the limitations on optical design. Furthermore, when the camera focal ratio is larger than those given by Equation 13, an increase in collimator aperture by a given factor, as permitted by larger gratings, produces at least as large an increase in speed as does an increase in the aperture of the telescope by the same factor. In order to take advantage of the large gains that could be obtained by doubling the collimator aperture, a design having a composite of four gratings was used for the coudé spectrograph of the 200-inch telescope. By this procedure the collimator aperture was pushed up to 30 cm.

Recent developments in the ruling of gratings hold forth the probability that single gratings of this 30-cm size may soon be available. It is therefore very desirable that future large telescopes be designed with space available for collimator focal lengths great enough to accommodate these apertures.

Echelle gratings are another development that may be of importance in connection with the use of image-intensifier tubes for high-dispersion spectroscopy. Thus, an echelle grating with appropriate cross-dispersion can be designed to give up to a total of 500 to 800 mm of spectra in parallel strips on an area 30×30 mm, which can be accommodated on the cathodes of present tubes. The chief disadvantages of such an echelle-grating spectrograph are the large variations in dispersion (up to ± 10 per cent) and in intensity (up to ± 35 per cent) along each strip of the spectra.

LITERATURE CITED

Baker, J. G. 1940a, *Proc. Am. Phil. Soc.*, 82, 323

Baker, J. G. 1940b, *ibid.*, 339

Baker, J. G. 1953, *Amateur Telescope Making*, Book 3, 1 (Sci. Am. Publ. Co., New York)

Baranne, A. 1966, *J. Observateurs* (Marseille), 49, 75

Bouwers, A. 1946, *Achievements in Optics* (Elsevier, New York and Amsterdam)

Bowen, I. S. 1938, *Ap. J.*, 88, 113

Bowen, I. S. 1952, *ibid.*, 116, 1

Bowen, I. S. 1960, *Stars and Stellar Systems*, I, *Telescopes*, 43 (Kuiper, G. P., Middlehurst, B. M., Eds., Univ. of Chicago Press, Chicago)

Bowen, I. S. 1962, *Stars and Stellar Systems*, II, *Astronomical Techniques*, 34 (Hiltner, W. A., Ed., Univ. of Chicago Press, Chicago)

Burch, C. R. 1942, *Monthly Notices Roy. Astron. Soc.*, 102, 159

Burch, C. R. 1947, *Proc. Phys. Soc. London*, 59, 41

Carathéodory, C. 1940, *Hamburger Math. Einzelschriften*, No. 28

Chrétien, M. H. 1922, *Rev. Opt.*, 1, 13, 49

Code, A. D., Liller, W. C. 1962, *Stars and Stellar Systems*, II, *Astronomical Techniques*, 281 (Hiltner, W. A., Ed., Univ. of Chicago Press, Chicago)

Dunn, R. B. 1964, *Appl. Opt.*, 3, 1353

Fehrenbach, C. 1947, *Ann. Ap.*, 10, 306 [also Duftot, M., *J. Observateurs* (Marseille), 44, 97, 1961]

Gutzeit, G., Mapp, E. T. 1956, *Corrosion Technology*, 3, No. 10, 1

Hendrix, D. O. 1939, *Publ. Astron. Soc. Pacific*, 51, 158

Hendrix, D. O., Christie, W. H. 1939, *Sci. Am.*, 161, 118

Ingrao, H. C. 1964, *Trans. I.A.U.*, XIIA, *Reports on Astronomy*, 59 (Pecker, J. C., Ed., Academic Press, London)

Köhler, H. 1966, *I.A.U. Symp. No. 27, The Construction of Large Telescopes* (Crawford, D. L., Ed.) (In press)

Kuiper, G. P. 1964, *Sky Telescope*, 27, 4

Linfoot, E. H. 1944, *Monthly Notices*,*Roy. Astron. Soc.*, 104, 48

Linfoot, E. H. 1949, *ibid.*, 109, 279

Linfoot, E. H. 1951, *ibid.*, 111, 75

Linfoot, E. H. 1955, *Recent Advances in Optics*, Chap. 4 (Oxford Univ. Press)

Maksutov, D. D. 1944, *J. Opt. Soc. Am.*, 34, 270

Malacara, D. 1965, *Bol. Obs. Tonantzintla Tacubaya*, 4, 64

Meinel, A. B. 1953, *Ap. J.*, 118, 335

Neugebauer, G., Martz, D. E., Leighton, R. B. 1965, *Ap. J.*, 142, 399

Pierce, A. K. 1964, *Appl. Opt.*, 3, 1337

Rosin, S. 1961, *J. Opt. Soc. Am.*, 51, 331

Rosin, S. 1966, *Appl. Opt.*, 5, 675

Ross, F. E. 1935, *Ap. J.*, 81, 156

Schmidt, B. 1932, *Mitt. Hamburgischen Sternw.*, 7, 15

Schulte, D. H. 1966, *Appl. Opt.*, 5, 309

Schwarzschild, K. 1905, *Abhandl. Kgl. Ges. Wiss. Gottingen Math. Phys.*, 4, 3

Slevogt, H. 1942, *Z. Instrumentenk.*, 62, 312

Strand, K. Aa. 1962, *Astron. J.*, 67, 706

Strand, K. Aa. 1963, *Appl. Opt.*, 2, 1

Strömgren, B. 1935, *Vierteljahrsschr. Astron. Ges.*, 70, 65

Theissing, H., Zinke, O. 1948, *Optik*, 3, 451

Wayman, P. A. 1950, *Proc. Phys. Soc. London, B*, 63, 553

Wynne, C. G. 1965, *Appl. Opt.*, 4, 1185

Zirkind, R. 1965, *Appl. Opt.*, 4, 1077

IMAGE-INTENSIFIER TUBES

Progress in development of these tubes has been summarized as follows:

Symposium on Photoelectric Image Devices. 1962 (McGee, J. D., Wilcock, W. L., Mandel L., Eds., Academic Press, London)

Transactions of the I.A.U. 1955, IX, 673

Transactions of the I.A.U. 1958, X, 143

Transactions of the I.A.U. 1961, XIA, 34; XIB, 180

Transactions of the I.A.U. 1964, XIIA, 65; XIIB, 135

Year Book of the Carnegie Institution of Washington, Report of the Committee on Image Tubes for Telescopes. 1954–1965, 53–64

WAVES IN THE SOLAR ATMOSPHERE[1]

By E. Schatzman and P. Souffrin

Faculté des Sciences and Observatoire de Paris
Institut d'Astrophysique de Paris

Statement of the Problem

In the following we shall consider, from a theoretical viewpoint, the hydrodynamics of the atmosphere of nonvariable stars, with special reference to the Sun. We shall ignore the magnetohydrodynamic (MHD) modes and other effects of the magnetic field, although we know quite well that in the plage areas, and probably in the chromosphere, this is a very crude assumption. We shall also disregard the dynamics of convective zones, because for the Sun this is not properly an atmospheric problem. As far as the Sun is concerned, we shall restrict our discussion to a rather thin layer, the high photosphere and the low chromosphere. The relevance of such a restricted study to astrophysics is twofold: first, the dynamics of atmospheres free from magnetic fields presents such a variety of physical problems, some of which have not been studied, that it is worth while to discuss the work that has been done in the field and to situate the yet-to-be-discovered areas with respect to the explored field; second, the above-mentioned thin layer of the solar atmosphere is a region where many absorption lines are formed, and information such as Doppler fields and brightness fields in monochromatic light is now available. The observational knowledge has been reviewed by Leighton (1963), and since further work has not essentially modified the picture he presented (see, however, Frazier 1966 and Roddier 1966), we shall not state it anew, but rather focus our attention on the theoretical problems.

As we are concerned with small-scale motions (as opposed to motions involving phase coherence over a significant fraction of the star's surface), we shall also ignore work done on spherical waves such as those observed in Cepheids. Some of the techniques devised by Skalafuris & Whitney (1961) for solving the problem of the structure of the shock front can also be applied to the case of solar and stellar atmospheres.

To work out a theoretical model of the hydrodynamics of a stellar atmosphere, one has to consider the following problems:

(*a*) How does one describe an atmosphere in which motions occur, i.e. choose a model for the atmospheric structure?

(*b*) What kind of motions are present? This involves decision about the physics of the motion, including thermodynamics and, to some extent, boundary conditions.

If (*a*) and (*b*) can be answered, then one is usually able to write down a complete set of equations and he is faced with the following problem:

[1] The survey of literature for this review was concluded in October 1966

67

(c) How does one solve the equations of motions?

Because the motions affect the atmospheric structure, questions (a) and (b) are intricately connected and cannot be answered separately. The techniques of solution nevertheless depend on either neglecting the relation between motion and structure so that (a) is answered first and then (b), or considering that relation in a simplified way that allows an iterative treatment.

Problems (a) and (b) are mainly physical, while problem (c) is mainly mathematical. But it must be stressed that the need to overcome the mathematical difficulties of problem (c) influences the other steps, so that physical and mathematical assumptions interact in a way that is often difficult to clarify.

MATHEMATICAL TREATMENT

Regarding the mathematical formulation, most of the work is characterized by the use of linearized equations of motion, a safe approximation in a stable atmosphere as long as the Mach number is small enough. It is nevertheless not self-consistent, as wave amplitudes increase with height in an atmosphere of decreasing density, so that ultimately the linear theory breaks down. Only in exceptional cases have the full nonlinear equations of motions been solved. Except for the work of Whitney (1955) on the development of waves induced by piston motions, it seems that the nonlinear problem in its full extent has not been considered. Most frequently, attempts to study the nonlinear equations of motion have been restricted to the shock-wave problem.

Mention must be made here of an original way to deal with nonlinear terms devised by Lighthill (1952). In a paper on noise generated by turbulence, Lighthill considered the asymptotic acoustic field induced at large distance in a stable fluid by a turbulent flow. Assuming that the outflowing pressure waves do not react on the motion in the turbulent region, he showed that the amplitude in the asymptotic field can be determined by linearizing the equations of motion outside the turbulent region, considering the nonlinear terms inside this region as known source terms [for a review of the current state of the method and results, see for example Ribner (1964)]. Schatzman (1953) first related Lighthill's mechanism to the solar granulation, and recently many efforts have been devoted in the astrophysical literature to extending Lighthill's treatment to stellar conditions (Moore & Spiegel 1964, Unno 1964, Kato 1966a, Stein 1966).

ATMOSPHERIC STRUCTURE

As a first approximation we shall consider plane-stratified atmospheres. Several simple models, each exhibiting some but not all of the qualitative properties of the motion which coexist in the actual atmosphere, can be considered.

(a) The *isothermal atmosphere* lends itself to a description of a large variety of hydrodynamical properties. As is well known, the linearized equations of motion can be solved exactly in this case when an equation of state is given for the fluid. The isothermal model is characterized by the existence of a frequency cutoff due to gravity. As concerns kinematical properties, the model is homogeneous, nonisotropic, and dispersive. The isothermal atmosphere mainly shows all the dynamical properties of an atmosphere which are independent of reflection and refraction effects. This restriction results in an absence of true mechanical resonance, although a divergent reponse can occur as a result of the infinite horizontal extension.

(b) *Models having a variation of kinematical properties with height* can also be investigated. The very best procedure would be to take into account the most up-to-date data for the structure of the photosphere and the chromosphere (see Heintze, Hubenet & de Jager 1964), but it is also of great interest to study highly simplified models which allow analytic treatment. Such models show eigenfrequencies and waveguide properties which are completely missing in the isothermal case. Even oversimplified models such as those in which gravitational fields are absent are worth investigating, and can throw light on the atmospheric problem.

THERMODYNAMICS

The use of an adiabatic relation implies the assumption of nondissipative motions. This assumption is undoubtedly too rough to be used without special care. Some aspects of the motion, such as reflections or refractions, are not qualitatively altered in using the equation of energy without dissipation; however, the quantitative results can change appreciably when the proper thermodynamics are taken into account. On the other hand, some properties depend critically on the actual energy balance, for instance the phase lag between the temperature fluctuations and the vertical velocity, the mechanical energy flux, and even the very nature of the modes of motion.

The most frequently considered problem is the effect of the radiative transfer on the (p, ρ, T) relation. Spiegel (1957) has achieved an explicit solution of the integrodifferential equation describing the radiative smoothing of temperature fluctuations in a gray homogeneous fluid. His solution clearly shows the difference between the optically thin and the optically thick cases, and shows that optically thin perturbations of temperature follow Newton's cooling law with the time of radiative relaxation given by:

$$T_R = \frac{\rho c_v}{16 \kappa \sigma T^3}$$

where ρ is the density, T is the temperature, σ is Stephan's constant, c_v is the specific heat per gram at constant volume, and κ^{-1} is the mean free path of photons. Spiegel's theory has been extended to the nonhomogeneous case (Whitney 1958) but the nonlocal character of radiative transfer allows only a numerical solution even in the one-dimensional problem. Nevertheless schematic estimates of the effect of the radiative transfer on motions in the solar atmosphere have been made with Newton's law of cooling and Spiegel's time of radiative relaxation (Noyes & Leighton 1963, Souffrin 1966, Mein 1966). Figure 1 gives an illustration of the values of the time of relaxation relevant in the solar atmosphere as obtained by Mein using Spiegel's result.

Further progress can be made by using an approximate local description of the radiative smoothing of temperature devised by Unno & Spiegel (1966) on the basis of Eddington's approximation. Nonthermodynamical equilibrium is not likely to modify wave propagation in the observable regions, but has to be considered in the problem of the heating of the chromosphere (Oster 1966).

It should be noticed that the damping effects can be divided into two classes:

1) Relaxation processes, in which we include the radiative processes as

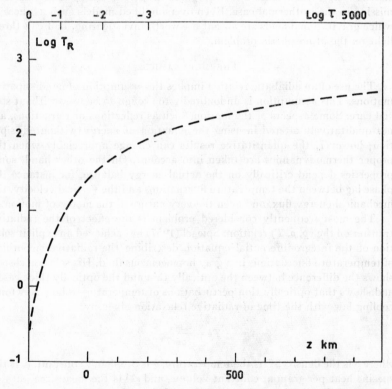

FIG. 1. Spiegel's time of radiative relaxation of temperature for optically thin perturbations in the solar atmosphere (Mein 1966). The model used for the solar atmosphere is basically the Utrecht model (Heintze, Hubenet & de Jager 1966).

mentioned above, and chemical reactions like ionization, excitation, and molecular dissociation. In the stellar case the chemical reactions have hardly been considered. Osterbrock (1961) considers charge exchange, but only in connection with the exchange of momentum in Alfvén waves. This is just one of the cases beyond the scope of this review.

2) Viscous and other diffusion effects, including the internal viscosity of each component of the fluid and friction between the electrons and the heavy particles, ions, and atoms. An exhaustive review of the problem of absorption of sound in homogeneous fluids, covering the methodological aspects as well as the experimental results, can be found in a paper by Markham, Beyer & Lindsay (1951) (see also Lighthill 1956). An account of more recent contributions has been given by Clarke & McChesney (1964).

The effects of damping are twofold: One effect is the heating of the medium by viscous dissipation, Joule heating, thermal conductivity, and radiative damping. The other effect is the limitation of the growth of the amplitude with height which can prevent the appearance of nonlinear effects. Little attention has been paid to the second effect in the astrophysical literature.

Modes of Motion

In a nonrotating, magnetic-free, inviscid atmosphere two kinds of forces act on every fluid element: the buoyancy force due to gravitational stratification, and the pressure force due to compressibility. In a stable atmosphere, both forces are restoring forces which induce a harmonic oscillation of a slightly displaced fluid element. The collective modes associated with this oscillation are the so-called gravity waves and pressure, or acoustic, waves. It is clear that neutral atmospheres (atmospheres with an adiabatic temperature gradient) cannot propagate gravity waves since buoyancy forces are absent. The bubble model makes it clear that, in general, gravity waves cannot propagate vertically, as the buoyancy force is an interaction between elements of different density situated at the same level, a situation precluded by vertical propagation. A more important insight which comes out of the consideration of the discrete model relates to the influence of dissipative processes on the modes of motion: if dissipation occurs during the motion of the bubble (viscous stress, radiative relaxation etc. . . .), it will behave as a damped oscillator, and will not oscillate at all if the damping is strong enough. This shows that even in a stable atmosphere the propagation of gravity modes might be prevented.

The different modes show qualitatively different behavior with respect to important properties: ability to transfer mechanical energy, direction of propagation, phase and amplitude relations between observable quantities such as temperature and velocity. In these differences lies the main interest of the mode analysis of the actual motion in the solar atmosphere. By mode analysis we mean the semiquantitative classification of the motion into the classes defined above, but it turns out that the quantitative analysis leading to such a classification is in fact the normal-mode analysis.

THE ISOTHERMAL ATMOSPHERE

A clear classification of the modes can be achieved only in an atmosphere in which the kinematical properties are independent of position. It is best illustrated by considering small adiabatic perturbations in a compressible isothermal atmosphere.

It is well known that in such an atmosphere the dispersion equation which relates the vertical wavenumber k_z to the pulsation ω and the horizontal wavenumber $k_\perp$ is (see for example Eckart 1960):

$$k_z{}^2 + k_\perp{}^2 \left(1 - \frac{N^2}{\omega^2}\right) - \frac{\omega^2 - \Omega_0{}^2}{c^2} = 0$$

where the Brunt-Vaïsälä frequency N and the cutoff frequency Ω_0 are defined by

$$N^2 = \frac{\gamma - 1}{\gamma} \frac{g}{H}, \qquad \Omega_0{}^2 = \frac{c^2}{4H^2}.$$

The other symbols have their usual meanings.

This equation is classically discussed with the help of the diagnostic diagram where $k_\perp$ is plotted against ω for constant k_z. The limit $k_z = 0$ separates the (k, ω) plane into two unconnected regions with $k_z{}^2 > 0$ which correspond to propagating modes, and a region with $k_z{}^2 < 0$ which corresponds to evanescent modes (see the curves labeled ∞ on Figure 2). The two regions of propagating modes can be identified in the following way. If one lets g go to zero, the region of low-frequency propagating modes vanishes and the diagnostic diagram ultimately reduces to the diagnostic diagram of the homogeneous fluid in which the high-frequency propagating modes are pure compression (or acoustic) modes. The other interesting limiting case is that of an incompressible fluid stratified according to the Laplacian density profile. The pertinent dispersion equation is obtained by letting the compressibility γ^{-1} go to zero. As the compressibility decreases, the high-frequency region vanishes, and the limiting diagram consists of a high-frequency region of evanescent modes and a low-frequency region of pure gravity modes. It is then seen that in the three-region diagram of Figure 2 the high-frequency region corresponds to compression modes (modified by gravity) and the low-frequency region corresponds to gravity modes (modified by compressibility). The evanescent modes of the intermediate region are often neglected; they are automatically left out in all methods considering asymptotic fields, such as the geometrical acoustic approximation or ray-tracing methods. However, for $k_\perp H \lesssim 1$, the minimum value of $k_z{}^2$ is not so large that the scale height can be neglected compared to $|k_z|^{-1}$. Therefore if such a mode is generated inside the convective zone it can reach observable regions. The properties of the resulting filtering have been investigated by Souffrin (1965, 1966). The importance of these nonpropagating modes in the solar atmosphere was first pointed out by Moore & Spiegel (1964), and developed further by Kato (1966a, 1966b). In the same paper Moore & Spiegel discuss the classification of the modes of motion in the isothermal atmosphere following a very general method of Lighthill (1960). Using the theory of the asymptotic evaluation of Fourier integral together with the principle of stationary phase, Lighthill obtained a compact expression for the response of a kinematically homogeneous atmosphere to a harmonic force. A byproduct of his study was to show that the surfaces of constant phase in the asymptotic field are the reciprocal polars of the slowness surface defined by the dispersion relation. This is not a new result (see Brandstatter 1959) but Lighthill's work gave it a new and more precise physical meaning. The geometry of the slowness surface is, then, a possible way to characterize the propagation. The dispersion equation shows that the slowness surface is an ellipsoid for high frequencies ($\omega > \Omega_0$) and a hyperboloid for low frequencies ($\omega < N$). The surface is a sphere in a homogeneous fluid ($g \to 0$) and a hyperboloid in the incompressible atmosphere ($\gamma \to \infty$). This gives a physical basis to the classification of the propagating modes into pressure (or acoustic) modes and gravity modes. A very thorough discussion of the physics and the kinematics of the different modes can be found in Eckart's treatise (Eckart 1960). Lamb waves, which propagate exactly horizontally, occur only in an isothermal atmosphere, because the sound velocity is independent of height. They do not exist in nonisothermal atmospheres.

The effect of radiative damping has been considered, for the isothermal atmosphere, only in the case of an optically thin atmosphere with constant damping time. When damping is present the vertical wavenumber k_z is no longer purely real or imaginary, but complex, as is shown by the dispersion relation, which becomes

$$k_z{}^2 + k_\perp{}^2 \left(1 - \frac{N^2}{\omega^2}\right) - \frac{\omega^2 - \Omega_0{}^2}{c^2} = \frac{i}{\omega\tau}\left\{k_z{}^2 + k_\perp{}^2 - \frac{\gamma_\omega{}^2 - \Omega_0{}^2}{c^2}\right\}$$

where τ is the time of relaxation. The physical interpretation is that the modes can no longer be purely progressive or stationary. The modes which were formerly progressive now show space damping due to the conversion of mechanical energy into heat, and the modes which were nonprogressive are modified in such a way that they can carry enough energy to balance the dissipation. The diagnostic diagram can still be defined through a slight generalization (Souffrin 1966), the modes being identified as mainly damped or mainly progressive and in the latter case as compression- or gravity-like modes (Figure 2). The new feature is the disappearance of the gravity modes when the damping constant is less than a finite critical value, as was expected on the basis of the phenomenological discussion given above.

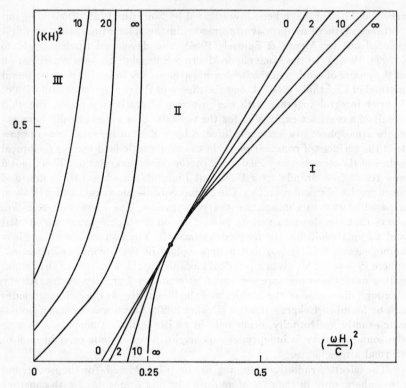

FIG. 2. Diagnostic diagrams for an isothermal atmosphere radiating according to Newton's law. The diagrams are labeled by the value of the time of radiative relaxation, times the Brunt-Vaïsälä frequency. The curves separate the horizontal wavenumber-frequency plane into regions where the vertical space damping of the density of kinetic energy is smaller (I and III) or larger (II) than the vertical wavenumber.

THE OSCILLATORY CHARACTER OF THE VELOCITY FIELD

In the last few years, a great deal of information has been obtained [see Leighton 1963; Edmonds, Michard & Servajean 1965; Mein 1966 (further reference will be found in this paper)] on the spacetime structure of the Doppler-shift field for a number of Fraunhofer lines. One of the most striking properties of the Doppler shifts of photospheric and low chromospheric lines is the damped oscillatory character of the time autocorrelation function, with a pseudoperiod of the order of 300 sec, slightly decreasing with height, the damping time being about 400 sec. The corresponding power spectrum, i.e. the Fourier transform of the autocorrelation function, is characterized by a maximum which defines the pseudoperiod.

The preceding discussion of the modes of motion clearly shows that a

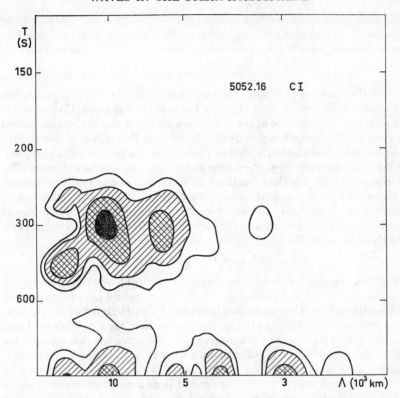

FIG. 3. Spectral density of the vertical velocity derived from Doppler shifts observed in the line C I 5052 (Mein 1966). The curves correspond to 0.1, 0.2, 0.4, and 0.8 times the maximum value.

time analysis of the field is insufficient to show the physical properties of the observed field. Actually, two-dimensional (space-time) power spectra of the Doppler shifts of several Fraunhofer lines are now available (Figure 3) but it is still very difficult to recover the velocity field because of uncertainties in the model of the solar atmosphere and in the depth of formation of the lines. Analysis and discussions of the observational material have been given recently by Edmonds et al. (1965) and by Mein (1966). The pseudoperiodic character of the motion is easily seen in Figure 3, the time power spectra being the projected density on the time axis.

Most of the efforts of theoreticians studying the motions in the solar atmosphere for the last years aimed at a satisfactory interpretation of the oscillatory property of the velocity field.

A detailed understanding of the dynamics of the lower parts of the solar atmosphere is a necessary first step toward the theory of the heating of the chromosphere and of the corona, which we shall not consider here (see Kuperus 1965, de Jager & Kuperus 1961, Osterbrock 1961, Weymann 1960).

Linear Filtering

We shall assume that the convection zone acts as a source of noise, but we shall not discuss the way in which the noise is generated. Consequently we represent the behavior of the convection zone by a stochastic distribution of velocities or pressure imposed on a given layer. This makes it easy to take into account the statistically stationary character (in time as well as in space) of the motions in the convection zone and its large horizontal extension. The motion at any level is then described by spectral densities in the frequency-wavenumber plane, i.e. in the diagnostic diagram, if such a diagram can be ascribed to the model. In the linear approximation, the spectral densities at any level are related to those at the source level through the relation characteristic of linear filtering

$$F(z, k, \omega) = T(z, k, \omega)F(0, k, \omega)$$

where F is the spectral density at the distance z of the source and where the transmission factor T depends only on the structure and physical properties of the atmosphere. The qualitative behavior of T can be inferred simply from the physics of the different kind of modes of motion, the z dependence being mainly determined by the progressive or nonprogressive character of the mode. Evans & Michard (1962) noticed a decrease with height in the pseudo-period of the solar atmospheric oscillation. This decrease was confirmed by Noyes & Leighton (1963) who interpreted it as a result of the filtering of purely vertical motions by an isothermal atmosphere. It is well known that for such motions the isothermal atmosphere behaves as a high-pass filter so that the power spectra is increasingly depleted at low frequencies as the altitude increases. The picture is modified if nonvertical motions are allowed for, because low-frequency modes may then propagate as gravity waves. Nevertheless, in the solar photosphere the radiative relaxation times estimated by Mein (see Figure 1) seem to be small enough to prevent the existence of gravity waves in the lower atmosphere which then acts as a high-pass filter for nonvertical motions also. The filtering by such an atmosphere has been considered in some detail by Souffrin (1966). Figure 4 gives an example of his results and shows the change in the power spectrum as a function of height in an isothermal atmosphere without temperature fluctuations. The figure shows that the filtering can cause a pseudoperiodicity at some distance from the source, but in the solar case the velocity field is observed to be already pseudoperiodic at the upper boundary of the convection zone, i.e. where the linear approximation breaks down, and the extrapolation of the observed spectrum inside the convection zone to a smoothed one of the type of Figure

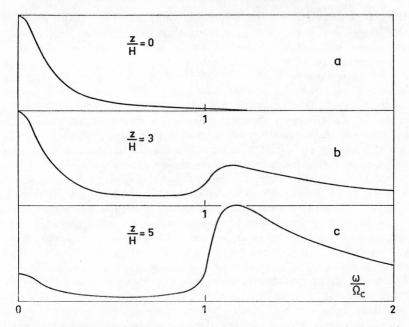

FIG. 4. Variation with height of the shape of the power spectra of vertical velocities as a result of linear filtering in an isothermal atmosphere with isothermal perturbations. At level $z=0$ (curve a) the autocorrelation function has been assumed to be $\exp -(\tau/\theta + r^2/2L^2)$ with a width L such that $H/L=0.15$ and a lifetime θ such that $(\theta\Omega_c)^2 = 40$. The normalization of each curve is arbitrary.

4a is very questionable. The model can only account for a modification of the spectrum from the type shown in Figure 4b to the type shown in Figure 4c. The observed attenuation at the low frequencies and the increase with height of the frequency of the maximum can actually be understood within such a description. The isothermal model also gives a qualitative interpretation of the observed increase with height of the size of the horizontal structure, the space damping being larger for the modes with larger horizontal wavenumber, but it does not provide any kinematical explanation for the most striking property of the velocity field in the solar atmosphere, i.e. its oscillatory character at *all* observed levels.

LAMB'S RESONANCE FOR THE ISOTHERMAL MODEL

The isothermal model gives no satisfactory kinematical interpretation for the peak of the velocity power spectrum, essentially because this model has no eigenmodes, so that no true resonance phenomena are to be expected.

It has been known since Lamb's investigation (Lamb 1909) that resonance-like divergences are found even for the isothermal atmosphere when applied forces are considered. Lamb showed that the velocity response of the atmosphere to an external force acting uniformly on a horizontal plane is infinite if the force is harmonic in time with the frequency $\Omega_c = (g/4H)^{1/2}$ (assuming isothermal perturbations). Noyes & Leighton (1963) stated that the atmosphere will preferentially amplify the velocity response to those components of the pressure-frequency spectrum near this critical frequency, which, with the solar values, is actually of the order of magnitude ($\simeq 260$ sec) of the observed frequency peak ($\simeq 300$ sec). Lamb's mechanism, then, seems to provide a consistent interpretation for the chromospheric oscillation, but further work is needed to clarify the physical meaning of this mechanism which is closely related to Lamb's wave. As we mentioned earlier, Lamb's wave is a singular mode of the isothermal atmosphere which has no counterpart if the sound velocity is not exactly constant within the atmosphere. If point sources are considered instead of the source of infinite extent of Lamb's model, there is no longer a divergence of the velocity reponse for $\omega = \Omega_c$, but rather an enchancement at the lower Brunt-Vaïsälä frequency (Kato 1966a). This clearly shows that the singular character of Lamb's waves and the very existence of the associated dynamical divergence are still questions to be investigated.

It is of interest that if the force does not act uniformly on a horizontal plane, but has a smooth spectrum of horizontal wavenumbers, Lamb's divergence is found for all modes in the diagnostic diagram lying on the frontier between the acoustic region and the stationary evanescent region (regions I and II on Figure 2). This frontier is also found to be the location of singular modes for the homogeneous model (i.e. without gravity) which has no characteristic frequencies; the divergence appearing for these modes, then, clearly has nothing to do with a mechanical resonance, as already noticed by Lamb.

THE ROLE OF THE LARGE-SCALE HORIZONTAL FLOW

A mixed interpretation, considering dynamical and kinematical properties, has been proposed by Moore & Spiegel (1963). As has been discussed above, frequencies in the range $N < \omega < \Omega_0$ cannot propagate in an isothermal atmosphere. Moore & Spiegel extended this result to the case of variable temperature by the $JWKB$ method, and showed that, at any level, propagation is not possible if the frequency is in the range $\omega_2 < \omega < \omega_1$ with

$$\omega_1^2 = \frac{\gamma^2 g^2}{4c^2}$$

$$\omega_2^2 = \frac{g^2(\gamma - 1)}{c^2}\left(1 - \frac{\beta}{\beta_c}\right)$$

where

$$\beta = \frac{dT}{dz}$$

and

$$\beta_a = \frac{g}{c_p}$$

Here β_a is the adiabatic temperature gradient. Moore & Spiegel suggest that if turbulent "sound" generation occurs in the convectively stable upper layers of the Sun, waves with frequencies outside the interval (ω_2, ω_1) will propagate upward to the corona, and the local motion will be built up mainly by oscillations in the nonpropagating frequencies $\omega_2 < \omega < \omega_1$. They argue, moreover, that such a local turbulent generation is likely to occur as a result of the shear in the large-scale horizontal flow observed by Leighton, Noyes & Simon (1962). This large-scale horizontal field can be expected in the Sun, according to Moore & Spiegel, because the edge of the Sun may be quite well defined for large convective elements originating deep in the convective zone and may stop their ascending motion abruptly. If the horizontal flow is confined in a layer of a few scale heights, there must be a vertical shear, and the associated Richardson number estimated by Moore & Spiegel does not contradict the possibility of turbulent instability. These authors, then, understood the observed motions as forced oscillations excited *in situ* by turbulent shear-flow. In terms of the diagnostic diagrams for the isothermal atmosphere, the picture is modified as follows: in the $JWKB$ approximation, a diagnostic diagram can be defined at any level in the atmosphere, and Moore & Spiegel's suggestion implies that motions are confined in the nonpropagating domain (region II of Figure 2). This is actually in fairly good agreement with the two-dimensional analysis of the observation in terms of a diagnostic diagram (see Figure 1) which locates the main energy of the oscillations in the nonpropagating modes of the local diagram. The assumption of Moore & Spiegel is very attractive, but it must be stressed that, as they mention, no quantitative analysis is given of the response of the atmosphere to excitation of nonpropagating frequencies. This problem has been investigated by Kato (1966a, 1966b), in the spirit of Lighthill's method, but the relevance of this method to the study of the motion inside the turbulent region is questionable and further work is needed to clarify this mechanism.

Nonisothermal Models

The idea that the observed oscillation of the solar atmosphere might result from a true mechanical resonance leads to the study of completely different models.

Basically, if the structure of the atmosphere, together with the boundary conditions, is such that there is a discrete set of eigenoscillations, these eigenmodes, along with the corresponding frequencies, will be preferentially selected by the response of the atmosphere. If there is an acoustic energy leak by means of outward-moving waves, the free oscillations will be damped so that the eigenfrequencies will be complex (Schatzman 1956). Mathematically these complex eigenfrequencies are the poles of the transmission factor, and one sees, on performing the integrations occurring in the computation of the response by the method of residues, that these complex frequencies will be selected by the atmosphere from any input of wide spectrum.

The existence of eigenproperties of an atmosphere is ultimately related to the reflection of waves, a phenomenon that was explicitly excluded by the semi-infinite isothermal model. In the outer layers of the Sun, reflection occurs upward because of the increase of the velocity of sound with temperature, and downward because of the very steep density gradient near the photosphere, and resonance phenomena can then be expected. This remark led Bahng & Schwarzschild (1963) to study nonisothermal models of the solar atmosphere. They considered models made up of superimposed layers in each of which an analytical solution can be obtained for horizontal waves, namely, isothermal layers and layers with a constant temperature gradient. As purely vertical motions cannot suffer total reflection, all the models are damped by acoustic leakage, and Bahng & Schwarzschild were able to specify three-layer models for the chromosphere and the transition to the corona whose eigenfrequencies and damping rates were in reasonable agreement with the observations. Schatzman (1964a) also considered vertical motions in models made of the same types of layers used by Bahng & Schwarzschild, but with a temperature minimum to mimic the temperature profile of the chromosphere, and was also able to obtain approximate agreement with the observed oscillation. It must be noticed that when the models are determined so as to give the correct pseudoperiod, the damping rate turns out to be somewhat too low. This discrepancy is not likely to be removed by taking horizontal motions into account: these will merely increase the reflections and decrease the leakage and hence the damping rate. This can be shown quantitatively by the study of the behavior of a three-layer model without gravity including oblique waves (Schatzman 1964b). Uchida (1965) worked out a numerical study of the eigenfrequencies of a model, including gravity with a parabolic temperature profile and with a horizontal wavenumber corresponding to the horizontal extent of the photospheric granules. His fundamental mode has a period near the observed pseudoperiod, but his model does not allow for damping and little can be said about what might happen if waves with different wavenumbers are superposed. It can be expected that the linear filtering would then combine with the resonance; the analysis would be quite tricky. The dissipation of kinetic energy into heat in the higher levels must increase the damping rate, and this effect may balance the decrease due to the reflection of nearly horizontal rays. This could be decided only by a quantitative analysis.

Considering the different models described in the two preceding paragraphs, one feels deceived on realizing that it is very difficult to escape obtaining roughly the observed period whatever the model, or that at least it is easy to adjust the models to it. This situation makes it impossible to use this characteristic of the observed motions to discriminate between the models. The problem is nevertheless not hopeless; a consideration of other characteristics of the motion, such as the phase lag between the velocity components and the thermodynamical quantities, or the anisotropy of the velocity field, must ultimately clarify the problem. To this end, improvement of the theoretical models, as well as improvement of the observational knowledge, is needed.

FORMATION OF SHOCK WAVES

It is not necessary to mention in great detail the problem of the development of acoustic waves into shock waves. The distance in which an acoustic wave propagating in a nonuniform fluid becomes a shock wave has been estimated by Osterbrock (1961), Schatzman (1961), and Bird (1964). Two aspects of the problem deserve special mention, however.

1) The growth of any wave is limited by dissipation; consequently the higher-order terms responsible for the formation of the shock become important later than they would in the absence of dissipation.

2) The waves are generally not purely progressive and the increase in amplitude of velocity with height is not given by the simple $\rho^{-1/2}$ law appropriate to a monochromatic wave in an isothermal atmosphere.

Theory of shock waves in a nonuniform medium.—There is no exact analytical solution for the propagation of a shock wave in a nonuniform medium. For cases with special symmetry the homology method, developed especially by Sedov (1959), can be applied (Hazlehurst 1962). However, this essentially applies to a problem similar to the Cepheid problem or to the nova problem, which we have excluded from our review. It should be noticed that the numerical computation of Colgate & White (1966) for the outer layers of a star lead to the same result as the similarity method used by Ono et al. (1960)

Several approximate methods have been devised which we shall briefly analyze in the following. It should be strongly emphasized that there is no guarantee that these methods are more than qualitatively valid. For example, the similarity method gives the same rate of growth of the spherical shock in the outerlayers of a star as the numerical method of Colgate; but the numerical method gives a shock velocity U growing like $\rho^{-1/6}$ whereas the approximate method gives instead of the power $-1/6$, respectively a power $-5/18$ for $\gamma = 4/3$, and $-3/11$ for $\gamma = 5/3$.

It is likely that the approximate methods are valid for weak shocks, which is just the case we need for the solar atmosphere.

There are in the literature three approaches to the problem of shock-wave propagation. One is derived from the paper of Brinkley & Kirkwood (1947).

Essentially, in the Brinkley-Kirkwood method, the energy losses through the irreversible process which takes place across the shock front are taken into account approximately. This gives the possibility of describing, by a set of ordinary differential equations, the rate of change of the energy available behind the shock and of the velocity of the shock. The approximation includes two assumptions. One concerns the thermodynamic path followed by the gas from the state in front of the shock front, to the state far behind the shock front. The other approximation consists in assuming that the expansion behind the front behaves similarly all during the propagation of the shock. The extension of this method to a horizontal shock wave in a nonuniform fluid has been made by Ono, Sakashita & Ohyama (1961), and the case of the oblique shock has been obtained by Saito (1964).

The second approach is derived from the Chisnell (1955) method which consists in dividing the region of varying density into layers of infinitesimal density jumps. It is easy to write the conditions of reflection and transmission across a discontinuity. It is then possible to obtain the rate of change of the shock strength during the propagation. The extension of this method to the case of an atmosphere has been made by Ono, Sakashita & Yamasaki (1961). The expression for the rate of change of the pressure behind the shock as a function of the path includes a dissipative term, a term depending on the pressure gradient, and a term depending on the density gradient. The Chisnell method does not allow correctly for dissipation due to irreversible processes. These correspond to a shock of infinite energy.

The third approach is an extension of Whitham's method (1958). His method consists in extending the equations of the characteristics by introducing in these equations the shock quantities. For the case of the propagation of a shock in a pipe of variable section, this happens to give satisfactory results because, for a reason which is not clear, the derivative with respect to time of a certain quantity is vanishingly small. The method of Whitham has been applied by Bird (1964) to horizontal shock waves in the solar atmosphere but we meet the same difficulty as with the Chisnell method, that the dissipation term is not included, and furthermore the approximation made by using the equation of characteristics is unclear. Ono et al. have found a term depending on the pressure gradient which differs at most by 10 per cent from the pressure term obtained by Whitham's method. Similarly, Saito (1964) has compared his method to Chisnell's method, for the case of no magnetic field, and showed that the pressure term differs at most by 14 per cent.

Shock Waves and the Heating of the Chromosphere

Owing to dissipation, the shock-wave strength never becomes large in the chromosphere. It seems likely that the approximate method based on the Brinkley-Kirkwood method is valid as long as the assumption of small shock strength is valid. However, it seems that because of shock-wave refraction not enough energy can reach the high chromosphere to match the radiation and conductive losses. As already suggested by Osterbrock (1961), it is necessary to consider interactions between the longitudinal waves and the transverse Alfvén waves. We refer to the papers of Frisch (1964) and Kahalas & Mc Neill (1964). Further work is needed in that direction but the subject is beyond the scope of the present review paper.

LITERATURE CITED

Bahng, J., Schwarzschild, M. 1963, *Ap. J.*, **137**, 901

Bird, G. A. 1964, *Ap. J.*, **140**, 288

Brandstatter, J. J. 1959, *SRI Project 2241*

Brinkley, S. R., Kirkwood, J. G. 1947, *Phys. Rev.*, **71**, 606

Chisnell, R. E. 1955, *Proc. Roy. Soc. London A*, **232**, 350

Clarke, J. F., McChesney, M. 1964, *The Dynamics of Real Gases* (Butterworths, London)

Colgate, S. A., White, R. H. 1966, *Ap. J.*, **143**, 626

Eckart, C. 1960, *Hydrodynamics of Oceans and Atmospheres* (Pergamon, London)

Edmonds, F. N., Jr., Michard, R., Servajean, R. 1965, *Ann. Ap.*, **28**, 534

Evans, J. W., Michard, R. 1962, *Ap. J.*, **136**, 483

Frazier, E. N. 1966, *An Observational Study of the Lower Solar Photosphere* (Thesis, Univ. of California)

Frisch, U. 1964, *Ann. Ap.*, **27**, 224

Hazlehurst, J. 1962, *Adv. Astron. Ap.*, **1**, 1

Heintze, J. R. W., Hubenet, H., de Jager, C. 1964, *Bull. Astron. Inst. Neth.*, **17**, 442

de Jager, C., Kuperus, M. 1961, *Bull. Astron. Inst. Neth.*, **16**, 71

Kahalas, S. L., McNeill, D. A. 1964, *Phys. Fluid*, **7**, 1321

Kato, S. 1966a, *Ap. J.*, **143**, 372

Kato, S. 1966b, *ibid.*, 893

Kuperus, M. 1965, *Rech. Astron. Obs. Utrecht*, **17**, 1

Lamb, H. 1909, *Proc. London Math. Soc.*, **7**, 122

Leighton, R. B. 1963, *Ann. Rev. Astron. Ap.*, **1**, 13

Leighton, R. B., Noyes, R. W., Simon, G. W. 1962, *Ap. J.* **135**, 474

Lighthill, M. J. 1952, *Proc. Roy. Soc. London A*, **211**, 564

Lighthill, M. J. 1956, *Surveys in Mechanics*, 250 (Batchelor, G. K., Ed., Cambridge Univ. Press)

Lighthill, M. J. 1960, *Phil. Trans. Roy. Soc. London A*, **252**, 397

Markham, J. J., Beyer, R. T., Lindsay, R. B. 1951, *Rev. Mod. Phys.*, **23**, 353

Mein, P. 1966, *Ann. Ap.*, **29**, 153

Moore, D. W., Speigel, E. A. 1964, *Ap. J.*, **139**, 48

Noyes, R. W., Leighton, R. B. 1963, *Ap. J.*, **136**, 631

Ono, Y., Sakashita, S., Ohyama, N. 1961, *Progr. Theoret. Phys.* (*Kyoto*), *Suppl. 20*, 85

Ono, Y., Sakashita, S., Yamasaki, H. 1960, *Progr. Theoret. Phys.*, (*Kyoto*), **23**, 294

Oster, L. 1966, *Ap. J.*, **143**, 929

Osterbrock, D. E. 1961, *Ap. J.*, **134**, 347

Ribner, H. S. 1964, *Advan. Appl. Mech.*, **8**, 103

Roddier, F. 1966, *Ann. Ap.*, **29**, 633

Saito, M. 1964, *Publ. Astron. Soc. Japan*, **16**, 179

Schatzman, E. 1953, *Bull. Acad. Roy. Belg.*, **39**, 960

Schatzman, E. 1956, *Ann. Ap.*, **19**, 45

Schatzman, E. 1961, *Nuovo Cimento Suppl.* **22**, 1047

Schatzman, E. 1964a, Preliminary report at 12th I.A.U. Symp. (Unpublished)

Schatzman, E. 1964b, *Astron. Norveg.*, **9**, 283

Sedov, L. I. 1959, *Similarity and Dimensional Methods in Mechanics* (Engl. transl., London)

Skalafuris, A., Whitney, C. 1961, *Ann. Ap.*, **24**, 420

Souffrin, P. 1965, *ibid.*, **260**, 2135

Souffrin, P. 1966, *Ann. Ap.*, **29**, 55

Spiegel, E. A. 1957, *Ap. J.*, **126**, 202

Stein, R. F. 1966, *Astron. J.*, **71**, 181

Uchida, Y. 1965, *Ap. J.*, **142**, 335

Unno, W. 1964, *Trans. 12th I.A.U. Symp. B*, 555

Unno, W., Spiegel, E. A. 1966, *Publ. Astron. Soc. Japan*, **18**, 85

Weymann, R. 1960, *Ap. J.*, **132**, 452

Whitham, G. B. 1958, *J. Fluid Mech.*, **4**, 337

Whitney, C. A. 1955 (Thesis, Harvard Univ.)

Whitney, C. A. 1958, *Smithsonian Contrib. Ap.*, **2**, 365

DETERMINATION OF MASSES OF ECLIPSING BINARY STARS[1]

By Daniel M. Popper

Department of Astronomy, University of California, Los Angeles

METHODS OF MASS DETERMINATION

Before discussing in some detail problems associated with determining stellar masses by analysis of data for eclipsing binaries, I comment briefly on methods of mass determination in general.

Direct methods of determining stellar masses are available in three cases. 1) Visual binaries with visual orbits analyzed, mass ratios evaluated, and parallaxes determined by geometrical means. 2) Double-lined eclipsing binaries with light- and velocity-curves analyzed. 3) Visual binaries with visual orbits analyzed and the radial velocities of the components measured. The first two cases provide the great majority of our information on this basic subject. Visual binaries furnish masses principally of main-sequence stars of type F and later, while main-sequence eclipsing binaries are mainly of type F and earlier. Subgiants are found in both groups, while more highly evolved stars are very rare among binaries of either type that are amenable to analysis. Insofar as mass determination itself is concerned, visual binaries suffer from the difficulties inherent in the determination of small angular quantities. The values of the masses are very sensitive to systematic errors in the separations of the components and in the parallaxes. Mass ratios of close visual components are subject to similar difficulties. Masses of eclipsing binaries do not involve angular measurements for their evaluation. If the spectrograms are properly interpreted and the measurements of velocity are made on spectrograms of adequate dispersion, the errors of mass determination may be quite small and free of systematic effects. On the other hand, the luminosity and position in the H-R diagram, essential for interpretation of the masses, are more readily obtained for visual than for eclipsing binaries.

In the remainder of this section brief comments will be made on methods other than the two principal methods, which are discussed later in this review and in the discussion of visual binaries by Eggen in an accompanying review.

The great importance of the third direct method, spectroscopic observations of a visual binary, is that one may obviate the need for an accurate trigonometric parallax. The conditions for useful application of this method are fulfilled in relatively few cases. One may refer to α Aur (1), δ Equ (2, 3), and ADS 10598 (4). The same precautions necessary in interpreting and measuring the spectrograms must be employed as for eclipsing binaries (see below) in order to avoid systematic effects. Difficulties with velocities of the

---

[1] The survey of literature for this review was concluded in December 1966.

hotter component of α Aur are well known. Because of the usually small difference in velocity of the components, the highest possible dispersion should be used in this method. Even if velocity differences are obtained with high precision, the masses will not be well determined unless the elements of the visual orbit are adequately known. In addition to P, a, and i, the velocity difference may depend critically on T and ω. A case in point is δ Equ, for which the difference of velocity has been carefully measured, but the elements T and ω are not well enough known for substantial improvement of the masses.

The use of group motions for evaluating parallaxes of visual binaries may or may not be considered a "direct" method, depending upon one's point of view. Groups used for this purpose by Eggen (5, 6) vary in character from the well-established Hyades cluster to much more widely scattered groups with loosely coupled motions. A cautious view is that the use of such groups may be of considerable importance in establishing correlations between the characteristics of galactic motion and distribution in the H-R diagram, for example. But for critical problems in mass determination, the parallaxes so evaluated should perhaps be considered provisional. In the case of the Hyades cluster itself, the determination of the parallaxes is subject to less skepticism. But as is well known, a decrease of only $0\overset{\prime\prime}{.}004$ in the parallax of the Hyades would place the Hyades binaries on the solar mass-luminosity relation, instead of their being appreciably more luminous for a given mass.

There are several indirect methods that have been employed for mass determination. For visual binaries with orbits analyzed there are moving-group parallaxes, photometric parallaxes, and parallaxes from H and K linewidths (7). For single-lined eclipsing binaries, the mass of the more luminous component may be assumed from its spectral type in order to evaluate the mass of the secondary component. Analysis of the spectrum of a single star may furnish its surface gravity and effective temperature, from which the mass may be evaluated if the luminosity is known [e.g. Bless (8)]. The virial theorem has been employed for evaluating masses in star clusters (9).

Every application of an indirect method of mass determination needs to be critically examined from three standpoints. 1) How valid is the hypothesis upon which the calculation of mass is based? For example, in the use of H and K linewidths for parallax, is the relation between linewidth and visual luminosity unique, independent of all other properties? 2) How well determined are the numerical parameters of the method? For example, how precisely is the relation between linewidth and luminosity known? 3) How well determined are the observed quantities required for the particular application? For example, how well determined are a^3/P^2 of the binary orbit and the H and K linewidths?

For some critical problems in the interpretation of stellar masses, parallaxes of visual binaries must be accurate to about 5 per cent. Since relatively few trigonometric parallaxes are determined this well, the use of indirect methods is attractive. But 5 per cent in parallax corresponds to 0.1 mag in absolute magnitude, the precision required in photometric and H and K line parallaxes. It is open to serious question whether the uniqueness and observational accuracy of the calibrating relations (color-luminosity and line-width-luminosity) are established more precisely than 0.2 mag. As an example of the uniqueness problem, one may refer to calculations of the effects of chemical composition on position in the H-R diagram (10). A change in the helium abundance by a factor of two corresponds, in the calculations cited, to a change of about 1 mag for a given color. Similarly each parallax determined by the moving-group method and used for mass determination must be critically examined to evaluate its uncertainty and validity. The use of these indirect methods of parallax determination for studying problems other than mass determination is subject to less stringent demands.

The masses determined for the fainter component of a single-lined eclipsing binary on the assumption that the mass of the more luminous component may be estimated from its spectral type must likewise be considered uncertain. In cases where the fainter component is a star of a kind for which masses are not readily obtainable, the results may be of considerable value. Determination of surface gravity, and indirectly the mass, by spectroscopic analyses of single stars would be of particular importance for evolved stars, for which our knowledge of masses is very incomplete. But it is just for such stars that our knowledge is also incomplete concerning the structure of the outer layers where the observed spectrum is formed. With respect to use of the virial theorem, hypotheses include the assumption that the velocity dispersion may be obtained from a few of the brightest stars, the absence of appreciable invisible matter, and the mass distribution in the cluster.

In each of the indirect methods, the validity of hypotheses and the uniqueness and determinacy of parameters are likely to be matters of opinion. The usefulness in the eyes of a reader of masses determined by these means will depend on his opinions and on the purpose for which the masses are to be employed.

Only masses determined by direct methods will be reliably enough known to be useful in elucidating some of the more critical problems of stellar structure and evolution, particularly for stars on or near the main sequence. The position of a star in a theoretical H-R diagram is very sensitive to its mass. Likewise the delineation of fine structure in the mass-luminosity diagram requires masses of high precision. Unless a valid estimate of the uncertainty in the mass of a star is available, its usefulness for studying such critical problems will be limited. The linear scale of the orbit of a binary star is the critical quantity for which estimates of systematic and accidental errors are essential. As discussed below, these estimates are obtained readily for the eclipsing binaries, in which the only critical quantities are the amplitudes of velocity variation. For the visual binaries, the orbital elements, parallaxes, and mass ratios for unequal components enter. Evaluations of the systematic and accidental uncertainties in the major axes based upon these quantities are not straightforward and are usually not attempted. Several important contributions have been made by Eggen (11), who has shown that the quantity a^3/P^2 may, under favorable circumstances, be determined with greater precision than would be expected from the separate uncertainties in a'' and P. He has also attempted to assess the determinacy of the orbital elements for a number of systems. But for the visual binaries in general, because of the complexity of the way in which the different kinds of observation enter the determinations of mass, valid estimates of uncertainty have not been carried out. This lack is an important weakness in the available data on masses from visual binaries.

MASS DETERMINATION OF ECLIPSING BINARIES
INTRODUCTION

Since the discoveries by Pickering of double lines in the spectrum of ζ UMa (12) and by Vogel of the velocity variation of β Per (13), the principles of mass determination in spectroscopic binaries have been clearly understood, the quantities required being radial velocities of the components and the orbital inclination. With respect to the latter, after increasingly valid attempts by Myers (14), Roberts (15), Schlesinger & Baker (16, 17), and Stebbins (18), a general solution for the inclination and other elements of the light-curve was given by Russell (19). Numerous modifications and improvements of the Russell scheme [e.g. Irwin (20)] have been proposed and employed. But if we restrict our discussion to eclipsing binaries for which the radii may be evaluated from the light-curve, uncertainty in the orbital inclination is generally a minor cause of uncertainty in evaluating the masses of the components, as is uncertainty in the orbital period or eccentricity. Another minor, although systematic effect is correction for the effect of reflection in a close binary (21).

Thus the central problem in determining the masses of the components of eclipsing binaries is, as it has been since the beginning, simply that of evaluating the amplitudes of velocity variation of the two components.

PROBLEMS IN MEASURING THE RADIAL VELOCITIES

Techniques of measuring spectrograms for radial velocity remain essentially as they have over the years. The use of projection-measuring devices and oscilloscopic scanning comparators (22), while convenient and efficient, does not introduce any essentially new factor and, in my as yet limited experience, does not increase the accidental or systematic precision of the results.[2] In some cases microphotometer tracings may be useful in estimating the positions of the centers of broad features.

Improvement in our ability to evaluate masses of eclipsing binaries has resulted primarily from advances in two directions: first, improvement in

[2] The Hartmann comparator method of radial-velocity measurement, insofar as it involves an attempt to obtain a simultaneous setting for a group of neighboring lines, is in my opinion not a valid method of precise measurement. Precision of 1 or 2 μ can only be obtained by concentration on a single feature at a time.

observational equipment and, second, improvement in interpretation of the spectrograms. To these endeavors should be added discoveries of eclipsing binaries, exemplified in recent years by the numerous relatively bright systems found in both hemispheres by Strohmeier and his colleagues. Such discoveries play an indispensable role in providing new possibilities for mass determination.

Among the improvements in observing equipment are larger telescopes, more sensitive photographic materials (including hypersensitization), and efficient grating spectrographs. As a consequence of these developments, spectrographic observations can be carried out on a much larger number of systems with adequate dispersion (see below) over a wider range of wavelengths than in earlier years. Image tubes have not yet been applied to this problem.

We turn next to the question of interpreting the spectrograms. "Interpretation" may appear an inappropriate concept when one is concerned with measuring the positions of spectrum lines in order to obtain the velocities of the components. But two qualitative decisions are required for each line before a valid measurement can be made. The first decision, which may be a difficult one for weak lines, concerns the reality of the feature. When one is seeking lines of a faint component, it is possible to confuse a pattern of grains in the photographic emulsion with a line. The possibility of confusion is great on weakly exposed and narrow spectrograms, particularly since one knows where on the spectrogram such a line should occur. The second decision concerns the assignment of a line to one or the other component. A line which is, in reality, an intrinsically weak line in the more luminous component may mistakenly be assumed an intrinsically strong line in the fainter component. Improvement in the spectrograms over the years, particularly in the dispersion available, has made it possible to make these decisions with greater certainty. In the past, numerous misinterpretations have resulted from the difficulties mentioned. The *Third Catalogue of Spectroscopic Binaries* of 1924 (23) lists 13 eclipsing binaries with masses determined by a combination of photometric and spectrographic data. It is now clear that for only six of these were the lines of the fainter component measured unambiguously. A recent compilation (24) lists 42 eclipsing binaries "with well-determined absolute dimensions." Brief comments on each of them have appeared elsewhere (25). Of these 42 at least six are systems for which the velocity variation of the fainter component is either unknown or very poorly determined. A still more recent discussion of eclipsing binaries (26) contains an even larger proportion of undetermined or poorly determined absolute dimensions resulting from difficulties in interpreting the spectrograms. Thus its appearance in the literature does not always provide assurance that a radial velocity results from measurement of real spectrum lines belonging unambiguously to the component in question.

Because of the difficulties of interpretation, it would be helpful to the interested reader if high-quality reproductions of all spectrograms of double-lined binaries were published along with the results of measurement of the spectrograms. For a number of good reasons such reproductions are not generally published. If one has questions concerning the interpretation, he must usually examine the original material himself or obtain new spectrograms. Both procedures have been employed in my own studies.

A difficulty of interpretation of a different kind, which shall not be referred to further, is that the absorption lines measured may be produced in part by atoms having mass motions systematically different from that of the stellar photosphere, as discussed by Struve and others (27). These effects reveal themselves by inconsistencies in the results that can usually be recognized. The associated problems are discussed elsewhere (28).

To illustrate some of the problems of interpreting the spectrograms, it may be instructive to examine a particular system. TV Cet is a nine-day eclipsing binary, magnitude 9.0 photographic, of sepctral type F2 approximately on the main sequence. The secondary component is clearly present in the spectrum, but its lines are weak, so that the difficulties of interpretation referred to are quite severe. In the cases of Z Her (29) and RS CVn (30) similar difficulties were circumvented by observations in the visual region of the spectrum. TV Cet is less favorable for effective use of this technique because of its faintness, because its secondary is of earlier type (G0?) than in the other systems so that lines in the visual region are not strong, and because interstellar components cause difficulty in measuring the weak D lines of the secondary. Hence the photographic region must be relied upon.

Figure 1 shows a spectrogram, original dispersion 21.5 Å/mm, obtained at the phase of maximum velocity of recession of the more luminous component. While effects of line doubling are clearly evident for a number of the lines, examination of spectrograms of higher dispersion, 10.2 Å/mm, shows that it is only with difficulty that unblended lines of the fainter component suitable for measurement can be found. Figures 2 and 3 show enlargements of two 10.2 Å/mm spectrograms of TV Cet, one obtained near each node of the orbit plane, together with a spectrogram of ι Peg (F5 V). The three spectrograms are aligned so that a feature of the stronger component of TV Cet on both spectrograms is in line with the corresponding feature in the spectrum of ι Peg. The lines of the weaker component are displaced to the red (right) in the upper spectrogram (phase $0.^P32$ after primary minimum) and to the violet in the lower ($0.^P88$). In Figure 2, only the following lines in the spectrum of the fainter component were considered suitable for measurement (numerous lines of the primary component are satisfactory):

λ 4045 (A). Good when displaced to the red (above), but blends badly when displaced shortward (below).

λ 4063 (B). Blends with a faint line when displaced to the red, but probably all right to measure. The line can be measured just shortward of λ 4062 of the brighter component on plates of excellent definition when displaced to the violet sufficiently, as at 0.88 phase.

λ 4071 (C) Blends when displaced longward. Similar to λ 4063 when dis-

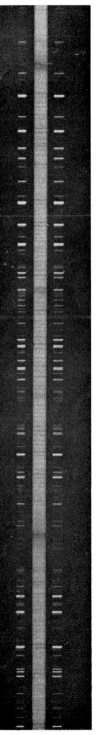

Fig. 1. TV Ceti, λλ 4000–4500, 0.74 phase. Original dispersion 21 Å/mm.

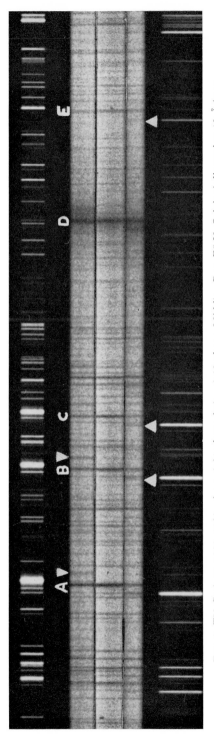

Fig. 2. TV Ceti, λλ 4020–4130, above 0.31 phase, below 0.89 phase. Middle ι Peg, F5 V. Original dispersion 10 Å/mm. Arrows show features of the fainter component described in the text.

placed shortward, being separable to the violet of λ 4070 of the stronger component.

Hδ (D). Not considered suitable because of the effects of overlapping wings.

λ 4118 (E). Blends badly when displaced longward. Free of blends when displaced shortward, though rather weak.

Similarly in Figure 3 we find the following features of the fainter component:

λ 4215 (A). Blends badly when displaced longward. Measurable as the longward line of the weak pair when displaced shortward.

λ 4226 (B). The line of the secondary appears unblended only when displaced longward.

λ 4254 (C). Good when displaced longward. The weak line at λ 4252 of ι Peg appears to lie sufficiently shortward of the weaker component of λ 4254 when that component is approaching to permit measurement.

λ 4271.2, 4271.8 (D). Both lines blend with lines of the brighter component when displaced longward, but are suitable for measurement when displaced shortward.

λ 4315 (E). Satisfactory only when displaced longward.

A total of 13 lines of the weaker component per plate was found suitable for measurement when displaced shortward and eight when displaced longward in the wavelength region λλ 4000-4600. Use of these lines should produce velocities free of systematic error. Although spectrographic investigation of this system is not yet completed, about 12 spectrograms obtained at suitably distributed phases will determine the amplitudes of velocity variation to a precision of 1 or 2 km/sec. Preliminary values are $K_1 = 69$ km/sec, $K_2 = 74$ km/sec, leading to minimum masses of 1.45 and 1.35⊙, somewhat larger than the values given for this system on the basis of a less satisfactory analysis (31). A definitive light-curve is badly needed.

Another example is V805 Aql (32), A2, Figure 4 (10.2 Å/mm). While the hydrogen lines are clearly double, it may not be assumed that they can be measured without systematic effects resulting from overlapping wings. Unblended lines clearly belonging to the fainter component have not been found in the photographic region. The D lines, Figure 5, are each triple, the central component being interstellar. The dispersion shown, 20 Å/mm in the visual, may not be adequate, and the star is too faint (7.8 mag) for much higher dispersion. Use of an image tube may be advantageous in such a case.

A third example is the 6th mag B star, U Oph, Figure 6. While the He lines are suitable for measurement, the H lines are not. Even with such poor lines as the He lines (as a matter of fact, they are unusually well defined in U Oph insofar as B-type spectroscopic binaries are concerned), the relatively high dispersion of 10 Å/mm is probably preferable to lower dispersions. When the lines are not very deep, irregularities in the grain pattern of the spectrograms have less influence on the measurements with higher dispersion. In a system of this character, the number of lines per spectrogram satisfac-

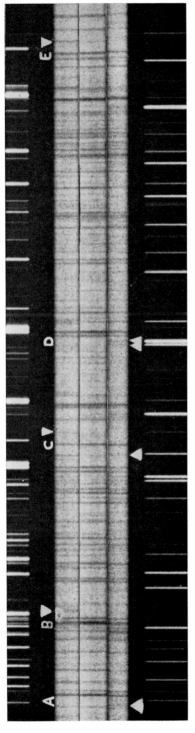

FIG. 3. Same as FIG. 2, λλ 4210–4320.

FIG. 4. V805 Aql, λλ 4020–4500, 0.15 phase, original dispersion 10 Å/mm.

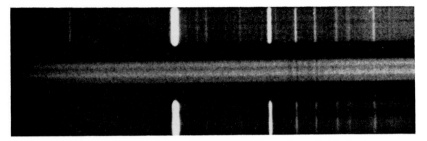

FIG. 5. V805 Aql, D lines, original dispersion 20 Å/mm. The sharp
central component of each D line is interstellar.

tory for measurement is small and varies appreciably from plate to plate.
Furthermore the plate error is small compared to the uncertainty of the position of a line. Under such circumstances, it is probably preferable to compute
the amplitude of velocity variation from measures of the individual lines
rather than from the mean velocity for each of the plates The spectrograms
reproduced were all obtained at the Mount Wilson Observatory.

While the systems illustrated here are representative, every system has
its individual problems that must be attacked in an appropriate manner.
Reference is made to the literature for other examples (29, 30, 33–37).

Even when the lines of both components are of comparable strength and
may be clearly visible, so that there is no ambiguity of interpretation, systematic effects in the measured velocities will result if inadequate dispersion
is used. The usual effect (36), well known to workers in this field, is that the
separation of the lines as measured is greater on lower dispersion plates. In
order to avoid such effects, dispersions of 10 Å/mm (or higher) are desirable,
while dispersions appreciably lower than 20 Å/mm are probably inadequate
except for B stars or perhaps for systems with stars of greatly different spectral types, so that the two components have essentially different sets of lines.
The hydrogen lines can be measured safely only in stars of types about F5 and
later, where their wings are sensibly absent.

The procedure used in computing the amplitudes of velocity variation as
well as the other elements are of relatively minor importance if the measurements are valid and free of systematic effects and if the observations are

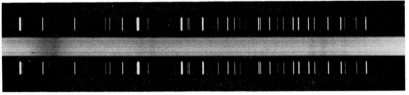

FIG. 6. U Oph, λλ 4380–4500, 0.72 phase, original dispersion 10 Å/mm.

suitably distributed in phase. The method of least squares has the advantages of being well defined and of yielding internal uncertainties of the elements.[3] In order to conserve observing time, which is severely limited, observations need to be concentrated near the times of nodal passage. Then the amplitudes will be determined with greatest weight. In such a case, the weight of determination of e and ω is generally low. While uncertainty in e and ω has a negligible effect in evaluating the masses, it does reduce the value of the observations for studying other problems, such as apsidal rotation. In any event these problems are generally better studied in single-lined binaries where the effects of blending of the lines of the components do not influence the measurements at the critical phases of rapid velocity variation.

A PROGRAM OF INVESTIGATION

By 1934, sufficient photometric and spectrographic work had been carried out on eclipsing binaries for A. B. Wyse (38) to undertake a survey of the results. A difficulty that he emphasized was that in many systems the spectral types published for the two components were much more nearly equal than would be predicted from the depths of the minima of the light-curve. It was this difficulty that first led me to investigate several well-known systems. It was found that in a number of cases (29, 33, 37, 39) the discrepancy was not a physical one but resulted from misinterpretations of the spectrograms of the kind referred to above. That is, the lines used for estimating the spectral type of the fainter component are too weak for one to be certain of their presence or are badly blended with lines of the stronger component. This is not to imply that there may not be real discrepancies between spectral types and surface brightnesses, particularly among the B stars. One may refer as an example to the discussion of α Vir by Struve and others (40).

Once the reasons for the incorrect assignments of spectral types were understood, it followed that the masses of the components in such systems had not been soundly based. Hence a re-examination of the masses was undertaken. In some cases it was concluded that the systems were not suitable for mass determination. In others improved material led to re-evaluation of the masses, with results sometimes in agreement and sometimes in significant disagreement with the earlier ones. During the first stages of this work a little more than ten years ago, theories of stellar interiors and stellar evolution were developing to the degree that, if significant tests of theory were to be supplied, existing data on the masses of eclipsing binaries needed to be assessed critically and increased. In the ensuing expansion of the observational program it was found that other effects, not associated with difficulties in the spectral types, could lead to significant errors in the masses. These include various effects due to the use of inadequate dispersion (34, 36) and of

[3] The fractional mean error of the masses is $\sqrt{5}$ times that of the amplitudes of radial velocity if the amplitudes and mean errors are the same for the velocities of both components.

inadequate measuring techniques (41). Some of these effects have been illustrated above. A number of results from this program, which is continuing, have been published in the references cited.

It is now clear, as discussed elsewhere (25), that in order to provide significant information in the discussion of such critical problems as the dichotomy of the mass-luminosity relation for main-sequence stars, the amplitudes of velocity variation must be known with accidental and systematic errors of 5 per cent or less. In reinvestigations to date of stars of type A0 or later, systematic corrections of the earlier observations of considerably more than this amount have been required in a number of cases.

It should not be inferred that the only systems to be studied are those previously investigated by others. A number of promising new systems that are being worked on will be referred to later. But most of the known double-lined binaries brighter than about 9th mag were observed spectrographically in earlier decades, and even with the most powerful equipment now available it is not possible to obtain adequate spectrograms for stars much fainter. As discussed above, grating dispersions of 10 Å/mm or higher are required in many cases, and dispersions lower than 20 Å/mm (except perhaps for some of the B-type systems) are inadequate in most cases. With the coudé spectrographs of the 100- and 120-inch telescopes the useful limiting magnitude is about 10.5 at 20 Å/mm and 9.0 at 10 Å/mm in the blue and violet region of spectrum. If one must employ the visual region to observe the lines of the fainter component, the useful limiting magnitude is 8.0 to 8.5, except for systems of long period.

A Survey of Systems

In this survey reference is made to double-lined eclipsing systems that are or may become useful in providing definitive determinations of stellar mass. If the system is to be of value, the radii and colors (on a standard system) of the components must be known in order to place the stars on the H-R diagram or its equivalent. Systems such as the W UMa stars and other "contact" systems are not included in the discussion. References to the literature for the systems mentioned may be found in such works as *Basic Astronomical Data* (24, 42), the *Finding List for Observers of Eclipsing Variables* (43), Martynov's bibliography (44), the *Bibliography and Program Notes on Eclipsing Binaries* (45), and the *Information Bulletins on Variable Stars* (46).

Main-sequence stars of type G-M.—The only well-observed systems of this group, particularly valuable for comparison with visual binaries, are YY Gem (M1) and UV Leo (G2). BH Vir has properties much like those of UV Leo, but with the components more unequal in temperature. Abt (47) has obtained radial-velocity curves using a dispersion of 62 Å/mm. Abt's masses of BH Vir may suffer from the same systematic effects found at 40 Å/mm with UV Leo (36), which has nearly the same velocity amplitude as BH Vir. It will be very difficult to investigate the system with higher dispersion because of its faintness and short period. ER Vul may also be similar to UV

Leo, but the inclination is probably too small and the ellipticity of the components too great to admit a definitive solution. Possibly promising new double-lined systems are PW Her and +47°781. Approximate types are K0 V and G5 IV–V respectively. Photometry is required in nearly all cases to establish the position of the stars relative to the main sequence.

Main-sequence stars of type F.—There is a considerable number of systems with approximately equal components, most as yet inadequately investigated either spectrographically or photometrically or both, that should yield masses of good precision. Of these, ZZ Boo, VZ Hya, WZ Oph, and probably CD Tau have adequate spectrographic orbits. Among the more promising additional double-lined systems are EI Cep, HD 90242, HD 123423, HD 185912, HD 190020, and in the Southern Hemisphere HD 93486. Careful photometry is needed for all these systems, not only to determine the geometric elements and the position of the components in the H-R diagram, but also to establish the magnitude and sign of the ultraviolet excess, since one of the outstanding current problems is the question of the dichotomy of the mass-luminosity relation discussed by Eggen (5, 6, 11), the discriminant between the two groups of stars being ultraviolet excess. Systems with unequal components, such as TV Cet described above, may also furnish significant data.

Main-sequence stars, A5-F0.—This is the realm of the metallic-line stars. The masses of WW Aur are probably well determined. This star was classified as A7, not as a metallic-line star, by Miss Roman (48), although the HD type is A0. TX Her and RR Lyn are being re-examined. The light-curves of both these systems appear to show abnormal effects. MY Cyg also has double lines. According to Abt (49), all eclipsing binaries in the range A4-F2 are expected to be metallic-line stars. The southern double-lined binaries HD 75747 (F0) and SZ Cen (A7) (50) fall in this range. Their classification as metallic-line stars requires verification, as does that of ZZ Boo (F2). HD 204038 probably has too great ellipticity in the light-curve for a definitive solution.

Early A stars.—Analyses of two of the best systems with nearly equal components, RX Her and AR Aur, have been published (34). V451 Oph is an additional A0 system with spectral features essentially identical with those of RX Her. As in the latter case, the published masses (2.4, 2.0 ⊙) based on lower-dispersion spectrograms are too small because of blending of stellar and interstellar K lines with inadequate dispersion. Revised preliminary values are 3.0 and 2.6 ⊙. Other systems with published masses, V805 Aql, SW CMa, WX Cep, and HD 205372, are less satisfactory systems because of blending of the lines. The hydrogen lines are not suitable for measurement because of their broad wings, and other lines may be exceedingly weak.

Systems among the A stars with unequal components include CM Lac and V477 Cyg. EE Peg may be a similar system. As in all cases of appreciable magnitude difference, measurement of the secondary requires great care. The published masses for CM Lac do not require modification to any extent, while those for V477 Cyg need to be decreased from 2.6 and 1.7 ⊙ to 2.1 and 1.5 ⊙. Another group of systems has A-type primaries with very small mass functions. Lines of the secondary component have been observed in AS Eri (unpublished), DN Ori [Smak (51)], AW Peg [Hilton & McNamara (52)], and XZ Sgr [Smak (53)]. The primaries have masses near 2.0 ☉ while the subgiant secondaries have masses of 0.2 or 0.3 ☉.

The B stars.—In the program of investigation, B stars have been studied relatively little as yet. Demands upon the precision of mass determination that are made by theory may not be so great as for stars of smaller mass. The best cases among the B stars are characterized by light-curves with two relatively deep minima and small ellipticity (radii and inclination determinate) and by spectra with well-defined helium lines. As in the A stars, the H lines are rarely suitable for velocity determination. Among the more favorable systems, both spectrographically and photometrically, are Y Cyg and U Oph. Nevertheless, conflicting values of the masses have been published for U Oph by Plaskett (54) and by Abrami (55). Measurement of a new series of spectrograms of U Oph (see Figure 6) confirms Plaskett's results. The smaller values of the velocity amplitudes obtained by Abrami may have resulted from inclusion of the H lines, which I find to give $K_1 + K_2$ 15 per cent smaller than the value from the He lines. A similar effect may be found in the measures of Y Cyg (56). Riggs' measurements of the separation of the lines of the components are about 4 per cent greater than Mrs. Zebergs'. Our measurements of the same Mount Wilson spectrograms, omitting the H lines, agree with Riggs'. It is not clear to what extent the masses of B stars appearing in the literature are influenced by such effects. The studies of line blending by Petrie (57) are pertinent.

Examples of systems with favorable spectra but light-curves not suitable for determinate solution are AH Cep, u Her, and V Pup. Double-lined systems in which both the light-curves and the spectra are difficult to analyze include σ Aql, EO Aur, CC Cas, V448 Cyg, and IM Mon.

Among systems with the primary star of type B and the secondary of type A or later, very few have lines of the cooler component measurable as they are in ZZ Cep and in the semidetached systems V356 Sgr and Z Vul. Other systems in this category for which the lines of the secondary may be measurable, but only with difficulty, are TU Mon, AU Mon, RY Per, and RS Vul.

Estimates of mass are available for the systems with Wolf-Rayet components, V444 Cyg and CV Ser.

Subgiants of types G and K.—These systems may be divided for convenience into several groups.

(a) Systems such as U Cep, U Sge, Z Her, and RS CVn, with normal mass functions and the primaries of type B to F on or near the main sequence. While such systems are numerous, reliable masses are available only for Z Her and RS CVn with F-type primaries, the secondaries being measurable only in the visual region. With more luminous primaries, the lines of the secondaries are too faint to be measured except briefly for some systems at mid-eclipse. Systems with unknown mass functions having D lines of the secondary faintly visible outside eclipse are RX Gem, RY Gem, and probably XY Pup. Mass determination will be difficult.

(b) A or early F-type primaries with very small mass functions. The secondaries are subgiants of very small mass. These systems are referred to under the A stars. It has been proposed (58) that some of the primaries (e.g. R CMa) may also have very small masses. The systems are all difficult because of the faintness of the secondaries.

(c) Systems in which both components are subgiants of type G or K. The one case with well-determined masses is AR Lac. Other systems are SS Boo and WW Dra, for both of which revisions of the published masses are under discussion. Provisional values of the minimum masses are 0.9 and 0.8 ⊙ for SS Boo and 1.4 and 1.3 ⊙ for WW Dra. The published masses of RT Lac also require rediscussion, and SZ Psc has not been re-examined. The large mass of 1.7 ⊙ obtained for the subgiant K star of SZ Psc (59) requires confirmation. Other systems of this group, which may yield masses with difficulty because of their faintness in addition to the spectrographic problems of blending of the lines and inequality of the components, are AW Her, MM Her, and +47°781. It is not clear whether this last system should be classed in this group or among the main-sequence systems. All these systems appear to have H and K emission associated with a component near type K0, as do Z Her and RS CVn. Each system with such emission lines that has been adequately studied photometrically shows irregular intrinsic variations in light.

Cool giants and supergiants.—The very meager information available for systems of this category has been reviewed elsewhere (25): one K-type supergiant (ζ Aur) of mass 8 ⊙ and three presumably normal giants, only one of which (RZ Cnc) has an unambiguous determination, with masses in the range 2 to 4 ⊙. It is not clear that other systems of this type are available that can yield reliable information on masses. The supergiants 31 Cyg and 32 Cyg are very poor from the spectrographic standpoint, although the unpublished work of Wright and Huffman leads to a value of 10 ⊙ for the mass of the supergiant K star in the former system. In the case of RZ Oph the velocity variation of the faint component through the long total eclipse does not appear sufficiently great to be measurable with adequate accuracy with the low dispersion that must be used. The faint supergiant systems V477 Sgr and AZ Cas should be studied, but they do not appear promising. Among the normal giants, AR Mon may be similar to RZ Cnc, but it is fainter. AL Vel does not appear to have measurable lines of the hotter component. UU Cnc has a large mass function, but the lines of the fainter component have not been found. The system is faint for adequate dispersion.

Peculiar systems.—This survey has been concerned with systems for which the masses may be determined unambiguously, without difficulties caused by absorption in gaseous envelopes and streams. Estimates of masses are available for a few of the numerous peculiar systems discussed by Struve and others. The very important work of Kraft and others (60) on cataclysmic variables as binaries should also be referred to.

Disk and halo stars.—I am unaware of any study of the characteristics of the motions of eclipsing binaries in space that might reveal systems not belonging to the spiral-arm population. It is not known whether the eclipsing binaries reported in globular clusters (61) are foreground stars or cluster members.

SOME OBSERVATIONAL RESULTS

As discussed here, it has been found desirable to re-examine the spectrographic observations of all eclipsing binaries in order to evaluate their suitability for mass determination. Because of the relatively small amount of observing time available for the spectrographic program and because of the vagaries of observing weather and orbital motions, an interval of from five to ten years is usually required to obtain adequate spectrographic material. Consequently the list of well-determined masses published several years ago (42) can be extended by only a few systems at this time. Furthermore, two of the stars on that list, TX Her and ζ Phe, are being re-examined, and the last word may not yet have been said on Y Cyg.

In Table I are listed values of masses of eclipsing binaries considered to be known without appreciable systematic error. Then the mean errors given, obtained from the internal agreement of the radial velocities, should be significant values. The letter p following the mass in place of a mean error indicates that the mass is known without appreciable scale error, but that the analysis of the data is as yet provisional. Some of these values are from unpublished material. Estimates of masses of systems such as those with Wolf-Rayet stars, U Gem Stars, and novae may be found in the literature cited in preceding paragraphs.

AS Eri is included in the table as the best determined of the A-type systems with very small mass functions.

As discussed in the next section, additional information concerning the stars is required in order to interpret the masses. Critical analyses giving the radii, colors, and luminosities of the components are not available for many of the systems. Values of the radii are listed when these have been obtained in a reasonably satisfactory manner. Less-well-determined values are followed by colons (: or ::). Spectral types are from a variety of sources, including color measurements. Spectral types of the secondaries in parentheses are adjusted in accord with the photometry.

OTHER PROPERTIES OF THE STARS

This discussion has concentrated on the determination of masses. It is only by correlation of the masses with other properties of the stars that they may furnish observational tests of structural and evolutionary theory. It is customary to base such tests on positions in the mass-luminosity and H-R diagrams. Since the parallax is required for determining the masses of a visual binary, its luminosities are readily evaluated. Then color and luminosity

TABLE I

MASSES OF ECLIPSING BINARIES

Star	Masses (⊙)	m.e.	Radii (⊙)	Spectra
Main-sequence systems with similar components				
YY Gem	0.58	0.02	0.60	M1
	0.58	0.02	0.60	
UV Leo	1.02	0.04	1.09	G2
	0.95	0.04	1.05	
WZ Oph	1.13	0.04	1.33	F8
	1.11	0.04	1.36	
VZ Hya	1.23	0.03	1.25	F5
	1.12	0.03	1.05	
CD Tau	1.33	0.05	—	F5
	1.40	0.05	—	
ZZ Boo	1.75p	—	1.75:	F2
	1.68p	—	1.70:	
WW Aur	1.81	0.10	1.9	A7
	1.75	0.10	1.9	
RX Her	2.75	0.06	2.4	A0
	2.33	0.03	2.0	
V451 Oph	3.0p	—	2.7	A0
	2.6p	—	2.2	
AR Aur	2.55	0.19	1.9	B9
	2.30	0.19	1.7	
U Oph	5.30	0.36	3.4	B5
	4.65	0.34	3.1	
Y Cyg	17.4	0.8	5.9	O9.5
	17.2	0.8	5.9	

p—Mass is known without appreciable scale error, but analysis of data is as yet provisional.   : or : :   Less-well-determined value.

TABLE I (*continued*)

Star	Masses ($\odot$)	m.e.	Radii ($\odot$)	Spectra
Systems with unequal components, primary on or near main sequence				
TV Cet	1.5p	—	—	F2
	1.4p	—	—	—
V477 Cyg	2.8p	—	1.4	A3
	1.4p	—	1.1	(F5)
CM Lac	1.88	0.09	1.6	A2
	1.47	0.04	1.4	(F2)
AS Eri	1.6p	—	1.6	A0
	0.2p	—	2.0	(K0)
Z Vul	5.4	0.3	4.7	B4
	2.3	0.1	2.0	A2
Subgiant systems				
SS Boo	0.91p	—	2: :	(G5)
	0.84p	—	2: :	(G8)
WW Dra	1.4p	—	2.3:	G2
	1.4p	—	3.9:	(K0)
AR Lac	1.32	0.06	1.8:	G2
	1.31	0.07	3.0:	(K0)
RS CVn	1.35	0.04	1.7:	F4
	1.40	0.03	4:	(K0)
Z Her	1.22	0.06	1.6	F4
	1.10	0.03	2.6	(K0)
Systems with giant and supergiant components				
RZ Cnc	3.1	0.2	11:	K1 III
	0.55	0.05	13:	K4 III
ζ Aur	8.3	1.6	160:	K4 II
	5.6	0.6	—	B7
31 Cyg	10	4	200:	K4 Ib
	6.6	0.9	—	B4

give position in the H-R diagram. The systematic photometric work and analysis by Eggen (5, 6, 11) have provided the relevant information for many of the important visual binaries.

For an eclipsing binary it is the radius rather than the luminosity that is usually well determined if the masses are known. The bolometric luminosities may then be computed if the effective temperatures are known. For stars near the main sequence of spectral type about A5 and later, the correlations of effective temperature with color $(B - V)$ and spectral type are well established. For the hotter as well as for the more evolved stars, the temperature relationships are less certain, and alternative methods of evaluating luminosities may be useful. Such methods include measurement of the equivalent widths of hydrogen lines (hotter stars) and use of some of the criteria of narrow- and intermediate-band photometry (62, 63). An example of application of the latter to eclipsing binaries is the work of McNamara (64). Blending of the light of the two components causes some difficulty in these evaluations of luminosity except in the method of effective temperatures, which has been the most widely used for eclipsing binaries.

In interpreting the positions of stars in a mass-luminosity diagram, it is essential to know their evolutionary states, which are normally evaluated by position in the H-R diagram. Since the positions of eclipsing binaries in the H-R diagram are determined only with some difficulty, evaluation of their evolutionary states by other methods may be desirable. One such method would be the use of intermediate-band photometry (62, 63). Another approach is to use a color-radius diagram, since these two parameters are usually available. Departure of a star from the lower envelope of such a diagram would presumably show its degree of departure from the main sequence.

ACKNOWLEDGMENT

Most of the spectrograms upon which the discussion in this review is based I obtained at the Mount Wilson Observatory as a guest investigator and at the Lick Observatory. The cooperation of the directors and staffs of these institutions is gratefully acknowledged. This observing program has been supported by the Office of Naval Research.

LITERATURE CITED

1. Struve, O., Kung, S.-M., *Ap. J.*, **117**, 1 (1953)
2. Wehlau, W. H., *Ap. J.*, **121**, 77 (1955)
3. Dworetsky, M. M., *Los Angeles Meeting of Am. Astron. Soc.* (1966)
4. West, F. R., *Astron. J.*, **71**, 186 (1966)
5. Eggen, O. J., *Ap. J. Suppl. 8*, 125 (1963)
6. Eggen, O. J., *Astron. J.*, **70**, 19 (1965)
7. Wilson, O. C. (In press)
8. Bless, R. C., *Ap. J.*, **132**, 532 (1960)
9. Wilson, O. C., Coffeen, M. F., *Ap. J.*, **119**, 197 (1954)
10. Bodenheimer, P., *Ap. J.*, **142**, 451 (1965)
11. Eggen, O. J., *Ann. Rev. Astron. Ap.*, **5**, 105 (1967)
12. Pickering, E. C., *Harvard Circ. No. 11* (1889)
13. Vogel, H. C., *Astron. Nachr.*, **121**, 241 (1889)
14. Myers, G. W., *Ap. J.*, **7**, 1 (1898)

15. Roberts, A. W., *Ap. J.*, **13**, 177 (1901)
16. Schlesinger, F., *Publ. Allegheny Obs.*, **1**, 123 (1909)
17. Schlesinger, F., Baker, R. H., *Publ. Allegheny Obs.*, **2**, 51 (1910)
18. Stebbins, J., *Ap. J.*, **34**, 112 (1911)
19. Russell, H. N., *Ap. J.*, **35**, 315; **36**, 34 (1912)
20. Irwin, J. B., *Astronomical Techniques: Stars and Stellar Systems*, **II**, 584 (Hiltner, W. A., Ed., Univ. of Chicago Press, Chicago, Ill., 1962)
21. Kitamura, M., *Publ. Astron. Soc. Japan*, **6**, 217 (1954)
22. Petrie, R. M., *Astronomical Techniques: Stars and Stellar Systems*, **II**, 63 (Hiltner, W. A., Ed., Univ. of Chicago Press, Chicago, Ill., 1962)
23. Moore, J. H., *Lick Obs. Bull. 11*, 141 (1924)
24. Wood, F. B., *Basic Astronomical Data: Stars and Stellar Systems*, **III**, 370 (Strand, K. A., Ed., Univ. of Chicago Press, Chicago, Ill., 1963)
25. Popper, D. M., *Trans. Intern. Astron. Union*, **XIIB**, 485 (1966)
26. Cester, B., *Z. Ap.*, **62**, 191 (1965)
27. Sahade, J., *Stellar Atmospheres: Stars and Stellar Systems*, **VI**, 466 (Greenstein, J. L., Ed., Univ. of Chicago Press, Chicago, Ill., 1960)
28. Popper, D. M., *Otto Struve Memorial Volume* (In press)
29. Popper, D. M., *Ap. J.*, **124**, 196 (1956)
30. Popper, D. M., *ibid.*, **133**, 148 (1961)
31. Popper, D. M., *Publ. Astron. Soc. Pacific*, **74**, 129 (1962)
32. Heard, J. F., Morton, D. C., *Publ. David Dunlap Obs.*, **2**, 255 (1962)
33. Popper, D. M., *Ap. J.*, **126**, 53 (1957)
34. Popper, D. M., *ibid.*, **129**, 659 (1959)
35. Popper, D. M., *ibid.*, **131**, 828 (1961)
36. Popper, D. M., *ibid.*, **141**, 126 (1965)
37. Popper, D. M., *Ap. J. Suppl. 3*, 107 (1957)
38. Wyse, A. B., *Lick Obs. Bull. 17*, 37 (1934)
39. Popper, D. M., *Ap. J.*, **124**, 208 (1956)
40. Struve, O., Sahade, J., Huang, S.-S., Zebergs, V., *Ap. J.*, **128**, 310 (1958)
41. Popper, D. M., *Ap. J.*, **134**, 828 (1961)
42. Harris, D. L., Strand, K. A., Worley, C. E., *Basic Astronomical Data: Stars and Stellar Systems*, **III**, 273 (Strand, K. A., Ed., Univ. of Chicago Press, Chicago, Ill., 1960)
43. Koch, R. H., Sobieski, S., Wood, F. B., *Publ. Univ. Pennsylvania, Astron. Ser.*, **IX** (1963)
44. Martynov, D. Ya., *Bibliografia Spektralno-dvoĭnykh Zvezd* (Izd. Astron. Sovetom Akad. Nauk SSSR, Moskva, 1961–1963)
45. International Astronomical Union, Commission 42
46. International Astronomical Union, Commission 27
47. Abt, H. A., *Publ. Astron. Soc. Pacific*, **77**, 367 (1965)
48. Roman, N. G., *Ap. J.*, **123**, 246 (1956)
49. Abt, H. A., *Vistas Astron.*, **8**, 75 (1966)
50. Popper, D. M., *Astron. J.*, **71**, 175 (1966)
51. Smak, J., *Publ. Astron. Soc. Pacific*, **76**, 210 (1964)
52. Hilton, W. B., McNamara, D. H. *Ap. J.*, **134**, 839 (1961)
53. Smak, J., *Acta Astron.*, **11**, 171 (1965)
54. Plaskett, J. S., *Publ. Dominion Ap. Obs.*, **1**, 138 (1919)
55. Abrami, A., *Mem. Soc. Astron. Ital.*, **29**, 381 (1958)
56. Struve, O., Sahade, J., Zebergs, V., *Ap. J.*, **129**, 59 (1959)
57. Petrie, R. M., Andrews, D. H., *Astron. J.*, **71**, 175 (1966)
58. Kopal, Z., *Ann. Ap.*, **19**, 299 (1956)
59. Bakos, G. A., Heard, J. F., *Astron. J.*, **63**, 302 (1958)
60. Kraft, R. P., *Advan. Astron. Ap.*, **2**, 43 (1963)
61. Sawyer, H. B., *Publ. David Dunlap Obs.*, **2**, No. 2 (1955)
62. Strömgren, B., *Basic Astronomical Data: Stars and Stellar Systems*, **III**, 123 (Strand, K. A., Ed., Univ. of Chicago Press, Chicago, Ill., 1963)
63. Strömgren, B., *Ann. Rev. Astron. Ap.*, **4**, 433 (1966)
64. McNamara, D. H., *I.A.U. Symp. 24*, 190 (1966)

MASSES OF VISUAL BINARY STARS[1]

By O. J. Eggen

Mount Stromlo Observatory, Canberra, Australia

INTRODUCTION

If determinations of stellar masses are as important astrophysically as one might believe, then the science of astrophysics owes a great debt to a succession of dedicated artists in double-star observation—R. G. Aitken (30 years with the Lick 36-inch refractor), G. van Biesbroeck (45 years with the Yerkes 40-inch refractor), W. H. van den Bos (40 years with the Johannesburg 26-inch refractor), and W. S. Finsen (25 years with an eyepiece interferometer on the 26-inch refractor). These long, consistent, and overlapping series of visual measures, together with an earlier series of nearly 40 years duration with a 9-inch refractor by O. Struve and supplemented by some 20 years of observations by Baize with the Paris Observatory refractors, provide the "standard observations" from which nearly all accurate mass determinations are derived.

Ironically this wealth of accurate measures has been grossly misused. The computed orbital elements have been revised and re-revised, in some cases almost yearly as these reliable measures became available and the literature clogged with what van den Bos ["Is This Orbit Really Necessary?" (1)] has compared with the publication of solutions to crossword puzzles. The aim seems to be to test, again and again, the accuracy of the relation $F = GM/R^2$.

The important result from the double-star orbits is the relation a^3/P^2, where a = the semimajor axis in seconds of arc and P = the period in years. Then the relation

$$\text{Total mass (unit, solar mass)/distance}^3 \text{ (pc)} = a^3/P^2$$

gives the mass.

The struggle for definitive values of the relatively unimportant orbital elements is often a long one. In 1835 Maedler produced the following elements for the equal components of 70 Ophiuchi only ten years after the first systematic observations by W. Struve:

	P	a	a^3/P^2	$m_1 + m_2$ ($\pi = 0''.188$)
1835	80.6 yr	4''.32	1.24×10^{-2}	$1.87 \odot$ (39)
1937	87.85 yr	4''.56	1.23	$1.85 \odot$ (40)

One hundred years and over 100 published orbits later, Strand computed a set of "definitive" elements based largely on photographic observations. However, the definitive value of a^3/P^2 differs insignificantly from that known 100 years previously.

[1] The survey of literature for this review was concluded in November 1966.

Other examples of this early determination of accurate values of a^3/P^2 are given elsewhere (2). The determinacy of the mass parameter a^3/P^2 essentially reflects the determinacy of the orbital inclination which, in turn, depends upon the accuracy with which the points of intersection of the orbital and sky planes (nodes) can be determined (2, 3). The criterion for an accurate determination of a^3/P^2 is the fact that one of the nodes is covered by the observations. The fairly reliable estimates of the mass parameter obtained in this way from poor orbits are especially valuable for types of stars for which masses are not otherwise available. For example, the following orbits are available for the old disk-type giants, γ Leo (ADS 7744):

	P	a	a^3/P^2	m_1+m_2 $(\pi=0\rlap{.}''025)$
1879	407.0 yr	$1\rlap{.}''98$	4.99×10^{-5}	$3.2\odot$ (41)
1956	701.4 yr	$2\rlap{.}''742$	4.19	$2.7\odot$ (42)
1958	618.6 yr	$2\rlap{.}''505$	4.11	$2.6\odot$ (43)

The group parallax (4) of $0\rlap{.}''025$ then gives a mean mass near $1.5 \odot$ and luminosities of $M_V = 0\rlap{.}^m8$ and $+0\rlap{.}^m4$ for the K0 III and G7 III components respectively. The colour-luminosity array of the group, like that for M67, shows a breakaway from the main sequence near $M_V = +3\rlap{.}^m5$ (mass $\sim1.5 \odot$).

ORBITAL ELEMENTS

The process of determining the orbital elements of a visual binary simply consists of (a) fitting the apparent ellipse (projected orbit) to the observations of separation (ρ, in seconds of arc) and position angle (θ), and (b) deprojecting this ellipse to obtain the true orbit. The means by which these operations are carried out are nearly as numerous as orbit computers. In recent years, however, the fundamental method used is that described by van den Bos (Thiele-Innes method) (5). More important than the method is the assessment and weighing of the individual observations. In general the orbits derived by observers of visual binaries are more reliable because the observers are more capable of assessing the value of the available observations. Most of the published orbits have been laboriously hand computed, the only check on the result being the comparison with the observations. In cases where an entire orbital period had not been covered by observations, the solution may not be unique but because of the labour involved, multiple solutions are seldom made.

Most of the orbits discussed here (except when available orbits appear to be definitive) have been computed with a 7094 computer using a programme written by John Castor of the Astronomy Department, California Institute of Technology. In general only the observations of the "Standard Observers" (6) already mentioned were used with weight dependent upon the number of observations; single observations were given very low weight. The programme called for a "basic solution," using the Thiele-Innes method, based on all observations. An additional 10 to 15 solutions were then made, each emphasizing a separate section of the observed arc. In the following, only solutions representing the extreme range found will be reproduced.

For convenience the following notation will be used:

A = Aitken
B = van den Bos
VB = van Biesbroeck
Fi = Finsen (interferometer)
JI = Jeffries (interferometer)
BZ = Baize
Bur = Burham
D = Dembrowski
S = W. Struve
OS = O. Struve

T = epoch of periastron passage
P = period in years
e = eccentricity of the orbit
a = semimajor axis in seconds of arc
i = inclination of orbit at place in degrees
A
$\dfrac{B}{F}$ = Thiele-Innes constants (5) in seconds of arc
G
θ = position angle in degrees
$\Delta\theta$ = observed minus computed position angle
$\Delta\rho$ = observed minus computed separation
ρ = separation in seconds of arc
t = epoch of observation
t' = epoch of observations $\pm P$

Three examples of the nonuniqueness of orbital elements are given below.

1. Twelve orbits were computed for ADS 3098. The range of solutions found is represented by the following:

	P	T	e	A	B	F	G	a	i	a^3/P^2
I	198.8	1969.47	0.29	−0.173	+0.402	+0.434	−0.012	0.52	130	3.55×10⁻⁶
II	228.5	1966.39	0.24	−0.270	+0.390	+0.375	+0.131	0.50	136	2.43
III	182.3	1971.07	0.34	−0.093	+0.387	+0.444	−0.128	0.53	125	4.48

The comparison with the observations (Table 1) indicates no preference for any of these orbits. Baize (7) has published an orbit similar to orbit II ($P = 223.^{y}4$, $a = 0.^{''}48$). The observed magnitude and colours for the combined light of the nearly equal components are $V_E = 7.^{m}05$, $B - V = +0.^{m}12$, $U - B = 0.^{m}08$. The photometric parallax (i.e. the parallax obtained by fitting the mean component to the Pleiades-Hyades main sequence) is $0.^{''}008$. The mean mass $m_1 = m_2$ then ranges as follows:

	m	M_V
I	3.5⊙	+2.^{m}2
II	2.4⊙	
III	4.4⊙	

2. Fourteen orbits were computed for ADS 9186. The range of solutions is represented by the following:

	P	T	e	A	B	F	G	a	i	a^3/P^2
I	148.2	1937.99	0.60	−0.278	−0.182	+0.078	−0.346	0.40	46	2.91×10⁻⁶
II	177.3	1938.18	0.64	−0.300	−0.215	+0.191	−0.366	0.435	47	2.64
Cou	204.6	1937.5	0.70	−0.394	−0.174	+0.073	−0.467	0.51	41	3.18

The orbit by Couteau (8), computed from essentially the same observations, is probably not acceptable (Table 2). A short ephemeris from orbit no. I (which results from a least-squares fit to all of the observations) is as follows:

t	θ	ρ	t	θ	ρ
1963.00	328	0.295	1985.24	8	0.415
1970.41	345	0.335	1992.65	16	0.455
1977.82	357	0.375			

3. Twelve orbits were computed for ADS 9982. The extreme range of elements is represented by the following:

	P	T	e	A	B	F	G	a	i	a^3/P^2
I	536.1	1907.43	0.83	−2.538	−0.019	+0.250	+1.869	2.565	135	58.9×10⁻⁶
II	391.1	1907.16	0.79	−2.155	−0.087	+0.153	+1.514	2.175	133	67.5
III	761.6	1907.15	0.87	−3.100	−0.151	+0.251	+2.189	3.145	133	53.6
Heintz	473.5	1906.9	0.82	⋯	⋯	⋯	⋯	2.455	133	65.7

The observations in Table 3 are satisfied equally well by all the orbits and by that determined by Heintz (9). The observed magnitude and colours of the combined light of the equal components (dK2) are $V_E = 8^m.74$, $B − V = +1^m.10$, $U − B = +1^m.05$. The photometric parallax of $0''.035$ gives the following mean masses:

	m	M_V
I	0.66⊙	$+7^m.2$
II	0.75⊙	
III	0.60⊙	
Heintz	0.73⊙	

A short ephemeris from orbit no. I is as follows:

t	θ	ρ
1959.00	28	2.165
1985.80	21	2.875
2012.60	16	3.425

An example of a stable value of a^3/P^2 for a large range of orbital elements is ADS 7846 for which 15 orbits were computed. The extreme range of results obtained is represented by the following:

	P	T	e	A	B	F	G	a	i	a^3/P^2
I	243.1	1948.38	0.82	−0.468	+0.812	+0.786	+0.058	1.045	127	19.4×10⁻⁶
II	157.3	1948.39	0.75	−0.346	+0.623	+0.646	+0.053	0.815	130	21.7
III	218.5	1948.57	0.81	−0.394	+0.793	+0.741	+0.023	0.98	128	19.8

The observations are in Table 4.

The photometric parallax of this mild subdwarf, based on the values of $V_E = 6^m.30$, $B − V = +0^m.48$, $U − B = −0^m.07$, is $0''.031$, giving the following mean masses for the nearly equal components:

	m	M_V
I	0.32⊙	+4.5
II	0.36⊙	
III	0.33⊙	

The stars are evidently evolved and the parallax is less than 0″031. A short ephemeris from orbit no. I is as follows:

t	θ	ρ
1964.00	334	0.675
1976.15	324	1.015
1988.31	318	1.255
2000.40	314	1.43

Two examples of definitive orbits are:

1. Fourteen orbits were determined for HR 3786 (ψ Vel) on the basis of van den Bos' remarkable series of observations that cover one orbital period (34 years). All orbits showed the small dispersion indicated for the following elements:

	P	T	e	A	B	F	G	a	i	a^3/P^2
I	33.99	1935.75	0.44	+0.453	−0.413	+0.088	+0.650	0.795	58.5	43.64×10⁻⁶
σ	0.15	0.12	0.00	0.004	0.006	0.004	0.006	0.005	0.5	0.65

The agreement with the observations in Table 5 shows the accuracy that can be obtained by an experienced observer. The photometric parallax obtained from the values $V_E = 3^m57$, $B-V = +0^m36$, $U-B = +0^m02$ is 0″059 compared to the trigonometric determinations of 0.059 Yale (wt. 16) and 0″058 Cape (wt. 8). The resulting mean mass of the equal components is $m_1 = m_2$ = 1.05 ⊙.

A short ephemeris is as follows:

t	θ	ρ	t	θ	ρ
1964.00	37	0.34	1979.62	116	0.205
1969.21	55	0.285	1984.83	159	0.20
1974.41	80	0.24	1990.04	202	0.21

2. ADS 7545 (ϕ UMa) consists of two, equal main-sequence A-type stars. The observations in Table 6 are well satisfied by the following mean elements, derived from 15 orbits;

P	T	e	A	B	F	G	a	i	a^3/P^2
104.1	1880.98	0.42	−0.327	+0.019	−0.103	−0.291	0.345	26.5	3.79×10⁻⁶
0.8	1.36	.02	.008	.008	.015	.010	.005	5.0	.19

Ambiguous Periods

When the separation of the nearly equal components of a close pair falls, temporarily, below the optical power of available telescopes, the observers may mistake the identification of the individual components when they again become visible and, as a result, misread the position angle by 180°. The result is the possibility of a double solution: a high eccentricity orbit with the companion, visible only in one or two quadrants, and a nearly circular orbit of twice the period. van den Bos (10) believes that the shorter-period orbit is in most cases the correct one. When accurate estimates for the parallaxes are available, it should be possible in some cases to differentiate between the two possibilities. Five cases in point are the following:

1. Observations of the equal components of ADS 9185 AB are given in Table 7. Twelve orbits (11) showed very small dispersion about the following elements:

	P	T	e	a	i	a^3/P^2
I	73.0	1930.85	0.06	0.30	110	5.18×10^{-6}
Baize	36.0	1931.8	0.90	0.28	124	16.93

Baize (12) has published an orbit with approximately half the period and a very high eccentricity. The residuals in Table 7 show no basis for a preference for either orbit. There is a third member of the system, separated from ADS 9185 AB by $4''$ and nearly identical to the equal components of the close pair:

	V_E	$B-V$	$U-B$	$R-I$	M_V	V_E-M_V	π(pt)
AB	$9^m.20$	$+0^m.87$	$+0^m.60$	$+0^m.28$	$+6^m.1$	$3^m.85$	$0''.017$
C	$9^m.75$	$+0^m.84$	—	$+0^m.28$	$+6^m.1$	$3^m.65$	$0''.0185$

The mean mass of the equal components of the close pair is then 0.44 ⊙ [not 0.55 ⊙ as given in (11)] from the long period and 1.45 ⊙ from the short period. Apparently the short period is incorrect.

2. Observations of the equal components of ADS 9397 are listed in Table 8. Two orbits (11), one with the starred values of θ in Table 8 changed by 180°, have been computed:

	P	T	e	a	i	a^3/P^2	$m_1 = m_2$
							($\pi = 0''.026$)
I	19.8	1935.17	0.05	0.155	36	9.05×10^{-6}	0.3⊙
B	9.85	1938.18	0.67	0.17	51	50.61	1.4⊙
II	9.92	1936.88	0.61	0.165	57	44.56	1.3⊙

van den Bos (10) has stressed van Biesbroeck's (VB) negative observation in 1946 and his own in 1958 in deciding that the short-period orbit is the correct one. However, his own orbit (13), indicated above and in Table 8 by B, represents these observations no better than does the long-period orbit no. I.

The photometric results for these late-type dwarfs (K2 V) are as follows:

V_E	$B-V$	$U-B$	$R-I$	M_V	π(pt)
$8^m.40$	$+0^m.88$	$+0^m.60$	$+0^m.34$	$+6^m.2$	$0''.026$

The resulting mean mass of 0.3 ⊙ for the long-period orbit is expected for a star $1^m.5$ fainter than the Sun whereas that from the short-period orbit, 1.4 ⊙, would appear to rule out this interpretation.

3. van den Bos (14) has derived the following sets of elements for ADS 9744

	P	e	a	i	a^3/P^2	$m_1=m_2$ $(\pi=0".016)$
B I	11.07	0.80	0.117	69	1.31×10^{-5}	1.60⊙
B II	22.14	0.00	0.210	84	1.89	2.3 ⊙

The comparison with the observation (15) gives no basis for deciding between the two alternatives. The equal components of the system are probably members of the Pleiades group:

Star	M_V	$B-V$	U	V	W	π
Pleiades star	$+1^m.4$	$+0^m.02$	$+9$	-27	-12	$0".008$
ADS 9744	$+1^m.3$	$+0^m.02$	$+8$	-27	-6	$0".016$

The resulting mean mass for the long-period orbit, 2.3 ⊙, is consistent with that for other Pleiades-group stars, as discussed below, but the smaller mass of 1.6 ⊙ is not clearly excluded. Also, if the pair is not a member of the Pleiades group and the components have evolved from the main sequence, both values of the mass given above will be increased.

4. Fifteen orbits have been computed for ADS 6314 on the assumption that the starred values of θ in Table 9 should be increased by 180°. The entire range of solutions is represented by the following:

	P	T	e	A	B	F	G	a	i	a^3/P^2	$m_2=m_1$ $(\pi=0".010)$
Cou	102.3	1931.33	0.10	$+0.153$	-0.165	-0.155	-0.141	0.225	159	1.33×10^{-6}	0.65⊙
I	55.8	1912.56	0.80	$+0.089$	-0.093	-0.237	-0.187	0.30	115	8.87	4.4⊙
II	55.3	1912.83	0.88	$+0.084$	-0.085	-0.314	-0.228	0.39	108	19.16	9.1⊙
III	57.1	1912.27	0.77	$+0.088$	-0.096	-0.224	-0.170	0.28	117	6.83	3.4⊙

Couteau (15), taking the values of θ as they stand in Table 9, derived a nearly circular orbit with about twice the period. The photometric results for this close pair and a distant (51″) companion are as follows:

	V_E	$B-V$	$U-B$	$R-I$	M_V	π(pt)
AB	$6^m.96$	$+0^m.235$	$+0^m.115$		$+2^m.0$	$0".010$
C	$10^m.88$	$+1^m.63$	$+2^m.00$	$+0".77$	$+9^m.1$	$0".044$

Unfortunately the distant companion is probably an optical one only. The photometric parallax of $0".01$ for the close pair leads to the unacceptable low mass of 0.6 ⊙ for the mean component, whereas that derived from orbits I and III is somewhat larger than expected. A parallax of $0".005$ would place the mean component 1^m above the main sequence and give the reasonable mass of 5 ⊙ from the long-period orbit, but unacceptably large masses from the short-period orbit. A luminosity estimate from spectra or narrowband photometry is needed to eliminate one of the two alternatives.

5. Several published orbits for the equal components of ADS 450 AB give

a period near 11 years. However, the well-determined photometric parallax obtained from the recently discovered proper-motion companion gives too large a mass for these late-type dwarfs (11), and the probability that the period should be doubled was suggested. Mr. L. S. T. Symms (16) has computed the following orbits:

	P	T	e	a	i	a^3/P^2	$m_1 = m_2$
							$(\pi = 0\overset{''}{.}023)$
Symms A	10.73	1940.50	0.58	0.155	140.4	3.23×10^{-5}	$1.3 \odot$
Symms B	21.2	1950.70	0.01	0.187	139.2	1.46	$0.6 \odot$

The photometric results are as follows:

	V_E	$B-V$	$U-B$	$R-I$	M_V	π(pt)
ADS 450 AB	$8\overset{m}{.}58$	$+0\overset{m}{.}85$	$+0\overset{m}{.}50$		$+ 6\overset{m}{.}3$	$0\overset{m}{.}025$
G 158–97	$14\overset{m}{.}59$			$+1\overset{m}{.}09$	$+11\overset{m}{.}2$	$0\overset{m}{.}021$
						$0\overset{m}{.}023$

The short-period orbit gives much too large a mean mass (1.3 $\odot$) and the long-period orbit is apparently the correct one.

HYADES CLUSTER AND GROUP

Some visual binaries that are also members of the Hyades cluster or group are listed in Table 10 (which contains the systems with reliable elements) and Table 11 (which contains pairs for which either the orbital elements are not considered reliable enough or the system is complicated by additional components). The orbits are thoroughly discussed and compared with the observations elsewhere (17).

The positions in the (M_V, log m) plane of the stars in Table 10 are shown in Figure 1: cluster members are indicated by the open circles, and group members by crosses. These stars are also shown in the (M_V, $B-V$) plane in Figure 2. The unit mass of the Sun ($M_V = 4\overset{m}{.}8$) is shown as a Sun symbol ($\odot$) in Figure 1. The group parallaxes for the group members, ADS 10140 AB and 10 UMa, are confirmed by other evidence; the physical, D component in the system of ADS 10140 falls in with the other stars of the same colour in the Hyades cluster, and the trigonometric parallax determinations for 10 UMa are $0\overset{''}{.}070$ Allegheny (wt. 16), $0\overset{''}{.}074$ McCormick (wt. 8), and $0\overset{''}{.}069$ Sproul (wt. 20), compared with the group value of $0\overset{''}{.}066$.

The stars in Figure 1 present the difficulty that although the Hyades cluster and group main-sequence stars follow a mean (M_V, log m) relation with a dispersion of about $0\overset{m}{.}1$, the Sun departs from this relation by $1\overset{m}{.}5$.

UMa CLUSTER AND THE SIRIUS GROUP

The binaries in the Sirius group (and UMa cluster), yielding accurate values of the mass, are listed in Table 12. The orbits are discussed elsewhere

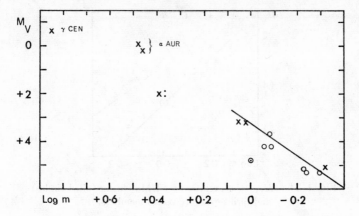

FIG. 1. The $(M_V, \log m)$ relation for Hyades cluster (open circles)
and group (crosses) members.

(11, 17). The resulting $(M_V, \log m)$ relation is in Figure 3 where the circles
represent cluster members and the crosses, group members. The $(M_V, \log m)$
relation for Hyades stars fainter than $M_V = +3^m$ is reproduced from Figure 1.
The $(M_V, B-V)$ relation for the stars in Figure 3 is shown in Figure 4.

Although the UMa cluster is closer to the Sun than is the Hyades cluster,
special problems are connected with determining the distance of individual
members (18) and there is a resulting uncertainty of $0^m.1$ or $0^m.2$ in the luminosities of the cluster members. The exact position of the cluster stars in the
$(M_V, B-V)$ plane is of some importance and accurate radial-velocity determinations of the members, which would help in determining accurate distances, are urgently needed (18).

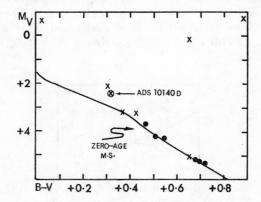

FIG. 2. The $(M_V, B-V)$ relation for the stars in Figure 1.

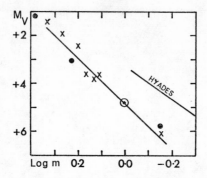

FIG. 3. The (M_V, log m) relation for UMa cluster (filled circles)
and Sirius group (crosses) members.

MULTIPLE SYSTEMS

Seven multiple systems, each consisting of a close pair of main-sequence
stars for which reliable values of a^3/P^2 are available and a distant, main-
sequence companion, are listed in Table 13. The system of α Cen (HD 126820/
1), for which the trigonometric parallax is 0 ."760, is also listed. The photometry and the orbital elements for the multiple system are discussed elsewhere (11, 17) except for new photometric observations of ADS 3159 AB, D.
The previously published photometry for this southern object ($\delta = -25°$)
was based on two observations in 1962 with the 100-inch reflector. Subsequent observations with the same instruments showed inconsistencies probably connected with observational difficulties. The photometric results given
in Table 12 are based on four observations with the 40-inch reflector at Siding
Spring Mountain.

The positions in the (M_V, $B-V$) and (M_V, log m) planes of the components of the close pairs in Table 12 are shown in Figure 5. The systems with
$\delta(U-B) < 0^{m}015$ are indicated by filled circles and those with $\delta(U-B)$
$\geq 0^{m}02$ (compared with Hyades cluster main-sequence stars) are indicated
by open circles.

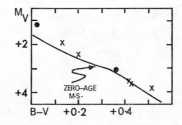

FIG. 4. The (M_V, $B-V$) relations for stars in Figure 3.

Figures 1, 3, and 5 show the fundamental data upon which is based the belief that two mass-luminosity relations exist for stars with M_V between $+3^m$ and $+6^m$. The stars populating the Hyades mass-luminosity relation are also distinguished from those populating the Sun-Sirius relation by at least two other characteristics.

1. The Hyades cluster and group members in Figure 1 and the three systems represented by open circles in Figure 5 populate the same ($U-B, B-V$) relation, whereas the stars shown as filled circles in Figure 5 and the members of the Sirius group in Figure 3 have values of $\delta(U-B)$, with respect to the ($U-B, B-V$) relation for the Hyades stars (19), of $0^m\!.02$ or more.

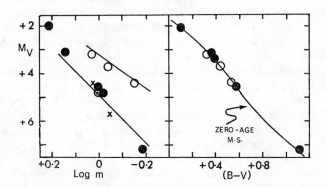

FIG. 5. The (M_V, $B-V$) and (M_V, log m) relation for the multiple system in Table 12. The systems with $\delta(U-B)<0^m\!.015$ are indicated by filled circles and those with $\delta(U-B)\geq0^m\!.02$ by open circles.

2. The results from a larger, and less select sample of stars (11, Figures 10 and 11) showed that the distribution in the (U, V) plane of those pairs populating the Sun-Sirius mass-luminosity relation is different from the distribution of those populating the Hyades relation and that this difference is the same as that found for single stars which (a) show an ultraviolet excess with respect to the Hyades stars (as do the binaries populating the Sun-Sirius mass-luminosity relation), and (b) populate the ($U-B, B-V$) relation defined by the Hyades cluster stars.

PLEIADES GROUP

Unfortunately an accurate value of a^3/P^2 is not available for any main-sequence binary brighter than about $M_V=+1^m$, for which an accurate determination of the parallax is also known. In an attempt to overcome this difficulty, the few systems containing relatively high-luminosity stars for which reliable values of a^3/P^2 are available have been examined for possible members of the Pleiades group (11). A selected list of candidates for group membership is given in Table 14 which contains the luminosity, mass, and space motion vectors (U, V, W) derived from the group parallaxes. The group parallax is that value making $V = -27$ km/s. The positions of these stars in the (M_V, log m) and (M_V, $B - V$) planes are shown in Figure 6 where the Pleiades main sequence, and a short segment of the mass-luminosity relation for Hyades stars in Figure 1, are shown as continuous curves. Two Hyades group members, γ Cen and ADS 16497, from Figure 1, are represented in Figure 6 by crosses.

The accuracy of the mass determination used to construct Figure 6 is considerably lower than is that of the stars in Figures 1, 3, and 5. This uncertainty not only arises in the correctness of the group assignment, but the group itself is probably part of the extensive "local association" (Gould Belt) of early-type stars in the solar neighbourhood (20, 21) and the degree of separation, in velocity, from that association is not known, so that the uniqueness of $V = -27$ km/s is questionable. Also, the spectral lines in most of these early-type systems are of poor quality and the computed values of V, insofar as they depend on the radial velocity, suffer in accuracy on this account. Accurate values of the mass for the stars of higher luminosity will depend almost entirely on the results from spectroscopic binaries.

A partial confirmation of the small slope shown in Figure 6 for the mass-luminosity relation for stars brighter than $M_V = +2^m$ should be mentioned. A discussion of evolved stars in well-observed binary systems, given in Volume 3 of these *Reviews* (21, Figure 8), showed that most of these evolved objects fall near the mass-luminosity relation shown as a broken line in Figure 6 when their original, main-sequence luminosity is considered.

RED DWARFS

The obvious dichotomy in the mass-luminosity relation, shown in Figures 1, 3, and 5, can be tested with the relatively large number of binaries fainter than $M_V = +6^m$. Twenty-four of these pairs, for all of which accurate values of a^3/P^2 can be derived, are listed in Table 15. New orbits have been derived for all except ADS 7284, 9031, 9716, 10075, 11046 and $-8°$ 4352, for which the definitive orbits in the literature, referenced in the notes to Table 15, have been used; the orbital elements, parallax, and mass ratio of the components of Ross 614 are by Lippincott (22). The new orbit determinations are based on 10 to 15 orbits computed for each system and, with two or three sets of elements, representing the entire range of solutions found, or a mean set with the small dispersion (σ) for the various parameters, are listed in Table 16. The representation of the observations is given in Tables 17 to 27, except in the few cases where this has been published elsewhere (11). The elements given in Table 16 show clearly the accuracy with which a^3/P^2 is known for these systems. Because accurate masses depend also on accurate distance determinations, most of these systems, which consist of nearly equal components, were observed in the ($R - I$) system of Kron, White &

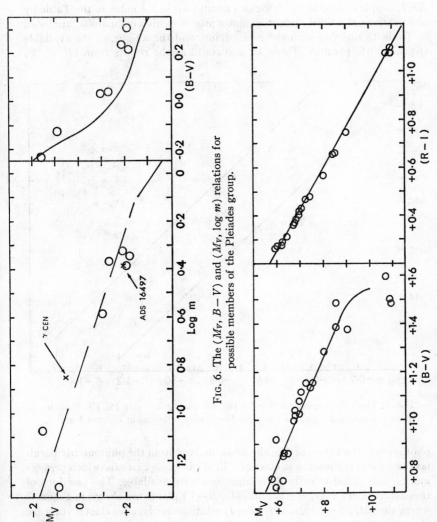

FIG. 6. The (M_V, $B-V$) and (M_V, log m) relations for possible members of the Pleiades group.

FIG. 7. The (M_V, $B-V$) and (M_V, $R-I$) relations for the red dwarfs in Table 15.

Gascoigne (23). The calibration of the $(M_V, R-I)$ relation for late-type dwarfs, based on stars with trigonometric parallax greater than 0″125, is given elsewhere (24). The photometric parallaxes obtained from the values of $R-I$, supplemented in the few cases mentioned in the notes to the Table by observations of a third distant common proper-motion companion, are listed in Table 15 together with the mean value, and the weight, of the available trigonometric parallax. Table 15 also contains the results from (U, B, V)

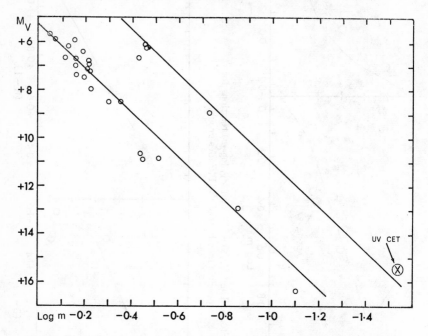

FIG. 8. The $(M_V, \log m)$ relation for the red dwarfs in Table 15. The straight lines are extrapolations for the mass-luminosity relations in Figures 3 and 5.

photometry, the values of M_V, the mass, derived from the photometric parallax, and the space motion vectors (U, V, W) in those cases for which proper-motion and radial-velocity determinations were available. The positions of these stars in the $(M_V, R-I)$ and $(M_V, B-V)$ planes are shown in Figure 7 where the $(M_V, B-V)$ and $(M_V, R-I)$ relations for Hyades cluster stars are also shown. Figure 8 contains the $(M_V, \log m)$ relation. The two straight lines in Figure 8 represent an extrapolation, to $M_V = +16^m$, of the Hyades (Figure 1) and Sun-Sirius (Figures 2 and 5) mass-luminosity relations. An additional, low-luminosity system, UV Cet (Table 11), which is a member of the Hyades group (17), is shown in Figure 8 by a circled cross.

The dichotomy in the mass-luminosity relations is obviously also present for stars with $M_V = +6^m$ to $+16^m$. It is surprising to find these relations to be approximately linear in M_V over the range $M_V = +6^m$ to $+16^m$ when the bolometric corrections may vary more than 2^m over the same range. However, the resulting curvature in the (M_{bol}, log m) relation is expected (25) as a result of the increasing depth of the convective envelope. The five stars in Figure 8 that populate the Hyades mass-luminosity relations are ADS 7114 BC, ADS 8048 BC, ADS 8166 AB, ADS 10158 AB, and UV Cet. The radial velocities are unknown for three of these and the other two have Hyades, or Hyades-like space motions.

Figures 1, 5, and 8 leave little doubt that the two nearly parallel mass-luminosity relations are populated by late-type, main-sequence stars.

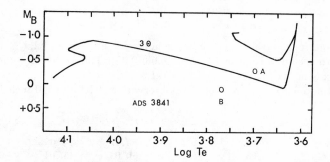

FIG. 9. The position of the components of α Aur in the (M_B, log m) plane. An evolutionary track for a 3 $\odot$ model, computed by Iben, is also shown.

LATE-TYPE GIANTS AND SUBGIANTS

Unfortunately there are no known halo stars among the few giants and subgiant binaries for which accurate values of a^3/P^2 are available. There are, however, a few late-type systems which probably populate subgiant sequences in the old disk (M67, NGC 188) population. Many of these pairs have attracted so little attention that not even radial-velocity determinations (by means of which they might be assigned to one of the groups among the old disk population) are available.

One red giant that has been thoroughly studied is ADS 3841 (α Aur). The orbital elements and the assignment to the Hyades group (17) give the results listed in Table 10. Adopting the bolometric corrections and temperature scale given elsewhere (17), the positions of these stars in the (M_{bol}, log T_e) plane are shown in Figure 9. The evolutionary track for a 3 $\odot$ model (26) is also shown.

Six systems that probably contain late-type subgiants are considered in the following.

1. The observations of ADS 12126 are listed in Table 28. Fifteen sets of orbital elements showed small dispersion from the following:

	P	T	e	A	B	F	G	a	i	a^3/P^2
I	110.5	1944.50	0.61	−0.092	−0.119	−0.134	+0.185	0.235	127	1.06×10⁻⁶
σ	4.4	0.35	0.01	0.005	0.003	0.003	0.005	0.005	2	0.05

The photometric results are $V_E = 6\overset{m}{.}76$, $B - V = +0\overset{m}{.}795$, and $U - B = +0\overset{m}{.}44$. The photometric parallax of $0\overset{.}{''}030$ gives the absurdly small mean mass of 0.04 ⊙ for the equal components. The following dependence of the mass on the parallax is shown in Figure 10:

M_V		$m_1 = m_2$
$0\overset{.}{''}010$	$+2\overset{m}{.}5$	0.5⊙
$0\overset{.}{''}008$	$+2\overset{m}{.}0$	1.0⊙
$0\overset{.}{''}007$	$+1\overset{m}{.}75$	1.5⊙
$0\overset{.}{''}006$	$+1\overset{m}{.}4$	2.4⊙

The evolutionary tracks shown in the figure for 1.5, 1.8, and 2.3 ⊙ (26) are based on a composition of $(X, Z) = (0.68, 0.03)$. The probable value of the parallax is near $0\overset{.}{''}0066$ ($M_B = 1\overset{m}{.}6$, $m = 1.8$ ⊙).

2. Photometric results from five observations with the 200-inch reflector of ADS 11579 are as follows:

	V_E	$B - V$	$U - B$	$\delta(U-B)$	C
AB	$6\overset{m}{.}94$	$+0\overset{m}{.}82$	$+0\overset{m}{.}41$	$+0\overset{m}{.}06$	$+0\overset{m}{.}88$
C	$8\overset{m}{.}83$	$+0\overset{m}{.}54$	$+0\overset{m}{.}03$	$+0\overset{m}{.}04$	$+0\overset{m}{.}575$

where $C = B - V + \triangle(B - V)$. The orbit of the close pair, which is discussed elsewhere (11), gives $P = 90$ years, $a = 0\overset{.}{''}225$ for $a^3/P^2 = 1.4 \times 10^{-6}$. Fitting

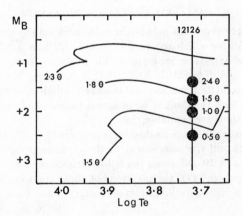

Fig. 10. Evolutionary tracks for 2.3, 1.8, and 1.5 ⊙ models. The values of M_B for ADS 12126, for various parallaxes, are shown as filled circles identified by the resulting mass.

the C component to the main sequence gives a photometric parallax of 0".0145, a mean mass of 0.23 ⊙, and $M_V = +3^m.5$ for the close pair. Apparently all three stars have evolved. However, the difficulty is illustrated in Figure 11 which contains the schematic colour-luminosity arrays for NGC 752 (28) and M67 (29). The values of $C = B - V + \triangle (B - V)$ for the mean AB component and the C component of ADS 11579 are indicated by vertical lines. Placing the mean component of the close pair on the subgiant sequence of M67 gives a parallax (0".014) similar to the photometric value derived by placing the C component on the main sequence. This value must be incorrect because (a) the resulting mass for the mean component of AB (0.3 ⊙) is much smaller than the value near 1 ⊙ expected (29, 30) for evolved stars in M67, and (b) the C component would be less evolved than other M67 stars of the same colour. A parallax of 0".0077 gives a mass of 1.5 ⊙ for the mean component and places the stars as shown in Figure 11. This solution would also

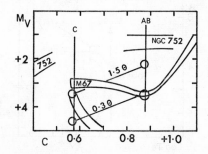

Fig. 11. The position of the component of the multiple system ADS 11579 in the (M_V, $B - V$) plane for various values of the parallax. The schematic arrays for NGC 752 and M67 are also shown.

place the mean components of the close pair on the evolutionary track for 1.5 ⊙ shown in Figure 10; the resulting age of the system is then about 2×10^9 years (26). However, the zero-age main-sequence luminosity for the mean component of the close pair was then about $M_V = +3^m$, or only 0.5 brighter than the present luminosity of component C. In the extreme, component C was at $M_V = +4.5$ at $t = 0$, so its mass is near 1 ⊙ and at $t = 2 \times 10^9$ years its luminosity should be considerably fainter than it now appears. An extremely important possibility is that the mean component of the close pair, at $M_V = +2^m$, is similar to the red "horizontal-branch" stars in M67 (29). The proper motion of ADS 11579 and the parallax of 0".0077 lead to (U, V, W) $= (-28 - 0.47\rho,\ -7 + 0.84\rho,\ -27 + 0.27\rho)$ and if the system has a radial velocity near $\rho = -30$ km/s, the resulting values of $(-14, -33, -35)$ would make it a good candidate for membership in the Wolf 630 group (31), which has a colour-luminosity array identical to that for M67.

3. The 11 orbits computed for ADS 14424 are represented by the following elements:

	P	T	e	A	B	F	G	a	i	a^3/P^2
I	113.5	1935.34	0.32	+0.229	−0.334	+0.102	+0.123	0.405	68	5.25×10^{-8}
II	97.2	1934.80	0.24	+0.194	−0.311	+0.102	+0.110	0.37	66	5.29

The comparison with the observations is given elsewhere (12). The photometric results are $V_E = 7^m.85$, $B - V = +0^m.63$, $U - B = +0^m.13$, $\delta(U - B) = +0^m.04$, $C = +0^m.66$. The photometric parallax of $0''.02$ gives a mean mass of 0.33 ⊙ for the equal components, and the stars have obviously evolved from the main sequence. The dependence of the mean mass on the parallax is as follows:

π	m	M_V
$0''.015$	0.75⊙	$+4^m.5$
$0''.013$	1.15⊙	$+4^m.1$

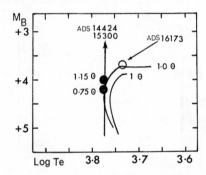

FIG. 12. The positions of ADS 14424 and 15300 in the (M_B, log m) plane for various parallaxes. The position of ADS 16173 is obtained from the common propermotion companion (see p. 123).

The resulting positions in the (M_{bol}, log T_e) plane are shown as filled circles in Figure 12 together with evolutionary tracks for 1 and 1.03 ⊙ models (31) with $(X, Z) = (0.67, 0.03)$ and $l = 1.6H$. The parallax of $0''.013$, which gives $(U, V, W) = (0, -3, -46)$, seems to be approximately correct.

4. The orbit of ADS 15300 is discussed elsewhere (11). The photometric results are $V_E = 8^m.81$, $B - V = +0^m.63$, $U - B = +0^m.13$, $\delta(U - B) = +0^m.04$, $C = +0^m.66$, and the equal components have the same temperature as those in ADS 14424 discussed above. Also, as in ADS 14424, the photometric parallax of $0''.0125$ yields the absurdly small mean mass of 0.37 ⊙. The dependence of the mean mass on the parallax is as follows:

π	m	M_V
$0''.010$	0.75⊙	$+4^m5$
$0''.0085$	1.75⊙	$+4^m2$

and the components are probably identical to those of ADS 14424 shown in Figure 12. The parallax of $0\overset{''}{.}0085$ gives $(U, V, W) = (-49, -42, +21)$.

5. The orbit of ADS 16173 AB is discussed elsewhere (11). The photometric results, including new observations of $R-I$, are as follows:

	V_E	$B-V$	$U-B$	$R-I$	$\delta(U-B)$	C
AB	$5\overset{m}{.}75$	$+0\overset{m}{.}72$	$+0\overset{m}{.}22$		$+0\overset{m}{.}06$	$+0\overset{m}{.}78$
C	$14\overset{m}{.}88$	$+1\overset{m}{.}70$		$+1\overset{m}{.}25$		

The value of $R-I$ for the C component leads to $M_V = +12\overset{m}{.}3$ or π (pt) $=0\overset{''}{.}030$. The orbital elements of $P = 20.93$, $a = 0\overset{''}{.}30$ give $a^3/P^2 = 61.6 \times 10^{-6}$ and a mean mass of 1.15 $\odot$, placing the star in the $(M_{bol}, \log T_e)$ plane as shown in Figure 12. The model computation (30) indicates that the components are nearly 10^{10} years old. When a grid of models, for different masses and composition, becomes available and the $(R-I, M_V)$ relation for such cool stars is more definitely calibrated, this system should prove of great importance for testing evolutionary theories. The space motion vectors are $(U, V, W) = (+48, -6, 0)$ and the stars are members of the old disk population.

6. The orbit of HD 117440 (See 179) is discussed elsewhere (11). The photometric results are $V_E = 3\overset{m}{.}88$, $B-V = +1\overset{m}{.}19$, $U-B = +1\overset{m}{.}03$. The orbital elements of $P = 78.\overset{y}{.}7$, $a = 0\overset{''}{.}165$ give $a^3/P^2 = 0.725 \times 10^{-6}$. The dependence of the mean mass of the equal components on the parallax is as follows:

π	m	M_V	M_V (Model)
0.010	0.36 $\odot$	-0.4	....
0.0075	0.85 $\odot$	-1.0	....
0.0062	1.5 $\odot$	-1.4	(0.0)
0.005	3.0 $\odot$	-1.9	-1.5
0.0042	5.0 $\odot$	-2.3	-2.3
0.0034	9.0 $\odot$	-2.7	-4.6

The parallax is probably not larger than $0\overset{''}{.}007$ despite the spectral type of G8 III (32). The comparison with evolutionary models (26) indicates a parallax near $0\overset{''}{.}0042$ and $m = 5$ $\odot$. The space motion vectors are then $(U, V, W) = (+3, -7, -12)$. The components are only slightly fainter and less massive than the K-type component of ζ Aur $(M_V \sim -3^m, m = 8 \odot)$.

SUMMARY

As it seems unlikely that the dichotomy in the mass-luminosity relations of Figures 1, 2, 3, and 8 is caused by errors in a^3/P^2, the only alternatives are (a) the Hyades, and Hyades-like main-sequence stars in the solar neighbourhood populate a different mass-luminosity relation from that of the other stars; or (b) the accepted distance of the Hyades cluster, and therefore the position in the $(M_B, \log T_e)$ plane of the Hyades main sequence, is incorrect. The supporting arguments for the first alternative—the differences in the

colours and kinematics of the two kinds of stars—are discussed elsewhere
(11). The second alternative has many ramifications.

Wayman, Symms & Blackwell (33) have recently completed a thorough
discussion of the distance of the Hyades cluster, as determined from the
convergent-point method applied to the proper motions. They found little
change from the previously accepted distance (34). They also concluded that:

(a) there is little evidence for a significant systematic rotation of the
cluster with respect to the standard frames of reference, and a small rotation
of this kind, if it did exist, would not affect the zero point of the absolute
magnitudes derived from the convergent-point method;

(b) the systematic differences of the proper motions of the Hyades stars
in the various fundamental systems (GC, N30, FK3) could lead to a systematic error in the distance moduli of only $0^{m}.06$, and an upper limit of 0.5 km/s
on the possible systematic error in the radial velocities could lead to only
$0^{m}.03$.

Wayman et al. find that the Hyades cluster can be represented by a mean
star at a distance of 40.6 pc. The cluster binaries represented in Figure 1 are
at a mean distance of 40 pc ($\pi = 0''.025$) and to make the resulting mass-
luminosity relation coincide with the Sun-Sirius relation would require an 11
per cent increase in the distance to 44.4 pc ($\pi = 0''.022$). The distance of the
"mean star" of Wayman et al. would then be at 45 pc and the main sequence of the Hyades cluster would be $0^{m}.23$ brighter than now believed.

Of course the consequences of this result would be wonderful and many.
They would not, however, solve the difficulty presented by Figures 5 and 8.
Although the "Hyades-like" stars in Figures 5 and 8 would be moved to the
Sun-Sirius mass-luminosity relation, those now populating the latter relation
will be similarly affected because the photometric parallaxes have been all
obtained from the same (M_V, $B-V$) relation. However, before discussing the
consequences of such a change in the Hyades main sequence, it is more germane to examine the probability that such a change is warranted. As mentioned above, the presently available astrometric results discussed by Wayman et al. give no basis for a change of this magnitude. The other direct
measurement of the cluster distance involves the trigonometric parallax
determinations. Those available for cluster members (33, Table 8) are listed
with their weights (35) in Table 29 where they are divided into determinations from the Allegheny Observatory (A) and the mean of results from all
observatories. The catalogue (35) precepts for systematic corrections, which
have been adapted in Table 29, have been challenged by several investigators (e.g. 36) who suggest that instead of these precepts, the published relative parallaxes should be used after correction, from standard tables (37, 38),
for the parallaxes of the comparison stars. For the Hyades the different procedures give essentially the same result: the values in Table 29 would be increased to $0''.0247$ for the Allegheny parallaxes and $0''.0236$ for the mean
values if the alternative procedure were adopted. Considering both the spread
of the individual parallaxes and the uncertainty of the correction from rela-

tive to absolute parallaxes, the trigonometric results give a mean parallax 0.ʺ0236 which agrees with that derived from the convergent-point method, 0.ʺ0246.

Note added in proof: After preparation of Table 29 the article by P. W. Hodge and G. Wallerstein (*Publ. Astron. Soc. Pacific*, **78**, 464, 1966) was received in Australia. The essential differences between Table 29 and their Table I result from their confusing *relative* parallaxes, taken for some Allegheny stars from the right-hand page of the parallax catalogue (35), and *absolute* values for the mean parallax on the left-hand page. Although Hodge & Wallerstein use three arguments to support their speculation that the Hyades distance, obtained from the convergent point of the proper motions, is in error by a large amount, one of these arguments—that the mass-luminosity relation does not fit with that of some field stars—is the *raison d'être* for the speculation, and a second argument—the questioning of the Wilson-Bappu effect—rests on one, controversial (11, ADS 1630) star. Since their remaining argument, the mean trigonometric parallax, is based on a partial misinterpretation of the available data, the challenge to the accepted Hyades distance appears unsupported.

TABLE 1. OBSERVATION OF ADS 3098

t	θ	ρ	n	Obs.	I		II		III	
					$\Delta\theta$	$\Delta\rho$	$\Delta\theta$	$\Delta\rho$	$\Delta\theta$	$\Delta\rho$
1829.52	320	0.54	4	S	−6	−0.06	−2	−0.07	−6	−0.02
1845.14	316	0.65	3	D	+2	+0.03	+6	+0.05	+3	+0.03
1904.83	250	0.36	2	Bur	+3	−0.04	+8	−0.03	+2	−0.03
1908.01	241	0.38	2	VB	+1	−0.01	+4	0.00	0	0.00
1913.10	229	0.36	4	VB	−1	−0.02	+2	−0.01	−2	−0.01
1920.92	216	0.44	4	VB	+3	+0.07	+5	+0.06	+3	+0.08
1927.41	200	0.40	4	VB	+1	+0.02	+2	+0.02	+2	+0.03
1943.77	164	0.39	3	VB	−4	−0.02	−3	0.00	−2	−0.02
1951.78	151	0.35	2	VB	−2	−0.06	−2	−0.04	0	−0.06
1958.02	141	0.36	5	VB	0	−0.03	0	−0.02	+1	−0.03
1960.97	136	0.39	5	VB	−1	+0.01	−2	+0.01	0	+0.01
1962.81	133	0.35	4	B	+2	−0.01	+1	−0.02	+3	0.00

TABLE 2. OBSERVATION OF ADS 9186

t	θ	ρ	n	Obs.	I		II		Couteau	
					$\Delta\theta$	$\Delta\rho$	$\Delta\theta$	$\Delta\rho$	$\Delta\theta$	$\Delta\rho$
1900.31	59	0.49	3	Hu	−1	−0.02	−4	−0.04	+2	−0.09
1916.23	68	0.43	3	Doo	−8	+0.05	−3	+0.01	−4	0.00
1923.27	90	0.34	5	VB	0	+0.04	−1	+0.05	+5	+0.02
1937.43	206	0.20	4	B	0	+0.07	0	+0.08	+4	+0.02
1944.27	262	0.24	2	VB	0	+0.04	0	+0.05	−3	+0.03
1950.33	292	0.24	2	VB	+2	0.00	+2	+0.01	+1	−0.03
1953.54	302	0.21	3	VB	+1	−0.04	+1	−0.03	+1	−0.09
1956.14	312	0.19	5	VB	+2	−0.07	+2	−0.06	+4	−0.13
1959.19	312	0.28	2	B	−7	0.00	−7	+0.01	−3	−0.06
1962.26	325	0.29	4	B	−2	0.00	−2	0.00	+4	−0.07

TABLE 3. OBSERVATION OF ADS 9982

t	θ	ρ	n	Obs.	I		II		III		Heintz	
					$\Delta\theta$	$\Delta\rho$	$\Delta\theta$	$\Delta\rho$	$\Delta\theta$	$\Delta\rho$	$\Delta\theta$	$\Delta\rho$
1830.94	346	2.54	4	S	+9	−0.03	+3	−0.02	+8	+0.01	+3	−0.04
1845.20	338	2.25	4	Ma	+5	+0.04	0	+0.04	+4	+0.09	0	0.00
1852.63	334	2.11	2	Ma	+4	+0.10	−1	+0.10	+3	+0.16	+1	+0.10
1867.64	325	1.45	4	D	+3	−0.10	−2	−0.10	+2	−0.04	−1	−0.10
1896.53	268	0.71	3	A	+4	+0.13	+3	+0.15	+3	+0.17	+6	+0.11
1897.25	257	0.59	3	A	−3	+0.02	−3	+0.04	−4	+0.07	−3	+0.04
1899.30	241	0.54	3	A	−7	+0.02	−6	+0.03	−7	+0.06	−5	+0.03
1903.38	218	0.44	2	A	+1	−0.02	+3	−0.03	+1	0.00	+3	−0.03
1905.43	198	0.40	3	A	−1	−0.04	0	−0.06	−1	−0.02	−1	−0.04
1906.39	191	0.38	3	A	0	−0.05	+2	−0.07	+1	−0.03	+4	−0.04
1911.44	133	0.38	2	A	−4	−0.02	−6	−0.03	−3	+0.01	−1	−0.03
1912.56	129	0.43	2	A	+4	+0.02	+2	+0.01	+5	+0.05	+6	+0.02
1914.44	106	0.51	2	A	−1	+0.06	−2	+0.06	0	+0.09	0	+0.06
1916.32	93	0.52	3	VB	0	+0.01	0	+0.01	+2	+0.04	+3	0.00
1916.41	92	0.50	2	A	0	−0.01	0	−0.01	+2	+0.01	+3	−0.02
1918.00	83	0.58	5	VB	+1	+0.01	0	+0.01	+2	+0.03	+1	−0.02
1920.39	69	0.64	2	A	−2	−0.04	−2	−0.04	−2	−0.02	−1	−0.06
1920.83	70	0.65	6	VB	0	−0.04	0	−0.04	+1	−0.04	+1	−0.07
1922.72	62	0.74	5	VB	−2	−0.04	−1	−0.04	−1	−0.02	−2	−0.07
1924.32	58	0.89	3	VB	−2	+0.04	−1	+0.04	−1	+0.06	−2	+0.01
1925.96	55	0.85	3	VB	−1	−0.08	0	−0.08	−1	−0.06	0	−0.10
1929.03	49	1.04	3	VB	−1	−0.02	0	−0.03	−2	0.00	−1	−0.06
1932.40	46	1.21	3	VB	0	0.00	+1	−0.01	−1	+0.02	−1	−0.04
1936.34	42	1.42	3	VB	0	+0.05	+1	+0.04	−1	+0.07	0	+0.01
1939.82	40	1.54	5	VB	+1	+0.03	+2	+0.02	0	+0.05	0	−0.02
1948.62	35	1.86	2	VB	+2	+0.03	+3	+0.02	0	+0.04	+1	−0.02
1958.33	31	2.12	4	B	+2	−0.02	+3	−0.02	0	−0.02	+1	−0.07

TABLE 4. OBSERVATION OF ADS 7846

t	θ	ρ	n	Obs.	I Δθ	I Δρ	II Δθ	II Δρ	III Δθ	III Δρ
1892.25	287	1.06	3	Bur	+4	−0.10	−2	−0.02	+6	−0.11
1926.25	282	0.55	1	B	−2	0.00	−4	−0.01	−1	−0.02
1928.70	254	0.56	6	B	+5	+0.05	+4	+0.04	+6	+0.04
1933.13	235	0.43	4	B	−2	+0.01	−3	0.00	−2	0.00
1935.92	224	0.37	4	B	−3	+0.01	−3	−0.01	−2	0.00
1938.21	215	0.37	4	B	−1	+0.05	0	+0.03	−1	+0.04
1940.75	202	0.30	2	B	+2	+0.01	+2	−0.01	+2	+0.01
1945.61	157	0.20	4	B	−1	−0.03	−2	−0.04	−1	−0.03
1947.32	137	0.18	4	B	0	−0.02	−1	−0.03	0	−0.02
1948.23	124	0.17	4	B	+1	0.00	+2	−0.01	+1	−0.01
1949.35	97	0.17	4	B	−1	+0.02	+1	+0.02	−2	+0.02
1951.30	39	0.16	2	B	−4	0.00	−3	0.00	−5	+0.01
1951.38	39	0.133	3	Fi (I)	−2	−0.023	−1	−0.036	−3	−0.020
1952.36	23	0.182	6	Fi (I)	+1	−0.009	+1	−0.025	+1	−0.004
1953.79	9	0.24	2	Fi (I)	+4	−0.015	+3	−0.033	+5	−0.009
1959.14	340	0.41	4	B	−2	−0.08	−2	−0.10	+1	−0.08
1960.20	341	0.56	4	B	+1	+0.02	+1	+0.01	+5	+0.03
1963.33	338	0.69	4	B	+4	+0.04	+4	+0.03	+7	+0.04

TABLE 5. OBSERVATION OF ψ VELORUM

t	θ	ρ	n	Obs.	Δθ	Δρ	t	θ	ρ	n	Obs.	Δθ	Δρ
1926.14	190	0.53	2	B	0	+0.01	1937.17	17	0.25	6	B	+2	0.00
1927.23	204	0.50	3	B	0	+0.01	1937.33	24	0.26	8	B	+1	0.00
1928.14	215	0.47	5	B	−1	−0.01	1938.20	55	0.33	6	B	−1	0.00
1929.25	231	0.50	4	B	0	+0.02	1939.30	77	0.47	6	B	−1	−0.01
1930.29	244	0.52	6	B	0	+0.02	1940.23	86	0.55	3	B	−2	−0.05
1931.43	257	0.47	4	B	−1	−0.06	1941.23	96	0.71	3	B	+1	0.00
1932.19	264	0.56	6	B	−2	+0.02	1945.04	113	0.98	4	B	+1	+0.02
1933.17	277	0.60	6	B	0	+0.07	1946.78	119	0.98	6	B	+2	−0.02
1934.18	290	0.50	6	B	+1	+0.01	1948.82	124	0.98	8	B	0	−0.02
1935.24	306	0.41	6	B	0	+0.01	1951.84	134	0.95	7	B	0	+0.03
1935.89	322	0.33	4	B	0	0.00	1956.26	157	0.70	4	B	+2	0.00
1936.05	327	0.32	4	B	+1	+0.01	1959.16	177	0.55	4	B	−2	0.00
1936.26	332	0.29	4	B	−2	0.00	1960.26	191	0.43	8	B	1	−0.08

TABLE 6. OBSERVATION OF ADS 7545

t	θ	ρ	n	Obs.	Δθ	Δρ	t	θ	ρ	n	Obs.	Δθ	Δρ
1845.38	9	0.43	8	S	+ 3	−0.02	1912.12	308	0.40	2	A	0	−0.02
1851.15	15	0.32	7	S	− 1	−0.09	1917.90	320	0.47	1	A	+1	+0.02
1856.22	27	0.37	9	S	0	0.00	1921.33	324	0.46	2	A	−1	−0.01
1863.58	48	0.32	5	S	− 1	+0.02	1924.22	327	0.55	3	VB	−2	+0.07
1872.42	78	0.23	2	S	−16	+0.01	1928.35	335	0.51	2	A	0	+0.02
1887.43	219	0.23	4	Sp	+ 6	+0.02	1931.22	337	0.56	2	VB	−2	+0.07
1892.13	251	0.24	3	Bur	+ 7	0.00	1935.02	342	0.50	8	BZ	−2	+0.01
1896.92	267	9.24	3	A	0	−0.04	1944.25	359	0.47	3	VB	+1	0.00
1902.47	288	0.32	3	Hu	+ 1	−0.02	1946.11	0	0.49	4	VB	0	+0.02
1903.38	286	0.30	2	A	− 3	−0.04	1958 29	22	0.40	4	B	0	+0.01
1905.35	294	0.32	2	A	0	−0.04	1963.06	36	0.35	4	B	+2	0.00

TABLE 7. OBSERVATION OF ADS 9185

t	θ	ρ	n	Obs.	I		BZ	
					Δθ	Δρ	Δθ	Δρ
1905.37	241	0.22	3	A	0	0.00	+1	−0.01
1914.48	220	0.28	2	A	− 1	−0.02	+1	−0.01
1917.36	218	0.33	2	A	+ 2	+0.04	+4	+0.04
1921.41	203	0.28	2	A	− 5	+0.02	−3	0.00
1926.36	192	0.27	2	A	0	+0.09	−2	+0.04
1933.56	too	close	2	A	$\rho_c =$	0.10	=	0.09
1936.42	89	0.16	2	A	+ 1	+0.04	+4	+0.02
1940.12	59	0.26	4	B	− 1	+0.05	−5	+0.03
1945.20	50	0.30	3	V	+ 3	+0.02	0	+0.03
1950.33	40	0.28	3	VB	+ 2	−0.02	0	−0.01
1955.53	25	0.19	4	VB	− 3	−0.09	−5	−0.09
1957.71	20	0.16	3	VB	− 4	−0.09	−6	−0.11
1962.24	19	0.19	4	B	+10	0.00	+7	−0.03

TABLE 8. OBSERVATION OF ADS 9397

t′	t	θ	ρ	n.	Obs.	I		B		II	
						Δθ	Δρ	Δθ	Δρ	Δθ	Δρ
1952.92	1933.12	304*	0.14	2	VB	+ 5	0.00	− 1	−0.04	0	−0.03
	1933.56	309*	0.18	2	A	+ 2	+0.04	− 1	0.00	− 1	0.00
1954.14	1934.34	317*	0.16	2	VB	− 2	+0.01	− 8	−0.02	− 3	−0.02
	1934.55	319*	0.20	2	A	− 4	+0.06	− 6	+0.02	− 5	+0.03
	1935.43	333	0.16	2	A	− 4	+0.02	− 5	−0.02	− 4	+0.01
1956.32	1935.52	16	0.095	2	VB	+20	−0.04	+14	−0.06	+ 2	+0.03
1916.40	1936.20	7	0.19	3	A	+15	+0.06	+18	+0.02	+ 8	+0.09
	1936.34		<0.15	1	VB	$\rho_c =$	0.13	=	0.16	=	0.09
	1936.98	346	0.165	4	B	−22	+0.03	−24	+0.03	+10	+0.10
1958.55	1939.75		<0.10	2	B	$\rho_c =$	0.13	=	0.10	=	0.12
1920.47	1940.27	80	0.16	2	A	0	+0.03	+ 4	0.00	+ 6	+0.03
1960.58	1940.78	65	0.13	5	VB	−20	0.00	−24	−0.03	−18	−0.01
1962.22	1942.42	113	0.18	6	B	− 3	+0.03	− 4	0.00	− 1	+0.02
1923.45	1943.25	128	0.16	3	A	0	0.00	+ 2	−0.02	0	−0.01
1963.40	1943.60	134	0.17	6	B	+ 2	+0.01	0	−0.01	+ 3	0.00
	1943.94	136	0.17	3	VB	− 1	+0.01	− 6	−0.01	0	−0.01
1924.54	1944.34	143	0.20	2	A	0	+0.04	+ 1	+0.02	+ 1	+0.03
	1945.32	163	0.16	2	VB	+ 7	0.00	+ 4	−0.02	+ 6	+0.01
1926.36	1946.16	168	0.16	2	A	− 1	+0.01	− 3	−1	−13	+0.07
	1946.52		<0.1	1	VB	$\rho_c =$	0.14	=	0.15	=	0.05
1927.46	1947.26	too	close	2	A	$\rho_c =$	0.14	=	0.12	=	0.09
	1950.33	248	0.13	2	VB	− 2	0.00	−14	−0.03	− 8	0.00

TABLE 9. OBSERVATION OF ADS 16314

t	θ	ρ	n	Obs.	Cou		I		II		III	
					Δθ	Δρ	Δθ	Δρ	Δθ	Δρ	Δθ	Δρ
1893.75	112*	0.25	3	Ho	−2	+0.05	+5	+0.02	+4	+0.03	+8	+0.03
1906.82	64*	0.20	1	A	+4	0.00	−1	+0.01	−3	0.00	+4	+0.02
1915.84	207	0.21	2	A	−2	+0.01	+3	+0.04	+8	+0.02	+3	+0.04
1918.11	194	0.22	2	A	−1	0.00	0	+0.02	+3	0.00	0	+0.02
1920.65	185	0.24	2	A	−4	+0.02	−1	+0.01	+1	0.00	−1	+0.02
1923.81	180	0.23	2	A	+1	0.00	+3	0.00	+3	−0.02	+3	0.00
1928.38	161	0.25	6	VB	+1	+0.01	−4	+0.01	−5	0.00	−4	+0.01
1928.71	168	0.24	2	A	+8	0.00	+4	0.00	+3	−0.01	+4	0.00
1933.12	151	0.24	5	VB	−1	0.00	−3	0.00	−3	0.00	−2	0.00
1940.40	135	0.23	3	VB	+3	−0.02	+2	0.00	0	0.00	+1	0.00
1941.10	132	0.20	4	J(I)	+2	−0.05	0	−0.03	−1	−0.03	0	−0.03
1943.73	126	0.21	3	VB	+3	−0.04	+2	−0.02	+1	−0.01	+2	−0.02
1945.58	120	0.18	2	J(I)	+2	−0.06	0	−0.04	−1	−0.04	−1	−0.04
1950.02	106	0.25	3	VB	0	+0.01	0	+0.02	0	+0.04	0	+0.03
1954.78	91	0.16	4	VB	−1	−0.07	0	−0.06	0	−0.05	−1	−0.06
1958.49	77	0.17	3	VB	−1	−0.05	−3	−0.04	−2	−0.04	−3	−0.04
1958.58	81	0.21	4	B	+3	−0.01	+2	0.00	+2	0.00	+1	0.00
1960.72	73	0.23	3	Wor	+1	+0.01	0	+0.03	0	+0.03	−1	+0.03
1961.76	68	0.24	4	B	−1	+0.02	−1	+0.04	0	+0.04	−1	+0.04

TABLE 10. HYADES CLUSTER AND GROUP BINARIES WITH WELL-DETERMINED ORBITS

ADS	M_V	$m(\odot)$	ADS	M_V	$m(\odot)$	U	V	W
Cluster			*Group*					
3135 A	+4.2	0.81	3841 A	−0.07	3.0	+37.8	−16.7	−10.0
3135 B	+5.2	0.59	B	+0.18	2.9			
3248 AB	+4.2	0.87	10 UMa A	+3.2	0.90	+40.3	−16.8	− 3.4
3248 B	+5.3	0.50	B	+5.1	0.48			
3210 AB	+5.2	0.59	γ Cen AB	−0.6	7.0	+40.4	−16.9	− 4.7
			ᵃ10140 AB	+3.2	1.04	+40.3	−16.8	+10.6
			16497 AB	+2.1	2.5±	+40.4	−16.4	− 4.4

ᵃ 10140D: $V_E = 8.02$, $B - V = +10.32$, $U - B = +0.06$, $M_V = +2^{m}.4$.

TABLE 11. HYADES CLUSTER AND GROUP BINARIES NOT USED IN THE DISCUSSION

ADS	M_V	$m(\odot)$	ADS	M_V	$m(\odot)$	U	V	W
Cluster			*Group*					
φ 342 AB	+4.1	—	UV Cet AB	+15.5	0.03	+39.5	−17.0	−20.4
				+15.4	0.04	+42.6	−18.7	−19.7
3017 AB	+4.9	0.8	15176 AB	+ 5.0	+0.6:	+38	−17	− 5
3169 AB	+5.1	—						
3475 AB	+4.4	0.88						
3730 BC	+5.5	—						
4617 AB	—	—						

TABLE 12. MEMBERS OF THE SIRIUS GROUP

ADS	M_V	$M(\odot)$	U	V	W		
UMa cluster							
8739 A	+3.ᵐ1	1.7	−14.2	+2.5	− 9.0	78 UMa	
8739 B	+5.8	0.7					
8891 Aa	+1.1	+2.45	−14.6	+2.6	− 8.6	ζ UMa	
Group							
1598 A	+1.9	1.8	−12	+3	−10		
1598 B	+3.6	1.5					
1709 AB	+3.6	1.3	−14	−2	−12		
5423 A	+1.4	2.2	−14.4	−0.4	−11.5	Sirius,	$\pi_t = 0.375$
9094 AB	+3.8	1.35	−12	−3	− 9		
9747 AB	+2.4	1.6	−12	+2	− 9		
17178 AB	+6.1	0.7	−12	0	−11		

TABLE 13. MULTIPLE SYSTEM

ADS/HD	V_E	$B-V$	$U-B$	$\delta(U-B)$	Sp	M_V	$m(\odot)$	U	V	W	π(pt)	π(tr) (0″001)
3159 AB	6.08	+0.36	+0.05	−0.02	dF2	+3.2	1.1	+ 6	−25	−11		
3159 D	8.16	+0.59	+0.10	0.00	G0	+4.7					20.5	49(10)
6650 AB	5.05	+0.54	+0.06	+0.01	F8 V	+4.4	0.7	−10	−10	− 2	52	40(98)
6650 C	6.20	+0.60	+0.13	0.00	dG2	+4.8					52	
6811 A	5.02	+0.32	+0.07	(−0.01)	dF1	+3.2		+16	−24	− 6		—
6811 BC	7.82	+0.52	−0.04	+0.09	dF6	+4.8	0.95				17	
9247 A	5.14	+0.01	−0.01	—	A0 V	+1.2		+24	−14	−10		
9247 BC	6.86	+0.43	−0.01	+0.01	A9 V	+3.7	0.9				16.5	8(48)
HD 126820	0.0	—	—	—	G2 V	+4.4	1.1	+29	+ 2	+13	—	760
HD 126821	1.0	—	—	—	dK5	+5.8	0.9					—
9689 A	6.98	+0.27	+0.04	+0.05	A2	+2.8		—	—	—	11	—
9689 BC	7.09	+0.34	+0.01	+0.03	A3	+3.1	1.40				11	
10660 A	5.33	+0.56	+0.07	+0.03	G1 V	+4.5	1.0	−37	− 1	−21		64(65)
10660 B	8.06	+1.10	+1.0			+7.2	0.65					
10660 C	9.95	(R-I) =	+0.78		M1 V	+9.1					68	
14775 AB	7.46	+0.11	+0.02	—	A0	+2.0	1.6	+12	− 2	− 7		—
14775 C	10.52	+0.56	+0.05	+0.03	—	+4.3		−.54ρ	+.65ρ	−.53ρ	7	

TABLE 14. POSSIBLE MEMBERS OF THE PLEIADES

ADS	M_V	$m(\odot)$	U	V	W
Pleiades	—	—	+ 9	−27	−12
2799	+1.9	2.1	+ 5	−27	+25
4265	−1.6	11.7	+15	−27	−15
7555	+1.0	3.8	+13	−27	−23
HD 114529	−0.9	20.2	+10	−27	+ 1
9301	+2.2	2.2	+10	−27	+ 4
9744	+1.3	2.3	+ 8	−27	− 6
10360	+2.1	2.4	+10	−27	−12

TABLE 15, RED DWARFS

ADS	V_B	$B-V$	$U-B$	$R-I$	M_V	Sp	$m(\odot)$	U	V	W	π_{pt}	$\pi_{tr(wt.)}$
												(0″001)
1729 AB	9.00				+7.0	K8	0.7	−20	+11	−13	28	46 (8)
1865 AB	8.68	+1.39	+1.11	+0.42	+8.5	dM2	0.45	+20	+7	+5	65	65 (20)
Ross 614 A	10.9			+1.38	+12.9	dM4e	0.14	+11	+28	+3		248 (77)
614 B	14.4				+16.4		0.08					
[a]6554 AB	7.90	+0.85	+0.48	+0.28	+6.2	dK2	0.75	−23	−14	0	32.5	40 (57)
[a]6664 AB	9.34	+1.49	+1.06	+0.66	+8.5	dM0	0.50	−11	−29	−37	50	53 (25)
[a]7114 BC	7.1	+1.02	+0.81	+0.75	+6.1	dM1	0.35	+28	−16	−20		66 (32)
[a]7284 AB	7.22	+1.38	+1.15	+0.36	+6.7	dK4	0.70	+24	−39	−20	56	56 (72)
[a]8048 BC	10.25	+1.08	+0.22	+0.405	+9.0	M0 V	0.19	+11 ±	−14 ±	−14 ±	40	—
[a]8166 AB	7.63	+1.02	+1.01	+0.385	+6.9	K0	0.35				37	—
[a]8635 AB	9.27	+0.72	+0.24	+0.27	+6.8	K8	0.60	+4	+15	+ρ^a	23	—
[a]8680 AB	9.20	+1.12	+1.04	+0.425	+5.9	K5	0.6	+20	−18	+ρ^a	23.5	—
8901 AB	8.28	+1.29	+1.27	+0.57	+7.0	dK6	0.85				22.5	62 (31)
[a]9031 AB	7.04	+0.92	+0.70	+0.275	+8.0	dM0	0.6	+32	−9	−10	78	40 (28)
[a]9352 AB	9.12	+0.86	+0.46	+0.315	+5.9	dK4	0.6	+15	−40	−14	41	52 (24)
[a]9716 AB	6.80	+1.00	+0.87	+0.37	+6.7	dK2	0.7	+43	−65	−31	48	57 (27)
[a]10075 AB	7.01	+1.04	+0.92	+0.38	+6.7	dK6	0.65	+53	−13	+5	58	—
[a]10158 AB	9.96	+1.60	+1.06	+1.08	+10.7	dM3e	0.38				17.5	12 (7)
[a]10188 AB	8.36	+1.49	+0.99	+1.08	+10.9	dM4	0.77	−4	−14	+3	33	155 (81)
[a]−8°4352 AB	8.98	+1.14	+1.14	+0.48	+7.4	dM0	0.37	−23	−33	+12		151 (48)
+45°2505 AB	9.38	+1.16	+1.05	+0.47	+7.2	K5	0.31	+50	−23	−21	151	151 (23)
10585 AB	9.01	+1.05	+0.33	+1.10	+10.9	dM4	0.70	−28	−40	+10	34	
[a]+27°2853 AB	9.20	+1.50			+5.7	K0 V	0.60	+53	−15	+9	29.5	
[a]10786 BC	9.80	+0.76			+7.5	dK6	0.36	−17	−32	−5	117	118 (48)
[a]11046 A	4.20	+1.15					0.90	−9	−22	−22		119
11046 B	6.00						0.65					

to (U, V, W) respectively.

[a] ADS 6554 C: $V_B=11^m40$, $B-V=+1^m30$, $M_V=+8^m9$.

ADS 6664 AB: The values of $m(\odot)$ range from 0.42 $\odot$ to 0.58 $\odot$ for the orbits in Table 16. The radial velocity may be variable.

7114 A: UMa $V_B=3^m12$, $B-V=+0^m18$, $U-B=+0^m07$ A7E.

7284: Definitive orbit by van den Bos (Union Obs. Contrib. No. 99, 448, 1938). +12%, −64%, and +76% of the unknown radial velocity, should be applied to (U, V, W) respectively.

8048 A: $V_B=7.64$, $B-V=+0.33$, $U-B=+0.77$, $U-B=+0.33$, $R-I=+0.26$.

8166: Comparison with the observations is given elsewhere (12).

8635, 8680: Comparison with the observations is given elsewhere. The W motion reflects only the unknown radial velocity.

9031: The identical orbits of Rabe (Astron. Nachr., 231, 121, 1927), from the visual observations, and those of Strand (Astron. J., 60, 42, 1955), from the photographic observations, have been used.

9352 C: $V_B=10.10$, $B-V=+1.32$, $U-B=+1.30$, $R-I=+1.30$, $R-I=+10.59$. The values of $m(\odot)$ range from 0.52 to 0.64 from the orbits in Table 16.

9716 C: $V_B=7.62$, $B-V=+0.95$, $U-B=+0.76$, $R-I=+0.29$. The definitive orbit by Young-Stephens (Publ. Am. Astron. Soc., 9, 170, 1939) has been used.

10075: The definitive orbit by Siegrist (Univ. Madrid, Fac. Sci. Publ. No. II, 1952) with $a=2.″415$ (12) was used.

10158: The values of $m(\odot)$ range from 0.36 $\odot$ to 0.40 $\odot$ from the orbits listed in Table 16. The comparison with the observations is given elsewhere (12).

10188: The values of $m(\odot)$ range from 0.70 to 0.85 from the orbits in Table 16.

−8°4352: C component $V_B=11.70$, $B-V=+1.70$, $U-B=+1.35$, $R-I=+1.22$; D component $V_B=16.66$, $B-V=+2.05$. The definitive orbit by Wieth-Knudsen (Lund Ann., 12, 1953) has been used.

+27°2853 C: $V_B=11.85$, $B-V=+1.44$, $U-B=+1.29$, $R-I=+0.78$. The comparison with the observations is given elsewhere (12): the third orbit listed in that reference has too long a period.

10786 BC = μ Her BC. The definitive orbit by Silbernagel (Astron. Nachr., 233, 257, 1928) has been used.

11046: The definitive orbit by Strand (Astron. J., 57, 97, 1952) and the trigonometric parallax have been used.

TABLE 16. RED-DWARF ORBITAL ELEMENTS

ADS	Orbit	P	T	e	A	B	F	G	a	i	a^3/P^2 $\times 10^6$	$m_1=m$
1729	I	35.52	1936.33	0.66	+0.086	−0.217	−0.073	−0.255	0.335	110	29.73	0.67
	σ	0.40	0.18	0.04	0.003	0.004	0.015	0.015	0.010	2	2.80	0.04
1865	I	25.32	1937.63	0.21	+0.218	−0.255	−0.065	+0.451	0.545	74	254.0	0.46
	σ	0.09	0.18	0.02	0.010	0.011	0.008	0.011	0.010	1	20.0	0.04
6554	I	44.58	1909.75	0.41	+0.159	+0.272	−0.261	+0.350	0.465	54	50.27	0.73
	σ	0.21	0.18	0.01	0.007	0.003	0.008	0.004	0.005	1	1.70	0.02
6664	I	53.85	1935.90	0.65	+0.503	−0.555	+0.392	+0.288	0.75	50	146.45	0.58
	II	49.25	1935.83	0.62	+0.452	−0.542	+0.381	+0.287	0.705	48	144.24	0.57
	III	63.56	1935.91	0.68	+0.528	−0.526	+0.384	+0.281	0.75	52	104.75	0.44
7114	I	39.69	1918.58	0.32	+0.562	+0.301	+0.307	−0.096	0.68	108	200.2	0.35
	σ	0.53	0.66	0.02	0.020	0.012	0.050	0.030	0.01	1	4.9	0.01
8048	I	23.71	1920.33	0.05	−0.055	+0.180	−0.182	−0.132	0.24	47	24.6	0.19
	σ	0.20	0.40	0.02	0.005	0.010	0.008	0.008	0.01	2	1.5	0.01
8166	I	49.0	1959.60	0.11	−0.349	+0.256	+0.254	+0.331	0.435	164	34.0	0.33
	σ	0.2	0.43	0.00	0.018	0.019	0.019	0.025	0.005	3	1.4	0.02
8635	I	62.5	1958.44	0.25	+0.025	+0.345	−0.374	+0.066	0.39	30	15.0	0.62
	II	61.3	1959.00	0.26	−0.005	+0.356	−0.367	+0.026	0.39	24	14.4	0.60
8680	I	124.1	1911.85	0.42	+0.368	−0.317	+0.097	+0.533	0.635	56	16.4	0.63
	II	136.3	1912.50	0.46	+0.394	−0.246	+0.110	+0.584	0.635	51	13.9	0.53
8901	I	43.65	1936.80	0.52	−0.297	−0.091	+0.053	−0.308	0.33	28	19.1	0.84
	σ	0.30	0.25	0.02	0.010	0.006	0.009	0.009	0.01	3	1.5	0.05
9352	I	52.04	1929.45	0.09	+0.579	−0.102	−0.192	−0.296	0.61	121	83.73	0.60
	II	51.33	1928.69	0.10	+0.604	−0.076	−0.116	−0.292	0.615	119	88.50	0.64
	III	52.20	1929.94	0.10	+0.55	−0.093	−0.184	−0.290	0.58	122	71.90	0.52
10158	I	133.8	1950.13	0.55	+0.241	−0.287	−0.307	−0.271	0.41	156	3.85	0.36
	II	122.8	1951.82	0.53	+0.208	−0.335	−0.330	−0.164	0.405	153	4.34	0.40
	III	111.9	1951.83	0.50	+0.163	−0.335	−0.297	−0.164	0.375	154	4.19	0.39
10188	I	131.86	1895.65	0.46	+0.838	−0.305	+0.143	−0.578	0.97	118	52.77	0.73
	II	121.24	1894.79	0.43	+0.829	−0.251	+0.214	−0.578	0.965	117	61.57	0.85
	III	138.31	1896.07	0.47	+0.857	−0.033	+0.119	−0.571	0.99	117	50.66	0.70
+45° 2505	I	12.98	1952.20	0.75	+0.315	+0.545	+0.636	−0.294	0.71	150	21.40	0.31
		0.03	0.25	0.01	0.005	0.005	0.003	0.004	0.01	2	70	0.01
10585	I	62.04	1974.80	0.17	−0.005	+0.191	−0.258	−0.512	0.60	82	55.52	0.70
	σ	1.00	0.98	0.02	0.004	0.010	0.006	0.002	0.01	1	2.50	0.04
+27° 2853	I	26.0	1960.25	0.21	+0.222	−0.119	−0.147	−0.233	0.28	153	31.7	0.61
	II	24.4	1961.06	0.21	+0.180	−0.180	−0.196	−0.176	0.265	160	31.7	0.61

TABLE 17. OBSERVATIONS OF ADS 1729

t'	t	θ	ρ	$n.$	Obs.	$\Delta\theta$	$\Delta\rho$
1944.66	1909.14	138	0.18	4	V	−12	+0.01
	1909.84	146	0.23	2	A	0	+0.05
1949.91	1914.39	122	0.30	6	VB	0	0.00
	1918.06	119	0.32	1	A	+ 6	−0.06
	1919.39	110	0.43	2	A	0	+0.03
	1921.36	106	0.43	3	A	− 1	0.00
1957.81	1922.29	106	0.34	2	VB	+ 1	−0.09
1958.38	1922.86	103	0.45	4	B	− 2	+0.01
1961.67	1926.15	100	0.42	4	B	0	−0.02
	1930.44	90	0.34	3	VB	− 2	−0.02
	1933.88	84	0.20	2	A	+ 5	+0.01
	1936.96	274	0.18	3	B	+ 1	+0.06
	1938.12	261	0.21	9	V	+ 3	+0.06

TABLE 18. OBSERVATION OF ADS 1865

t'	t	θ	ρ	$n.$	Obs.	$\Delta\theta$	$\Delta\rho$
1929.21	1903.89	236	0.25	2	A	+13	+0.06
1955.79	1905.15	253	0.19	4	VB	− 1	−0.07
1957.85	1907.21	274	0.27	2	VB	− 1	−0.14
1959.70	1909.08	284	0.49	4	Wor	− 3	+0.02
1959.92	1909.28	286	0.47	2	B	− 1	0.00
1934.99	1909.67	285	0.43	2	B	− 4	−0.04
1961.67	1911.03	298	0.37	4	B	+ 1	−0.03
	1911.38	300	0.38	3	A	0	+0.02
1936.94	1911.62	303	0.36	4	B	+ 1	+0.02
1937.93	1912.61	311	0.26	3	B	− 5	+0.03
1940.00	1914.68	40	0.15	1	B	0	−0.01
	1919.33	108	0.60	2	A	0	+0.01
1945.63	1920.31	114	0.60	3	B	+ 3	0.00
1945.74	1920.42	114	0.62	5	VB	+ 2	+0.02
	1921.22	113	0.60	4	A	− 1	0.00
1948.72	1923.40	122	0.58	3	VB	0	+0.06
1948.73	1923.41	122	0.49	4	B	0	−0.02
1950.99	1925.67	130	0.42	2	VB	− 6	+0.07

TABLE 19. OBSERVATION OF ADS 6554

t'	t	θ	ρ	$n.$	Obs.	$\Delta\theta$	$\Delta\rho$
	1906.11	324	0.35	3	A	− 1	+0.08
	1909.27	46	0.20	2	A	0	+0.03
	1912.25	101	0.30	2	A	− 1	0.00
1958.08	1913.49	114	0.31	3	B	+ 1	−0.03
	1916.16	136	0.40	2	A	+ 3	+0.03
1961.88	1917.29	139	0.37	4	B	− 1	0.00
	1919.14	157	0.39	2	VB	+ 4	+0.03
	1920.97	166	0.39	2	VB	− 1	+0.04
	1921.87	172	0.35	3	A	− 1	0.00
	1924.17	190	0.34	3	A	− 1	−0.01
	1924.69	192	0.37	2	VB	− 2	+0.02
	1926.62	208	0.42	2	VB	0	+0.05
	1927.71	231	0.45	2	B	+15	+0.07
	1932.89	244	0.46	5	B	0	0.00
	1933.28	246	0.49	4	B	+ 1	+0.03
	1934.25	250	0.47	4	B	+ 1	−0.01
	1938.17	264	0.51	4	B	0	−0.01
	1942.37	277	0.47	4	VB	− 1	−0.06
	1945.94	290	0.44	2	VB	− 1	−0.03
	1947.12	298	0.43	3	VB	+ 1	0.00
	1948.24	302	0.36	2	B	− 1	−0.03

TABLE 20. OBSERVATION OF ADS 6664

t	θ	ρ	$n.$	Obs.	I		II		III	
					$\Delta\theta$	$\Delta\rho$	$\Delta\theta$	$\Delta\rho$	$\Delta\theta$	$\Delta\rho$
1900.30	128	1.02	3	Hu	0	+0.02	−1	+0.05	− 1	+0.03
1911.26	144	0.86	6	Doo	0	+0.01	−2	+0.01	− 3	+0.01
1923.66	159	0.93	3	VB	− 1	0.00	−2	+0.02	− 3	0.00
1932.30	208	0.33	2	B	+13	+0.07	+5	+0.08	+15	+0.08
1933.30	229	0.31	2	B	− 4	−0.04	−4	−0.01	− 4	−0.03
1934.28	266	0.27	2	B	+ 1	+0.02	+1	+0.06	+ 2	+0.02
1935.24	300	0.20	2	B	− 5	+0.03	−7	+0.08	− 3	+0.03
1937.18	345	0.23	5	B	+ 3	0.00	+2	0.00	+ 1	0.00
1938.21	15	0.24	4	B	+ 2	+0.02	+2	+0.03	+ 1	+0.02
1939.24	36	0.28	2	B	0	+0.06	+1	+0.08	− 1	+0.06
1940.74	65	0.32	4	B	0	−0.02	+1	0.00	− 1	−0.03
1943.64	91	0.59	2	B	− 6	+0.07	−4	+0.09	− 5	+0.06
1947.33	106	0.86	4	B	− 1	−0.07	−2	−0.05	+ 3	−0.04
1952.33	119	0.98	3	B	− 1	0.00	0	+0.02	0	+0.01
1959.62	127	1.24	7	B	+ 1	−0.03	+3	−0.01	+ 3	−0.04

TABLE 21. OBSERVATION OF ADS 7114 BC

t'	t	θ	ρ	$n.$	Obs.	$\Delta\theta$	$\Delta\rho$
	1924.04	348	0.30	4	VB	− 1	−0.02
	1925.12	330	0.22	4	VB	− 2	−0.04
	1926.27	304	0.19	2	VB	− 1	−0.03
	1927.98	270	0.26	2	VB	+ 6	0.00
	1931.09	239	0.42	3	VB	+ 9	−0.04
	1932.22	231	0.49	2	VB	+ 7	−0.05
	1933.69	219	0.55	6	VB	0	−0.08
	1935.22	215	0.69	2	VB	0	−0.03
	1936.83	210	0.79	6	VB	− 1	0.00
	1938.19	207	0.76	3	VB	− 2	−0.08
	1940.28	205	0.90	6	VB	0	+0.02
	1942.64	199	0.95	6	VB	− 2	+0.08
1903.38	1943.07	203	0.93	2	A	+ 3	+0.06
1905.05	1944.74	196	0.88	2	Hu	− 1	+0.06
	1944.76	196	0.93	7	VB	− 1	+0.11
	1946.06	191	0.87	3	VB	− 4	+0.01
	1947.18	190	0.70	2	VB	− 2	−0.01
1909.22	1948.91	187	0.63	2	Bur	0	+0.04
	1950.69	183	0.40	6	VB	+ 4	−0.04
	1952.67	154	0.18	5	VB	− 3	−0.08
	1954.12	113	0.15	3	VB	0	−0.02
	1956.35	58	0.17	3	VB	+12	−0.12
	1957.28	36	0.38	3	VB	0	+0.01
1920.86	1960.55	15	0.42	5	VB	+ 1	−0.05
	1962.56	4	0.36	4	B	+ 3	−0.02

TABLE 22. OBSERVATION OF ADS 8048

t'	t	θ	ρ	$n.$	Obs.	$\Delta\theta$	$\Delta\rho$
	1925.85	215	0.21	2	A	0	−0.02
1904.18	1927.89	237	0.22	2	A	0	−0.03
1956.18	1932.47	300	0.12	2	VB	+ 8	−0.06
	1933.34	295	0.16	2	B	−12	−0.05
	1933.39	310	0.19	1	A	+ 2	−0.01
	1934.35	17	0.20	2	B	+50	+0.02
1960.26	1936.55	1	0.21	4	B	− 8	+0.01
	1937.26	19	0.20	2	B	− 1	−0.01
1962.30	1938.59	41	0.22	2	B	+ 4	−0.01
	1940.79	64	0.23	2	B	+ 2	0.00
1917.79	1941.50	70	0.25	3	A	0	+0.02

TABLE 23. OBSERVATION OF ADS 8901

t'	t	θ	ρ	$n.$	Obs.	$\Delta\theta$	$\Delta\rho$
1962.34	1918.69	30	0.52	4	B	+2	+0.04
	1919.15	28	0.44	3	A	−2	−0.03
	1921.31	35	0.44	2	A	−1	−0.02
	1924.46	43	0.48	2	A	−4	+0.05
	1928.27	63	0.44	2	VB	0	+0.08
	1933.52	115	0.22	2	A	+5	+0.02
	1934.33	123	0.30	4	VB	−2	+0.12
	1935.49	156	0.16	2	A	0	0.00
	1940.74	281	0.20	2	VB	+2	−0.03
	1943.30	307	0.26	2	VB	0	−0.02
	1945.32	322	0.32	2	VB	0	0.00
	1950.30	349	0.42	3	VB	+1	+0.02
1907.54	1951.19	355	0.40	3	A	+3	−0.01
	1953.03	357	0.40	7	VB	−2	−0.04

TABLE 24. OBSERVATION OF ADS 9352

t'	t	θ	ρ	$n.$	Obs.	I		II		III	
						$\Delta\theta$	$\Delta\rho$	$\Delta\theta$	$\Delta\rho$	$\Delta\theta$	$\Delta\rho$
	1912.42	128	0.40	3	F0X	+5	+0.03	−2	+0.01	+6	+0.04
	1920.95	31	0.48	4	VB	0	+0.07	−1	+0.07	−1	+0.06
	1923.70	12	0.55	4	VB	−4	+0.03	−3	+0.05	−4	+0.06
	1927.34	0	0.47	3	VB	+1	−9.08	+1	−0.09	0	−0.06
	1929.42	351	0.48	3	VB	+1	−0.05	+1	−0.06	0	−0.02
	1930.28	347	0.45	3	VB	+1	−0.06	+1	−0.07	+1	−0.04
	1933.48	328	0.44	2	VB	+1	+0.03	+1	+0.02	+2	+0.05
	1934.88	316	0.33	2	VB	+2	−0.04	0	−0.04	+2	−0.02
	1936.40	298	0.33	3	VB	+1	+0.01	−1	+0.01	+2	+0.02
	1936.57	291	0.44	2	B	−4	+0.12	−6	+0.12	−3	+0.13
	1937.32	292	0.35	2	VB	+7	+0.04	+5	+0.05	+8	+0.06
	1932.42	283	0.32	4	B	−1	+0.02	−2	+0.02	0	+0.03
	1938.34	278	0.30	2	VB	+8	0.00	+6	+0.01	+8	+0.01
	1939.31	256	0.31	3	VB	−1	0.00	−1	+0.01	0	+0.01
	1940.36	241	0.36	2	VB	−2	+0.03	−2	+0.04	−2	+0.03
	1940.74	238	0.35	3	B	−1	0.00	0	+0.02	−1	+0.01
	1942.28	220	0.38	2	VB	−4	−0.02	−3	−0.01	−5	−0.01
	1943.33	217	0.45	2	VB	+1	+0.01	+2	+0.02	0	+0.02
	1944.26	209	0.47	2	VB	−1	−0.01	0	0.00	−2	+0.01
	1945.36	201	0.51	2	VB	−3	−0.01	−2	0.00	−4	+0.01
	1949.23	188	0.62	3	VB	−1	−0.01	0	−0.01	−2	+0.02
	1950.33	186	0.64	3	VB	0	−0.01	+2	−0.02	0	+0.02
	1953.52	177	0.62	3	VB	+1	−0.04	+2	−0.05	0	−0.01
1902.51	1954.55	171	0.68	3	Hu	−2	+0.03	−3	+0.01	−2	+0.06
1907.22	1959.26	164	0.63	2	Doo	+8	+0.08	+5	+0.05	+8	+0.11
	1961.52	146	0.48	3	B	+1	+0.01	0	0.00	+1	+0.02
	1962.23	140	0.45	4	B	−1	0.00	−1	0.00	−1	+0.02

TABLE 25. OBSERVATION OF ADS 10188

t'	θ	ρ	$n.$	Obs.	I		II		III	
					$\Delta\theta$	ρ	$\Delta\theta$	ρ	$\Delta\theta$	ρ
1870.62	131	0.93	5	D	0	+0.04	−3	+0.01	+2	+0.05
1881.42	105	0.46	3	Bur	+2	0.00	−1	0.00	+3	0.00
1883.73	89	0.42	6	sp	0	+0.05	−1	+0.05	+1	+0.05
1887.11	50	0.32	6	sp	−4	+0.02	−2	+0.03	−4	+0.03
1892.56	356	0.40	3	Com	−1	+0.01	+1	−0.02	−2	+0.02
1895.54	341	0.43	3	Com	0	−0.05	+1	−0.09	0	−0.04
1897.36	329	0.50	4	A	−4	−0.02	−4	−0.06	−4	−0.01
1899.35	328	0.55	2	A	+3	0.00	+2	0.00	+2	+0.01
1905.53	312	0.50	3	A	+9	−0.03	+7	−0.05	+8	−0.02
1906.22	301	0.52	2	A	0	0.00	−1	−0.02	0	+0.01
1912.58	270	0.42	2	A	−1	−0.02	−3	−0.04	−2	−0.01
1914.50	257	0.44	2	A	−3	0.00	−5	0.00	−4	+0.02
1917.07	247	0.47	2	A	+1	+0.03	0	+0.02	+1	+0.04
1917.60	243	0.42	3	VB	0	−0.02	−1	−0.03	0	−0.01
1920.43	230	0.56	2	A	+2	+0.09	+1	+0.08	+2	+0.10
1920.48	229	0.50	2	VB	+1	+0.03	0	+0.02	+1	+0.04
1924.33	218	0.51	2	VB	+6	+0.03	+6	−0.04	+6	−0.02
1926.93	205	0.56	2	VB	+2	−0.04	+2	−0.05	+2	−0.03
1929.41	197	0.65	3	VB	0	0.00	+1	−0.03	+1	0.00
1930.33	193	0.66	3	VB	−2	−0.03	−1	−0.04	−1	−0.02
1931.34	190	0.73	2	VB	−2	−0.02	−2	0.00	−2	+0.03
1944.27	174	1.06	2	VB	0	+0.04	+1	+0.01	0	+0.04
1945.34	173	1.10	3	VB	+1	+0.05	+1	+0.03	0	+0.05
1947.50	170	1.08	3	VB	0	−0.01	0	−0.03	−1	−0.01
1952.29	164	1.23	3	VB	−2	+0.05	−2	+0.04	−3	+0.05
1958.41	162	1.26	5	B	0	0.00	+1	−0.01	−1	−0.02
1962.98	159	1.22	7	VB	+1	−0.08	+2	−0.09	0	−0.09

TABLE 26. OBSERVATION OF +45°2505

t'	t	θ	ρ	$n.$	Obs.	$\Delta\theta$	$\Delta\rho$
	1939.49	357	0.23	7	VB	−1	0.00
	1940.39	301	0.46	4	VB	+1	−0.05
1953.54	1940.56	294	0.48	3	VB	−1	−0.07
	1941.40	280	0.68	3	VB	+1	−0.04
1954.77	1941.79	274	0.98	3	VB	+1	+0.19
	1942.40	265	0.77	3	VB	−1	−0.10
1955.76	1942.78	262	1.14	2	VB	0	+0.22
	1943.59	257	1.00	6	VB	+2	0.00
	1944.51	249	1.04	5	VB	+1	−0.02
1957.60	1944.62	250	1.14	3	VB	+3	+0.07
	1944.62	248	1.08	3	B	0	+0.01
	1945.31	244	1.05	2	VB	+1	−0.04
1958.57	1945.59	245	1.17	2	VB	+4	+0.07
1958.64	1945.66	240	1.10	3	B	0	0.00
	1946.44	234	1.10	5	VB	−1	0.00
1935.43	1948.41	217	1.01	4	VB	−4	+0.01
	1948.88	222	0.98	4	VB	+5	+0.02
	1949.35	210	0.78	3	VB	−3	−0.12
1936.40	1999.38	210	0.94	3	VB	−3	+0.05
1962.65	1999.67	209	0.83	4	B	0	−0.02
1962.71	1949.73	208	0.79	2	VB	−1	−0.04
1937.30	1950.28	199	0.77	3	VB	−3	+0.03
1938.40	1951.38	175	0.43	4	VB	−1	−0.02
	1951.78	152	0.30	4	VB	+1	+0.01
	1952.30	30	0.15	3	VB	0	−0.03

TABLE 27. OBSERVATION OF ADS 10585

t'	t	θ	ρ	$n.$	Obs.	$\Delta\theta$	$\Delta\rho$
	1917.67	227	0.28	2	A	0	+0.07
	1920.14	234	0.38	5	A	− 2	+0.02
	1924.62	242	0.54	4	A	− 1	−0.01
	1930.12	248	0.62	3	A	0	0.00
	1932.75	248	0.59	3	A	− 1	0.00
	1934.15	252	0.56	2	A	+ 1	−0.01
	1936.48	257	0.51	2	A	+ 4	0.00
	1937.19	257	0.47	4	VB	+ 3	−0.02
	1940.02	262	0.38	5	VB	+ 3	0.00
	1942.02	267	0.33	2	VB	+ 3	+0.03
	1945.00	280	0.21	3	VB	− 1	+0.04
	1946.52	287	0.13	3	VB	−16	+0.01
	1948.69	22	0.14	3	VB	+21	+0.03
	1951.79	40	0.20	4	VB	− 1	−0.02
	1953.56	50	0.30	2	VB	+ 1	0.00
	1954.77	52	0.30	2	VB	0	−0.05
	1957.63	56	0.48	3	B	− 2	+0.03
	1959.65	63	0.58	3	VB	+ 3	+0.07
	1962.38	64	0.57	4	B	+ 1	+0.02
	1962.46	67	0.51	2	VB	+ 5	−0.04
1902.48	1964.52	65	0.55	4	A	0	−0.01
1910.31	1972.35	76	0.38	2	A	− 1	+0.07

TABLE 28. OBSERVATION OF ADS 12126

t	θ	ρ	$n.$	Obs.	$\Delta\theta$	$\Delta\rho$
1900.46	37	0.23	3	A	0	0.00
1901.71	34	0.20	4	A	−1	−0.03
1910.74	20	0.19	1	A	0	−0.02
1915.05	16	0.22	4	A	+5	+0.02
1918.89	4	0.22	3	A	0	+0.03
1922.69	353	0.18	5	A	−2	0.00
1925.67	350	0.17	2	A	+2	−0.01
1927.76	346	0.17	2	A	+4	0.00
1928.60	343	0.18	2	A	+3	+0 01
1930.55	327	0.18	4	B	−8	+0.01
1931.62	329	0.16	3	A	−3	0.00
1933.62	329	0.15	2	A	+4	0.00
1937.57	312	0.16	4	B	+2	+0.03
1944.83	213	0.13	2	VB	−8	+0.08
1946.54	173	0.12	2	VB	+5	+0.06
1948.65	140	0.12	2	VB	+2	+0.02
1952.70	114	0.155	3	Fi(I)	−3	0.000
1954.76	112	0.190	2	Fi(I)	+1	+0.012
1959.67	102	0.23	3	B	+1	+0.01
1959.76	103	0.210	2	Fi(I)	+2	−0.007
1960.73	99	0.21	2	B	0	−0.01
1961.39	100	0.25	3	VB	+2	+0.03
1962.53	96	0.24	4	B	−1	+0.01

TABLE 29. TRIGONOMETRIC PARALLAXES (0".001) OF HYADES CLUSTER MEMBERS DETERMINED AT THE ALLEGHENY OBSERVATORY (A) AND THE MEAN OF ALL DETERMINATIONS WITH THEIR WEIGHTS

HD	Y	A	wt.	mean	wt.
G 43272	551.0			53	10
18404	615.0	30	28	30	28
21663	731.0	35	28	35	28
23050	769.0	29	16	29	16
+23°571	842.0			38	4
24357	854.0	30	28	24	30
26162	916.0			22	6
27383	955.0	46	20	28	27
27691	964.0	31	20	29	32
27697	965.0	16	20	16	28
27819	970.0			16	8
27991	974.1			41	3
28024	977.0			33	7
28052	979.0			3	7
28305	987.0	26	20	18	25
28307	988.0			33	7
28319	989.0			25	7
28363	991.0			8	8
28485	995.0			22	10
28527	996.0			50	10
28910	1003.0	23	20	22	28
29388	1022.0	18	28	18	36
29488	1027.1	20	28	20	28
30210	1059.1	3	28	3	28
+25°733	1064.0			26	10
30780	1083.0	9	28	9	28
30810	1084.0	39	20	28	46
30869	1087.0			17	14
32301	1132.0	22	20	9	28
33254	1162.0			6	10
Mean		0.0245(352)		0.0221(557)	

LITERATURE CITED

1. van den Bos, W. H., *Publ. Astron. Soc. Pacific*, **74**, 297 (1962)
2. Eggen, O. J., *Quart. J. Roy. Astron. Soc.*, **3**, 259 (1962)
3. van Albada, B. B., *Bull. Astron. Inst. Neth.*, **16**, 178 (1962)
4. Eggen, O. J., *Roy. Obs. Bull. No. 51* (1962)
5. van den Bos, W. H., *Union Obs. Circ. No. 68* (1926); *No. 86* (1932)
6. Eggen, O. J., *Astron. J.*, **61**, 361 (1956)
7. Baize, P., *J. Observateurs*, **41**, 170 (1958)
8. Couteau, P., *J. Observateurs*, **44**, 57 (1961)
9. *Circ. Inf. I.A.U. No. 29* (1963)
10. van den Bos, W. H., *J. Observateurs*, **44**, 76 (1961)
11. Eggen, O. J., *Astron. J.*, **70**, 19 (1965)
12. Baize, P., *J. Observateurs*, **36**, 2 (1953)
13. van den Bos, W. H., *Republic Obs. Circ. No. 123* (1964)
14. van den Bos, W. H., *J. Observateurs*, **47**, 34 (1964)
15. *Circ. Inf. I.A.U. No. 30* (1963)
16. *Roy. Obs. Bull.* (In press)
17. Eggen, O. J., *Ap. J. Suppl. No. 76* (1963)
18. Eggen, O. J., *Observatory*, **85**, 104 (1965)
19. Sandage, A. R., Eggen, O. J., *Monthly Notices Roy. Astron. Soc.*, **119**, 278 (1959)
20. Eggen, O. J., *Roy. Obs. Bull. No. 41* (1961)
21. Eggen, O. J., *Ann. Rev. Astron. Ap.*, **3**, 235 (1965)
22. Lippincott, S. L., *Astron. J.*, **60**, 379 (1955)
23. Kron, G. E., White, H. S., Gascoigne, S. C. B., *Ap. J.*, **118**, 502 (1953)
24. Eggen, O. J., Greenstein, J. L., *Ap. J.*, **141**, 83 (1965)
25. Iben, I., Jr., *Ap. J.*, **142**, 1447 (1965)
26. Iben, I., Jr., *ibid.*, **140**, 1631 (1964)
27. Eggen, O. J., *Ap. J.*, **138**, 356 (1963)
28. Eggen, O. J., Sandage, A. R., *Ap. J.*, **140**, 120 (1964)
29. Sandage, A. R., *Ap. J.*, **135**, 349 (1962)
30. Demarque, P. R., Larsen, R. B., *Ap. J.*, **140**, 544 (1964)
31. Eggen, O. J., *Observatory*, **85**, 191 (1965)
32. Evans, D. S., Menzies, A., Stoy, R. H., *Monthly Notices Roy. Astron. Soc.*, **117**, 534 (1957)
33. Wayman, P. A., Symms, L. S. T., Blackwell, K. C., *Roy. Obs. Bull. No. 98* (1965)
34. van Bueren, H. G., *Bull. Astron. Inst. Neth.*, **11**, 385 (1952)
35. Jenkins, L. F., *General Catalogue of Trigonometric Stellar Parallaxes* (Yale Univ. Obs., 1963)
36. Strand, K. Aa., *Basic Astronomical Data* (Univ. of Chicago Press, 1963)
37. Vyssotsky, A. N., Williams, E. T. R., *Astron. J.*, **53**, 78 (1948)
38. Binnendijk, L., *Bull. Astron. Inst. Neth.*, **10**, 9 (1943)
39. Maedler, J. A., *Astron. Nachr. No. 289* (1835)
40. Strand, K. Aa., *Leiden Ann.*, **18** (1937)
41. Doberck, W., *Astron. Nachr.*, **94**, 255 (1879)
42. Guntzel-Lingner, U., *Astron. Nachr.*, **283**, 73 (1956)
43. Rabe, W., *Astron. Nachr.*, **284**, 97 (1958)

ASTRONOMICAL FABRY-PEROT INTERFERENCE SPECTROSCOPY[1]

By Arthur H. Vaughan, Jr.

Mount Wilson and Palomar Observatories, Carnegie Institution of Washington
California Institute of Technology, Pasadena, California

INTRODUCTION

During recent years the need for high-resolution spectrophotometry of faint astronomical sources has been increasingly emphasized. The study of H_2O and CO_2 molecular lines in the spectrum of Mars (Kaplan, Münch & Spinrad 1964; Spinrad et al. 1966), the study of weak CN interstellar line absorption (Field & Hitchcock 1966), and the determination of stellar isotope abundances Li^6/Li^7 (Herbig 1964) have required spectroscopic resolving powers close to the limit of the largest coudé spectrographs. On the other hand, detection of interstellar hydroxyl absorption would apparently require a detection limit of less than 6 mÅ equivalent width in stars of sixth magnitude or fainter (Goss & Spinrad 1966) while for the determination of the degree of saturation of interstellar CN lines, a resolving power as high as 10^6 might profitably be used. Although it is possible to produce fairly large diffraction gratings or echelles giving resolving powers as high as 10^6 (Stroke 1963), such gratings have not been made sufficiently large to avoid severe loss of light at the slit in even moderately high-resolution spectroscopy with large telescopes. At a resolving power of 80,000 with the 12-inch mosaic grating in the coudé spectrograph of the 200-inch telescope, for example, the slit must be only 0.07 arc-sec in width; with typical 1-sec atmospheric blurring only a few per cent of the starlight delivered by the telescope are actually utilized. This is equivalent to a reduction of the effective aperture from 200 to less than 50 inches!

By the use of optical interferometers in conjunction with physical quantum detectors, it is now possible to overcome this limitation of spectrographs for certain classes of problems. The essential feature of interferometers is the ability to accept 100 times more light from an extended source of given brightness or surface radiance (a seeing disk, for example) than is geometrically possible with a slit spectrograph of equal size and resolving power. Thus, even a comparatively small Fabry-Perot (F-P) interferometer can outperform the largest gratings and reach a resolution suitable for the narrowest spectral lines, yet utilize the entire stellar seeing disk of a large telescope.

In the thirteen years since Jacquinot (1954) first pointed out this fundamental advantage of F-P interferometers, much progress has been made in the development and use of laboratory instruments for high-resolution interference spectroscopy. Quite early, Dunham (1956) strongly emphasized the importance of utilizing such techniques for high-resolution spectroscopy with large astronomical telescopes. The feasibility of such an application was ex-

---

[1] The survey of literature for this review was concluded in January 1967.

plored and demonstrated in practice by Geake, Ring & Woolf (1959). However, partly because of the admittedly distressing practical difficulties involved, the implementation of F-P techniques by observing astronomers has proceeded slowly, and reported successes in achieving truly high resolution—in excess of 10^5—are only a development of the past three or four years.

The usefulness of the fundamental advantages offered by F-P interferometers has depended significantly upon technological progress in several fields in the past decade or two. Techniques for producing semitransparent dielectric multilayer mirror coatings having very low absorption have improved markedly. Sensitive quantum detectors covering a wide spectral range have become available. Refined electronic pulse-counting and data-recording techniques have improved the accuracy of intensity measurements in faint astronomical spectra and have greatly simplified the problem of seeing-compensation. Quite recently the precision, stability, and convenience of operation of F-P interferometers have gained substantially through the introduction of modern techniques of automatic control in optics.

In recent years F-P scanners have been increasingly used for the study at moderate spectroscopic resolution of extended emission-line sources such as H II regions, and the classical photographic methods first introduced by Fabry & Buisson (1911) for the same purpose have been resurrected and considerably improved. In such specialized studies (which can be carried out with small telescopes) the greatest gains over conventional spectroscopy are possible.

Jacquinot, in a major article on interference spectroscopy, has discussed in detail the theory of F-P techniques and their state of development as of 1960. Several review articles of a more summary nature concerning F-P instrumentation have appeared (Code & Liller 1962, Courtès 1964, Connes 1964). It seemed, therefore, that a somewhat detailed discussion of the design of specifically astronomical instrumentation involving F-P interferometers would be worth while, and the present paper is devoted largely to this purpose.

Although two-beam interferometry forms an important part of what is now customarily called interference spectroscopy, it is outside the scope of this article. The special techniques and problems of this field have been reviewed expertly, and with emphasis on astrophysical implications, by Mertz (1965). The comparison between two-beam or Fourier transform devices and F-P interferometers is elaborated in the reviews mentioned and by Gebbie & Twiss (1966).

THEORY OF THE FABRY-PEROT INTERFEROMETER

The principle of the F-P etalon and more recent theoretical refinements for describing the properties of real (imperfect) etalons have been discussed in the reviews listed and will only be summarized in this section with a few additions.

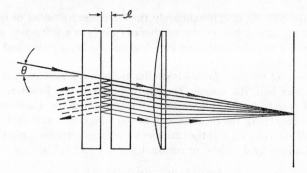

FIG. 1. The Fabry-Perot (F-P) interferometer.

THE IDEAL ETALON: DEFINITIONS

The ideal etalon consists of a pair of identical plane-parallel semireflecting mirrors (Figure 1). The transmission characteristics of such an etalon are given approximately by the Airy formula

$$A(\lambda) = \tau_A \{1 + [4r/(1-r)^2] \sin^2 \pi x\}^{-1} \qquad 1.$$

where

$x = (2\mu l/\lambda) \cos \theta$

$\tau_A = [1 - k/(1-r)]^2$

$\lambda =$ vacuum wavelength

$\mu =$ index of refraction of dielectric spacer medium

$\theta =$ angle of incidence

$r =$ reflectance of each mirror

$k =$ absorptance of each mirror

$l =$ effective mirror separation (geometrical distance plus geometrical equivalent of phase shifts at the interfaces)

The peak transmittance τ_A is limited by absorption in the semireflecting layers but may be quite high ($\tau_A \sim 0.9$) for high-reflecting multilayer stacks. Transmittance maxima occur at integral interference orders

$$m = (2\mu l/\lambda) \cos \theta \qquad 2.$$

Typically m is a number of order 10^3–10^4. Control of the wavelength is available by varying l or μ or by tilting. At normal incidence the spectral range (distance between maxima) is

$$Q = \lambda^2/2\mu l = \lambda/m \qquad 3.$$

The half-intensity width w_A (full width at one-half peak transmittance) is given approximately by $(\lambda/m)(1-r)/r^{1/2}$. The ratio Q/w_A, called the reflectance finesse, depends only upon the reflectance of the coatings. It is given by

$$N_R = \pi r^{1/2}/(1-r) \qquad 4.$$

The quantity N_R is approximately the equivalent number of interfering beams of equal amplitude, or the number of rulings in a diffraction grating or echelon, which would give the same theoretical resolving power and the same order of interference.

The shape of the Airy profile is similar to that of a Lorentz profile and is such that one-half the energy transmitted within one interference order ($\Delta m = 1$) lies outside the half-width of the transmittance maximum. When illuminated normally the ideal etalon has an integral transmittance or *filtrage*, $\mathfrak{F} = (1-r)/(1+r)$, and the ratio between transmitted intensities at integral and half-integral orders, or contrast is $\mathfrak{C} = 1 + \pi^2 N_R^2/4$.

ACCEPTABLE SOLID ANGLE

$A(\lambda)$ is the instrumental function of an F-P spectrometer such as the one shown in Figure 2 when the entrance diaphragm has zero area and the light

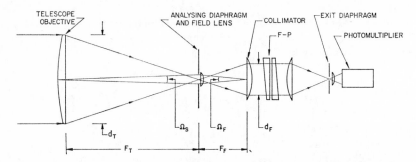

FIG. 2. Principle of the F-P interference spectrometer.

striking the etalon is fully parallel. When light is incident over a finite range of angles, the instrumental function can be expressed by a convolution of $A(\lambda)$ with a distribution function $F(\cos \theta) = d\Omega/d(\cos \theta)$ of elemental solid angle with respect to angles of incidence. This function may be written in the form $F(\lambda)$ by means of Equation 2. As $r \to 1$, $N_R \to \infty$, $A(\lambda)$ approaches a Dirac delta function and the instrumental function becomes $F(\lambda)$ itself. In the most important case of an axially symmetrical diaphragm, $F((\lambda)$ is a periodic rectangular function equal to 2π when the inequality $(2\mu l/m)$ $\cos \theta_2 \leq \lambda < (2\mu l/m) \cos \theta_1$ is satisfied, and zero otherwise. Accordingly, a spectral width $w_F = (2\mu l/m)(\cos \theta_2 - \cos \theta_1) = \lambda \Omega_F/2\pi$ is associated with the diaphragm. The ratio $N_F = Q/w_F$, where

$$N_F = \pi\lambda(2\mu l\Omega_F)^{-1} \qquad 5.$$

is the maximum attainable finesse with a diaphragm that gives an acceptable solid angle Ω_F as viewed from the collimator. N_F is called the diaphragm finesse. Similarly the maximum resolving power $R_F = \lambda/w_F$ satisfies the basic relation

$$R_F \Omega_F = 2\pi \qquad 6.$$

which also governs the Michelson interferometer.

Notice that in addition to broadening the instrumental function, a finite axial diaphragm also shifts the transmission maxima toward longer wavelengths from their positions for purely axial illumination. The new positions can be calculated from Equation 1 and are, approximately, $(\lambda/\mu)_{\theta=0}[1+\Omega_F/\pi]$. The diaphragm-effect may have to be considered when carrying out precise wavelength measurements. A detailed discussion is given by Bruce (1966).

The flux entering the spectrometer from an extended source of unit surface radiance is equal to the product of the aperture area and the acceptable solid angle. In accordance with common practice we denote this product by $\mathcal{L}$ (abbreviation of the French *Luminosité*) but use the name throughput to avoid conflict with the photometric meaning of luminosity. The terms radiance response, luminous efficiency, and light grasp are also used by some authors. Our definition of $\mathcal{L}$ is the same as that of Mertz (1965) but differs slightly from that of Jacquinot (1960) who includes peak transmittance in the product.

DIFFRACTION LIMIT

The theoretical resolving limit R_0 of a perfect diffraction grating is $R_0 = \Delta/\lambda$ where Δ is the maximum optical retardation or path difference between interfering beams. However, it is easily verified that R_0 is also essentially equal to the reciprocal angular dispersion $d\lambda/d\theta$ of the grating multiplied by the angular deviation associated with diffraction by an aperture having the cross-sectional area of the beam.

It is, therefore, appropriate to consider that the theoretical resolving limit R_0 of a perfect F-P etalon (as $r \rightarrow 1$) is given by Equation 6 when Ω_F is the spread introduced by diffraction. Taking the first zero of the Airy diffraction disk as defining the spread when the diameter of the etalon is d_e, we obtain $R_0 \sim 0.9 \, (d_f/\lambda)^2$.

This is a very large number. For example, when $d_F = 1$ cm and $\lambda = 5000$ Å, then $R_0 = 3.6 \times 10^8$. The diffraction limit of an etalon is indeed sometimes reached, but only rarely (e.g. in lasers, when l is very large and d_F small). By contrast the diffraction limit R_0 of a grating is proportional to only the first power of d/λ with the result that it does intervene in many instances.

Let us assume that the etalon is to be employed for stellar observations with a telescope of area $\mathcal{C}_T$ in the configuration shown in Figure 2. If the telescope is diffraction-limited in angular resolution, it has a throughput (cf. Mertz 1965) $\mathcal{C}_T\Omega_T = (1.22\pi/2)^2\lambda^2$. From the geometry in Figure 1, it can readily be shown that if the etalon were operated at the diffraction-limited resolving power, its throughput $\mathcal{C}_F\Omega_F$ would precisely equal that of the diffraction-limited telescope. It could, in this case, just accept all the light flux delivered to it by the telescope. The throughput of a diffraction-limited grating spectrograph in a similar configuration is about one-half this value, as may be shown from the analysis given by Jacquinot (1954). In either case, since the Airy disk is exceedingly small for a large telescope (0.06 arc-sec for a 100-inch telescope), it is evident that R_0 cannot be used without wasting an enormous amount of light if the source has the 1 arc-sec angular extent associated with "seeing."

DESIGN CONSIDERATIONS

Because multiple reflections are involved, the optical tolerances required of the F-P etalon are much more critical than is the case in most other instruments. The distortion of a wavefront after many reflections from a non-flat surface greatly exceeds the distortion characterizing the surface itself. The main effects of this are to reduce the spectral resolving power or broaden the transmission maxima or fringes, and to reduce the peak transmission. A secondary effect is the introduction of angular aberrations, resulting in limitations described later.

To the extent that the total response of an interferometer with flatness imperfections can be taken as the sum of intensities transmitted by different elementary areas of the aperture, each element being plane-parallel, the instrumental function can be written as the convolution (Chabbal 1953, 1958)

$$I(\lambda) = A(\lambda) * F(\lambda) * D(\lambda) \qquad\qquad 7.$$

The distribution function $D(l) = d\Sigma/dl$ represents the fractional surface area $d\Sigma$ over which the plate separation lies between l and $l+dl$. The function $D(l)$ can be found experimentally from measurements carried out in the laboratory. The function $D(\lambda)$ can be deduced by means of Equation 2. It is a rectangular function in case one of the plates contains a slight spherical curvature. Explicit or numerical evaluations of Equation 7 for errors of specified form (sinusoidal or Gaussian errors, or errors associated with a wedge angle between plates) have been studied by Hill (1963) and Hernandez (1966).

For design purposes it is sufficient, following Jacquinot (1960), to consider $D(l)$ as an essentially Gaussian function characterized by a half-intensity width λ/z_λ. The concomitant half-intensity width of $D(\lambda)$ is then $w_D = \lambda^2/z_\lambda l$. The ratio

$$N_D = Q/w_D = z_\lambda/2 \qquad\qquad 8.$$

is frequently called the limiting finesse. One plate may be regarded as perfectly flat while the other has a flatness tolerance of λ/z_λ. For random errors we have approximately $N_D = \zeta_\lambda/2\sqrt{2}$, where λ/ζ_λ refers to the tolerances of the individual plates. To obtain a limiting finesse as high as 50 (a good value) at λ 5000 Å would require a flatness tolerance of $\lambda/140 = 35$ Å.

Flatness defects whose typical size h (lateral extent) is small compared to the diffraction spread produced at a distance l, that is, $\lambda/h \ll h/l$, cannot be analysed by means of Equation 7. In the case of such microscopic defects the limiting finesse due to variations in separation is of order z_λ^2 (Stoner 1966) and may be ignored. The loss of finesse resulting from the angular deviations due to diffraction or scattering may not be negligible, however.

The finesse of an instrument cannot exceed the lowest of the three values N_R, N_F, and N_D. The actual finesse and also the peak transmittance depend upon the ratios N_R/N_F, N_R/N_D, in virtue of the convolutions performed in Equation 7. Chabbal (1953) has studied and Jacquinot (1960) has summarized the variation of finesse and transmittance as functions of these ratios. The main conclusion is that for design purposes it is sufficient to regard the functions $A(\lambda)$, $F(\lambda)$, $D(\lambda)$ as being approximately Gaussian functions, in which case the resultant finesse N is the root-mean-square value

$$N \sim (N_R{}^2 + N_F{}^2 + N_D{}^2)^{1/2}/3^{1/2} \qquad 9.$$

The peak transmittance, in the same approximation, is maximum if

$$N_R \sim N_F \sim N_D \qquad 10.$$

If this latter condition is satisfied by the design, the peak transmittance will be ~ 0.6, apart from absorption losses in the plates or coatings. If N_R is chosen to be $1.5N_F$, the resultant finesse will be very little increased but the transmittance will be reduced to 0.4–0.5. Further departure from Equation 9 results in still more loss of light. With multilayer dielectric reflective coatings the reflectance and hence N_R are wavelength dependent. Furthermore N_F varies proportionally with the wavelength, with the result that a given arrangement can, in general, be used efficiently only in the limited region of the spectrum for which it has been designed.

The precise value of N_D plays a less critical role, provided it is sufficiently large, but there is seldom a reason to choose a larger value than is suggested by Equation 10. In some circumstances the actual form of the function $D(l)$ may significantly influence the transmission, particularly where chains of etalons are involved (Mack et al. 1963).

PHOTOGRAPHIC APPLICATIONS

Classical Method

Direct photography of F-P interference fringes in a plane coincident with an image of the sky provides a powerful and well-known method of detecting faint H II regions (because of the contrast of the fringes), as well as of measuring Doppler shifts (because of the high angular dispersion of the F-P). The method, first applied by Fabry & Buisson in 1911, entails isolating the desired line with an auxiliary filter and employing an optical system that feeds the F-P with collimated light in a cone large enough to illuminate the desired fringes. Recently this method has been further developed for astronomical purposes by Courtès (1960), who has constructed an optical system (Figure 3) consisting of field lens, collimator, and f/1 Schmidt camera, the ensemble being given the name focal length reducer (*reducteur focal*). The field lens provides an exit pupil of suitable size in the collimated beam where the etalon is located. The focal length reducer is designed for the best utilization of the resolving properties of the etalon, and the fast camera permits high-definition photography of Hα in nebulae whose surface radiances are 10^3 to 10^4 times fainter than that of the Orion Nebula.

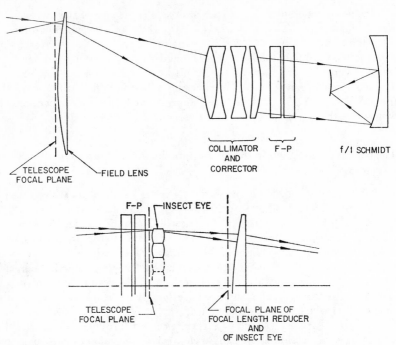

FIG. 3. *Top:* Focal length reducer for photographing interference fringes in H II regions (schematic, after Courtès 1960). *Bottom:* The insect eye method (after Courtès, Fehrenbach et al. 1966).

In an instrument of this kind there is no focal plane diaphragm, but the solid angle Ω_F is in effect determined by the angular resolution on the photographic plate. This resolution may, in turn, be determined by the granularity of the emulsion, by aberrations of the optical system, or by seeing. In any case the design considerations of the preceding section are relevant to this use of the F-P.

To analyze the interference pattern obtained, assume $\theta \ll 1$, so that Equation 2 can be written in the form $(2\mu l/\lambda)(1-\rho_m^2/2L^2) = m$, where ρ_m is the radius on the photograph of the ring of order m, and L is the focal length of the camera. To the center of the ring system ($\rho = 0$) is assigned an interference order $m_0 = 2\mu l/\lambda = m+q+\epsilon-1$, where m (an integer) is the order of the qth bright ring, counted outward from the center, and ϵ ($0 \leq \epsilon \leq 1$) is called the fractional order at the center. The number q is called the rank of the ring. In the focal plane there exists a dispersion $d\rho_m/d\lambda = L(m_0/2)^{1/2}\lambda^{-1}$ $(q-1+\epsilon)$, where the condition $1-q-\epsilon \ll m_0$ has been assumed. The radial dispersion is infinite (and therefore useless in the present application) at the center of the ring pattern and it decreases outward for the off-axis lower-order rings. For the ring of rank $q = 1$, in particular, the dispersion becomes

$$dp/d\lambda = \begin{cases} (L/\lambda)(m_0/2)^{1/2}, & \epsilon = 1 \text{ (bright center)} \\ (L/2\lambda)(m_0)^{1/2}, & \epsilon = \tfrac{1}{2} \text{ (dark center)} \end{cases}$$

It is of interest that the dispersion depends only upon the order (or free range) of the etalon. The resolution of finesse may thus be given any desired value by an appropriate choice of other parameters of the etalon.

If the highest speed (or throughput) compatible with the resolving power is to be reached, the geometrical width of a ring should equal the limit of definition ω of the photographic plate. This condition is satisfied if $L = \omega$ $[2RN_R(q-1+\epsilon)/\sqrt{3}]^{1/3}$. It is evident that both the photographic speed and the useful dispersion are highest if the first ring is used. To give a numerical example, with $\epsilon = \tfrac{1}{2}$, $q = 1$, $\lambda = 4861$ Å, $\omega = 30\,\mu$, and $N_R = 30$, a resolving power $R = 60,000$ (5 km/sec resolution) can be obtained with $L = 3$ cm.

In practice there may be advantages in using the sectors of fringes of higher rank, obtained by tilting the etalon to give more regular coverage of the field. If the rank is too high, however, the throughput falls below what can be achieved more simply with other methods, such as the multislit spectrograph (cf. Jacquinot 1960).

INSECT EYE

A novel alternative method introduced by Courtès (Courtès, Fehrenbach et al. 1966) is illustrated in Figure 3. In front of the field lens of the focal length reducer is placed an F-P etalon of large size, followed by a closely packed array of small lenses (an "insect eye") of numerical aperture equal to that of the telescope. The insect eye and the telescope are focused so that the interference rings are in the focal plane of the focal length reducer; the final image of the field with its superimposed first rings is then photographed as before. This method has the advantage over the one first described that the entire field can be covered with the maximum possible photographic speed and sensitivity to Doppler shifts.

To assure that only the first rings are actually illuminated entails designing the etalon so that the outer diameter of the first ring subtends an angle slightly smaller than the telescope entrance pupil. At the same time, the inner diameter of the ring must be slightly larger than the cone obscured by the telescope secondary mirror if one exists. The design is thus somewhat circumscribed by the telescope used. Instruments of this type, and of the type described previously, have recently been employed for survey studies of galactic H II regions in the Northern and Southern Hemispheres by Courtès, Cruvellier & Georgelin (1966).

THE F-P AS A TUNABLE FILTER

When only the axial fringe is illuminated the etalon acts as an interference filter, giving a channeled spectrum of which the wavelengths of the passbands may, for example, be varied at will by variation of l or μ. A feasible arrangement permitting two-dimensional filter photography with this apparatus is one in which the etalon is placed immediately in front of the image plane of a telescope, the entrance pupil of the telescope serving to isolate the axial fringe. Velocity maps of H II regions, in which an emission line can be isolated by an auxiliary filter, can be made simply in this way (Courtès 1958).

Study of continuous sources, especially at high resolution, involves the more troublesome problem of spectral isolation of a single F-P order. However, where little or no tuning of the auxiliary isolator is needed, a narrowband F-P type all-dielectric interference filter, a Lyot filter, or a Hallé filter with bandwidth N times that of the F-P might be used. Since Ω_F is determined by the focal ratio of the telescope, the maximum resolving power of a correctly designed F-P is, according to Equations 5 and 10, $R = 8f^2/\sqrt{3}$. This is equal to $\sim 10^5$ in the case of a telescope with a focal ratio of f/150. The field is determined by the diameter of the etalon; for the f/150 telescope, for example, the entire Sun could be photographed, provided $d_F/d_T \geq 0.86$. The largest etalons now available with high finesse have diameters of ~ 60 mm. In this case, the telescope aperture must be ~ 70 mm or smaller, resulting in a focal length of ~ 10.5 m. Such a telescope would have a scale of some 50 μ/sec of arc in the focal plane, so that an angle of $\sim .5$ sec of arc could be resolved photographically (in the absence of aberrations). An arrangement of this kind could thus in theory serve as a compact spectroheliograph or spectrocinematograph with many valuable applications to studies of solar activity. Although simple in principle, such a device would entail several rather serious practical difficulties. Among these, in addition to spectral isolation, is the need for very accurate parallelism of the F-P plates both to avoid photometric variation across the field and to minimize the loss of angular resolution due to aberrations introduced and amplified by the multiple reflections.

MEASUREMENT OF THE ZEEMAN EFFECT

The recent development of a tunable birefringent F-P interferometer suitable for the direct recording in two dimensions of solar magnetic fields using the longitudinal Zeeman effect has been described by Ramsey & Smartt (1966). A thin sheet of optically uniform mica is placed in the cavity of the etalon, splitting the F-P transmission into two orthogonally plane-polarized channeled spectra. The displacement $\Delta\lambda_{12}(=\lambda^2[4t\Delta\mu_{12}])$ between polarizations for which the two indices are μ_1, μ_2 is determined by the mica thickness t. A quarter-wave plate placed in front of the etalon converts the circularly polarized longitudinal Zeeman components into plane-polarized ones. A 90° rotation of the angle ψ ($= \pm 45°$) between "fast" directions of the two birefringent plates exchanges the senses of circular polarization in each pair of oppositely polarized F-P passbands. Unpolarized or plane-polarized light is not modulated by rotation of the quarter-wave plate.

For study of solar magnetic fields this instrument is arranged so that a pair of oppositely polarized passbands are spectrally isolated and lie in the wings of an absorption line sensitive to the Zeeman effect. If $A(x, y)$ and

$B(x, y)$ are the two-dimensional intensity distributions or filtograms corresponding to $\psi = \pm 45°$, their difference $B - A$ is a magnetic map or magnetogram. All the light is continuously utilized since no polaroids are employed. To first order, $B - A$ is independent of Doppler shifts. An experimental magnetograph employing this principle has been constructed by Ramsey and Alan Young using an etalon of 60 mm diameter and a Vidicon system enabling images to be electronically recorded, stored, subtracted, and displayed on a monitor screen. The etalon gives a bandwidth of about 0.05 Å at λ 6103. Magnetograms having 5″ angular resolution, 2-min time resolution, and 20-G threshold sensitivity have been obtained (Ramsey 1966, private communication).

The method has a number of attractive advantages over conventional methods of mapping the solar magnetic field with a slit spectrograph (Babcock 1953) in that the equipment is compact, the transmission is fairly high, all the light of the solar disk is utilized simultaneously, and a direct display is obtained, greatly simplifying the reduction of the data.

Somewhat simpler arrangements could be tolerated in the measurement of stellar magnetic fields: no imaging is then needed; a smaller etalon could be used; the $\lambda/4$ plate could be replaced by an "ADP" plate; and spectral isolation could be accomplished with a grating spectrograph as the premonochromator. To date, however, such an application of the principle of the birefringent interferometer has not been made.

APPLICATION TO SPECTROMETRY

In this section we shall discuss applications of the F-P etalon in series with a photometric receiver. The assembly is then an instrument capable of spectrum scanning, combining the advantages of high throughput of the F-P with the advantages of photoelectric recording: high quantum efficiency, linearity, large dynamic range, and sensitivity in the infrared.

Some features of the optical layout are shown in Figure 2. If the entrance diaphragm is very large, a field lens is desirable to image the objective onto the F-P, thus assuring a fixed instrumental profile, regardless of how the diaphragm is illuminated. An exit diaphragm can be employed to prevent transmission of spuriously reflected rays or ghosts such as are shown by Candler (1951). For spectral isolation a monochromator may be used, either before or after the etalon, the exit pupil of either one forming the entrance pupil of the other. Since premonochromator requirements differ greatly for emission- and absorption-line spectrometry, these two applications will be considered separately.

EMISSION-LINE SPECTROMETRY

In this simplest case the premonochromator need only be a filter suitable for isolating the desired emission line. For astronomical purposes the first apparatus of this kind was constructed by Geake & Wilcock (1957) and further used experimentally by Geake, Ring & Woolf (1959). The technique was utilized for measurement of $[\text{O II}]\,\lambda\,3726/\lambda\,3729$ line-intensity ratios in diffuse and planetary nebulae by Davies, Ring & Selby (1964). A similar instrument, constructed at the University of Rochester by Vaughan and M. Savedoff (Vaughan 1964), has been used for study of profiles and intensities of He I λ 10830 emission in gaseous nebulae (Figure 4).

Let us briefly consider the design of an F-P interferometer for emission-line spectrophotometry suited to a specific telescope and a specific observational task. We suppose that the focal ratio of the collimating lens is equal to the focal ratio of the telescope so that no light is lost. Let Ω_s be the acceptable solid angle in the plane of the sky. The required etalon diameter is

$$d_F = d_T[\Omega_s/\Omega_F]^{1/2} = d_T[\sqrt{3}\,R\Omega_s/2\pi]^{1/2} \qquad 11.$$

For example, with a 24-inch telescope, a resolving power of 40,000 and a field 1' in diameter can be achieved with an etalon of 16-mm diameter. The throughput, $\pi d_T{}^2\Omega_s/4$, of such an instrument is approximately 10 times that of the coudé spectrograph of the 200-inch telescope for the same resolving power in the third-order spectrum (dispersion $= 2.3$ Å/mm; entrance slit width $= 100\mu$) if the maximum slit length of 2 cm is used. Increasing the telescope size will necessitate a proportionately larger etalon for a given field and resolving power. Alternatively, if d_F is given, increasing the telescope size will allow studies with higher angular resolution.

It is worth pointing out that if λ, R, Ω_s, Q, and d_T are specified, and d_F is chosen in accordance with Equation 11, all the other parameters of the F-P follow uniquely from Equation 10, the condition for maximum peak transmission.

In an instrument of the kind shown in Figure 4 many F-P passbands are present in the spectral region in which the transmission of the isolating filter is nonzero. This means that Q must be chosen at least as large as the spectral interval that one hopes to scan. On the other hand, a fixed filter of much narrower bandwidth than $10Q$ might introduce difficulties because of the wavelength dependence of its transmissivity.

If a continuous source is observed with the instrument, all the exposed passbands are illuminated, with the result that absorption-line observations are not feasible. However, stellar observations can be used for making photometric calibrations of nebular emission lines, provided the effects of stellar absorption lines may be ignored.

To show that this is possible in principle, we shall first show that the effective instrumental equivalent width for starlight (a continuous spectrum) is independent of the size of the star image. For this purpose we first suppose the diaphragm to be uniformly illuminated by starlight. Let the diaphragm be divided into annular zones of equal solid angle. The angle of incidence θ at the F-P is different for each zone. However, this angle merely determines the wavelength at which the F-P will have a transmission maximum. Thus, if the incident flux is the same at all wavelengths, corresponding to the different values of θ, the same amount of light will be received from each zone.

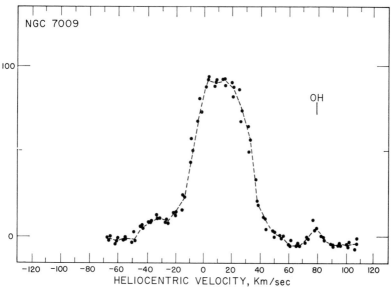

FIG. 4. Cassegrain F-P interference photometer, Mees Observatory, The University of Rochester. *Top:* Instrument. The ruler is 30 cm. *Bottom:* He I λ 10830 emission profile in the planetary nebula NGC 7009 scanned with the Mount Wilson 60-inch telescope. Approximate resolving power 30,000, 1 arc-min analysing diaphragm. Plotted points are individual measurements of 60-sec duration.

We may therefore, without changing the total light received, superpose all zones at the center zone. Furthermore, since an arbitrarily large number of zones could be chosen, the center zone may be arbitrarily small. Thus it is evident that the total signal from a star is independent of its image size, provided, of course, no vignetting occurs.

By an analogous argument, one can show that the integral over wavelength of an observed nebular line is independent of the actual distribution of surface radiance within the field of view subtended by the diaphragm. These conclusions are valuable in showing that the observed ratio of integrated line intensity to stellar intensity can be calibrated empirically through observation of a line of known integrated intensity and a star of known monochromatic magnitude. Absolute calibration through laboratory measurements on the F-P and isolating filter is also possible although more detailed considerations than are given here would be necessary.

Various modifications of the basic configuration of the emission-line scanner have been proposed. Instead of a single circular entrance diaphragm, a series of concentric rings matching the interference rings and comprising essentially a Fresnel zone plate has been suggested (Jacquinot & Dufour 1948) as a way of increasing the acceptable solid angle by a factor equal to the number of rings used. The light transmitted by all the rings goes to a single detector. Shepherd et al. (1965) have investigated the use of optical fiber bundles to intercept the light from several concentric rings in the focal plane of the collecting lines. The bundles from different rings may be arranged to conduct the light to different detectors, which results in a multichannel instrument. If the rings are chosen to match the interference pattern of monochromatic light, each detector observes the same line as it appears in different areas of the extended source. Kinematical studies of planetary nebulae might, for example, be done this way. Alternatively, if the telescope objective rather than the sky is imaged on the fiber ends, all detectors see the entire field equally.

Another interesting approach suggested by Shepherd et al. is the use of an expandable annular diaphragm (in practice, a stepping system presenting a sequence of annular diaphragms of equal area, but different radii, was actually constructed). The solid angle subtended by each annulus in turn is the same as that of an axial one which would give the same resolution. It is significant that a solid, rather than an air-spaced, etalon may be "scanned" in wavelength using a system of this kind. Thus a mechanically very stable instrument could be achieved. Another advantage of a solid etalon which they point out is an increase in the $R\mathcal{L}$ product by approximately the factor μ^2 due to refraction in the spacer medium (ignored in Equation 6).

ABSORPTION-LINE SPECTROMETRY

Exploration of an absorption-line spectrum imposes fairly serious requirements upon the premonochromator in that (a) it must (usually) be capable of isolating a single F-P passband; (b) the premonochromator and F-P etalon

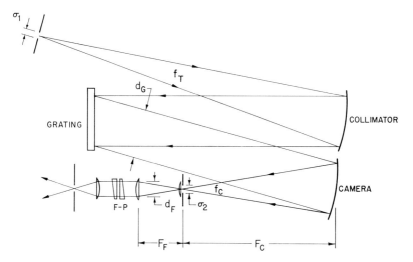

FIG. 5. Parameters used in matching F-P and grating premonochromator.

must scan the spectrum synchronously; and (c) if the spectrometer is intended for stellar observations and is used under conditions where the entrance diaphragm and stellar image are comparable in size (implying an efficiently designed system), adequate compensation for fluctuations introduced by "seeing" and guiding errors becomes essential.

The necessary spectral isolation can be accomplished by several methods.

Grating spectrograph as the premonochromator.—The basic optical layout of an F-P and grating spectrograph system is indicated in Figure 5. In the following analysis, the focal ratios of the F-P collimator and spectrograph camera are assumed equal so that the available aperture of the etalon is filled with light. We further assume that the exit slit of the spectrograph is a square whose circumscribed circle satisfies the first equality of Equation 10. Accordingly, we have

$$\sigma_2 = F_F \left(\frac{2}{\pi} \Omega_F \right)^{1/2} \qquad\qquad 12.$$

where F_F is the focal length of the F-P collimator, σ_2 the width of the slit (see Figure 5), and Ω_F the half-intensity solid angle of the central fringe from Equation 6.

In order to assure that a single F-P passband is isolated by the spectrograph, we must also require

$$Q = c\sigma_2 D, \qquad c \geq 1 \qquad\qquad 13.$$

where D is the reciprocal dispersion of the spectrograph in $\text{Å}/\text{mm}$. The constant c should be chosen large enough to reduce light leakage through adjacent F-P passbands in the wings of the transmission profile of the spectrograph, but not large enough to reduce the throughput of the equipment unnecessarily. In the following we assume the value $c = 1.5$, which has been found satisfactory for the coudé spectrograph of the 100-inch telescope on Mount Wilson. The best value, however, will depend essentially upon the quality of the spectrograph and grating used. Specification of the design problem is completed by requiring the instrumental parameters to satisfy the condition for maximum peak transmittance, given by Equation 10.

It can be ascertained from the equations governing this design (Equations 10, 12, 13) that of the ten parameters at our disposal (R, D, F_F, r, l, λ, Q, z_λ, σ_2, and Ω_F), four in fact are independent parameters. It is convenient to regard λ, D, and F_F as three of these; the remaining one adjustable parameter may be R, σ_2, or z_λ.

As R is increased, practical limits are met when z_λ reaches the maximum flatness tolerance to which interferometer plates of the required aperture can be figured or, alternatively, when σ_2 becomes unacceptably small (much smaller than a star image, for instance). In either event the resolving power then cannot exceed a limiting value R_z or R_σ and yet satisfy the matching conditions outlined above.

Formulae for R_z and R_σ are derived readily from the foregoing consideration. Since $\sigma_1 = \sigma_2 f_T / f_c = \phi f_T d_T$ where f_T and f_c are the focal ratios of the telescope and spectrograph camera, d_T is the diameter of the telescope, and ϕ is the angular slit width in the plane of the sky, we may write

$$\sqrt{R_\sigma} = \frac{2}{(\sqrt{3})^{1/2}} \cdot \frac{d_F}{d_T \phi} \qquad 14.$$

and

$$\sqrt{R_z} = \frac{1}{4c(\sqrt{3})^{1/2}} \cdot \frac{\lambda z_\lambda}{d_F f_c D} \qquad 15a.$$

Equation 14 shows that when the slit width ϕ is specified, the resolving power can be increased by using either larger interferometer plates or a smaller telescope; for example, when $\phi = 1$ arc-sec, the size of a star image in average seeing, a resolving power of 3×10^5 can be obtained with plates 1/570th the diameter of the telescope or larger.

In the case of a Littrow spectrograph, if λ is the blaze wavelength, Equation 15a can be rewritten in the form

$$\sqrt{R_z} = \frac{1}{2c(\sqrt{3})^{1/2}} \cdot \left(\frac{d_G}{d_F}\right) z_\lambda \tan \beta \qquad 15b.$$

where d_G and β are the collimator diameter and blaze angle of the grating and we have used the relation $\lambda / f_c D = 2 d_G \tan \beta$. From Equation 15b it follows, for example, that with $z_\lambda = 200$, $c = 1.5$, and $\beta = 30°$, a resolving power of 3×10^5 can be obtained with plates 1/19th the diameter of the grating or smaller. In this particular numerical example it is interesting that an etalon diameter can be selected which satisfies both Equation 14 and 15b only if

$d_G/d_T \geq 1/30$. Otherwise either a smaller resolving power or a smaller slit width would have to be used.

It is instructive to analyse here the inverse relation between resolving power and F-P diameters found in Equation 15b. This argument for the use of a small illuminated area of the F-P arises as a general consequence of the facts that (*a*) R_G, the effective resolving power of the isolating spectrograph, varies inversely as the first power of the slit width σ_2; (*b*) the resolving power R of the F-P varies inversely as the square of σ_2/d_F, whereas (*c*) under the constraint of constant finesse (because z_λ is fixed), the ratio R/R_G must remain invariant.

We note that Equation 15b is not the only consideration favoring the use of a small illuminated plate area in the application under discussion, since in

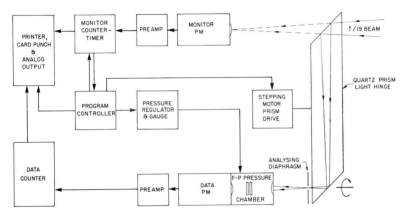

FIG. 6. Mount Wilson 100-inch coudé scanning F-P interferometer.

addition, the value of z_λ may well be much greater over a smaller area than over a larger one. In practice it can be advantageous to employ physically larger plates of which the flattest partial area only is illuminated.

Use of the F-P in conjunction with a spectrograph for stellar spectroscopy was first investigated in practice by Geake & Wilcock (1957) and Geake et al. (1959). More recently, a pressure-scanning F-P interferometer employing interchangeable etalons of 1 cm aperture has been used in conjunction with the 114-inch coudé spectrograph camera of the 100-inch telescope on Mount Wilson for high-resolution observations ($R \sim 2.4 \times 10^5$) of interstellar Na I D and Ca II K lines (Vaughan & Münch 1966). This equipment has also been employed by Vaughan & Zirin (1966) for study of chromospheric He I λ 10830 absorption in late-type stars at $R \sim 30,000$.

The optical arrangement is essentially as shown in Figure 5. The scanning system and electronics are indicated schematically in Figure 6. Some results are illustrated in Figure 7. The premonochromator is scanned by a stepping motor which rotates a double quartz prism in front of the exit slit as described

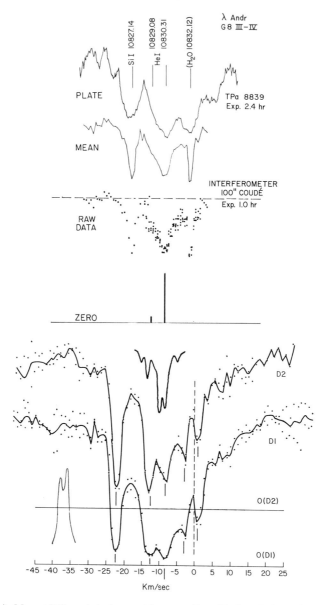

FIG. 7. Mount Wilson interferometric scans. *Top:* He I λ 10830 region in λ Andr. Approximate resolution 30,000; 1-hour exposure. Lower two curves refer to interferometric observations. For comparison the upper curve is a densitometer tracing of a spectrogram, taken with an RCA C70071 image tube at the 200-inch coudé; original dispersion 8.4 Å/mm. *Bottom:* Interstellar Na I lines D1 and D2 in α Cygni, estimated resolving power 240,000. Upper inset shows scan of iodine (I_2) line absorption at room temperature. Lower inset is scan of a Gates sodium lamp.

by Oke & Greenstein (1961). The F-P is pressure-scanned synchronously (in steps) with air to within a precision of 0.001 Å. "Seeing" compensation is provided by a digital ratio recorder (Oke & Dennison 1965). The signal-to-noise ratio in the compensated output is very close to the limit set by photoelectron counting statistics.

At the D lines, with $R \sim 2.4 \times 10^5$ the observed photoelectron counting rate equals the "dark" emission (of 60 counts/min) for a seventh-magnitude star. At this effective magnitude limit, 5 per cent photometric accuracy over a scanning range of 2 Å is feasible with an integration of seven hours. The observed counting rate is about one-third the value predicted by Dunham (1956) for an assumed detector quantum efficiency of 0.1, grating efficiency of 0.5, and peak transmission of 0.5. However, the additional rather serious losses of light in the telescope, spectrograph, lenses, and windows also associated with the F-P would appear likely to account for this discrepancy.

In the violet, preliminary observations by Münch (1966) of interstellar ground-state rotational lines of CN at λ 3874 in ζ Oph indicate that it will be possible to measure equivalent widths as low as 0.004 Å in stars of fifth magnitude.

An F-P spectrometer of similar type has been constructed for the 200-inch telescope (Figure 8). It is designed to give a resolving power of 1.5×10^5. Stars of magnitude 9.5 or 10 should be observable with a signal-to-dark emission ratio of unity. A similar coudé spectrometer is being constructed by Courtès & Cruvellier at Haute Provence.

The chief advantage of using a grating as premonochromator is the simplicity of the method. The resolving power and throughput of an existing spectrograph can be greatly increased for specialized purposes with relative ease. The method has some important shortcomings which are emphasized when construction of a complete instrument is contemplated: (a) the losses of light in a spectrograph can be serious; (b) a large grating is needed to utilize the capabilities of even a small F-P fully, so that the grating is used at only a small fraction of its theoretical resolution limit—this last is, however, also true of the F-P; (c) to provide the required high spectral purity or sharp cutoff outside the premonochromator passband, freedom from grating ghosts is necessary, and in general the spectrograph must be considerably overdesigned in quality compared to the actual resolving power at which it is used.

Multietalon spectrometers.—Another method of isolating a single F-P passband is to use a series or chain of etalons of successively decreasing finesse, with spacers so selected as to avoid exact coincidences between all but a few widely spaced passbands. This results in increased finesse with no loss of resolving power. The maximum value of the latter is given essentially by Equation 10. If enough etalons are used, no premonochromator other than a filter may be required, so that the peak transmission may be quite high. Application of the multietalon principle to photoelectric spectrometry was first proposed and studied theoretically and experimentally by Chabbal (1953).

FIG. 8. High-resolution interferometer scanner, Palomar 200-inch telescope. *Inset:*
Scanner (*A*) is located near focus of the on-axis 144-inch (*f/12*) camera in the coudé
spectrograph, used as a predisperser. (*B*) Spectrograph entrance slit. (*C*) 12-inch
mosaic grating. Dashed lines indicate path of light from *f/12* camera mirror (not
shown). North is at top left. *Upper photo:* (*D*) 45° plane mirror. (*E*) Quartz prism light
hinge and stepping motor prism drive assembly. (*F*) Analysing diaphragm. (*G*) Interferometer compartment, containing F-P pressure cell, interchangeable etalons,
aligning microscope, and collimating lens. (*H*) Fixed monitor beam-defining slots.
(*I, J*) Monitor and scanner photomultipliers. Scanner can be used in a direct scanning
mode at lower resolution by removing the etalon. The instrument and its associated
electronic data system were developed by Dennison, Münch, Oke, and Vaughan under
a grant from NASA.

Much of the recent progress of F-P interference spectroscopy has been concerned with the development of instruments of this kind.

Perhaps the best known of these multietalon instruments is the pressure-scanning PEPSIOS (trade mark) spectrometer (Mack et al. 1963, McNutt 1965, Stoner 1966). A pilot model of PEPSIOS containing three etalons, and yielding $R \sim 5 \times 10^5$, has been used successfully for observations of interstellar sodium D lines with the Lick Observatory 120-inch telescope (Hobbs 1965). The instrument is fully described in the literature cited.

Design considerations are more critical for multietalon scanning systems than for systems using one etalon and a spectrograph. The total transmissivity for wavelengths outside the isolated passband (i.e., the parasitic light transmission) depends strongly and in a complicated manner upon the ratios of spacer thicknesses used in the etalons. For more than two etalons, no way has been found to locate spacer ratios other than by numerically evaluating the parasitic light for many different ratios. It can be shown that "vernier" spacer ratios (ratios near unity) are always suitable; in the 3-etalon PEPSIOS the McNutt (1965) ratio 1.000:0.8831:0.7244 has been used. The minimum total parasitic light of this instrument is reported to be about 4 per cent of the observed level of the continuum, whereas the best value obtained so far with the Mount Wilson instrument is 7 per cent under observing conditions.

Special precautions are necessary to avoid unwanted interference (resulting in increased parasitic light) caused by reflections between the etalons. In PEPSIOS it is done by tilting the axes slightly, throwing the spuriously reflected beams outside the exit diaphragm. Other methods are discussed by Chabbal (1953), Bens, Cogger & Shepherd (1965), and Schwider (1965).

S.I.S.A.M. method.—Graner (1965) describes another type of instrument which has been developed for high-resolution infrared absorption spectroscopy. In this method an "Interferometric Spectrometer with Selection by Amplitude of Modulation" (French abbreviation: S.I.S.A.M.) of the type invented by Connes (1958) plays the role of "premonochromator" for an F-P. Essentially, the S.I.S.A.M. is a Michelson interferometer in which the two mirrors have been replaced by diffraction gratings. If it is illuminated by monochromatic light having a wavelength λ_0 such that the incident and diffracted rays from each grating coincide (so that the gratings operate in autocollimation), then the two recombining beams are parallel and interfere. As the path difference between them is modulated linearly with time (by rotating the compensating plate in one beam), the outgoing flux is modulated sinusoidally in a narrow band around λ_0, the amplitude of modulation decreasing rapidly at neighboring wavelengths. Thus, measurement of the amplitude of modulation is equivalent to a measure of the light intensity at λ_0. Use as a premonochromator for the F-P takes advantage of the fact that the revolving power of an S.I.S.A.M. is close to the theoretical limit R_0 of the gratings, while the acceptable solid angle (in contrast to an ordinary spectograph) is governed by the same relation as for the F-P: $R_0\Omega = 2\pi$.

As a result, an S.I.S.A.M. optimally matched to a given F-P can be much smaller than an isolating spectrograph giving comparable performance. In particular we have, according to Graner,

$$\frac{d_F}{d_G} = \frac{\sigma_F}{\sigma_s}\left(\frac{R}{R_0}\right)^{1/2}$$

where d_G is the height of the gratings and σ_s and σ_F are the diameters of the exit diaphragm of S.I.S.A.M. and the entrance diaphragm of the F-P. If the latter are equal it is seen that the S.I.S.A.M. may actually be smaller than the F-P, since one could in principle use $R/R_0 \sim N$. In practice it is probably best to consider an F-P and an S.I.S.A.M. of about equal size so as to obtain $R \sim R_0$, giving more nearly complete isolation of the F-P passband and minimum modulation of sidebands (the analog of parasitic light in the instrument).

Since the S.I.S.A.M. is a selective modulator rather than a monochromator, light is transmitted to the detector by many F-P passbands in addition to the one being modulated. This unmodulated light carries no useful information but does increase the photon noise, which is proportional to the square root of the flux. The method is, thus, inappropriate for observations in spectral regions in which the desired output signal-to-noise ratio is photon-noise-limited.

In the infrared, where the detector itself is the dominant noise source, the unmodulated photon flux can be ignored, hence the method is most efficient in this region. The S.I.S.A.M., however, does not provide the advantage of multiplexing (Fellgett 1958, Code & Liller 1962) given by other methods of two-beam interference spectroscopy.

TECHNICAL CONSIDERATIONS

Because of its advantageous properties, optical-quality fused silica, such as "Spectrasil" or "Utrasil," is the material usually employed for the interferometer plates. One surface is figured to attain the required flatness finesse, the other is often figured to about $\lambda/20$ and is antireflection-coated. Figured plates and, for that matter, ready-made etalons, are now obtainable from a number of competent commercial suppliers. Small plates giving flatness finesse up to $N_D = 100$ (measured at λ 5500) are readily obtained. The ability to produce much flatter plates, giving up to $N_D = 200$, is claimed by some makers. With increase in size the attainable flatness is greatly decreased. Etalons with $N_D = 100$ over a useful diameter of up to 60 mm have been produced in some laboratories.

Whether or not a given pair of plates meets critical finesse requirements can only be determined by laboratory measurements on the assembled etalon. Furthermore the most critical requirements can be met only if the plates are matched, since it is the variations in l which must be minimized. With unmatched plates the achieved finesse may be much less than the value suggested by the tolerances of the individual plates. The precision optical testing of assembled uncoated etalons has been considered in detail by Roesler & Traub (1966).

The choice of semitransparent coatings is discussed in some detail by Jacquinot (1960), Davis (1963), and others. Whenever possible, multilayer dielectric coatings are desirable because of their low absorption losses. A major discussion of the properties of dielectric coatings and their method of manufacture is given by Heavens (1960). Such coatings can be made to yield any required reflectance up to 99.5 per cent at any specified wavelength within essentially the entire optical and near-infrared spectrum, and are available commercially through most of the makers who supply plates. The wavelength range over which any given coatings have substantially constant reflectance depends upon the number of layers used and is unfortunately limited to only a few hundred angstroms for common (seven-layer) coatings. In the ultraviolet the materials often used tend to be water soluble and must be protected from moisture and rough handling. For precise wavelength determinations over a substantial wavelength interval, the dispersion in phase change in the coatings must be taken into account (Baumeister & Jenkins 1957).

A serious limitation to the use of dielectric coatings occurs when the greatest possible finesse is required, regardless of aperture and substrate flatness. This is true not only because the coatings, with a thickness of several λ, cannot be put on perfectly evenly, but also because they inevitably contain scattering centers that feed some of the light into off-axis modes, thus causing a reduction of effective finesse. Metallic coatings, properly made, are much less subject to this limitation and may be used where dielectric coatings fail, provided the increased absorption losses can be tolerated and the required reflectance achieved. The absorption losses can be minimized and the rate of deterioration of the coatings reduced through use of a protective dielectric layer on the outside (Bradley 1963).

Below 2400 Å there is no transparent material of high refractive index for multilayer dielectric films. However, Bradley et al. (1964) have been able to extend the spectral region accessible to F-P interferometers into the vacuum ultraviolet ($\lambda\lambda$ 1600–2400 Å) through the use of semitransparent aluminum films overcoated with magnesium fluoride. A reflective finesse of 30 and a transmissivity of 25 per cent have been achieved. The authors note that since the flatness finesse also becomes worse with decreasing wavelength, there seems to be little reason to use a reflectance finesse in excess of 30.

The mounting of plates depends upon the scanning method envisaged. Whatever mounting is used must assure that the plates are parallel and that they maintain the desired separation. Use of a spacer is the classical method and the simplest one in case of pressure-scanning. Quartz and Invar are the commonly emphasized materials because of their low coefficients of expansion. However, it might be noted that if air at a pressure near 1 atm is the scanning gas, Invar is the better material because its slightly larger coefficient of expansion very nearly compensates that of air. With gases of much larger refractive indices than that of air, like freon, sometimes used to increase the scanning range, an even larger coefficient of expansion may be desirable.

Methods of making spacers have been described by Phelps (1965) and Saksena (1966). The spacer should preferably not rest upon the reflective coating but upon the substrate and may even be optically contacted to it (cf. Smartt & Ramsey 1964). For less critical applications, mass-produced precision steel bearing-balls positioned with a lightweight retainer will work if temperature variations are not too severe. For very thin spacers, not a usual requirement except in filters, a buildup of evaporated coatings or sheets of mica has been used. Etalons constructed with spacers require mechanical force to achieve the parallelism adjustment, through compression of the spacer by means of an adjustable spring, as described by many authors. The thermoelastic modulus of the spring should be about the same as that of the spacer, although good results are obtained if this is not so.

The simplest and most frequently used method of varying the wavelength is to place the F-P etalon in an air-tight chamber in which the pressure of a gas is varied. This provides for a very straightforward design and a high degree of rigidity. A grating spectrograph can also be scanned in this way at the same rate (Hirschberg & Kadesch 1958). The variation of wavelength with pressure is independent of l and is given by

$$\Delta\lambda = \frac{\lambda}{\mu} \frac{d\mu}{dP} \Delta P$$

If μ_0 is the refractivity at $P = 1$ atm, $t = 0°$ C, then for a perfect gas, $d\mu/dP = (\mu_0 - 1) [1 + t/273]^{-1}$. For air at $t = 20°$ C, the value of $(\lambda/\mu)d\mu/dP$ goes from 1.2 Å/atm at 4000 Å to 3.0 Å/atm at 10,000 Å. Thus a pressure change of only 1–2 atm air gives a scanning range great enough for most investigations of interstellar absorption and nebular emission-line profiles. Sometimes (as in the investigation of stellar Ca II K absorption) the air pressure change required becomes excessive unless the sawtooth scanning procedure proposed by Chabbal and Jacquinot in 1955 is used. In this procedure the spectrum is scanned in sections, the premonochromator isolating F-P passbands of successively higher (or lower) order than in the previous scan, while the pressure is varied repeatedly over the same interval (corresponding to the free range). Some workers have used pressure changes as great as 100 atm (Beer & Ring 1961) or gases with very high refractivities, such as freon or propane (Davies, Ring & Selby 1964). If air is used, a pressure change of 7×10^{-4} atm, corresponding approximately to the limit of resolution of a good-quality regulator or Bourdon-type pressure gage, gives a wavelength shift of about 1.0 mÅ, surely adequate for even the most accurate astronomical work.

Despite the simplicity, precision, and elegance of pressure scanning, recourse to more sophisticated control methods is dictated when the need exists for extremely accurate parallelism control, long-period stability, remote operation, stability against environmental disturbances such as vibration, ability to scan rapidly, or ability to maintain a precisely determined spacer ratio between two etalons. For such purposes, methods making use of mechanical displacements of the F-P plates through deformation of piezoelectric or magnetostrictive transducer elements have been successful. The earliest investigation of such methods is that of Dupeyrat (1958); much new progress has since been made. In the future of interference spectroscopy these techniques seem likely to play an important role.

Arrangements permitting a degree of technical simplicity and involving the magnetostrictive distortion of Invar or Nilvar have been described by Bennett & Kindlmann (1962), who were primarily concerned with laser stabilization, and by Slater, Betz & Henderson (1965). The latter describe a sturdy magnetostrictively scanned F-P interferometer, with a finesse of 30, for geophysical and solar observations from airplanes, balloons, and rockets. A re-entrant design was used to achieve temperature compensation. The magnetostrictive properties of Invar depend upon its precise constitution and annealing history: in general the effect is nonlinear and saturates at a field strength of 6000–8000 A-turns/m, at which $\delta l/l \sim 10^{-5}$.

The manufacture and properties of piezoelectric transducer substances applicable to interferometers have been studied by Crawford (1961) and Ramsey & Mugridge (1962). With suitable preparation including prepolarization, barium titanate ceramic transducers give a linear movement of about 2500 Å/kV applied voltage up to a maximum working voltage of about 3.5 kV/mm, at which $\delta l/l = 9 \times 10^{-3}$. Consequently the element may be small. A slightly larger movement is obtained with zirconium titanate. The two principal advantages of piezoelectric transducers are their high-frequency responses and low power consumption.

Ramsey (1962, 1966) has described an ingenious and sophisticated application of piezoelectric transducers to achieve automatic control of parallelism and spacing of F-P interferometers. Of particular interest is the control technique. The basic scheme takes advantage of the white light or superposition fringes produced when a beam passes through two or more interferometers in series. A transmission maximum for white light exists only when the optical spacings μl are identical independently of wavelength (because the channeled spectra produced by the separate interferometers do not fully coincide otherwise). For parallelism detection a beam is passed first through a small area near one edge of the aperture of an etalon, then returned through a similar area near the diametrically opposite edge and received by a photomultiplier. Another system operates in a plane normal to this one, so that a maximum response received simultaneously from two photomultipliers indicates parallelism of the plates. The four small areas around the rim of the aperture are used for control purposes, leaving a clear working aperture in the center.

The control signal used is generated by the photomultipliers when one of the plates is oscillated about its center by means of oscillatory voltages applied with 120° phase differences to the three identical barium titanate supports on which the plate is mounted. When the mean plate separations at two diametrically opposed control apertures are equal during a cycle, the corresponding photomultiplier output contains no oscillatory component at the driving frequency.

When such a component appears, a phase-sensitive amplifier applies the necessary correcting bias to the barium titanate supports, which returns the plates to parallelism. At a driving frequency of 9 kc/s, the automatic parallelism control will operate reliably with an oscillatory amplitude of as little as 5 Å at the edge of the plates, which is much less than the flatness errors of the plates. Since the response time of the servo system is about 5 msec, the servo can compensate to some degree for vibrations up to about 200 c/s.

The plate separation can be controlled, varied slowly, or oscillated by applying equal bias voltages to the three barium titanate supports. Alternatively the separation can be maintained precisely constant if monochromatic instead of white light is used for the control beam; in this case the possible separations depend upon the wavelength used. More generally, one of the control beams may be white light, while the other is the channeled spectrum produced by an auxiliary reference interferometer; the plate separation of the main interferometer automatically will follow that of the reference, which may have any desired value.

LITERATURE CITED

Babcock, H. W. 1953, *Ap. J.*, **118**, 387

Baumeister, P. W., Jenkins, F. A. 1957, *J. Opt. Soc. Am.*, **47**, 57

Beer, R., Ring, J. 1961, *Infrared Phys.*, **1**, 94

Bennett, W. R., Jr., Kindlmann, P. J. 1962, *Rev. Sci. Instr.*, **33**, 601

Bens, A. R., Cogger, L. L., Shepherd, G. G. 1965, *Planetary Space Sci.*, **13**, 551

Bradley, D. J. 1963, *Appl. Opt.*, **2**, 539

Bradley, D. J., Bates, B., Juulman, C. O. L., Majumdar, S. 1964, *Nature*, **202**, 579

Bruce, C. F. 1966, *Rev. Sci. Instr.*, **37**, 349

Candler, C. 1951, *Modern Interferometers*, 296–99 (Hilger & Watts, Ltd., London)

Chabbal, R. 1953, *J. Rech. Centre Natl. Rech. Sci. Lab. Bellevue (Paris)*, No. 24, 138 (Engl. transl. by Jacobi, R. B., AERE, Harwell, *AERE Lib. Trans. 778*, 1958)

Chabbal, R. 1958, *Rev. Opt.*, **37**, 2, 336, 501

Code, A. D., Liller, W. C. 1962, *Astronomical Techniques, Stars and Stellar Systems*, **2**, 281 (Hiltner, W. A., Ed., Univ. of Chicago Press)

Connes, P. 1958, *J. Phys. Radium*, **19**, 197

Connes, P. 1964, *Quantum Electronics and Coherent Light*, 207 (Academic Press, New York)

Courtès, G. 1958, *J. Phys. Radium*, **19**, 342

Courtès, G. 1960, *Ann. Ap.*, **23**, 115

Courtès, G. 1964, *Astron. J.*, **69**, 325

Courtès, G., Cruvellier, P., Georgelin, R. 1966, *J. Observateurs*, **49**, 329

Courtès, G., Fehrenbach, C., Hughes, E., Romand, J. 1966, *Appl. Opt.*, **5**, 1349

Crawford, A. E. 1961, *Brit. J. Appl. Phys.*, **12**, 529

Davies, L. B., Ring, J., Selby, M. J. 1964, *Monthly Notices Roy. Astron. Soc.*, **128**, 399

Davis, S. P. 1963, *Appl. Opt.*, **2**, 727

Dunham, T. 1956, *Vistas Astron.*, **2**, 1223

Dupeyrat, R. 1958, *J. Phys. Radium*, **19**, 290

Fabry, C., Buisson, H. 1911, *Ap. J.*, **33**, 406

Fellgett, P. 1958, *J. Phys. Radium*, **19**, 237

Field, G. B., Hitchcock, J. L. 1966, *Ap. J.*, **146**, 1

Geake, J. E., Ring, J., Woolf, N. J. 1959, *Monthly Notices Roy. Astron. Soc.*, **119**, 42

Geake, J. E., Wilcock, W. L. 1957, *Monthly Notices Roy. Astron. Soc.*, **117**, 380

Gebbie, H. A., Twiss, R. Q. 1966, *Rept. Progr. Phys.*, **29**, Part II, 729

Goss, W. M., Spinrad, H. 1966, *Ap. J.*, **143**, 989

Graner, G. 1965, *Appl. Opt.*, **4**, 1620

Heavens, O. S. 1960, *Rept. Progr. Phys.*, **23**, 1

Herbig, G. H. 1964, *Ap. J.*, **140**, 702

Hernandez, G. 1966, *Appl. Opt.*, **5**, 1745

Hill, R. M. 1963, *Opt. Acta*, **10**, 141

Hirschberg, J. G., Kadesch, R. R. 1958, *J. Opt. Soc. Am.*, **48**, 177

Hobbs, L. M. 1965, *Ap. J.*, **142**, 160

Jacquinot, P. 1954, *J. Opt. Soc. Am.*, **44**, 761

Jacquinot, P. 1960, *Rept. Progr. Phys.*, **23**, 267

Jacquinot, P., Dufour, C. 1948, *J. Rech. Centre. Natl. Sci. Lab. Bellevue (Paris)*, **6**, 91

Kaplan, L. D., Münch, G., Spinrad, H. 1964, *Ap. J.*, **139**, 1

Mack, J. E., McNutt, D. P., Roesler, F. L., Chabbal, R. 1963, *Appl. Opt.*, **2**, 873

McNutt, D. P. 1965, *J. Opt. Soc. Am.*, **55**, 288

Mertz, L. 1965, *Transformations in Optics* (Wiley, New York)

Münch, G. 1966, *Year Book 65*, 153 (Carnegie Inst. of Washington)

Oke, J. B., Dennison, E. W. 1965, *Year Book 64*, 49 (Carnegie Inst. of Washington)

Oke, J. B., Greenstein, J. L. 1961, *Ap. J.*, **133**, 349

Phelps, F. M. III, 1965, *J. Opt. Soc. Am.*, **55**, 293; see also Phelps, F. M. III, Newbound, K. B. 1966, *J. Opt. Soc. Am.*, **56**, 831

Roesler, F. L., Traub, W. 1966, *Appl. Opt.*, **5**, 463

Ramsey, J. V. 1962, *Appl. Opt.*, **1**, 411

Ramsey, J. V. 1966, *ibid.*, **5**, 1297

Ramsey, J. V., Mugridge, E. G. V. 1962, *J. Sci. Instr.*, **39**, 636

Ramsey, J. V., Smartt, R. N. 1966, *Appl. Opt.*, **5**, 1341

Saksena, G. D. 1966, *J. Opt. Soc. Am.*, **56**, 256

Schwider, J. 1965, *Opt. Acta*, **12**, 65

Shepherd, G. G., Lake, C. W., Müller, J. R., Cogger, L. L. 1965, *Appl. Opt.*, **4**, 267

Slater, P. N., Betz, H. T., Henderson, G. 1965, *Japan. J. Appl. Phys.*, **4**, *Suppl. I*, 440

Smartt, R. N., Ramsey, J. V. 1964, *J. Sci. Instr.*, **41**, 514

Spinrad, H., Schorn, R. A., Moore, R., Giver, L. P., Smith, H. J. 1966, *Ap. J.*, **146**, 331

Stoner, J. O., Jr. 1966, *J. Opt. Soc. Am.*, 56, 370

Stroke, G. W. 1963, *Progr. Opt.*, 2, 1

Vaughan, A. H., Jr. 1964, *An Investigation of the He I λ 10830 Emission Line in the Orion Nebula* (Doctoral thesis, Univ. of Rochester, Rochester, N. Y.)

Vaughan, A. H., Jr., Münch, G. 1966, *Astron. J.*, **71,** 184; *Year Book 65*, 153, 175 (Carnegie Inst. of Washington)

Vaughan, A. H., Jr., Zirin, H. 1966, *Astron. J.*, **71,** 188 (Submitted to *Ap. J.*, 1967)

OBSERVING THE GALACTIC MAGNETIC FIELD[1,2]

BY H. C. VAN DE HULST

Sterrewacht, Leiden, Netherlands

HISTORY

Some highlights of the subject in my opinion are: the symposium on cosmical aerodynamics in Paris 1949, where, owing to the surprise discovery of optical interstellar polarization by Hall and Hiltner a few months earlier, magnetic fields formed a main theme in the discussion. Then the exciting year 1955 in which optical polarization measurements of the Crab Nebula sparked from the USSR through Leiden to Mount Palomar and provided good proof that at least in one object the synchrotron emission is at work and the magnetic field can be mapped. From that time on, measuring the polarization and thus mapping the magnetic field became a prime desideratum in galactic radio astronomy. The first trustworthy results were available by the time of the Princeton symposium in the spring of 1961, where champions of strong fields ($\sim 2 \times 10^{-5}$ G) and of weak fields ($\sim 0.2 \times 10^{-5}$ G) waged a heavy theoretical battle. At the present symposium[2] the theoretical problems seem to have been somewhat eased but not solved. On the other hand, the amount and the quality of observational data already available is surprising. Some of them are very recent and should await a more thorough discussion by the authors themselves.

Traditionally, polarization phenomena have been discovered too late. There was a delay of about 200 years between the presence of adequate instruments (a wine glass and a birefringent crystal) and the discovery of polarization by reflection. For the optical interstellar polarization the delay was about 40 years. But for the radiopolarization of the galactic synchrotron emission it was something like minus 1 or 2 years; so keen was everybody to get this phenomenon established.

REVIEW OF "OTHER METHODS"

Table I presents a list of observed effects from which we may infer something about the magnitude of the magnetic field, or its direction or topology, or both. The comments are conservative and, like many other statements in this review, subjective. I shall first review the "other methods," then the radio methods given in the last two lines.

Optical interstellar polarization.—Active programs of observation in the years after the discovery of the effect led to maps showing the direction and degree of polarization for over 3000 stars (Hall 1958, Behr 1959). There is no

---

[1] The survey of literature for this review was concluded in September 1966.

[2] Review paper, with minor additions, presented at the Symposium on Radio Astronomy of the Galactic System, organized by the International Astronomical Union, Noordwijk, August 25 to September 1, 1966.

reasonable doubt that a magnetic field is the basic reason for the partial alignment of the interstellar grains that gives rise to the optical polarization. The mechanism probably is the one first proposed by Davis and Greenstein, which means that the predominant electric vector of the observed starlight is parallel to the magnetic field projected on the sky. Hence the maps give some idea of the topology of the magnetic field. The alignment parallel to the galactic equator is most perfect near $l = 140°$ and completely lacking near $l = 80°$ (all longitudes in this paper are on the new scale). This led Chandrasekhar and Fermi to propose that these are the respective directions where we look across and along the local spiral arms. The conclusion is approximately correct, but the picture of the arm as a continuous tube of force

TABLE I

OBSERVATIONAL DATA ABOUT THE GALACTIC MAGNETIC FIELD

	Magnitude	Direction topology
Optical interstellar polarization	q	f
Shapes of filamentary nebulae	—	q
Cosmic-ray energy density and confinement	q	q
Cosmic-ray anisotropy	—	q
Cosmic-ray electrons plus nonthermal radio emission	f	—
Zeeman effect, H	q	—
Zeeman effect, OH	—	—
Polarization of nonthermal radio emission	—	f
Faraday effect	f	f

Key: — no data or don't believe,
q questionable or marginal,
f fair or fine.

with wiggles does not necessarily follow. An arbitrary magnetic configuration stretched by differential rotation would give much the same observed effect.

Further detailed studies can give a great deal more information. The study of nearby stars (Behr 1959), for which very small degrees of polarization have to be measured, is of special interest. Hiltner's new rotating telescope will offer possibilities for continuing this work with great precision. Many further statistical studies can be made, to find the field topology or to determine scale parameters, which can then be used in a theoretical discussion, or to examine correlation with other properties of stars and nearby nebulae. Many papers have been devoted to these topics in the past ten years.[3] We note, in particular, correlation studies of polarization in clusters,

[3] Where no detailed references are given, we suggest the triannual Reports on Astronomy of the International Astronomical Union for fairly complete but not necessarily critical reviews.

from which Serkowski (1965 and unpublished work) determines a micro-scale in the magnetic field of the order of 1 pc.

The magnitude of the field needed to produce enough alignment of the interstellar grains to explain the observed polarization was initially a worry. Thanks to the work of Greenberg (Greenberg & Shah 1966) on the extinction by nonspherical grains, it is now possible to compute the polarization for a given field strength and for a given shape and composition of the grains (including their complex paramagnetic permeability coefficients). Reasonable assumptions, given a reasonable field. Unfortunately our poor knowledge of the grains still leaves this a questionable method for determining the field strength.

Shapes of filamentary nebulae.—Field-aligned irregularities occur in the solar corona (where the polar plumes were the first evidence for a general solar magnetic field), in the Earth's magnetosphere, in aurorae, and in laboratory plasmas, so there is good reason to expect them also in interstellar space. At the Noordwijk Symposium Pikelner sketched a mechanism that could lead to a gaseous nebula stretched along a magnetic-field line. Shajn argued many years ago that many filamentary emission nebulae are oriented along the magnetic field. My problem is only: which nebulae? For it is clear that filaments could also be curtains or shells or shock fronts seen edgewise and that many such features could be explained by nonmagnetic gas dynamics. For this reason it may be difficult to make much progress with this method.

Cosmic-ray energy density.—A 20-year-long discussion must here be compressed into a few lines. The argument can best be discussed in the form of an application of the virial theorem (Biermann & Davis 1960)

$$2T + 3P + M = -\Omega$$

All terms are positive. T is the total kinetic energy, P the pressure integrated over the entire volume, M the total magnetic energy, and Ω the (negative) total gravitational energy. If the system is to be maintained, this equality must hold. The large contribution to $2T$ arising from the galactic rotation can be estimated fairly well. Upon subtracting this, it appears already difficult to accommodate the term $3P$, which arises mostly from cosmic-ray pressure. The traditional argument, therefore, is that we cannot accept a value M that is substantially higher than $3P$. On the other hand, if the magnetic fields confine the cosmic rays, it would also be surprising to find M much smaller than than $3P$. Taking the same volume (presumably the halo volume) and precise equality, we would find $3P = 1.6 \times 10^{-12}$ erg cm^{-3}, $B = 7 \times 10^{-6}$ G.

Obviously, this is at most a vague estimate. A questionable point in this argument is that it is not clear yet to what extent the cosmic rays really are confined. Also, as Puppi, Setti, & Woltjer (1966) have pointed out, confinement may be helped by clouds impinging upon the Galaxy from outside.

Cosmic-ray anisotropy—Dozens of positive results have been announced but have not stood up under subsequent examination. This makes us suspicious of further claims. The results announced by Jacklyn (1966) seem above suspicion, however. Data from underground counters in Hobarth, Budapest, and London at depths equivalent to 35–60 m of water gave amplitudes and phases of the diurnal and semidiurnal variation. The errors are typically 0.01 per cent, i.e. about a factor 10 smaller than the maximum variations. The results are consistent with a slight preference for small pitch angles with respect to a field direction to or from $1 = 62°$, $b = +12°$. The uncertainty is $\pm 5°$. The relatively good agreement with the field direction found in other ways lends support to this determination.

Cosmic-ray electrons combined with nonthermal radio emission.—The principle of this determination is straightforward. Synchrotron emission comes from fast electrons in a magnetic field. If we can measure the emission by observing the nonthermal radio continuum, and the electrons by detecting them as cosmic-ray electrons near the Earth, then we can calculate the field strength.

The theory underlying this determination has been well reviewed by Biermann & Davis (1960) and by Ginzburg & Syrovatskii (1964, 1965) and presents no hazards. But the practical execution involves a number of uncertainties. The main questions are:

(*a*) Can we reliably convert the observed radio brightness into a volume emissivity arising from synchrotron emission?

(*b*) Can we reliably measure the cosmic-ray electrons among the hundred-times-more-abundant protons, other nuclei, and their secondary products?

(*c*) Is the electron density measured near the Earth typical of the density in interstellar space?

With the accumulation of observational data at more frequencies and with better angular resolution, question (*a*) has become more difficult to answer than 10 years ago, when the separation between a relatively smooth disk and smooth halo seemed rather evident. The extreme assumption that most of the nonthermal continuum at low latitudes is due to unresolved sources can pretty well be excluded, so the emissivity in the disk remains about what it was. But the high-latitude distribution shows so many details that some authors prefer to describe it as a collection of shells (Quigley & Haslam 1965). The concept of a large halo with fairly uniform radio emission has almost vanished from the literature. This may be an overcorrection.

Until recently, question (*b*) seemed by far the most difficult one. Several satellite experiments to measure the cosmic-ray electrons near the Earth are in preparation. But the preparatory balloon flights have already given rather convincing results. Figure 1 shows the cosmic-ray electron spectrum as we know it (Tanaka 1966). In the interpretation of these data, naturally, question (*c*) arises. Several arguments are in favor of blaming the change of slope near 1 GeV in Figure 1 on solar modulation. Electrons of higher energy would be unaffected by such modulation. A more secure answer may have to await observations during a solar cycle.

Altogether it appears that the numbers change very little from those

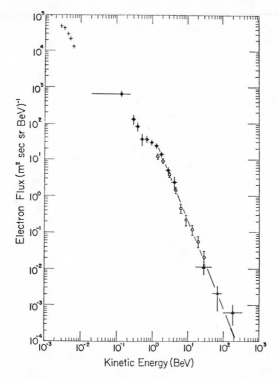

FIG. 1. Energy spectrum of cosmic-ray electrons. This is a composite diagram prepared by Dr. Y. Tanaka (October 1966) from data by Cline et al. (+, 1964), Meyer et al. (●, unpublished), Bleeker et al. (o, unpublished), and Daniel & Stephens (Δ, unpublished).

given by Biermann & Davis (1960), but that their lower limit of the electron density and hence the higher limit of the field strength can now (tentatively) be regarded as actually measured values. This would make the disk field 2×10^{-5} G. An independent estimate by Sironi (1965) gave 0.9×10^{-5} G, by Tanaka (1966) 1.6×10^{-5} G. The slope $\gamma = 2.4$ of the observed electron spectrum in the range 2–30 GeV would give a radio spectral index $\alpha = \frac{1}{2}(\gamma - 1) = 0.7$, well within the range of values determined by direct observation.

Zeeman effect.—The atomic hydrogen line at 21 cm has a Zeeman effect, well established in the laboratory and by theory. The two circularly polarized components in a longitudinal field are separated by 28 c/s per 10^{-5} G. Unfortunately, this separation is so small that the best efforts, devoted to the sharpest absorption peaks available, still give only an upper limit for the field strength. The values quoted for two such clouds are $(-2 \pm 5) \times 10^{-6}$ G and $(-3 \pm 3) \times 10^{-6}$ G (Verschuur 1966). In interpreting these data it should be noted that the field in such a dense cloud may actually be smaller than it generally is in the disk.

Some polarization phenomena in the OH lines have been interpreted as Zeeman effect in fields of the order of 10^{-3} G. As long as the conditions of excitation of these lines remain enigmatic (they probably involve some chance maser effect), it is hard to take this quantitative result seriously. The qualitative argument that circular polarization can be produced only in the presence of a magnetic field may be incorrect. Heer (1966) has shown that

TABLE II

CONTINUUM POLARIZATION SURVEYS

Observatory	Authors	Year of publication	Fre- quency	Beam- width
Parkes	Mathewson & Milne	1965	408	48'
Cambridge	Wielebinsky & Shakeshaft	1964	408	8°
Dwingeloo	Westerhout, Seeger, Brouw & Tinbergen	1962	408	2°
Dwingeloo	Brouw, Muller & Tinbergen	1962	408	2°
Dwingeloo	Berkhuijsen & Brouw	1963	408	2°
Dwingeloo	Brouw	Unpublished	465	1°8
Dwingeloo	Berkhuijsen, Brouw, Muller & Tinbergen	1965	610	1°3
Parkes	Mathewson, Broten & Cole	1966	620	32'
Dwingeloo	Brouw	Unpublished	820	1°
Cambridge	Bingham	Unpublished	1407	2°
Dwingeloo	Brouw	Unpublished	1411	30°
Parkes	Mathewson, Broten & Cole	1966	1410	14'
Parkes	Högböm	Unpublished	1410	14'

circular polarization could result from saturation effects in a maser amplifier with an energy level structure similar to that of the OH molecule.

THE RADIO-POLARIZATION DATA

The effects noted in the last two lines of Table I are: the polarization of synchrotron emission, which shows the existence of a magnetic field at the source of radiation; and the Faraday effect, which shows the existence of a field along the line of sight. The discussion of these topics cannot quite be separated. We shall place the main emphasis on the first one.

Continuum polarization surveys.—By now we have a number of reliable continuum polarization surveys, and have passed from the stage when instrumental corrections formed the main topic to the early stages of astronomical interpretation. Table II lists all surveys available to date.

Since no circular polarization has been found, the quantities measured, in principle, are three Stokes parameters for any point on the sky. They can be separated into a polarized and an unpolarized component as follows

$$\begin{bmatrix} I(l, b, \nu) \\ Q(l, b, \nu) \\ U(l, b, \nu) \end{bmatrix} = \begin{bmatrix} I_u \\ 0 \\ 0 \end{bmatrix} + \begin{bmatrix} I_p \\ I_p \cos 2\theta \\ I_p \sin 2\theta \end{bmatrix}$$

Here l, b = galactic longitude and latitude, ν = frequency, I_u = unpolarized intensity, I_p = polarized intensity, and θ = position angle. The degree of polarization is $p = I_p/(I_u + I_p)$.

The instrumentation usually is designed to measure the polarized component with great accuracy at one frequency. It is therefore natural to display the results as maps in which each observed point (l, b) shows a dash with length I_p and direction θ. We may call I_p the polarization brightness and convert it in the usual manner into a brightness temperature.

Figures 3 and 4 show such maps for two regions in the sky that have attracted special interest. The complete maps would require too much space to be included in this review. Taking the 408 Mc/s as an example, we observe that in most points of the sky the polarization brightness rises just above the instrumental errors. The amount and direction in adjacent points usually are similar and sometimes create a coherent pattern over 10° or more on the sky. This shows at once that there is no typical magnetoturbulence with fields tangled on a small scale. If 200 pc is adopted as a representative distance, the coherence scale is about 40 pc.

Spectra, depolarization.—Before entering into a more detailed discussion of these maps, it should be noted that this presentation misses some important points: the maps do not show the unpolarized component or the degree of polarization, nor do they give ready information about the spectrum. Figure 2 provides this information in the form of a sketch of some typical spectra.

This illustration, made from eye estimates on the preliminary maps of Brouw, will need revision in detail, but suffices to show the general trend. The ordinate is proportional to brightness I_p for the polarized part, and to $I_p + I_u$ for the total brightness. Since temperature units are more important in the discussion than brightness units, the lines of constant brightness temperature have been drawn in for reference. The polarization temperatures in the four regions shown are low, but remain well above the internal mean errors shown by crosses. The total brightness is rather similar in these four regions and is shown in the top part of the figure by one solid line with adopted slope -0.65. In this Figure the published polarization temperatures at 408 Mc/s and 620 Mc/s have been revised upwards by the factors 1.4 and 1.5, respectively, on the basis of a new calibration by Brouw.

The most striking feature of these spectra is the existence of a strong depolarization. The theoretical synchrotron radiation of electrons with isotropic velocities in a homogeneous magnetic field has the degree of polarization

$$p = (\gamma + 1) \bigg/ \left(\gamma + \frac{7}{3}\right)$$

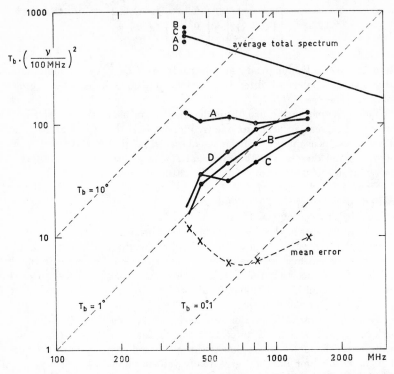

FIG. 2. Spectra of polarized and unpolarized radiation in four regions of the sky (rough sketch based on unpublished data by Brouw). Region $A:l=140°$, $b=5°$; Region $B:l=125°$, $b=5°$; Region $C:l=30°$, $b=30°$; Region $D:l=30°$, $b=50°$.

where γ is the exponent of the electron energy spectrum. The value $\gamma = 2.4$ gives $p = 0.72$. The regions shown in Figure 2, however, have degrees of polarization from 3 to 50 per cent, much lower than the theoretical 72 per cent. In most regions of the sky the polarization is even weaker.

Since no physical effect will destroy the polarization, its weakness must be attributed to superposition of some kind. Many possibilities are open: superposition of thermal emission, which is unpolarized; superposition of different intrinsic directions of polarization within the beam; different intrinsic directions of polarization along the line of sight. Further, the Faraday effect may rotate by different amounts the radiation emitted at different distances along the line of sight, or at different directions within the beam, or even at different frequencies within the band. The positive slope of the spectra in Figure 2, i.e. the rapid (but perhaps not quite smooth) decrease of polarization with increasing frequency, points very clearly to the Faraday effect. The currently popular explanation is that only the relatively nearby regions contribute to the observed polarization. The emission at greater distances arrives with so many different angles of rotation that it is virtually unpolarized. It is, so to say, covered by a "Faraday fog." This fog gradually lifts when we go to higher frequencies. We may recall that the Faraday rotation turns the plane by $R\lambda^2$ radians where R, the rotation measure, is

$$R = 0.81 \int N_e(B \cos \theta)dl$$

with N_e = electron density in cm^{-3}, $B \cos \theta$ = longitudinal component of field strength in 10^{-6} G, dl = element of line of sight in parsecs.

Mathewson's belt.—Among the details shown by the polarization maps, the "fan region" in the northern sky near $l = 140°$ is most striking. Figure 3 shows maps of the region at two frequencies. This region is in no other way peculiar, and it seems likely that it is just a rather prominent hole in the Faraday fog. It fans out more widely at 465 Mc/s than at 1411 Mc/s. If the rotation is taken out by its proportionality to ν^{-2}, the dashes become vertical, showing a magnetic field parallel to the galactic plane. The rotation is 0 at $l = 140°$, showing that at that longitude the field is perpendicular to the line of sight, which fixes its direction (including the sign) in space.

An analysis of this type is permissible if a strict separation between a fully depolarized background and a polarized but rotated foreground emission can be made. Consistent with this extreme assumption would be a polarization spectrum with the same slope as the normal synchrotron spectrum. We see in Figure 2 that the fan region (region A) indeed comes closest to this assumption. And only with this assumption is it possible to define a unique rotation measure for the continuum polarization.

Closer examination shows that the simple picture just sketched does not explain all details of the fan region. For one thing, a second zero in the rotation occurs near $l = 160°$. Both Hornby (1966) and Bingham (unpublished) have fitted more complicated field models to this region. More generally, models computed by Komesaroff (unpublished) for emission along the line of sight, with various values of the rotation measure, have shown that the direction of polarization may suggest a unique rotation measure, but when the degree of polarization is plotted against frequency, prominent fluctuations show that this simple interpretation is incorrect.

Polarization maps of the entire sky reveal the existence of a large-scale feature sometimes called "Mathewson's belt." Mathewson & Milne (1965) marked all places where, as in the fan region, the observed polarization is relatively strong. With hardly any exceptions these fall in a belt, about 60° wide, cutting the galactic equator at $l = 320°$-20° and at $l = 120°$-180°, and perpendicular to it. This belt contains the fan region just described, and Mathewson's explanation is similar to the one given above: the belt is the locus of directions perpendicular to the local magnetic field, where Faraday rotation is smallest and synchrotron polarization strongest. The deviation by about 10° of the belt from a great circle may be due to the magnetic-field lines expanding outwards in the direction of $l = 250°$.

Rotation measures from extragalactic sources.—At this point a comparison

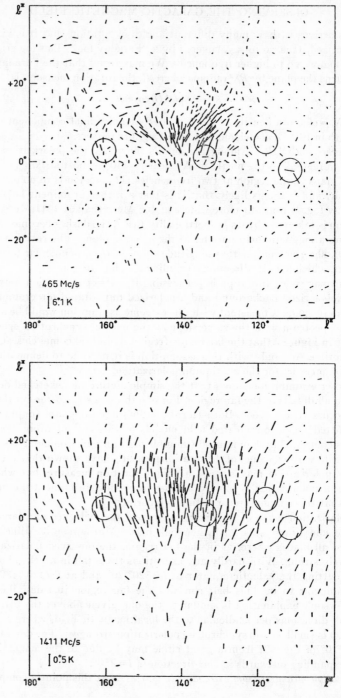

FIG. 3. Polarization maps of the "fan" region near $l = 140°$ at two frequencies, showing different Faraday effect (Brouw, unpublished).

with rotation measures derived from the observation of extragalactic radio sources suggests itself. This has been a very active field of research in the past few years, including many sets of observations and statistical and theoretical studies (e.g. Bologna et al. 1965, Maltby & Seielstad 1965, Gardner & Davies 1966a, Gardner & Whiteoak 1966).

We shall not review this entire field. Rotation measures thus found typically are 10–100, or even larger, whereas the galactic polarization studies just discussed give typical values 0–5. At least three independent arguments for such a difference can be advanced: Faraday rotation in the extragalactic source; Faraday rotation in the more distant parts of the Galaxy, which are virtually depolarized in the continuum studies; the fact that determinations from the observed continuum polarization constitute a selection of regions where the rotation measure is small. Different authors assess these explanations with different weights. Probably further model calculations and statistical studies will be necessary before a conclusion can be reached. That is why we are not quite ready to discuss this important subject in full in the present review.

A few results may be noted. Gardner & Davies (1966b) have constructed a map in which the rotation measures of extragalactic sources, with their proper sign, are plotted as a function of galactic coordinates. They have tentatively drawn iso-R contours, which presumably give information about the large-scale field of the Galaxy. The sign changes not only with longitude, at about 200° and 340°, but also with latitude, thus requiring rather complicated field models for its explanation. Seymour (unpublished) has made model studies using Legendre expansion. Mathewson & Milne, disregarding the signs, note from the same map that large rotation measures systematically occur outside Mathewson's belt; which is understandable if this belt is more than a local phenomenon. Most statistical studies correlating the rotation measure with other parameters have been disappointing (Maltby 1966). However, Bologna, McClain & Sloanaker (unpublished) find a significant absence of high degrees of polarization at low latitudes only in the longitude quadrants towards the center, where we look through much galactic gas. The explanation would be that the solid angle subtended by the source is wide enough to permit different values of R, thus leading to partial depolarization. With source diameters of the order of 1′ and reasonable distances of the gas this would point to a fine structure in the magnetic field of the order of 1 pc.

The north-galactic spur.—In all continuum surveys the north-galactic spur forms the most prominent feature outside the galactic plane. It was already visible on Reber's old maps. The spur can be traced from the galactic equator near $l = 40°$ to the galactic pole and beyond; according to some authors it extends all around a great circle. Nobody knows what it is, although there are plenty of speculations. There are a few fainter arcs which may have a similar character (Quigley & Haslam 1965).

Naturally, when polarization methods became feasible, the spur was one

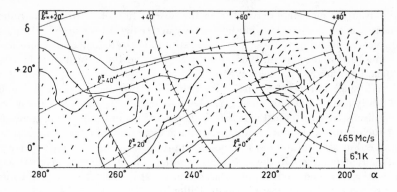

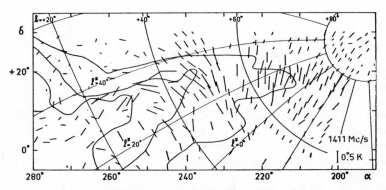

Fig. 4. Polarization maps of the north-galactic spur at two frequencies,
from data obtained at Dwingeloo (Brouw, unpublished).

of the first objects to be studied. Figure 4 shows its polarization maps at two
frequencies with the Dwingeloo 25-m telescope. A full discussion of these
data is not yet available. Measurable polarization gradually appears at lower
latitudes as we let the frequency increase. This fact, which is also shown by
the strong positive slopes of curves C and D in Figure 2, points to a distant
object that gradually becomes more visible through the Faraday fog. If the
peculiar convergence of optical polarization towards the point $l = 37°$, $b = 0°$
noted by Hall (1958) has anything to do with the spur, this also would indicate that it cannot be very near. On the other hand, Bingham (unpublished),
in correlating his polarization measurements with the optical polarization
data of Behr (1959), finds a fair correlation with stars at a distance of only
100 pc and suggests that the spur may be at a distance of that order. Finally,
the few polarization scans made by Högböm (unpublished) with a beam of

11' show examples of marked changes of direction, which would have been obliterated by a larger beam (Figure 5). The obvious conclusion is that any explanation based on low-resolution maps may be subject to drastic revision when better data become available.

CONCLUSIONS

The review in this paper has been conservative, dealing mostly with the observations and adding only relatively unassailable theoretical interpretations. This attitude can be justified by the fact that I have been permitted to use much unpublished observational material that has not yet been fully discussed by the authors themselves.

For a full understanding of the galactic magnetic field a more aggressive theoretical approach, exploring not only the geometry but also the dynamics and stability of all kinds of configurations, is certainly necessary. Important facts entering into such an approach are the existence of differential galactic rotation and that of spiral arms. Recent studies of these subjects have been presented by Wentzel (1963), Woltjer (1965), and Parker (1966).

Returning to the data directly inferred from observation, we find, in summary, that a fair measure of agreement exists about the direction of the magnetic field in our neighborhood. The points 90° or 270° away from the directions where we look perpendicularly to the field, or the points 0° or 180° away from the directions where we look along it, are:

$$
\begin{array}{ll}
\text{from optical polarization} & l = 50°\text{--}80° \\
\text{from cosmic-ray anisotropy} & l = 62° \\
\text{from the polarization ``fan''} & l = 50° \\
\text{from Mathewson's belt} & l = 70° \\
\text{from rotation measures} & \\
\text{of extragalactic sources} & l = 70°\text{--}110°
\end{array}
$$

For comparison, the direction of the local spiral arm (Sharpless 1965) is

$$
\begin{array}{ll}
\text{as outlined by O associations} & l = 50° \\
\text{as outlined by H II regions} & l = 60° \\
\text{as outlined by H I gas} & l = 70°
\end{array}
$$

These values deserve to be more than quoted and averaged. Each of them is uncertain by at least 10°, not only in determination but also in definition. Much depends on the definition of "local" in this context. The spiral-arm studies refer to objects at distances up to several kiloparsecs, the radio continuum polarization surveys effectively to objects at several hundred pc, and the cosmic-ray anisotropy to a minute fraction of 1 pc. The best this comparison does is to relieve us somewhat from the worry that the astronomical studies of magnetic fields might not be relevant at all on the small scale covered by cosmic-ray studies.

The question of topology, whether the direction just quoted is the direction of a wiggly but continuous tube of force, or just the predominant direc-

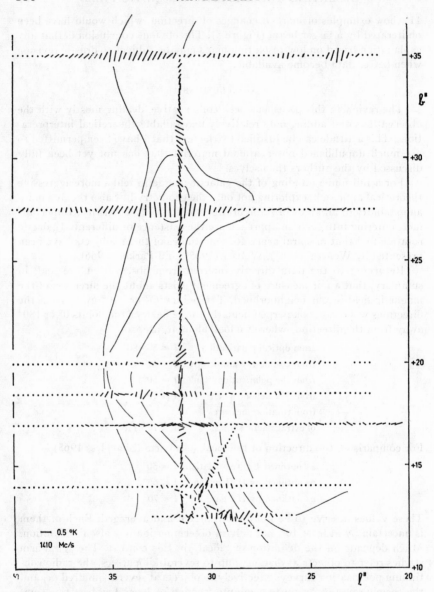

FIG. 5. Polarization scans through the north-galactic spur made with an angular resolution of 14′ with the Parkes telescope (Högböm, unpublished). This Figure covers about ⅛ of the area of Figure 4.

tion of a more tangled field, remains open. The microscales of the order of 1 pc suggested by some optical and radio studies should warn us not to take the simplest picture for granted. The question of magnitude remains very much up in the air. The estimate "of the order of 10^{-5} G" seems all right. But factors of the order of 3 in the field, and hence 10 in the pressure—which would make an enormous difference to the dynamical picture—cannot yet be firmly decided by direct observation.

LITERATURE CITED

Behr, A. 1959, *Nachr. Akad. Wiss. Göttingen, Math.-Phys. Kl. Nr.* 7, 185–240

Berkhuijsen, E. M., Brouw, W. N. 1963, *Bull. Astron. Inst. Neth.*, **17**, 185–202

Berkhuijsen, E. M., Brouw, W. N., Muller, C. A., Tinbergen, J. 1965, *Bull. Astron. Inst. Neth.*, **17**, 465–94

Biermann, K., Davis, L. 1960, *Z. Ap.*, **51**, 19–31

Bologna, J. M., McClain, E. F., Rose, W. K., Sloanaker, R. M. 1965, *Ap. J.*, **42**, 106–21

Brouw, W. N., Muller, C. A., Tinbergen, J. 1962, *Bull. Astron. Inst. Neth.*, **16**, 213–23

Cline, T. L., Ludwig, G. H., McDonald, F. B. 1964, *Phys. Rev. Letters*, **13**, 786–89

Gardner, F. F., Davies, R. D. 1966a, *Australian J. Phys.*, **19**, 129–39

Gardner, F. F., Davies, R. D. 1966b, *ibid.*, 441–59

Gardner, F. F., Whiteoak, J. B., 1966, *Ann. Rev. Astron. Ap.*, **4**, 245

Ginzburg, V. L., Syrovatskii, S. I. 1964, *The Origin of Cosmic Rays* (Pergamon, Oxford)

Ginzburg, V. L., Syrovatskii, S. I. 1965, *Ann. Rev. Astron. Ap.*, **3**, 297–350

Greenberg, J. M., Shah, G. 1966, *Ap. J.*, **145**, 63–74

Hall, John S. 1958, *Publ. U.S. Naval Obs.*, **17**, 273–342

Heer, C. V. 1966, *Phys. Rev. Letters*, **17**, 774

Hornby, J. M. 1966, *Monthly Notices Roy. Astron. Soc.*, **133**, 213–24

Jacklyn, R. M. 1966, *Symposium on radio and optical studies of the galaxy, held at Mount Stromlo Observatory*, 8–13 (Mimeographed rept., Hindman, J. V., Westerlund, B. E., Eds.)

Maltby, P. 1966, *Ap. J.*, **144**, 219–32

Maltby, P., Seielstad, G. A. 1965, *Ap. J.*, **144**, 216–18

Mathewson, D. S., Broten, N. W., Cole, D. J. 1966, *Australian J. Phys.*, **19**, 93–109

Mathewson, D. S., Milne, D. K. 1965, *Australian J. Phys.*, **18**, 635–53

Parker, E. N. 1966, *Ap. J.*, **145**, 811–33

Puppi, G., Setti, G., Woltjer, L. 1966, *Nuovo Cimento*, **45**, 252–53

Quigley, M. J. S., Haslam, C. G. T. 1965, *Nature*, **208**, 741–43

Serkowski, K. 1965, *Ap. J.*, **141**, 1340–61

Sharpless, S. 1965, *Stars and Stellar Systems*, **V**, *Galactic Structure*, 131 (Blaauw, A., Schmidt, M., Eds., Univ. of Chicago Press)

Sironi, G. 1965, *Nuovo Cimento*, **39**, 372–76

Tanaka, Y. 1966, Communication at I.A.U. Symp. Noordwijk

Verschuur, G. L. 1966, Communication at I.A.U. Symp. Noordwijk. Also in British Natl. Rept. to URSI

Wentzel, D. G. 1963, *Ann. Rev. Astron. Ap.*, **1**, 195–218

Westerhout, G., Seeger, C. L., Brouw, W. N., Tinbergen, J. 1962, *Bull. Astron. Inst. Neth.*, **16**, 187–224

Wielebinsky, R., Shakeshaft, J. R. 1964, *Monthly Notices Roy. Astron. Soc.*, **128**, 19–32

Woltjer, L. 1965, *Stars and Stellar Systems*, **V**, *Galactic Structure*, 531–88 (Blaauw, A., Schmidt, M., Eds., Univ. of Chicago Press)

OH MOLECULES IN THE INTERSTELLAR MEDIUM[1]

By B. J. Robinson and R. X. McGee

Radiophysics Laboratory, CSIRO, Sydney, Australia

Introduction

Since the discovery of the first microwave lines of the hydroxyl radical, in October 1963, a series of most surprising observations have been made. A number of processes have been suggested to account for them, but as of the end of 1966 there is no sound theoretical framework on which to fit the results. It has accordingly been difficult to find a satisfactory arrangement of the material for this *Review*. There appears to be no observation which is completely consistent with our prior knowledge of conditions in the interstellar medium. Thus the investigations of OH must contain most pregnant clues to the behaviour of the interstellar gas, but these continue to elude our understanding.

The historical development of OH investigations has been covered in a number of articles (1–6). In this review we shall arrange the observational material in an order that is essentially one of increasing perplexity. There has been great activity at a number of observatories during 1966, and many of the results are available only in preprint form.

OH has been observed in absorption, or emission, or both, in the spectra of about 50 radio sources close to the galactic plane. The great proportion of these are thermal sources (H II regions). The OH profiles display a wide variety of characteristics, and our discussion will mainly be based on the results for a few particular sources which demonstrate this variety. Information as to the distribution of OH in the Galaxy is incomplete and heavily biased by observational selection. Only near the galactic centre is there a clear picture of the distribution and motions of OH. The processes which populate the energy levels are not yet known with any certainty, and it is not possible to give even order-of-magnitude estimates of the OH densities. We have therefore omitted discussion of the processes which might form OH; various possibilities for molecular formation have recently been reviewed by Salpeter (43).

Lambda Doubling in the Hydroxyl Radical

The $^2\Pi_{3/2}$, $J = 3/2$ ground state of OH is split by lambda doubling—an interaction between the rotation of the nuclei and the motion of the unpaired electron in its orbit. The nature of the interaction has been discussed by Barrett (1). Two different states of electronic motion are possible—with the electron distribution along the axis of rotation or in the plane of rotation. Hyperfine interaction with the unpaired spin of the proton further splits the levels.

[1] The survey of literature for this review was concluded in December 1966.

The transition frequencies have been determined by both astronomical and laboratory methods. The most precise values available are Radford's laboratory determinations (7), given in Table I. The errors were taken as 4 times the standard deviation of the means, to allow for possible systematic errors. The rest frequencies given are mutually inconsistent (8) since they should obey the sum rule

$$1612,231 + 1720,533 (=) 1665,401 + 1667,358$$

and fail to do so by 5 kHz. This introduces an uncertainty of nearly 1 km/s in radial velocity, and makes it difficult to match velocity components between any two OH lines when the profiles are complex.

The relative intensities for the four lines in Table I are 1:5:9:1. The lines are electric-dipole transitions, about 10^4 times stronger than the magnetic-dipole transitions producing the hydrogen and deuterium lines. For the 1667-MHz line the absolute value of the Einstein coefficient A_{22}

TABLE I

LABORATORY DETERMINATIONS OF OH REST FREQUENCIES[a]

Transition	Frequency (kHz)	Relative intensity
$F=1\rightarrow2$	$1612,231 \pm 2$	1
$F=1\rightarrow1$	$1665,401 \pm 2$	5
$F=2\rightarrow2$	$1667,358 \pm 2$	9
$F=2\rightarrow1$	$1720,533 \pm 2$	1

[a] See reference (7).

has been computed (9) as 2.86×10^{-11} sec^{-1}, revised (10) to 9.64×10^{-12} sec^{-1}, and most recently revised[2] (11) to 7.7×10^{-11} sec^{-1}. For 1665 MHz the corresponding revised value is 7.1×10^{-11} sec^{-1}.

OH IN ABSORPTION

The two main lines of OH were first observed (12) in absorption in the spectrum of Cassiopeia A. More recent measurements (13) are shown in Figure 1, with the intensity scale for 1665 MHz scaled by 9/5. The shapes of the two profiles agree to within the noise. The radial-velocity scale is based on the rest frequencies of Table I. Absorption features at -0.8 and -48.2 km/s correspond closely to features in the 21-cm absorption spectrum of Cas A. This correspondence provides independent confirmation for the rest frequencies adopted. It also establishes that this OH is located in the H I 'clouds' producing the 21-cm absorption.

[2] Lide (65) has again corrected A_{22} to 7.68×10^{-11} sec^{-1}. His numerical values for the four transitions are inconsistent (Goss, private communication). The most probable values for A_{FF} are:

$$A_{12} = 1.29 \times 10^{-11} \qquad A_{22} = 7.71 \times 10^{-11}$$
$$A_{11} = 7.11 \times 10^{-11} \qquad A_{21} = 0.94 \times 10^{-11}$$

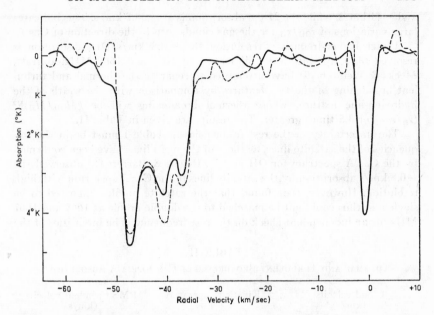

FIG. 1. Absorption profiles of Cassiopeia A observed (13) at 1665.401 (---) and 1667.358 MHz (——), the former being scaled by 9/5. (Green Bank 140-ft telescope, bandwidth 10 kHz.)

The ratio of optical depths τ_{OH}/τ_H is 8.7×10^{-3} and 4.8×10^{-3} at -0.8 and -48.2 km/s, with linewidths of 2.8 and 3.7 km/s (15 and 21 kHz). We know that

$$\int \tau(\nu) d\nu = \frac{hc^2 A}{8\pi k \nu_0} \cdot \frac{g_i}{\Sigma g_i} \frac{N}{T_s} \qquad 1.$$

where T_s is the excitation temperature, ν_0 is the line frequency, g_i is the statistical weight of level i, and N is the number of absorbers per unit cross section. Substitution of the numerical values gives

$$N_{OH}/T_s = 8 \times 10^{12} \text{ per cm}^2 \text{ for } V = -0.8 \text{ km/s}$$

and

$$N_{OH}/T_s = 1.8 \times 10^{13} \text{ per cm}^2 \text{ for } V = -48.2 \text{ km/s}$$

The value of N_{OH} cannot be determined until we know the value of T_s. There is no observational or theoretical material which can give more than a rough limit to T_s. Thus we have no indication yet of the abundance of OH relative to H, nor can we say whether the variation of N_{OH}/T_s for the two absorption features discussed is produced by a variation in N_{OH} or T_s, or both.

The two weaker OH absorptions in Figure 1 appear at velocities of -37

and −41 km/s, while the 21-cm absorption is at −38.2 km/s. This indicates large variations of τ_{OH}/τ_H for the gas clouds seen in the direction of Cas A.

Under higher-frequency resolution the −0.8 km/s OH absorption is resolved (8, 9, 14) into a double feature at 1665 and 1667 MHz. Barrett, Meeks & Weinreb (9) have attempted to separate the thermal and turbulent broadening of the two features by comparison with the width of the hydrogen-line feature, whose thermal broadening will be $(M_{OH}/M_H)^{1/2}$ (ν_H/ν_{OH}) or 3.5 times greater. The results are given in Table II.

The uncertainty in the rest frequencies in Table I must be in the frequencies of the satellite lines, as those of the main lines have been confirmed by the Cas A spectrum for OH and H. Rogers & Barrett (8) observed the −0.8 km/s absorption by the satellite lines in the Cas A spectrum with high resolution. However, they found that the satellite profiles appeared to be single and thus could not be matched to the double profile at 1665 and 1667 MHz for an independent check on the rest frequency. The intensities of the

TABLE II

THERMAL AND TURBULENT BROADENING OF CASSIOPEIA A ABSORPTION[a]

Cloud velocity (km/s)	Kinetic temperature (°K)	RMS turbulent velocity (km/s)
−0.1	120°	0.27
−1.5	90°	0.24

[a] See reference (9).

two satellite lines differed by a factor of 2, despite the equality of the transition probabilities. The 1720-MHz profile also showed some evidence for anomalous emission on its positive velocity side.

Before proceeding further it is important to resolve the uncertainty in the rest frequencies of the satellite lines. Seventeen sources have been observed on all four lines, and in only one case—W12—are the shapes of all profiles closely similar. Measurements of W12 by Goss (14) are reproduced in Figure 2. The profiles for the satellite lines have been matched to those of the main lines to determine the rest frequencies. That of the 1612-MHz line (Table I) is found to be correct, that of the 1720-MHz line to be 6 kHz high. Thus we shall adopt the rest frequencies: 1612,231 ± 2 kHz; 1720,527 ± 3 kHz.

Although the shapes of the W12 satellite profiles are closely similar, the absorption on 1720 MHz is 50 per cent stronger than on 1612 MHz. If the populations of the energy levels were in thermal equilibrium, with the same excitation temperature T_s for each transition, the satellite lines would have identical optical depths. In nearly every case of OH absorption where all four lines have been measured, the populations of the levels depart from thermal equilibrium. This anomaly was first found in absorption in the

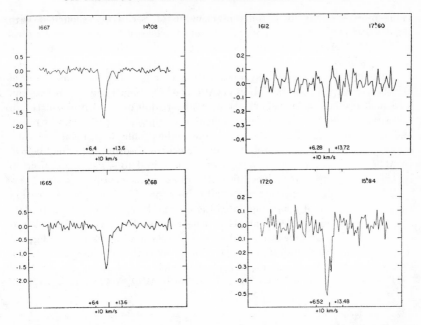

FIG. 2. Absorption profiles of W12 observed (14) at 1667.358, 1665.401, 1612.231, and 1720.527 MHz. The rest frequency of the last transition has been modified to align the absorption with that on the other lines. (Hat Creek 85-ft telescope, bandwidth 2 kHz, integration lines as marked.)

galactic centre region (15, 16), but is found as well in M17 and NGC 6357 (14).

The W12 absorption provides an example of yet another puzzle. Whatever the population distribution, Rogers & Barrett (8) have pointed out that the excitation temperatures T_{ij} corresponding to each transition must obey the equation:

$$(\nu/T)_{1612} + (\nu/T)_{1720} = (\nu/T)_{1665} + (\tau/T)_{1667} \qquad 2a.$$

which can be written in terms of optical depths τ_{ij} (V) at velocity V as

$$\tau(V)_{1612} + \tau(V)_{1720} = \tau(V)_{1665}/5 + \tau(V)_{1667}/9 \qquad 2b.$$

and when $\tau \ll 1$, the observed absorptions $\Delta T(V)_{ij}$ obey

$$\Delta T(V)_{1612} + \Delta T(V)_{1720} = \Delta T(V)_{1665}/5 + \Delta T(V)_{1667}/9 \qquad 2c.$$

Now for W12 the observed opacity at 1667 MHz is 0.44, so we must sum the observed $\tau(V)$ as in Equation 2b. However, for the peak τ, we find

$$\tau_{1612} + \tau_{1720} = 0.21$$
$$\tau_{1665}/5 + \tau_{1667}/9 = 0.13$$

which are far from equal. To satisfy the sum rule we must substitute a value of τ_{1667} equal to 2. This high value of τ can be reconciled with the observed absorption of only 44 per cent if we postulate that the OH does not absorb over the full area of the continuum source, but is concentrated in a blob (or blobs) with $\tau \sim 2$ covering half the source.

The postulate of a patchy distribution of absorbing OH was required (15, 17), to explain the initial observations of the Sagittarius A absorption. The measurements showed 60 per cent absorption of the 'core' source, an intensity ratio for the lines which indicated an optical depth of 3.5 or more, and no sign of saturation of the strong lines—the profile shape being the same for each line. We shall discuss the galactic centre absorption further below. In M17 and NGC 6357 the optical depth required to satisfy Equation 2b is also many times that deduced from the depth of absorption.

The departure from thermal equilibrium found in the OH absorption poses the problem of what populates the energy levels. When the main OH lines were first discovered, Barrett & Rogers (18) discussed collisional and radiative excitation mechanisms, but the population anomalies highlighted by the satellite lines require an excitation process that is F-dependent. We shall return to a discussion of such mechanisms.

It might be objected that the absorption in W12, M17, and NGC 6357

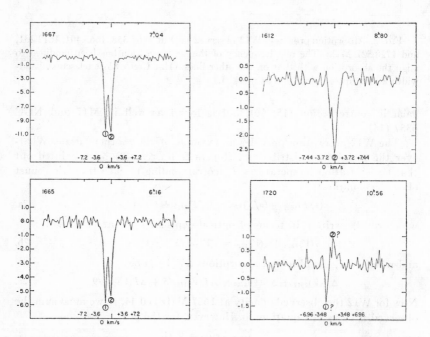

FIG. 3. OH profiles for Cassiopeia A near −0.8 km/s (14). The rest frequency of the 1720-MHz line is the same as in Figure 2. (Hat Creek 85-ft telescope, bandwidth 2 kHz, integration times as marked.)

could be occurring close to the H II region itself and that the population anomaly is only a minor case of the exotic departures from equilibrium found near other H II regions. It is therefore instructive to reconsider the Cas A −0.8 km/s absorption reported by Rogers & Barrett (8). Independent measurements of this absorption by Goss (14) are shown in Figure 3, with the corrected rest frequency used for 1720 MHz. For the lines at 1612, 1665, 1667, and 1720 MHz the intensity ratios of component [1] are 1.16:5.1:9:1.18, the closest case known to the expected 1:5:9:1 for low τ and thermal equilibrium. However, the strength of the satellite lines indicates a τ of 0.30 at 1667 MHz, while the observed absorption is only 4.8 per cent. For component [2] the emission at 1720 MHz is confirmed, and indicates an inversion of the populations for this transition. Also $\Delta T_{1612} + \Delta T_{1720} = 1.70$ while $\Delta T_{1665}/5 + \Delta T_{1667}/9 = 1.23$, and the failure to satisfy Equation 2c again indicates that the actual values of $|\tau|$ are much higher than the apparent values (5.2 per cent opacity at 1667 MHz).

The examples quoted suffice to indicate that the absorbing OH is located in condensations in H I clouds and is perturbed from thermal equilibrium. The size of the small-scale OH structure is completely unknown. There is evidence that the large-scale distribution is similar to the extent of H I regions. In the case of NGC 6334 and NGC 6357, separated by 2°, there are closely similar absorption features in each for both OH (19) and H (20). If we adopt a distance of 600 pc for the absorbing gas complex (19), its diameter would be of the order of 50 pc. Many OH clouds are also seen near the galactic centre (6, 16), projected against the extended group of continuum sources, and the OH absorption again extends over distances of up to 50 pc.

OH in Emission

No thermal emission from OH molecules has been detected. If a uniform distribution of OH with an excitation temperature T_s were present, we would see an emission line with brightness temperature $T_s(1 - e^{-\tau})$ adjacent to a continuum source which is absorbed by $e^{-\tau}$. No such emission has been found, even near sources where τ is observed to be near unity. Searches have shown that the emission does not exceed 1° K adjacent to Cas A (12), 0.5° K near the galactic centre (17), 0.1° K at the anticentre (21), and 0.05° K on the galactic plane at $l = 58°$ (22). These measurements set a limit to T_s of about 5° K if the OH is present and is distributed uniformly. The general association of OH absorption with H I complexes leads us to expect OH to have a widespread distribution. However, if the OH is distributed in very small clouds of high τ (as the absorption results suggest), their emission will suffer considerable dilution in the antenna beam and lead us to underestimate T_s.

Nonthermal emission, with most curious properties, is seen in the direction of a score of continuum sources. We have already noted the anomalous emission at 1720 MHz in the direction of Cas A (Figure 3). There is a continuous progression of observations where there is emission on one to four of the lines. Figure 4 shows the spectra of the supernova remnant W44, where

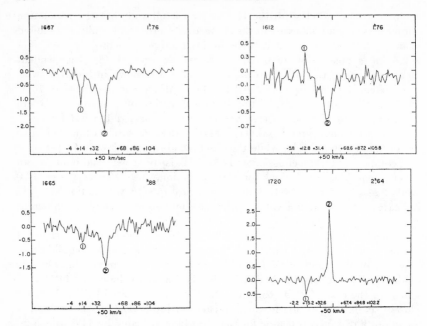

FIG. 4. OH profiles for W44 showing anomalous emission on the satellite lines (14). (Hat Creek 85-ft telescope, bandwidth 10 kHz, integration times as marked.)

the satellites appear alternately in emission at the velocities where absorption is observed for the main lines. Figure 5 shows the spectra of the thermal source W43, where emission appears for all lines on at least one velocity.

The above examples of weak emission for one or more lines seem to be directly related to the cases of spectacular emission: W3, W49 (Figure 6), W51, W75, NGC 6334, Sgr B2, 1608-51, RCW 74, and 1617 50. Details of all sources where OH has been detected are given in Table III. Emission is seen on at least one line in 20 of the 55 sources.

In general the emission profiles show very little similarity for any of the four OH lines. The profiles appear to be composed of a number of narrow components, and their relative intensities differ widely for the four lines. The emission is usually stronger on 1665 MHz than on 1667 MHz, and tends to be stronger on 1720 MHz than on 1612 MHz. But the examples given show that this is not always the case. In W49 (Figure 6) there are two groups of velocities for the main lines near $+5$ and $+16$ km/s, one of which is strongest at 1665 MHz and the other at 1667 MHz.

The emission components have typical widths of 2 kHz. If this were entirely thermal broadening, the kinetic temperature would be less than 50°K. Barrett & Rogers (23) have found that the 1667-MHz line at -43.1 km/s in W3 has a width of approximately 600 Hz; the corresponding kinetic temperature would be 4°K. The lowest temperatures found for H I regions

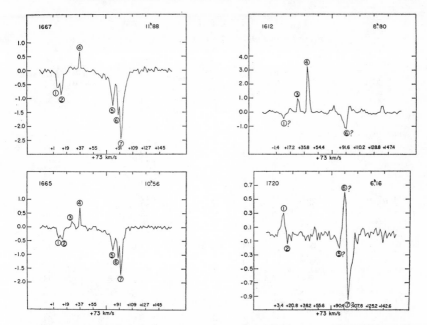

FIG. 5. OH profiles for W43 (14). Emission is seen from several different spiral arms. The 109α recombination line for W43 has a velocity of +97 km/s. (Hat Creek telescope, bandwidth 10 kHz, integration times as marked.)

in absorption are about 60°K. However, it is widely held that the OH linewidth does not give a measure of the kinetic temperature, but is narrowed by maser amplification. It would be valuable to have a determination of the linewidths of the emission features in Figures 4 and 5, where the maser action appears to be weak.[3]

Most of the OH emission has been found to be associated with large H II regions, but this may be due to observational selection: the emission was found (25, 26) during searches for absorption in Westerhout's catalogue of galactic sources, which are predominantly thermal. In many cases in Table III there is a close correspondence between the velocity of one of the OH-emission components and the velocity of the H II region determined from hydrogen-recombination lines (27, 28). But there are some interesting exceptions. In the direction of Cas A, emission at 1720 MHz is seen from somewhere in the local spiral arm, where there is no recorded H II region. Also for W43 (which has a recombination line velocity of +97 km/s) there is emission at velocities of +7, +27, +37, and +90 km/s. The last of these is located in the Scutum arm close to W43, while the first lies in the Sagittarius arm; W43 shows 21-cm absorption at both these velocities (20). The

---

[3] For W43, component 6 at 1720 MHz has a width of 26 kHz (14).

TABLE III

Sources with OH in Emission or Absorption

Source catalog no. — Westerhout	NGC	Other	l^{II}	b^{II}	Lines observed	Absorption and/or emission	Polarization	OH velocity (km/s)	Recombn. line velocity	References
W1	7822		118.3	+ 4.8	1	A		− 14 to − 7		14
W3		IC 1795	133.8	+ 1.2	4	E	c, e	− 41 to −48	−43, −50	23
W7			170.6	−11.7	1	A		5		14
W9		Tau A	184.5	− 5.8	4	A		+ 2 to +13		14
W10		Ori A	209.0	−19.4	4	A, E	c	+ 3 to +21	− 2	30
W12	2024	IC 434	206.4	−16.5	4	A		10	+ 4, +6	14
W14		IC 443	189.1	+ 3.0	1			− 12 to − 3		14
W22	6357		353.2	+ 0.7	4	A	−	− 6 to + 5	− 2 to − 6	14, 24
W28	6514	M20	6.6	− 0.2	4	A, E		+ 7		14
W29		M8	6.0	− 1.2	1	A		+ 11		14
W30			8.6	− 0.2	1	A		+ 17 to + 33		14
W31			10.2	− 0.2	2	A		+ 12 to + 28		14
W33		IC 4701	12.6	− 0.1	4	A, E		+ 35 to + 63		14
W35	6604		19.4	+ 1.9	1	A		+ 28		14
W38	6618	M17	15.1	− 0.7	4	A		+ 20	+17	14, 29
W41			23.2	− 0.3	4	A, E		+ 4 to +77		14, 29
W42			25.5	− 0.1	4	A, E		+ 6 to +81		14
W43			30.8	− 0.1	4	A, E		+ 7 to +94	+97	14, 29
W44			34.6	− 0.4	4	A, E		+ 12 to +42		14
W47			37.7	− 0.1	1	A		+ 17 to +83		14
W49			44.2	− 0.1	4	E	c	+ 3 to +22	+ 7	30
W51			49.2	− 0.4	4	A, E		+ 58 to +62	+59	4, 29
W57		Cyg A	76.2	+ 5.7	2	A		+ 3	—	29
W66		Cyg X	78.3	+ 1.9	1	A		0		14
W67		IC 1318	78.5	+ 0.9	1	A		+ 1		14
W72			81.5	+ 1.3	1	A		+ 5		14
W73			80.6	+ 0.4	1	A		+ 4		14
W75			81.7	+ 0.2	2	E	c, e	− 1 to + 6		30
W80	7000		84.9	− 0.7	2	A		− 1		14
—		Sgr B2	0.7	0	4	E	c	+ 61 to +74	+74	13, 24, 30
W81	—	Cas A	111.7	− 2.1	4	A, E		− 48 to 0	—	8, 14
—		Sgr A	0	− 0.1	4	A		−160 to +100		16
—	6334	—	351.3	+ 0.7	4	A, E	c	− 13 to + 5	− 4	13, 19, 41
		RCW 36	265.1	+ 1.5	1	A, E		+ 11, +27		24
		RCW 38	267.9	− 1.1	3	A		+ 1, + 3	+ 4	24
		RCW 46	282.0	− 1.2	1	A, E		− 7 to + 7		24
		RCW 49	284.3	− 0.3	1	A		− 20		24
		RCW 57	291.6	− 0.5	1	A		− 8	−12	24
		RCW 74a	305.3	+ 0.2	2	A, E		− 44 to −36	−35	24
		η Car	287.5	− 0.6	2	A		− 21	−20	24
		1207−62	298.2	− 0.3	1	A		− 52, + 4		24
		1404−61	311.9	+ 0.1	1	E		− 8		24
		1437−59	316.3	− 0.0	1	A		− 9, − 4		24
		1441−59	316.8	− 0.1	1	A		− 37, −26		24
		1442−59	317.0	+ 0.3	1	A		− 19		24
		1541−53	326.7	+ 0.6	1	A, E		− 31, −15		24
		1548−56	326.2	+ 1.7	1	A		− 94 to −25		24
		1548−54	327.3	− 0.5	1	A		−126 to −42		24
		1608−51b	331.5	− 0.1	1	A, E		−101 to −90	−87	24
		1617−50	332.2	− 0.4	2	A, E		− 59 to −45	−49	24
		1618−49	333.6	− 0.2	1	A, E		−114 to + 8	−112, −50	24
		1630−47	336.8	+ 0.1	1	A		−120, −53		24
		1636−46	338.4	+ 0.1	1	A		− 92, −24		24
		1716−38	348.7	− 1.0	1	A		+ 31		24
W37		M16	17.1	+ 0.8	2	A		+ 20		14

a OH Source is 8′ arc from 1308−62.
b Previously referred to as MHR 49.

OH emission at $+27$ and $+37$ km/s has not been located with certainty, but it probably lies beyond W43, where the line of sight cuts the Sagittarius arm on the far side of the Galaxy. The $+7$, $+27$, and $+37$ km/s emissions are not associated with any well-known H II regions [although Dieter and Goss (private communication) have found a weak recombination line at $+44$ km/s]. There thus appears to be insufficient observational evidence to isolate those objects with which OH emission is associated.

Polarization of the OH emission.—The OH-emission components are highly polarized, most with a high degree of circular polarization. The sources for which polarization has been observed are noted in Table III, where *c* denotes circular and *e* elliptical. The emission from W49 observed (30) with left- and right-hand circular polarization[4] is shown in Figure 6, and is typical of the results found in other sources. Many of the emission components are more than 90 per cent circularly polarized on all four lines; there is in general a predominance (about 30 per cent) of the left- over the right-hand sense (30).

Circular-polarization observations are available for only a limited number of sources. These are W3 (23, 31, 32), W49 (30, 31), W75 (30), NGC 6334 (19, 41), Orion A (30), Sgr B2 (13, 24, 30), RCW 74 (24), 1608–51 (24), and 1617–50 (24)

Elliptical polarization is found in a number of sources. Table IV gives the complete polarization parameters for W3 at 1665 MHz (32); for the central velocities the components overlap and a model has been used to separate them. The component at -45.13 km/s has an ellipticity b/a of 0.65, while all the others have b/a greater than 0.75. The feature at -49.13 km/s is the only one for which the percentage of polarization is below 86 per cent.

The sources which show a significant amount of linear polarization (and so a high degree of ellipticity) are W3 (32, 33), W51 (4), W75 (4), and RCW 74 (24).

One of the few physical mechanisms which will give circular polarization is the Zeeman effect. For a longitudinal magnetic field, the 1665- and 1667-MHz lines split into a normal Zeeman doublet, while the satellite lines split into six components. The polarization of the profiles in Figure 6 cannot readily be interpreted as Zeeman patterns. For the two main lines there is no obvious pairing of the senses, and the left- and right-hand components also have widely different intensities. For the satellite lines we find isolated, highly polarized components rather than a complex Zeeman multiplet. This would suggest that we have emission with one sense of polarization from isolated OH clouds with characteristic velocities.

More extensive polarization observations are required to determine the

---

[4] The optical convention for sense of polarization is used in (30). This is the opposite sense of rotation to the radio convention used in all other papers cited (compare Figure 6 and Table V).

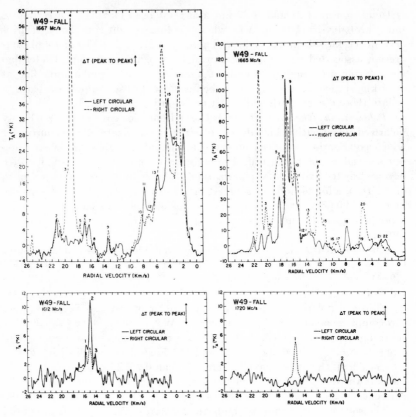

FIG. 6. OH-emission profiles for W49 (30) observed with opposite circular
polarizations. (Green Bank 140-ft telescope, bandwidth 1.25 kHz.)

four Stokes parameters for many sources, including particularly the polarization of the satellite lines. At present this important information is
being collected at a very slow rate.

No polarization of the OH absorption features has been detected.

Sizes of OH-emission sources.—Attempts to resolve the OH-emission
sources with interferometers at 1665 and 1667 MHz have been made at
MIT (Massachusetts Institute of Technology) Lincoln Laboratory with a
maximum spacing of 3800 λ (34, 35), at Owens Valley with 2700 λ (36), at
Jodrell Bank with 2200 λ (37), and at Parkes with 2300 λ (38). At these
spacings the emission sources are completely unresolved, and size limits of
15″ to 20″ arc have been set. At the time of writing, an interferometer of
43,000 λ (Millstone Hill 84-ft telescope and Agassiz 60-ft telescope) has produced fringes on W3 with an amplitude of 1.0, which reduces the size limit
to 1″ arc (39). (See note at end of text.)

The most complete results are those of the MIT Lincoln Laboratory group (35) on W3, W49, Sgr B2, and NGC 6334 reproduced in Table V.

W3 and Sgr B2 are found to be single sources. W3 has been observed at 1665 and 1667 MHz, and the positions at these two frequencies agree to within 15″ arc. Perhaps the most remarkable fact is that components with a velocity spread of 5 km/s agree in position to within 3″ arc for W3 (7″ arc for Sgr B2). If we adopt 1700 pc as the distance of W3, then 3″ arc corresponds to less than 0.025 pc.

The correspondence in position for different velocities suggests that the sources are themselves smaller than 0.025 pc. If the observed radial velocities are components of a space motion of individual OH concentrations, we can imagine that they are expanding outwards from their original source. For a constant expansion velocity of 5 km/s the duration of the expansion is less than 5×10^3 years.

TABLE IV

POLARIZATION OF 1665-MHz OH EMISSION FROM W3[a]

Radial velocity (km/s)	Maximum T_A (°K)	Width (km/s)	Ellipticity ($\pm b/a$)	Ellipse position angle (degrees)	Polarization (per cent)	Comments
−49.13	11.0	0.27	−0.75 (± 0.1)	170 (± 30)	51 (± 3)	Isolated feature
−46.39	28.0	0.68	−0.93 (± 0.07)	10 (± 10)	100 (−6)	Slightly overlapped feature
−45.42	35.0	0.55	−0.84	65	100	Parameters from model
−45.13	50.0	0.45	+0.65	55	100	Parameters from model
−44.55	16.0	0.50	−0.70	105	100	Parameters from model
−43.73	30.0	0.35	+0.78 (± 0.06)	165 (± 10)	86 (± 6)	Slightly overlapped feature
−43.08	8.0	0.26	+0.78 (± 0.2)	90 (± 14)	100 (−20)	Isolated feature
−41.73	14.0	0.36	+0.96 (± 0.04)	110 (± 30)	100 (−12)	Isolated feature

[a] See reference (32).

W49 and NGC 6334 are found to be double sources, each smaller than 15″ arc. In the continuum the thermal source of W49 is resolved into a double source with separation 68″ arc (40, 41), while the two discrete sources of OH emission are separated by 120″ ± 4″ arc (35, 36, 38). Each appears to be displaced by a comparable amount from the associated continuum maximum. W49 is believed to be 15 kpc from the Sun, which would make the maximum size of the emitting regions about 1 pc, and their separation 9 pc. At some velocities (near 16 to 18 km/s) there is emission at 1665 MHz from both positions 1 and 2. Most other velocities come from position 1 only, the velocity spread being 15 km/s at 1665 MHz and 17 km/s at 1667 MHz. For each velocity the positions agree to within 7″ arc. The left-hand polarized 1667-MHz emission at a velocity of +19.0 km/s appears to come from *both* sources, with 37 per cent of the flux from position 2. The positions of the weaker components of the profiles are not yet known. Nor has any interferometry been carried out on the satellite lines.

In NGC 6334 the separation of the two emission centres is 16′ arc, and they had been resolved by single dish measurements (19, 35). The northern source has been studied with the MIT Lincoln Laboratory interferometer

TABLE V

Positions and Diameters of OH-Emission Regions

Radial velocity (km/s)	Polari-zation	Effective source diameter	Separation from line with position listed	Position epoch 1950 α	Position epoch 1950 δ
W3	**1665 MHz**				
−45.1	right	<15″		$02^h23^m16.3 \pm 1^s$	$61°38'57 \pm 5''$
−43.7	right	<20″	<3″		
−41.7	right	<25″	<3″		
−45.4	left	<20″	<3″		
−46.4	left	<20″	<3″		
W3	**1667 MHz**				
−42.3	right	<30″	<7″		
−44.8	left	<30″		<15″ from 1665 position	
W49	**1665 MHz**	*Position 1*			
+17.0	right		<7″		
+ 5.5	left	<25″	<15″		
+12.0	left	<25″	<10″		
+16.8	left		<7″		
+20.9	left	<20″		$19^h7^m49.7 \pm 1^s$	$9°1'12 \pm 5''$
W49	**1667 MHz**	*Position 1*			
+ 2.0	right	<30″	<7″		
+ 5.0	right	<25″		<7″ from 1665 position 1	
+ 3.0	left	<25″	<7″		
+ 5.0	left	<20″	<7″		
+19.0	left		<7″		
W49	**1665 MHz**	*Position 2*			
+16.0	right		<7″		
+15.7	left	<25″			+8.5^s in R.A.
+17.9	left		<7″		−68″ in Decl.
					from Position 1
W49	**1667 MHz**	*Position 2*			
+19.0	left			<10″ from 1665 Position 2	
NGC 6334	**1665 MHz**				
−12.4	right	<25″		$17^h17^m33.5 \pm 2^s$	$−35°45'35 \pm 10''$
− 9.1	left	<25°	<7″		
Sgr B2	**1665 MHz**				
+74.0	right	<25″	<7″		
+67.5	left	<20″		$17^h44^m11 \pm 2^s$	$−28°23'29 \pm 10''$

[a] See reference (35).

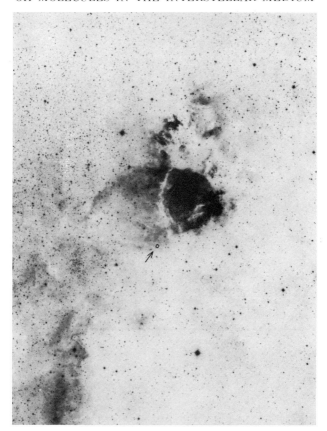

Fig. 7. Location of the OH-emission source near IC 1795 (W3). The circle
indicates the uncertainty in the position. (Palomar Sky Survey print.)

(35), and found to have an angular size of less than 25″ arc (0.1 pc at a distance of 1 kpc). The position of the emission at different velocities agrees to
within 7″ arc (0.03 pc at 1 kpc), and suggests that the source size is smaller
than this. A limit to the size of the southern source of 5′ arc has been set
by single dish measurements (19). A similar range of velocities is covered by
the profiles from each of the two sources. The northern source is strongest
at 1665 MHz, the southern source at 1667 MHz.

   Identification of the OH-emission sources.—The source positions determined by interferometry (Table V) have errors of only one or two seconds in
R.A., 5″ to 10″ in declination. This accuracy is sufficient to show for W3
and NGC 6334 that no familiar object is responsible for the emission.

   An enlargement of a Sky Survey print of IC 1795 (W3) is shown in
Figure 7, with a circle indicating the errors of the OH position. It is about
15′ away from the arc of nebulosity IC 1795, and from the centres of con-

FIG. 8. Location of the two OH-emission sources near NGC 6334. The circles indicate the uncertainties in the positions. (Uppsala Schmidt telescope, Mount Stromlo.)

tinuum emission. However, a weak continuum source has been detected (40) at 6 cm near the OH position.

An Uppsala Schmidt photograph of NGC 6334 is shown in Figure 8. The circles indicate the positions and uncertainty of the OH sources. The northern position lies in a region of faint nebulosity (in line with a lane of dark matter) where there are a number of stellar images. It would be valuable to have colours and spectra for these stars. For the position of the southern source (19, 35) the errors are greater, but the emission comes from a region adjacent to a bright patch of nebulosity.

Both W49 and Sgr B2 lie at great distances from the Sun and cannot be seen optically. In each case the OH emission is displaced from the centres of the continuum emission at centimetre wavelengths (35). Orion A presents the only case where the position of the OH emission (4) coincides with that of the continuum radiation.

On the data presently available we are unable to specify the conditions which lead to OH emission. It is frequently found near H II regions, but not associated with any marked peculiarity in the visible nebulosity. It may also be associated with regions where there is much absorbing dust, but nebulae like IC 1795 or NGC 6334 are crossed by many dark lanes with no trace of OH emission. In the cases of Cas A, W41, W43, and W44 there is emission at velocities very different from those of the source, apparently in a different spiral arm (or arms). Some of it arises close to the Sun, but has not yet been associated with an H II region.

Better radio-position data and further infrared observations are obvi-

TABLE VI

LIMITS TO ANGULAR SIZES AND BRIGHTNESS TEMPERATURES
OF OH-EMISSION SOURCES

Source	Line frequency (MHz)	Radial velocity (km/s)	Flux (10^{-26} Wm^{-2} Hz^{-1})	Assumed size " arc	Brightness temperature (°K)
W3	1665	−45.1	300	<1″	>5×10⁷
	1667	−44.8	25	<1″	>4×10⁶
W49	1665	+20.9	230	<7″	>2.5×10⁶
	1667	+19.0	94	<7″	>1×10⁶
Sgr B2	1665	+67.5	170	<7″	>1.8×10⁶
NGC 6334	1665	−12.4	107	<7″	>1×10⁶

ously needed to enable us to narrow the range of conditions leading to OH emission.

Brightness temperatures for OH emission.—The brightness temperatures T_B for many of the OH-emission components are very high. Only lower limits can be set until the sources have been resolved. For the strongest emission components the present upper limits to source size lead to the T_B limits in Table VI. As W3 proves to be unresolved by the Millstone Hill-Agassiz interferometer, the peak brightness temperature must exceed 10^8°K. The range of brightness temperatures is not known. If we assume that the weakest features detected have the same angular size as the sources in Table VI, they will have brightness temperatures of more than 10^4°K.

Time variations of the OH emission.—Emission profiles of NGC 6334 and Orion A recorded at the Hat Creek Observatory have shown changes in the intensity of some components with time (4). Major changes in NGC 6334 were noted in profiles taken three months apart, and closely spaced observations suggested that the variations have a period as short as ten days.

No long-term time variations have been seen at other observatories. But the observations have been made sporadically and do not rule out the variations reported. Palmer & Zuckerman (30) have seen no significant change in the Orion profile recorded in October 1965 and May 1966. Gardner et al. (19) have seen no change in the emission from the northern source in NGC 6334 recorded in August 1965, February 1966, and March 1966 nor did Meeks & Ball (41) detect any variation in the NGC 6334 profile recorded at the position of the continuum maximum (and so mainly from the northern source) during the first six months of 1966. Neither of these groups has made any extended observations of the southern source of NGC 6334. Since both sources were included in the wider Hat Creek beam, the time variations would appear to be restricted to the southern source. Extended observations of this source are urgently required (preferably by two observatories simultaneously), but none have been made during 1966.

Time variations within ten days would imply a very small source—less than 0.01 pc on a light travel time argument. The longest interferometer baseline used (39) shows that the (nonvariable) source in W3 is smaller than this.

Maser Mechanism for OH Emission

The high brightness temperatures and the high degree of polarization preclude any thermal explanation of the OH emission. Also the high values of T_B contrast with the kinetic temperatures of tens of degrees that we would deduce if the linewidth were determined by thermal broadening.

The most plausible explanation proposed is that the populations of the levels have been inverted, and that maser amplification then occurs (16, 33). However, we cannot specify what is being amplified. The emission has been seen so far only in the direction of continuum sources, but in W3 and NGC 6334 the emission comes from a position where the continuum is weak (40, 41). Even if there were no nearby source of continuum to serve as an input to the maser, the background radiation from the Galaxy (typically 5° to 20°K in the plane) or from the cosmic fireball (3.5°K) would be an effective stimulus. If the absolute value of the excitation temperature, $|T_s|$, is greater than the background temperature of about 10°K, spontaneous transitions will be amplified. A value $|T_s|$ greater than 10°K corresponds to an inversion of less than 1 per cent. Our interpretation of the size limits of the emission sources depends on the input assumed for the maser. If the continuum background is dominant, the source size is then that of the OH condensation. If the maser is emitting spontaneously (inversion less than 0.01), the size is simply that of the triggering fluctuation. The observations do not exclude the further possibility that there is a small continuum source being amplified by a more extensive OH cloud.

The rate of stimulated transitions corresponding to the high values of T_B exceeds the rate of inversion by the pumping processes so far proposed. The maser may thus be saturated, the output being limited by the pumping rate. Some evidence for saturation is provided by the relative intensities of the OH lines. If the level increases exponentially, with $\exp|\tau_{1665}|$ greater than 10^6, then $|\tau_{1665}|$ is about 14. If the inversion were the same for the satellite transitions, $|\tau_{1612}|$ would be 2.75 and T_{1665}/T_{1612} would be 6.4×10^4. However, the observed ratio T_{1665}/T_{1612} is typically 50/1. Thus we must conclude either that the maser is strongly saturated or that the inversion of the satellite lines is nearly 4 times stronger than that of the 1665-MHz line. The latter possibility cannot be excluded, as cases such as W28, W38, W41, W43, W44, and Cas A demonstrate that under some conditions one or both of the satellites can be inverted while the main lines remain in absorption. Also saturation will not directly explain the many cases where T_{1665} exceeds T_{1667}.

The maser hypothesis provides an explanation for the narrowness of the emission lines. At an exponential gain of 10^6 the lines would be narrowed by a factor of $\tau^{1/2}$, about 4 at 1665 MHz. The width of the unamplified line would then typically be about 8 kHz; if the line were broadened by thermal motions only, the corresponding temperature would be near 700°K. Saturation of the maser gain would reduce the amount of line narrowing which occurs.

Although we cannot yet specify a model for the OH maser, it is of interest to compute order-of-magnitude limits to the OH masses necessary to give the observed gain. We shall first consider the case of exponential gain. The numerical values assumed will be the limits for W3:

T_B = peak brightness temperature = 5×10^7K

T_C = background continuum temperature = 10°K

$\Delta\nu$ = linewidth before amplification = 8 kHz

Now the equation of transfer, neglecting spontaneous emission, gives

$$\ln(T_B/T_C) = c^2 \cdot A_{21} \cdot \int \Delta n dl/8\pi\nu^2\Delta\nu$$

$$= 9.1 \times 10^{-10} \cdot (\Delta\nu)^{-1} \int \Delta n dl \quad \text{at 1665 MHz}$$

where Δn is the population difference (per cm³) produced by pumping and dl is an increment of path length through the OH cloud (cm). Inserting $\ln(T_B/T_C) = 15.4$ and $\Delta\nu = 8$ kHz, we find:

$$\int \Delta n \, dl = 1.3 \times 10^{14} \text{ per cm}^2$$

The value of T_B assumed corresponds (through the measured flux) to a linear size of 0.008 pc = 2.6×10^{16} cm at the distance of W3. If the value of $\int\Delta n \, dl$ is constant over the surface of an amplifying cloud of volume V, it will then have a total population difference

$$\iiint \Delta n dV \doteqdot 10^{47}$$

If we arbitrarily set the inversion at 10 per cent ($T_s \doteqdot -1$°K), the cloud will contain 10^{48} OH molecules. We can estimate the mass of such a cloud by assuming that the abundance O:H is normal (about 1:1500), and that a fraction α of oxygen atoms have formed OH molecules. The cloud then contains $1.5 \times 10^{51}/\alpha$ hydrogen atoms (80/α per cm³) and has a mass of $10^{-6}/\alpha M_{\odot}$. No estimate of the value of α is available (43), but we can note that the cloud will contain $1 M_{\odot}$ if $\alpha = 10^{-6}$.

For the case of a saturated maser the density and mass are considerably increased. If the induced-transition rate W_P of the pumping process is much less than the stimulated-transition rate, it can be shown (51) that the flux density received is

$$S = h\nu \cdot W_P \cdot \iiint \Delta n dV / R^2 \cdot \Omega \cdot \Delta \nu$$

where Δn is the population difference prior to saturation; R is the distance to the emitting volume; and Ω is the solid angle of the emission, and depends on the geometry of the pumped OH cloud. For the most efficient of the pumping processes discussed in the next section, W_P is of the order of 10^{-8} per sec. Then for W3 we have (taking $\Delta \nu = 2$ kHz):

$$\iiint \Delta n \cdot dV = 3 \times 10^{51} \Omega$$

If we again assume $\Delta n = 0.1 n_{\mathrm{OH}}$; $\iiint n_{\mathrm{OH}} dV = 3 \times 10^{52}$ Ω; $\iint n_{\mathrm{H}} dV \div 5 \times 10^{55}$ $\cdot \Omega/\alpha$; and the cloud has a mass of $4 \times 10^{-2} \Omega/\alpha$ $M_{\odot}$. For a spherical cloud the mass will attain its maximum value of $1/2\alpha$ $M_{\odot}$. Since $\alpha \ll 1$, these clouds would be quite massive objects, with densities of $5 \times 10^4/\alpha$ hydrogen atoms per cm³.

Pumping processes for OH.—A number of processes have been proposed to account for the inversion of the populations of the Λ-doublet levels. Each process has been able to account for some of the characteristics of the observed emission. Other characteristics have presented difficulties—particularly the high degree of circular polarization. All the theoretical work has been based on the early observation that the emission came from near the edge of H II regions. Accordingly the models are based on the conditions close to the ionization shock front at the boundary of the H II region, where energetic particles and protons are present but the kinetic temperature is low. The precise location of two of the emission sources by interferometry has, however, indicated no clear connection with the ionization fronts. The situation is further confused by the controversy between the observers about possible time variations in the emission, and by our having only upper limits to the source sizes and lower limits to the brightness temperatures.

The initial association of the emission with ionized regions led Symonds (44, 45) to propose that the OH molecules were formed in excited states by the exothermic association of protons and negative oxygen ions, both of which should be abundant in the ionization front. No quantitative estimates of the degree of population inversion have been made.

Shklovsky (46) has suggested that the compact and rather dense sources producing the emission are "protostars" which condense where the expanding ionization front compresses heterogeneities in the surrounding H I region. The molecules are then excited to higher vibrational states by the far-infrared radiation from the core of the protostar, and inversion takes place through the subsequent cascade transitions. No quantitative analysis of the cascade has been made, but it appears (42) that only the satellite lines would be inverted.

Johnston (47) has shown that collisions with electrons or ions streaming radially outwards in the ionization front can affect the OH energy levels asymmetrically, and that this can lead to a population inversion. However, at an OH density N_{OH} of 10^{-2} per cm^3, a stream of electrons would become completely uncollimated by collisions with OH molecules alone before it had penetrated more than 10^{15} cm. Greater penetration would occur if the OH molecules were continually formed throughout the front. Calculations indicate that only the 1665- and 1667-MHz transitions would be inverted, the former more strongly. The maximum brightness temperature predicted is $2 \times 10^{4} °K$ over an amplification path of 3 pc, the radiation being plane polarized.

Several authors have discussed pumping of OH by ultraviolet radiation from the H II region or the exciting star, followed by a cascade which over-populates the upper levels of the ground-state Λ doublet. Cook (48) noted that there is probably a close coincidence between the wavelengths for Lyman α and the transition $X^2\Pi_i$, $v=2$, $J=5/2$ to $C^2\Sigma^+$, $v=$o, $J=5/2$, $7/2$ in OH. The symmetry selection rules permit a decay from the $^2\Sigma^+$ state to the upper, but not to the lower, of the ground-state Λ-doublet levels. In H I regions the population of the $v=2$ state of $^2\Pi_i$ would be negligible. If the OH were formed in the ionization front, a significant proportion of the molecules might be in the $v=2$ state. However, OH is dissociated by photons with energy greater than Ly-α at about the same rate as the absorption of Ly-α into the $^2\Sigma^+$ state.

Another close resonance (49) is that of N II 3063 Å with the 3062.7 Å OH transition between $^2\Pi_{3/2}$ and $^2\Sigma^+$. No details of the pumping process have been published.

Pumping of OH by the continuum near 3080 Å, from the $^2\Pi_{3/2}$ ground state to $^2\Sigma_{1/2}^+$, is the most efficient process yet suggested. For anisotropic 3080 Å radiation incident on an OH cloud Perkins et al. (50) found that the $P_1(1)$ transition (Figure 9) was followed by a cascade that inverted the ground-state Λ doublet. The 18-cm background would be then amplified, with linear polarization, for propagation normal to the direction of the ultraviolet. Litvak et al. (51) have examined the 3080 Å pumping in detail and find that, of the six transitions from the ground state (Figure 9), $P_1(1)$, $R_1(1)$, and $R_{21}(1)$ lead to inversion of the Λ doublet while $Q_1(1)$, $Q_{21}(1)$, and $S_{21}(1)$ are anti-inverting. The UV pumping initially leads to an anti-inversion of the Λ doublet (region 1 of Figure 10). But as the UV penetrates into the OH cloud, the wavelengths corresponding to $Q_1(1)$, $P_1(1)$, and $Q_{21}(1)$ are preferentially absorbed. The $R_1(1)$ and $R_{21}(1)$ transitions then become domi-

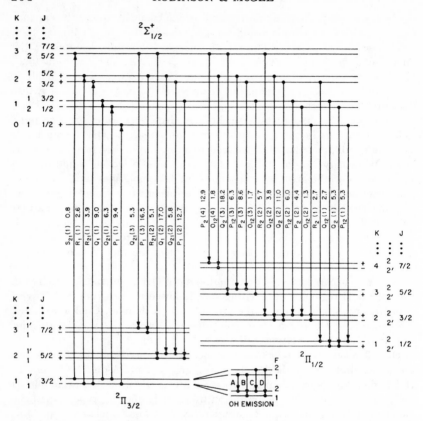

FIG. 9. Energy level diagram (51) of OH molecule showing rotational states of the lowest vibrational level for the $^2\Pi_{3/2}$, $^2\Pi_{1/2}$, and $^2\Sigma_{1/2}^+$ electronic states (not to scale). Hyperfine splitting is indicated only for the ground-state Λ doublet. The relative strengths of the UV transitions are marked.

nant (in region 2) and invert strongly until they too are absorbed. The remaining UV then produces a further region of enhanced absorption (region 3).

Litvak et al. consider an OH cloud at two light years from an O5 star, the UV flux being 10^{-17} watts m^{-2} Hz^{-1}. The maximum induced transition rate W_{uv} between the Λ-doublet states (right-hand scale of Figure 10) is then about 3×10^{-8} per sec. For the sources of high brightness temperature the stimulated transition rate is greater than 10^{-5} per sec, so that the maser would be strongly saturated. Electron collisions will induce transitions at a rate W_e of about $7 \times 10^{-5} n_e \cdot T_e^{-1/2}$, which for n_e equal to 1 per cm^3 and an electron temperature T_e of 100°K will give $W_e \doteq 10^{-5}$ per sec. Thus electron collisions will seriously modify the inversion unless $n_e \ll 1$ per cm^3.

The UV-pumping model with differential absorption has been extended to consider the four 18-cm transitions separately (52). The degree of inversion

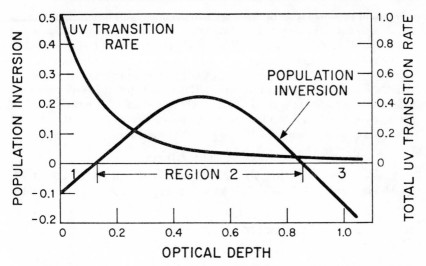

FIG. 10. Population inversion of OH Λ doublet (51) by isotropic UV pumping, as a function of the UV penetration. The effects of selective absorption are clearly seen, producing inversion in region 2. Unit optical depth for the UV is about $1.5 \times 10^{15}/n_{\mathrm{OH}}$ cm, n_{OH} being the molecular density per cm³.

decreases in the order 1667, 1665, and 1720 MHz; 1612 MHz is anti-inverted. Thus the model does not account for the relative intensities observed. However, the results will be significantly modified by the inclusion of: (a) differential absorption as the amplified waves leave the inverted region; (b) different beam divergence for the four transitions, associated with differences in the saturation path length; (c) the effect of microwave saturation on the population distribution; (d) the effect of electron collisions on the inversion; (e) various orientations of the pumping direction relative to the magnetic field; (f) the effect of re-radiated UV and IR from the cascade transitions.

The predominantly circular polarization of the emission and the preference for one sense of polarization have proved difficult to explain. With directional pumping by unpolarized UV the $|M_F|$ populations are unequal, but there is no skewing to positive or negative M_F that would result in large gain differences for the left- and right-hand polarizations. Most of the models predict strong linear polarization. Because the maser would be strongly saturated (for any of the pumping mechanisms proposed), there could well be nonlinear suppression of one polarization sense over the other (53–55). The dominance of one sense could be initiated (51) by an asymmetry of the Doppler features combined with the gyrotropic effect of the plasma electrons.

The UV pumping accounts qualitatively for the anomalous absorption which is generally observed, as well as for the emission. However, there is a marked difference in scale between the two phenomena. The emission sources are observed to have sizes of less than 0.01 pc, and the depth of UV penetration into a dense OH cloud is smaller still. On the other hand, the absorption extends over dimensions comparable with those of H I complexes, typically tens of pc. There is evidence to suggest that the absorbing OH occurs as small, dense concentrations throughout H I clouds, but it is difficult to see what pumping process (except Shklovsky's IR pumping of protostars) can act on condensations over distances up to 10 pc. There are also the cases of anomalous emission in nearby H I clouds with no associated H II region.

OTHER OH TRANSITIONS

Lambda doubling of $^2\Pi_{1/2}$, $J = 1/2$.—If the inversion of the $^2\Pi_{3/2}$, $J = 3/2$ Λ doublet takes place by a selective cascade, the population of the metastable

TABLE VII
LIMITS TO LINE STRENGTH OF $^2\Pi_{1/2}$, $J = 1/2$ LAMBDA DOUBLET[a]

Source	Transition	Computed frequency (MHz)	Search range (MHz)	$\Delta T/T_{1665}$
W3	$F = 1 \rightarrow 1$	4717.2	-7.1 to $+6.1$	<0.013
	$F = 1 \rightarrow 0$	4702.3	-8.1 to $+4.5$	<0.020
	$F = 0 \rightarrow 1$	4807.5	-8.9 to $+9.7$	<0.020
W49	$F = 1 \rightarrow 1$	4717.2	-6.6 to $+6.6$	<0.010
	$F = 1 \rightarrow 0$	4702.3	-6.1 to $+4.7$	<0.016
	$F = 0 \rightarrow 1$	4807.5	-9.9 to $+3.3$	<0.016

[a] See reference (56).

$^2\Pi_{1/2}$, $J = 1/2$ state (Figure 9) might be sufficient to make its Λ-doubling transitions detectable. A search for these lines in W3 and W49 has been made by Zuckerman et al. (56) with the Green Bank 140-ft telescope. Table VII gives the frequency range searched, and the upper limits to the $^2\Pi_{1/2}$ line strengths relative to the peak 1665-MHz emission (both averaged over a velocity range of 1.8 km/s). The transition frequencies are not known precisely, but were computed from Radford's (57) hyperfine coupling constants; the values differ slightly from those computed by Barrett & Rogers (58).

Lambda doubling of $^{18}O^1H$.—The frequencies of the ground-state Λ-doublet transitions of the isotopic species $^{18}O^1H$ were computed by Barrett & Rogers (58) from the measured frequencies of $^{16}O^1H$ (Table I). The $F = 2 \rightarrow 2$ transition has been detected in absorption in the spectrum of Sagittarius A by the Sydney (6) and MIT (59) observers. From a comparison with the 1667-MHz absorption the rest frequency was determined to be 1639·460 MHz. The MIT results indicate an isotopic abundance ration of 1/450 for the gas absorbing at $+40$ km/s and 1/750 for that at -135 km/s, in good agreement with the terrestrial abundance of 1/490. These observations yield the first measure of an isotopic abundance for the interstellar medium.

The $F = 1 \rightarrow 1$ transition of $^{18}O^{1}H$ at 1637.4 MHz has not yet been detected for Sgr A (6). It would be of interest to search for this line in emission in sources such as W3 and W49.

OH Clouds Near the Galactic Centre

So far we have concentrated mainly on the physics of the interstellar OH molecules, discussing the profiles of individual sources. The OH lying in directions near the galactic centre is relatively extended, and has been surveyed in detail at 1667 MHz. The survey has brought to light some results of considerable astronomical significance. This work, begun in 1964 (60), was interrupted by a shift of emphasis onto observations of the OH-emission line, but has recently been completed.

The survey at 1667 MHz was made with the 12′.2 beam of the Parkes radio telescope from $l^{II} = 358°$ to $l^{II} = 3°$ and $b^{II} = +30′$ to $b^{II} = -30′$. Profiles have been observed over a 5′ grid in the inner regions and a 10′ grid in the outer. About 320 points have been included. In general, all the fine structure of the absorption features is also found in the other three lines (1612, 1665, 1720 MHz) observed at about 20 sample points spread throughout the region.

Figure 11 gives contours of intensity of the 1667-MHz absorption on coordinates of galactic longitude and radial velocity (LSR) for a latitude of $-0°05′$. It is one of ten such plots which summarize all the line profiles ob-

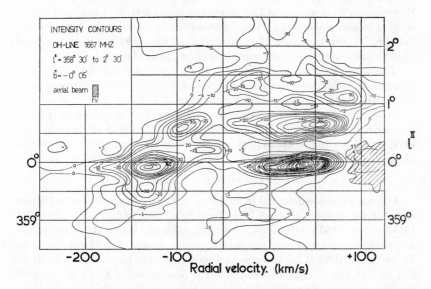

Fig. 11. Contours of 1667-MHz absorption near the Galactic centre as a function of galactic longitude and radial velocity, for galactic latitude $-0°05′$. Contour interval is 0.34° K in T_a. (Parkes 210-ft telescope, bandwidth 37 kHz.)

served. The absorption in the nearer spiral arms (Sagittarius and 3 kpc arms) is weak, and can barely be recognized in the figure.

One of the controversial issues when the preliminary results were announced (60) was unexpected constancy of the radial velocity for the partially resolved minima, which extended over ranges of l^{II} of one or two degrees. This contrasted with the neutral hydrogen results (62) which indicated rapid changes of velocity with l^{II} and suggested the rotating disk model of the galactic nucleus. Robinson & McGee (61) have shown that, even if the OH data be degraded in velocity resolution so that the discrete nature of the velocity distribution disappears, only a few features of the H I model can be recognized. The neutral hydrogen observations are handicapped by the presence of the very large contributions of nearby gas in the range ± 40 km/s, and the galactic centre profiles have to be interpolated across this range.

As can be seen in Figure 11, the observed OH velocities extend from -200 to $+160$ km/s. In the hydrogen case positive velocities extend up to $+250$ km/s. Thus the gas with velocities greater than $+160$ km/s must lie on the far side of the nuclear sources and not contribute to the absorption.

OH clouds may be delineated by converting the observed absorption to apparent opacity (defined as $T_{line}/T_{continuum}$). About 35 clouds of apparent opacity 0.2 to 0.7 are revealed with radial velocities ranging from -165 to $+160$ km/s. (Opacities less than 0.2 were excluded in determining clouds.) A typical cloud size is 50 pc for distances near R_0 (10 kpc) and in many cases these appear to combine into complexes of about 200 pc in extent. It is possible that many more clouds are unresolved in view of the large-velocity half-widths seen in Figure 11 compared with the average widths for clouds observed in absorption in other parts of the galactic plane (14, 24).

The ratio τ_{OH}/τ_H varies throughout the central region by an order of magnitude. Differences in concentrations of OH and H are illustrated in the comparison of H and OH profiles for Sagittarius A (3, 61).

It has not been possible to construct a dynamical model of the galactic nuclear region, since no radial distances of the OH clouds are known. However, radial motion is at least as important as any component of rotation about the galactic centre.

OH in Other Galaxies

OH has not been detected in any other galactic system. Roberts (63) has searched for emission at 1665 MHz from NGC 224 (M31), 2403, 2903, 3034 (M82), 5194 (M51), 5195, 5457, and 6822, using the Green Bank 140-ft telescope. When averaged over the velocity range observed at 21 cm, the upper limits to the 1665-MHz antenna temperatures were less than 0.1°K. If the OH and H had similar distributions the abundance of OH relative to H would then be in the range 10^{-5} to 10^{-6}. However, the distribution of OH is likely to be very patchy (as in the Galaxy), and the emission integrated over galactic dimensions and velocity ranges exceeding 100 km/s would be most difficult to detect.

Radhakrishnan (64) has made a preliminary search for anomalous emission at 1665 and 1667 MHz from the 20 brightest H II regions in the Magellanic Clouds, using the Parkes 210-ft telescope. Any source comparable in strength with the powerful galactic emitters W49 and Sgr B2 would have been detected at the distance of the Clouds. No signal above the noise limit of 1°K (in a 37-kHz band) was found for any of the 20 regions.

Concluding Remarks

The OH which we see in absorption appears to be related generally to H I regions. The velocity dispersion in the profiles has values that would be expected from our knowledge of the temperature and turbulent motion in hydrogen clouds; the radial velocities of the absorbing OH agree closely with those of the 21-cm absorption profiles. However, the relative intensities of the OH lines indicate marked departures from thermal equilibrium, and suggest that the radical occurs in small condensations whose optical depth is an order of magnitude greater than the measured opacity. The size of such condensations should be measurable by interferometry, and we may then be able to set better limits to the excitation temperatures. Measurements of the absorption of all lines on adjacent continuum sources are required to indicate the distribution and excitation of the OH through H I clouds or complexes.

The emission profiles are observed to have peak brightness temperatures of more than $10^{7}°$K, yet to have linewidths of 2 kHz or less. The components of the profiles are highly polarized, predominantly circular. Interferometry has indicated that the emitting sources have sizes of less than 0.01 pc. Most of the sources are associated with H II regions—about a third of the catalogued regions have shown the emission on one or more of the OH lines. However, a careful survey of parts of the galactic plane is required to rule out an association with other objects. There are several cases where emission occurs from a source in a spiral arm different from that in which the catalogued H II region is located. Emission has also been seen on the satellite lines in the direction of two supernova remnants. Precise positions for the emitting sources have been determined near two H II regions. In neither case is there any unusual object at the position, nor any clear association with an ionization front or obscuring matter. For two objects the OH emission comes from a pair of sources having some correlation of velocities and anticorrelation of polarization; further measurements are required to determine whether the sources are independent or not.

Maser action with a very high gain is the most plausible explanation for the anomalous emission. Ultraviolet pumping is the most efficient process proposed to invert the populations. However, the pumping efficiency seems to be rather low to account for the high brightness temperatures. This would suggest that the maser is highly saturated, and would involve quite massive condensations. A high degree of saturation might also account for the emission of isolated, circularly polarized Doppler components.

It is clear that the investigations of OH are still in an exploratory phase. Those engaged in the field expect to find still more that is surprising. Identification of the objects producing the anomalous emission may be the next significant step. Or theoretical developments may succeed in unifying the discordant results and suggest further areas of observation.

Note added in proof: Observations made during January 1967 complement or modify some of the remarks on the sizes of OH-emission sources, the data in Table VI, and the calculations of OH masses.

Several OH-emission sources have been observed with an interferometer between Jodrell Bank and Malvern (66) with a separation of 127 km (equivalent to 7×10^5 wavelengths at 18 cm). The fringe visibilities, size limits, and brightness temperatures were found to be:

Source	Line frequency (MHz)	Radial velocity (km/s)	Fringe visibility	Angular size	Linear size (pc)	Brightness temperature (°K)
W3	1665	−45	>0.85	<0.05″	<5×10⁻⁴	>1.0×10¹¹
W49	1665	+21	>0.6	<0.1″	<7×10⁻³	>1.6×10¹⁰
	1667	+19	>0.6	<0.1″	<7×10⁻³	>6.5×10⁹
W75	1665	+0.5	>0.6	>0.1″	<7×10⁻⁴	>6.0×10⁹
Sgr B2	1665	+67	>0.6ᵃ	<0.2″	<1×10⁻²	>3.5×10⁹

ᵃ Shorter projected baseline.

The separation of the −45 and −46 km/s components of W3 (R.H. circular polarization) was measured to be 1.4″ arc (equivalent to 0.014 pc), about 30 times the upper limit on the apparent size of the components. In W49 the separation of the strongest spectral features in the source at R.A. $19^h07^m50.0^s$, $\delta = 09°01'12''$, was found to be 0.1″ arc (equivalent to 0.007 pc).

The increased limits to the brightness temperatures and reduced sizes modify the estimated masses and densities of the OH clouds.

For the case of exponential gain $\ln(T_B/T_C)$ for W3 increases to 23, and we find

$$\int \Delta n dl = 2 \times 10^{14} \text{ per cm}^2$$

$$\iiint \Delta n dV \doteqdot 5 \times 10^{44}$$

The mass estimate will then be reduced to $5 \times 10^{-9}/\alpha \ M_\odot$ but the hydrogen density will increase to $2 \times 10^3/\alpha$ per cm³.

For a saturated maser the mass has been estimated from the pumping rate W_p. The value of W_p of 10^{-8} per sec used for the UV-pumping model appears to be quite inadequate to account for a brightness temperature of 10^{11} °K, for which the stimulated transition rate would be about 10^{-1} per sec.

LITERATURE CITED

1. Barrett, A. H., *IEEE Trans. Antennas Propagation*, 12, 822–31 (1964)
2. McNally, D., *Sci Progr.*, 53, 83–87 (1965)
3. Robinson, B. J., *Sci. Am.*, 213, 26–33 (July 1965)
4. Dieter, N. H., Weaver, H. F., Williams, D. R. W., *Sky Telescope*, 31, 132–36 (1966)
5. Bolton, J. G., *Discovery*, 27, 24–29 (May 1966)
6. Robinson, B. J., *Radio Astronomy and the Galactic System* (van Woerden, H., Ed., Academic Press, London, 1967, in press)
7. Radford, H. E., *Phys. Rev. Letters*, 13, 534–35 (1964)
8. Rogers, A. E. E., Barrett, A. H., *Radio Astronomy and the Galactic System* (Academic Press, London, 1967, in press)
9. Barrett, A. H., Meeks, M. L., Weinreb, S., *Nature*, 202, 475–76 (1964)
10. Goss, W. M., Weaver, H., *Astron. J.*, 71, 162–63 (1966)
11. Turner, B. E., *Nature*, 212, 184–85 (1966)
12. Weinreb, S., Barrett, A. H. Meeks, M. L., Henry, J. C., *Nature*, 200, 829–31 (1963)
13. Barrett, A. H., Rogers, A. E. E. (Private communication)
14. Goss, W. M., *OH Absorption in the Galaxy* (Ph.D. thesis, Univ. of California, Berkeley, 1967)
15. Gardner, F. F., Robinson, B. J., Bolton, J. G., van Damme, K. J., *Phys. Rev. Letters*, 13, 3–5 (1964)
16. McGee, R. X., Robinson, B. J., Gardner, F. F., Bolton, J. G., *Nature*, 208, 1193–95 (1965)
17. Robinson, B. J., Gardner, F. F., van Damme, K. J., Bolton, J. G., *Nature*, 202, 989–91 (1964)
18. Barrett, A. H., Rogers, A. E. E., *MIT Res. Lab. Electronics, Quart. Progr. Rept. No. 74*, 16–18 (1964)
19. Gardner, F. F., McGee, R. X., Robinson, B. J., *Australian J. Phys.*, 20 (June 1967, in press)
20. Clark, B. G., Radhakrishnan, V., Wilson, R. W., *Ap. J.*, 135, 151–74 (1962)
21. Weaver, H. F., Williams, D. R. W., *Nature*, 201, 380 (1964)
22. Penzias, A. A., *Astron. J.*, 69, 46 (1964)
23. Barrett, A. H., Rogers, A. E. E., *Nature*, 210, 188–90 (1966)
24. McGee, R. X., Gardner, F. F., Robinson, B. J., *Australian J. Phys.* (In press)
25. Gundermann, E. J., *Observations of the Interstellar Hydroxyl Radical* (Ph.D. thesis, Harvard, June 1965)
26. Weaver, H. F., Williams, D. R. W., Dieter, N. H., Lum, W. T., *Nature*, 208, 29–31 (1965)
27. Mezger, P. G., Höglund, B., *Ap. J.*, 147, 490–518 (1967)
28. Gardner, F. F., McGee, R. X., *Proc. Astron. Soc. Australia*, 1, 19 (1967)
29. Menon, K., Cunningham, A. A. (Private communication)
30. Palmer, P., Zuckerman, B., *Harvard Radio Astron. Preprint No. 124* (1966)
31. Davies, R. D., de Jager, G., Verschuur, G. L., *Nature*, 209, 974–77 (1966)
32. Meeks, M. L., Ball, J. A., Carter, J. C., Ingalls, R. P., *Science*, 153, 978–81 (1966)
33. Weinreb, S., Meeks, M. L., Carter, J. C., Barrett, A. H., Rogers, A. E. E., *Nature*, 208, 440–41 (1965)
34. Rogers, A. E. E., Moran, J. M., Crowther, P. P., Burke, B. F., Meeks, M. L., Ball, J. A., Hyde, G. M., *Phys. Rev. Letters*, 17, 450–52 (1966)
35. Rogers, A. E. E., Moran, J. M., Crowther, P. P., Burke, B. F., Meeks, M. L., Ball, J. A., Hyde, G. M., *Ap. J.*, 147, 369–77 (1967)
36. Cudaback, D. D., Read, R. B., Rougoor, G. W., *Phys. Rev. Letters*, 17, 452–55 (1966)
37. de Jager, G. (Private communication)
38. Radhakrishnan, V., Whiteoak, J. B., *Proc. Astron. Soc. Australia*, 1, 20–21 (1967)
39. Burke, B. F. (Private communciation)
40. Mezger, P. G. (Private communication)
41. Meeks, M. L., Ball, J. A. (Private communication)
42. Turner, B. E. (Private communication)
43. Salpeter, E. E., *Radio Astronomy and the Galactic System* (Academic Press, London, 1967, in press)
44. Symonds, J. L., *Nature*, 208, 1195–96 (1965)
45. Symonds, J. L., *Symposium on Radio*

and *Optical Studies of the Galaxy*,
89 (Mount Stromlo Obs., May
1966)

46. Shklovsky, I. S., *Astron. Tsirk. No.
372* (June 1966)

47. Johnston, I. D., *Cornell Univ. Preprint
CSUAC 46* (July 1966)

48. Cook, A. H., *Nature*, **210**, 611–12
(1966)

49. Shklovsky, I. S. (Private communication)

50. Perkins, F., Gold, T., Salpeter, E. E.,
Ap. J., **145**, 361–66 (1966)

51. Litvak, M. M. McWhorter, A. L.,
Meeks, M. L., Zeiger, H. J., *Phys.
Rev. Letters*, **17**, 821–26 (1966)

52. Litvak, M. M., McWhorter, A. L.,
Meeks, M. L., Zeiger, H. J., *Astron.
J.* (In press)

53. Heer, C. V., Graft, R. D., *Phys. Rev.*,
140, A1088 (1965)

54. Culshaw, W., Kannelaud, J., *Phys.
Rev.*, **145**, 257 (1966)

55. Heer, C. V., *Phys. Rev. Letters*, **17**, 774
(1966)

56. Zuckerman, B., Palmer, P., Penfield,
H., *Nature*, **213**, 1217–18 (1967)

57. Radford, H. E., *Phys. Rev.*, **126**, 1035–
45 (1962)

58. Barrett, A. H., Rogers, A. E. E., *Nature*, **204**, 62–63 (1964)

59. Rogers, A. E. E., Barrett, A. H., *Astron.
J.*, **71**, 868–69 (1966)

60. Bolton, J. G., Gardner, F., F. McGee,
R. X., Robinson, B. J., *Nature*, **204**,
30–31 (1964)

61. Robinson, B. J., McGee, R. X., *Determination of Radial Velocities and
Their Applications*, 133–37 (Academic Press, London, 1967)

62. Rougoor, G. W., *Bull. Astron. Inst.
Neth.*, **17**, 381–441 (1964)

63. Roberts, M. S., *Ap. J.* (In press)

64. Radhakrishnan, V., *Australian J. Phys.*,
20, 203–4 (1967)

65. Lide, D. R., *Nature*, **213**, 694–95 (1967)

66. Davies, R. D., Rowson, B., Booth,
R. S., Cooper, A. J., Gent, H.,
Adgie, R. L., Crowther, J. H., *Nature*, **213**, 1109–10 (1967)

STRUCTURE OF THE SOLAR CORONA[1]

By Gordon Newkirk, Jr.

High Altitude Observatory, Boulder, Colorado

Introduction

Recent years have seen varied and intense investigations of the solar corona. To a large extent these have been the result of techniques for observing the corona in ways which, only a short time ago, were unimagined or impractical. Radio telescopes have not only supplemented the more classical optical observations but have revealed the unsuspected existence of transient bursts of particles and magnetohydrodynamic waves. Radar probing of the corona allows direct measurement of the inner solar wind. X-ray and ultraviolet photographs taken above the Earth's atmosphere show the inner corona over the entire solar disk while space-probes outside the magnetosphere measure the density, velocity, and magnetic fields of the outermost regions. Balloon- and satellite-borne coronagraphs promise to yield the long-sought synoptic observation of the intermediate corona. Theoretical studies have described the general character of the solar wind and in doing so have uncovered such intriguing problems as the interaction of the wind with the magnetic field of the Sun. Along with these impressive advances other rather old problems remain as perplexing as ever, for example, the connection of structural features in the intermediate corona with surface features and geomagnetic activity as well as the mechanism of heating the corona. Although the long-unresolved question of the ionization of the corona appears to have been settled, the anomalously high abundance of the iron group of elements in the corona remains unexplained.

These many aspects are not to be reviewed in a single article. Discussions of the heating of the corona and the chromosphere-corona interface as well as the interplanetary plasma and high-energy particles from the Sun appear in other papers in this volume. Likewise, an examination of the local physics of the coronal gas is not included. Our concern will be with the density, temperature, and velocity structure of the corona considered from a phenomenological viewpoint. Even within the restricted scope our treatment may not be exhaustive but, we hope, will illuminate existing problems and areas of future advance.

Overall Density Structure

With the recent addition of space-probe and radio-star occultation measurements, coronal densities are now known from the interface with the chromosphere out to the orbit of Earth (Figure 1 and Table I). Since the merit of various theoretical models of the corona is often judged on their ability to predict the decrease of density with increasing distance from the

[1] The survey of literature for this review was concluded in November 1966.

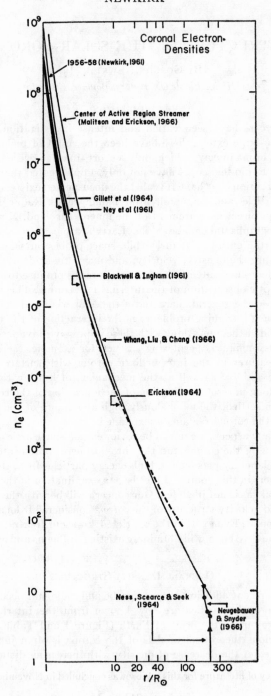

TABLE I

REPRESENTATIVE CORONAL ELECTRON DENSITIES

(Equator at sunspot minimum)

$R(R_\odot)$	$N_e(\text{cm}^{-3})$
1.02	4.0×10^8
1.1	1.4×10^8
1.2	7.0×10^7
1.3	3.7×10^7
1.4	2.3×10^7
1.6	1.0×10^7
2.0	2.8×10^6
2.5	9.0×10^5
3.0	4.0×10^5
4.0	1.2×10^5
6.0	3.1×10^4
8.0	1.3×10^4
10.0	9.8×10^3
15	2.5×10^3
20	1.3×10^3
30	4.6×10^2
50	$(1\ \times10^2)$
100	$(2\ \times10)$
215	2.5

Sunspot maximum densities can be expected to be $\simeq 2$ times higher. Polar densities are uncertain as are those between 30 and 215 $R_\odot$.

Sun, it is appropriate to examine the accuracy with which the density structure has been established.

Coronal photometry.—Photometric studies during eclipse (for a summary see Hata & Saito 1966) and outside of eclipse by means of sensitive polarimeters (Wlerick & Axtell 1957, Dollfus 1958, Newkirk 1961) yield values of the coronal radiance out to as far as 30 $R_\odot$. The technique for translating the observed radiance and polarization into $n_e(R)$ is quite standard (van de Hulst 1950a) and rests upon the assumption that the K corona is entirely due to the

FIG. 1. A compilation of electron densities in the equatorial solar corona derived from a variety of techniques. No attempt has been made to rectify data made during different portions of the sunspot cycle. A single theoretical model (Whang, Liu & Chang 1966) of the density structure in the solar wind appears for comparison.

Thompson scattering of photospheric radiation by free electrons.[2] Although the technique is straightforward in principle, it is fraught with practical difficulties. Eclipse observations are made against a background of sky radiation $B_{sky} \sim 6 \times 10^{-10} \, B_\odot$ to $2 \times 10^{-9} \, B_\odot$ ($B_\odot$ is the mean radiance of the entire solar disk), which is brighter than the K corona at distances > 6–$10 \, R_\odot$. Even when high-altitude aircraft, balloon, and satellite observations are employed to reduce the magnitude of this correction, the F corona or inner zodiacal light is still present as an unwanted background for this purpose. Separation of the contributions of K and F coronas is effected by assuming that the radiation of the F corona is unpolarized and symmetric about the plane of the ecliptic. In addition, one assumes the K corona to be the sole source of polarization and, as a first approximation, the polarization to be that expected from a corona symmetric along the line of sight. The effect of these assumptions on the density determinations must be examined in detail.

Separation of the contributions of the F and K coronas is relatively certain out to a distance of $\sim 10 \, R_\odot$. Beyond this point, the radiance of the K corona is such a small fraction of the total that even minute polarizations in the F corona, such as the recent calculations of Giese (1961) warn us to expect, would drastically modify the inferred K-component and electron densities. Since present models of the zodiacal light (e.g. Weinberg 1964) are not unique and so do not permit an unambiguous calculation of the polarization to be expected from the F corona, even the most carefully executed observations (e.g. Blackwell & Ingham 1961) cannot yield reliable electron densities beyond $\sim 10 \, R_\odot$ except in abnormally bright streamers. To our knowledge measurements in the centers of deep Fraunhofer lines have not been fully exploited to their full capacity to separate the two components.

The recent observation by space-probes (Neugebauer & Snyder 1966) that the solar wind is *always* present suggests that the assumption of symmetry along the line of sight in reducing photometric observations to electron densities is roughly valid for the equatorial regions. However, the well-known appearance of the corona during sunspot minimum indicates that such symmetry over the poles is certainly not present during that part of the solar cycle, if ever. Reductions of photometric measurements for electron densities at high latitude have generally followed the work of van de Hulst (1950a), who used microphotometer tracings made perpendicular to the poles on eclipse plates to estimate the contribution of the lower-latitude corona seen in projection. These analyses have led to the conclusion that electron densities are approximately half the equatorial values at the same height (Allen

---

[2] Although the presence of an optical continuum from synchrotron radiation from high-energy particles has been suggested (Kellogg & Ney 1959), careful polarization measurements (Ney et al. 1961) show that this mechanism can contribute only an insignificant fraction to the light of the corona. This negative result does not rule out the possibility that some of the optical continuum from dense sporadic condensations or solar flares may be synchrotron radiation; however, its existence has still to be demonstrated.

1963a). However, the fact that such an interpretation is not unique and may be incorrect was demonstrated when Ney and collaborators (1961) found that they could fit their photoelectric observations of both radiance and polarization at sunspot maximum (1959) with a corona absolutely free of electrons above latitude 70°. A similar absence of electrons above 65° latitude could have occurred at the sunspot minimum eclipses of 1952 and 1963 (Gillett et al. 1964). Although the presence of limb brightening over the poles at X-ray wavelengths (Figure 2) suggests that a polar corona does at times exist, we must conclude that quantitative information on high-latitude electron densities is gravely uncertain and that the corresponding density models must be used with great caution.

Radio-source occultations.—Knowledge of electron densities from 10 to 80 $R_\odot$ can be obtained from observation of the occultation of discrete radio sources by the solar corona. The pursuit of these measurements (Hewish 1958, Slee 1961, Vitkevitch 1961, Erickson 1964, Slee 1966) for a variety of radio

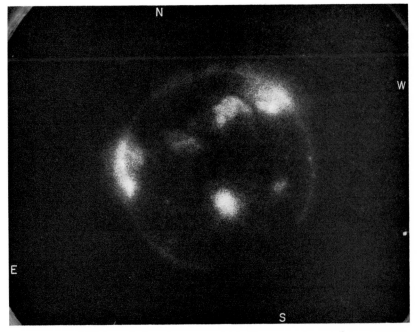

FIG. 2. A photograph of the corona in the wavelength range 27–43 Å, in which the permitted lines of highly ionized atoms contribute most of the radiation. Note the appearance of several X-ray "plages" due to coronal density enhancements, and the outstanding limb brightening. The absence of coronal emission over the southern polar cap compared to that in the north is presumably related to the weak activity which had prevailed in the south for many months prior to the date of observation (20 May 1966). (Courtesy J. Underwood, NASA Goddard Space Flight Center.)

sources over more than a solar cycle has revealed both the shape of the outer corona and its change with solar activity.

Rays from a discrete radio source will be refracted, scattered, and attenuated as they pass through the coronal gas which inevitably contains fluctuations in density about some mean local value. Although in principle the measurement of the amount of refraction would give valuable clues concerning the electron density along the optical path, the ray deviations are so small as to have eluded detection. Of the remaining effects—attentuation and scattering—only the latter as exemplified by the increase in apparent diameter of the radio source has yielded quantitative information concerning the structure of the corona.

The theory of the multiple scattering of radio waves by small density fluctuations in the corona has been developed by Hewish (1955) and Högbom (1960) and recently reviewed by Erickson (1964). In these analyses the outer corona is considered to contain density fluctuations of root-mean-square diameter l and amplitude Δ n_e. Although for a completely arbitrary collection of filaments it would be impossible to relate the half-width ϕ_0 of intensity in the scattered radio source to electron density, several simplifying assumptions are possible: (a) the effective region of scattering is a thin sector of the corona in the plane of the sky, (b) the fluctuation diameter l increases linearly with distance from the Sun, (c) the amplitude of the fluctuations and the fraction of space occupied by high and low densities remains constant throughout the corona. It is then possible to show that

$$\phi_0(R) \sim C\nu^{-2}\bar{n}_e(R_{\min}) \qquad\qquad 1.$$

in which C is a constant which depends upon such parameters as the amplitude of the fluctuations, ν is the frequency of the radiation, and $\bar{n}_e(R_{\min})$ is the mean electron density at the point of closest approach. Within the limits imposed by these assumptions, Equation 1 allows $\bar{n}_e$ to be estimated directly from measures of $\phi_0(R)$ once the constant of proportionality has been fixed. Thus, by scaling determinations of ϕ_0 at various frequencies according to ν^{-2} and by fitting $\bar{n}_e$ to the Blackwell & Ingham (1961) values in the region 8–10 $R_\odot$, Erickson (1964) has extended the estimates of electron density to 80 $R_\odot$ (Figure 3).

We see (Figure 1) that the densities inferred from this analysis of radio-star occultations appear to be somewhat too high in the outer regions when compared with those measured by interplanetary probes. Clearly, one or more of the assumptions of the model are unrealistic. Probably, most suspect is the assumption that the amplitudes and space-filling factor of the fluctuations remain constant in space. Parker (1964) has found that in a filamented corona with equilibrium between magnetic and gas pressure, the density fluctuations will tend to smooth out with distance. Also, the assumption of a single root-mean-square size is unreliable if a wide range of sizes is, in fact, present or if one scale of irregularities is randomly oriented while another is preferentially radial. We must await further refinement of the powerful tool of radio-source occultations before these ambiguities are resolved.

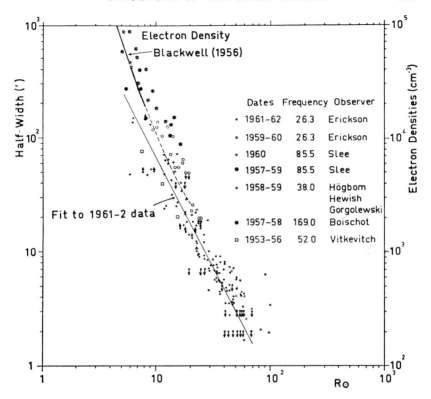

FIG. 3. Electron densities determined from the scattering of the radiation from dis-crete radio sources occulted by the corona. On the basis of several reasonable assump-tions, the scattering half-width is expected to be proportional to $\bar{n}_e$. The factor of proportionality is set by comparison with optically determined densities. (From Erickson 1964.)

Spacecraft measurements.—Although the state of the interplanetary plasma is reviewed elsewhere in this volume (Ness 1967), we include two recent determinations of the density of the corona in the region of 1 a.u. for comparison. During late 1963 (sunspot minimum) a mean value of 3 protons /cm³ was observed (Ness, Scearce & Seek 1963). Detectors aboard Mariner II showed a similar value as well as an increase somewhat more rapid than $1/R^2$ as the spacecraft approached the orbit of Venus (Neugebauer & Snyder 1966). Of course, the difficulty of distinguishing temporal and spatial varia-tions in the latter data prevents definite conclusions concerning the gradient of density of the interplanetary medium.

LOCAL DENSITY STRUCTURE

General morphology.—That the corona possesses an intricate morphology is obvious from its appearance during solar eclipse (Figure 4). Attempts have

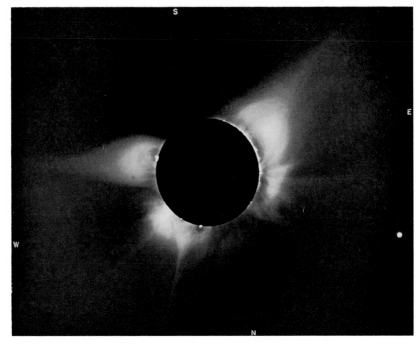

Fig. 4. The solar corona of 12 November 1966 displayed the shape typical of the "intermediate" years of the sunspot cycle. This photograph, made with a radially symmetric, neutral-density filter in the focal plane to compensate for the steep decline of coronal radiance with increasing distance, allows structural features to be traced from the chromosphere out to 4.5 $R_\odot$. The overexposed image of Venus appears in the NE quadrant. Typical "helmet" streamers overlie prominences in the SE and SW quadrants while another streamer at high latitude in the NW develops into a narrow ray at large distance. Arches and the absence of coronal material in the domes immediately above prominences are particularly striking at the bases of the NW and SW streamers. A coronal condensation is visible on the NW limb at a latitude of about 25°. Faint polar plumes over the South Pole represent the typical orientation along the lines of force of a magnetic dipole while those over the North Pole show a much more complex structure. (High Altitude Observatory photograph.)

been made to classify the various forms according to their shape (e.g. Bugoslavskaya 1949) and to associate particular structures with features visible on the solar disk. A classification based on inferences about the geomagnetic effects produced by different coronal structures has been devised by Mustel' (1962a,b; 1964). As one might expect with anything so complex and subtle as the form of the corona, the names used to describe particular features have varied from author to author. Table II displays the reviewer's attempt to form a classification based on earlier works but emphasizing the evolution of coronal forms.

TABLE II

SUMMARY OF CHARACTERISTICS OF CORONAL STRUCTURES

Structure	Diameter at 1 $R_\odot$	Extent in height	Typical density enhancement	Typical density at 1.1 $R_\odot$ (cm^{-2})	Typical temperature (10^6 °K)	Typical age of region	Lifetime	Associated surface feature	Associated coronal emission-line feature	Associated radio, X-ray, and UV feature
Active-region enhancement	200,000 km (5–7 min arc)	200,000 km (0.3 $R_\odot$)	up to 2×	4×10^8	1.5 to 2	2 weeks	2–3 weeks	young active-region plage	green and red line enhancement	"plage"
Active-region streamer	300,000 km (5–7 min arc)	many $R_\odot$	2 to 5×	10^9	2	3 weeks	2–3 weeks	most vigorous active-region, type E sunspot, plage		
"Permanent" condensation	50,000 to 130,000 km	80,000 km	5×7	10^9	2–3	3 weeks	several days		green and red line enhancement	small bright "button" within plage
Sporadic condensation	20,000 km	20,000 to 80,000 km (may appear as isolated cloud above surface)	50–500×	10^{10}–10^{11}	4–5	3 weeks	fraction of an hour to several hours	often associated with flares and loop prominences	yellow line (5674 Å Ca XIV), continuum	gradual rise at cm radio wavelengths, post-flare enhancements in X rays
Helmet streamer	300,000 km	many $R_\odot$	upper limit of 7–25×	[b]	1.5–2	8 weeks	many months	prominence, extended magnetic fields	?	?
Equatorial streamer	300,000 km	many $R_\odot$	—	2–4×10^8	1.5–2	?	months to years	15 hours	?	?
Polar plume	30,000 km	many $R_\odot$	4–8×[a]	5×10^8	1.2	?	~15 hours	bright polar faculae	?	?
Narrow ray	30,000 km	many $R_\odot$	?	?	?	?	?	?	?	?

[a] Because of uncertainty in true electron densities over the pole, this figure may represent a lower limit.

[b] Density enhancement seldom appears close to surface.

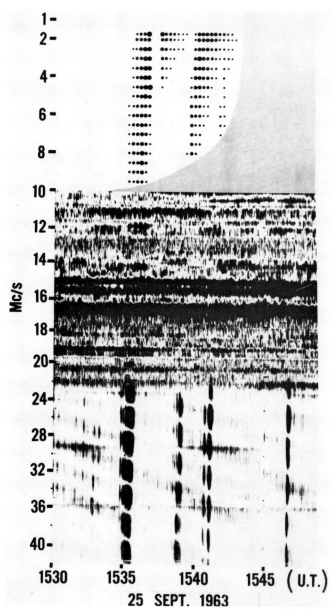

FIG. 5. Type III radio bursts observed from 40 to 22 Mc/s from the ground and from 10 to 1.6 Mc/s from the satellite Alouette I. The frequency range 10–22 Mc/s is obliterated by man-made interference. Periodic breaks in the lower record result from minima in the multilobed interferometer. The drift of the bursts toward lower frequency with time is interpreted as the excitation of plasma oscillations at progressively higher and higher levels in the corona although an independent measure of the heights is required to derive a density model. (From Hartz & Warwick 1966.)

The term "streamer" is taken to refer to the larger structures which appear brighter than the background corona and which extend beyond 0.5 to 1.0 $R_\odot$. Streamers seen over active regions, over the equator, and over prominences (helmets) appear to comprise distinct subclasses. "Enhancement" is reserved to describe the diffuse region of high density which surrounds an active region but which does not extend beyond 0.3 $R_\odot$ above the limb. "Permanent" and "sporadic condensations" were first recognized by Waldmeier & Müller (1950) as smaller-sized density enhancements above active regions while "equatorial streamers" and "polar plumes" have been described at length in the literature (Saito 1958, 1959, 1965a). "Narrow rays" occasionally appear as thin, slightly curved features extending great distances out into the corona.

Densities above active regions.—Densities in the various coronal structures tabulated above are most often expressed in terms of their enhancement above the background or smoothed-out coronal densities at the same height. Unfortunately, two difficulties in the concept remain unresolved: one is never quite sure that the "background" corona may not consist of a collection of unresolved structures such as streamers, thus special attention may be granted only to the truly abnormal features; the location and the extent of the structure along the line of sight are generally unknown so that densities are derived on the basis of some assumption about the three-dimensional shape.

Active-region enhancements appear above plage regions in the coronal emission lines 5303 Å and 6374 Å as well as on K-coronameter and eclipse observations. At X-ray wavelength (Figure 2) and in radioheliograms enhancements may be seen against the disk as well as projected against the sky. These and the more highly developed active-region streamers appear similar to the corresponding plage as seen in K_3 spectroheliograms except that the fine structure is less evident and the edges are more diffuse in the corona. Although little attempt appears to have been made to distinguish between enhancements and active-region streamers, which are in our opinion only two evolutionary stages of the same structure, density increases up to $2\times$ over the background corona appear to be typical (Christiansen et al. 1960, Newkirk 1961, Kundu 1965).

By the time an active region has reached the height of its development 2–4 weeks after its birth, it has produced numerous flares. Accompanying these flares are radio bursts which allow an independent assessment of the electron density. As is well known (Wild, Smerd & Weiss 1963; Maxwell 1965; Warwick 1965), the drift of the radio-noise bursts towards lower frequency with increasing time (Figure 5) is interpreted as the excitation of electromagnetic waves at the plasma frequency (and at times its harmonics) by a disturbance traveling upward through the corona. Type II (slow-frequency drift) bursts are believed to be associated with hydromagnetic shock waves while type III (fast-frequency drift) bursts owe their excitation to a pulse of highly energetic particles. Radio-interferometer measures of the

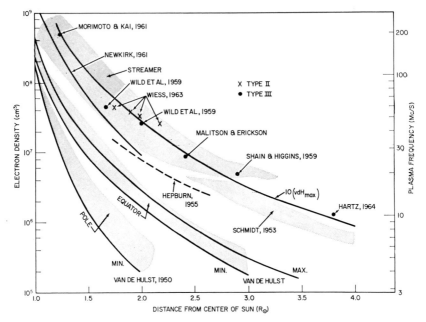

FIG. 6. Radio-burst measurements require electron densities over active regions to be enhanced $\sim$ 10 times those in the model derived for the equator by van de Hulst. The fact that optical observation of streamers (Newkirk, Hepburn & Schmidt) yields slightly lower densities than the radio data is probably caused by the preferential occurrence of bursts in the most active (and most dense) regions. (From Malitson & Erickson 1966.)

heights of the disturbances (Wild, Sheridan & Neylan 1959) give their spatial velocity so that the known-frequency drifts can yield a density model of the corona. Analyses of this type (Maxwell & Thompson 1962, Malitson & Erickson 1966) have been remarkably consistent during the last solar cycle in giving densities above active regions out to 4 $R_\odot$ of 10 times those suggested by van de Hulst (1950a) or Baumbach and Allen (Allen 1963a) for sunspot maximum (Figure 6). Since there is evidence (Newkirk 1961, Ney et al. 1961) that the background corona was some 2 times the van de Hulst values during the last maximum, densities above active regions must have been augmented by a factor of 5 times the background. The height of type III bursts inferred from the statistics of their visibility across the solar disk has been used by Morimoto (1964) to arrive at a similar conclusion.

Initially it might appear that this 5 times augmentation is in disagreement with the factors of 1.5 to 2.5 times found from K-coronameter observations (Newkirk 1961) and from the slowly varying radio component (Christiansen et al. 1960, Kakinuma & Swarup 1962). The difference is presumably real and the result of the biased selection of the most vigorous active regions, and presumably those with the higher densities, by the radio bursts. Channeling of the exciters of type II and type III bursts along the dense axis of the active-region streamers may also contribute to the higher densities suggested by these phenomena. Thus, we conclude that an active region during the stage of its development when flares are present always contains enhanced densities of 2 to 5 times the local background out to at least 4 $R_\odot$.

The structure of "permanent" and "sporadic" condensations was established by Waldmeier & Müller (1950) and has been confirmed on several occasions (Saito & Billings 1964, Kundu 1965) by optical, radio, and X-ray observations. As described schematically by Waldmeier & Müller, the permanent condensation represents a low mound of enhancement ~80,000 km in diameter with densities of ~10^9 cm^3 at its core. This feature often appears as a substructure within the more extensive active-region streamer. Permanent condensations often contain even denser knots of transitory nature in which the densities (10^{10} to 10^{11} cm^{-3}) are high enough to allow their continuum radiation to be detected on coronagraph spectrograms. These so-called sporadic condensations are intimately associated with loop prominences and the loops and arches seen in coronal emission lines (Kleczek 1963). As with all qualitative descriptions of complex phenomena, one should beware of attaching too much significance to the somewhat arbitrary classification of various types of condensations and to their "typical" characteristics. The intense coronal condensation (Figure 7) observed at the February 1962 eclipse and analyzed by Saito & Billings (1964) amply demonstrates that the distinction between active-region enhancement and permanent condensation is often extremely artificial.

Densities in streamers.—The densities in helmet streamers, which frequently appear above prominences, have been established only by means of optical photometry since, by our definition, these features occur away from active regions and the probing of radio-burst exciters. Although their association with prominences has long been known both statistically and by inference of their true position from the observed polarization (Bugoslavskaya 1949, Schmidt 1953), only a few attempts have been made to distinguish these streamers from others in estimates of density enhancement. Schmidt examined three streamers, two of which could be definitely associated with prominences and the third with a region that had displayed a prominence two months previously; he concluded that densities ~7 times the background occurred. A large streamer observed in 1952 and studied extensively by Hepburn (1955) and Michard (1954) showed negligible enhancement close to the Sun with a rapid increase to 25 or 30 times the background at a height of 3 $R_\odot$. Beyond 3 $R_\odot$ the enhancement above the background corona remained at this very high value. Both these analyses include the assumption that the streamer is roughly cylindrical in form. There is evidence (Saito 1959) that some streamers have considerable extent in longitude and that the inferred enhancements must be considered as upper limits only.

Densities in polar plumes.—Polar plumes can scarcely be discussed outside the context of the variation of the form of the corona with solar cycle. Their striking resemblance to the lines of force about a bar magnet has been

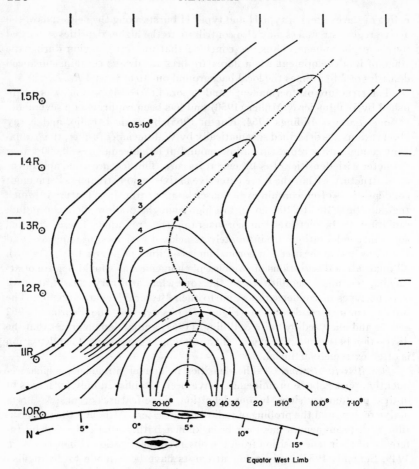

FIG. 7. Electron densities in a coronal condensation derived by Saito & Billings (1964) under the assumption of symmetry about the axis shown as a dot-dashed line. Several isolated loops (not shown) of high-density material appeared within the main structure of the condensation. (From Saito & Billings 1964.)

known for many years, as has the fact that the length of the dipole varies with solar cycle. Since the polar regions of the Sun represent more a patchy collection of magnetic regions than a classical dipole (Severny 1965), it is not surprising that the lines of force outlined by polar rays seldom if ever appear symmetrically about the rotation poles (Godoli 1965). The fact that discrete radio-source occultations show a dipole-like alignment of the scattering ir-regularities out of 5 to 10 $R_\odot$ (Wyndham 1966) suggests that polar rays maintain their identity far out into the interplanetary medium.

The same analytical tools used to derive coronal densities elsewhere in the

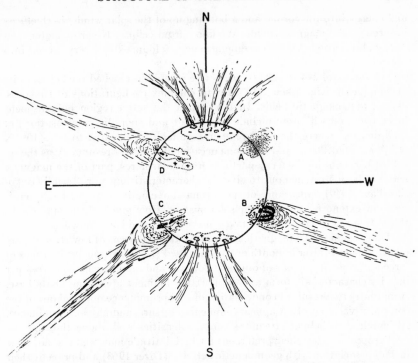

FIG. 8. A schematic representation of the evolution of a "typical" region in the corona from its early appearance as an enhancement (A) through the active-region streamer stage (B) to the long decline as a helmet streamer (C and D). Both the helmet and the accompanying filament are believed to require the presence of extended regions of opposite magnetic polarity (schematically shown).

corona yield densities in polar plumes. Although their contrast with the background corona is only 10–20 per cent (van de Hulst 1950b, Saito 1965a), their small line-of-sight extension requires densities a factor of 4 to 8 above the background. Such enhancements of density above the background corona, as well as the observation of Saito (1965a) that the density gradient in polar streamers is less steep than for the background, must always be regarded in the light of our previous discussion of the paucity of our knowledge of the density or even the existence of a general polar corona.

The evolution of coronal structures and their connection with surface features. —Although the necessary data—synoptic observations of the corona out to at least $6R_\odot$—are not available, we can attempt to construct a picture of the evolution of a "typical" active region as it appears in the corona (Figure 8). Immediately following the birth of an active region (A) with its plages and emergent magnetic fields, a density enhancement appears in the corona. Presumably the enhancement owes its origin to the increased flow of material and energy into the corona and a bottling up of the solar wind—both effects induced in the magnetic fields. We know from eclipse, K-coronameter, and X-ray observations that these enhancements seldom extend very far out into the corona.

At the age of 2–4 weeks (B) the active region has reached the height of its development (Kiepenheuer 1953) with complex configurations of sunspots, plages, and magnetic fields. In the corona the active-region enhancement has grown and will often include permanent and sporadic condensations of high density, where the local temperature may be elevated to $4-5 \times 10^6$ °K (Billings 1966). Loops in the corona occasionally form. Presumably as the result of bursts of high-energy particles from solar flares, part of the magnetic field has been stretched out to give the characteristic open or "bush" (Bugoslavskaya 1949) appearance of the corona above active regions. High densities now extend far out into the solar system. How much of the material in this active-region streamer partakes of the solar wind is unknown.

Dissipation of the magnetic fields into extended areas of low strength becomes apparent after a month or so has passed (C) (Leighton 1964). Lines of force still connect regions of opposite polarity and mold the helmet streamer into its characteristic form and support the filament appearing at its base. Apparently as a result of condensation of cool prominence material out of the corona, a vacancy often appears immediately surrounding the prominence. Although such helmet streamers contain densities well above those of the background corona, their preference for high latitude may well explain their lack of association with geomagnetic activity (Dizer 1963) and prevent their investigation by present space-probes. In the latter stages of development of such a region, differential rotation will have dragged out the underlying magnetic fields in longitude (D). The associated streamer will be expected to have a correspondingly great extent in longitude and would be called a fan by many authors. During the many months of lifetime of this stage, the underlying prominence may from time to time disappear without modifying the overlying streamer. As the field decays further, it loses its ability to channel the solar wind into a streamer, which gradually subsides into the background corona.

Little is known concerning the evolution of polar plumes although several recent researches have been devoted to establishing what if any connection exists between them and surface features. Using the statistical distribution of plumes across the polar cap as well as their observed polarizations to infer positions in three dimensions, Saito (1965a,b) concludes that plumes occurred with maximum frequency at latitudes 75°–80° during the sunspot minimum of 1965 and that the pole itself was bare of these features. Even though his polarization measures seem to lack the accuracy required for such a subtle inference of position, the idea is provocative. What connection exists between such a "plume zone" and the interface between the magnetic fields of the polar region and those migrating toward the pole from lower latitudes (Leighton 1964, Hyder 1965a) is unknown.

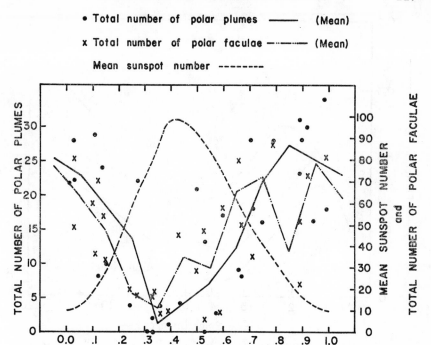

- Total number of polar plumes ———— (Mean)

x Total number of polar faculae —·——·— (Mean)

Mean sunspot number - - - - - - - -

FIG. 9. The number of polar plumes visible on published eclipse photographs (1889–1963) is compared with the number of bright polar faculae (Sheeley 1964) and average sunspot number. Phase in a given sunspot cycle is defined as

$$\phi = \frac{t - t_{\text{preceding min}}}{t_{\text{following min}} - t_{\text{preceding min}}}$$

This and other evidence suggest that plumes originate over the brightest polar faculae.

The association of polar plumes with surface features visible either in white light or in spectroheliograms is difficult to establish for obvious reasons. Bugoslavskaya (1949) states that plumes are rooted on "chromospheric prominences," which would be referred to as large polar spicules by most authors. A distinct correlation between the brightness variation of the north polar corona of 20 July 1963 and that in a polar strip of the K_3 spectroheliogram made the same day has been interpreted by Harvey (1965) as evidence that plumes originate over bright polar faculae. We also note (Figure 9) the similarity between the variation of the number of polar faculae (Sheeley 1964) with sunspot cycle and the number of polar plumes counted from published eclipse photographs extending from 1889 to 1963. The hazards of inferring causal relationships from such statistical evidence are well known; however, if we accept the counts at face value we conclude that, since only half

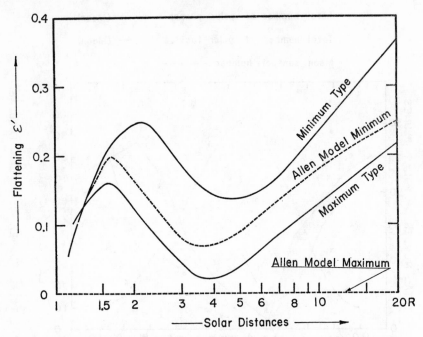

FIG. 10. Variation of the flattening index ϵ' of coronal isophotes with distance. Even at sunspot maximum the corona is not spherically symmetric and a strong concentration at the equator is always apparent. (From Hata & Saito 1966.)

the polar faculae are visible at any one time, only about one in seven faculae possesses a visible polar plume. Thus, these several pieces of data all support the plausible but hardly well-demonstrated conclusion that polar plumes overlie polar faculae, which represent regions of higher material flux into the corona.

Solar-cycle variations in the corona.—The familiar variation of the appearance of the corona with solar cycle has been well documented (van de Hulst 1953) and recently reviewed (Hata & Saito 1966) to include more recent data. A flattening index

$$\epsilon' = \frac{R_{eq}}{R_{pole}} - 1$$

for coronal isophotes allows a large quantity of data to be surveyed at a glance. We note in Figure 10 that the difference between the sunspot minimum and sunspot maximum is one of degree only—both are far from spherically symmetric. Evaluation of the scattering by the corona of the radiation from many discrete radio sources has led Slee (1966) to conclude that the flattening index continues to increase to a value of about 0.7 in the region 20–80 $R_\odot$. Although it is questionable that the shape of the ellipse of constant scattering can be directly equated to the shape of coronal isophotes, it seems clear that the concentration of the corona toward the equator becomes more exaggerated at large radial distances.

The variation of the innermost corona visible in the 5303 Å line has been described by Waldmeier (1957), who discovered three zones of emission: 1) a dominant maximum which progresses toward the equator with the sunspot zone, 2) a band which originated at about 50° latitude early in the sunspot cycle and reaches the pole at about the time of sunspot maximum, and 3) a zone which appears at high latitude late in the cycle and reaches 40° latitude at about the time the new cycle begins. The behavior of the active-region zone needs little explanation, except that we have only a qualitative explanation (Babcock 1961) for the fundamental processes underlying sunspot activity in general. The similarity between the poleward migrating zone and a similar class of prominences (Kiepenheuer 1953), as well as that of the migrating magnetic regions (Babcock 1961, Leighton 1964), suggests that the magnetic regions are responsible for both phenomena. The origin of the third zone remains an enigma.

Density variations in the corona throughout the last solar cycle have been documented by eclipse photometry [see Hata & Saito (1966) for a review], radio-interferometer observations of the thermal emission of the corona (Kundu 1965), and radio scattering data (Slee 1966, LeSqueren-Malinge 1964, Erickson 1964). A consensus of these various investigations appears to be that electron densities throughout the corona at sunspot maximum are about twice those at sunspot minimum. Comparison of observations made during the last sunspot maximum (Ney et al. 1961, Newkirk 1961) with those of previous maxima (van de Hulst 1953) indicates that an increased level of activity at a given maximum leads to abnormally high electron densities—during 1957–1959 densities of 1.4 to 2 times the van de Hulst model for sunspot maximum were present.

SMALL-SCALE DENSITY FLUCTUATIONS

There is overwhelming evidence that the corona possesses, in addition to the gross structural features associated with active regions and streamers, fine-scale fluctuations in density. High-resolution eclipse photographs (e.g., those made by the Swarthmore expedition in 1930) present a corona which appears to have been carefully groomed with a fine comb. Interplanetary probes of the solar wind show a wide spectrum of temporal variations of plasma density and velocity. Such effects as radio-source scattering, interplanetary radio scintillation, diffusion of solar cosmic rays, and radar Doppler displacements in the corona would be nonexistent or vastly different if the corona were a smooth plasma. The presence of such fluctuations also plays a vital role in explaining the excitation of type III radio bursts by high-speed particles (Ginzburg & Zhelezniakov 1958). Finally, comet tails show the results of sudden accelerations due to small-scale structures in the interplanetary medium (Antrack, Biermann & Lüst 1964). Recent investigations have yielded exciting details concerning the scale, amplitudes, and orientations of these fluctuations in the corona.

The inner corona.—Optical observations of coronal irregularities are intrinsically limited by the fact that the intensity fluctuations produced by small features become lost in the noise of the photographic or photoelectric detector as soon as more than a few features appear along any single line of sight. No quantitative analysis of the spectrum of structure sizes visible on high-resolution coronal photographs has been made although we have determined that the smallest features visible on the Swarthmore plates had a diameter 3×10^3 km—a figure suspiciously close to the resolution limit of the observations. Recent spectroscopic observation of the K line of Ca^+ in emission in the corona out to $0.3\ R_\odot$ has been interpreted by Deutsch & Righini (1964) as due to the presence of dense, cool regions in the corona. Although the necessary insulation of such regions (hypothesized to occupy only 10^{-4} of the coronal volume) by magnetic fields is not impossible, it appears that the basic observation of K line emission from the corona is erroneous—the result of scattered chromospheric radiation by the image-forming optics and the aircraft window.

Small-scale density fluctuations in the lower corona ($\sim$1.3 $R_\odot$) make their presence indirectly known by the diffusion of the radiation from radio-noise bursts. Using a Monte Carlo technique to trace rays from the burst through randomly oriented coronal irregularities, Fokker (1965) has concluded that such scattering can account for the general characteristics of the angular distribution of radiation emerging from bursts in the 200 Mc/sec range. His model does not predict the detailed appearance of these bursts as a bright core surrounded by an extensive halo (Weiss & Sheridan 1962), however. Fokker finds that the magnitude of the scattering in the inner corona is nearly an order of magnitude lower than that found from 10 to 80 $R_\odot$ by radio-source occultations and suggests that the amplitude of the fluctuations must increase in the outer parts of the corona. However, an alternate explanation for this difference must be considered. Erickson (1964) has noted that the observed angular distribution of radiation from an occulted radio source is consistent with scattering elements which are more or less randomly oriented but with a preference for radial alignment. Radiation from a source imbedded in such a corona could thus emerge nearly parallel to most of the density inequalities and experience less scattering than the radiation from an occulted radio star.

The outer corona.—In addition to the estimates of coronal electron densities described earlier, the analysis of discrete radio-source occultations reveals information on the scales and orientations of the density fluctuations. The fact that the corona acts as a polarizer suggests that the fibers over the poles are oriented along the magnetic dipole (Hewish & Wyndham 1963). However, Erickson (1964) finds that the generally radial elongation of scattering elements must still be mixed with a random component to explain the observations.

Using measures of the size ϕ_0 of a diffused radio source alone, one cannot discover the scale of the interplanetary density fluctuations. The same value of the proportionality constant C in Equation 1 may be produced by a few large fluctuations or a great number of small ones. An additional observation —the absence of seeing effects as the rays pass through first one and then another irregularity—is crucial for an estimation of the scale. In a series of papers Hewish (1958), Slee (1961), and Hewish & Wyndham (1963) have attempted to establish an upper limit for the scale of the irregularities from the lack of observed seeing effects with amplitudes larger than 6 sec of arc. They argue that cells of enhanced density of angular diameter larger than the interferometer beam carried through the line of sight by the solar wind would produce temporal variations in the apparent position of the centroid of the scattered source. Since such seeing effects are not observed, they conclude that $l < 5 \times 10^3$ km. Such structures have dimensions of a few hundred kilometers when projected back to the Sun and might be related to chromospheric spicules.

Critics [see discussion following Wyndham (1966)] of this interpretation of the radio-occultation data have noted that while a structure of scale $l \sim 5 \times 10^3$ km may be present and may even be required, the data do not truly rule out the presence of larger features. We note that the crucial absence of seeing effects larger than 6 sec of arc over intervals of several minutes is observed at a separation between the Sun and the source of 80 $R_\odot$. At this distance a ray from a discrete radio source would still intercept and be deflected through small angles by some 20 to 50 irregularities even if their scale were as large as 10^6 km, and the chance that seeing effects could be observed would still be small. Moreover, anticipating the discussion of the section to follow, we find that the larger coronal density irregularities detected by direct space-probe measurements can explain much of the observed radio scattering.

Recent observations of interplanetary scintillations of discrete sources (Hewish, Scott & Wills 1964; Cohen et al. 1966; Vitkevich, Antonova & Vlasov 1966) and of Jupiter (Slee & Higgins 1966) have been interpreted as of interplanetary origin and appear to give support to the presence of small-scale ($l \sim 5 \times 10^3$ km) irregularities in the interplanetary medium. Hewish and collaborators estimate that these small structures move across the line of sight with speeds 50–190 km/s and they conclude that the scintillations are produced as the filaments are swept by the Earth with the rotation of the interplanetary medium. The observed change of scintillation frequency with heliographic latitude lends credence to this argument.

Space-probe measurements.—Direct space-probe measurements provide, of course, the most unambiguous evaluation of density fluctuations in the outer solar corona. Examination of the data relayed by such satellites as Mariner II (Neugebauer & Snyder 1966) reveals long-period increases in density lasting for several days. These stable structures, in which the density may rise to 70 protons/cm³ as compared to the normal 3 protons/cm³, often recur at 27-day intervals and appear to be related to the sectored structure of the interplanetary magnetic field found by Wilcox & Ness (1965). It is not surprising that these large-density structures have appeared in radio-occultation records as abnormal changes in scattering and attenuation of several days' duration (Slee 1961).

Within these large regions of unusually high density, smaller-scale fluctuations appear (Figure 11) which are identified as filamentary structures imbedded in the solar wind. Even when the data are presented as 3-hour averages, density variations by factors of 5 or more frequently occur within 3 to 12 hours' time. Although the small-scale irregularities visualized by the radio observers would be completely masked in these averaged data, it is quite clear that large-amplitude density fluctuations of dimensions as large as 4×10^6 to 2×10^7 km (a mean velocity of 400 km/s is assumed) are quite common in the corona at 1 a.u.

Measurements of the interplanetary magnetic field also allow inferences concerning the detailed density structure of the medium although it must be kept in mind that field and plasma measurements are not one and the same thing. It is well known that the field is remarkably stable in value at 4–7 $\times 10^{-5}$ G and exhibits a streaming angle corresponding to an Archimedes spiral for a velocity of 300–700 km/s (Ness, Scearce & Seek 1963). Within this gross alignment the field is observed to reverse sign within periods of 10 minutes up to several hours. From this duration of fields of the same orientation we conclude that the width of a given magnetic bundle is between 3×10^5 and 3×10^6 km. Fluctuations in the field of even shorter duration found by Pioneer VI (Ness, Scearce & Cantarano 1966) are presumed to represent Alfvén waves.

Complementary information concerning the structure of the interplanetary magnetic field is obtained by satellites carrying cosmic-ray detectors outside of the Earth's magnetosphere. It has long been known that cosmic rays produced by solar flares initially arrive at the Earth from the direction of the Sun and only after considerable time do they become isotropic through diffusion within the solar system. Recent measures in the 13-MeV range (Bartley et al. 1966, Fan et al. 1966) show that although the average direction of arrival of the flare protons centers about the streaming angle of the overall interplanetary field, abrupt and large fluctuations in velocity vector occur every 0.5 to 4 hours. The strong correlation (McCracken & Ness 1966) found between the direction of the interplanetary field and that of arrival of the cosmic rays suggests that the particles are being conducted along magnetic tubes of force with overall dimensions of 7×10^5 km to 6×10^6 km. Moreover, the extreme anisotropy of the flux within a given tube as well as the abruptness of the boundaries observed between tubes requires that the magnetic field must be smooth on a scale of 10^5 km, the gyroradius of the particles.

In summarizing the observational aspects of fluctuations of density in the corona we are faced with apparently contradictory evidence. Radio-source occultations and interplanetary scintillations require the presence of struc-

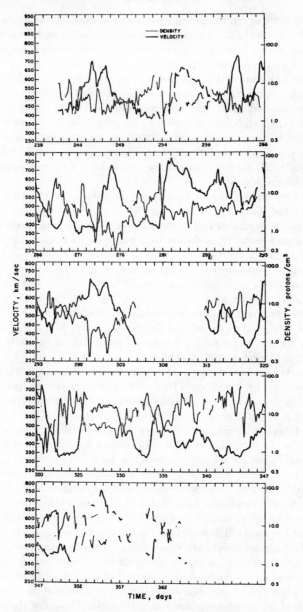

FIG. 11. Variation of the velocity and density of the interplanetary medium as measured by the Mariner II spacecraft and averaged over 3-hour intervals. The occurrence of high velocities with low densities and vice versa is apparent. Rapid changes in density (or velocity) over a few hours' time show that the interplanetary plasma contains density irregularities with a scale of some 10^6 km diameter. (From Neugebauer & Snyder 1966.)

tures with a scale of 5×10^3 km and rule out the existence of fluctuations as large as 10^6 km. On the other hand, direct plasma probes and the inference of density fluctuations from the structure of the interplanetary magnetic field demonstrate unquestionably the presence of structures of dimensions of the order of 10^6 km, and cosmic-ray measurements suggest that irregularities smaller than 10^5 km cannot occur in any great numbers in the entire space between the Sun and the Earth. The contradiction is only partly resolved by the fact that the radio observations are primarily sensitive to the very small and presumably numerous density fluctuations while present space detectors reveal only the larger structures. The cosmic-ray data appear to require that the small-scale density structures, if they exist, have no corresponding structure in the magnetic field.

Origin.—Several origins for density fluctuations in the outer solar corona can be imagined. Certainly some irregularities represent shock waves which develop when a stream of fast particles overtakes a slower stream (Parker 1963). Helmholz instabilities similar to the flapping of flags in the wind will develop at the interface between parallel streams with different velocities. Alfvén waves also undoubtedly occur. Although these various sources of density irregularities bear detailed investigation, the correspondence between solar and interplanetary magnetic fields on a large scale discovered by Wilcox & Ness (1965) suggests that photospheric, magnetic structures of still smaller dimensions may also be carried into the interplanetary gas. In the absence of a quantitative theory for such a conformal mapping, we can only assemble several pieces of circumstantial evidence.

At 1 a.u. the satellite data reveal density fluctuations of a factor of 5 or 10 with scales of 3×10^5 km $< l < 6 \times 10^6$ km. In the inner corona, polar plumes and what few rays can be occasionally discerned over faculae (van de Hulst 1953) have a characteristic diameter of approximately 3×10^4 km (Saito 1965a) and appear to contain densities approximately 5 times the background. We note that radial expansion of coronal irregularities of this size at the Sun would produce features rather close to the dimensions of those observed at 1 a.u. by space-probes. Thus we might envision a corona completely interlaced with tangled filaments which originate over the chromospheric network forming the boundaries of the supergranulation. Magnetic fields present in the network (Howard 1959, Simon & Leighton 1964) serve to constrain the material flowing into the bottom of the corona to individual tubes while the general outward streaming of the solar wind sweeps the filaments out into interplanetary space. Justification for considering the chromospheric network as the origin of these coronal filaments is offered by Kulsrud (1955), de Jager (1962), and Kuperus (1965) who note that the presence of a magnetic field in the photosphere will substantially increase the density as well as the energy flow into the overlying corona. Of course, this model of the detailed mapping of chromospheric magnetic structures out into the corona and interplanetary medium and the origin of whatever filamentary structure exists in the corona is only a plausible hypothesis. A self-consistent picture of the small-scale structure of the corona and interplanetary plasma will, it is to be hoped, result from future investigations.

DYNAMICS OF CORONAL STRUCTURES

The steady flux of coronal gas out into interplanetary space amply demonstrates that the corona is a dynamical medium in an approximately steady state rather than a static one. The solar wind with a flux $\sim 5 \times 10^8$ protons/cm²/s (Ness, Scearce & Seek 1963) demands that the lower corona be replaced about once a day. In addition to the general flow of the solar wind, occasional displacements and oscillations appear in localized regions of the corona. Moreover, at least the inner portion of the corona must share in the rotation of the Sun. We shall first examine the observational evidence for these various dynamic aspects of the corona. Where a well-developed theory exists, as for the solar wind, we shall compare the results with the observations wherever possible, although a complete review of the theory is not intended.

Local macroscopic motions.—Bugoslavskaya's (1949) classical study of the structure of the corona also included an attempt to determine what macroscopic motions occur in various features. Using eclipse photographs made at several stations along the totality path and covering a few hours of elapsed time, she inferred the following magnitudes for the motions:

arches over prominences—2 km/s
helmets— <4 km/s
archlike structures over sunspots—10 km/s
small coronal "clouds"—15 km/s
knots and "clouds" over sunspots—47 km/s
displacements in thin rays—45 km/s
displacements of polar plumes—0.6 km/s
expansion of outer limits of streamers (fans)—1 km/s

However, since position measurements of diffuse coronal features to an accuracy of better than several seconds of arc are extremely difficult, the velocities of only a few km/s must be regarded with caution. Motion pictures in the 5303 Å line show velocities of tens of km/s for the emergence of coronal loops above sunspots (Kleczek 1963), as well as speeds of 600 km/s in the violent release of one end of the loop, to create a coronal "whip" (Evans 1957). That such motions represent real displacements of the material was questioned by Lyot (1944), who concluded that most of the apparent transverse movement in emission-line regions was an illusion created by traveling zones of excitation or the successive growth and decay of adjacent structures.

Of course, no such ambiguity afflicts measurements of the Doppler displacement of coronal emission lines. Waldmeier (1947) and Dollfus (1957) have found that, except over active regions, these velocities seldom exceeded a few km/s. However, in the dense and hot sporadic condensations, which often appear as intense loops in 5303 Å, speeds of up to 150 km/s occur. A detailed examination (Newkirk 1957, Karimov 1961) of Doppler motions in

6374 Å (Fe X) and Hα from such regions shows that although the coronal ions generally travel in the same direction as the neutral hydrogen in the associated loop prominence, the average speed of the coronal material is much lower. Those few instances in which the velocities are equal appear to be associated with the rapid expansion of the coronal and prominence loop. For material streaming along the magnetic lines of force, the speeds of coronal ions and hydrogen were found to be in the ratio $v_C/v_H \sim 0.28$. Apparently, electrical forces, which impart accelerations proportional to the ratio of particle charge and mass, play a larger role than gravity in the flow along such loops.

Oscillations.—Observations of macroscopic oscillations in the solar corona are meager. Billings (1959a,b) interpreted a quasi-periodic variation of the width of the 5303 Å line with height above the chromosphere as evidence for the presence of a transverse hydromagnetic wave. More recently Lilliequist (1966) has analyzed a long time-series of spectrograms of an isolated coronal condensation. Genuine fluctuations in both the intensity and width of 5303 Å appear with periods of 150 and 300 sec. A similar periodicity at 490 sec appears in the 6374 Å line. Variations in the intensity of 5303 Å as far out as 1.25 $R_\odot$ have also been detected by Noxon (1966) at an estimated period of 500 sec. Although the cause of these oscillations is unknown, one might speculate that they represent a modulation of the corona driven by oscillations (Evans, Michard & Servajean 1963) of the underlying chromosphere.

Rotation.—Intuition suggests that the inner solar corona must rotate synchronously with the underlying surface; magnetic fields undoubtedly penetrate into the lower regions with sufficient strength to drag the plasma along. Naturally, a point must be reached further out in the corona where the fields are incapable of maintaining rigid rotation and where a given parcel of material begins to lag behind the solar surface. Still further out in the interplanetary medium the gas must follow the classical "garden-hose" trajectory, which closely resembles the Archimedes spiral. To avoid confusion we shall refer to rotation in the innermost region as *rigid rotation* while the term *co-rotation* is reserved to describe the Archimedes spiral region.

Since even the most accurate measures of coronal emission lines are ineffective in determining the rotation of the inner corona, our knowledge is based almost exclusively on the apparent motions of coronal features. Minute systematic shifts visible on her eclipse plates led Bugoslavskaya (1949) to conclude that the corona does rotate with the rest of the Sun. Waldmeier (1950) and Trellis (1957) followed the recurrent passage of regions of bright 5303 Å emission over the solar limbs to determine the rotation rate as a function of latitude. Cooper & Billings (1962) were able to follow one such high-latitude region for over two years to estimate the period at latitude 65°. These data and the more recent rotation rates determined by Hansen & Hansen (1966) from autocorrelation analyses of K-coronameter observations appear in Figure 12 together with estimates of the rotation rate of filaments (d'Azambuja & d'Azambuja 1948), sunspots (Allen 1963a) and polar faculae

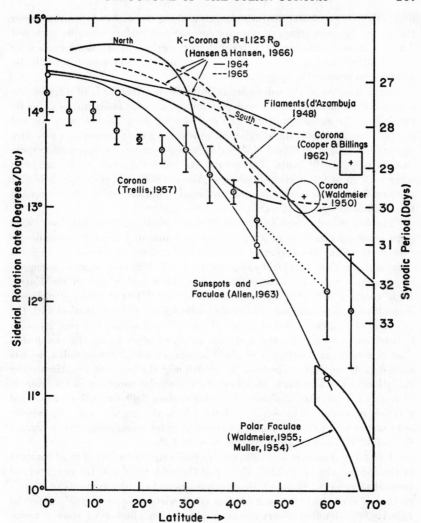

FIG. 12. A compilation of data on the rotation rate of the inner solar corona compared to that of filaments, polar faculae, and sunspots and faculae. (Courtesy R. T. Hansen & S. Hansen.)

(Waldmeier 1955, Müller 1954). While some of these observations appear to suggest that the corona below 1.25 $R_\odot$ rotates with the underlying magnetic fields as revealed by filaments, the Southern Hemisphere data of Hansen & Hansen and of Cooper & Billings require the corona to rotate *faster* than the surface. This apparently anomalous behavior may be produced by dominant magnetic fields at lower latitude dragging along the higher-latitude corona. However, the comparison of data on coronal and surface features gathered during different solar cycles or inconsistent portions of the same cycle may well yield misleading results. Concomitant information would actually be required to resolve the question.

Rigid rotation of the solar corona out to some distance is of interest not only for its implications concerning coronal magnetic fields but also for its bearing on the evolution of the Sun and similar stars. As has been pointed out by Brandt (1966), the slinging off of coronal gas at the present rate at a rigid radius arm of 30 $R_\odot$ could slow down solar rotation with a characteristic time of only 7×10^9 years. The torque exerted by the outflow of gas from stars with more active chromospheres (Wilson 1966), and presumably higher stellar-wind fluxes, could have a profound influence on the evolutionary track followed by such stars.

The first calculations of the limit of rigid rotation (Lüst & Schlüter 1955) were made on the basis of a static corona containing the imbedded dipole field of the Sun. Distortion of the field by streams of corpuscles to form "magnetic bottles" was later suggested by Gold (1959) as the mechanism by which solar cosmic rays are conducted to the neighborhood of the Earth. Rigid rotation in both models extended out to 20–100 $R_\odot$. Recently, Pneuman (1966a) has examined the detailed interaction of the solar wind with the dipole field to determine the expected variation of solar-wind velocity with latitude and the configuration of the distended field. Using the conclusion that beyond a few solar radii the field becomes approximately radial, he has extended the analysis to include the rotation of the Sun and to estimate the variation of the angular velocity of the equatorial corona as a function of distance. His calculation shows that for a surface field strength of 1 G, rigid rotation extends out to ~ 3 $R_\odot$, above which the angular velocity rapidly approaches that given by conservation of angular momentum. For a surface field of 0.1 G, rigid rotation reaches to only 2 $R_\odot$.

Two basic measurements allow us to estimate the radius of rigid rotation of the corona: the tangential velocity of the solar wind and the curvature of coronal streamers. Although in principle spacecraft observations of the direction of arrival of the solar plasma could yield a value of the tangential velocity, the detailed observations obtained with Pioneer VI show a variation of aberration angle from 8° west ecliptic longitude for particles of 1600 eV per unit charge to about 4° east ecliptic longitude for particles of 600 eV per unit charge. Wolfe et al. (1966) interpret these results as the sum of the *radial* motion of the plasma and a thermal anisotropy of the ions, which have a larger temperature along the magnetic field. They imply that the tangential motion imparted to the ions at the Sun has been masked by the channeling by the magnetic field.

Biermann and his collaborators (Biermann 1951, 1957; Antrack, Biermann & Lüst 1964; Biermann & Lüst 1966) have long recognized that the orientation of ionic tails of comets provides a valuable tool for study of the solar wind throughout the inner parts of the solar system. Since such tails are believed to be swept back from the head of the comet by the solar plasma and its entangled magnetic fields, the alignment of the tail simply represents the vector difference of the solar-wind velocity and the known velocity of the comet. Although the study of many such comet tails leads to reasonable values for the radial component of the solar wind, this analysis could establish only an upper limit of 20 km/s for the tangential component. Recently, Brandt, Belton & Stevens (1966) have completed an exhaustive study of the orientation of ionic tails of retrograde and direct comets at approximately the same distance from the Sun. This comparison yields a value of $\sim$10 km/s for the tangential velocity of the solar wind at 1 a.u. and requires that rigid rotation must occur out to 32 $R_\odot$. Unfortunately, this estimate is based on data from only a few paired comets with similar but opposing orbital velocities. The errors in the inferred tangential velocity are thus expected to be large although they are not stated by the authors.

Using a unique series of observations of the intermediate corona, from the eclipse of 30 May 1965, from two flights of a balloon-borne coronagraph (Newkirk & Bohlin 1963, 1964, 1965), and from nearly daily coverage by K coronameter, Bohlin, Hansen & Newkirk (1966) were able to estimate the three-dimensional structure of a single helmet streamer from immediately above the surface out to $\sim$5 $R_\odot$. As shown in Figure 13, the projection of this streamer on the plane of the solar equator exhibits the garden-hose curvature and leaves little doubt that rigid rotation dose not extend out as far as 5 $R_\odot$. An attempt to make a self-consistent model of this streamer, including the run of solar-wind velocity with distance, has led these authors to conclude that rigid rotation in this structure extended to between 2.2 and 2.6 $R_\odot$ in remarkable agreement with the calculations of Pneuman.

Certainly, the existence of estimates of the radius of rigid rotation differing by an order of magnitude does not lead to confidence that the problem is thoroughly understood. However, the above data and those presented in our review of coronal magnetic fields suggest that rigid rotation beyond a few solar radii does not occur. Of course, co-rotation of coronal structures with appropriate curvature extends out into interplanetary space until the solar wind becomes overpowered by interstellar magnetic fields at some 50 to 100 a.u. (Axford, Dessler & Gottlieb 1963).

The solar wind.—Scholars of the history of science cannot but be intrigued by the story of the solar wind. Although the presence of streams of corpuscles from the Sun had been known for many years (Chapman & Bartels 1940), the discovery of the solar wind as a continuous outflow of coronal gas is properly attributed to Biermann (1951, 1957), who found that radiation pressure alone could not account for the behavior of many comet tails. Soon after Chapman (1957) pointed out that conduction of energy from the lower corona maintains the interplanetary plasma at 1 a.u. at a temperature of $\sim$200,000 °K, Parker (1958) demonstrated the inevitability of a steady-state solar wind. Parker's ideas were challenged by Chamberlain (1960, 1961), who

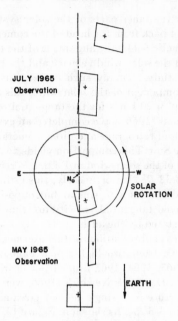

FIG. 13. Detection of a high-latitude helmet streamer over the pole during two meridian passages reveals the "garden-hose curvature" of the feature projected against the plane of the solar equator. Crosses mark the estimated location of the axis of the streamer while the boxes indicate the uncertainty. Such observations can yield data on the velocity structure of the inner solar wind and the height of rigid coronal rotation. (From Bohlin, Hansen & Newkirk 1966.)

claimed that subsonic solutions to the hydrodynamical equations could also satisfy the boundary condition of zero pressure at infinity and that a "solar breeze" was more representative of the corona.

The controversy was resolved in favor of the supersonic solar wind by the first direct space-probe measurements, which showed the velocity of the plasma to be ∼400 km/s. Recent studies (Noble & Scarf 1963; Scarf & Noble 1965; Whang, Liu & Chang 1966) have refined the theory of the solar wind while increasingly more sophisticated space-probes have revealed subtle (and often bewildering) details of its structure. Since excellent reviews (Lüst 1962, de Jager 1962, Parker 1965a) of the physics of the solar wind are available, this report will be largely restricted to a comparison of measurements of the solar wind with the various theoretical models rather than a discussion of the subtleties of the theory.

Unfortunately, determinations of the velocity of the inner solar wind are few and often ambiguous. One of the earliest attempts to describe the outflow of coronal gas from the Sun was that of van de Hulst (1950b), who examined the density distribution in polar plumes. Using dynamical equations including only the initial upward velocity of the material and gravity, he concluded that a significant flow did not exist. Measurements of the outward motion of distinct coronal features by Bugoslavskaya (1949) cannot be readily interpreted in terms of the velocity of the solar wind. One might expect to observe transient changes in the flow of gas in isolated tubes such as polar plumes. However, a recent series of photographs covering nearly two hours of elapsed time from the High Altitude Observatory expeditions at the 1963 eclipse failed to reveal any such events. Several attempts (e.g. Saito 1959) have been made to discover the variation with height of solar-wind velocity using the observed density distribution, cross-section area of coronal streamers, and equation of continuity. The velocity scale is fixed by comparison with the measured value at 1 a.u. This approach is invalid since the shape of the streamer cannot be identified with the flow lines of the material (see later discussion of interaction of magnetic field and solar wind).

Indirect inference of the velocity of the inner solar wind may be made from determinations of the temperature and temperature gradient in the corona and the solution of the appropriate hydrodynamic equations. Billings & Lilliequist (1963) employed measurements of the profile of the 5303 Å line in stable coronal features to estimate a temperature gradient of 3°K/km between 1.03 and 1.82 $R_\odot$. Using the base temperature of 2.6×10^6 °K at $1.03 R_\odot$, they derive the velocities shown in Figure 14.

In spite of their low contrast, polar plumes, considered as magnetically isolated tubes of gas, appear to present an ideal opportunity for study of the solar wind since the density gradient can be established unambiguously. Unfortunately, as was shown by Saito (1965a), an independent estimate of temperature gradient is required for numerical solution of the dynamical equations for the velocity. His upper limits (Figure 14) were obtained using a gradient of 3°K/km.

Although radar measurements of the solar corona at present allow the only direct determination of the velocity of the inner solar wind, these reflections are so complex that a simple interpretation of the results is impossible. Chisholm, James, and their collaborators (Chisholm & James 1964, James 1966) find an average returned radar signal with a mean frequency shift of ~4 kc/s (16 km/s) and a spread between half-power points of 25–40 kc/s. The time delay between transmitted and reflected signals leads to a calculated reflection level between 1.05 and 1.25 $R_\odot$ although a spread of nearly 0.6 $R_\odot$ appears always to be present. Often reflections are received with frequency shifts of 60 kc/s (240 km/s) at ranges corresponding to heights of several solar radii. The above observers conclude that the corona at a radar frequency of 38 mc/s must resemble a spiny Christmas tree ornament from which occasional specular reflections with various velocities are seen.

Unfortunately, a consistent model of the corona capable of explaining the complex array of observed Doppler displacements, ranges, and cross sections does not exist. Thus the interpretation to be placed on such a parameter as the mean displacement of 4 kc/s is uncertain. Brandt (1964) has noted that,

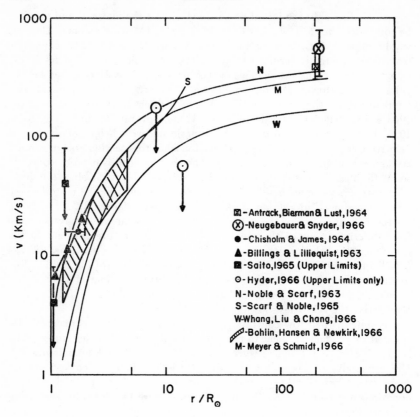

FIG. 14. Estimates of the velocity of the solar wind at various distances
compared with several theoretical models.

even if we take this velocity at its face value, lack of knowledge as to how
many streamers may be present and whether the streamers contain higher or
lower than average solar-wind velocities only allows the reflection level to be
placed between 1.7 and 2.0 $R_\odot$. Using a Baumbach-Allen model containing
random density irregularities, Chisholm & James suggest a mean reflection
level of 1.5 $R_\odot$. The mean outward velocity is represented in Figure 14 with
range lines in height to represent these uncertainties. Until it is known how
much of the spread in Doppler displacements is due to actual dispersion in
the velocity of the solar wind and how much is due to the presence of plasma
waves and anomalous reflection effects at density irregularities, corresponding range lines cannot be placed in the velocity coordinate.

Another indirect estimate of the distribution of velocity in the solar wind
close to the Sun has been made by Bohlin, Hansen & Newkirk (1966) from
the garden-hose curvature observed in a single high-latitude streamer (Figure

13). The requirement that the velocity structure be consistent with both the observed curvature and the fluxes detected at 1 a.u. allows velocities in the cross-hatched area of Figure 14. This estimate is subject to the criticism that an initially radial orientation for the base of the streamer must be assumed and that the outermost portion of the streamer admittedly is poorly located. Moreover, we are still uncertain of how representative the velocities in such a streamer are of the general corona.

As mentioned earlier, study of the interaction of the solar wind with ionic comet tails not only brought about the discovery of the solar wind, but still provides valuable information in the interplanetary medium. Extensive analysis of the aberration of type I tails has led Antrack, Biermann & Lüst (1964) and Brandt, Belton & Stevens (1966) to conclude that at 1 a.u. the mean velocity of the solar wind is ~375 km/s. Individual comets yield values ranging between 300 and 500 km/s with the higher velocities associated with periods of high geomagnetic activity in a manner similar to that observed by space-probes (Neugebauer & Snyder 1966). Moreover, the observed accelerations of material in the tail are frequently found to fluctuate considerably as a result of the patchy structure of the interplanetary plasma. Although comets often penetrate the inner reaches of the solar system, presently published analyses have not been able to describe the variation of the solar wind with distance. The Sun-grazing comet Ikeya-Seki (1965f) appeared to hold promise as a close-in solar probe until it failed to develop an ion tail. However, using measurements of the polarization in the sodium D lines, Hyder (1966a) has attempted to infer the orientation of the cometary magnetic field, which he presumes to be the interplanetary field swept up by the comet. The fact that the inferred field is apparently parallel to the velocity vector of the comet sets the upper limits for the solar-wind velocity at 8 and 14 $R_\odot$ at 175 and 56 km/s at high solar latitudes.

Direct measurements of the solar-wind velocity by space-probes are discussed in detail elsewhere in this volume (Ness 1967) and need not be repeated in detail. The mean velocity observed during the last few years (Coleman et al. 1962, Neugebauer & Snyder 1966) was 500 km/s with values between 300 and 800 km/s appearing frequently.

Theoretical descriptions of the stationary expansion of the corona into interplanetary space are all based on solutions of the equation of motion

$$\frac{1}{N}\frac{dN}{dR} + \frac{1}{T}\frac{dT}{dR} = -\frac{\mu}{kT}\left(g - v\frac{dv}{dR}\right) \qquad\qquad 2.$$

where

μ = mean molecular weight of the plasma

$g = g_0\left(\dfrac{R_\odot}{R}\right)^2$ = local gravitational acceleration

and the other symbols have their conventional meaning. Except for such special circumstances as the influence of the magnetic field on the solar wind

(Pneuman 1966a), authors have uniformly coupled Equation 2 with the equation of continuity for radial expansion. Parker's (1963) early solutions of this equation were made by replacing the temperature term by that for a polytropic gas, for which

$$T = T_0 \left(\frac{N}{N_0}\right)^{\alpha-1} \qquad 3.$$

in which T_0 and N_0 represent the conditions at some reference level low in the corona and α is the polytropic index. Although the use of the polytropic law is somewhat artificial, his solutions for a variety of indices contributed greatly to our insight into just how the solar wind responds to a variety of conditions in the lower corona. Noble & Scarf (1963) discarded the arbitrary assumption of a particular temperature variation by calculating the observed temperature distribution which would be produced by conduction of heat outward from a thin shell low in the corona. They later (Scarf & Nobel 1965) included the effects of kinematic viscosity in the equations, only to find that the added complexity produced instabilities in the numerical solution. A similar model was constructed by Whang, Liu & Chang (1966), who carried out the integration of the equations from far out in the solar system inward towards the Sun. Note that one basic criterion for all these models—their ability to reproduce the observed density distribution (Figure 1)—was particularly well met by the latter calculation. Unfortunately, Whang, Liu & Chang's computed mean velocity of 165 km/s at 1 a.u. fell far short of the observed value. Recently, Meyer & Schmidt (1966) have pointed out that although the model calculated by Whang and collaborators is basically correct, the presence of a radial magnetic field throughout the interplanetary gas so inhibits the transfer of ions from one radial line to another that the effects of viscosity are made negligible. This change scarcely influences the density distribution but brings the calculated velocities into closer agreement with those observed.

In summarizing present knowledge we note that although the broad outlines of the solar wind near the Sun may now be described, the data are scarcely accurate enough to select one of the theoretical models as preferable. Many other unsolved problems confront us. Although space-probes show the solar wind to exist over the entire Sun at low latitudes, we are ignorant of how the flow varies from streamer to interstreamer regions. The evolution of the flow from an active region as the latter develops and decays is unknown, as is the connection between the temperature and magnetic structure of such a region and the flux of the solar wind. Coronal streamers—perhaps the most characteristic structure of the corona—undoubtedly owe their form to the interaction of the solar wind and the underlying magnetic fields; however, we still lack a quantitative theory of their formation. The embarrassing riches provided by space-probe measurements are largely unexplained. The connection between the velocity, density, and magnetic features in the interplanetary medium and visible features in the corona excites the imagination with hypotheses that only the coming years of research will prove or disprove.

Temperature Structure

The quiet corona.—Although our major concern is with the structure of the corona, some discussion of its temperature is still required. Particular attention will be paid to the temperature gradient and its influence on the solar wind, as well as the characteristic temperatures of various morphological structures, rather than the details of different methods of estimating coronal temperatures. It is well to bear in mind that local thermodynamic equilibrium does not exist in the corona and that the entire concept of "temperature" is often illusive. Careful distinction between excitation, ionization, and kinetic temperature is required, although for lack of a complete theory we are forced to treat them as equivalent in our discussion. For a review of the problems attendant on determining the temperature of the corona, the reader is referred to the recent books of Billings (1966) and Zirin (1966) and the treatise of Shklovsky (1965).

The discrepancies which plagued the evaluation of temperature in the lower corona are familiar (Seaton 1962). Estimates from the observed density gradient with the assumption of hydrostatic equilibrium (Hepburn 1955, Pottasch 1960, Ney et al. 1961, Saito 1965a) yielded values of the kinetic temperature of electrons of $\sim 1.5 \times 10^6$; widths of the 5303 Å line led to a somewhat higher $T_i \sim 2 \times 10^6$ °K for the kinetic temperature of ions, while the equilibrium implied by the observed intensities of coronal emission lines demanded ionization-excitation temperatures in the 5×10^5 to 8×10^5 °K range. Some interpretations of the thermal radio emission of the corona (Allen 1963b) appeared to favor a lower electron temperature while detailed studies of the radio spectrum in combination with simultaneous optical measures of coronal density were consistent with kinetic temperatures of 1 to 2×10^6 °K (Christiansen et al. 1960, Newkirk 1961). Recent investigations have largely removed these discrepancies. Burgess (1964, 1965) has found that the mechanism of di-electronic recombination is vastly more important in the equilibrium of coronal ions than had previously been suspected, and that the new ionization cross sections lead to ionization temperatures of about 2×10^6 °K. Although the radio fluxes in the centimetric and decimetric range are roughly consistent with a similar value for the kinetic temperature of electrons (Kundu 1965), quiet-Sun fluxes at wavelengths of 1–10 m still appear lower than could be explained by reasonable density models. Recently, Oster & Sofia (1965) have pointed out that dispersion of radio waves near the plasma frequency has a profound effect on the opacity of the plasma and that brightness temperatures 0.3 to 0.6 as great as the electron temperatures must be expected at these longer wavelengths. The somewhat higher kinetic temperatures required by the widths of coronal lines have frequently been ascribed to the presence of macroscopic turbulence. Billings (1965) has pointed out that the turbulent radial velocities of 35 km/s apparent from Doppler radar observations of the Sun (Chisholm & James 1964) rather nicely account for the unusual width of the 5303 Å line seen at the limb. Thus, we must conclude that in general the lower corona can be regarded as maintaining a "temperature" of approximately 1.5×10^6 °K with, perhaps, an increase by 20 per cent during sunspot maximum (Waldmeier 1963). Electron temperatures in polar plumes (Saito 1965a) appear to be somewhat lower—1.2×10^6 °K. On the basis of the heating of the lower corona by shock waves, Kuperus (1965) finds roughly similar values.

Active regions.—Detailed temperature structure is evident in the corona above active regions (Table II). Linewidths in active-region enhancements suggest elevated kinetic temperatures of up to 2×10^6 °K (Billings 1966). The observed brightness temperatures associated with the slowly varying component of radio emission (Kundu 1965, Newkirk 1961, Kakinuma & Swarup 1962) yield similar values when account is taken of the increased opacity of the plasma in such regions due to the mechanism of magnetoionic resonance (Zhelezniakov 1962). Likewise the coronal "plages" visible at X-ray wavelengths suggest brightness temperatures of 1.8 to 2.3×10^6 °K (Acton 1966).

Even higher temperatures appear in permanent and sporadic condensations, which appear as small bright knots in both X-ray and decimetric radio observations and as the source of 5694 Å emission in the optical region. Indeed, such regions at brightness temperatures of from 3 to 5×10^6 °K account for a major fraction of the X-ray radiation of the Sun (Mandelshtam 1964, Acton 1966, Widing 1966). Although some authors have suggested temperatures higher than 10^7 °K to account for the bursts of microwave radiation received from sporadic condensations (Kawabata 1963), it appears likely that the concept of temperature is inappropriate for such phenomena. Suprathermal streams of particles occur sporadically in such regions with sufficient density (Jefferies & Orrall 1965) to explain the observed bursts of both microwave and X-ray radiation (Acton 1966).

Variation with height.—The temperature structure of the corona with height above the chromosphere naturally divides itself into two regimes. The chromosphere-corona interface in which the temperature rises abruptly has been discussed by Pottasch (1964) and by Athay (1966) and is more appropriately examined in the review of the solar ultraviolet spectrum in this volume (Goldberg 1967). As we proceed further out into the corona, the decrease in temperature will be determined by the supply of energy from below, thermal conduction, and the loss of energy produced by the escape of material in the solar wind. We shall first summarize the observational data for the temperature of the outer corona and later examine some of the implications of this structure.

Away from active regions, temperatures in the intermediate corona may be estimated from the general gradient of coronal electron density and the width of coronal lines. Although the former technique requires the assumption of hydrostatic equilibrium, which the presence of a solar wind demonstrates to be untrue, a simple evaluation of the magnitudes of the terms in Equation 2 shows that the departures from the hydrostatic equation caused by the existence of a solar wind are insignificant in the region of subsonic expansion, i.e., below $\sim 10\ R_\odot$. The electron temperature distribution inferred

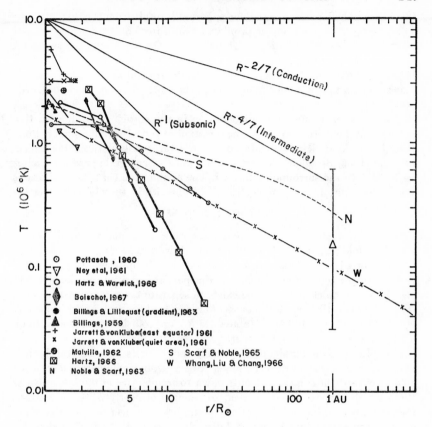

FIG. 15. The temperature structure of the solar corona. The temperature gradient in the quiet corona appears consistent with conduction from the base while that of active regions (heavy lines) appears to be so steep that only a subsonic solar wind can be maintained. Lines showing the $R^{-2/7}$, $R^{-4/7}$, and R^{-1} decay of temperature are included for comparison.

from the hydrostatic equation by Ney et al. (1961) from their observation of the 1959 eclipse and by Pottasch (1960) from a compilation of several density models thus are based on the assumption that both the influence of the solar wind and any transfer of momentum into the corona by shock waves (Kuperus 1965) are negligible (Figure 15)

Measurement of the Doppler widths of coronal emission lines allows the ion kinetic temperature and temperature gradient to be established free from any assumption of hydrostatic equilibrium although, as has already been mentioned, other uncertainties intrude. Interferometric observation of the widths of the 6370 Å and 5303 Å lines during eclipse led Jarrett & von Klüber (1961) to kinetic temperatures which even in the quiet corona exceed 3×10^6

°K. These unbelievably high values apparently result from the difficulties of determining the true Doppler profile with an interferometer whose instrumental half-width (0.6 Å) is a large fraction of the line itself. Conventional spectroscopic measurements outside of eclipse by Billings (1959), Billings & Lehman (1962), and Billings & Lilliequist (1963) yield kinetic temperatures that are consistently lower but still above those inferred from the hydrostatic equation.

The electron temperature distribution in the corona high above active regions may be discovered from the duration of type III radio bursts (Boischot, Lee & Warwick 1960, Malville 1962, Hartz & Warwick 1966, Hartz 1964 and 1966, Boischot 1967) in which a pulse of energetic particles excites oscillations in the surrounding plasma. The assumption that the oscillations are damped out by proton-electron collisions leads to a simple relation

$$T_e = 0.65 \times 10^{-4} \nu^{4/3} \tau^{2/3} \qquad\qquad 4.$$

between the electron temperature T_e, the radio (and plasma) frequency ν, and the e-folding duration τ of the burst. However, as Hughes & Harkness (1963) pointed out, the duration of a burst is actually determined both by the velocity dispersion in the spray of exciting particles and by collisional damping. Although no clear technique exists to separate these two contributions, these authors estimate, on the basis of an assumed coronal model, that only at frequencies below 100 Mc/s (above 1.5 $R_\odot$) are collisions sufficiently dominant to allow a meaningful temperature estimate.

In spite of these ambiguities type III radio bursts have several distinct advantages as a coronal thermometer. Space-probe radiometers have been able to extend these measurements down to frequencies of 0.6 Mc/s (Hartz 1966) which are taken as equivalent to a height of nearly 20 $R_\odot$. Moreover, the estimates are independent of any assumptions concerning the dynamics of the corona and may be compared directly with the theoretical models. Inspection of Figure 15 shows immediately that although the electron temperatures derived by this technique at 2–3 $R_\odot$ are roughly concordant with other measures, the indicated temperature gradients are steeper in the entire range 2.5 to 20 $R_\odot$ than either the theoretical models or the results determined from the hydrostatic equation. Remembering that the analysis of burst durations really results in a relation between temperature and density, we examine the validity of the density models chosen to establish the height scale.

First, we note that the combined observation of both frequency drift and burst position necessary for an *independent* determination of the electron density extends down to only 10 Mc/s ($\sim$4 $R_\odot$). The electron temperature estimates were made under the assumption that a particular model (e.g. 10 $\times$ the Baumbach-Allen densities or 5 $\times$ the background) extends out to $\sim$20 $R_\odot$ above active regions. This assumption appears unjustified; in fact, there is evidence (Michard 1954) that the enhancement of a streamer may increase at large distances to values of 20–30 times the background. We note that the perhaps more realistic assumption of a lower density gradient in the streamer would lead to a less drastic decrease of the electron temperature above $4R_\odot$. A more consistent picture of the corona high above an active region would presumably result from a combination of the observed T vs. n_e and the dynamical equations for the solar wind.

Space-probes allow direct measurement of the energy dispersion of the interplanetary gas and show an average proton temperature of 1.5×10^5 °K but with values between 3×10^4 °K and 6×10^5 °K appearing frequently. The higher temperatures generally occur with the highest bulk velocities of the plasma (Neugebauer & Snyder 1966). One interpretation of the direction of arrival of particles at 1 a.u. is that the proton temperature transverse to the magnetic field is smaller by a factor of 5 than the longitudinal temperature (Wolfe et al. 1966) although the use of temperature to describe such an aniso-tropic flow is questionable. Clearly, the fact that the range of temperatures experienced in interplanetary space far exceeds that found in the lower corona indicates that the conduction model is not completely adequate.

Implications for the solar wind —In comparing the above estimates with solar-wind models, we should remind ourselves of several conclusions regarding the relation between the temperature structure of the corona and the be-havior of the wind. In the idealized isothermal solar wind the interplanetary flux, velocity, and density are all monotonically increasing (and rather sensi-tive) functions of the temperature T_0 at some reference level in the corona. Changes in N_0, the density at the reference level, have no other effect than to modulate the density at 1 a.u. proportionately. Parker (1965b) has shown that the role of the base density in the purely conductive corona is more criti-cal since increasing N_0 furnishes more and more material which must be ex-panded and lifted out of the solar gravitational field by the limited amount of energy supplied by conduction from below. He identifies two temperature-dependent critical densities, N_A and N_B, to separate the solutions of the hydrodynamic equations into three regimes in which the flow and the tem-perature distribution differ vastly (Table IIIa). For $N_0 < N_A$, conduction overpowers the convective loss of energy and $T \propto R^{-2/7}$ as in the purely con-ductive corona (Chapman 1957). At an intermediate value of N_0 the tem-perature drops sharply as $R^{-4/7}$ out to some radius R_2 beyond which it follows the conductive law. For $N_0 > N_B$, convective losses choke off the flow of heat into the outer corona and the solar wind becomes subsonic with a fall of temperature as $\sim R^{-1}$ beyond the critical radius R_C.

In Table IIIb the values of N_A and N_B (Parker 1965b) for conditions taken to represent the quiet and active-region corona at a reference level $R = 2.4\ R_\odot$ show immediately that although the quiet corona should be con-sistent with the purely conductive model and an approximately $R^{-2/7}$ tem-perature law, we might expect active regions to be described by the intermedi-ate or even the completely subsonic cases. Comparison of the observed tem-perature structure with the theoretical models supports these conclusions. As well as can be determined, temperatures in the quiet corona follow that pre-dicted by the conductive models of Noble & Scarf (1963), Scarf & Noble

TABLE III

A. CHARACTER OF THE TEMPERATURE DISTRIBUTION AND SOLAR WIND

Condition	Temperature structure	Character of solar wind at 1 a.u.
$N_0 < N_A$	$T \propto R^{-2/7}$	supersonic
$N_A < N_0 < N_B$	$T \propto R^{-4/7}, R \leq R_2$ $T \propto R^{-2/7}, R > R_2$	supersonic
$N_0 > N_B$	$T \propto R^{-4/7}, R \leq R_c$ $T \propto \sim R^{-1}, R > R_c$	subsonic

B. REPRESENTATIVE DENSITIES AT 2.4 $R_\odot$

Quiet corona	Active region
$T_0 = 1.5 \times 10^6$ °K $N_{obs} = 1.3 \times 10^6$ cm^{-3} $N_A = 10^7$ cm^{-3} $N_B = 1.2 \times 10^8$ cm^{-3}	$T_0 = 2 \times 10^6$ °K $N_{obs} = 10^7$ cm^{-3} $N_A < 10^7$ cm^{-3} $N_B = 2 \times 10^7$

(1965), or Whang, Liu & Chang (1966) quite nicely. Moreover, if we accept the approximately R^{-1} decline found from the analysis of type III burst durations, we conclude that the density is, indeed, high enough to have choked off the heat flow and that only subsonic expansion occurs in the corona above active regions. The temperature in such a "cold streamer" might be expected to decrease as $\sim R^{-1}$ to about 10^4 °K and then become constant out to the limit of the Sun's H II or Strömgren sphere at several astronomical units. At approximately the location where the temperature has dropped significantly below the conduction values (i.e., at 5–10 $R_\odot$), the density gradient would be expected to become significantly less steep than in the background corona. Although present theories are adequate for the description of isolated subsonic and supersonic winds, little attention has been given to the problem of the interaction of adjacent subsonic and supersonic sectors. The fact that space-probes always detect a substantial supersonic wind suggests that the high-speed sectors must fold in upon whatever cool sectors may exist. Since present space-probes, because of their accumulation of a net charge of a few volts, may not be capable of detecting slow particles, the existence of cool, slow streams of interplanetary plasma remains hypothetical.

Finally, we note that the assembled temperature data give scant support to the proposal that conduction is insufficient to explain the temperature structure of the inner corona and that dissipation of the energy of shock waves must extend out ot two or so solar radii (Parker 1965a). Except for the

Pottasch (1960) model there is no evidence for the broad temperature plateau which would require such a mechanism.

MAGNETIC FIELDS

Perhaps, the most central agent in determining the morphological structure of the corona is the magnetic field. The field is clearly responsible for the shape of polar plumes as well as the appearance of loops and arches in the corona above magnetically active regions on the disk. Moreover, the presence of enhanced magnetic fields in particular regions at the surface can be expected (Kulsrud 1955, de Jager 1962, Kuperus 1965) to increase the rate of transfer of mass and energy from the chromosphere into the corona. Magnetic fields must also play a vital role in moulding coronal streamers into the shapes observed and determining the distribution of mass in coronal condensations. It is rather ironic that we know so little about coronal magnetic fields. Efforts to analyze these fields have yielded information on either the configuration or the gross magnitude but rarely on both simultaneously. The detailed connection between coronal and photospheric fields has yet to be established.

Geometrical configuration.—The simplest way to examine the configuration of coronal magnetic fields is to assume that an identifiable feature such as an arch represents a magnetic tube which contains an abnormal amount of material. Thus, the arrangement of polar plumes projected against the plane of the sky may be fitted to the lines of force of an immersed dipole and the length of the dipole derived (e.g. Saito 1965a). Coronal arches visible in photographs made in the 5303 Å line appear to connect regions of opposite magnetic polarity in the photosphere with the plane of the arch perpendicular to whatever filaments may be present (Bumba, Howard & Kleczek 1965), as would be required by the theory of prominence support proposed by Kippenhahn & Schlüter (1957). A detailed polarimetric analysis of an extensive coronal condensation allowed Saito & Billings (1964) to determine the three-dimensional distribution of material and to estimate the direction of the field at various positions in the region (Figure 16). Here again the field approximately parallels the density contours, which indicates that the field is capable of containing the material.

Magnitude.—The fact that occasionally the field is of such a configuration as to support the relatively cool material of a prominence against gravity permits the conventional technique of Zeeman splitting in Hα to be used to estimate the projected magnitude and orientation of magnetic fields in coronal space (Zirin & Severny 1961). A detailed investigation of prominence fields led Rust (1966) to conclude that, indeed, prominences occur only where the field is horizontal and that the lines of force run approximately perpendicular to the long axis of a filament. The magnitudes of the fields ranged from ∼50 G for active-region prominences, through ∼15 G for prominences associated with young bipolar groups, to ∼5 G in the polar crown of filaments. Although the polarities of the fields found in prominences agreed with that

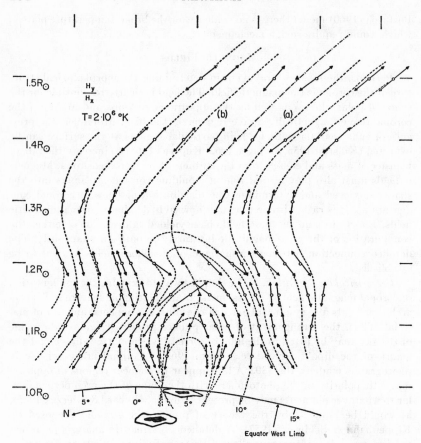

FIG. 16. The projected orientations of magnetic fields in a coronal condensation inferred from the departures of the observed density structure from hydrostatic equilibrium. (From Saito & Billings 1964.)

calculated by potential theory from the underlying photospheric fields, Rust discovered that the observed magnitudes were far in excess of those calculated. Whether this discrepancy is due to the neglect of weak photospheric fields outside the region considered to be the source, or to the increase of the coronal field strength by twisting is unknown.

Measurement of the polarization of emission lines is also capable of yielding estimates of the magnetic field since the plane and magnitude of polarization are governed by competition between resonance scattering of photospheric light and the perturbation of the atomic energy levels by the field (Warwick & Hyder 1965, Hyder 1965b). Field strengths determined from polarimetry of prominence lines (Brückner 1963, Hyder 1964, 1965c) appear in agreement with the corresponding Zeeman estimates. The magnetic perturbation for forbidden lines such as 5303 Å is so great that studies of their polarization (Charvin 1964, Hyder 1965c, Eddy & Malville 1966) have been unable to yield data on coronal magnetic fields.

A unique estimate of the magnetic field in a filament and the surrounding corona has been made by Hyder (1966b) from the decay of the oscillatory motion of a winking filament. The frequency of the oscillation required a field in the range 2–30 G in the prominences while the viscosity needed to produce the observed damping suggested fields greater than 0.1–0.2 G in the surrounding corona. It is of interest that all these estimates give fields in prominences much smaller than those required by Warwick (1957) to account for the observed curvature in the trajectories of flare-associated spray prominences. Apparently the sprays represent events in which the ejected material stretches abnormally intense fields high into coronal space. Under the bold assumption that all these estimates of prominence magnetic fields are roughly appropriate to the corona, we place them in Figure 17, in which the field magnitude is plotted against height above the surface.

The propagation of radio waves through the coronal plasma is profoundly influenced by the presence of a magnetic field. Not only is the opacity increased by the mechanism of magnetoionic resonance (Zhelezniakov 1962, Kundu 1965) but also the refractive index and opacity become dependent upon the polarization and direction of propagation. Since the height in the corona corresponding to optical depth unity is then different for extraordinary and ordinary modes of propagation, the resultant thermal radiation from coronal condensations and active-region enhancements is polarized. Following earlier investigators (Gelfreich et al. 1958, Christiansen et al. 1960, Korol'kov & Soboleva 1962), Kakinuma & Swarup (1962) have used the observed variation of polarization of the slowly varying component with frequency and position of the region on the disk to derive a self-consistent model of the magnetic field above active regions. Although the method requires the assumption of a rather idealized configuration of the field in order to be mathematically tractable, it appears to be the least uncertain of all the techniques for estimating coronal magnetism.

The characteristics of radio bursts also permit the estimation of coronal magnetic fields, although the relatively straightforward observations are often subject to a variety of interpretations. For example, the sense of circular polarization of microwave bursts is found to reverse at a frequency of $\sim 4 \times 10^3$ Mc/s and the profile of emitted flux about the reversal frequency is remarkably similar from burst to burst. Takakura (1960) attributes the microwave bursts to synchrotron radiation and the reversal of polarization to resonance absorption in the extraordinary mode at the gyrofrequency in the source region. Conversely, Cohen (1961) takes the site of the anomalous absorption and the polarization reversal to be at a location where the field is approximately perpendicular to the escaping radiation at an altitude of some 10^5 km. Thus, the basic datum appears to be explained by either a field of ~ 600 G at 3×10^4 km or one of ~ 4 G at 10^5 km.

The interpretation of other radio-burst phenomena is subject to similar

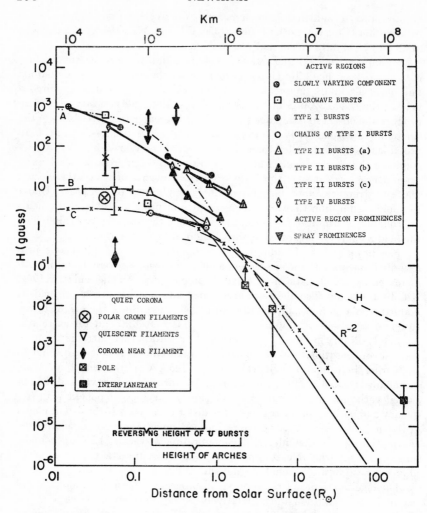

FIG. 17. The strength of coronal magnetic fields as a function of height above the photosphere as estimated by various techniques. Values referring to active regions appear as heavy lines. Simple models for the potential field above an active region (A), an extended dipole region (B), and the general solar field at the equator (C) appear for comparison. The curve H represents the field required for pressure equilibrium with the solar wind.

uncertainties. Takakura (1966) regards type I bursts as the result of coherent plasma oscillations excited by electrons traveling at a few times the mean thermal speed. The upper limit of field required to accelerate these electrons by the Fermi mechanism and the lower limit needed to explain their polarization establish the field strength at a few tens of gauss in the frequency range 50–200 Mc/s (height range 10^5 to 2×10^6 km). Examination of chains of type I bursts, on the other hand, led Wild & Tlamicha (1965) to conclude that the field strengths are at least an order of magnitude lower. They make no assumption concerning the details of the excitation of the bursts but interpret the frequency drift observed in a chain as an upward-propagating Alfvén wave.

Type II bursts have given rise to no less than three rather distinct estimates of coronal magnetic fields. From their time-frequency behavior and the assumption that the burst exciter is a hydromagnetic shock wave (Takakura 1966, Fomichev & Chertok 1966), it is possible to set limits for the field at a given frequency or height [type II bursts (a), Figure 17]. Field strengths may also be inferred from the splitting of these bursts although there appears to be no unanimity as to just how the data are to be interpreted. Considered as a magnetoionic phenomenon similar to the Zeeman effect, the observed splittings (Maxwell & Thompson 1962, Takakura 1966) lead to values somewhat in excess of the limits imposed by the shock-wave model [type II bursts (b), Figure 17]. However, as pointed out by Weiss (1965), there is even some question whether the splitting should be considered a true magnetoionic effect with a splitting half the gyrofrequency, or a simple beat phenomenon, with a splitting equal to the gyrofrequency. Other models of the splitting formulated by Sturrock (1961), in which two exciter waves propagate at different angle to the magnetic field, and by Tidman, Birmingham & Stainer (1966), in which the exciter is a cluster of electrons with an anisotropic velocity distribution, require still higher values of the field [type II bursts (c), Figure 17]. Even the basic premise that the splitting is due to the magnetic field has been questioned by Roberts (1959), who points out that the plausible formation of a double front in the exciting shock wave could also explain the observations.

The analysis of type III bursts, believed to be excited by a pulse of relativistic particles, has yielded still less information on coronal magnetic fields. The fact that some bursts are polarized while others appear unpolarized allows the field strength to be estimated only between limits of nearly two orders of magnitude.

The frequency distribution of the radiation of type IV bursts can give an idea of the field since the maximum flux is expected at twice the gyrofrequency. However, resonance absorption at the gyrofrequency and its harmonics will shift the maximum by an uncertain amount; and the fields recorded in Figure 17 for type IV bursts were derived by Takakura (1966) under the assumption that the maximum occurs at 4 times the gyrofrequency. It is quite clear that the majority of estimates of coronal magnetic fields in the height range 10^5 to 10^6 km from radio bursts all suffer from a common difficulty—the unknown validity of the assumed excitation mechanism. An order-of-magnitude uncertainty in the field strength results.

Beyond one solar radius from the surface only limits are available for coronal magnetic fields. Högbom's (1960) earlier mentioned study of coronal irregularities suggested that polar plumes are maintained out to at least 3 $R_\odot$. Using a standard model for polar densities, he thus concluded that a field of at least 3×10^{-2} G was required to contain the plumes against lateral expansion at that height An upper limit to the field of 8×10^{-3} G at 5–6 $R_\odot$ was found by Golnev, Parijstky & Soboleva (1964) from the absence of any detectable Faraday rotation in the radiation from the Crab Nebula as it was occulted by the corona

The measurement of interplanetary magnetic fields has been discussed in an earlier section and reviewed in detail elsewhere in this volume (Ness 1967). For comparison with the other estimates in Figures 17 we note that the magnitude of the field varies between 2×10^{-5} G and 10^{-4} G with an average of about 5×10^{-5} G (Greenstadt 1966). The direction of the field is generally along the Archimedes spiral (Ness, Scearce & Seek 1963) although wide variations in direction show that the field is often kinked and twisted (McCracken & Ness 1966). The close connection of the direction of the interplanetary field and that present on the Sun demonstrated by Wilcox & Ness (1965) suggests that the solar fields are indeed swept into space by the solar wind.

Interaction of the magnetic field and the solar wind.—In the attempt to integrate the various measures of coronal magnetic fields, we have added several simple calculations to Figure 17. Curves A, B, and C represent the decrease in field strength above the equator of simple dipoles taken to represent an active region, an extended magnetic-dipole region such as visualized by Leighton (1964), and the general magnetic field of the Sun, respectively. The parameters of these oversimplified models appear in Table IV. Two other curves show the R^{-2} dependence fitted to the interplanetary value and H, the field necessary to give a magnetic pressure just equal to the dynamic pressure of the solar wind taken from the isothermal model with $T_0 = 2 \times 10^6$ °K. Although no particular merit is claimed for these calculations, they suggest the following conclusions. First, the potential dipole model appears capable of yielding at least a crude representation of the magnetic field above active regions and the more quiet portions of the Sun. Second, comparison of these fields with the curve H implies that below a

TABLE IV

PARAMETERS OF THE POTENTIAL FIELD MODELS

Model	Flux at surface (maxwells)	Separation between poles
A (active region)	3×10^{22}	2×10^4 km
B (extended dipole)	3×10^{21}	3×10^5 km
C (general)	3×10^{22}	5×10^5 km (0.7 $R_\odot$)

height of $\sim 3\ R_\odot$ the coronal material is controlled by the magnetic field. It is equally clear that above this height the coronal magnetic fields are being strung out into interplanetary space by the solar wind. This conclusion is supported by the observation that the highest closed arches visible in the eclipse data assembled by Bugoslavskaya from 1887 to 1945 have a mean height of only 0.6 $R_\odot$ above the limb and that the maximum height of U bursts is some 4×10^5 km (0.57 $R_\odot$) (Takakura 1966). The fact that the R^{-2} curve intersects the others in the same region implies that the measured interplanetary fields are simply those which have been unraveled from the corona at this level. Finally we note that the dominance of the solar wind above $\sim 3\ R_\odot$ precludes rigid rotation of the corona with the Sun further out into the solar system.

The detailed theoretical investigation of the interaction of the solar wind and coronal magnetic fields is one of great complexity. A complete solution of the problem would have as its boundary conditions the distribution of magnetic fields as well as the variation of flux into the corona from place to place on the surface. The flow of the material will, of course, distort the field and conversely the presence of the field will modify the flow. The resultant distribution of material should then be that of the corona we see with its streamers and other features. Parker (1958) has described the basic pattern of the dipole field extended by the solar wind while both he (Parker 1963) and Pneuman (1966a, b) have attempted to estimate the detailed density distribution and field configuration near the Sun. Parker reasoned that the escape of the solar wind must be primarily at the equator and formally calculated the configuration of the field with a conducting sheet interposed at the equator. Material flowing out along the lines of force produces a distribution of density reminiscent of an equatorial coronal streamer (Figure 18). Pneuman divided the flow into two regimes in which the stream lines follow the dipole field close to the Sun and are radial far out, but was unable to calculate their form in the crucial intermediate region (Figure 19).

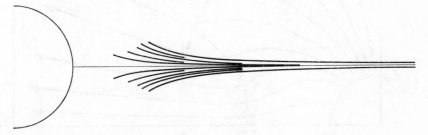

FIG. 18. Contours of equal density in the corona are distorted into a "streamer" by the presence of a dipole field in the solar wind. This caluated structure, although made for a flow model slightly different from that shown in Figure 19, demonstrates that the apparent shape of a streamer cannot be taken as representing the flow lines of the material. (From Parker 1963.)

One consequence of the magnetic field is to reduce the velocity of the solar wind at the equator compared to that at high latitudes. Comparison of Figures 18 and 19 shows rather dramatically that the apparent shape of a coronal streamer can no-wise be equated with the flow lines of the material

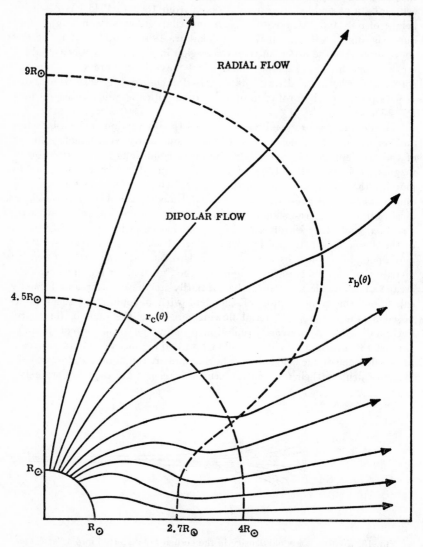

FIG. 19. Stream lines in the solar wind are distorted toward the equator by the presence of the general, dipole magnetic field during sunspot minimum. The velocity of the wind is also modified so that the equatorial speeds are slower than those at the pole. (From Pneuman 1966a.)

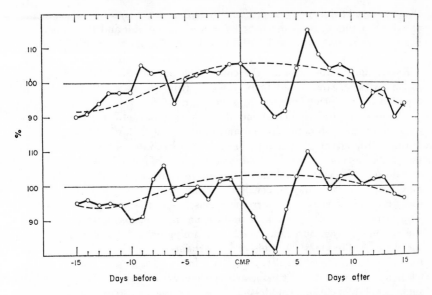

Fɪɢ. 20. The superposed variation of the geomagnetic character figure K_p with respect to the central meridian passage of plages. Only recurrent or M-region magnetic storms are considered. The top graph contains data for the brightest plages occurring at all phases of the sunspot cycle while the lower one considers only the declining portion of the cycle. (From C. W. Allen 1964.)

except far from the Sun where radial, supersonic expansion obtains. Although calculations of the interaction of the solar wind with the more complex fields in active regions or in the neighborhood of filaments have not been attempted, we can be confident that a similar distortion of the flow and a "coronal streamer" will result.

The Corona and Geomagnetic Activity

The basic question "What is the connection between visible forms in the corona and geomagnetic activity?" still has no uniformly accepted and satisfactory answer. Since geomagnetic storms are one of the most spectacular features of geomagnetism, many of the early investigations (see the reviews of Chapman & Bartels 1940, Mustel' 1964, Allen 1964) attempted to relate the occurrence of storms to particular solar surface features. These statistical studies were handicapped in several ways. First, as a detector of the state of the interplanetary medium, the geomagnetic field at the surface is unsatisfactory since it is uncertain whether the "storm" is caused by an increase in the interplanetary density, velocity, flux, energy density, magnetic field, or in the time fluctuation of these parameters. [Recent space-probe measurements suggest that the K_P index is controlled by the velocity of the interplanetary stream (Neugebauer & Snyder 1966) although it appears that the intensity of the magnetic field is also an important factor (Wilcox, Schatten & Ness 1966)]. Second, both the source on the Sun and the velocity of particles were unknown and had to be inferred from the statistical data. Third, synoptic observations of the intermediate and outer corona were, and still are, unavailable so that the connection of, for example, geomagnetic storms with streamers had to be inferred from the correlation of both of these with an intermediate feature such as prominences.

These statistical studies frequently divide geomagnetic disturbances into large storms, which occur around and immediately following sunspot maximum, and moderate and small storms, which continue to be found through the declining portion of the solar cycle into minimum. The large storms seldom recur for more than one or two solar rotations while the small storms may reappear at 27-day intervals for many months or even years. It was early recognized that nonrecurrent storms follow the passage of active regions across the disk by $\sim$3 days, and the natural conclusion was drawn that the region was the source of particles streaming outward from the Sun at a speed of $\sim$600 km/s. [Impulsive disturbances in the geomagnetic field following solar flares are attributed to blast waves (Parker 1963) in the corona and are not directly associated with any permanent solar feature.] No such simple conclusion could be reached for the recurrent storms, which show a minimum in K_P $\sim$3 days after the central meridian passage (CMP) of an active region and a maximum $\sim$6 days after CMP (Figure 20). The lack of any obvious candidate for the source of these storms has led to the hypothesis of the "M region" which has been variously, and not very successfully, identified with plages, the space between plages, prominences, unipolar magnetic regions, and streamers (Allen 1964).

Two schools of thought have developed around the controversy of whether the maximum or the minimum is the principal event. The "cone of avoidance" school regards the minimum at CMP+3 days as evidence for a vacancy (Pecker & Roberts 1955) above the active region caused by the magnetic steering of particles toward the boundary of the region. A later modification of this picture (Billings & Roberts 1964) suggests that the material immediately over the region is contained by the magnetic field while that at the border is allowed to escape as the solar wind. The presence of the minimum at CMP+3 days with regions bright in the 5303 Å line (Bell & Glaser 1957) rather than active regions appears to support this idea. Outside the "cone of avoidance" the solar wind is considered to have a speed of $\sim$600 km/s. The "R-ray" school, championed for many years by Mustel' (1961, 1962a, 1962b, 1964), considers the minimum of completely secondary importance and the maximum at CMP+6 days as evidence for the arrival of particles at a speed of $\sim$300 km/s from the center of the active region. Support for this theory is claimed from the fact that active regions appear to have high coronal densities extending far out into the corona.

Unfortunately, the advent of space-probes from which the speed of the interplanetary particles may be measured has done little to resolve the dilemma. Using the plasma velocities found by Mariner II, Snyder & Neugebauer (1966) attempted to locate the sources of several well-defined magnetic storms and could find no consistent connection of the storms with either active-region plages or, as the "cone of avoidance" hypothesis would dictate, the interplage areas. The satellite observations also show (Wilcox & Ness 1965) that the structure of the solar wind within a given sector is extremely complex. Particles with the highest speeds at 1 a.u. do not, as would be suggested by Mustel''s (1964) models, arrive at the beginning of the sector but several days later.

Returning to the corona, we note that streamers are known to occur over active regions, over prominences, and occasionally over areas which have no distinctive appearance whatsoever (Bohlin, Hansen & Newkirk 1966). Both from our earlier discussion of the temperature profile over active regions and from Pneuman's (1966a) study of the solar wind above the equator of a dipole, we should expect that the solar wind from active regions shoud be reduced or completely choked off to create a "cone of avoidance" in the *flux*. The lower field intensities and density in streamers over prominence would suggest that such features could be a likely source of geomagnetic storm particles. However, Dizer (1963) finds no correlation of geomagnetic activity over the period 1920–1954 with long-lived filaments. Mustel' (1964) interprets this as evidence that such streamers do not penetrate more than $\sim$30 $R_\odot$ into the solar system although an equally satisfactory explanation is that their generally high latitude causes them to miss the Earth. Certainly, there is much reason to believe that there are large variations in the speed of the solar wind from place to place on the Sun. How these variations compare with visible structures in the corona and how streams of different velocity interact in interplanetary space to produce the wind observed at the orbit of Earth is presently unknown.

LITERATURE CITED

Acton, L. W. 1966, *Ap. J.* (Submitted August 1966)

Allen, C. W. 1963a, *Astrophysical Quantities* (2nd ed., Athlone Press, London)

Allen, C. W. 1963b, in *The Solar Corona*, *IAU Symp. No.* 16 (Academic Press, New York)

Allen, C. W. 1964, *Planetary Space Sci.*, 12, 487–94

Antrack, D., Biermann, L., Lüst, R. 1964, *Ann. Rev. Astron. Ap.*, 2, 327

Athay, R. G. 1966, *Ap. J.*, 146, 223

Axford, W. I., Dessler, A. J. Gottlieb, B. 1963, *Ap. J.*, 137, 1268

d'Azambuja, M., d'Azambuja, L. 1948, *Ann. Obs. Paris*, 6

Babcock, H. W. 1961, *Ap. J.*, 133, 572

Bartley, W. C., Bukata, R. P., McCracken, K. G., Rao, U. R. 1966, *J. Geophys. Res.*, 71, 3297

Bell, B., Glaser, H. 1957, *Smithsonian Contrib. Ap.*, 2, 51

Biermann, L. 1951, *Z. Ap.*, 29, 274

Biermann, L. 1957, *Observatory*, 107, 109

Biermann, L., Lüst, R. 1966, in *The Solar Wind*, 3–23 (Jet Propulsion Lab. and California Inst. of Technol., Pasadena)

Billings, D. E. 1959a, *Ap. J.*, 130, 215–20

Billings, D. E. 1959b, *Solar Res. Memo No. 137* (28 December)

Billings, D. E. 1965, *Ap. J.*, 141, 325

Billings, D. E. 1966, *A Guide to the Solar Corona* (1st ed., Academic Press, New York)

Billings, D. E., Lehman, R. C. 1962, *Ap. J.*, 136, 258–65

Billings, D. E., Lilliequist, C. G. 1963, *Ap. J.*, 137, 16

Billings, D. E., Roberts, W. O. 1964, *Astron. Norveg.*, 9, 147–50

Blackwell, D. E., Ingham, M. F. 1961, *Monthly Notices Roy. Astron. Soc.*, 122, 129

Bohlin, J. D., Hansen, R. T., Newkirk, G. A. 1966 (Paper presented at October meeting A.A.S.)

Boischot, A. 1967, *Ann. Ap.*, 30, 85

Boischot, A., Lee, R. H., Warwick, J. W. 1960, *Ap. J.*, 131, 61–67

Brandt, J. C. 1964, *Science*, 146, 1671

Brandt, J. C. 1966, *Ap. J.*, 144, 1221

Brandt, J. C., Belton, M. J. S., Stevens, M. W. 1966, *Astron. J.*, 71, 157

Brückner, G. 1963, *Z. Ap.*, 58, 73

Bugoslavskaya, E. Y. 1949, *Publ. Sternberg Inst.*, 19

Bumba, V., Howard, R., Kleczek, J. 1965, *Publ. Astron. Soc. Pacific*, 77, 55

Burgess, A. 1964, *Ap. J.*, 139, 776

Burgess, A. 1965, *IAU Symp. No. 23*, 95–99

Chamberlain, J. W. 1960, *Ap. J.*, 131, 47

Chamberlain, J. W. 1961, *ibid.*, 133, 675

Chapman, S. 1957, *Smithson. Contrib. Ap.*, 2, 1

Chapman, S., Bartels, J. 1940, *Geomagnetism* (Oxford Univ. Press, New York)

Charvin, P. 1964, *Compt. Rend.*, 259, 733

Chisholm, J. H., James, J. C. 1964, *Ap. J.*, 140, 377–79

Christiansen, W. N., Mathewson, D. S., Pawsy, J. L., Smerd, S. F., Boischot, A., Denisse, J. F., Simon, P., Kakinuma, T., Dodson-Prince, H., Firor, J. 1960, *Ann. Ap.*, 23, 75

Cohen, M. H. 1961, *Ap. J.*, 133, 978

Cohen, M. H., Gundermann, E. J., Hardebeck, H. E., Harris, D. E., Salpeter, E. E., Sharp, L. E. 1966, *Science*, 153, 745

Coleman, P. J., Davis, L., Smith, E. J., Sonett, C. P. 1962, *Science*, 138, 1100

Cooper, R. H., Billings, D. E. 1962, *Z. Ap.*, 55, 24–28

Deutsch, A. J., Righini, G. 1964, *Ap. J.*, 140, 313

Dizer, M. 1963, *Observatory*, 82, 250

Dollfus, A. 1957, *Compt. Rend.*, 244, 1880–83

Dollfus, A. 1958, *ibid.*, 246, 2345

Eddy, J. A., Malville, J. M. 1966 (Paper presented at October meeting A.A.S.)

Erickson, W. C. 1964, *Ap. J.*, 139, 1290–1311

Evans, J. W. 1957, *Publ. Astron. Soc. Pacific*, 69, 421

Evans, J. W., Michard, R., Servajean, R. 1963, *Ann. Ap.*, 26, 368

Fan, C. Y., Lamport, J. E., Simpson, J. A., Smith, D. R. 1966, *J. Geophys. Res.*, 71, 3289

Fokker, A. D. 1965, *Bull. Astron. Inst. Neth.*, 18, 111–24

Fomichev, V. V., Chertok, J. M. 1966, *Soviet Astron.*, 9, 976

Gelfreich, G., Korol'kov, D., Rishkov, N., Soboleva, N. 1958, *Paris Symp. Radio Astron.*, 125–28

Giese, R. H. 1961, *Z. Ap.*, 51, 119

Gillett, F. C., Stein, W. A., Ney, E. P. 1964, *Ap. J.*, 140, 292–305

Ginzburg, V. L., Zhelezniakov, V. V. 1958, *Soviet Astron.—AJ*, 2, 653

Godoli, G. 1965, in *Stellar and Solar Magnetic Fields*, 149 (Lüst R., Ed., North-Holland, Amsterdam)

Gold, F. 1959, *J. Geophys. Res.*, 64, 1665

Goldberg, L. 1967, *Ann. Rev. Astron. Ap.*, 5

Golnev, V. J., Parijstky, Y. N., Soboleva, N. S. 1964, *Izv. Glavnoi Astron. Obs.*, 23, 22–24

Greenstadt, E. W. 1966, *Ap. J.*, 145, 270–95

Hansen, R. T., Hansen, S. 1966 (Privately circulated manuscript)

Hartz, T. R. 1964, *Ann. Ap.*, 27, 831

Hartz, T. R. 1966 (Privately circulated manuscript)

Hartz, T. R., Warwick, J. W. 1966 (Privately circulated manuscript)

Harvey, J. W. 1965, *Ap. J.*, 141, 832–34

Hata, S., Saito, K. 1966, *Ann. Tokyo Astron. Obs., 2nd Ser.*, 10, 16

Hepburn, N. 1955, *Ap. J.*, 122, 445–59

Hewish, A. 1955, *Proc. Roy. Soc. A*, 228, 238

Hewish, A. 1958, *Monthly Notices Roy. Astron. Soc.*, 118, 534

Hewish, A., Scott, P. F., Wills, D. 1964, *Nature*, 203, 1214–17

Hewish, A., Wyndham, J. D. 1963, *Monthly Notices Roy. Astron. Soc.*, 126, 469

Högbom, J. A. 1960, *Monthly Notices Roy. Astron. Soc.*, 120, 530

Howard, R. 1959, *Ap. J.*, 130, 193

Hughes, M. P., Harkness, R. L. 1963, *Ap. J.*, 138, 239

Hulst, H. C. van de. 1950a, *Bull. Astron. Inst. Neth.*, 11, 135–50

Hulst, H. C. van de. 1950b, *ibid.*, 150–60

Hulst, H. C. van de. 1953, in *The Sun* (Univ. of Chicago Press)

Hyder, C. L. 1964, *Ap. J.*, 140, 817

Hyder, C. L. 1965a, *Ap. J.*, 141, 272

Hyder, C. L. 1965b, *Ap. J.*, 141, 1374

Hyder, C. L. 1965c, *ibid.*, 1382

Hyder, C. L. 1966a, *ibid.*, 146, 748

Hyder, C. L. 1966b, *Z. Ap.*, 63, 78–84

Jager, C. de. 1962, *Space Sci. Rev.*, 1, 487–521

James, J. C. 1966, *Ap. J.*, 146, 356

Jarrett, A. H., Klüber, H. von. 1961, *Monthly Notices Roy. Astron. Soc.*, 122, 223

Jefferies, J. T., Orrall, F. Q. 1965, *Ap. J.*, 141, 505

Kakinuma, T., Swarup, G. 1962, *Ap. J.*, 136, 975–94

Karimov, M. G. 1961, *IAU Symp. No. 16*, 297–300

Kawabata, K. 1963, in *The Solar Corona*, 143 (Academic Press, New York)

Kellogg, P. J., Ney, E. P. 1959, *Nature*, 183, 1297

Kiepenheuer, K. O. 1953, in *The Sun* (Univ. of Chicago Press)

Kippenhahn, R., Schlüter, A. 1957, *Z. Ap.*, 43, 36

Kleczek, J. 1963, *Publ. Astron. Soc. Pacific*, 75, 9–14

Korol'kov, D. V., Soboleva, N. S. 1962, *Soviet Astron.—AJ*, 5, 491

Kulsrud, R. M. 1955, *Ap. J.*, 121, 461

Kundu, M. R. 1965, *Solar Radio Astronomy* (Interscience, New York)

Kuperus, M. 1965, *Rech. Astron. Obs. Urecht*, 17(1), 1–69

Leighton, R. B. 1964, *Ap. J.*, 140, 1547

LeSqueren-Malinge, A. M. 1964, *Ann. Ap.* 27, 183

Lilliequist, C. G. 1966 (Paper presented at October meeting A.A.S.)

Lüst, R. 1962, *Space Sci. Rev.*, 1, 522–52

Lüst, R., Schlüter, A. 1955, *Z. Ap.*, 38, 190

Lyot, B. 1944, *Ann. Ap.*, 1, 41–44

McCraken, K. G., Ness, N. F. 1966, *J. Geophys. Res.*, 71, 13

Malitson, H. H., Erickson, W. C. 1966, *Ap. J.*, 144, 337–51

Malville, J. M. 1962, *Ap. J.*, 136, 266–75

Mandelshtam, S. L. 1964, *IAU Symp. No. 23*, 81–89

Maxwell, A. 1965, in *The Solar Spectrum*, 342–97 (Reidel Publ. Co., Dordrecht)

Maxwell, A., Thompson, A. R. 1962, *Ap. J.*, 135, 138–50

Meyer, F., Schmidt, H. 1966, *Mitt. Astron. Ges.*, 21

Michard, R. 1954, *Ann. Ap.*, 17, 429

Morimoto, M. 1964, *Publ. Astron. Soc. Pacific* 16, 163–69

Müller, R. 1954, *Z Ap.*, 35, 61

Mustel', E. R. 1961, *IAU Symp. No. 16* (Academic Press, New York)

Mustel', E. R. 1962a, *Astron. Zh.*, 39, 418

Mustel', E. R. 1962b, *ibid.*, 619

Mustel', E. R. 1964, *Space Sci. Rev.*, 3, 139–231

Ness, N. F. 1968, *Ann. Rev. Astron. Ap.* (To be published)

Ness, N. F., Scearce, C. S., Cantarano, S. 1966, *J. Geophys. Res.*, 71, 3305

Ness, N. F., Scearce, C. S., Seek, J. B. 1963, *J. Geophys. Res.*, 69, 3531

Neugebauer, M., Snyder, C. W. 1966, in *Solar Wind*, 3–23 (Jet Propulsion Lab. and California Inst. of Technol., Pasadena)

Newkirk, G. A. 1957, *Ann Ap.*, 20, 127–36

Newkirk, G. A. 1961, *Ap. J.*, 133, 983–1013

Newkirk, G. A., Bohlin, J. D. 1963, *Appl. Opt.*, 2, 131

Newkirk, G. A., Bohlin, J. D. 1964, *ibid.*, 3, 543

Newkirk, G. A., Bohlin, J. D. 1965, *Ann. Ap.*, 28, 234

Ney, E. P., Huch, W. F., Kellogg, P. J., Stein, W., Gillett, F. 1961, *Ap. J.*, 133, 616

Noble, L. M., Scarf, F. L. 1963, *Ap. J.*, **138**, 1169

Noxon, J. F. 1966, *Ap. J.*, **145**, 400–10

Oster, L., Sofia, S. 1965, *Ap. J.* **141**, 1139

Parker, E. N. 1958, *Ap. J.*, **128**, 664

Parker, E. N. 1963, *Interplanetary Dynamical Processes* (Interscience, New York)

Parker, E. N. 1964, *Ap. J.*, **139**, 690

Parker, E. N. 1965a, *Space Sci. Rev.*, **4**, 666–708

Parker, E. N. 1965b, *Ap. J.*, **141**, 1463–78

Pecker, J.-C., Roberts, W. O. 1955, *J. Geophys. Res.*, **60**, 33

Pneuman, G. W. 1966a, *Ap. J.*, **145**, 242–54

Pneuman, G. W. 1966b, *ibid.*, 800

Pottasch, S. R. 1960, *Ap. J.*, **131**, 68–74

Pottasch, S. R. 1964, *Space Sci. Rev.*, **3**, 816

Roberts, J. A. 1959, *Australian J. Phys.*, **12**, 327

Rust, D. M. 1966, *Measurements of the Magnetic Fields in Quiescent Solar Prominences* (Thesis, Univ. of Colorado, Boulder, Colo.)

Saito, K. 1958, *Publ. Astron. Soc. Japan*, **10**, 49

Saito, K. 1959, *ibid.*, **11**, 234–52

Saito, K. 1965a, *ibid.*, **17**, 1

Saito, K. 1965b, *ibid.*, **17**, 421

Saito, K., Billings, D. E. 1964, *Ap. J.*, **140**, 760

Scarf, F. L., Noble, L. M. 1965, *Ap. J.*, **141**, 1479

Schmidt, M. 1953, *Bull. Astron. Soc. Neth.*, **12**, 61–67

Seaton, M. J. 1962, *Observatory*, **82**, 111

Severny, A. B. 1965, *Soviet Astron.*, **9**, 171

Sheeley, N. R. 1964, *Ap. J.*, **140**, 731

Shklovsky, I. S. 1965, in *Physics of the Solar Corona* (2nd ed., Pergamon Press, New York)

Simon, G. W., Leighton, R. B. 1964, *Ap. J.*, **140**, 1120

Slee, O. B. 1961, *Monthly Notices Roy. Astron. Soc.*, **123**, 223

Slee, O. B. 1966, *Planetary Space Sci.*, **14**, 255–67

Slee, O. B., Higgins, C. S. 1966, *Australian J. Phys.*, **19**, 167–80

Snyder, C. W., Neugebauer, M. 1966, in *The Solar Wind*, 25 (Jet Propulsion Lab. and California Inst. of Technol., Pasadena)

Sturrock, P. A. 1961, *Nature*, **192**, 58

Takakura, T. 1960, *Publ. Astron. Soc. Japan*, **12**, 325

Takakura, T. 1966, *Space Sci. Rev.*, **5**, 80

Tidman, D. A., Birmingham, T. J., Stainer, H. M. 1966, *Ap. J.*, **146**, 207

Trellis, M. 1957, *Suppl. Ann. Ap.*, 5

Vitkevitch, V. V. 1961, *Soviet Astron.*, **4**, 897

Vitkevitch, V. V., Antonova, T. D., Vlasov, V. I. 1966, *Soviet Phys. "Doklady,"* **11**, 369

Waldmeier, M. 1947, *Astron. Mitt. Zurich*, *No. 151*

Waldmeier, M. 1950, *Z. Ap.*, **21**, 24

Waldmeier, M. 1955, *ibid.*, **38**, 37

Waldmeier, M. 1957, *Die Sonnenkorona II* (Birkhauser Verlag, Basel)

Waldmeier, M. 1963, in *The Solar Corona*, 129 (Academic Press, New York)

Waldmeier, M., Müller, H. 1950, *Z. Ap.*, **27**, 58

Warwick, J. W. 1957, *Ap. J.*, **125**, 811

Warwick, J. W. 1965, in *Solar System Radio Astronomy*, 131–70 (Plenum Press, New York)

Warwick, J. W., Hyder, C. L. 1965, *Ap. J.*, **141**, 1362

Weinberg, J. L. 1964, *Ann. Ap.*, **27**, 718

Weiss, A. A. 1965, *Australian J. Phys.*, **18**, 167

Weiss, A. A., Sheridan, K. V. 1962, *J. Phys. Soc. Japan*, **17**, *Suppl. A-II*, 223

Whang, Y. C., Liu, C. K., Chang, C. C. 1966, *Ap. J.*, **145**, 255

Widing, K. G. 1966, *Ap. J.*, **145**, 380–99

Wilcox, J. M., Ness, N. F. 1965, *J. Geophys. Res.*, **70**, 5793

Wilcox, J. M., Schatten, K. H., Ness, N. F. 1966, *Space Science Laboratory Report*, Ser. 7, No. 29

Wild, J. P., Sheridan, K. V., Neylan, A. A. 1959, *Australian J. Phys.*, **12**, 369

Wild, J. P., Smerd, S. F., Weiss, A. A. 1963, *Ann. Rev. Astron. Ap.*, **1**, 291

Wild, J. P., Tlamicha, A. 1965, *B.A.C.*, **16**, 73

Wilson, O. C. 1966, *Science*, **151**, 1487

Wlerick, G., Axtell, J. 1957, *Ap. J.*, **126**, 253

Wolfe, J. H., Silva, R. W., McKibbin, D. D., Mason, R. H. 1966, *J. Geophys. Res.*, **71**, 3329

Wyndham, J. E. 1966, in *The Solar Wind*, 109–22 (Jet Propulsion Lab. and California Inst. Technol., Pasadena)

Zhelezniakov, V. V. 1962, *Astron. Zh.*, **39**, 5–14

Zirin, H. 1966, *The Solar Atmosphere* (Blaisdell, Waltham, Mass.)

Zirin, H., Severny, A. B. 1961, *Observatory*, **81**, 155

ON THE ORIGIN OF THE SOLAR SYSTEM

By D. ter Haar

Magdalen College, Oxford

Introduction

In 1963 Cameron and the present author (ter Haar & Cameron 1963) puslished a survey of various theories of the origin of the solar system. This review was based upon an unpublished manuscript of the present author which was finished in 1950. As a result, apart from a few references to later papers, theories published after 1949 were not covered in any depth. Since that time a fair amount of work has been done, but the problem seems as far from a definite solution as it was in 1950. Lack of space makes it necessary to restrict the discussion to relatively few, relatively arbitrarily chosen, developments, but a more detailed and comprehensive discussion will be given in a forthcoming monograph by Cameron and the present author (ter Haar & Cameron 1968). Recently, Suess (1965) has discussed chemical evidence bearing on the origin of the solar system, and we shall not discuss this aspect, although we are fully aware of the importance and relevance of such evidence. In that respect we must also mention a recent paper by Urey (1966; see also Ostic 1965) as well as Urey's Silliman Lectures (1952).

Our main discussion will be related to some papers of Berlage (1953a,b; 1954; 1957; 1959a,b; 1962a,b; 1964; 1967), Alfvén (1954, 1964a,b), McCrea (1960, 1963), Hoyle (1960), Cameron (1962, 1963a,b), and Whipple (1964). Before entering into a detailed consideration of the various problems, we shall briefly list some of the data which a satisfactory theory must explain. Our list is not complete, and there are other facets which have to be covered before a theory is really satisfactory.

The main features we want to stress are:

(*a*) the regularity of the planetary orbits, that is, their approximate coplanarity, their small eccentricities, and the fact that the direction of motion of the planets around the Sun is the same for all of them and the same as the direction of rotation of the Sun itself;

(*b*) the Titius-Bode law which gives the following relation for the mean distance r_n of the nth planet from the Sun:

$$r_n = r_0 \beta^n \qquad\qquad 1.1.$$

where $\beta = 1.9$;

(*c*) the difference between the small, heavy, slowly rotating terrestrial planets (Mercury, Venus, Mars, and the Earth) which are near to the Sun and have few satellites, on the one hand, and the large, light, fast-rotating giant planets (Jupiter, Saturn, Uranus, and Neptune) which are further away from the Sun and have extensive satellite systems, on the other; and

(*d*) the similarity between the planetary system and the systems of regular satellites which show features (*a*) and (*b*).

We must draw attention to the fact that many authors feel that the Titius-Bode law is an accidental relation; but we feel strongly that this is not the case, especially as it seems to be repeated in the satellite systems. To some extent the division between regular and irregular satellites is an artificial and tautological one, in that the regular satellites are those for which (d) holds. However, they still comprise a large fraction of the satellites and we feel that the division is a real one.

We have not included in our list the problem of the solar angular momentum. As is well known, the Sun, while possessing about 99.9 per cent of the total mass of the solar systems, possesses only about 1 per cent of its total angular momentum. This, however, seems to us to be connected with the general problem of why early-type stars rotate faster, on the average, than late-type stars and therefore to be a problem which can be divorced from the other problems connected with the origin of the solar systems. It is likely that the solution of the angular momentum puzzle must be looked for in electromagnetic processes (see, e.g., ter Haar 1949, Hoyle 1960) as was first suggested by Alfvén.

In earlier theories, both dualistic and monistic—to use Belot's terminology—one often considered what would happen to either a filament or a gaseous envelope which was in the neighborhood of the Sun, and the Sun was assumed to be essentially the same star as it is now. Many aspects of the planetary system can be explained in such theories, for instance, the general mass distribution and the regularity of the orbits (see e.g. ter Haar 1948, 1950 or Berlage 1962a,b); but it seems very difficult to reconcile the lifetime of such an envelope with the time needed for condensation into planets and satellites. Recent work has therefore increasingly concentrated on theories in which the solar system originated—at least in some protoform—at the same time as the Sun itself. A corollary to such theories is that in such a case it seems likely that planetary systems are rather common and this has led to many discussions about the possibility of life outside our Earth (for a very readable account of such topics see, for instance, Sullivan 1964). However, one of the most persistent of modern cosmologists, the Dutch meteorologist Berlage, has continued to study the evolution of a solar envelope at a stage where the Sun was essentially in its present state. In the next section we shall discuss his recent work because there are many features which we feel are worth studying in more detail. Another point which we feel should be studied is the possible stability of a regular system of vortices such as the one proposed by von Weizsäcker (1944) and the stabilizing influence of the solar magnetic field—especially in the later stages of the envelope. After having discussed Berlage's work, we shall consider some recent work where the solar envelope develops at the same time as the Sun itself.

BERLAGE'S THEORY

Berlage's basic idea is that a gaseous disk around the Sun might spontaneously develop into a system of concentric rings. This idea was the basis of his 1940 and 1948 papers (Berlage 1940a,b; 1948a,b). The state of the disk is supposed to be governed by the requirement that the dissipation of energy due to viscosity be a minimum. Berlage's most detailed papers are his 1957 paper which was corrected in the 1959 papers (Berlage 1957, 1959a,b), while his latest papers (Berlage 1962a,b; 1964; for a semipopular account see Berlage 1967) have filled in some quantitative details.

If the axis of rotation is taken to be the z axis and s the distance from the z axis, Berlage derives from the condition of minimum energy dissipation the following expressions for the density ρ in the disk:

$$\rho = \rho_0 \exp\left[-a\sqrt{s} - \frac{GM_0}{2RT}\frac{z^2}{s^3}\right] \qquad 2.1.$$

and

$$\rho = \rho_0 \left(\frac{r}{s}\right)^{n/2}\left(\frac{s}{r_0}\right)^{n-\kappa} \exp\left[-as^{1/2+n} + k(s^{n-1} - r^{n-1})\right] \qquad 2.2.$$

Expression 2.1 was derived (Berlage 1957), assuming absence of turbulence and the relation

$$T = \text{constant} \times \mu \qquad 2.3.$$

where T is the temperature and μ the mean molecular weight. In Equations 2.1 and 2.2, G is the gravitational constant; R is the gas constant; M_0 is the mass of the central body (Sun); ρ_0, r_0, k, and a are constants; n is a constant, occurring in the expression for the pressure

$$P = P_0 \left(\frac{r_0}{r}\right)^n \qquad 2.4.$$

where r is the distance from the centre of the system; and finally κ is a constant measuring the anisotropy of the pressure:

$$\kappa = \frac{2(P_1 - P_2)}{P_1} \qquad 2.5.$$

where P_1 and P_2 are, respectively, the radial and tangential pressures. Berlage (1962a,b) considers $n = 0$, $\frac{1}{2}$, 1, 2, while κ varies from $2(P_2 = 0)$ through $0(P_1 = P_2)$ to $-\infty(P_1 = 0)$.

The derivation of Equations 2.1 and 2.2 is far from convincing, but it is interesting to note that ter Haar (1950; see also 1948), on the basis of the mechanical equations of motion, assuming the pressure to be the turbulent pressure and introducing an *ad hoc* cutoff following von Weizsäcker (1944), derived for the density the expression (γ and δ are constants)

$$\rho = \rho_0 \frac{r}{r_0}\left(\frac{s}{r}\right)^\gamma e^{-\delta_s} \qquad 2.4.$$

which has a qualitative similarity to Equation 2.2.

In his earlier paper Berlage (1957) now considers a variation in ρ such that in the equatorial plane, the density ρ_{eq} is given by

$$\rho_{eq} = \rho_{eq}{}^0 + \delta\rho_{eq} = \rho_0(1 + \phi) \exp\left[- a(1 + \psi)\sqrt{r} + \epsilon \cos\left(p \ln \frac{r}{r_m}\right)\right] \qquad 2.5.$$

where ψ, φ, and ϵ are all $\ll 1$, and such that the variations in total mass, total angular momentum, and total energy vanish, while the total kinetic energy increases at the cost of a decrease in potential energy.

We still fail to see both the basis of the particular form of Equation 2.5 and the consistency of requiring the total mass and the total energy to be constant while a dissipation process is taking place, so that the basis of Berlage's work remains questionable. However, once Berlage's premises have been accepted, the conclusions are rather striking.

Berlage now proceeds as follows. The three conditions of zero variations in the total mass, total angular momentum, and total energy lead to three homogeneous linear equations in φ, ψ, and ϵ, and the requirement that these equations have a nontrivial solution leads to an equation giving a functional relation between $\cos(p \ln a^2 r_m)$ and $\sin(p \ln a^2 r_m)$ and p. For any value of p one can thus get a solution r_m. However, the solution is not unique, as with r_0 also all r_k satisfying the relation

$$p \ln r_k = p \ln r_0 + 2k\pi \qquad 2.6.$$

are solutions.

From Equation 2.5 we see that there will be density maxima and minima. Condensation will presumably take place at a maximum. Once we accept this, we see that condensation may be expected to occur also at other distances which are interrelated through Equation 2.6. This then leads immediately to the Titius-Bode law (Equation 1.1) with

$$\beta = e^{2\pi/p} \qquad 2.7.$$

Berlage (1948b) had earlier shown that if there is just one condensation product, it will occur at a distance r_0 from the primary, satisfying the relation

$$a^2 r_0 \sim 50 \qquad 2.8.$$

From Equation 2.6 Berlage finds for a given p the possible values of r_k and then from Equation 2.5 the corresponding maxima and minima in the density. Plotting in a p–r_k diagram the loci of density maxima and intersecting those with the straight line 2.8, Berlage finds possible positions where primary condensations can start. Secondary condensation can then presumably start at positions governed by Equation 2.6 for given p. In this way, Berlage is able to reproduce with remarkable accuracy the sequence of the mean distances from the primary of the terrestrial planets, the giant planets, and the satellite systems of Mars, Jupiter, Neptune, Uranus, and Saturn. The division of the planets into two groups had also been suggested by Schmidt (1944a,b; 1945a,b; 1946).

We do not have the space to discuss in depth Berlage's detailed consideration of the satellite systems and their development (Berlage 1953a,b; 1954) or his consideration of the mass and density distribution in the solar system (Berlage 1953a,b; 1959a,b; 1962b). We shall just make a few remarks.

First of all, we note that it was shown by ter Haar (1948, 1950) that a density distribution such as Equations 2.1, 2.2, or 2.4, together with a consideration of possible condensation mechanisms, will lead automatically to roughly the same mass distribution as is observed. Secondly, we think that Berlage gilds the lily in superimposing upon the density distribution 2.5 further oscillating components There are then a sufficient number of adjustable parameters so that agreement between observed and calculated masses can be obtained in even the smallest detail. Thirdly, we mention that in order to explain the particular variations in the specific density of the various planets, Berlage makes the plausible assumption that the planets were formed from rings consisting of both gas and solid particles, whereas the satellites were formed from rings with a negligible gas content.

Finally, we note that Berlage (1962b) finds that depending on the parameter n, he finds different relations between r_k and r_{k-1}. For $n = 1$, the Titius-Bode law 1.1 is recovered, but $n = \frac{1}{2}$, for instance, leads to

$$\sqrt{r_k} - \sqrt{r_{k-1}} = \text{constant} \qquad 2.9.$$

a relation suggested both by Berlage in one of his earliest papers (Berlage 1935) and by Schmidt (1945a,b) to represent the observations as well as Equation 1.1.

We have dealt with Berlage's work in great detail because it contains a wealth of ideas which unfortunately have not always received the attention they deserve. We must refer the reader to the original papers for details and especially for Berlage's discussion of the accretion process.

ALFVÉN'S THEORY

Another author whose work does not seem to have had as much acclaim as it deserves is Alfvén, who has persistently advocated and stressed the importance of electromagnetic processes. We shall briefly review his ideas about the important processes involved in the formation of planetary systems, as given in his monograph on the subject (Alfvén 1954), before discussing his recent work on the origin of the asteroids.

First of all, we note that we have seen that theories which assume the Sun to have been in essentially its present state during the formation of the planets seem to run into difficulties of finding enough time to produce the planets, especially if we neglect electromagnetic effects. In a solar envelope of a density necessary to produce the planets, the solar radiation field will be insufficient to produce any appreciable ionization. Alfvén has pointed out, however, that even in such circumstances the various magnetohydrodynamic processes which are responsible for the complicated behavior of the corona and the chromosphere may well produce ionization. Moreover, once ionization occurs, the currents produced in the ionized medium will themselves tend to produce more ionization. Alfvén therefore states that a "cold" theory—that is, a theory proceeding from an un-ionized envelope—is probably a much more specialized assumption than a "hot" theory. This point of view is, of course, compatible with the idea that the formation of the planets is connected with the formation of the central star, as T Tauri stars, which according to Struve (1950) represent the state of formation of stars from interstellar matter, are surrounded by ionized material as shown by the emission lines in their spectra.

Alfvén points out that in our solar system there are six (or seven) families of celestial bodies: the terrestrial planets, the giant planets, two groups of Saturnian satellites, the Jovian satellites, and the Uranian satellites (and the Martian satellites), and he emphasizes that these groups should all be treated in essentially the same way. Moreover, Alfvén assumes that the only important features of the central body are its mass, its axial rotation, and the fact that it possesses a magnetic moment. We shall follow the outline of Alfvén's theory given in the first chapter of his book (Alfvén 1954).

His theory starts with a central mass surrounded by an ionized gas cloud, the so-called "initial cloud" which stretches over a distance large compared with the final dimensions of the system it produces. This initial cloud may be the cloud surrounding the protosun when we are discussing the formation of the planets, but in the case of the formation of the satellites, the formation process of the primary body itself will lead to the existence of the relevant initial cloud. The ionized gas in the cloud is acted upon by both the magnetic field and the gravitational field. When atoms become de-ionized the gravitational field may attract them and they will tend to fall towards the central body. Depending on their ionization potentials, different elements will become de-ionized at different times and will fall towards the central body at different rates, and eventually will be stopped at different distances from the central body. Alfvén distinguishes an A-cloud containing mainly helium together with some solid-particle impurities ("meteor rain"), a B-cloud containing mainly hydrogen, a C-cloud containing mainly carbon, and a D-cloud containing mainly silicon and iron. In the planetary systems, the impurities in the A-clouds are supposed to condense into Mars and the Moon (later captured by the Earth), and the B-cloud impurities condense into the Earth, Venus, and Mercury. The C-cloud condenses into the giant planets. Inside each cloud, the mass distribution is governed by the angular momentum of the cloud, and Alfvén is able to explain in this way the mass distribution in the different groups of planets. Pluto and Triton may have been formed in the D-cloud.

The satellite systems are formed in a similar way. The initial cloud around the giant planets will be mainly C- and D-cloud material, and Alfvén suggests that the Galilean satellites of Jupiter, Saturn's rings, and its inner satellites are C-cloud satellites, while the outer Saturnian and the Uranian satellites are D-cloud condensation products.

On Alfvén's theory the conditions for planet formation are the existence of an initial cloud which falls towards a star with sufficiently high magnetic and gravitational fields. Of those conditions, the only ones which need investigating are the presence of a magnetic moment and the requirement of non-ionized material near the star. As long as we know so little about the origin and evolution of stellar magnetic fields, it is very difficult to make any definite statements about the presence or absence of a magnetic field in the early stages of stellar formation, but Alfvén's estimates seem to indicate that most stars would probably have a magnetic moment sufficient for the requirements of his present theory. The requirement of the presence of non-ionized material rules out early-type stars, whose H II regions (and even He II regions) would be too extensive. In fact, one probably would expect that only for stars of types later than F would we expect planet formation to take place. Whether Alfvén's suggestion that the limit of planet formation and the limit between fast and slow rotation are the same is plausible, seems rather open to doubt, since it appears to us that the question of stellar rotation is divorced from that of planet formation.

In two recent papers Alfvén (1964a,b) discusses the origin of the asteroids. He points out that as there seem to be two groups of planets (compare the discussion in Section 2) there is no reason to look for a "missing" planet at the position of the asteroids and that, moreover, any attempts to account for the existence of the asteroids on the basis of explosion, rotational instability, tidal disruption, or collision of such a missing planet look very far fetched. Moreover, recent measurements of the rotational periods of several asteroids (see Alfvén 1964a for references) have shown that, although there is a range of a factor 10^5 in the masses of the asteroids, their rotational periods vary less than a factor 10 (between 3 and 17 hours), and in fact are also—within that factor—equal to the rotational periods of the Earth, Mars, Jupiter, Saturn, Uranus, and Neptune. This suggests most strongly, as Alfvén emphasized, that a common condensation process must have been at work. Alfvén shows that in a rotating plasma, condensation products will appear on a central mass in such a way that the ultimate rotational velocity, which will be reached asymptotically, will be independent of the rotational velocity of the plasma. The reason for this is that when the plasma is rotating fast, only matter close to the central mass will be able to strike the central mass while it describes a Kepler orbit, while matter further out can condense onto the central mass, if the plasma rotational velocity is low. Moreover, the limiting period is of the order of a few hours, in good agreement with the observed rotational periods.

McCrea's Theory

McCrea (1960, 1963) has looked at various aspects of the origin of the solar system. In his theory no electromagnetic phenomena are taken into account. McCrea starts from the idea that stars cannot be formed singly, but only in clusters, because otherwise there would be no mechanism by which the angular momentum in the protostar could be dissipated. He thus assumes that a star cannot be formed by the condensation of all the material originally within any one particular region of the interstellar medium. This has a corollary that a star formed in the medium can be formed only by the aggregation of parts of the material from various regions, these being parts that in the aggregate do not possess undue proper angular momentum, i.e., angular momentum about its centre of mass. It seems to us that McCrea is begging the important question of the angular momentum distribution in the solar system, and on this ground his theory needs other advantages before it can be taken seriously. McCrea pushes this even further in assuming that a particular portion of material goes into a particular condensation —which is to form a star—for the most elementary reason possible, simply that it chances to be moving towards that particular condensation. This ensures that it will contribute only a negligible amount to the angular momentum.

Let us follow McCrea in spelling out in more detail the evolution of a cloud of material with a total mass several orders of magnitude larger than that of the Sun and in a highly turbulent state. The turbulent state will be represented, according to McCrea, by the presence of many cloudlets or "floccules" which to a first approximation are independent, randomly moving entities. When two floccules collide, they will partly coalesce and so, through collisions, fluctuations in the size of the floccules will occur. The larger ones will have a better chance to capture more material until they are large enough to start contracting under the influence of their own gravitational field. Having assumed that when part of a large floccule which condenses into a star has practically no angular momentum, in order to account for planets with a much larger angular momentum, McCrea suggests that the gravitational field of a major condensation can capture into a closed orbit a minor condensation which is then not available for incorporation into a major condensation.

In order to obtain agreement with certain observational data, McCrea starts from a system with a mean density in the cloud of 4.10^{-12} g cm^{-3}, a floccule density of 8.10^{-40} cm^{-3}, and a floccule radius of 9.10^{11} cm (to be precise, McCrea chooses a value for the floccule mean free path λ rather than their radius, but as the floccules are supposed to be spherical, the one quantity determines the other). He assumes on some general grounds that the mean random speed of a floccule is 10^5 cm sec^{-1} and that the mean kinetic temperature at the stage considered is 50° K. From these data it follows that within a sphere of radius λ, just over 1 solar mass is present, that the mass of one floccule is about three times the mass of the Earth, that the density in a floccule is about 7.10^{-9} g cm^{-3}, that at such a density one needs to combine about 20 floccules to produce a gravitationally unstable mass (at the given temperature), and that the total angular momentum of a cloud of radius λ—assuming the motion of the floccules to be random—is about half of the angular momentum of Jupiter's orbital motion, i.e., about 100 times the angular momentum of the Sun. The reason McCrea considers what happens inside a sphere of radius λ is essentially that the mean free path of the floccules determines the scale of length for what happens in the cloud—it should be of the order of magnitude of the turbulent mixing length.

Consider now a sphere with radius λ around the position of the protosun. In this sphere there is about the right amount of mass and of angular momentum to produce the solar system. The number N of floccules inside this sphere is 10^5. On McCrea's picture, the early stages of the formation of the Sun are the combination of about 20 floccules to produce a gravitationally unstable mass. On the assumption of random motion of the floccules, this stage of the Sun would have an angular momentum of the order of $20^{1/2}/N^{1/2} \sim 1$ per cent of the total angular momentum within the sphere, that is, about the present solar angular momentum. In the later stages, when the protosun has a radius a of something like 0.05 λ, only those floccules which will come as close as a to the protosun will be captured. Those have an angular momentum about 5 per cent of the mean angular momentum, resulting in a total angular momentum of the sun of about 5 per cent of the angular momentum of the cloud. It seems to us that, even on McCrea's simple hypothesis, the Sun acquires more angular momentum than it has at present. However, it might be possible, by suitably rearranging the various parameters, to get a lower angular momentum. As the mass left behind and subsequently condensed into the planets will possess practically all the angular momentum, which on the assumption of random motion of the floccules is of the order of $N^{1/2}$ times the angular momentum of a floccule, while in the motion of a planet the angular momentum vectors of all the material add up, we would expect that a fraction of only about $N^{-1/2}$ of the total mass will be used to produce the planets, which is in reasonable agreement with observational data.

McCrea shows that the formation of the Sun will take about 10^5 years—which may be too long, as the lifetime of a turbulent medium of dimensions of the order of λ may well be a few orders of magnitude less than that period. As McCrea points out himself, his theory does not give any positive suggestions regarding the occurrence of satellites. The rotational periods of the outer planets come out more or less correctly, which is not surprising as his theory in this respect resembles the calculations of Alfvén mentioned in the preceding section.

HOYLE'S THEORY

We next want to discuss Hoyle's recent work (Hoyle 1960, 1963; see also comments by Öpik 1963 and Whipple 1964). In discussing the crucial angular momentum problem, Hoyle objects to McCrea's ideas on statistical grounds: it is known that most dwarf stars rotate slowly and it seems extremely unlikely that condensation always takes place in regions of quite exceptionally small differential motion. To solve the angular momentum problem, Hoyle suggests that star formation takes place in three stages. There is an initial phase during which the condensation loses angular momentum to its surroundings; in this phase matter does not slip across the magnetic lines of force. A second phase follows during which matter slips across the magnetic lines of force and angular momentum is conserved. During this phase the condensation contracts. The final phase is a slow contraction during which the star becomes a main-sequence star. Even so, Hoyle indicates that the loss of angular momentum in the first stage cannot be anything like enough to explain the slow rotation of dwarf stars.

Another difficulty, and one which Gold (1963; see also Öpik 1963) mentions as one of the most important unsolved problems, is the deficiency of hydrogen and helium in Uranus and Neptune. Correcting for this deficiency, that is, assuming that the original mass and angular momentum were originally about 100 times as large as they are now, one finds that practically all of the total angular momentum of the initial condensation is now present in the outer planets. This means that one must find a process whereby the angular momentum of the initial protosun was transferred from the protosun to the planetary material. Hoyle then concludes that a similar process must also have been operative in the great majority of dwarf stars and hence that such stars must possess planetary systems. We do not see the logic of the second conclusion, especially since, while Hoyle's discussion of the transfer of angular momentum seems reasonable, his discussion of the planet formation is far from convincing.

Hoyle sees the evolution after stage two as follows. During the contraction of the protosun, it becomes rotationally unstable when its diameter becomes of the order of magnitude of the radius of Mercury's orbit. If the rotational instability evolves without any concomitant electromagnetic processes, the material is shown by Hoyle to condense into one or more bodies of stellar mass which will orbit around the Sun at a very small distance. Hoyle suggests that this is the origin of W Ursae Majoris stars. Turbulent processes do not help either, and Hoyle concludes that no purely hydrodynamic process can explain the very slow rotation of the Sun (see also ter Haar 1949). He therefore looks for a magnetic torque transmission, the main features of which are:

(a) the protosun became rotationally unstable when its radius was about 3.10^{12} cm;

(b) magnetic coupling between the central condensation and the Laplacian disk produced by the rotational instability maintains equal angular velocities of the central condensation and of the inner edge of the disk—which is supposed to stay at a distance of about 3.10^{12} cm, while its material spirals outwards at the same time.

Hoyle shows that a relatively low degree of ionization will suffice to produce a conductivity high enough to have the more or less rigid torque coupling. As far as the anchoring of the lines of force at the protosun is concerned, there must be a deep convection zone, thus early-type main-sequence stars are excluded from the present process.

The temperatures in the inner regions of the disk are sufficiently low for condensation to take place. However, it is necessary to obtain the condensation products at distances which are further away from the Sun—at the present distances of the giant planets from the Sun. Hoyle assumes that the condensation products are swept along by the gas which is spiralling outwards. If this were, indeed, the case one could understand some of the aspects of the formation of the giant planets—apart from the H and He deficiency already mentioned. However, Whipple (1964) has made it very plausible that Hoyle's process will not take place. In the magnetic-field configuration envisaged by Hoyle, the magnetic lines of force exert a pressure on one another and tend to separate. The forces on the gas are an outward pressure from the Sun and a smaller component in the forward sense of rotation due to the lines of force. The last component is the one removing the gas. The outward pressure, however, partially counteracts the solar gravitational attraction, leading to a slower rotational velocity of the gas, as it moves essentially in a weaker gravitational field. Hence the gas will move more slowly than the solid particles which move with Keplerian velocities: The solid particles move effectively in a resistive medium, leading to a reduction of their angular momentum and thus to a motion of them inwards rather than outwards.

CAMERON'S THEORY

To conclude, we shall briefly mention some aspects of Cameron's work on the formation of the solar nebula and of the Sun and planets (Cameron 1962, 1963a,b). We do not have the space to consider the detailed physico-chemical considerations given by Cameron, but we shall briefly discuss the general outline of a theory of the origin of the solar system which he gives, and we shall lean heavily upon Whipple's account (1964). Cameron's theory starts with a protosun of about 1–2 solar masses extending over a volume of dimensions of about 10^5 astronomical units which is gravitationally unstable, collapses, and breaks up into smaller subunits. The magnetic field in this protosun will be of the order of 10^{-5} gauss. During the collapse the magnetic lines of force are twisted. The collapse is quite fast, lasting not much more than the time for free fall to the centre, provided the gravitational potential energy can be removed. Cameron suggests that this is done by the dissociation of hydrogen molecules, followed by the ionization of hydrogen and the double ionization of helium. The collapse is finally governed by the difficulty of disposing of the angular momentum. The angular momentum present leads to rotational instability of the protostar which will thus produce a Laplacian disk. At this stage, radiation will remove excess energy and the disk will be quite cool in a relatively short period (of the order of 10^6 years), and condensation into what Whipple calls cometesimals takes place. Aggregation of such cometesimals produces the giant planets which during their formation were surrounded themselves by Laplacian disks from which the satellites—or more likely some of the satellites—were formed. The formation of the terrestrial planets, the comets, and the asteroids was probably complicated, involving formation, disintegration, heating, melting, solidifying, and so on, but most of these processes involve chemical and metallurgical considerations outside the scope of the present article.

CONCLUSION

We have covered very little ground in the present article and, unfortunately, definite conclusions can be reached on only very few points. It seems likely that magnetohydrodynamics must be looked to for explaining the low rotational velocity of the Sun. How much magnetohydrodynamics is involved in the condensation processes leading to the planets is still undecided. Although, generally speaking, we are getting closer to an understanding of the different chemical composition of the various bodies in the solar system, many details are still baffling—such as the escape of H and He from the Uranus-Neptune regions. The existence of a regular law for the mean distances from the primary object is doubted by some authors, and in general the origin of satellite systems is still far from being explained satisfactorily. Altogether, there seems to be still much to be done, even though the increasing store of observational data is helping to discard many theories more easily than used to be the case.

LITERATURE CITED

Alfvén, H. 1954, *On the Origin of the Solar System* (Oxford Univ. Press)

Alfvén, H. 1964a,b, *Icarus*, **3**, 52, 57

Berlage, H. P. 1935, *Proc. Roy. Dutch Acad. Sci.*, **38**, 857

Berlage, H. P. 1940a,b, *ibid.*, **43**, 532, 557

Berlage, H. P. 1948a,b, *ibid.*, **51**, 796, 965

Berlage, H. P. 1953a,b, *ibid.*, **B56**, 45, 56

Berlage, H. P. 1954, *ibid.*, **B57**, 452

Berlage, H. P. 1957, *ibid.*, **B60**, 75

Berlage, H. P. 1959a,b, *ibid.*, **B62**, 63, 73

Berlage, H. P. 1962a,b, *ibid.*, **B65**, 199, 211

Berlage, H. P. 1964, *Verslag. Roy. Dutch Acad. Sci., Phys. Sec.*, **73**, 8

Berlage, H. P. 1967, *The Origin of the Solar System* (Pergamon, Oxford)

Cameron, A. G. W. 1962, *Icarus*, **1**, 13

Cameron, A. G. W. 1963a, *ibid.*, **1**, 339

Cameron, A. G. W. 1963b; Jastrow & Cameron 1963, p. 85

Gold, T. 1963; Jastrow & Cameron 1963, p. 171

ter Haar, D. 1948, *Proc. Roy. Danish Acad. Sci.*, **25**, No. 3

ter Haar, D. 1949, *Ap. J.*, **110**, 321

ter Haar, D. 1950, *ibid.*, **111**, 179

ter Haar, D., Cameron, A. G. W. 1963, Jastrow & Cameron 1963, p. 1

ter Haar, D., Cameron, A. G. W. 1968, *Origin of the Solar System* (Pergamon, Oxford)

Hoyle, F. 1960, *Quart. J. Roy. Astron. Soc.*, **1**, 28

Hoyle, F. 1963, Jastrow & Cameron 1963, p. 63

Jastrow, R., Cameron, A. G. W., Eds. 1963, *Origin of the Solar System* (Academic Press, New York)

McCrea, W. H. 1960, *Proc. Roy. Soc. London A*, **256**, 245

McCrea, W. H. 1963, *Contemp. Phys.*, **4**, 278

Öpik, E. 1963, Jastrow & Cameron 1963, p. 73

Ostic, R. G. 1965, *Monthly Notices Roy. Astron. Soc.*, **131**, 191

Schmidt, O. J. 1944a, *Dokl. Akad. Nauk SSSR*, **45**, 229

Schmidt, O. J. 1944b, *ibid.*, **46**, 355

Schmidt, O. J. 1945a,b, *ibid.*, **52**, 577, 667

Schmidt, O. J. 1946, *ibid.*, **54**, 15

Struve, O. 1950, *Stellar Evolution* (Princeton Univ. Press, Princeton, N. J.)

Suess, H. E. 1965, *Ann. Rev. Astron. Ap.*, **3**, 217

Sullivan, W. 1964, *We Are not Alone* (McGraw-Hill, New York)

Urey, H. C. 1952, *The Planets* (Yale Univ. Press, New Haven, Conn.)

Urey, H. C. 1966, *Monthly Notices. Roy. Astron. Soc.*, **131**, 191

von Weizsäcker, C. F. 1944, *Z. Ap.*, **22**, 319

Whipple, F. L. 1964, *Proc. Natl. Acad. Sci. U. S.*, **52**, 565

ULTRAVIOLET AND X RAYS FROM THE SUN[1]

By Leo Goldberg

Harvard College Observatory, Cambridge, Massachusetts

INTRODUCTION

Progress in the observation of ultraviolet radiation and X rays from the Sun has been very rapid since the subject was last reviewed in these pages (Friedman 1963). Most of the advances reported have been achieved with sounding rockets, of which well over a dozen have been flown successfully, but results of major importance have also been obtained with satellites, notably OSO-1, Ariel I and II, and Electron II and IV. Major gains have been made in extending the solar emission-line spectrum to a new short-wavelength limit of 6 Å, and in improving the spectroscopic resolution at all wavelengths below 2100 Å, by as much as a factor of 10 in some spectral regions. A great deal has been learned about solar X-ray emission, particularly with respect to its spatial and spectral distribution and its variation with time. Various aspects of solar XUV radiation have been treated in considerable detail in a number of review articles, some reporting results (Tousey 1963, 1964), others emphasizing interpretation (Pagel 1963, Pottasch 1964b, Allen, C. W., 1965, Mandel'shtam 1965a,b), and still others dealing with techniques (Boyd 1965, Hinteregger 1965). In this review, we survey the latest observational results and discuss the current status of certain of the more interesting problems of interpretation.

SOLAR XUV SPECTRUM

OBSERVATIONS

Table I summarizes the principal observations of the solar XUV spectrum which have been performed from rockets and from the OSO-1 spacecraft since 1962, and for which some results have been published. The first column gives the date on which observations were made, the second column the spectral region, the third column the laboratory at which the instrument was prepared, and the fourth column the observational technique. The fifth column gives the precision of the wavelength measurement and the sixth column the approximate resolving power $\lambda/\Delta\lambda$. A table giving absolute fluxes of emission lines in the spectral region from 1750 to 1 Å under relatively quiet solar conditions has been published by Hinteregger, Hall & Schmidtke (1965).

For purposes of discussion, the XUV solar spectrum may be divided into three spectral regions: 3000–1200 Å, 1200–500 Å, and $\lambda < 500$ Å. The lower wavelength limit of the first region, approximately 1200 Å, represents the limit of usefulness of aluminum as a reflecting surface, whereas 500 Å is

---

[1] The survey of literature for this review was concluded in January 1967.

TABLE I

Recent Observations of the Solar XUV Spectrum

Date	λ (Å)	Laboratory	Technique	Wavelength precision (Å)	Resolving power λ/Δλ
19 November 1964	3000–2085	NRL	Photography disk spectrum	0.01–0.02	10⁶
1959	3100–2470	USSR	Photography disk spectrum	0.04	3 ×10⁴
17 December 1964	2950–930	Culham	Photography integrated disk spectrum		6 ×10³
9 April 1965	2950–930	Culham	Photography limb spectrum		
22 August 1962	2085–1200	NRL	Nearly stigmatic photography disk spectrum	0.05	10⁴
22 August 1962	1200–500	NRL	Nearly stigmatic spectra	0.03	10⁴
10 May 1963	500–149	NRL	Grazing incidence spectra	0.05–0.1	10³
2 May 1963	310–55	AFCRL ⎫	Photoelectric recording	0.2	2–5 ×10²
March 1964	310–55	AFCRL ⎭	Grazing incidence		
3 November 1965	128–30	AFCRL	Photoelectric recording, grazing incidence	0.1	2–6 ×10²
10 March– 26 May 1962	340–170	GSFC	Grazing incidence, photoelectric recording (OSO-1)	0.3	2 ×10²
20 September 1963	80–33	NRL	Grazing incidence, photographic	0.05	2 ×05
25 July 1963 4 October 1966	25–13⎫ 13–6 ⎭	NRL	Bragg spectrometer	0.1	10²

roughly the boundary between the useful domains of normal-incidence and grazing-incidence optics. These divisions also have considerable physical significance, the more so when the first spectral region is further subdivided at 2085 Å. Thus, between 3000 and 2085 Å the spectrum is essentially an extension of the visible Fraunhofer spectrum. The region 2085–1200 Å is noteworthy because it contains the radiation emitted by the solar atmosphere in the region of the temperature minimum and therefore exhibits the transition from limb darkening to limb brightening and from a dark-line or Fraunhofer spectrum to a bright-line spectrum. A sudden drop in the intensity of the continuous spectrum at 2085 Å signals the onset of the transition region. Below 1200 Å, the spectrum contains radiation emitted by the chromosphere and corona. On the average, there is an inverse correlation between the wavelength of an emission line and the temperature of the region that emits it. Hence, the quiet chromosphere and corona are best studied in the wavelength region 1200–500 Å, where normal-incidence techniques are applicable, whereas the most information about transient solar activity and active regions in the chromosphere and corona is provided by the wavelengths below 500 Å.

Newly reported spectra in the region between 3000 and 1200 Å have been photographed from rockets with spectrographs prepared by groups in the United States, in the United Kingdom, and in the Soviet Union (Kachalov & Yakovleva 1962, Black et al. 1965, Burton & Wilson 1965, Tousey et al. 1965). The echelle spectrograph originally flown on August 21, 1961, by Purcell, Garrett & Tousey (1963), was reflown on November 19, 1964 and the solar spectrum extended with very high resolution to 2100 Å.

Photographs of the ultraviolet solar spectrum are usually made with the spectrograph slit placed across the solar disk. In 1964 and 1965, the spectrum between 3000 and 950 Å was photographed for the first time with the slit placed 10 sec of arc outside the solar limb. This development was made possible by the perfection of a triaxial attitude control unit attached to a Skylark rocket which, together with a servo-controlled optical alignment system, achieved an overall root-mean-square pointing accuracy of a few seconds of arc. Scattering by dust on the collector mirror during the first flight gave spectra characteristic of the sky spectrum (Black et al. 1965), but the second flight of April 9, 1965 was completely successful (Burton & Wilson 1965). As shown in Figure 1, the limb spectrum extended to 950 Å, the measured resolving power being about 5000. Analysis (Burton, Ridgely & Wilson 1966) shows many previously unreported emission lines including forbidden lines of highly ionized coronal atoms similar to those which have been observed in the visible spectrum. Three such lines were observed: [Fe XI] $3p^4\,{}^3P_1-3p^4\,{}^1S_0$ at 1445.89 Å, [Fe XII] $3p^3\,{}^4S^0-3p^3\,{}^2P^0$ at 1242.15 and 1349.57 Å. The emission-line spectrum displays many other interesting features, e.g., intersystem transitions in C III, N IV, and O IV, more than 100 permitted lines of Fe II and abnormal enhancement of Si I multiplets arising from transitions between high levels with $n = 6$, 7, 8 and the ground state. The absence of transitions to the ground state from upper levels more than 0.4 eV from the series limit is quantitatively unexplanied.

The spectral region from 2085 to 1200 Å has also been photographed by the Naval Research Laboratory with a nearly stigmatic double-dispersion grating spectrograph having a resolving power of 0.2 Å (Tousey et al. 1965). The spectrograph slit was placed along a diameter of the solar image, which is closely stigmatic near 1650 Å but deviates on either side until at the extremes, 1200 and 2000 Å, there is ~1 min of uncorrected astigmatism. The continuum from 2000 to 1530 Å shows no change in intensity from the center of the disk to within 20 sec of the limb, the approximate limit set by the jitter of the pointing control. Below 1530 Å, the continuum shows conspicuous limb brightening which is attributed to radiation from the ionization continuum of Si I. The spatial resolution is not sufficient to establish the precise shape of the temperature minimum at the photosphere-chromosphere

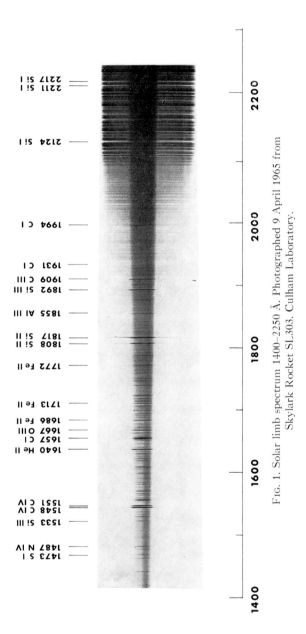

FIG. 1. Solar limb spectrum 1400–2250 Å. Photographed 9 April 1965 from Skylark Rocket SL303. Culham Laboratory.

interface. The extreme weakness of Fraunhofer lines between 2000 and 1700 Å makes identification difficult without detailed comparison with laboratory absorption spectra produced at approximately the same temperature.

Following the appearance of the emission line of longest wavelength, C I, 1993.65 Å, the number of emission lines increases as the number of absorption lines diminishes, until the two appear in equal numbers between 1725 to 1750 Å. About one half of the emission lines from 1725 to 1200 Å have been identified. The emission line at 1241.9 Å is seen to extend across the disk and beyond the Sun's limb, which supports its identification as a forbidden line of [Fe XII] in the Culham limb spectrum. Most of the chromospheric emission lines show pronounced limb brightening, which is generally greater the higher the temperature required to produce the line. Thus, the lines of C I show the least limb brightening and those of C IV, near 1550 Å, and N V, near 1240 Å, show the most. The intensities of the emission lines seen on the disk are well correlated with those observed above the limb at Culham, but the ratio of limb to disk intensities increases with the ionization potential of the emitting ion (Burton, Ridgely & Wilson 1966), as shown in Table II.

On the same rocket flight of August 22, 1962, the Naval Research Laboratory (NRL) observers also obtained a high-resolution spectrum of the region 1250–800 Å (Tousey et al. 1964), the major improvement over the earlier observations being the replacement of a 600 line/mm grating by one having 2400 lines. This spectrum and its companion at longer wavelengths are typical of the quiet Sun, as contrasted with a spectrum made in the same wavelength region in 1959 at solar maximum. The gain in resolution over

TABLE II

LIMB-TO-DISK INTENSITY RATIO (IN RELATIVE UNITS)
vs. IONIZATION POTENTIAL

Ion	Ionization potential (eV)	Limb-disk ratio
C I	11.3	1
O I	13.6	1
H I	13.6	1
Si II	16.3	2
C II	24.4	3
Al III	28.4	5
Si III	33.5	2
Si IV	45.1	8
C III	47.9	3
He II	54.4	6
C IV	64.5	11
N V	97.9	7
O VI	138.1	6

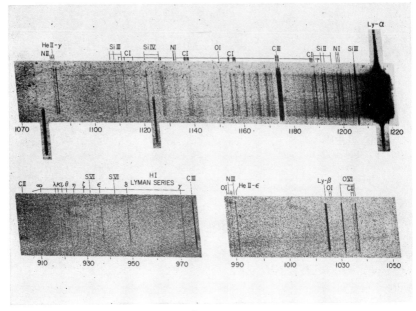

Fig. 2. Solar spectrum 900-1220 Å. Photographed 22 August 1962 from an Aerobee-Hi rocket. The two lines projecting below the spectrum near 1080 and 1120 Å are Lyman-α ghosts. Naval Research Laboratory.

earlier spectra is such that many multiplets are seen to be resolved into lines, notably those of neutral carbon between 1100 and 1200 Å (see Figure 2). The Lyman lines are observed through Lyman λ 918.13 Å, their widths being about 0.3 Å. Both Lyman α and Lyman β are strongly self reversed, the peak to peak separation in Lyman α being 50 per cent greater than in Lyman β. The most prominent emission lines arise from screening-type transitions in which the principal quantum number does not change for such ions as C II–IV, N I–V, O II–VI, Si III and IV, and S II–V. Members of the Balmer series of He II are also present as are several multiplets of O I.

Much improved grazing-incidence spectra at wavelengths below 500 Å have been obtained in a series of rocket flights in 1963 and 1964 by the NRL (Tousey et al. 1965, Austin et al. 1966) and the Air Force Cambridge Research Laboratory (AFCRL) (Hinteregger, Hall & Schweizer 1964; Hall et al. 1965; Manson 1967), the former employing photography and the latter photoelectric recording. The photographic spectra give the most accurate wavelengths and the highest resolution, whereas the photoelectric spectra provide the most accurate intensities and also fill in the yet unphotographed region between 148 and 80 Å. The gap occurs because all spectra photographed below 500 Å are taken with an aluminum filter, 1000 Å in thickness, placed just in front of the entrance slit. The filter eliminates stray light by absorbing all wavelengths greater than 837 Å, but it is also opaque from 148 to 80 Å.

Figure 3 shows the solar spectrum between 500 and 149 Å as photographed by NRL. A section of the spectrum between 160 and 304 Å is also shown, together with the photoelectrically scanned spectrum recorded by AFCRL in March 1964. The qualitative agreement is excellent. Most of the lines between 171 and 220 Å have been shown to belong to iron in various high stages of ionization. The exact assignments will be discussed later. The second-order images of these lines account for most of the strong lines between 340 and 400 Å. Other identifications are shown in the Figure.

Figure 4 shows the AFCRL spectrum of 3 November 1965 covering the wavelength region from 30 to 128 Å with a resolution of 0.2 Å (Manson 1967). The original AFCRL telemetering monochromator was modified by the in-

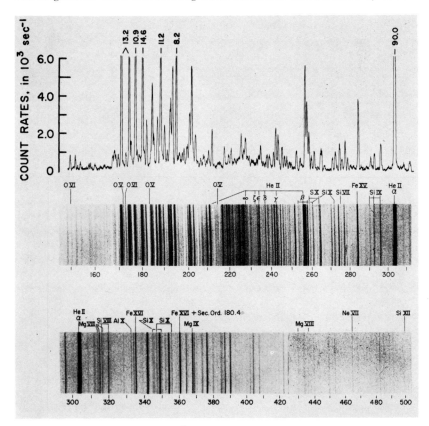

FIG. 3. Solar spectrum 149–500 Å. The photographic record obtained by the Naval Research Laboratory on 10 May 1963 is compared with a photoelectric scan by the Air Force Cambridge Research Laboratory on 30 March 1964.

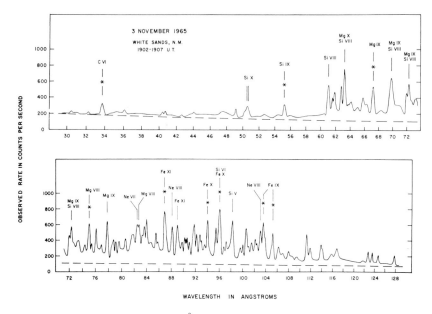

FIG. 4. Solar spectrum 30–128 Å. Recorded photoelectrically by the Air Force Cambridge Research Laboratory on 30 November 1965. Asterisks denote lines used to establish the wavelength scale.

troduction of a 2400 line/mm gold replica grating, instead of the 1200 line grating used previously, and a thin-window flow Geiger detector replaced the windowless photomultiplier with a LiF photocathode used in 1963 and 1964. The instrument was calibrated by careful measurements of the efficiencies of the grating and detector, so that absolute intensities derived from the data should be accurate to within a factor of 1.5. Manson (1967) lists absolute intensities for well over 100 lines. They are in good agreement with fluxes derived from the NRL spectrum of 20 September 1963 for wavelengths greater than 61 Å.

At still shorter wavelengths, Bragg X-ray crystal spectrometers are more effective than grating instruments. Results have been obtained from two NRL flights with crystal spectrometers in Aerobee rockets. The first, on July 25, 1963, recorded the spectrum from 13 to 25 Å, and revealed line emissions from O VII and VIII, N VII, and Fe XVII (Blake et al. 1965). Since the spectra are slitless, the width of the emission line is a measure of the angular size of the source; and in this way it could be shown that the O VII line emission came mainly from the solar disk as a whole, whereas the Fe XVII emissions were localized in a single active region, as would be expected from the respective ionization potentials of O VI (138 eV) and Fe XVI (489 eV). The second flight, on October 4, 1966 (Friedman 1967), scanned the spectrum in the two ranges 1–8 and 9–25 Å. The shorter-wavelength region is still undergoing analysis. The region from 6.6 to 25 Å contains 38 lines, almost all of which can be identified with N VII, O VII and VIII, Ne IX and X, Mg XI, Al XII and XIII, Si XIV, S XVI, Fe XVII, and Ni XIX. The 11–25 Å band has also been measured with a pair of Bragg crystal spectrometers on a stabilized Skylark rocket at the University of Leicester (Pounds 1966). According to Pounds, "the sun was somewhat more active on the day of flight (5 May 1966) than for the previous NRL observation in 1963 and all the lines they reported below 20 Å are seen again, but considerably enhanced. In addition, four shorter wavelength lines now appear and these are of particular interest since three of them come from adjacent stages of ionization of Ne."

IDENTIFICATIONS

The very large number of unidentified lines in the solar spectrum is a continuing challenge to astrophysicists and laboratory spectroscopists. Even in that portion of the spectrum observable from the ground, i.e., between 3000 and 8000 Å, recent estimates by Mrs. Moore-Sitterly (1965) placed the number of unidentified lines in the Fraunhofer spectrum of the solar disk at about 7000. Many of these lines are quite strong (Merrill 1961). Identifications in the XUV solar spectrum are even less complete than in the visible. According to a recent estimate by Tousey (1966a), about one half of all observed Fraunhofer lines between 3000 and 2085 Å are unidentified. The status of identifications below 2000 Å has recently been reviewed by Jordan (1965b), who tabulates about 750 solar lines from published data in the region between 1994 and 13.7 Å. Although nearly 600 of the lines are assigned suggested identifications, many are based on extrapolated wavelengths rather than on direct laboratory data, and hence are subject to a great deal of uncertainty. The solar data are themselves incomplete in certain spectral regions, notably in the region 1750–1550 Å, which contains a large number of unpublished lines. Additional high-resolution solar spectra and continued efforts in the identification and classification of laboratory spectra are still badly needed. The suggestion by Sinanoğlu, Skutnik & Tousey (1966) that a number of far-ultraviolet solar-emission lines may be attributed to quarks has been criticized as unlikely by Bennett (1966).

Since the publication of Miss Jordan's table, the fourth-positive band system of CO has been identified as a very important contributor to the solar spectrum in the region 1550–1850 Å (Goldberg, Parkinson & Reeves 1965) and the classification of strong solar lines between 167 and 220 Å has been completed (Fawcett, Gabriel & Saunders 1967). Figure 5 shows the coincidence of CO bands photographed in absorption with a shock tube and spectrograph at the Harvard College Observatory and features in the NRL rocket spectrum of August 1962. It is likely that much fine detail in this region of the solar spectrum can be accounted for by the rotational structure of the CO bands (Porter, Tilford & Widing 1967), as is apparent from Figure 6.

For well over a year after they were first observed in the solar spectrum,

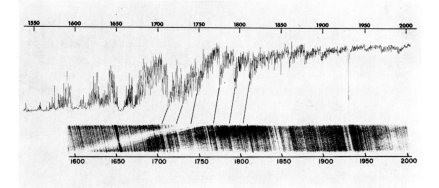

FIG. 5. *Lower:* The solar spectrum 1600–2000 Å. Photographed by the Naval Research Laboratory 22 August 1962. *Upper:* Densitometer tracing of CO shock-tube absorption spectrum. Photographed at the Harvard College Observatory. The connecting lines denote bands of the fourth-positive system of CO identified in the solar spectrum.

almost all of the very strong lines in the region 169–220 Å were unidentified, with no clues as to their origin. Their importance may be judged by the fact that their estimated total intensity exceeds that of He II 304 Å by more than a factor of 3 (Hinteregger, Hall & Schmidtke 1965). The first clue to their identification came when their presence was noted in the spectrum of the Zeta high-current toroidal discharge (Fawcett et al. 1963). Since the walls of the discharge chamber were lined with stainless steel, the lines were ascribed to iron, presumably in intermediate stages of ionization. The enhancement of these lines when iron was added to Theta-pinch discharges confirmed the identifications (Elton et al. 1964; House, Deutschman & Sawyer 1964). High-excitation laboratory sources excite several stages of ionization simultaneously and therefore special methods have to be devised for the assignment of lines to the correct ion and for their classification. Fawcett & Gabriel (1965) introduced iron into a thetatron by using a high-current spark across the ends of a metallic-iron strap shaped like a letter C. By using C straps constructed of other metallic elements, the same system of lines (8 strong lines and 20–30 weaker ones) could be produced and the wavelengths of lines studied in isoelectronic sequences. Alexander, Feldman & Fraenkel (1964) (see also Feldman, Fraenkel & Hoory 1965; Alexander, Feldman & Fraenkel 1965; Alexander et al. 1965) devised a method of differentiating between atomic spectra in high stages of ionization by photographing the spectra separately with a three-electrode spark and with a sliding spark arrangement. The ratio of the intensities of lines in the two sources is roughly constant and depends on the stage of ionization. Cowan & Peacock (1965) also have studied the iron lines in the range 167–220 Å using two sources of considerably dif-

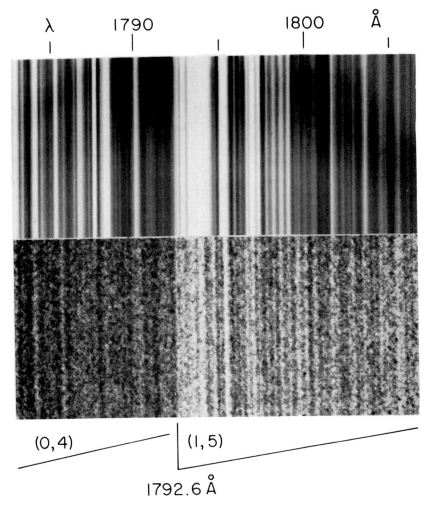

λ 1790   1800   Å

(0,4)   (I,5)

1792.6 Å

FIG. 6. Coincidences between rotational structure of 0, 4 and 1, 5 bands of the fourth-positive system of CO observed in the laboratory at the Harvard College Observatory at $T = 6000°K$, and lines in the solar spectrum photographed by the Naval Research Laboratory 22 August 1962.

ferent excitation and comparing the observed spectra with those predicted by atomic theory. Since most of the solar lines are considerably reduced in intensity in the low-excitation source, in which lines known to arise from Fe VII and VIII appear strongly, they concluded that most of the strong iron lines in this region of the spectrum belong to Fe IX and higher ionization stages.

The newest and definitive work by the Culham group (Gabriel & Fawcett 1965; Gabriel, Fawcett & Jordan 1965, 1966; Fawcett, Gabriel & Saunders 1967) is based on: (a) the form of the wavelength variation along the iso-electronic sequence; (b) previous knowledge in some cases of the levels at the beginning of the sequence; (c) recognition of the effects of configuration interaction; and (d) the use of a time-resolved spark to obtain separation of spectra from adjacent stages of ionization (Fawcett et al. 1967). All of the intense lines from 167 to 188 Å are accounted for by transitions of the form $3p^n3d - 3p^{n-1}3d^2$ and $3p^n - 3p^{n-1}3d$ in the ions Fe VIII-Fe XII (Gabriel et al. 1966). Intense lines between 188 and 220 Å are shown to be due to similar transitions in Fe XII-Fe XIV (Fawcett et al. 1967). Gabriel & Fawcett (1965) and Gabriel et al. (1966) have also observed several strong lines in the Zeta spectrum between 140 and 160 Å which have been classified as iso-electronic transitions in nickel similar to those of iron at 167–220 Å. These lines are also present in the solar spectrum, and together with one line of O VI account for all the strong solar lines between 140 and 160 Å. The majority of the emission lines in the solar spectrum in the range 60–170 Å are found to be due to iron and nickel, mostly the former. Finally, the coincidences between the laboratory spectra and the extremely weak features in the solar spectra (Hall et al. 1965), for example, at 103.6 Å, suggest to Gabriel & Fawcett that the solar lines represent genuine signals and are not noise pulses. Fawcett et al. (1967) have also produced spectra of Cr, Mn, and Fe in the Fe XIV–XVI sequences and of Mn, Fe, and Co in the Fe XVIII sequence from the plasma produced when a laser beam is focused on the solid surface of the element. Identifications of other solar emission lines in the region from 310 to 55 Å have been given by Zirin (1964) and are based on term differences from values listed in the tables of atomic energy levels.

Until recently, the permitted lines of Fe XIV had not been observed in the laboratory, and discussion of their identification in the solar spectrum was based on wavelengths that were either calculated or extrapolated from isoelectronic sequences beyond Sc IX (Tousey et al. 1965). Thus, multiplets terminating in the ground state and arising from the upper terms $3d\ ^2D$, $3p^2\ ^2P$, and $3p^2\ ^2S$ were expected to fall in the wavelength range 210–300 Å. One aid to identification in the absence of laboratory wavelengths is the requirement that two lines having a common upper state must be separated by an amount corresponding to the wavenumber of the green line at 5303 Å. The lines in this spectral region have also been observed from the OSO-1 satellite, and their intensity variations provide added clues to the identifications. Thus, Neupert and Smith (Neupert 1965) have grouped possible lines of Fe XIV into three multiplets by requiring that lines of a given multiplet all have the same time dependence, while Tousey et al. (1965) gave greater weight to the coincidences in wavelength and to the required wavelength separation of line pairs. The identifications proposed by Neupert and Smith for the P-S and P-P multiplets and by Tousey for the P-D multiplet are shown in Table III. After the multiplet and line designations, the third

TABLE III

IDENTIFICATION OF FE XIV LINES IN THE SOLAR SPECTRUM

Transition	ΔJ	λ (Å)	I (10^{-3} erg cm^{-2} sec^{-1})	I (calc.)	Notes
$3s^2\ 3p\ ^2P^0 - 3s3p^2\ ^2P$	1/2–1/2	257.32	6.5	0.32	Masked
	3/2–1/2	270.57	2.2	0.51	
	1/2–3/2	251.96	3.3	0.36	Blend
	3/2–3/2	264.29	3.1	1.60	Blend S X
$3s^2\ 3p\ ^2P^0 - 3s3p^2\ ^2S$	1/2–1/2	274.23	3.2	0.43	
	3/2–1/2	289.31	0.8	0.02	Blend S XI
$3s^2\ 3p\ ^2P^0 - 3s^23d\ ^2D$	1/2–3/2	211.32	8.0	1.10	Masked
	3/2–3/2	220.16	3.2	0.29	
	3/2–5/2	219.04	1.5	2.00	

and fourth columns give, respectively, wavelengths and intensities observed in the solar spectrum, the latter in 1963 (Hinteregger et al. 1964). Column 5 gives the calculated relative intensity in each multiplet. Tousey thought that the discrepancy between calculated and observed intensities was evidence that the lines of Fe XIV are all blended in the solar spectrum and hence that their presence had not yet been proved. Neupert and Smith were persuaded by the intensity anomaly to assign the solar lines at 203.8 and 211.3 Å as the leading lines of the P-D multiplet. However, the lines at 211.32 and 219.04 and 220.16 Å have now been observed in the laboratory at Culham (Gabriel et al. 1966, Fawcett et al. 1967) with the predicted theoretical relative intensities. The explanation of the anomaly seems to be that in the Sun the electron density is so low that the $^2P_{3/2}$ level is greatly underpopulated (Stockhausen 1965, Jordan 1965a). Consequently, the $^2D_{5/2}$ level is also underpopulated relative to $^2D_{3/2}$, since it is fed primarily by electron collision from $^2P_{3/2}$. Thus the 1/2–3/2 line at 211.32 Å is much enhanced relative to the 3/2–5/2 line at 219.13 Å.

Table IV, adapted from Tousey et al. (1965), summarizes the elements and stages of ionization, denoted by solid bars, that have been observed in the XUV spectrum. The solid dots also mark the ions observed at longer wavelengths above λ 2500 Å and represented by permitted lines in the case of neutral and singly ionized atoms and by forbidden lines in the case of highly ionized atoms. The numbers above each element are the logarithms of the abundance numbers relative to hydrogen for which log N is arbitrarily set equal to 12.0. The numbers in parentheses refer to those elements which are not observed in the Fraunhofer spectrum; their abundances are therefore estimated from the intensities of their lines in the XUV spectrum. Tousey et al. (1965) point out that the absence of almost all lines in the fluorine and neon isoelectronic sequences is surprising. In reality, several lines in both sequences have already been observed, although their intensities are low. Pottasch (1964b) suggested that two resonance lines of Si VI in the fluorine sequence were weakly present in the Sun at 246.0 and 249.1 Å, together with

TABLE IV

Elements and Ions Observed in the Solar Spectrum[a]

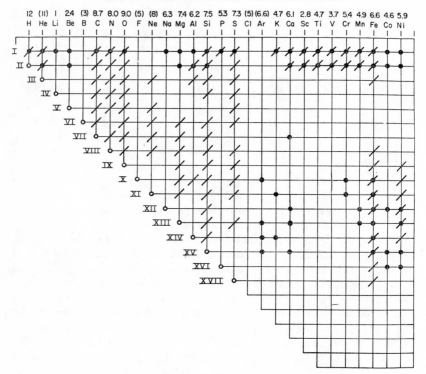

[a] Slant lines denote ions observed in the XUV spectrum, filled circles those observed at $\lambda > 2500$ Å. Numbers above each element are the logarithms of the abundance numbers relative to hydrogen. The abundances are usually taken from analyses of the Fraunhofer spectrum, but numbers in parentheses refer to elements absent in the Fraunhofer spectrum, whose abundances are estimated from line intensities in the XUV spectrum or in other stars or nebulae.

the corresponding resonance transition of S VIII at 198.6 Å. The newer spectra shown in Figure 3 strengthen the case. Similarly, Manson (1967) has identified a line at 97.1 Å as the resonance line $2p^6\,{}^1S_0$-$2p^5 3d\,{}^1P_1$ of Si V in the neon sequence. The weakness of the lines in the neon sequence may be connected with the fact that the ground configuration is a closed shell and hence the energy required to excite the resonance lines is relatively large in relation to the ionization energy. Table IV shows no dearth of lines from ions requiring between 100 and 300 eV for formation, which had been inferred earlier by Zirin & Dietz (1963) and attributed to a steep rise of temperature with height in the chromosphere.

This very brief review of accomplishments in XUV spectroscopy shows that while much progress has been made during the past few years, a great deal remains to be done before the data can be considered adequate for the solution of basic problems in solar physics. In particular, the spectral resolution and purity are not good enough to permit the determination of accurate line profiles, with the possible exception of such strong emission lines as Lyman α and the Mg II lines at 2800 Å. Furthermore, very meager information has been acquired on the spatial dependence of spectra—the manner in which they vary from center to limb and above active regions. Probably the most important need is for accurate intensity calibration of spectral data, both relative and absolute. The problem of calibration is particularly acute at wavelengths below 500 Å, where most workers agree that absolute calibrations may be in error by as much as a factor of 4, although the most recent work, by Manson (1967), is a refreshing exception.

INTERPRETATION

Photosphere-chromosphere.—Throughout most of the visible spectrum, the equivalent black-body temperature, or brightness temperature of the solar disk is nearly independent of wavelength, reflecting the slow variation with wavelength of the continuous absorption coefficient of the negative hydrogen ion. Below 4000 Å, however, photoionization continua of abundant neutral atoms prevail over H^- and consequently, as the absorption coefficient per gram increases with decreasing wavelength, the brightness temperature begins to diminish rapidly, as the observed radiation is emitted by higher and cooler layers of the photosphere. Measurements made on rocket spectra (Detwiler et al. 1961, Tousey 1964) show that the brightness temperature shows a rather steady decrease, punctuated by dips and depressions caused by the enhanced absorption of various individual atoms and molecules, until a minimum is reached near 1500 Å where the brightness temperature is about 4700°K (see Figure 7). The minimum in brightness temperature is caused by a corresponding minimum in the solar temperature at the interface between the photosphere and the chromosphere. Shortward of 1500 Å, the brightness temperature increases with decreasing wavelength as the radiation emanates from progressively higher layers in the chromosphere. Series limits are now marked by local increases in emission, e.g., those attributed to Si I, 1525 Å, C I, 1101 Å, and H I, 912 Å.

At λ 2085 Å, a sudden increase in opacity causes the brightness temperature to fall abruptly from 5500°K at $\lambda > 2100$ Å to 5050°K near 2050 Å (Tousey et al. 1965). At the same time, the Fraunhofer lines display a corresponding drop in intensity and their number decreases very rapidly. According to Tousey et al. (1965) "the Fraunhofer lines all but vanish by 1690 Å and none have been detected below 1530 Å." At approximately the same wavelengths that show a minimum in the brightness temperature, the solar disk should be uniform in brightness and should exhibit limb brightening at still shorter wavelengths. The observations show that between 2000 and

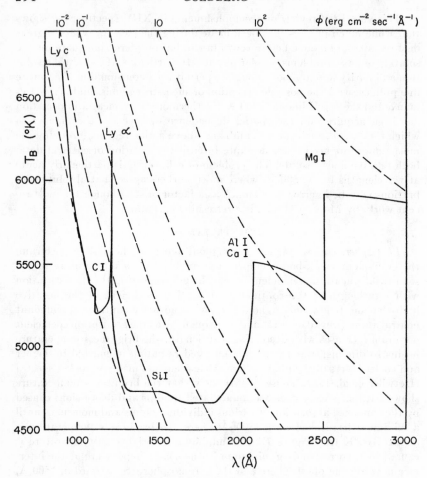

FIG. 7. Variation of brightness temperature of the solar continuum with wavelength. The dashes denote lines of constant flux in ergs cm^{-2} sec^{-1} Å^{-1} incident at the top of the Earth's atmosphere.

1530 Å there is no change in the intensity of the continuum to within 20 sec of the limb, which is about the limit set by the accuracy of the pointing control. Below 1530 Å, both the continuum and lines show marked limb brightening.

Several attempts have been made to calculate the flux of the ultraviolet continuum according to various assumptions as to the solar model and sources of opacity (Widing 1961, de Jager 1963, Matsushima 1964). Kodaira (1965) has confirmed the suggestion by Tousey (1964) that the depression in the solar spectrum at 2085 Å may be attributed to the photoionization of Al I from the ground level $3p\ ^2P^0$, on the basis of theoretical calculations with a model atmosphere and calculated values of the absorption cross section (Vainstein & Norman 1960, Peach 1962). The calculated cross section is in agreement with recent laboratory measurements with the shock tube (Parkinson & Reeves 1966).

The most thorough calculations of the solar continuum have recently been made by Gingerich & Rich (1966) in the wavelength region 2500–1400 Å. Figure 8 shows the results of computations performed using the Utrecht reference model of the solar atmosphere (Heintze, Hubenet & de Jager 1964) together with recently measured experimental values of silicon absorption coefficients (Rich 1966a). Also shown is the flux observed by Detwiler et al. (1961). The drop in intensity at 2508 Å seems to be satisfactorily accounted for by the absorption edge due to Mg I, as had already been shown by Kodaira (1965). The most striking result, however, is the very large value of the absorption cross section for photoionization from the first excited term ^1D of Si I at 1682 Å, which suggests that all continuum radia-

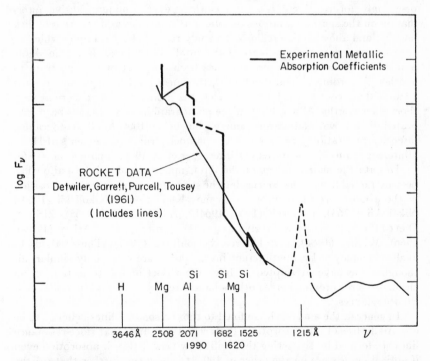

FIG. 8. Comparison between observed and calculated solar fluxes in the ultraviolet solar spectrum. Light full line: observed flux (Detwiler et al. 1961). Heavy full line: calculated flux using Utrecht reference model and experimental metallic absorption coefficients. The dashed portion of the curve was derived using an absorption coefficient calculated for the Si I series limit by the quantum defect method.

tion shortward of 1682 Å originates in the low chromosphere and hence that no Fraunhofer lines should be visible below this wavelength. The excess absorption between 1525 and 1680 Å is probably caused by overlapping lines of the fourth-positive system of CO, which was identified in that region of the solar spectrum by comparison with shock-excited laboratory absorption spectra (Goldberg et al. 1965). Careful examination of the solar spectra (Porter et al. 1967) also reveals that several of the CO bands appear in absorption below 1682 Å, including the (0, 0) and (1, 1) bands near 1550 Å. This implies that the Utrecht model employed by Gingerich & Rich needs revision at the temperature minimum. Possible modifications have been discussed in detail by Rich (1966b).

Chromosphere-corona.—In principle, the accurate measurement of emission-line profiles and their center-to-limb variations should elucidate the structure of the chromosphere-corona. The fact is, however, that except for Lyman α and Lyman β, and the emission cores of the H and K lines of Mg II, nothing is known about the shapes of far-ultraviolet emission lines. Moreover, only for relatively few lines is there even semiquantitative information on the spatial variations of total line intensities as a function of disk position and above active regions. The only really solid data upon which to base analyses are measurements of the total line emission from the whole Sun and even here the accuracy varies from one part of the spectrum to another. According to Tousey (1966a), the uncertainties vary from between 10 and 30 per cent at the longer wavelengths to factors of 4 or more at very short wavelengths. Above 1000 Å, the photographic spectrum has been calibrated directly with laboratory sources, but below 1000 Å, all photographic intensity calibrations are based on the photoelectric measurements of H. E. Hinteregger and his co-workers (Hinteregger et al. 1964, Austin et al. 1966).

Despite the shortcomings of the data, much valuable information has been extracted from them concerning the structure and chemical composition of the atmosphere in analyses by Ivanov-Kholodnyi & Nikol'skii (1963), Nikol'skii (1963), Pottasch (1963, 1964b), Allen (C. W., 1965), Zirin & Dietz (1963), Suemoto & Moriyama (1964), Jordan (1966), Athay (1965b, 1966), Widing (1966), and Dupree & Goldberg (1967). The methods of analysis employed by the different investigators are essentially similar, although some have attempted to infer the model of the transition-region chromosphere-corona whereas others have stressed the derivation of chemical abundances.

In general, the analysis is confined to those resonance lines believed to be essentially free of self absorption. Pottasch (1963) finds that since resonance lines are formed by scattering they will show very little self absorption even if optical depths are on the order of 100. It is assumed further that (a) the upper level of the transition is populated by electron impact and depopulated by spontaneous line emission and (b) the ionization equilibrium is determined by the balance between collisional ionization by electron impact and radiative recombination. It has recently been established (Burgess

1965a, b) that dielectronic recombination is more important under coronal conditions of high temperature and low density than radiative recombination, and the newest calculations (Burgess & Seaton 1964; Allen, J. W. 1965; Jordan 1966; Widing 1966) take account of this process.

With the foregoing assumptions, the flux E in ergs cm^{-2}sec^{-1} incident on the top of the Earth's atmosphere from the whole Sun may be written as follows (Pottasch 1964b):

$$E = 2.4 \times 10^{-20} g f_{\mathrm{lu}} A_{\mathrm{el}} \int \frac{N_i}{N_{\mathrm{tot}}} N_e^2 T_e^{-1/2} 10^{-5040W/T_e} dh \qquad 1.$$

where the integration is taken over the region of the atmosphere emitting the line. In Equation 1, T_e and N_e are the temperature and electron density, g is the statistical weight of the ground level, f_{lu} is the absorption oscillator strength of the line, W is the excitation potential in volts. The ratio N_i/N_{tot} is the fractional concentration of the element in a given stage of ionization while A_{el} is the abundance of the element relative to hydrogen. Although the use of h, the height above the Sun's limb, as the independent variable implies spherical symmetry, the integration may just as well be taken over the emitting volume, in which case spherical symmetry need not be assumed. In practice, values of the integral are derived empirically from observations of E and then used either to obtain relative values of A_{el}, in which case no assumption need be made about spherical symmetry, or to infer a model of the transition-region chromosphere-corona, which is possible only if spherical symmetry or some other assumption about the angular temperature and density distribution is adopted. It is not correct (Nikol'skii 1963, Ivanov-Kholodnyi & Nikol'skii 1963), however, when deriving a model of the transition-region chromosphere-corona, to apply Equation 1 with its implication of spherical symmetry to such lines as those of Mg X, Si XII, Fe XV, Fe XVI, etc., which are radiated almost entirely by localized active regions.

Equation 1 is further simplified by assuming that within the emitting region, the function

$$g(T_e) = \frac{N_i}{N_{\mathrm{tot}}} T_e^{-1/2} 10^{-5040W/T_e} \qquad 2.$$

has a constant value equal to 0.7 times its maximum value:

$$\langle g(T_e) \rangle = 0.7 g(T_{\mathrm{max}}) \qquad 3.$$

Thus the final simplified expression for E becomes

$$E = 2.4 \times 10^{-20} g \cdot f_{\mathrm{lu}} \langle g(T_e) \rangle A_{\mathrm{el}} \int_R N_e^2 dh \qquad 4.$$

where the integral is understood to be limited to the region R of line formation.

Calculation of the function $g(T_e)$ gives for each line the range ΔT_e in temperature over which most of the radiation is emitted as well as the value of T_e for which the emission is a maximum. If the abundance of the element relative to hydrogen is known, observed values of E give a relation between the quantity

$$\int_R N_e^2 dh \qquad\qquad\qquad 5.$$

and the mean temperature of line formation. This relation may be used to check the consistency of atmospheric models derived from other data. It was shown by Ivanov-Kholodnyi & Nikol'skii (1961), who first constructed such a curve of "emission measure" from observational data, that systematic deviation of the lines of a given element from the curve might be due to errors in the assumed abundance. Pottasch (1963, 1964b) has used the method to derive the chemical abundances of the heavy elements by plotting the product of the integral and the abundance ratio in Equation 4 against the temperature separately for the different ions of a given element and then determining the constant factor required to bring the separate curves into coincidence. It is assumed that the integral of the square of the electron density, taken over the entire emitting volume, will on any given day always vary smoothly with the mean temperature of line formation, although the shape of the curve must depend on the degree of solar activity. This procedure gives the relative abundances of the heavy elements. The abundance relative to hydrogen is then obtained by making use of values of $\int N_e^2 dh$ derived from radio-frequency measurements made at approximately the same phase of the solar cycle as the ultraviolet measurements.

The most recent results obtained in this way by Pottasch (1964b, 1965) are shown in the second column of Table V together with photospheric val-

TABLE V

SOLAR ABUNDANCES DERIVED FROM PERMITTED XUV LINES[a]

Element	Pottasch (20)	Athay (131)	Photosphere (139)
H	1,000,000		1,000,000
C	600	610	520
N	60	110	95
O	450	420	910
Ne	50	32	
Mg	90	32	25
Al	5	2	1.6
Si	100	32	32
P	0.8	.22	0.22
S	14	20	20
Ca	3		1.4
Fe	40	3.8	3.7
Ni			0.8

[a] Values by Athay are given relative to the photospheric value of the Si abundance. Other values are relative to that of $H = 10^6$.

ues derived from observations of the Fraunhofer spectrum (Goldberg, Müller & Aller 1960), which appear in the fourth column. For the elements C, N, O, S, and Ca, the differences are well within the uncertainties of the values. For Mg, Al, and Si, the photospheric abundances are smaller than the coronal values by about a factor of 3 while for Fe the discrepancy is a factor of 10. The relative abundance O to Si is found to be about 30 in the solar photosphere but only 4.5 in the corona. It is doubtful whether a discrepancy of even a factor of 3 is outside the error of the derived abundances, and hence the disagreement in the O:Si ratio is not particularly disturbing. But the relatively large coronal abundance found for Fe cannot be explained away so easily, particularly since an abnormally high Fe:H ratio is also obtained from the intensities of forbidden coronal lines in the visible spectrum (Pottasch 1964a).

In Athay's (1966) reformulation of Pottasch's analysis, the temperature gradient is introduced explicitly by changing the independent variable in Equation 1 from h to T_e, as follows:

$$dh = (dT_e/dh)^{-1} \cdot dT_e \qquad\qquad 6.$$

The electron pressure is then assumed to be constant in the transition region, so that

$$P_0 = N_e T_e = \text{const} \qquad\qquad 7.$$

On the basis of a model by Pottasch (1964b), Athay finds that the value $P_0 = 6 \times 10^{14}$ is a reasonable one for the chromosphere-corona. Equation 7 may be used to eliminate the electron density from Equation 1. If the temperature gradient is assumed constant in the region of line formation for each ion, we may compute the quantity

$$A_{\text{ol}} \left(\frac{dT_e}{dh}\right)^{-1} = E\left(2.4 \times 10^{-20} g f_{1u} P_0^2 \int \frac{N_i}{N_{\text{tot}}} T_e^{-5/2} 10^{-5040W/T_e} dT_e\right)^{-1} \qquad\qquad 8.$$

as a function of the mean temperature for each emitting ion. The relative abundances may then be derived, as in the Pottasch analysis, by superposition of the curves for different elements.

Athay notes that the temperature gradient probably does not have a simple monotonic form below $T_e = 10^5$ °K, e.g., the temperature distribution may exhibit plateaus. Therefore, the abundances should be derived only from the high-temperature portion of the curve, which characterizes the transition-region chromosphere-corona. In this region, where T_e varies between 10^5 and about 10^6, Equation 8 is found to fit the relation

$$T_e^{5/2} \cdot (dT_e/dh) = \text{const} \qquad\qquad 9.$$

which corresponds to a constant conductive energy flux of 5×10^5 ergs cm^{-2} sec^{-1}, assuming $A_{\text{Si}} = 3.2 \times 10^{-5}$ and no magnetic field (Athay 1966). The relative abundances derived by Athay, using essentially the same data as Pottasch, are given in the third column of Table V, where the abundance for Si has arbitrarily been set equal to its photospheric value. With the exception of O, which differs by a factor of 2, all elements are found to have abundances within 30 per cent of the photospheric values.

Neither Pottasch nor Athay has taken account of dielectronic recombination in his calculations. Its influence on the ionization equilibrium was first investigated in the case of Fe by Burgess & Seaton (1964), who found that it increases the temperature at which each ion is in maximum abundance by about a factor of 2 and also extends the range of temperature over which the ion emits. They also found that the ratio N_i/N_{tot} is not greatly affected by the change in temperature. Pottasch (1964b) points out that the two other temperature-dependent factors entering into the expression for the line intensity, namely, $T_e^{-1/2}$ and $10^{-5040W/T_e}$, change in opposite directions and hence that the abundance results are not likely to be affected in any major way by the introduction of dielectronic recombination. It is by no means clear that other elements will show the same behavior as Fe and in fact there is considerable evidence that they do not. For example, Allen's (J. W., 1965) investigation of oxygen shows that the fractional concentrations of O VII and O VIII as a function of T_e are very little affected by dielectronic recombination, and also that for O VI the maximum value of the concentration is abnormally low.

The reason why dielectronic recombination does not always act to increase the temperature at which ionization equilibrium curves maximize is readily apparent. Owing to the fact that the rate coefficient for dielectronic recombination depends on the product of $T^{-3/2}$ and an exponential in $1/T$, its value goes through a maximum at some value of the temperature and then declines as $T_e^{-3/2}$. In general, dielectronic recombination will lead to a substantial change in the ionization equilibrium curves only when the electron temperature is greater than the temperature corresponding to the maximum value of the rate coefficient. The recombination coefficient will have its maximum value at a temperature such that the rms velocity of the free electrons corresponds to the energy of the resonance transition of the recombining ion. By a coincidence, the temperature is approximately equal to the wavenumber in cm^{-1} of the resonance line.

Analyses of solar ultraviolet emission intensities with dielectronic recombination included have recently been carried out by Widing (1966), Jordan (1966), and Dupree & Goldberg (1967). Widing finds that silicon line intensities in the XUV spectrum are consistent with a temperature of 1.6–1.8×10⁶ °K, whereas the intensities of Fe XV and Fe XVI lines in the spectrum of an active center require a temperature in the range 3–5×10⁶ °K and values of $\int N_e^2 \, dh$ 50–80 times greater than in quiet coronal regions.

Jordan (1966) has applied the Pottasch method of analysis to the newly identified lines of Fe IX–XI and XIV, and Ni X–XIII in the spectral region 100–274 Å and has also included Fe XV 284 Å and lines of Si VI–X in the region 215–296 Å in deriving the abundance of Ni and Fe relative to Si. The Fe:Ni ratio of 12.6 is not greatly different from that in the photosphere, but the Fe:Si ratio of 1.3 is about an order of magnitude larger in the corona. The principal advantage of the new data is that the Fe and Ni ions together define the shape of the function $\int N_e^2 \, dh$ at high temperatures between 6×10^5 and $2.5 \times 10^{6\circ}$K. Thus the curve shows a maximum at $T_e = 1.4 \times 10^{6\circ}$K, which Jordan interprets as the temperature of the "normal" corona, the higher temperatures corresponding to those of active regions. The observations were taken near solar minimum, in March 1964 (Hall et al. 1965), and it may be presumed that at solar maximum the greatly enhanced emission from active regions would cause the maximum in the curve to disappear.

Although in principle the use of intermediate stages of ionization of Fe and Ni to define the curve of emission measure is sound, the relatively low accuracy of the data is troublesome. Thus, the published line intensities for wavelengths shorter than 256 Å are judged only to be "correct to an order of magnitude down to 100 Å" (Hall et al. 1965). The silicon lines lie between 220 and 296 Å, but the iron lines, with two exceptions, are in the region below 180 Å, where the calibration error may be substantial.

Dupree & Goldberg (1967) limit their analysis to lines of Si, O, and Fe, lying above 246 Å, for which reliable intensities have been compiled by Pottasch (1964b). Whenever required, new f values were calculated by use of a Hartree-Fock approximation, with configuration interaction included when necessary. The results are illustrated in Figures 9–11. In Figure 9 log A_{Si} $\int N_e^2 \, dh$ is plotted against log T_e with relative abundances Fe:Si and O:Si as derived by Pottasch (1964b). The scatter is undesirably high for $T_e > 10^5$°K. A much smoother curve is obtained with the photospheric abundances, as shown in Figure 10. Finally, in Figure 11 the reciprocal temperature gradient, derived with $P_0 = 6 \times 10^{14}$, is plotted against temperature, following Athay (1966). Again, the assumption of photospheric abundances brings the three elements together on a common curve above $T = 10^5$°K. The dashed line corresponds to a constant conductive flux of 6×10^5 ergs cm^{-2} sec^{-1}, which is in good agreement with Athay's (1966) value.

As first shown by Pottasch (1963), the relative abundances derived from ultraviolet emission lines may be related to the abundance of hydrogen by making use of measurements of undisturbed solar radio emission at centimeter-meter wavelengths. For each of several assumed values of the abundance ratio A_{Si}, Figure 10 gives a relation between $\int N_e^2 \, dh$ and T_e. With this relation and the absorption coefficient for bremsstrahlung emission, the solar radio flux as a function of wavelength may be predicted for each assumed value of A_{Si}. In this way, Dupree & Goldberg (1967) found that the photospheric abundance value $A_{Si} = 3.2 \times 10^{-5}$ gave predicted brightness temperatures for wavelengths between 10 and 300 cm that fell midway between those observed at sunspot maximum and sunspot minimum. The relevant ultraviolet emission-line intensities were measured principally during 1960–61, approximately at the halfway point in the cycle. The conclusion is that the correctness of the photospheric abundances is supported by the observed radio emission of the undisturbed Sun.

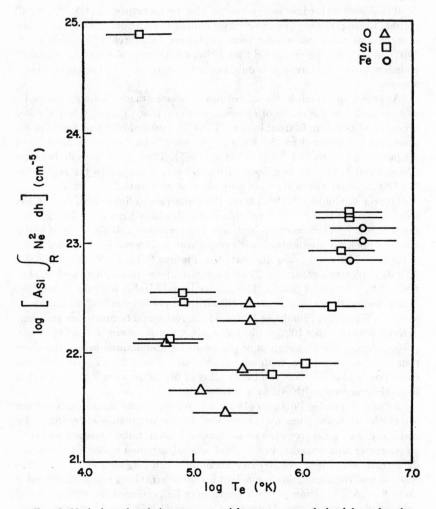

FIG. 9. Variation of emission measure with temperature derived from far-ultra-violet line intensities. Assumed relative abundances Fe:Si and O:Si according to Pottasch (1964b).

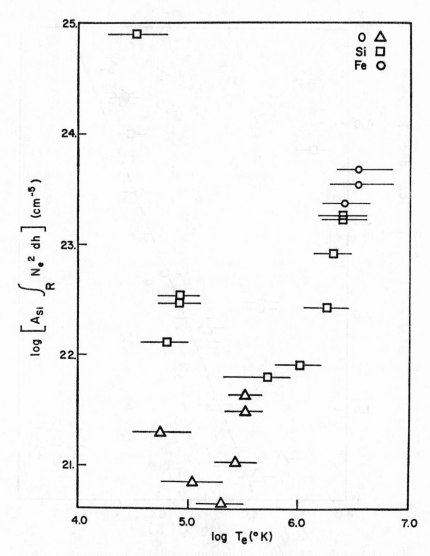

FIG. 10. Variation of emission measure with temperature derived from far-ultraviolet line intensities. Assumed relative abundances Fe:Si and O:Si according to Goldberg et al. (1960).

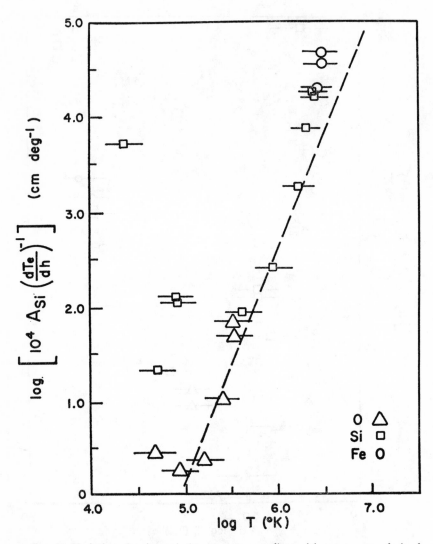

F𝚒𝚐. 11. Variation of reciprocal of temperature gradient with temperature derived from far-ultraviolet line intensities. Assumed relative abundances Fe:Si and O:Si according to Goldberg et al. (1960).

Finally, the relation between the temperature gradient and the temperature may be used to derive a temperature model of the transition region, provided that a value of the temperature is assumed at some arbitrary height. Athay adopts $T = 5 \times 10^{4}°K$ at $h = 2000$ km from consideration of the intensities and heights of hydrogen and helium lines in the flash spectrum. A similar model derived from Figure 11 is shown in Figure 12. The model is characterized by an extremely steep rise from chromospheric to coronal temperatures. The dashed portion of the curve is derived from line emission originating principally from active regions and therefore should not be regarded as real. The steep rise in temperature is in agreement with predictions by Athay (1965b) based on a new method of calculating the total intensities of strong emission lines in solar and stellar chromospheres (1965a). Athay finds that the observed intensities of resonance and subordinate lines of He I and He II and of Lyman α and Lyman β of H are consistent with their formation in a layer no greater than 2000–3000 km thick with $T_e \approx 25,000$–$30,000°K$ at its

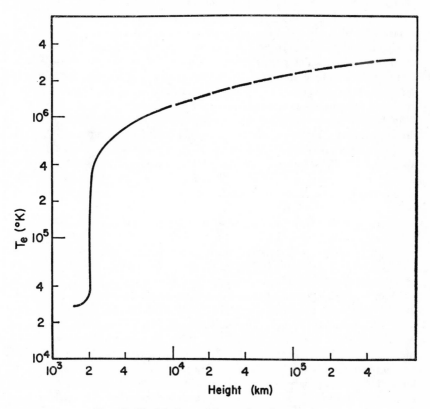

FIG. 12. Model of transition-region chromosphere-corona from far-ultraviolet emission lines.

base at about $h = 1000$ km and with $T_e \approx 50,000–60,000°K$ at its top. It is also required that the temperature immediately above the layer rise sharply to coronal values.

SOLAR IMAGING

ULTRAVIOLET IMAGES

Monochromatic images of the Sun in such lines as Hα of hydrogen and the K line of ionized calcium have provided basic information on velocity and magnetic fields and on transient phenomena in the low to middle chromosphere. Similarly, high-resolution spectroheliograms made in various XUV line radiations extend these studies to the high chromosphere and corona and thereby help to define the mechanisms that heat the chromosphere-corona as well as to clarify the physical nature and height of origin of solar activity. The first high-quality ultraviolet spectroheliogram was made in the light of Lyman α of hydrogen in 1959 by the Naval Research Laboratory (Tousey 1963). Progress in the technique of recording in other lines has been quite slow but is beginning to accelerate. The difficulty is that narrowband interference filters like those in use for Hα are not feasible in the extreme ultraviolet, and true spectroheliographs with moving slits are not practical, at least in rockets. The procedure followed by NRL is simply to use a wide entrance slit or to eliminate it from the spectrograph. This technique is satisfactory only when the spectral lines are rather widely separated. The first such spectroheliograms in lines other than Lyman α were obtained on May 10 and September 20, 1963 (Tousey 1964). The principal conclusions drawn from these rather rough photographs (Tousey et al. 1965) are that the solar disk shows little or no limb brightening in such lines as Lyman α and β and C III 977.02 Å, but quite pronounced limb brightening in O VI 1031.91 Å, 1037.61 Å, and Mg IX, 368 Å. The lines of very highly ionized atoms, such as Fe XV 284 Å and Fe XVI 335 Å, seem to emit radiation only from plage regions.

Ultraviolet images of much improved quality, showing a wealth of new information, have been obtained in a number of new NRL flights in 1966, made with a single-pass concave-grating spectrograph at normal incidence (Tousey 1966b). As shown in Figure 13, the objective-grating spectra made on 28 April 1966 extend in wavelength from 150 to 700 Å. The resolution is ∼10 arc sec. Also shown are the Fraunhofer Institute maps for the period 26–30 April 1966. The arrows at the extreme right-hand side of the spectra mark the location of plages, the numbers being those assigned to the plages by the McMath-Hulbert Observatory. With the exception of the He II image 304 Å, all lines in the region 171–400 Å appear as bright coronal rings and in plage areas. Note that lines arising from successively higher stages of ionization form a progression with increasing wavelength and that the plage intensities also increase in the same direction. There are also striking differences in the appearance of the limb in different lines, some of which show more or less uniform brightening all around the limb (O IV 555 Å) while

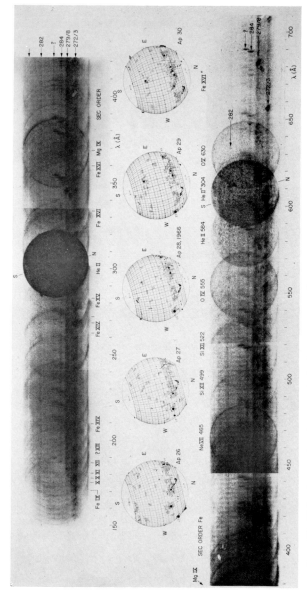

Fig. 13. Objective-grating images of the Sun 150–700 Å. Photographed by the Naval Research Laboratory 28 April 1966. Fraunhofer Institute maps for the period 26–30 April 1966 are shown for comparison.

others show gaps in brightness at the poles (Mg IX 368 Å) or in local regions elsewhere on the limb (Ne VIII 465 Å).

The emission in lines of very highly ionized atoms (Mg IX, Fe XII–XVI) extends well into the corona above the limb at 70° latitude, as may be seen in Figure 13. In a later flight on 27 July 1966, a heliograph consisting of an off-axis paraboloidal mirror, with filters transmitting the entire band from 171 to 400 Å, produced photographs which revealed emission above active regions extending at least to 30 arc min above the limb (Purcell & Tousey 1966). Figure 14 shows a reproduction of three exposures with the heliograph together with a K-line spectroheliogram taken on the same day at the Mc-Math-Hulbert Observatory. Note the gap in limb brightening at the south limb and the strong enhancement of the plage areas. The resolution of the images is ∼1 arc min. Images in the same band of wavelengths have also been photographed with pinhole cameras (Zhitnik et al. 1964).

Referring back to Figure 13, we note that both He II 304 Å and He I 584 Å show the chromospheric network, but that it is weak or absent in O IV 554.5 Å, O V 629.7 Å, and lines of still higher ionization. Figure 15 compares two He images with two Ca–K spectroheliograms of the same date, the upper from Mt. Wilson and the lower from McMath-Hulbert. The He and Ca images resemble one another in considerable detail, but also exhibit important differences. For example, the He images are uniformly bright, except for a small enhancement in a narrow zone at the very limb and pronounced weakening over the polar regions, while the Ca images show uniform limb darkening. The prominence on the southwest limb is real, but the feature in the northeast is an overlapping plage from the adjacent image in Fe XVI. In general, plages in He closely resemble those in Ca, but the correspondence is not exact. Tousey notes a small plage in Ca that is completely absent in He, as well as other smaller differences.

The Culham Laboratory in the United Kingdom has developed a compact extreme-ultraviolet spectroheliograph in which a pinhole camera is combined with a plane diffraction grating used at grazing incidence (Burton & Wilson 1965). The instrument was flown in a stabilized Skylark rocket on 9 April 1965 and it recorded monochromatic images at 304 Å and at 171 Å (Fe IX). The emission in the wavelength band between 60 and 150 Å was found to be strongly limb brightened. More conventional techniques employing transmission optics may be used in the near ultraviolet. K. Fredga (1966) has obtained the first monochromatic photographs of the Sun in the Mg II line at 2803 Å, with a Cassegrain-Maksutov telescope and a Solc-type birefringent filter flown in Aerobee rockets on 12 April and 2 December 1965.

X-Ray Images

The distribution of X-ray-emitting sources over the solar disk has been studied in four different ways: 1. From the widths of X-ray emission lines observed with a Bragg crystal spectrometer (Blake et al. 1965). 2. By means of slit scans of the solar disk in broad X-ray wavelength bands (Blake et al.

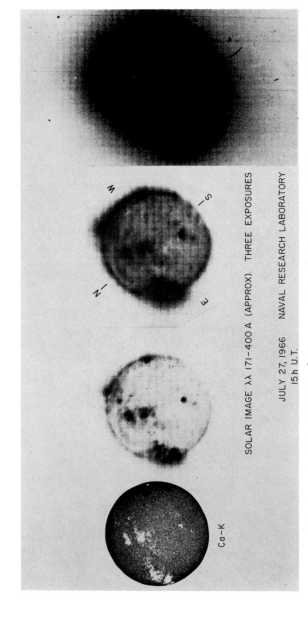

FIG. 14. Solar images in the band 171–400 Å. Photographed by the Naval Research Laboratory on 27 July 1966, at three different exposures. A Ca-K spectroheliogram photographed on the same day at the McMath-Hulbert Observatory is shown for comparison at the left.

1964, 1965). 3. By pinhole photography (Burton & Wilson 1965; Russell 1965a, b; Russell & Pounds 1966). 4. By photography with a grazing-incidence imaging telescope (Giacconi et al. 1965, Underwood & Muney 1966). The principal conclusions derived from these observations are as follows: 1. Relatively soft X radiation in the band 44–60 Å seems to be distributed uniformly over the quiet solar disk and is also enhanced over active regions. It is also more intense at the limb except near the poles. 2. Sources of harder X rays, in the wavelength band 8–12 Å, are geometrically smaller than 44–60 Å sources, some having dimensions of the order of 1 arc min (Blake et al. 1965); virtually all of the radiation observed in the 8–15 Å band originates in active regions (Blake et al. 1965, Giacconi et al. 1965). 3. Large regions of low X-ray intensity are found to be distributed across the solar disk (Giacconi et al. 1965). 4. The X-ray intensity in the band 8–20 Å correlates well with the intensity of coronal green-line emission indicated by Fraunhofer maps (Giacconi et al. 1965). 5. Coronal X-ray emission at the limb appears to be considerably softer than that associated with plage regions (Giacconi et al. 1965). From the observation that the softer radiation sources seem to have dimensions comparable with those of large plage regions whereas the 8–12 Å

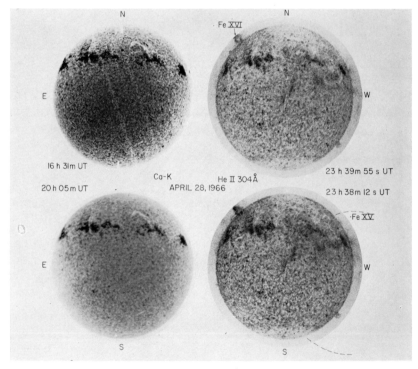

FIG. 15. Solar images in He II 304 Å (right) and Ca-K (left).

sources have much smaller dimensions, it is suggested that coronal condensations are large regions of relatively low temperature in which small, hotter centers are imbedded (Blake et al. 1965).

Most of the conclusions above are readily apparent from the reproductions in Figure 16, which are the newest and best photographs made to date (Underwood & Muney 1966). The photographs were taken from an Aerobee rocket on 20 May 1966, 15^h 12^m UT, with a grazing-incidence telescope of the Giacconi-Rossi type through a series of filters transmitting different wavelength bands. Thus, the upper right was taken through a Mylar-Al filter passing the two bands 3–13 and 44–60 Å, the lower left with an Al filter transmitting 8–20 Å and the lower right Be passing 3–11 Å. For comparison, an Hα filtergram taken on the same day at the Goddard Space Flight Center is shown at the upper left.

SOLAR MONITORING

EXTREME-ULTRAVIOLET SPECTRA

Up until now, monitoring of extreme-ultraviolet solar spectra from satellites has been limited to the observation of radiation from the whole disk. It is possible nevertheless to learn a good deal about the spatial distribution of the emission from such observations, as Neupert (1965) has shown from analysis of photoelectric measurements of the solar spectrum between 170 and 400 Å, obtained from the first Orbiting Solar Observatory (OSO-1) during March, April, and May 1962. Since the flux in most extreme-ultraviolet emission lines is greatly enhanced over active regions, the line intensities display a periodic variation in phase with the solar rotation. The total variation in the line 304 Å He II is ∼33 per cent, whereas in the line 284 Å Fe XV, the variation is more than a factor of 3. These examples serve to illustrate the general rule that the intensity variations of lines are greater the greater the electron temperature required to produce the ion. This rule is illustrated in Figure 17, in which the relative increase in flux due to the presence of an active region on March 22, 1962 is plotted against the electron temperature at which each ion has its maximum concentration. The curve given originally by Neupert (1965) has been corrected by Mrs. A. K. Dupree for the effects of dielectronic recombination. The form of the curve varies from day to day depending on the amount of solar activity, but as a fairly general rule, those ions that exist at electron temperatures below about 1,600,000° K (e.g. Si VIII–X, and Mg VIII–IX) seem to show little association with active regions as contrasted with those at higher temperatures. All lines show an increase in activity during the first two weeks in March, which Neupert suggests is due to an increase in density in and around the active region. It is not clear whether the large increases in the intensities of the lines of Fe XV and XVI are due to a combined increase of electron temperature and density over plages, or whether localized regions in which these emissions occur merely increase in number over plages.

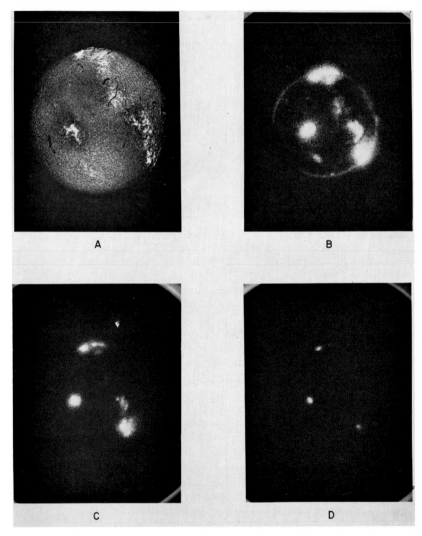

FIG. 16. X-ray photographs of the Sun by the Goddard Space Flight Center of NASA.

A: Hα filtergram.
B: wavelength bands 44–60 and 3–13 Å.
C: wavelength band 8–20 Å.
D: wavelength band 3–13 Å.

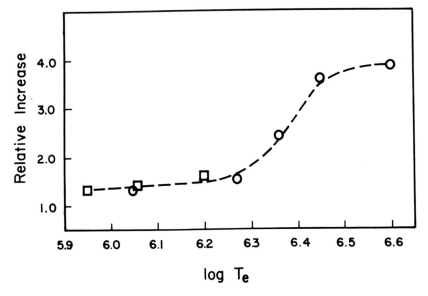

FIG. 17. Relative increase in flux of far-UV emission lines during passage of active region on 22 March 1962. The abscissa is the electron temperature at which the emitting ion is at maximum concentration.

The intensity variations in solar emission lines are closely correlated both with the flux of 2800 MHz radiation and with the presence of plage regions in Ca+. The correlations are not exact and do not extend to small secondary maxima and minima. According to Neupert (1965), the 284 Å radiation increases more slowly than does the microwave radiation as the active center develops, but remains intense even after the sunspots and flare activity have disappeared and the microwave radiation is decreasing.

NONFLARE X RAYS

Measurements of solar X-ray emission have been accumulating at a rapid rate during the past few years. Sounding rockets continue to play a useful role but since X-ray emission is a much more sensitive indicator of solar activity than radiation at longer wavelengths, it is not surprising that the most interesting results have been obtained from continuously monitoring satellites, such as OSO-1, the NRL solar-radiation satellites, the Injun I satellite (Van Allen et al. 1965), the British Ariel I satellite, and the USSR Electron-II and IV stations. In most cases, the total flux from the Sun is monitored in one or more fairly standard energy ranges such as 20–100 keV, 2–20 keV, 1–10 Å, 8–12 Å, 8–18 Å, and 44–60 Å; the use of a low-resolution proportional-counter spectrometer for the region 4–14 Å on board Ariel I and similarly the measurement of the spectral energy distribution of solar X rays

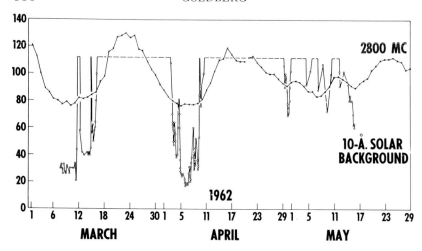

FIG. 18. Time variation of 2–8 Å X-ray Flux, 7 March–
29 May 1962. Goddard Space Flight Center.

from 2–20 keV, also with proportional counters, are notable exceptions. Observations of solar X rays in wavelength bands peaked at about 2 Å and at
about 4 Å have also been made from a Vela satellite during the period from
October 18 through November 1, 1963 (Conner et al. 1965). The NRL solar
radiation-monitoring satellite launched in January 1964 (1964–01-D) transmitted real time data which were successfully recorded by several European
observatories (Landini et al. 1964, Pounds 1965). A detailed review of all
aspects of solar X rays has been given by Mandel'shtam (1965).

In discussing solar X rays, we must recognize that more than one mechanism is responsible for their production. Following de Jager (1965), we distinguish between quasi-thermal X rays, which may be emitted by (*a*) the
quiet corona, (*b*) activity centers without flares, and (*c*) X-ray flares, and
nonthermal X-ray bursts, which are always associated with flares. Since the
Sun is almost never completely quiet, and active regions are sources of enhanced X radiation, the total X-ray flux from the Sun in the absence of flares
exhibits variations correlated with the solar rotation and with plage activity.
The variability is well shown by the measurements of 2–8 Å X-ray flux made
from OSO-1 between March 7 and May 29, 1962 and plotted in Figure 18.
Although the detector became saturated at readings greater than 120, the
record clearly shows both the slowly varying component, which also correlates well with 2800 MHz radio emission, and a number of additional variations in time periods ranging from one second to several hours (White 1964).
Similar results were obtained from continuous observations of fluxes in the
2–10 and 8–18 Å ranges between January 30 and March 16, 1964 from the
Electron-II station (Tindo 1965, Mandel'shtam 1965). Measurements with
the NRL Solrad series of satellites have been reported for three satellites,

1963-21-C (Thomas 1964), 1964-01-D (Landini et al. 1964; Thomas 1964; Kreplin 1965; Thomas, Venables & Williams 1965), and 1965-16-D (Friedman 1966) in the ranges 2–8, 8–20, and 44–60 Å. The spectral energy distribution of X rays in the 4–14 Å region under "quiet" Sun conditions was measured with proportional counters from Ariel-1 (Bowen et al. 1964, Pounds 1965). The corresponding grey-body "temperatures" were found to vary with wavelength, decreasing from 1.8×10^{6}°K for 7–9 Å to 1.3×10^{6}°K for 13–15 Å.

From a careful study of all available satellite measurements, Mandel' shtam (1965b) summarizes the present knowledge of nonflare X-ray emission from the Sun as follows.

1. In the absence of flares, the total flux below 100 Å may remain practically constant for periods on the order of 24 hours.

2. Over time periods between a few days and several months, the flux below 10 Å may change by a factor of 10–100 and in the 10–20 Å range by a factor of 10. The intensity in the 44–100 Å range does not change by more than a factor of about 3 in these time periods.

3. During the period 1960–1964, when solar activity varied from near maximum to minimum, the intensities in ergs cm^{-2} sec^{-1} at the Earth's distance in the absence of flares were in the following ranges, which agree with the results from rocket measurements:

Range	Flux (minimum)	Flux (maximum)
0–10 Å	10^{-5}	$2–3 \times 10^{-3}$
0–8 Å	5×10^{-6}	$1–1.5 \times 10^{-3}$
10–20 Å	10^{-4}	$1–2 \times 10^{-2}$
44–60 Å	10^{-2}	5×10^{-2}

4. Variations in the X-ray flux are closely correlated with other kinds of solar activity, such as centimeter-decimeter radio emission, relative sunspot number, Ca+ plage area on the visible disk.

X-Ray Flares

X-ray emission from solar flares was first observed in 1956 (Byram, Chubb & Friedman 1956). Measurements made in three wavelength bands, 2–8, 8–20, and 44–60 Å, first from Aerobee rockets and later from the NRL solar-radiation satellites, revealed the great intensification and hardening of the X-ray spectrum during solar flares and the correlation between X-ray flare emission and sudden ionospheric disturbances. During bright flares, radiation can be detected at wavelengths as short as 0.1 Å or less and the total flux below 10 Å can easily increase by an order of magnitude (Mandel' shtam 1965b). The total flux in the band 0–100 Å may increase by a factor of 2 over that of the undisturbed Sun. Subsequent measurements from rockets and particularly from satellites have further confirmed these relationships and have also revealed the presence of short-lived bursts often superimposed upon a less intense burst of shorter duration.

According to the classification of de Jager (1965), solar X-ray events may be divided into two main groups: I, quasi-thermal X rays and II, nonthermal X-ray bursts. Quasi-thermal emission in flares occurs over a period of many minutes, comparable with the lifetime of the associated optical flare and of the associated radio emission in centimeter or decimeter waves. Most of the radiation is at wavelengths longer than 0.5 or 1 Å. Class I events produce SID's (Sudden Ionospheric Disturbances). The radiation is assumed to be emitted by a hot plasma at a temperature of several million degrees; the relatively long lifetime suggests confinement by a magnetic field.

Nonthermal bursts of class II X rays have lifetimes ranging between a few seconds and a few minutes, are in the energy range 10^4–10^6 eV (1–0.01 Å), and are not clearly related to SID's. They evidently signal the presence of high-speed streams of electrons radiating as a result of deceleration either in a magnetic field or by interaction with the surrounding dense medium. The nonthermal bursts themselves fall into three subcategories: class IIa, in which the bursts are coincident with impulsive radio bursts at centimeter or short decimeter waves and with the flash phase of large optical flares (80 per cent of class II X-ray bursts fall into this category); class IIb, hard X-ray bursts occurring simultaneously with type III radio bursts, accounting for 10 per cent of the class II bursts; and class IIc, which consists of hard X-ray bursts occurring without any well-defined radio effects at the very beginning of minor optical flares.

Many dozens of solar X-ray events have been observed in various wavelength bands from monitoring satellites, including eleven with a five-channel proportional-counter spectrometer on the satellite Ariel I (Bowen et al. 1964, Culhane et al. 1964, Pounds 1965). The wavelength bands are centered near 5, 6, 7, 9, and 12 Å. Figure 19 shows the flux measurements on all five channels during the flare of importance 2, on April 27, 1962, beginning at 13^h 46^m, peaking at 14^h 13^m, and terminating at 14^h 40^m. During the early stages of the flare, the shape of the spectral curve is hardly different from that of the preflare spectrum (lowest curve), but there is a marked hardening of the spectrum at 14^h 10^m, just before the beginning of the flash phase, which is marked by an impulsive burst in both X rays and 2800 MHz radio noise between 14^h 11^m and 14^h 12^m. These and other data suggest that flare X-ray emission has both a quasi-thermal (de Jager class I) and nonthermal (class II) component (Bowen et al. 1964). Pounds (1965) believes that the most likely nonthermal mechanism is electron bremsstrahlung. Numerous transient events connected with solar flares were also observed during the life of OSO-1, but the usefulness of the data is limited by the low time resolution of 8 min. The largest increase during flares is associated with the ion of highest ionization potential, Fe XVI. For the flare of March 13, 1962 the line at 335 Å increased in intensity by 60 per cent. Hard X rays in the band 20–100

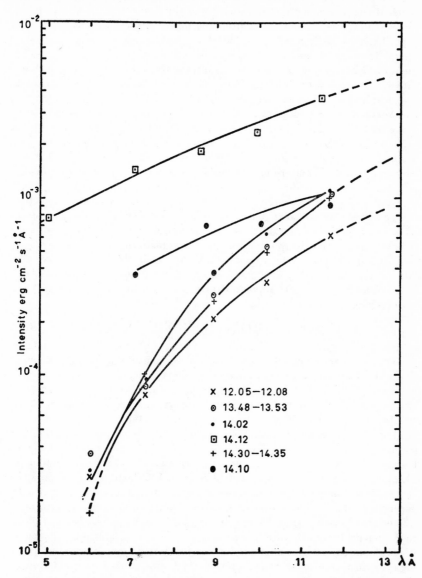

FIG. 19. X-ray spectra during solar flare of importance 2, 27 April 1962. Recorded on Ariel I satellite with five-channel proportional-counter spectrometer.

keV were monitored on board OSO-1 over a period of $\sim$3 months during which the satellite saw the Sun about two thirds of the time (Frost 1965, Lindsay 1965). No definite flux from the quiet Sun was detected, but the upper limit was estimated to be 3.40 ± 0.95 photons cm^{-2} sec^{-1}. Eight high-energy X-ray bursts were observed to be associated with solar flares. The event of 17 March 1962, which had a duration of about 1-1/2 min, coincided with an impulsive radio burst at 2800 MHz lasting for 3 min. All eight events observed by OSO-1 appear to fit de Jager's class IIa classification. Especially noteworthy is the double peak in the X-ray flux measurements shown by two of the eight events. A similar double peak is shown by radio measurements at 3750 MHz (see Figure 20). The search for gamma rays from the quiet Sun has now been extended from balloons up to energies of 10 MeV and again only an upper limit may be estimated (Frost, Roth & Peterson 1966; Peterson et al. 1966).

Sounding rocket measurements with proportional counters of the spectral energy distribution of solar X rays from 2–20 keV show that the flare spectrum is markedly enhanced at the higher energies during subflare activity (Chubb, Friedman & Kreplin 1964). The spectrum of the quiet Sun decreases rapidly and monotonically from 2 to 11 keV, but during two subflares the flux was observed to display a maximum near 10 keV in one case and 12 keV in the other, which suggests emission associated with Fe XXV and XXVI, respectively.

INTERPRETATION OF X–RAY INTENSITIES

Early attempts to derive information about conditions in the quiet solar corona and in active regions were based on comparisons between measured X-ray fluxes at various times during the solar cycle and theoretical predictions of solar X radiation, e.g. by Elwert (1961). Elwert's prediction that the X-ray spectrum of the quiet Sun in the wavelength region above 33 Å should be almost wholly a line spectrum has been fully confirmed (Tousey et al. 1965, Hinteregger et al. 1964). Conversely, his expectation that the spectrum at wavelengths below 30 Å should be mostly continuous has not been borne out by the observations, since the region 14–25 Å has been found to consist entirely of spectral lines (Blake et al. 1965), although it is possible that at least some of the line emission could arise from active regions (de Jager 1965). Calculations similar to Elwert's have recently been performed by Fetisov (1963, 1964) for the spectral region 1–10 Å, and extended to 10 Å, by Mandel'shtam (1965a). For each of the 15 most abundant solar elements, they calculate the degree of ionization as a function of temperature by equating the rate of collisional ionization by electron impact with the rate of radiative recombination. Approximate cross sections are then employed to calculate the flux emitted (a) in individual lines, (b) in recombination, and (c) in free-free emission or bremsstrahlung. The results for the spectral region 2–20 Å are shown in Figure 21 for a temperature of $2 \times 10^{6\circ}$K and an average model of the corona with $y = \int N_e^2 \, dV = 3.2 \times 10^{49}$. It is seen that below 15 Å,

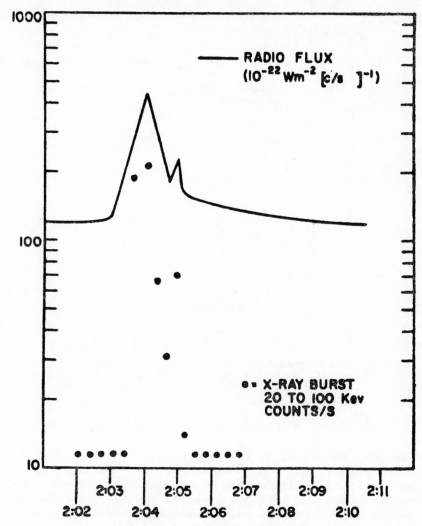

FIG. 20. Hard X-ray burst and coincident radio burst
at 2800 MHz. Goddard Space Flight Center.

the main contribution to the emission is by recombination of electrons with heavy ions. The predominant role of recombination emission in this region is due to the small number of lines in that spectral interval.

Calculations made for a quiet corona at a temperature of about one million degrees give for the total flux in the band 2–10 Å about 10^{-7} erg cm^{-2} sec^{-1}, whereas the minimum flux observed in the band 1–8 Å in early 1964

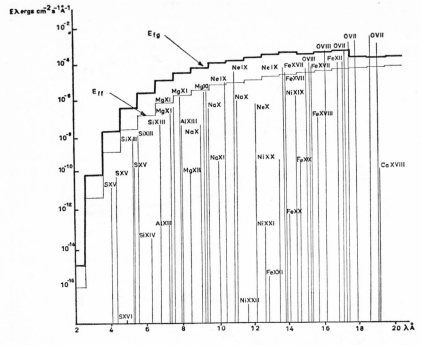

FIG. 21. X-ray spectrum of quiet Sun calculated for a temperature of $2 \times 10^{6\circ}$K (Mandel'shtam 1965). The heavy black curve labeled E_{fg} refers to emission by recombination to heavy ions, whereas E_{ff} denotes the corresponding emission by free-free transitions. The light vertical lines refer to emission by individual lines of the various ions.

was 5×10^{-5} erg cm^{-2} sec^{-1} (Pounds 1965). Experimental fluxes on five dates in 1962 quoted by Mandel'shtam are still greater, hence it is assumed that most of the flux comes from active regions or coronal condensations. To calculate the contribution from active regions, 9.1-cm radio spectroheliograms from Stanford University are used to derive values of the electron temperature and the integral of the square of the electron density over the active region. In order to calculate the contribution of active regions to the X-ray flux, one requires a knowledge of the distribution of electron temperature T_e and of the emission measure $Y_R = \int_R N_e{}^2 \, dV$ over the solar disk, where both T_e and N_e vary from one active region to another. The Stanford data give the distribution of brightness temperature or flux over the disk from which the emission measure can be derived if the temperature is known. Since there is no way of obtaining the temperature distribution independently, Mandel'shtam assumes that there is a correlation between electron density and temperature and adopts the following rather arbitrary relation between brightness temperature T_B at 9.1 cm and electron temperature T_e:

		Undisturbed corona	Active regions		Condensations
T_B	$(10^3 \, °\mathrm{K})$	$T_B < 30$	$30 \leq T_B < 60$	$60 \leq T_B < 150$	$T_B \geq 150$
T_E	$(10^6 \, °\mathrm{K})$	1.0	1.5	1.75	2.5

In this fashion, agreement within a factor of 2 is obtained for six different dates and hence it is concluded that X-ray emission from the quiet Sun (in the absence of flares) is thermal. The emission is thus comprised of two components, an undisturbed component and a slowly varying component contributed by active regions.

White (1964a) attempts to account for the slowly varying component of the X-ray flux in the 1–10 Å band observed from OSO-1 with a model consisting of localized sources having diameters equal to those of Ca+ plages observed on the same day and thicknesses equal to half the diameters. On any given day, the value $\int N_e^2 \, dV$ required to fit the observed X-ray flux may then be calculated for different assumed temperatures, and also assuming that line emission is absent. If the temperature is taken to be $2.8 \times 10^{6} \, °\mathrm{K}$, as indicated by the shape of the spectrum between 7 and 11 Å observed from Ariel I (Bowen et al. 1964), the required electron density is about 2×10^{10} cm^{-3}. From an optical observation of one of the Ca+ plages, Billings estimated the electron density to be $0.5 \times 10^{10} \, \mathrm{cm}^{-3}$, which requires, according to Elwert's (1961) calculations, that the ratio of line emission to continuum emission be 30:1. In that case, however, the measurement of temperature from the shape of the spectrum would be vitiated. If the emission from active regions is thermal, the conclusion is that the densities of these regions are $5 - 10$ times higher than average (Culhane et al. 1964, White 1964), since otherwise the required high temperatures would be inconsistent with observed UV line intensities.

The mechanisms responsible for X-ray production in flares are not yet well understood, apart from evidence that they are both thermal and nonthermal and that the latter are produced by streams of electrons giving off energy by bremsstrahlung and synchrotron radiation. The close correlation between X rays and radio radiation of centimeter and decimeter wavelengths suggests that the same physical processes are responsible and helps to identify them. A careful and detailed review of the observed correlations has been given by Kundu (1963). They are also an important feature of de Jager's (1965) classification of X-ray flares. The process of Compton scattering of thermal photons by relativistic electrons is favored as a source of γ rays by Zheleznyakov (1965) and of X rays by Shklovsky (1965), while thermal synchrotron radiation of accelerated electrons has been investigated by Korchat (1965). Acton (1964) has criticized the inverse Compton effect, by pointing out that any quantity of relativistic electrons producing a measurable amount of inverse Compton photons will always produce a greater flux of higher-energy bremsstrahlung photons than has been observed. He suggests (Acton 1965) that electrons with energy in the range 10–30 keV, which are produced during solar flares, may cause K-shell ionization, and that the resulting X rays would contribute to the total X-ray emission of flares.

The author wishes to thank Mrs. A. K. Dupree of the Harvard College Observatory for a number of valuable suggestions.

LITERATURE CITED

Acton, L. W. 1964, *Nature*, **204**, 64

Acton, L. W. 1965, *ibid.*, **207**, 737

Alexander, E., Feldman, U., Fraenkel, B. S. 1964, *J. Quant. Spectry. Radiative Transfer*, **4**, 501

Alexander, E., Feldman, U., Fraenkel, B. S. 1965, *Phys. Letters*, **14**, 40

Alexander, E., Feldman, U., Fraenkel, B. S., Hoory, S. 1965, *Nature*, **206**, 176

Allen, C. W. 1965, *Space Sci. Rev.*, **4**, 91

Allen, J. W. 1965, *Sci. Rept. No. 7* (Shock-Tube Spectry. Lab., Harvard Coll. Obs.)

Athay, R. G. 1965a, *Ap. J.*, **142**, 724

Athay, R. G. 1965b, *ibid.*, 755

Athay, R. G. 1966, *ibid.*, **145**, 784

Austin, W. E., Purcell, J. D., Tousey, R., Widing, K. G. 1966, *Ap. J.*, **145**, 373

Bennett, W. R., Jr. 1966, *Phys. Rev. Letters*, **17**, 1196

Black, W. S., Booker, D., Burton, W. M., Jones, B. B., Shenton, D. B., Wilson, R. 1965, *Nature*, **206**, 654

Blake, R. L., Chubb, T. A., Friedman, H., Unzicker, A. E. 1964, *Space Res.*, **4**, 785

Blake, R. L., Chubb, T. A., Friedman, H., Unzicker, A. E. 1965, *Ap. J.*, **142**, 1

Bowen, P. J., Norman, K., Pounds, K. A., Sanford, P. W., Willmore, A. P. 1964, *Proc. Roy. Soc. A*, **281**, 538

Boyd, R. L. F. 1965, *Space Sci. Rev.*, **4**, 35

Burgess, A. 1965a, *Ann. Ap.*, **28**, 774 (also in 1964, *IAU Symp. No. 23*, Liège)

Burgess, A. 1965b, *Proc. 2nd Harvard-Smithsonian Conf. Stellar Atmospheres*, 47

Burgess, A., Seaton, M. J. 1964, *Monthly Notices Roy. Astron. Soc.*, **127**, 355

Burton, W. M., Ridgely, A., Wilson, R. 1966, *Preprint CLM-P109*, Culham Lab. (Submitted to *Monthly Notices Roy. Astron. Soc.*)

Burton, W. M., Wilson, R. 1965, *Nature*, **207**, 61

Byram, E. T., Chubb, T. A., Friedman, H. 1956, *J. Geophys. Res.*, **61**, 251

Chubb, T.A., Friedman, H., Kreplin, R.W. 1964, *Space Res.*, **4**, 759

Conner, J. P., Evans, W. D., Montgomery, M. D., Singer, S., Stogsdill, E. E. 1965, *Space Res.*, **5**, 546

Cowan, R. D., Peacock, N. J. 1965, *Ap. J.*, **142**, 390; 1966, *ibid.*, **143**, 283

Culhane, J. L., Willmore, A. P., Pounds, K. A., Sanford, P. W. 1964, *Space Res.*, **4**, 741

de Jager, C. 1963, *Bull. Astron. Inst. Neth.*, **17**, 209

de Jager, C. 1965, *Ann. Ap.*, **28**, 125

Detwiler, C. R., Garrett, D. L., Purcell, J. D., Tousey, R. 1961, *Ann. Geophys.*, **17**, 263

Dupree, A. K., Goldberg, L., 1967, *Solar Phys.*, **1**, 229

Elton, R. C., Kolb, A. C., Austin, W. E., Tousey, R., Widing, K. 1964, *Ap. J.*, **140**, 390

Elwert, G. 1961, *J. Geophys. Res.*, **66**, 391

Fawcett, B. C., Gabriel, A. H. 1965, *Ap. J.*, **141**, 343

Fawcett, B. C., Gabriel, A. H., Griffin, W. G., Jones, B. B., Wilson, R. 1963, *Nature*, **200**, 1303

Fawcett, B. C., Gabriel, A. H., Saunders, P. A. H. 1967, *New Spectra of Fe XII to Fe XVIII and their Isoelectronic Sequences* (Submitted to *Proc. Phys. Soc.*)

Feldman, U., Fraenkel, B. S., Hoory, S. 1965, *Ap. J.*, **142**, 719

Fetisov, E. P. 1963, *Cosmic Res.*, **1**, 171 (*Kosmich. Issled.*, **1**, 209)

Fetisov, E. P. 1964, *Soviet Astron.—AJ*, **8**, 231 (*Astron. Zh.*, **41**, 299)

Fredga, K. 1966, *Ap. J.*, **144**, 854

Friedman, H. 1963, *Ann. Rev. Astron. Ap.*, **1**, 59

Friedman, H. 1966, *Montioring of X-Rays by Solrad-8 (1965-16-D)*, *Rept. 7th COSPAR Space Sci. Symp.*, *Vienna, 1966*

Friedman, H. 1967 (Private communication)

Frost, K. J. 1965, *Space Res.*, **5**, 513

Frost, K. J., Rothe, E. D., Peterson, L. E. 1966, *J. Geophys. Res.*, **71**, 4079

Gabriel, A. H., Fawcett, B. C. 1965, *Nature*, **206**, 808

Gabriel, A. H., Fawcett, B. C., Jordan, C. 1965, *Nature*, **206**, 390

Gabriel, A. H., Fawcett, B. C., Jordan, C. 1966, *Proc. Phys. Soc.*, **87**, 825

Giacconi, R., Reidy, W. P., Zehnpfennig, T., Lindsay, J. C., Muney, W. S. 1965, *Ap. J.*, **142**, 1274

Gingerich, O., Rich, J. C. 1966, *Astron. J.*, **71**, 161

Goldberg, L., Müller, E. A., Aller, L. H. 1960, *Ap. J. Suppl. 5*, 1

Goldberg, L., Parkinson, W. H., Reeves, E. M. 1965, *Ap. J.*, **141**, 1293

Hall, L. A., Schweizer, W., Heroux, L., Hinteregger, H. E. 1965, *Ap. J.*, **142**, 13

Heintze, J. R. W., Hubenet, H., de Jager, C. 1964, *Bull. Astron. Soc. Neth.*, **17**, 442

Hinteregger, H. E. 1965, *Space Sci. Rev.*, **4**, 461

Hinteregger, H. E., Hall, L. A., Schmidtke, G. 1965, *Space Res.*, **5**, 1175

Hinteregger, H. E., Hall, L. A., Schweizer, W. 1964, *Ap. J.*, **140**, 319

House, L. L., Deutschman, W. A., Sawyer, G. A. 1964, *Ap. J.*, **140**, 814

Ivanov-Kholodnyi, G. S., Nikol'skii, G. M. 1961, *Soviet Astron.—AJ*, **5**, 31

Ivanov-Kholodnyi, G. S., Nikol'skii, G. M. 1963, *ibid.*, **5**, 31 (Orig: 1962, *Astron. Zh.*, **39**, 777)

Jordan, C. 1965a, *Phys. Letters*, **18**, 259

Jordan, C. 1965b, *Commun. Univ. London Obs.*, No. 68

Jordan, C. 1966, *Monthly Notices Roy. Astron. Soc.*, **132**, 463

Kachalov, V. P., Yakovleva, A. V. 1962, *Izv. Krymsk. Astrofiz. Obs.*, **27**, 5

Kodaira, K. 1965, *Z. Ap.*, **60**, 240

Korchat, A. A. 1965, *Cosmic Res.*, **3**, 613 (*Kosmich. Issled.*, **3**, 751)

Kreplin, R. W. 1965, *Space Res.*, **5**, 951

Kundu, M. R. 1963, *Space Sci. Rev.*, **2**, 438

Landini, M., Piatelli, M., Righini, G., Russo, D., Tagliaferri, G. L. 1964, *Ann. Ap.*, **27**, 765

Lindsay, J. C. 1965, *Ann. Ap.*, **28**, 586

Mandel'shtam, S. L. 1965a, *Ann. Ap.*, **28**, 614

Mandel'shtam, S. L. 1965b, *Space Sci. Rev.*, **4**, 587

Manson, J. E., 1967, *Ap. J.*, **147**, 703

Matsushima, S. 1964, *Proc. 1st Harvard-Smithsonian Conf. Stellar Atm.*, 246

Merrill, P. W. 1961, *Ap. J.*, **134**, 556

Moore-Sitterly, C. E. 1965, *The Solar Spectrum*, 89 (D. Reidel, Dordrecht, Holland)

Neupert, W. M. 1965, *Ann. Ap.*, **28**, 446

Nikol'skii, G. M. 1963, *Geomagnetism Aeronomy*, **3**, 643 (*Geomagnetizm Aeronomiya*, **3**, 793)

Pagel, B. E. J. 1963, *Planetary Space Sci.*, **11**, 333

Parkinson, W. H., Reeves, E. M. 1966 (To be published)

Peach, G. 1962, *Monthly Notices Roy. Astron. Soc.*, **124**, 371

Peterson, L. E., Schwartz, D. A., Pelling, R. M., McKenzie, D. 1966, *J. Geophys. Res.*, **71**, 5778

Porter, J. R., Tilford, S. G., Widing, K. G. 1967, *Ap. J.*, **147**, 172

Pottasch, S. R. 1963, *Ap. J.*, **137**, 945

Pottasch, S. R. 1964a, *Montlhy Notices Roy. Astron. Soc.*, **128**, 73

Pottasch, S. R. 1964b, *Space Sci. Rev.*, **3**, 816

Pottasch, S. R. 1965, *Ann. Ap.*, **28**, 148

Pounds, K. A. 1965, *Ann. Ap.*, **28**, 132

Pounds, K. A. 1966 (To be published)

Purcell, J. D., Garrett, D. L., Tousey, R. 1963, *Space Res.*, **3**, 781

Purcell, J. D., Tousey, R., *XUV Heliograms* (Paper presented at A. A. S. Special Meeting on Solar Astron., Boulder, Colo., 3–5 Oct. 1966)

Rich, J. C. 1966a, *Silicon and Carbon Monoxide Absorption in the Ultraviolet Solar Spectrum* (Dissertation, Harvard Univ. Dept. of Astron.)

Rich, J. C. 1966b, *Sci. Rept. No. 10* (Shock-Tube Spectry. Lab., Harvard Coll. Obs.)

Russell, P. C. 1965a, *Nature*, **205**, 684

Russell, P. C. 1965b, *ibid.*, **208**, 281

Russell, P. C., Pounds, K. A. 1966, *Nature*, **209**, 490

Shklovsky, I. S. 1965, *Soviet Astron—AJ*, **8**, 538 (1964, *Astron. Zh.*, **41**, 676)

Sinanoğlu, O., Skutnik, B., Tousey, R. 1966, *Phys. Rev. Letters*, **17**, 785

Stockhausen, R. 1965, *Ap. J.*, **141**, 277

Suemoto, Z., Moriyama, F. 1964, *Ann. Ap.*, **27**, 775

Thomas, L. 1964, *Nature*, **203**, 962

Thomas, L., Venables, F. H., Williams, K. M. 1965, *Planetary Space Sci.*, **13**, 807

Tindo, I. P. 1965, *Investigation of Cosmic Space* (Nauka, Moscow)

Tousey, R. 1963, *Space Sci. Rev.*, **2**, 3

Tousey, R. 1964, *Quart. J. Roy. Astron. Soc.*, **5**, 123

Tousey, R. 1966a, *Space Res.*, *Directions for the Future*, Part 2, *Solar Astronomy* (Natl. Acad. Sci.–Natl. Res. Council, Washington, D.C.)

Tousey, R. 1966b, *Henry Norris Russell Lecture, 122nd Meeting, A.A.S., Cornell Univ.* (To be publ. in *Ap. J.*)

Tousey, R., Austin, W. E., Purcell, J. D., Widing, K. G. 1965, *Ann. Ap.*, **28**, 755

Tousey, R., Purcell, J. D., Austin, W. E., Garrett, D. L., Widing, K. G. 1964, *Space Res.*, **4**, 703

Underwood, J. H., Muney, W. S. 1967, *Solar Phys.*, **1**, 129.

Vainstein, L. A., Norman, G. E. 1960, *Opt. Spectry.*, **8**, 79 (*Opt. Specktroskopiya*, **8**, 149)

Van Allen, J. A., Frank, L. A., Maehlum, B., Acton, L. W. 1965, *J. Geophys. Res.*, **70**, 1639

White, W. A. 1964, *Space Res.*, **4**, 771

Widing, K. G. 1961, *Astron. J.*, **66**, 298

Widing, K. G. 1966, *Ap. J.*, **145**, 380

Zheleznyakov, V. V. 1965, *Soviet Astron.-AJ*, **9**, 73

Zhitnik, I. A., Krutov, V. V., Malyavkin, L. P., Mandel'shtam, S. L. 1964, *Kosmich. Issled.*, **2**, 920 (*Cosmic Res.*, **2**, 801)

Zirin, H. 1964, *Ap. J.*, **140**, 1332

Zirin, H., Dietz, R. D. 1963, *Ap.J.*, **138**, 664

EXTRASOLAR X-RAY SOURCES[1]

By Philip Morrison

Department of Physics and Center for Space Research
Massachusetts Institute of Technology
Cambridge, Massachusetts

The high excitation temperatures of the solar corona were established a generation ago by optical spectroscopy. Atomic lines in the soft X-ray region were among the first results sought—and found—once detectors were sent beyond the atmosphere in sounding rockets. Continuum radiation in the range of tens or hundreds of Rydbergs was earlier predicted from the observed ionosphere disturbances accompanying solar activity, and it too has been studied, directly and indirectly. The Sun, then, has been known as an X-ray source for a long time (1). But cosmic X rays from sources farther than the Sun were found for the first time in the summer of 1963 (2). Three years later (September 1966) the astronomy of such X-ray sources is a strong if immature branch of science. How young it is can be realized from the fact that a total successful exposure to the sky of about an hour from rockets and of perhaps 150 hours from balloons sums up the world's entire positive experience of extrasolar X-ray astronomical observation. Yet a plausible case can be made for the statement that most of the matter in the universe is observable only by means of X rays, a channel whose exploitation has obviously only just begun. That is its interest and its challenge.

THE X-RAY WINDOW

The X-ray region of the electromagnetic spectrum is here taken to extend from a photon energy well above the ionization energy of hydrogen, say 20 or 30 eV/photon, up the region of nuclear gamma rays, about 100 keV/photon. The upper limit is, of course, largely arbitrary; the lower limit is set to exclude the discrete levels of the most abundant atomic species. Ever since Röntgen looked through black paper, we have come to think of X rays as penetrating. This can be justified in an elementary way by an estimate of the cross section for a single dominant resonance based on the dipole sum rule. For a purely radiative width $\Delta\omega$:

$$\int \sigma(\omega)d\omega = (e^2/mc^2)c = r_0 c$$

and

$$\sigma_{\text{res}} = (r_0 \lambdabar_{\text{res}})/(1/137)(a/\lambdabar_{\text{res}})^2$$

with (ea) the dipole matrix element. All this is for hydrogenic atoms (3). Thus, in more detail, the X-ray cross section at 1 keV is something like 10^{-19} cm², while that for visible light in resonance is a millionfold greater, some 10^{-12} cm². Yet the optical astronomers work more or less unimpeded

[1] The survey of literature for this review was concluded in December 1966.

under a meter of lead equivalent, while X-ray astronomy cannot be ground based. This paradox is wholly a resonance effect; if their energies lie between resonance levels, photons travel unabsorbed. Once levels are continuous, chemical state ceases to have importance, and numbers of atoms alone are significant. Thus the X rays which in the mean penetrate splendidly cannot pierce the thick atmosphere, while for that select octave we call the optical region the air is magically transparent. Of course, the importance for X-ray absorption of proximity to the atomic K edge or to analogous threshold energies is familiar. The consequence of continuum absorption is a heavy experimental burden: X rays are attenuated by air in the amounts plotted in Figure 1. It is plain that X-ray astronomers must use rockets, or place detectors in extraterrestrial orbit, to receive photons of energies up to some 10 keV, beyond which balloons floating at a few grams per square centimeter become acceptable observatories. Aircraft at extreme altitudes are usable for photons around 100 keV and up, but Mount Everest lies under ten mean free paths of air, even as high as 1 MeV.

Once the atmospheric shutter is opened, what absorption remains in interstellar space? The answer is contained in Figure 2. The nearest stars might become visible at 80–100 eV, but truly galactic distances can be

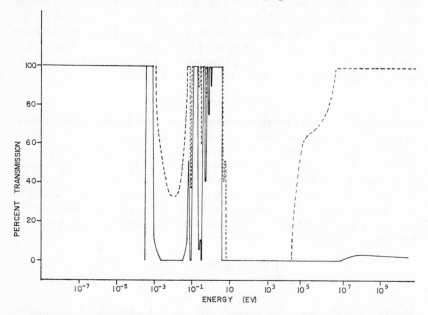

FIG. 1. The atmospheric filtration of electromagnetic radiation. The solid curve shows absorption from mountain stations (computed for Mount Everest!); the dotted curve, from balloon altitude (below 3 g/cm²). Note that the kilovolt region requires a rocket or an orbital station, and that balloon detectors become competent beyond some 20 keV. Adapted from Rossi (4).

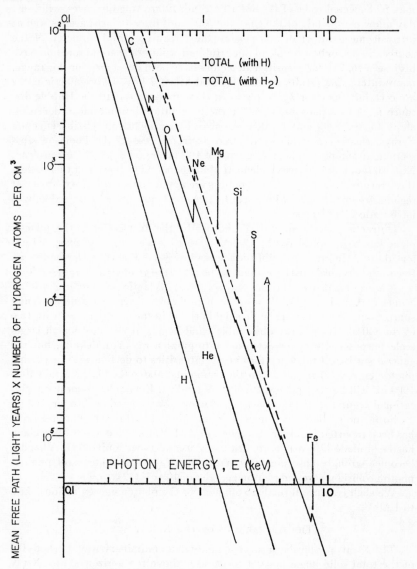

FIG. 2. The photon absorption in interstellar space. Various K edges are noted. Adapted from Felten & Gould (5), who estimated the effect of hydrogen both in atomic and in molecular form (H₂).

traversed only beyond 1.2–1.6 keV. This energy region, where optical depths may be expected to be of the order unity for interesting distances, will some day allow close study of the gas composition of interstellar space, as well as intrinsic measures of distance using the X-ray spectrophotometry of the source. So far, only a couple of doubtful indications of space absorption have been seen (6, 7). Once sensitive detectors have been developed for this range, many interesting results can be expected below 2 keV. Extragalactic paths are certainly open for 2-keV photons; they are open out to the Hubble distance for 8-keV photons, even for the maximum plausible mean densities, about 10^{-3} protons/cm^3. Evidently ionized hydrogen has a negligible effect on X rays; elastic scattering by the free electrons presents the Thomson cross section, and implies a mean free path of 10^{10} lyears for $\frac{1}{6} \cdot 10^{-3}$ electrons/cm^3. Note that anomalous heavy-element abundances might here and there affect these results; but if cosmic abundances are about as we think (perhaps most regions are low in *neutral* hydrogen), elements beyond $Z \simeq 10$ are unlikely to have noticeable effects.

Diffraction scattering of keV photons by the dust grains of the galactic plane has been considered by Overbeck (8); it seems to be of marginal observability. He predicts a diffusion spread of $\sim \frac{1}{2} - 3'$ at 10 Å for one mean free path in typical dust (reckoning in terms of the geometrical cross section).

A line of sight from the Sun out toward the galactic pole crosses $\simeq 10^{21}$ H atoms/cm^2. It is patent that we shall probably never study directly any sources outside our near galactic neighborhood in the energy range from 13.6 eV to 800 eV. The X-ray observable band begins at ~ 1 keV for all large-scale purposes. Its upper extension into gamma rays is indefinite, but the general statistical tendency for overall intensities to decline with increasing photon energy (9) suggests that the region just above the black curtain 100–1000 eV will be the first to be fruitfully studied. Events have borne out this rational argument, though the actual history owes a good deal to fortune (2, 4). So far no statistically compelling detection of extrasolar photon sources has been reported for energies above 1 or 2 MeV, though one tentative claim has been made (10) and there are many useful upper limits (11). All space becomes strongly absorbing to gamma rays (as a consequence of photon-photon collisions) around photon energies of 10^{14} eV (12).

We shall confine all further comment to the photon energy region 1 keV to 1 MeV.

THE APPEARANCE OF THE X-RAY SKY

The X-ray sky has been surveyed only in a preliminary way. Perhaps 0.3 of the total solid angle has yet to be seen above the horizon of any X-ray detector. In the rest of the sky, a score of discrete sources have been reported, and there is also an unresolved quasi-isotropic background, over the whole energy range. Excluding the Sun, all the discrete sources so far detected lie within an intensity range of only 3 or 4 mag (5 mag = 10^2). The Sun is an intense and highly variable source at the lowest X-ray energies, though its dominance is modest. The quiet Sun at 1 keV exceeds the brightest extrasolar source only by two or three decades, instead of the eight decades of its optical splendor. In the 10-keV range, the quiet Sun is actually less bright than the brightest stellar X-ray source, in spite of its huge advantage of an inverse-square distance factor of at least 10^{14}. This is exactly the situation already familiar in the meter band.

In Figure 3 the strongest known radio sources are plotted in order of intensity, and, for comparison, the X-ray sources. The similarity of the radio and X-ray intensity-rank distributions is marked. It is clear that the meter-band emission of main-sequence stars like the Sun is very inefficient, whereas other objects, large or small, have considerably more effective means of emitting radio waves. The same statement evidently applies to X rays. But there is one remarkable dynamical difference. The meter-band flux from even a very strong radio source, like Cas A, is about 10^{-14} W/m². Cas A radio emission is a few times $L_\odot$. But the Scorpius X-ray source is 10,000 times stronger in energy flux received on Earth. If X-ray sources are at galactic distances, they must have X-ray luminosities some 10^4 larger than the total power output of the Sun in all frequencies. If any distant extragalactic X-ray source is confirmed at something like the present detectable level of received flux, its X-ray luminosity must be 1000 or more times the total output of our Galaxy in the visible region. Such X-ray sources could not simply represent, as do radio sources, objects with an unusual mechanism for a specialized emission, which need have no major influence on the dynamical history of the emitting object. Their X-ray power must arise as a major energy-transfer mode during the history of the emitting source, whereas radio emission is only a marker of the presence of relativistic magnetic plasma. Such X-ray sources would be X-ray objects; their visible emission is energetically secondary, or at best only comparable. X-ray astronomy may therefore be a new branch of the subject in a way radio astronomy is not. The great physical diversity of the brightest radio sources is evident; the X-ray sources, however, are still mostly unidentified, though hints of their diversity, too, are already present. The suggestion that the extension of X-ray observations over a dynamic range of three more decades will disclose hundreds of sources, just as happened in the radio region, is strong. Extended exposure time, larger detector area, instruments of higher angular gain, and possible focusing devices are all planned (4). Spectral resolution too is to be expected to increase; work in the rocket region is often rather analogous to two-or-three-color optical photometry, and at best yields 5 or 10 per cent energy resolution. No atomic spectral lines are yet seen either in emission or in absorption; indeed, in the region of a few keV and above we shall see that they are not likely to be found with sizable contrast.

THE DISTRIBUTION AND IDENTIFICATION OF SOURCES

In Figure 4 the current map of the X-ray sky is presented, with a source list. Since the sources are not equally resolved by different observers viewing

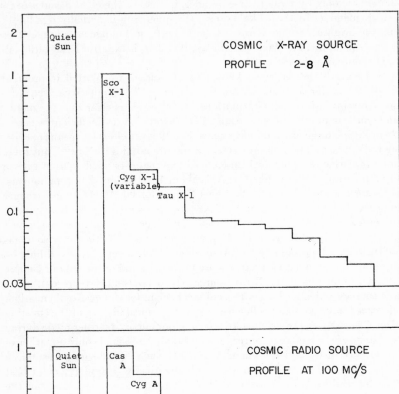

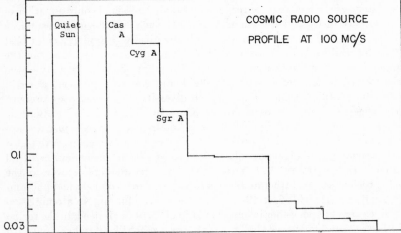

FIG. 3. Relative intensities for the brightest sources in the X-ray and in the meter band. References: Table I and Allen (43), Friedman (1).

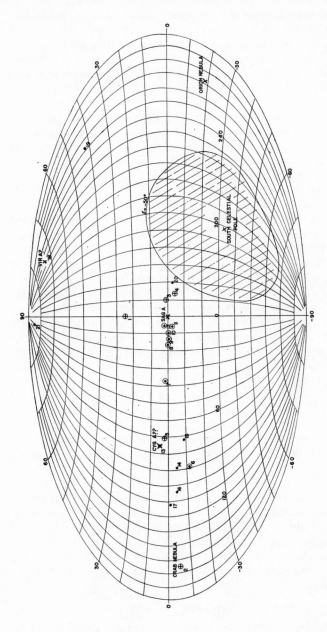

X-RAY SKY 1966.9

FIG. 4. The X-ray sky as it is known for epoch 1966.9. Galactic coordinates (l^{II}, b^{II}) on Aitoff projection. Three classes of sources are shown: confirmed, marked by circled cross; probable, marked by circled dot; unconfirmed, marked by solid dot. A few radio or optical objects are indicated with an X. The shaded region has not been observed at all. See Table I for source particulars.

the same directions, not even the count of sources is unambiguous, let alone their intensities or angular positions. It seems proper to count a source as confirmed only if it has been seen on at least two attempts. A reasonable judgment of conclusions to be drawn from this work follows, neither accepting every indication of a source as real, nor demanding statistically rigorous evidence of its existence. Naturally, such a judgment remains subjective in part. Some of these conclusions may be in error; I believe they are broadly correct.

(*a*) Two sources are surely identified by optical means, from precise X-ray positions. These are the strongest source, called Scorpius X-1, and a weaker one in NGC 1952, the Crab Nebula. Scorpius X-1 has no detectable radio emission; the Crab Nebula is the strong wideband radio source Tau A, known to be some 3000 to 5000 lyears distant. The SN remnant Cas A is a third possible identification (17, 20).

(*b*) The other known galactic Type I supernova remnants—SN 1604, in Ophiuchus, and SN 1572, in Cassiopeia (Cas B radio source)—are not X-ray sources at the levels so far detectable. There is no X-ray source in the Cygnus Loop (19).

(*c*) Most of the rest of the discrete sources lie near the galactic plane, with a noticeable concentration in longitude within ± 20 degrees of the galactic center. This seems to safely establish the bulk of the sources as galactic objects, and probably also as including a number of sources which lie within a few kiloparsecs of the center of the Galaxy, to account for the cluster near the galactic center. The distribution resembles that of novae, or other objects of an intermediate population (22). The physical center of the Galaxy, very near in direction to the radio source Sgr A, is not an X-ray source at presently detectable levels.

Notice that the first source to be found, Sco X-1, lies at galactic coordinates $b^{II} = 23.6°$, $l^{II} = 359.7°$. This close agreement in longitude with the galactic center is entirely accidental, for Sco X-1 is an optical stellar source. It almost surely belongs to the galactic disk, more than 8 kpc radially out from the galactic center (13). Beware of statistical conclusions from one example!

(*d*) There is still no confirmed source at high galactic latitudes. A case has been made for an extended source in the direction of the Coma cluster of galaxies (distance 300 million lyears (21). A claim for coincidence of an X-ray source with the direction of Cyg A is disputed (19), and a weaker claim for M87 is still unconfirmed (17). Analogy and theoretical speculations suggest that extragalactic sources will be found. Establishment or exclusion of such extragalactic sources is perhaps the most urgently needed experimental result at this time.

Angular Size and Precise Positions

Most of the sources observed are known only not to exceed several degrees in angular diameter. The unconfirmed Coma cluster source, 21 in Table I, is reported as definitely an extended source, perhaps 3 deg in diameter. Sources 5 and 6 in Cygnus are less than a few minutes across (19). The NRL group, in a *tour de force* of controlled rocketry, observed in the energy range of 1.5 to 6 keV a rare lunar occultation of the Crab Nebula. They established that the X-ray source lay on a line through the center of the optical Crab Nebula and that it had a diameter of 1 or 2 min of arc (14). Recently, the American Science and Engineering-MIT group, using the ingenious modulation collimator of M. Oda (23), located the Sco X-1 source within an error rectangle (actually, a pair of them) a fraction of a square minute of arc in area. Optical detection at this position of an unusually blue star which exhibits a flickering spectral variability over a range of 1 mag, continued for a century without important secular change, seems to clinch the stellar identification (Figure 9; see pp. 346–47). The same measures show that the X-ray source in Scorpius is less than 10 sec of arc across (24). The ASE-MIT group has also confirmed the size of the X-ray source in the Crab Nebula in the few-keV range; they place it within 15 or 20 sec of arc from the center of the source of the optical continuum, and give its dimensions as 110 ± 20 sec along the long axis of the Crab, and 60 ± 60 sec in the transverse direction (25) (Figure 10; see p 348).

Physically, this implies that the kilovolt X rays from the Crab originate from a region 2 lyears across, while the Scorpius source can hardly be larger than 0.01 lyears, and may very well be much smaller. It seems appropriate to refer to Scorpius as a point source. The distance of the Sco X-1 source is not well known. The star is of 13th magnitude, shows a radial velocity of -100 km/s, and no proper motion (26). The strong, complex, and variable emission spectrum (lines of H I; He I and II; and probably C III, N III, and O II), with a strong violet continuum, resembles that of an old nova. The visible output varies irregularly by as much as a factor of 2 in a day, and 1 per cent/min. It is seen unchanged on plates back to 1896. This physical identification is not secure, for the typical periodic spectral shifts of the close binary star which most old novae exhibit are not surely present. Interstellar absorption lines of Ca II are found, and the most likely distance seems to be many hundreds of parsecs. [On the other hand, the X-ray measurements of Friedman (7) indicate, rather indirectly to be sure, the presence of a large component of X rays below 300 eV, which would imply a distance of under 100 pc.] The object can be placed at 1000 lyears, with a possible error either way of a factor of $\sim$3.

No other sources have yet been given positions in the galactic plane sufficiently accurate to fix their coincidence with any of the many known very blue stars of the general type of the Scorpius X-1 variable star.[2] No observations have yet placed sufficiently small limits to angular size to establish whether any source more closely resembles the extended Crab source, or the point source in Scorpius.

---

[2] *Note added in proof:* Source 6 of Table II, Cyg X-2, has a position very close to that of a star of unusual blue excess. This is probably another example of the class of Sco X-1 (19, 74).

TABLE I
LIST OF X-RAY SOURCES MAPPED IN FIGURE 4

Source number and name	Position R.A.	Dec.	Approx. intensity[a] keV/cm²-sec	Remarks
1 Sco X-1	16h15m	−15.2°	∼100	Found by all groups. Identified 1966.6 (13).
2 Tau X-1	5h31.5m	+22°	∼15	Also found generally. Identified as Crab Nebula, M1 (14).
3 Sco X-2	17h08m	−36.4°	∼10	NRL, Lockheed give this position (15, 16).
4 Sco X-3	17h23m	−42.2°	∼5	Confirmed several times.
5 Cyg X-1	19h53m	+34.6°	∼5 to 15	Intensity variable by factor of 2 to 4 (17).
6 Cyg X-2	21h43m	+38.8°	∼5	NRL on several flights (16, 17).
7 Ser X-1	18h45m	+ 5.3°	∼5	NRL.
8 Sgr X-2 (?) (L5)	18h03m	−20.7°	∼5	The source earlier named Sgr X-2 has been resolved by Lockheed group (15) into three sources, here labeled L5, L6, L7. The first two are in Sgr, the third in Ser.
9 L6	18h11m	−17.2°	∼5	
10 L7	18h14m	−14.3°	∼5	
11 Sgr X-1 (?)	18h02m	−24.9°	∼5	No agreement on position; this value from (18).
12 Sgr	17h44m	−23.2°	≤5	A weaker member of a rather uncertainly resolved pair. Probably the source earlier called Oph X-1, not confirmed, was this one.
13 Cyg X-3	19h58m	+40.6°	∼5	Position that of Cyg A, probably premature identification (19).
14 Cyg X-4	21h21m	+43.7°	∼5	Suggested by NRL data (17).
15 Cyg X-5	20h40m	+29°	∼5	Tentative fit to NRL data (17), possibly Cygnus Loop?
16 Lac X-1	22h40m	+54°	∼5	Another tentative position (16).
17 Cas X-1	23h21m	+58.5°	∼5	Position given that of Cas A, identification uncertain (17, 20).
18 Vir X-1	12h28m	+12.7°	∼5	Position given that of M87, postulated identification, NRL (17).
19 Leo X-1	9h35m	+ 8.6°	∼5	Reported only once (17).
20 Ara X-1	16h52m	−46.6°	∼5	Reported only once (17).
21 Coma X-1	13h	+28°	∼10	Reported as extended over several degrees by Boldt et al. (21) and identified tentatively with Coma cluster of galaxies, distance ca. 200 million lyears (21).

[a] All intensities are rough integrals over rocket energy range, except for source 21, seen (?) so far only in the 20–50 keV range. Sources 1 and 2, sources 5 and 6, and perhaps sources 13 and 14, unresolved, are seen also at balloon energies.

The Diffuse Background

In addition to the discrete sources which stand out against the sky, there is good evidence for a diffuse background. The background can be due to randomly distributed sources of any angular size, even sources so small as not to overlap, but spaced closer than the few degrees resolution of the typical detectors. There are not enough data to give anything but the roughest limits on the isotropy, but perhaps one can say that neither the Sun nor the Milky Way plane nor any other directions stand out by a factor of as much as ~1.5–2 in intensity. The background flux is rather consistently observed over a wide range of energies, as is shown in Figure 5. Note that the Arnold-Metzger data, statistically highly precise, form the entire high-energy end of the spectrum. They were obtained on two cislunar probes, Rangers 3 and 5, and have been recently confirmed on ORS-3 (36). It is possible to argue that all or most of the general isotropic background is instrumental in origin, but this view disregards a whole series of apparently credible and consistent experiments. We propose instead to accept the plot shown at face value, at least for the present. The isotropic X rays are then a phenomenon of large scale and considerable energetic importance, a weaker analogue to the still more impressive isotropic energy flux in the microwave region, which has been interpreted as a pervasive black-body flux at 3°K. The X-ray flux corresponds to a total energy density of about 5×10^{-4} eV/cm^3 while the 3°K radiation amounts to ~0.8 eV/cm^3, galactic visible starlight to ~0.5 eV/cm^3, and starlight averaged over the whole of space to ~10^{-2} eV/cm^3. The spectrum can be represented by the relation

$$I(\epsilon) = dN(\epsilon)/dEd\Omega da$$
$$= 7(E_0/E)^{1.3}, \quad \text{for } 2 \text{ keV} < E < 1 \text{ MeV}$$

where $I(\epsilon)$ is the specific intensity in eV/eV-cm^2-sec-sr, and $E_0 = 40$ keV. The single power-law shape suggests a simple origin, so far unknown.

Time Variations

Secular changes.—The statistics of the X-ray sources within the Galaxy allow some lifetime estimates. If supernova remnants are typically X-ray sources—and the Crab Nebula is one example—then the lifetime of a source must be long enough so that a dozen or so with measurable flux exist in the galactic plane at one time. The estimate depends upon the square of the mean distance to the sources. If we assume that the sources are intrinsically not very different from the Crab Nebula, they must have an active X-ray lifetime of some 10^4 years if SN explosions occur every century or so in the Galaxy, since not every known SN is a source, but only one among a few. The Scorpius source might be an old nova, but the class of old novae includes scores not known to be X-ray sources. On this basis, taking the distances of the dozen sources to be 1 kpc, we need to accumulate a few tens of thousands of novae in the whole Galaxy. At the expected rate of a few tens per year, this

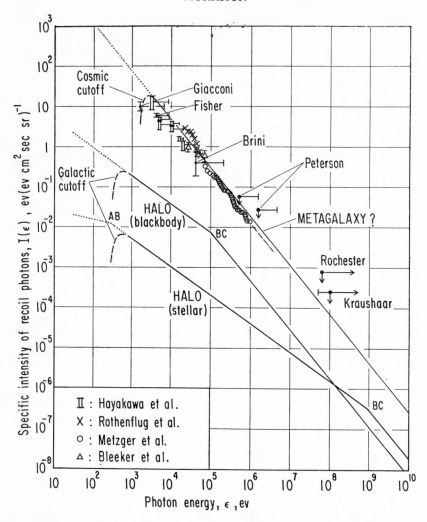

FIG. 5. The diffuse background spectrum in the keV to MeV range. The solid curves are theoretical models irrelevant here (and wrong) (27). References: Hayakawa et al. (28), Rothenflug et al. (29), Metzger et al. (30), Bleeker (31), Giacconi & Gursky (32), Fisher et al. (33), Brini et al. (34), Peterson (35, 36), Duthie et al. (37), Kraushaar et al. (38).

implies a lifetime of 1000 years in the X-ray emitting state. The object has not changed visibly in 70 years (13).

Short-term changes.—There is good evidence, perhaps not quite certain, that Cygnus X-1 sharply declined in X-ray emission in the 2–6 keV region during the year 1965. It was detected on two rocket flights (16, 17), both times with detectors which in the same flight viewed Scorpius and a couple of weaker sources. The ratio of Cygnus X-1 to Scorpius decreased by about a factor of 4 between the flight of June 1964 and that of April 1965. Other observations seem to be consistent with this remarkable result.

The inconsistencies of the balloon experiments for spectral measurement (39–41) have led to the suggestion of time variations of the Crab Nebula in the region above 20 keV. Such variations would have to be sporadic changes of factors of 2 to 5 in a month's time. At the lower energies, where the Crab source almost surely occupies a volume of many cubic light years, such variations seem impossible. If they were to turn out to be real at the higher energies, they would imply a source of much smaller volume for the higher-energy photons, one or a few hot spots within the low-energy sources. It is more probable that the systematic errors of the various experiments have merely simulated a changing source.

POLARIZATION

This diagnostic parameter has not been measured at all in the X-ray region. Atomic X-ray scattering is strongly polarized, so that in principle a way to measure the incoming radiation is at hand. Plans for exploiting this at balloon altitudes have been made (42).

THE ATOMIC MECHANISMS OF X-RAY EMISSION

All sources of radio continuum (apart from the 3°K microwave background) appear to emit bremsstrahlung, either magnetic or ordinary Coulomb bremsstrahlung. X-ray continuum as well can arise from those two mechanisms, and from black-body radiation, inverse Compton scattering of lower-energy photons by relativistic electrons, knock-on collisions of fast protons with atomic electrons, and atomic recombination, with or without additional electron transitions. Lines, too, can arise, either from atomic or nuclear level decays; these can be produced in absorption together with continuum. Since the atomic levels of all the elements of abundance above a part in 1000, except perhaps for iron, lie below a kilovolt, lines cannot be expected except under very unusual circumstances until the region under 1 or 2 keV is studied, either in emission or in absorption. Nuclear lines are rare, too, for light nuclei; they can be expected only in the energy region above many tens of keV. Significant stimulated emission is plainly unlikely because of the short lifetimes in this spectral region. We shall therefore say no more about line spectra; the couple of tentative reports of line structure in any spectrum are still far from convincing.

The first four processes listed above are those of chief interest. Their distinguishing features are clear. *Magnetic bremsstrahlung* requires electrons of extremely high energy, compared to the energy of the emitted photon. Thus for X rays in magnetic fields of the usual cosmic intensities of microgauss or tens of microgauss, X-ray emission means electron energies of 10^{14} eV at least. The radiative lifetimes are cosmically brief. Thermal excitation of *Coulomb bremsstrahlung*, by contrast, implies that the electron energies are comparable with the photon energies emitted, therefore around 10 keV. The radiative lifetimes from this process can be very long. *Black-body emission*—to be distinguished sharply from thermal bremsstrahlung emitted by dilute gaseous structures, which are almost transparent to these photons—is spatially so efficient as an emitter that it implies sources of small physical area, and hence of small angular dimensions at a distance. Here lifetimes again are short. *Inverse Compton processes* produce X rays from the relativistic electrons of some objects without the need for magnetic fields, and therefore without the simultaneous emission of detectable radio-frequency radiation. If one recalls that there are both point and extended sources, that most X-ray sources are not radio sources at observed levels, and that both statistically and by observation, X-ray sources seem of quite long life, one sees the possible utility of these four mechanisms. The others have for one or another reason been less attractive.

We do not intend to present a detailed summary of the relations describing all these processes. They are handled in many references (44, 45). Rather, we shall quote a few mainly approximate formulae which enable construction of rough models for the various sources so far observed.

Black-body radiation.—This mechanism is characterized by the familiar Planck spectrum and by very high surface emissivity. For a black body at T degrees K, the spectrum is:

$$j(\nu)d\nu = 2\pi(\nu^2/c^2)(e^{h\nu/kT} - 1)^{-1}d\nu \text{ photons/cm}^2\text{-sec in } d\nu$$

and the total power

$$F = \int_0^\infty j(\nu)h\nu d\nu = \sigma T^4$$

where $\sigma = 5.67 \cdot 10^{-5}$ erg/cm^2-sec deg^{-4} (43). The emission is unpolarized. One may estimate the lifetime of a spherical black body for emission of the thermal energy content:

$$\tau \simeq (R\bar{\rho}/T_s^3)(T_{av}/T_s) \times 10^{13} \text{ sec}$$

where

T_s = surface temperature; T_{av} = mean temperature;
R = radius; $\bar{\rho}$ = mean density (all c.g.s. and °K)

The small size of the source implied by the modest luminosities (for the galactic sources so far, only 10^3 to $10^4 L_\odot$) at these very high temperatures implies a lifetime rather too brief to be believable for any numerous class of sources. Observe that sufficiently small R leads to nuclear density.

Synchrotron radiation.—The electron energy and field for synchrotron radiation with maximum in the kilovolt region are governed by the relation:

$$h\nu_{\text{Max}} = 5.8 \cdot 10^{-9} B_\perp \gamma^2 \text{ eV}$$

with

$$E = \gamma m_0 c^2, \quad B \text{ in G}$$

so that 10-keV radiation means $\gamma = 1.3 \cdot 10^8$ for 100 μG. The orbit size is given by the Larmor radius:

$$R = E/300B\perp, \quad E \text{ in eV}, \quad B \text{ in G}$$

which is $R = 2.2 \cdot 10^{15}$ cm for 100 μG. The lifetime for a single electron is given by:

$$\tau_{1/2} = 5.1 \cdot 10^8/B\perp^2\gamma, \quad B \text{ in G}$$

and for our example is $3.9 \cdot 10^8$ sec. The short lifetime allows the possibility of short-term variations, especially above 10 keV. At the same time, it requires a mechanism for production of these very energetic electrons, which cannot easily be pushed back into a more explosive past. The spectral shape and total power emitted are given by these formulae [see Ginzburg & Syrovatskii (46)]:

$$P(E, \nu_m) = 2.16 \cdot 10^{-22}B\perp \text{ erg/sec} - \text{(c/s) per electron}$$

$$P(\gamma m_0 c^2, B\perp) \simeq 9.9 \cdot 10^{-16}\gamma^2 |B\perp \ (\mu G) \ |^2 \text{ eV/s per electron}$$

and roughly an isotropic electron distribution of number (cm^{-3}):

$$n(E)dE = KdE/E^m$$

$$P(\nu)d\nu = 1.35 \cdot 10^{-23}A(m)KLB^{(m+1)/2}(6.3 \cdot 10^{18}/\nu)^{(m-1)/2} \text{ erg/cm}^2\text{-sec-sr-(c/s)}$$

for a line-of-sight diameter L cm, B in G (randomly oriented). $A(m)$ varies slowly from $\sim$3 for $m = 1$ to $\sim$1 for $m = 5$.

The polarization is strong, elliptical in general. The observed synchrotron emission of the Crab in the optical spectrum is the best possible model for such a process, extended into the X-ray region. The familiar power-law spectrum is given above:

$$P(\nu)d\nu \propto d\nu/\nu^\alpha, \quad \alpha = (m - 1)/2$$

Inverse Compton effect.—Like the synchrotron process, this process requires relativistic electrons and an electromagnetic field. Instead of a dc magnetic field, given by B in the synchrotron case, here a photon field is the collision partner. The processes are related closely, as the formulae show:

$$P_c(\gamma m_0 c^2, \rho_{ph}) \simeq 2.66 \cdot 10^{-14}\gamma^2\rho_{ph} \text{ eV/sec}$$

$$\rho_{ph} = \text{isotropic photon energy density in eV/cm}^3$$

$$\tau_{1/2} \text{ (Compton)} \sim 2.10^{19}/\gamma\rho_{ph} \text{ sec}$$

$$\text{and } h\nu_{Max} \simeq \gamma^2(h\nu_{ph})$$

The spectral shape is broadly the same for inverse Compton and for synchrotron emission, though in detail this is true only for electrons with a smooth power-law distribution in energy (47). The main difference is, of course, that the inverse Compton effect can produce X rays using electrons of much lower γ than synchrotron radiation needs, provided $h\nu_{ph} \geq 1$ eV. The source must be spatially extended if for no other reason than that the magnetic field which presumably confines such electrons has to be weak enough not to rob the electron energy too rapidly by synchrotron loss. If the X-ray source is not a radio source, then strong conditions are placed on its minimum extension in size.

Thermal emission from "thin" sources.—If a source contains a near-Maxwellian distribution of electrons, they will collide with the ions present to emit thermal bremsstrahlung, or free-free emission. Integrating the bremsstrahlung formula over the Maxwell distribution of electron velocities, one obtains in a good approximation (43):

$$j(\nu) = 5.44 \cdot 10^{-39} \bar{g}(\nu, T) \exp\left(-h\nu/kT\right) T^{-1/2} \sum n_e n_z Z^2 \text{ erg/cm}^3\text{-sec-sr},$$
$$T \text{ in } °K, \quad n\text{'s in } cc^{-1}$$

for the power radiated from unit volume of a thin source with number densities: n_e for electrons; n_z for ions, charge ze. The effective Gaunt factor $\bar{g}(\nu, T)$ is a slowly varying function, which approaches unity for photon energies high compared to kT, and has been carefully studied (48, 49). At low frequencies

$$g \simeq \frac{\sqrt{3}}{\pi} \ln\left[4kT/(0.6h\nu)\right]$$

is an estimate good to a factor of better than 2. The total isotropic emission integrated over all frequencies is given well by

$$P = 1.44 \times 10^{-27} T^{1/2} g \sum n_e n_z Z^2 \text{ erg/cm}^3\text{-sec}, \quad T \text{ in } °K$$

Bremsstrahlung dominates for $kT \geq 2$ keV (48). With ionization of H and He complete in the X-ray region (kT above 15 eV) and the other abundant elements well ionized, it seems good enough to estimate the sum, for normal chemical composition, by noting that $n_e = n_H \sum (n_z/n_H)z = 1.34 n_H$; and $\sum n_e n_z z^2 = 1.8 n_e n_H$ (48).

The absorption mean free path in the dilute gas for photons of energy $h\nu$ is given well enough by the relation:

$$\text{m.f.p.} \simeq 4 \times 10^{34} T^{1/2} (h\nu/1 \text{ eV})^3/z^2 g n_e n_z \text{ cm}, \quad T \text{ in } °K$$

which emphasizes how thin the medium is for the energies we are considering and for number densities possible for a source extended over spheres of radius r with stellar mass:

$$n_e \sim \frac{10^{2-3}}{(r_{1\text{year}})^3}$$

The medium is thick in the keV range only for $r \stackrel{<}{\sim} 1 - 0.1$ a.u. Such radiation has been discussed by Tucker & Gould (48) [and Hovenier (50)] in the temperature range between 10^6 and 10^8 degrees, or 1–100 keV. Tucker & Gould consider, besides the free-free transitions described above, the radiative recombination, the line emission induced by Coulomb collisions, and the so-called dielectronic recombination, in which an ion captures an electron into an excited state, while at the same time another ionic electron is excited to a state below the continuum. Thus the ion has picked up enough energy to enter its continuum, but splits it two ways, so that both electrons are bound without photon emission. A radiative transition to a lower level by one electron then completes the recombination. The dielectronic process is more important than normal recombination at temperatures high compared to ionization temperature.

At temperatures above 1 keV, free-free processes dominate in emission, and ionization is more or less complete. The line-emission recombination edges of elements like oxygen and neon, which are significant below 1 keV, fade out. Structure in these thin thermal spectra will appear more or less in the same region where interstellar absorption is important; the result will be complex but plainly of real content once it can be observed. Possible edges for nearly stripped iron will be of low contrast at the energies involved—under 1 keV for the L shell, but near 7.1 keV for the K shell—though it is a real possibility at the higher energies and temperatures. Low-lying nuclear states are also uncommon in the abundant light elements, but they can be Coulomb excited, and represent a possible source for high-energy structure in the spectrum.

The Observed Spectra

Four spectra can be presented from cosmic X-ray sources. They are given in Figures 5 through 8, combining the results of many investigations. Not all experiments are plotted, but the omitted data do not contradict the data shown. In Figures 6 and 7 rough fits have been shown for various simple emission mechanisms: the power-law spectrum of the synchrotron theory, and the thermal free-free spectrum from dilute plasma. Isothermal plasma will fit none of the sources; power-law spectra will not fit the Scorpius source very well. But it appears rather naive to expect that in this new regime the familiar synchrotron power-law spectrum will work unmodified, in spite of the very great radiative lifetime change between the electrons emitting meter-band radiation, and those at 50 keV, a factor of $>10^5$, applying to a large fraction of the total relativistic electron energy present. At least a sharp break in the spectrum is to be expected. On the other hand, it is unlikely that the plasma would be isothermal. Any other distribution of temperature and density is sure to produce a spectrum more complex than the simple exponential. Thermal gradients are not difficult to maintain in this very dilute material, and density gradients of course will make important differences in intensity. No map of any source seen in different X-ray energies is yet available; all present spectra sum over the entire source. Here is evidently a refinement much to be sought after.

Toward Astronomical Models

The Scorpius source.—This source is likely to emit via a thermal-plasma mechanism (58). For there is no radio emission (13), and the visible continuum is roughly what would be expected from a hot plasma fitting the observed X-ray flux at a few keV. Synchrotron emission ought to increase greatly as the emitted photon energy decreases. Moreover, the source is small in physical extent, observed to be less than some 100 a.u. and possibly much less. From the integrated emission, we have:

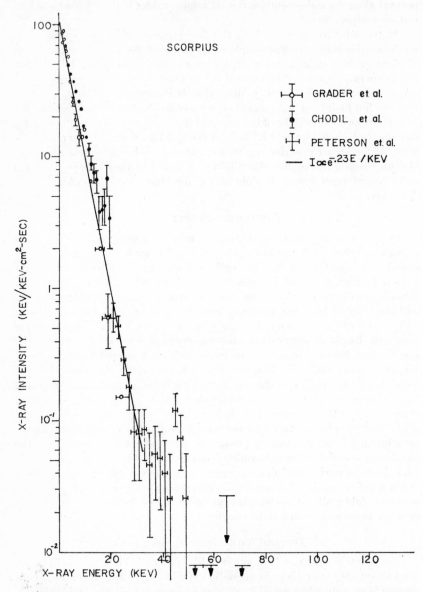

FIG. 6. X-ray spectrum of the Scorpius source. In this and Figures 7 and 8, not all experimental results are plotted. The omitted data serve as confirmation in most cases. References: Grader et al. (69), Chodil et al. (6), Peterson & Jacobson (71).

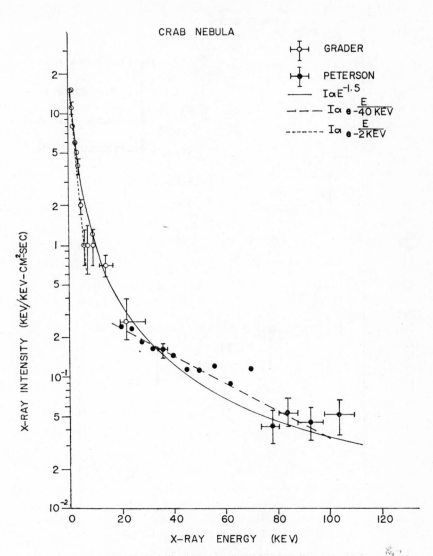

FIG. 7. X-ray spectrum of the Crab Nebula. References:
Grader (69), Peterson et al. (40).

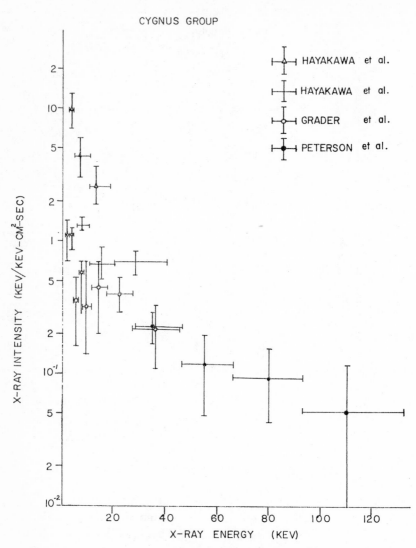

FIG. 8. X-ray spectrum of the group of sources in Cygnus. These sources are not uniformly resolved. Presumably the spectrum is dominated by the brightest source nearby, Cyg X-1. References: Hayakawa et al. (28), Grader et al. (69), Peterson & Jacobson (71).

$$\phi \simeq K \langle n^2 V \rangle \sqrt{T} \propto (M^2/V) \sqrt{T}$$

and thus we obtain a value for $\langle n^2 V \rangle$, or more directly for M^2/V. The plasma cannot be gravitationally self confined. It could be a corona to a central object, provided that object is more massive than $1\ M_\odot$, for we require then that $kT/m_H \simeq GM/r$. The radius is then between 0.01 and a few a.u. for $M_\odot < M < 10^3 M_\odot$. A smaller radius will make the source too opaque; therefore the large masses are favored. This situation is fully possible (51). More attractive perhaps is the idea of a transient plasma, whose mean particle lifetime is $r/\sqrt{kT/m}$. For this we get r about 0.1 a.u., and a mass-loss rate of a few $\times 10^{-6} M_\odot$/year. This is a quite acceptable plasma model. The energy source is unknown, but could be related to the idea of a close binary (52–54) whose streaming plasma transfers are accompanied by shock collisions (60) or the like, drawing energy essentially from the gravitational and thermal store of the stars. The optical emission would on this view come from a cooler plasma volume, only partly ionized (55). On this sort of model, the visible light variations would be accompanied by a similar X-ray variation, which might be slower, if the plasma volume is large, or faster, in the reverse case. The model is obviously preliminary. The lifetime of such objects could be large enough, say 10^4–10^6 years, to furnish most of the discrete sources in the galactic plane. Such special conditions for a close-binary remnant would explain the low yield among old novae.

The Crab Nebula.—The Crab Nebula is distinguished from Scorpius X-1 by three features: its spectrum is much harder, its source diameter is light years, not astronomical units, and its radio and indeed its optical synchrotron emission are strong. Still, it remains an X-ray object predominantly. Its overall X-ray luminosity rivals the optical emission, and requires more overall electron energy, in a much smaller volume. Given strong acceleration, or very special circumstances to preserve fast electrons very copiously made in early stages (51, 56), which may or may not be possible if leakage diffusion across the lines of force is at all effective, the X rays may be synchrotron emitted (61–64). This gives a simple, if not compelling, model for the power spectrum. But, admitting that the spectrum may be merely that of a nonisothermal plasma, another model is possible (57). Here the X-ray source is a central freely expanding ball of dilute plasma. [The radiopolarization data relied upon to exclude this in (59) are in error.] The energy content of this ball is 100 times the total energy content of the large visible and radio portion of the nebula. It obtained its energy from the same source—still unknown—whence the observed radio and optical synchrotron electrons got it, the acceleration mechanism. But the plasma gained heat energy from the inefficiencies of the acceleration; what energy did not get stored in fast electrons went to heat the dilute expanding plasma, with something less than a solar mass. This suggests a very general result; most synchrotron radio sources, probably explosion aftermaths on many scales, may in the end turn out to be primarily X-ray objects, still undetected because at the moment L_X/L_{radio}, the ratio of detectable luminosities, is much larger than 10^3.

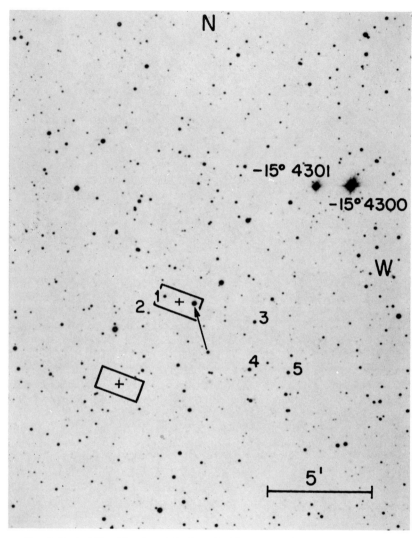

FIG. 9. Optical identification of the Scorpius X-1 source. *Above:* we show the optical fields singled out (within the rectangles) as possible positions of the source by the modulation collimator study (13). *Opposite:* we show the spectrum obtained in the optical region for the star indicated by the arrow.

Whether the other sources, like the Cygnus source, will resemble the Crab, or Sco X-1, or neither, is now the issue. *If* the rapid time variation reported for it is real, Cyg X-1 must turn out to be a "point" source, like Sco X-1.

There still remains the open question of the neutron star. The original idea that such a dense body would emit kilovolt radiation by black-body thermal emission no longer seems natural. The diameters of the sources, their spectrum, the presence of photons at tens of kilovolts, and various lifetime arguments now stand against, rather than for, the neutron star (65–67). But a more complex model of such an object, perhaps bearing large magnetic "handles," or holding a thin gravitationally bound corona, can then emit by synchrotron or by thermal bremsstrahlung processes, respectively. The neutron star then enters as a store of energy and momentum, perhaps vibrationally or in some other way feeding the radiating region it confines. Cameron, in particular, has presented the properties of such a model in a series of papers, especially those cited in (68).

Extragalactic sources.—No extragalactic source has yet been firmly identified. If the strong radio sources do turn out to possess strong thermal X-ray emission, the arguments of the explosion aftermath will be supported. If, like the Coma galaxy cluster, the sources are not strong radio emitters, a thermal plasma, perhaps gravitationally bound, heated by gravitational interaction, may be involved (70). A large fraction of all the mass in space might possibly take part in such X-ray emission, otherwise to be invisible. The question is open, but exciting. How much such great hot coronas might contribute to the diffuse background and how much of that is merely secondary to cosmic-ray meson and electron production over all of space is another open question.

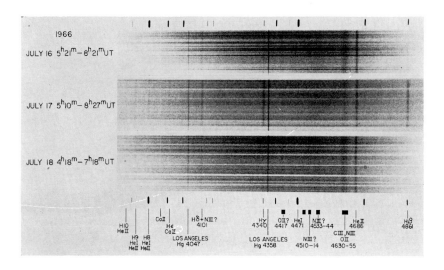

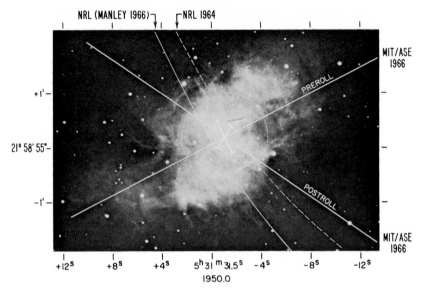

FIG. 10. Optical identification of the Taurus source, and its X-ray size. The photograph shows the Crab Nebula in the continuous synchrotron light. On it are marked two large arcs, one (NRL 1964) being the quoted position locus of the NRL group. The arc (Manley 1966) is the same result corrected by Manley (72) for rocket motion. The two lines mark the position set by the modulation collimator in two directions of scan, between which the rocket rolled to a new position (25). The rms error bars are shown normal to each line. The solid circle shows the best rms fit for a circular source. Its best position is $\alpha = 5^h31^m\ 30.3$ s, and $\delta = +21°59'9''$. The measured widths are $110'' \pm 25''$ postroll, and $60'' \pm 60''$ preroll. This is consistent with a source elongated along the longer optical axis, but is too rough to be a strong argument for elongation.

The next review of X-ray astronomy is likely either to set it out as one of the most fruitful channels of data about the largest scales of the universe, or to relegate it to the detailed history of certain unusual stars or binaries. Experiment alone can tell. Plenty of exposure at the top of the air ocean or beyond is the essential for the future; the ingenuity and energy of the experimenters are already certain.

I am grateful to Mr. B. Skeist for his indispensable assistance.

LITERATURE CITED

1. Friedman, H., *Ann. Rev. Astron. Ap.*, **1**, 59 (1963)
2. Giacconi, R., Gursky, H., Paolini, F., Rossi, B., *Phys. Rev. Letters*, **9**, 439 (1962); **11**, 530 (1963)
3. Bethe, H., Salpeter, E. E., *Quantum Mechanics of One- and Two-Electron Atoms* (Academic Press, New York, 1957)
4. Rossi, B., in *Perspectives in Modern Physics* [Interscience (Wiley), New York, 1966]
5. Felten, J. E., Gould, R. J., *Phys. Rev. Letters*, **17**, 401 (1966)
6. Chodil, C., Jopson, R., Mark, H., Seward, F., Swift, C., *Phys. Rev. Letters*, **15**, 605 (1965)
7. Friedman, H., Byram, E. T., Chubb, T. A., *Science*, **153**, 1527 (1966)
8. Overbeck, J., *Ap. J.*, **141**, 864 (1965)
9. Cocconi, G., *Ap. J. Suppl. 4*, 417 (1960)
10. Duthie, J. G., Cobb, R., Stewart, J., *Phys. Rev. Letters*, **17**, 263 (1966)
11. Greisen, K., *Perspectives in Modern Physics* [Interscience (Wiley), New York, 1966]
12. Gould, R. J., Schreder, G., *Phys. Rev. Letters*, **16**, 253 (1966)
13. Sandage, A. R., Osmer, P., Giacconi, R., Gorenstein, P., Gursky, H., Waters, T., Bradt, H., Garmire, G., Sreekantan, B. V., Oda, M., Osawa, K., Jugaku, J., *Ap. J.*, **146**, 316 (1966)
14. Bowyer, S., Byram, E. T., Chubb, T. A., Friedman, H., *Science*, **146**, 912 (1964)
15. Fisher, P. C., Jordan, W. C., Meyerott, A. J., Acton, L. W., Roethig, D. T., *Nature* (Preprint, 1966)
16. Bowyer, S., Byram, E. T., Chubb, T. A., Friedman, H., *Science*, **147**, 394 (1965)
17. Byram, E. T., Chubb, T. A., Friedman, H., *Science*, **148**, 152 (1966)
18. Giacconi, R., Gursky, H., Waters, J. R., Clark, G., Rossi, B., *Nature*, **204**, 981 (1964)
19. Gursky, H. (Private communication)
20. Friedman, H. L. (Address to AIAA Convention, Boston, 1966)
21. Boldt, E., McDonald, F. B., Riegler, G., Serlemitsos, P., *Phys. Rev. Letters*, **17**, 447 (1966)
22. Plaut, L., in *Galactic Structure* (Univ. of Chicago Press, 1965)
23. Oda, M., *J. Appl. Opt.*, **4**, 143 (1965)
24. Gursky, H., Giacconi, R., Gorenstein, P., Waters, J. R., Oda, M., Bradt, H., Garmire, G., Sreekantan, B. V., *Ap. J.*, **146**, 310 (1966)
25. Same as ref. 73.
26. Luyten, W. J., *Circ. No. 1980* (Central Bureau for Astron. Telegrams, 1966)
27. Felten, J. E., Morrison, P., *Phys. Rev. Letters*, **10**, 453 (1963)
28. Hayakawa, S., Matsuoka, M., Yamashita, K. (Preprint, 1966)
29. Rothenflug, R., Rocchia, R., Koch, L., *Proc. Intern. Conf. Cosmic Rays, 9th, London, 1965*, 446
30. Metzger, A. E., Anderson, E. C., Vandilla, M. A., Arnold, J. R., *Nature*, **204**, 766 (1964)
31. Bleeker, J. A. M., Burger, J. J., Scheepmaker, A., Swanenburg, B. N., Tanaka, Y., *Proc. Intern. Conf. Cosmic Rays, 9th, London, 1965*
32. Giacconi, R., Gursky, H., *Space Sci. Rev.*, **4**, 151 (1965)
33. Fisher, P., Johnson, H., Jordan, W., Meyerott, A., Acton, L., *Ap. J.*, **143**, 203 (1966)
34. Brini, D., Ciriegi, V., Fuligni, F., Gandolfi, A., Moretti, E., *Nuovo Cimento*, **38**, 130 (1965)
35. Peterson, L. E., *Space Research VI, Proc. Intern. Space Sci. Symp., 6th, Amsterdam, 1965*
36. Peterson, L. E. (Rept. to Am. Geophys. Union Meeting, Washington, D.C., 1966)
37. Duthie, J. G., Hafner, E. M., Kaplon, M. F., Fazio, G. G., *Phys. Rev. Letters*, **10**, 364 (1963)
38. Kraushaar, W., Clark, G. W., Garmire, G., Helmken, H., Higbie, P., Agogino, M., *Ap. J.*, **141**, 845 (1965)
39. Clark, G., *Phys. Rev. Letters*, **14**, 91 (1965)
40. Peterson, L., Jacobson, A., Pelling, R., *Phys. Rev. Letters*, **16**, 142 (1966)
41. Haymes, R. C., Craddock, W. L., Jr., *J. Geophys. Res.*, **71**, 3261 (1966)
42. Novick, R. (Private communication)
43. Allen, C. W., *Astrophysical Quantities* (Univ. of London, 1963)
44. Gould, R. J., Burbidge, G. R., in *Handbuch der Physik*, **46/II** (Preprint, 1966)
45. Hayakawa, S., Matsuoka, M., Yamashita, K., *Space Sci. Rev.*, **5**, 109 (1966)
46. Ginzburg, V. L., Syrovatskii, S. I., *Ann. Rev. Astron. Ap.*, **3**, 297 (1965)
47. Jones, F. C., *Phys. Rev.*, **137**, B1306 (1965); and in press

48. Tucker, W., Gould, R., *Ap. J.*, **144**, 244 (1966)
49. Karzas, W. J., Latter, R., *Ap. J. Suppl.* **6**, 167 (1961)
50. Hovenier, J. W., *Bull. Astron. Inst. Neth.*, **18**, 185 (1966)
51. Tucker, W., *Cosmic X-ray Sources* (Thesis, Univ. of California, San Diego, 1966)
52. Hayakawa, S., Matsuoka, M., *Progr. Theoret. Phys. Suppl. No. 30*, 204 (1964)
53. Kraft, R. P., *Ap. J.*, **135**, 408 (1962)
54. Nikolskii, G. M., *Soviet Phys. "Doklady,"* **8**, 646 (1964)
55. Matsuoka, M, Oda, M., Ogawara, Y. (Submitted to *Nature*)
56. Manley, O. P., *Nature*, **209**, 901 (1966)
57. Morrison, P., Sartori, L., *Phys. Rev. Letters*, **14**, 771 (1965)
58. Manley, O., *Ap. J.*, **144**, 628 (1966)
59. Shklovsky, I. S., *Soviet Astron.—AJ*, **10**, 6 (1966)
60. Heiles, C., *Ap. J.*, **140**, 420 (1964)
61. Woltjer, L., *Ap. J.*, **140**, 1309 (1964)
62. Ginzburg, V. L., Syrovatskii, S. I., *Space Sci. Rev.*, **4**, 267 (1965)
63. Ginzburg, V., *Soviet Phys. "Doklady,"* **9**, 831 (1965)
64. Shklovsky, I. S., *Soviet Astron.-AJ*, **9**, 224 (1965)
65. Chiu, H. Y., Salpeter, E. E., *Phys. Rev. Letters*, **12**, 413 (1964)
66. Bahcall, J. N., Wolf, R. A., *Phys. Rev. Letters*, **14**, 343 (1965)
67. Swedin, M., *Nature*, **209**, 62 (1966)
68. Cameron, A. G. W., *Nature*, **212**, 493 (1966)
69. Grader, R. J., Hill, R. W., Seward, F. D., Toor, A., *Science*, **152**, 1499 (1966)
70. Felten, J. E., Gould, R. J., Stein, W. A., Woolf, N. J. (Preprint Letter to *Ap. J.*, 1966)
71. Peterson, L. E., Jacobson, A. S., *Ap. J.*, **145**, 962 (1966)
72. Manley, O. (Private communication, 1965)
73. Oda, M., Bradt, H., Garmire, G., Spada, G., Sreekantan, B. V., Gursky, H., Giacconi, R., Gorenstein, P., Waters, J. R. (Submitted to *Ap. J.*, 1966)
74. Sandage, A. (Unpublished)

ENERGETIC PARTICLES FROM THE SUN[1]

By C. E. Fichtel and F. B. McDonald

Goddard Space Flight Center, Greenbelt, Maryland

INTRODUCTION

Nature seems to accelerate particles with surprising ease. On a galactic scale the cosmic radiation extends to energies of at least 10^{20} eV. On a local scale the data from radiation belt and auroral studies strongly suggest that there is also an accelerating mechanism operating within the confines of the Earth's magnetosphere. Intermediate in this hierarchy is the solar production of energetic particles, which usually seems to occur in discrete events. These events range from large flare-associated increases in which the maximum particle energy exceeds 10^{10} eV, and the maximum intensity of particles with energies greater than 20 MeV may exceed 10^3 particles/cm²-sec-sr at the orbit of Earth, down to barely detectable events with intensities of about 1 particle/cm²-sec-sr at energies above 1 MeV and with no identifiable flare association. At present there is no adequate theory for the acceleration either of galactic cosmic rays or of magnetospheric and solar energetic particles.

In the case of solar events, it is possible to study the development of the active region, the optical flare, and radio emission, and then to measure the charge and energy spectra of the particles which reach the vicinity of the Earth. Thus the solar-particle event appears to be one particular particle-accelerating phenomenon in which we can observe a number of the accompanying processes. Also, since it is one of the unique aspects of the active region where it occurs, this acceleration process is important in understanding solar phenomena. At the same time, the particle-propagation characteristics should give new information on the magnetic-field configuration in the vicinity of the Sun and in interplanetary space.

The first evidence for the production of solar cosmic rays was the February and March 1942 observations of Lange & Forbush (1942), using sea-level, shielded ion chambers. In the late 1940's, development of the sea-level neutron monitor provided a much more sensitive instrument than the ion chamber, but despite this significant improvement, solar cosmic-ray increases were still observed only at the rate of one every few years. The large increase of February 23, 1956 provided the first detailed study of the high-energy particles. Because a number of neutron monitors were in operation at that time, accurate energy spectra as a function of time could be obtained (Meyer, Parker & Simpson 1956). An important feature of this event was the correlation by Bailey (1957) of the intense absorption of cosmic radio noise in the polar-cap region with the solar flare. The IGY Balloon observations by Anderson (1958) in 1957 and 1958 furnished the first direct identification of

[1] The survey of literature for this review was concluded in September 1966.

low-energy solar proton events. Almost simultaneously, the riometer studies of Reid & Collins (1959), Leinbach & Reid (1959), and Hultqvist (1959) identified many polar-cap absorption events as being produced by low-energy solar cosmic-ray increases. Both the riometer and the forward-scatter techniques were immensely valuable in providing a more sensitive continuous monitor of solar-flare cosmic radiation. Their results revealed that the Sun is a far more frequent source of cosmic rays than had been realized; over the period 1957 to 1961, events occurred at an average rate of greater than 12 per year. The riometers also furnished information about the initiation of events, which made possible the timely launching of large skyhook balloons and research rockets that gathered direct information on the charge and energy spectra of the particles. The University of Minnesota balloon groups detected a helium nuclei component and made differential energy spectral measurements for both H and He nuclei down to approximately 100 MeV/nucleon (Freier, Ney & Winckler 1959). Research rockets launched by the group at Goddard Space Flight Center extended the spectral observations down to 15 MeV for a number of events in 1960 (Davis, Fichtel, G uss & Ogilvie 1961). These observers also found that heavier nuclei such as carbon, oxygen, and neon were being accelerated to velocities of the order of 0.3 of the speed of light, and the observed abundances appeared to agree with those observed in the photosphere (Fichtel & Guss 1961).

Since 1961, satellite and space-probe experiments have played a dominant role. The first satellite observations were from Explorer IV in August 1958 by Rothwell & McIlwain (1959) using integral detectors with a threshold of 30 MeV. In the years since Explorer IV, there has been a steady evolution of experimental techniques by a number of research groups. Continuous monitoring and precise proton differential energy spectra from 3 to 500 MeV were provided by Explorers XII and XIV (Bryant, Cline, Desai & McDonald 1965a). With Mariner II and IV, these measurements were extended down to proton energies of 0.5 MeV. The frequency of events increased steadily as the energy threshold was lowered and the detector sensitivity increased (Krimigis & Van Allen 1966, O'Gallagher & Simpson 1966). The angular distribution of low-energy protons were measured by the Pioneer VI experiments of Bartley et al. (1966) and Fan, Lamport, Simpson & Smith (1966). The long-term anisotropies which were observed argued convincingly for an intermediate storage mechanism at the Sun. This view is supported by the observation by Explorer XIV and IMP of recurrence events over many solar rotations (Bryant et al. 1965b; Fan, Gloeckler & Simpson 1966).

The close correlation of type IV radio emission and flare-particle increases strongly suggested that electrons must be accelerated along with protons and heavier nuclei. Recent observations by Van Allen & Krimigis (1965) and Anderson & Lin (1966) on Mariner IV and IMP-III have revealed the existence of electron events with large fluxes at energies above 40 keV.

Most of the detailed satellite measurements were made either during the period of decreasing solar activity or over solar minimum. Significant changes may be expected as these measurements are extended toward solar maximum.

Thus, in the last decade, with the advent of the riometer and forward-scatter networks, balloon patrols, research rockets, satellites, and space probes, our knowledge of solar cosmic-ray events has been greatly extended. It is now possible to estimate the frequency of occurrence, directional properties, characteristics of the source region, energy spectra, and composition for a number of events over the last solar cycle. We shall try to summarize the knowledge of these events and give a brief interpretation of their possible implications.

GENERAL CHARACTERISTICS OF SOLAR COSMIC RAYS

One of the most striking features of solar cosmic-ray events has been the tremendous variety observed. In the largest of the flare-associated events, the flux of protons with energies above 20 MeV has exceeded 10^3 particles /cm^2-sr-sec for more than a day, and the total energy arriving at the top of the atmosphere of the Earth for the whole event has been 10^4 erg/cm^2-sr, about the same order of magnitude as that for galactic cosmic rays for a year. The present intensity distribution extends down to events which are more than a thousand times smaller, with no observable associated solar flare. Events with large fluxes of electrons at energies above 40 keV are seen in conjunction with small flares. There are recurring solar proton events which appear over many solar cycles with an approximately 27-day periodicity. The frequency and character of all types of events are probably strongly associated with the 11-year cycle of solar activity. This is definitely true of the large flare-associated events. Observation of the small, low-energy events began in the declining phase of the last solar cycle, well after the 1957–59 maximum, so a definite conclusion about the solar-cycle variations of these events is not yet possible.

FLARE-ASSOCIATED EVENTS

The larger events which produce protons with energies greater than 20 MeV have displayed an excellent correlation with solar flare and solar radio-burst activity. For the purpose of classification, a threshold for these events has been defined such that the peak flux of protons with energies above 20 MeV must exceed 0.5 proton/cm^2-sec-sr. Although these energy and intensity thresholds are somewhat arbitrary, they give a reasonable separation of these events from the normal recurrence events and other events which cannot be directly associated with flares. A few events of moderate size which have exceeded these thresholds did not have an obvious parent flare. In general, however, the term "solar cosmic-ray events" will imply flare-associated events unless specifically noted otherwise.

The character of these events as defined by the energy spectrum of the particles and its time variation differs markedly from one occurrence to another. For example, at comparable times in the November 12, 1960 and

July 12, 1961 events, the integral fluxes at energies above 10 MeV were nearly the same, but above 100 MeV they differed by more than a factor of 200. In some events, the maximum low-energy intensity has occurred as early as 4 hours after the flare, whereas in others it has occurred as late as 30 or 40 hours after the flare. When events are examined in detail with regard to time histories, differences in the abundances of protons, helium nuclei, and electrons become very striking. Because of these variations, it is not possible to set down a single set of characteristics for each feature of the flare-associated events, and it will often be difficult even to speak in terms of a "typical" characteristic. With this in mind, we proceed to discussion of the properties of these events, making whatever generalizations are possible and describing the variations in as much detail as seems warranted.

Frequency of occurrence.—The frequency distribution is naturally one of the first characteristics of interest. Because of the steady evolution of experimental techniques, it is necessary to set definite criteria for the different classes of events. The sea-level neutron monitor and ionization chambers provide significant coverage over the longest period of time, but the detectors see only the secondaries of the primary particles with energies above 500 MeV. The second row of Table I shows the number of sea-level events which have been detected each year since 1942. Two other ground-level instruments have provided data for some time: the ionosonde and the riometer. Both give rough estimates of the flux of particles with energies above 20 MeV by noting the increased absorption of radio signals in the ionosphere when it is bombarded by the solar particles. The first measures the reflection of a signal transmitted from the ground, and the second, the absorption of cosmic radio noise. The early riometers were able to detect events with fluxes above 20 MeV of more than 10 particles/cm²-sr-sec; improved riometers are more sensitive, but their data have been excluded from Table I in order to provide a fair comparison.

Table I shows that there is a clear tendency for the number of low-energy events to follow the level of solar activity. There is some tendency for the high-energy events to fall on each side of the sunspot maximum; however, this feature, although possibly of considerable interest, is not yet considered to be statistically significant.

Table II is a catalogue of the solar minimum events covering the period from November 1963 through August 1966. The flare association is excellent. The occurrence of all the list flares to the west of the central meridian reflects the strong collimation of the particles along the more-ordered interplanetary field which exists over solar minimum. Furthermore, all the events are very small (see Figure 1 for the time history of the 1964–1965 events).

Intensity.—For the larger events it is possible to obtain a rough estimate of the size distribution. The distribution given in Table III is based on events occurring from 1956 through 1961 (Malitson & Webber 1963).

The data are not complete for smaller events, but they do suggest a steady increase for the 10^6–10^7 and 10^5–10^6 particle/cm²-sec regions. How-

TABLE I

COMPARISON OF NUMBER OF SOLAR PARTICLE EVENTS
PER YEAR AND SOLAR SUNSPOT CYCLE[a]

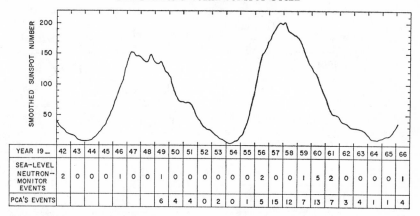

YEAR 19_	42	43	44	45	46	47	48	49	50	51	52	53	54	55	56	57	58	59	60	61	62	63	64	65	66
SEA–LEVEL NEUTRON– MONITOR EVENTS	2	0	0	0	1	0	0	1	0	0	0	0	0	0	2	0	0	1	5	2	0	0	0	0	1
PCA'S EVENTS								6	4	4	0	2	0	1	5	15	12	7	13	7	3	4	1	1	4

[a] In the above Table, the number of solar-particle events per year detected on sea-level neutron monitors and the number of polar-cap absorption events (PCA's) which are solar-particle events detected by ground level-radio absorption techniques, are shown and compared to the Zurich Smoothed Sunspot Number. In interpreting this Table several considerations should be noted: (a) The sensitivity of neutron monitors was increased generally around 1963. (b) Riometer sensitivity was increased somewhat in about 1962 or thereafter, but only events which would have been detected before this time have been included. (c) Before 1955 only ionosonde data exist, not riometer data, hence some events may have been missed. (d) The sunspot number is the index most commonly used to indicate overall solar activity, but it is not at all clear that it is the best indicator.

ever, for very small, low-energy events there appears to be a dramatic increase which will be discussed later.

The largest events observed have had peak fluxes above 20 MeV of about 3×10^4 particles/cm²-sec in the vicinity of the Earth. This number can be compared to the galactic cosmic-ray flux, which is about 2 particles/cm²-sec. There is some debate regarding which event was the largest observed to date; however, at least at moderate energies, it was probably the November 12, 1960 event, when the integrated flux for the whole event at the Earth for particles above 20 MeV was about 4×10^9 particles/cm². Again for comparison, the integrated galactic cosmic-ray flux for a whole year is about 6×10^7 particles/cm². Integrating over the last solar cycle, there were well over ten times as many solar particles in the vicinity of the Earth above 20 MeV as there were galactic cosmic rays. As will be discussed later, the energy spectra of solar cosmic rays are very steep, so that at about 1 BeV, the solar cosmic rays represent only a very small fraction of 1 per cent of the galactic particles when averaged over a solar cycle.

TABLE II

CATALOGUE OF PRINCIPAL SOLAR COSMIC-RAY EVENTS
NOVEMBER 1963–AUGUST 1966

Date	Time of flare	Importance	Location	Peak flux >20 MeV
Mar. 16, 1964	1550	2	N-05, W-75	.7
Feb. 5, 1965	1750	2	N-07, W-25	50
Oct. 4, 1965	0938	2	S-20, W-29	3
Mar. 24, 1966	0233	3B	N-18, W-37	15
May 2, 1966	Type II radio emission, 1222–1237, location unknown			
July 7, 1966	0022	2B	N-34, W-45	30
Aug. 28, 1966	1530	3B	N-08, W-03	15

For the period spanning the last solar maximum, 1956–1960, the integrated flux of particles at energies above 20 MeV is estimated to be 10^{10} particles/cm²-year. For 1964, the integrated intensity above 20 MeV was about 10^5 particles/cm²-year. For 1965, this had risen to 10^7 particles/cm²-year. Not only does the number of solar cosmic-ray events decrease at solar minimum, but the individual events are much smaller in intensity on the average. As indicated above, the net result is a variation of almost five orders of magnitude in the production of solar cosmic rays from solar maximum to solar minimum.

Intensity—time profiles.—The important parameters in describing the time history of an event are the onset characteristics of the particles, especially with respect to the main features of the flare, and the event rise and decay characteristics. These characteristics are strongly energy dependent. The higher-energy particles (i.e., above 200 MeV) display a faster rise and

TABLE III

SIZE DISTRIBUTION OF LARGE EVENTS

Integrated intensity for energies >30 MeV (particles/cm²-sec)	Number of events
10^7–10^8	10
10^8–10^9	10
$>10^9$	2

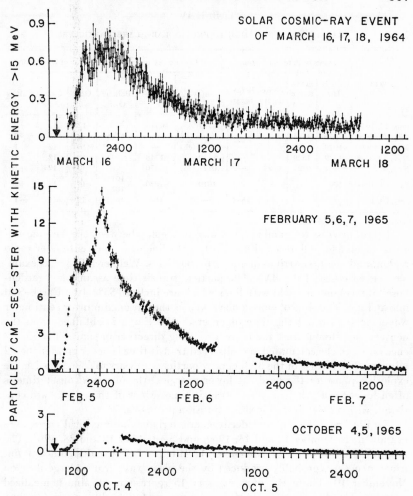

FIG. 1. The intensity profile of particles greater than 15 MeV for the three flare-associated events during the 1964–1965 solar minimum period. Note the enlarged scale for the March 16, 1964 event. These events are very small and display faster rise and decay times than the average event near solar maximum (McDonald 1966).

decay time than the particle population at lower energies. Generally, the higher-energy component has a relatively simple profile with no marked structure. The lower energies show a greater variety. The maximum intensity may occur a few hours after the flare, or several days later when the storm cloud reaches the Earth. There is often significant structure, and this will be discussed later.

TABLE IV

Transit Times for High-Energy Solar-Flare Particles

Event	Times of flare maximum				Cosmic-ray onset			Transit time (min)		
	Hα	cm Radio burst	White light	X-ray burst	$n(I)$	$n(0)$	Balloon OBS	Obs.	Calc.	Δt
Feb. 23, 1956	0340	0341	0340	—	0343	0348	—	12	12	0
Aug. 22, 1958	1456	1504	—	—	—	—	1524	28	30	−2
May 4, 1960	1023	—	—	—	1027 (ch)	—	—	13	12	+1
Nov. 12, 1960	1329	1329	—	—	~1330	1348	—	10	12	−2
Nov. 15, 1960	0221	0227	0223	—	0227 (M)	0250	—	13	12	+1
July 18, 1961	1005	—	—	—	1010	~1030	{1015 1026}	14	12	+2
July 20, 1961	—	—	—	1553	—	1605	1604	17	18	−1

It is of interest to examine the time relationship between the flare param-eters—such as the flare maximum in the Hα line or in the centimeter radio region, and the first arrival of energetic particles. Webber (1963) has exam-ined seven cases (Table IV). The particle transit time assumes a "garden-hose" interplanetary field with lines which are inclined 55° to the Earth–Sun line at 1 a.u. The time of emission is taken as the Hα maximum or centimeter wave maximum and the time of onset at the Earth is obtained from the neutron monitor nearest the correct viewing direction in space. The differ-ence between the calculated and observed transit time is never greater than 2 minutes. The release of the high-energy particles appears to coincide with the explosive phase of the flare. At lower energies, the observed onset time is often longer than the calculated transit time. Most of these cases are con-sistent with a delay introduced by diffusion processes.

To illustrate the rise and decay characteristics, four typical cases were examined: November 12 and 15, 1960, and September 10 and 28, 1961. The neutron-monitor data for November 12 (Figure 2a) reveal significant un-usual structure, probably produced by the blast wave from a large flare on November 10. The event of November 15 represents an almost idealized example. There is a very rapid rise to maximum, followed by a series of fluc-tuations, and then a smooth decline. The increase on May 4, 1960 (Figure 12) was similar to this. The neutron monitor responds to particles with energies greater than 500 MeV. The low-energy data above 20 MeV for the November 1960 events exhibit a very different behavior. The rise to maximum is much slower and at low energies the event is present over a period of days. Both the rise and decay characteristics are markedly different from the solar minimum events of Figure 1. The increase on September 10, 1961 is much more complex (Figure 3). The maximum is reached several days after the parent flare, and there are large variations in intensity throughout the event.

The increase of September 28, 1961 (Figure 4) possesses features common to both types discussed in the last paragraph. There is a simple rise to maxi-

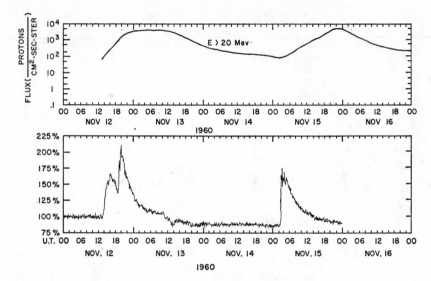

FIG. 2. *Above:* Neutron-monitor increase recorded for November 12 and 15, 1916 events. This instrument responds to nuclei above 500 MeV. The peak on November 12 and 15 represents a primary flux of ~8 protons/cm²-sec-ster (Steljes et al. 1961).

Below: Time history of particles >20 MeV for November 12 and 15, 1961 events. This represents a compilation of riometer, balloon, and rocket data (Fichtel, Guss & Ogilvie 1963).

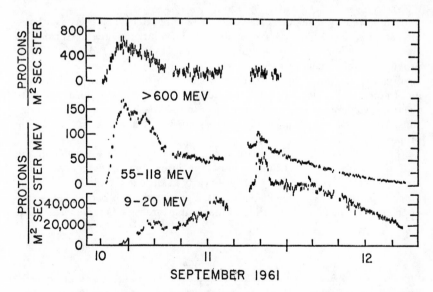

FIG. 3. The event of September 10, 1961. The intensities of relativistic protons and of two lower-energy groups indicate the complicated structure of this event (Bryant et al. 1965).

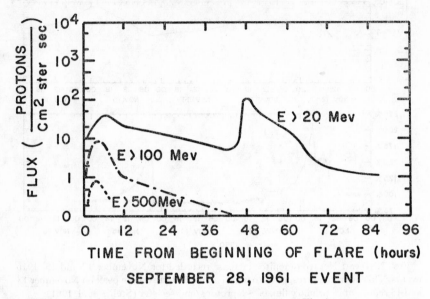

FIG. 4. Time history of the September 28, 1961 event. The increase at some 50 hours after the flare coincided with a large Forbush decrease and a great geomagnetic storm with mid-latitude auroral displays (Bryant et al. 1965).

mum followed by a smooth decay until 50 hours later when there is a large increase associated with the commencement of a large magnetic storm. The Goddard cosmic-ray experiment on Explorer XII (Bryant et al. 1962, 1965a) provided detailed energy spectra throughout the event (Figure 5). For each energy, the maximum intensity corresponds to a distance traveled of ~10 a.u. In fact, on a scale of distance traveled (or velocity × time) it is found that the time profiles for each energy have the same shape. The long rise and decay times suggest that a diffusion-like process must play a significant role in the propagation.

The intensity time profiles are also a function of solar longitude of the initiating flare. For west-limb events, the intensity curves generally are strongly peaked with a power-law decay early in the event and a transition to an exponential decay later in the event. For events with parent flares east of the central meridian, the time of maximum intensity is less marked and

FIG. 5. The differential intensities of solar protons for the event of September 28, 1961 plotted against time after the X-ray burst at the Sun. The data were interrupted when the satellite passed through the magnetosphere and when the delayed increase occurred on September 30, 1961 (Bryant et al. 1965a).

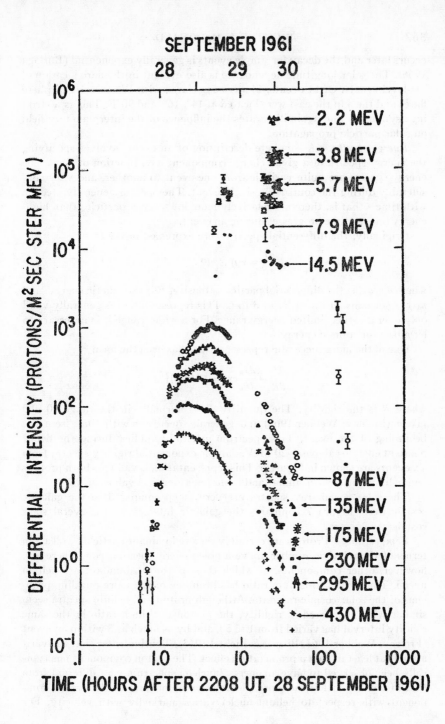

SEPTEMBER 1961

28   29   30

2.2 MEV

3.8 MEV

5.7 MEV

7.9 MEV

14.5 MEV

87 MEV

135 MEV

175 MEV

230 MEV

295 MEV

430 MEV

DIFFERENTIAL INTENSITY (PROTONS/M² SEC STER MEV)

TIME (HOURS AFTER 2208 UT, 28 SEPTEMBER 1961)

occurs later and the decay for simple events is generally exponential (Burlaga 1966). The solar-longitude dependence is also evident in the flare location of sea-level events. Of 14 events, 11 were associated with Western Hemisphere flares and the 3 in the east were located at 14°, 16°, and 90°E. This is a striking asymmetry and again illustrates the influence of the interplanetary field on solar-particle propagation.

Energy spectra.—A complete description of an event involves specifying the energy spectra of a given charge component as a function of time. The energy spectra are quite variable from one event to another, and vary considerably during the course of a single event. The spectra generally steepen with time—that is, there are relatively more low-energy particles than high-energy ones as time progresses during an event.

Originally, the differential spectra were expressed in the form

$$\frac{dJ}{dE} = c(t)\, E^{-n(t)} \qquad\qquad 1.$$

where dJ/dE is the differential particle intensity, E is the kinetic energy, and $n(t)$ is normally of the order of 3 to 6. This representation is generally valid only over a rather limited energy range. For a wider range it is necessary to let n be a function of energy.

One of the more successful representations has been the form

$$\frac{dJ}{dR} = \frac{dJ_0}{dR_0} \exp\, -[R/R_0(t)] \qquad\qquad 2.$$

where R is the rigidity. The quantity R_0 is normally in the range 40–400 mV/c (Freier & Webber 1963) and generally decreases with time from the beginning of the event. This spectrum generally applies during the decay phase at energies above $\sim$20 MeV. Sample exponential rigidity spectra from six events are shown in Figure 6. This representation is valid for both protons and helium nuclei with similar and sometimes identical values of R_0.

The solar cosmic-ray spectra are very steep compared to the galactic cosmic rays. Figure 7 compares the galactic intensity with several solar cosmic-ray spectra.

Composition.—Protons are certainly the predominant particles, at least in terms of the relative number above a given energy per nucleon. There are, however, other components, of which the next most abundant are helium nuclei. The proton-to-helium ratio has been one of the more puzzling problems of the solar cosmic-ray data. Although proton and helium spectra seem similar when expressed in rigidity, the proton-to-helium ratio in the same rigidity interval has varied from 1 to 50, and by as much as 5 within an event (Freier & Webber 1963, Biswas & Fichtel 1963). This indicates that the variation is at least in part a propagation effect. The proton component has typically been observed to have an energy/nucleon spectrum very different from that of helium nuclei; therefore, the relative abundance of the proton component with respect to helium nuclei varies markedly with velocity. For

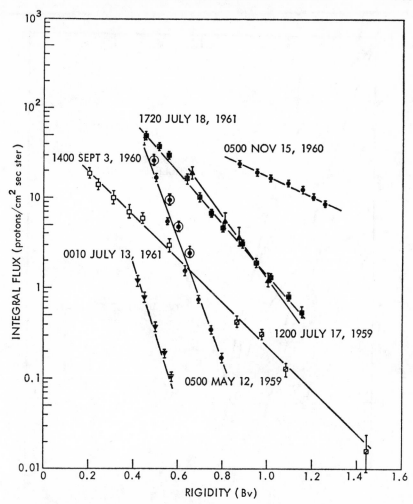

FIG. 6. Exponential spectra for solar cosmic rays observed at selected times in different events (Freier & Webber 1963).

example, in one event, the proton-to-helium ratio varied from about 20 at 40 MeV/nucleon to about 300 at 120 MeV/nucleon. The situation in other events has been similar, as shown in Figure 8 (Biswas & Fichtel 1963). Nuclei heavier than helium have also been observed; and being of considerable interest, they will be discussed separately below.

Present experimental results of the isotopic composition of hydrogen indicate that the deuteron-to-proton ratio for solar particles in the approximate interval from 10 to 10^2 MeV/nucleon is on the order of 10^{-3}, or less, and

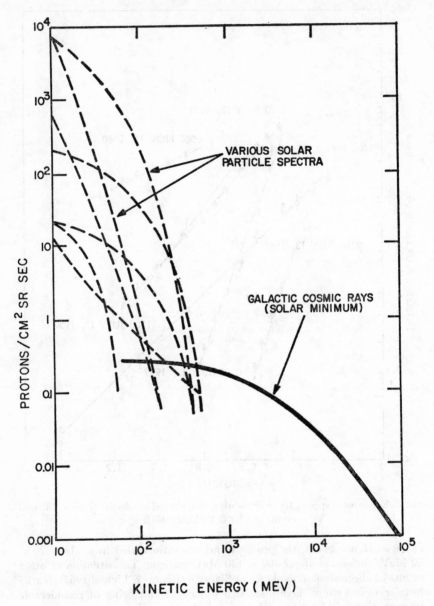

FIG. 7. Energy spectra from several moderate-size events compared with the galactic cosmic-ray spectrum (Fichtel, Guss & Ogilvie 1963).

the triton-to-proton ratio is probably less than 10^{-3}. Variations in these ratios from event to event almost certainly occur, and the ratio might be expected to vary with the velocity interval. The electron component will be discussed in a separate section.

The solar electron component.—The strong correlation between solar cosmic-ray events and types II, III, and IV radio emission suggests that energetic electrons are also accelerated in the flare region, and then lose much of their energy by synchrotron radiation near the source region. In fact, it has been shown that the synchrotron radiation from at least one flare can be explained in terms of electrons whose initial rigidity spectrum is similar to that of the protons and whose number is comparable to that of the protons (Stein & Ney 1963).

Solar electrons near the Earth were first observed by Meyer & Vogt (1962) three days after the large flare of July 20, 1961. They observed a significant increase in the flux of 100–1000 MeV electrons. More recently on Mariner IV, Van Allen & Krimigis (1965) have detected large electron increases with energies greater than 40 keV in interplanetary space following chromospheric flares. Anderson & Lin (1966) on IMP-III have also observed a large number of solar electron events. The data of Van Allen & Krimigis for an electron event on June 5–7, 1965 are shown in Figure 9. The electron identification is based on the failure to observe any increase in a low-energy solid-state detector D_1 which had a proton threshold of 0.5 MeV and negligible electron response, while Geiger counters A and B, which were sensitive to both protons above 0.5 MeV and electrons above 45 keV and 40 keV, displayed large increases. A similar particle identification was obtained by Anderson & Lin for this and other events, with a completely different experimental technique involving electron scattering. They observed 38 events in the 1964–1966 period. Of these, 28 had a clear association with a solar flare, 4 had no observed flare, and 6 were delayed events associated with the arrival of a storm cloud and a consequent magnetic storm observed at Earth. The associated flare is often accompanied by radio-burst events, and sometimes by X-ray emission. The appearance of the electrons is delayed from 21 to 75 minutes with respect to the radio burst or, in cases in which no radio burst has been reported, with respect to the flare maximum. These time delays are reasonable, in view of the fact that the travel time of an unscattered 50 keV electron with small pitch angle from Sun to Earth along the interplanetary field line is 24 minutes. Most of the prompt events are associated with flares that have a heliographic longitude in the interval $60°W \pm 16°$. Anderson & Lin found that events outside this region were generally accompanied by a type I radio-noise region covering a substantial fraction of the solar disk.

The observed electron fluxes are highly anisotropic and have a net streaming of electron flux away from the Sun even late in the event.

Anderson & Lin have also observed what appears to be a marked solar-cycle variation over a period of less than 3 years: in 1964, during 7.5 months coverage, there was one event; in 1965, with 100 per cent satellite and space

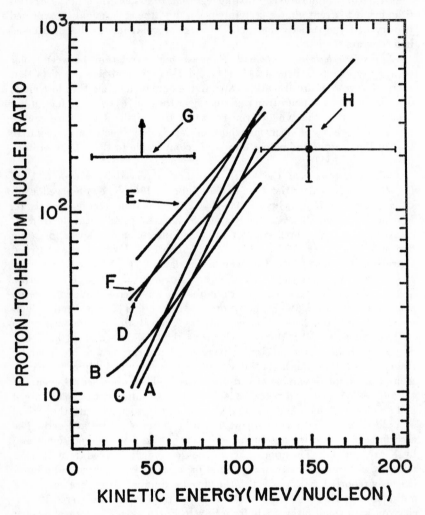

FIG. 8. The proton-to-helium nuclei ratio as a function of kinetic energy per nucleon at several different times. For curves A through F, the curves represent data taken from the work of Biswas et al. (1962), Biswas et al. (1963), and Biswas & Fichtel (1964). Uncertainties in the ratios range from 25 to 50 per cent. The data represented by G are the lower limits set by McDonald et al. (1965). The times at which the measurements were made are as follows: A-1840 UT, November 12, 1960; B-1603 UT, November 13, 1960; C-1961 UT, November 16, 1960; D-0600 UT, November 17, 1960; E-0339 UT, November 18, 1960; F-1408 UT, September 3, 1960; G-March 16, 1964 and February 5, 1965.

coverage, 8 events were observed; and in the first 8 months of 1966, 23 events were observed. Since the electron fluxes are strongly confined to the cones of propagation, this count of events probably represents about one-third the total number of events.

A number of the flare events produce both electrons and protons. Most of the proton events in Table II had accompanying electrons, and for the event of March 16, 1964, Anderson & Lin have pointed out that the data strongly

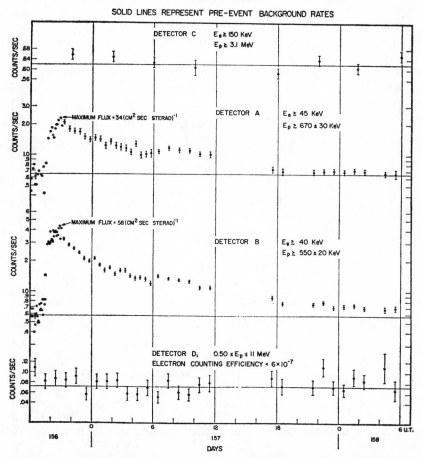

FIG. 9. Comprehensive plot of data from all relevant detectors of the University of Iowa, Mariner IV experiment for the June 5 to 7, 1965 solar electron event (Van Allen & Krimigis 1965).

suggest that the electrons were released at the Sun before the protons. There are, however, a number of electron events where no protons were observed above the detector threshold of 0.5 MeV.

The flare-associated event of July 7, 1966 had an unusually complex and intense type IV radio burst. This event also provided the first direct detection of electrons in the 3–10 MeV region (Cline & McDonald 1966). Figure 10 shows the integral counting rate of particles above 15 MeV. Figure 11 shows the initial phase, indicating the arrival of the electrons some 40 minutes after the flare, with the protons following 40 minutes later. Type IV emission had previously been explained in terms of synchrotron radiation from trapped relativistic electrons. This provided the first opportunity to observe these electrons. The intensity is low, which suggests that most of the electrons remained trapped at the Sun.

Anisotropy.—Early studies of the angular dependence of solar cosmic rays

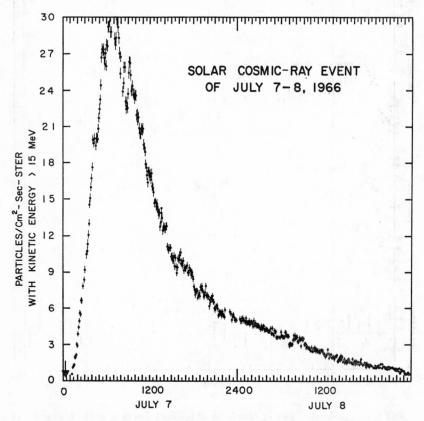

FIG. 10. The event of July 7, 1966. The curve represents the integral intensity of protons > 15 MeV and the flux of electrons > 3 MeV (McDonald 1966).

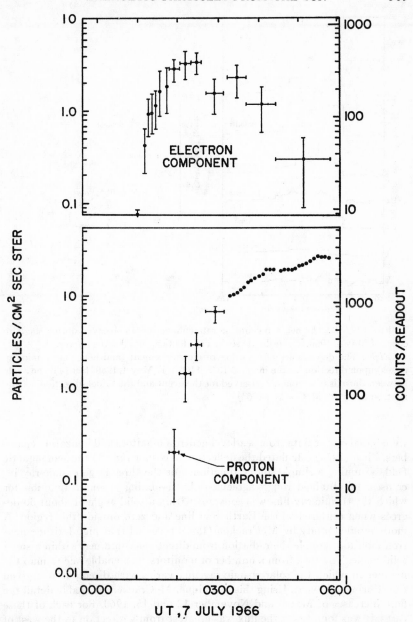

FIG. 11. The time history of 3–10 MeV electrons and protons > 15 MeV during the onset phase of the July 7, 1966 event. The higher-velocity electrons arrive almost 50 minutes before the slower protons (Cline & McDonald 1966).

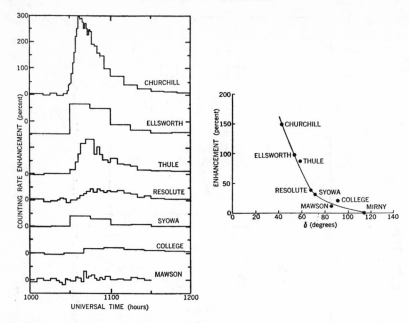

FIG. 12. *Left:* The neutron counting rate enhancements observed during the flare effect of May 4, 1960. The ordinate scales are identical for all seven curves.

Right: The dependence of counting rate enhancement (and hence of cosmic-ray fluxes) upon direction for the interval 1045–1100 UT, May 4, 1960. Here δ is the angle between the axis of symmetry observed for the event and the "Look-direction" of the neutron monitor (McCracken 1962).

were based on the data from sea-level neutron monitors and ionization chambers. Firor (1954) calculated the deflection of cosmic rays by the geomagnetic field, assuming a simple source function. For the three large cosmic-ray increases, he obtained rough agreement by assuming a source function for which the cosmic-ray flux was constant within a solid angle of about 30 degrees when centered on the Earth–Sun line and zero outside this region. A more detailed study by McCracken (1962) revealed that high-latitude neutron monitors sample the radiation from directions contained within a small solid angle. The data from a number of monitors then enable one to map the manner in which the solar cosmic-ray flux varies with direction at a given time during the event. Using this technique, McCracken studied in detail the flare increases of May 4 and November 12 and 15, 1960. For each of these cases, it was found that the flux was greatest from a direction to the west of the Sun. The anisotropy for the May 4, 1960 event was particularly strong. The neutron-monitor data in Figure 12a show this very clearly. All seven stations are at sufficiently high latitudes that the air cutoff is greater than the geomagnetic cutoff. The angular dependence during the onset from these data is seen in Figure 12b. During the entire event, the particle fluxes were symmetrical about a direction 55° west of the Sun and 10° north of the equatorial plane. The radiation was extremely well collimated initially; however, the angular distribution became progressively broader and tended toward isotropy later in the event.

Recent satellite measurements have shown that low-energy particles (0.6–45 MeV) display very strong collimation along the interplanetary magnetic-field lines (Fan, Lamport et al. 1966, Bartley et al. 1966). Figure 13 shows a typical plot of this data from Pioneer VI for an event commencing on December 30, 1965. This strong anisotropy persisted for at least 48 hours. During this period there were major changes in the direction of the anisotropy on an hour-to-hour basis.

Apparently, the correct explanation of the observed anisotropy is related to the spiral-arm concept of the magnetic field of the Sun suggested by the solar-wind theory of Parker (1958). The generally regular garden-hose field with its small-scale irregularities controls the arrival of the particles at the Earth once they leave the region close to the Sun. As the uniform field lines spread apart with increasing distance from the Sun, the particles tend to have smaller and smaller pitch angles with respect to the field, hence, except for scattering, particles would arrive essentially parallel to the field line. Early in an event, the high-energy particles seem to show this type of assymetry, but

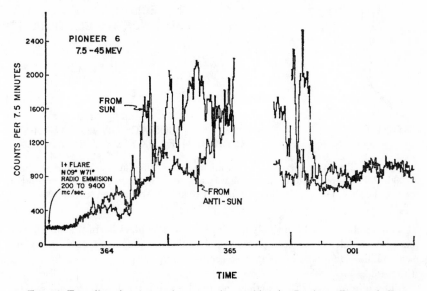

FIG. 13. Two-directional counting rates observed by the Graduate Research Center of the Southwest cosmic-ray experiment on Pioneer VI during the period December 30, 1965–January 1, 1966. The data are 7.5-minute samples (Bartley et al. 1966).

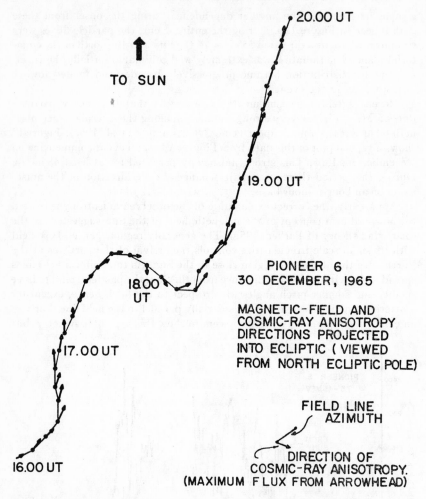

TO SUN

20.00 UT

19.00 UT

18.00 UT

17.00 UT

16.00 UT

**PIONEER 6
30 DECEMBER, 1965**

**MAGNETIC-FIELD AND
COSMIC-RAY ANISOTROPY
DIRECTIONS PROJECTED
INTO ECLIPTIC (VIEWED
FROM NORTH ECLIPTIC POLE)**

**FIELD LINE
AZIMUTH**

**DIRECTION OF
COSMIC-RAY ANISOTROPY.
(MAXIMUM F LUX FROM ARROWHEAD)**

Fig. 14. The magnetic-field and cosmic-ray azimuths during the interval 1600–2000 UT, December 30, 1965. Note the very close correspondence of the two azimuths during the abrupt change in direction around 1800 UT (McCracken & Ness 1966).

it seems to be diluted relatively quickly by the irregularities in the field. Low-energy particles seem to maintain their anisotropy over an extended time scale. This extended time scale also seems to require particle storage close to the Sun.

The alignment of the particle motion by the interplanetary field can be shown by plotting the direction of the solar cosmic-ray anisotropy and the magnetic-field direction (Figure 14) (McCracken & Ness 1966). The direction of arrival of these low-energy protons is completely controlled by the local direction of the interplanetary field.

This strong collimation of particles along field lines is the major factor producing the solar longitudinal distribution of events over solar minimum (Table II). This distribution at solar maximum is more symmetric across the visible disk, which suggests that diffusion across field lines will play a much more prominent role during active periods.

In addition, Figure 13 reveals large intensity variations. This has been explained by Bartley et al. in terms of a filamentary structure in interplanetary space. According to this view, the cosmic rays are being constrained to move through the interplanetary medium along well-defined "streams" by the interplanetary magnetic field, and consequently the cosmic radiation characteristics exhibit a similar filamentary structure. This structure, being frozen into the interplanetary plasma (the solar wind), sweeps radially outward past the Earth at the plasma velocity. The magnetic field itself behaves as if it were co-rotating with the Sun. In this model, the changing direction of the cosmic-ray anisotropy results from the fact that the co-rotating filamentary structure causes the spacecraft to sample the cosmic radiation in various filamentary members at various times, the magnetic-field vectors in the various neighboring filaments being nonparallel. The whole filamentary structure, despite the twisting and intertwining of the filaments, is proposed to retain the basic Archimedes spiral configuration.

Periodic intensity fluctuations.—For a number of events, it has been found that superimposed on the large-scale features of the intensity versus time profile, there is a series of quasi-periodic fluctuations. These have been observed both on a scale of 15–20 minutes and on a longer scale of 1–3 hours. The first report of periodic fluctuations (Steljes et al. 1961) was during the large increases of November 12 and 15, 1960. The neutron monitor showed a rise to maximum for the event on November 15, followed by four rounded peaks with a periodicity of about 20 minutes (Figure 2). Following several slower variations, a smooth decay was obtained. The Explorer XII and XIV cosmic-ray experiment (Bryant et al. 1965a) reported periodic fluctuations in a number of events with periods ranging from 1 to 1.5 hours. Figure 15 shows the integral intensity versus time profile on a linear scale of the 5.7 and 30 MeV proton components for the September 10, 1961 solar cosmic-ray event. These data indicate a series of 90-minute oscillations. To show the periodic nature more clearly, the lower frequencies have been removed by subtracting the running mean of one period length. The results for four different energies are shown in the lower half of Figure 15. Over the interval 6–90 MeV, the fluctuations are periodic, with the same frequency and phase at all energies. Thus, these particular variations must occur in a local region of interplanetary space and do not reflect the properties of the source region.

During the December 31, 1965 particle increase, Simpson and co-workers (Fan, Lamport et al. 1966) observed a superposition of intensity variations of about 15 minutes and a quasi-periodic fluctuation of 3.5–4 hours. These are

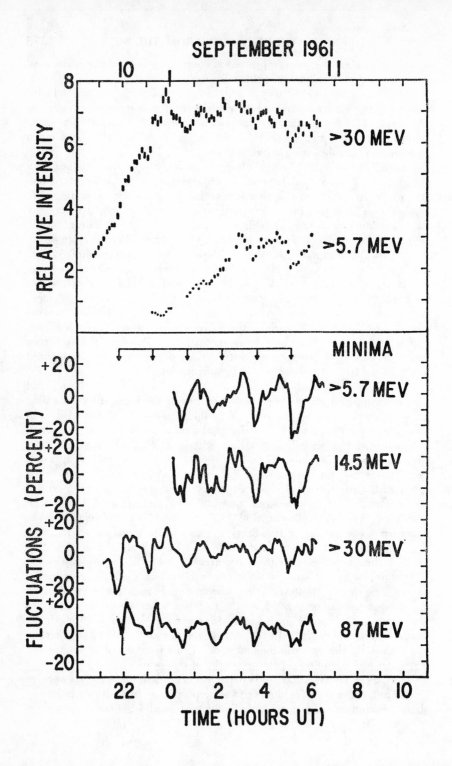

shown in Figure 16. The insert in this figure shows the short-term changes on an expanded scale, clearly revealing large-scale intensity changes with a period of some 15 minutes and extending over at least 7 hours. The characteristic rise and fall times are in the range 2–3 minutes. The arrows in Figure 16 indicate the quasi-periodic fluctuations of 3.5–4 hours.

Fan, Lamport et al. have suggested that periodic increases in field intensity across a magnetic region of 5×10^4 km scale size could momentarily control the flow of particles along the lines of force passing the spacecraft. The scale size is large, compared to the gyroradius of a 600 keV proton. Thus, magnetohydrodynamic waves in the interplanetary medium offer one possible explanation for the observed modulation. These authors assumed an Alfvén wave moving across the magnetic field with a velocity of 40–50 km/sec. From this, they obtained a modulating period of 1000 seconds for waves crossing the magnetic-field structure connecting with the point of observation. This is in agreement with the short-term fluctuation, but does not rule out the possibility that variations during the onset, such as were observed in the November 15, 1960 event, reflect changes in the source region.

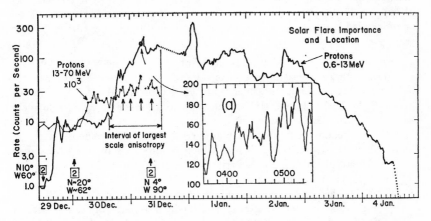

FIG. 16. The intensity-time distribution of protons of 0.6–13 MeV energy and protons 13–70 MeV energy. Anisotropies were observed for a period of approximately two days after the flare of December 30, 1965. The arrows refer to quasi-periodic bursts of period ∼4 hours. Insert (a) is an expansion of the region shown within the circle. Data points are ∼56 seconds apart. Note the quasi-periodic oscillations of ∼15 minutes (Fan, Lamport, Simpson & Smith 1966).

FIG. 15. The event of September 10, 1961. Several components of the event are plotted in the upper half of the figure to illustrate the energy fluctuations. In the bottom half of the figure the lower frequencies have been numerically filtered out to display the fluctuations more clearly. The modulation appears to have a period and a phase independent of energy (Bryant et al. 1965a).

The long-period variations of several hours observed by Fan, Lamport et al. (1966) and Bryant et al. (1965a) have time scales similar to that observed by Bartley et al. (1966) for the passage of a typical filament. Whether this reflects a quasi-periodicity in the filamentary structure is not known at present.

OTHER TYPES OF SOLAR-PARTICLE EVENTS

Recurrence events.—The increased sensitivity and the long-term monitoring capability of satellite-borne particle detectors has made possible the detailed observation of a new component of solar cosmic rays. In 1963 low-energy protons were observed by the Goddard Cosmic Ray Experiment to be contained in streams lasting for several days and corresponding to 60 to 120 degrees of solar rotation on at least seven consecutive solar rotations. In late 1963, and early 1964, the same sector was observed by the University of Chicago Experiment on IMP-I to contain enhanced proton fluxes. In the latter experiment, it was also possible to identify energetic helium nuclei as one of the stream constituents. The remarkable recurrence property can be seen from a compilation of the Chicago and Goddard data in Figure 17. Furthermore these events can be identified with a single long-lived series of M-type or 27-day recurrent, magnetic storms which began in August 1963 and extended through 1964. The salient features of these new particle events are:

(a) A very low intensity level (i.e., 1–5 protons/cm^2-sec-sr above 3 MeV).

(b) An energy spectrum of the form $dJ/dE = $ const $\exp(-E/E_0)$, where E is the proton kinetic energy and E_0 is typically about 2 MeV in the 3–20 MeV region. Unlike flare-associated events, E_0 is relatively constant through an event.

(c) No direct association with solar flares or type IV radio emission.

(d) A state of quasi-equilibrium which is indicated by the absence of velocity dispersion in the arrival of the particles.

(e) A 27-day recurrence.

(f) A close association with geomagnetic disturbance.

(g) A rise and decay characteristic markedly different from that observed for flare-associated events at this stage of the solar cycle.

Previous to these observations, there had been observed recurrent events associated with the central meridian passage of an active region on its next rotation after a flare-associated increase. The September 28, 1961 event which was produced by a flare at 30°E was followed 29 days later by a small but well-defined increase (Figure 18) (Bryant et al. 1963). In a similar manner the November 10, 1961 event was produced by a flare at 90°W and was followed by a proton increase 21 days later at the next central meridian passage of the active region producing the flare. In each case, there was a large magnetic storm and a Forbush decrease of galactic cosmic rays, but no associated flare or radio emission. These events would then seem to be short-lived recurrences or co-rotation events with properties similar to those of events associated with long-lived regions.

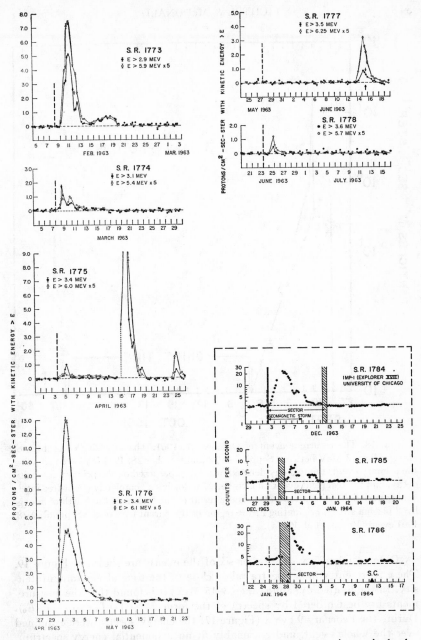

FIG. 17. The 27-day recurrence tendency of low-energy solar cosmic rays is clearly seen from a compilation of the Goddard Space Flight Center Explorer XIV (Bryant et al. 1965b) and University of Chicago IMP-I data (Fan et al. 1966) over nine solar rotations. For the last three recurrences the position of sector boundaries observed by Wilcox & Ness (1965) is indicated.

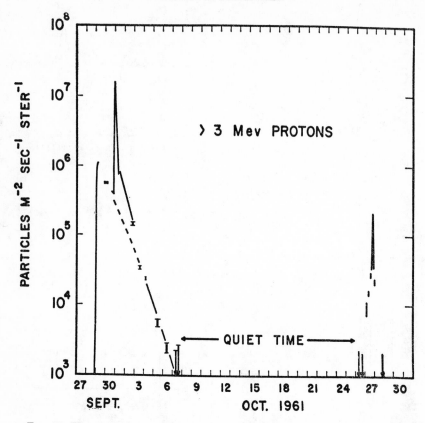

Fig. 18. The recurrent event of October 27, 1961. The intensity of protons of energy above 3 MeV between September 28 and October 28, 1961 is plotted showing the primary event, the two-day delayed increase superposed on the primary intensity decay, and the recurrent event on October 27 following a completely event-free intervening period. This recurrent event is the result of the arrival at the Earth of a long-lived plasma stream originating at the region of the parent flare of the September 28, 1961 event (Bryant et al. 1965a).

Differential energy spectra for six of the events are shown in Figure 19. These represent sample spectra taken close to the time of maximum intensity. In the limited dynamic range 3–15 MeV, it is found that the measurements are best ordered by spectra of the form $J(\geq E) = J_0 \exp(-E/E_0)$. During the February 9 event (Figure 17) the energy spectra were examined over the whole event, and reasonably fit an exponential energy spectrum. However, E_0 varied from 1.8 at the leading edge to a maximum of 3.3 on February 12, 1963, decreasing then to 1.4 in the tail of the event. In contrast to the conventional flare-associated event, the particle energy spectrum does not vary strongly as a function of time. Since the recurrent nature of the

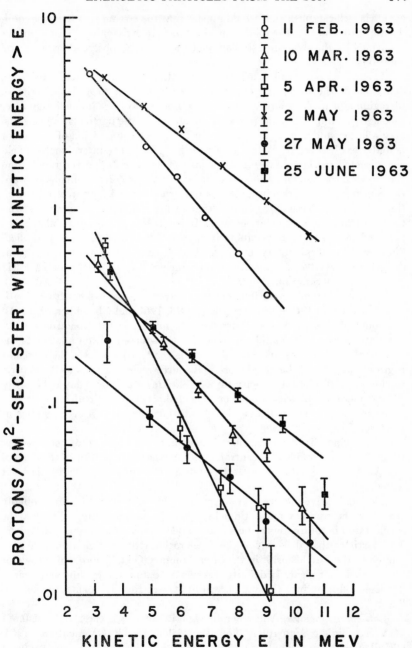

FIG. 19. Energy spectra from the six Explorer XIV recurrence events. These represent values close to the peak intensity of each event (Bryant 1965b).

events is well established, it would appear that the stream is in a quasi-equilibrium state and that the spatial distribution is being sampled. This is further confirmed by the structural similarity of the events as observed over several rotations.

This class of events has a strong correlation with magnetic disturbance, and each of the long-lived events is directly related to a single series of M-region storms. The larger events are observed to start immediately after the sudden-commencement phase of the storm. These particular storms display a large sudden commencement. In Figure 17 the position of the sector boundaries observed by Wilcox & Ness is indicated. The interplanetary field was directed predominantly away from the Sun for 2/7 of the revolution, then toward the Sun for 2/7 of a revolution, away for 2/7, and, finally, toward the Sun for 1/7 of a revolution. This pattern persisted for three solar rotations. Wilcox & Ness (1965) noted that in general, the particles were contained within one particular sector. Krimigis & Van Allen (1966) also noted the tendency of nonflare events to be associated with neutral sheets on field lines in the interplanetary medium. The co-rotation of these events was directly observed by observations on Mariner IV, OGO-I, and IMP-III (O'Gallagher & Simpson 1966).

The solar region responsible for the recurrent events has not been identified. For the two 1961 events (September 28, 1961 and November 10, 1961), previously discussed, which occurred on the next solar rotation after a flare-associated event, it is possible to determine the source position relative to the start of the event. The unexpected result is that for both cases the increases begin about one day before the original source region returns to the central meridian. With plasma velocity of 600–700 km/s, a 2- or 3-day delay after central meridian passage would be expected. It is possible that the region over the source is magnetically closed and that particles are seen leaking at the sides.

For the long-lived source, there is also an active region that persists over the entire period and has its central meridian passage just prior to the start of the events. These data are suggestive but a definitive identification has yet to be made.

The qualitative picture that emerges is as follows. Particles are probably stored in the complex magnetic field above the active center. Whether the acceleration is continuous or occurs in some intermediate way is not known. The particles then drift out of the acceleration-storage region and are conducted to the Earth's orbit in the interplanetary field. The role of the sector boundary is not clear. It definitely serves to define the leading edge, since particles diffusing into this region will drift very rapidly normal to the field lines.

Low-energy events.—A number of experiments with thresholds of 0.5–1 MeV (Krimigis & Van Allen 1966; Van Allen, Frank & Venkatesan 1964; O'Gallagher & Simpson 1966) have shown that the Sun is a very frequent source of low-energy protons. A typical plot from Mariner IV (Figure 20)

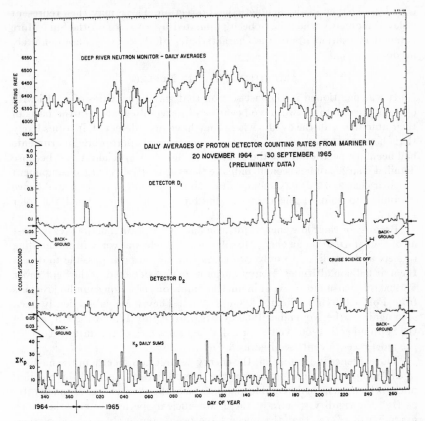

FIG. 20. Ten-month summary of daily averages of the counting rate of Mariner IV interplanetary proton detectors D_1 and D_2 (two center curves); cosmic-ray neutron-monitor rates at Deep River, Canada (top curve, courtesy of H. Carmichael) and daily sums of three-hour geomagnetic disturbances indices KP (Krimigis & Van Allen 1966).

shows the daily average counting rate of two channels of a solid-state detector with proton energy limits of 0.5–11 MeV for D_1 and 0.9–4 MeV for D_2. Krimigis & Van Allen found 20 resolvable proton events having clear statistical significance. In fact, during the period November 29, 1964–April 20, 1965, a detectable flux of solar protons was present for 10 per cent of the period of observation. For the time interval April 21–September 30, 1965, this increased to 32 per cent. For only one of these 20 events is there a convincing association with a specific solar flare. That is the increase of February 5, 1965, which had previously been listed in Table II with the flare-associated events. It seems probable that the number of these "microevents" will increase as the energy of the solar-wind ions is approached. There has been no identifiable solar source for these events. A number of them are bounded on the leading edge by a change in the magnetic-field direction. These may then represent isolated filaments which have been populated by a source on the Sun. More detailed measurements of the characteristics of these events are definitely needed.

MULTIPLY CHARGED NUCLEI

It was mentioned in the last section that small quantities of nuclei with charges greater than two have been seen in several events. On the basis of their abundance in the events where they have been detected, it seems likely that they would have been seen many other times if appropriate instruments had been in a position to see them. These nuclei are of special interest because detailed charge studies seem to indicate that the multiply charged component (nuclear charge $Z \geq 2$) has a composition similar to that of the Sun, and hence dissimilar to that of the cosmic radiation coming from beyond the solar system.

Because of their low energy/nucleon and hence their short ranges, almost all of the information on energetic solar heavy nuclei has come from sounding rockets and satellites. In only two cases has it even been possible to detect them at balloon altitudes. Heavy nuclei were first detected in the September 3, 1960 solar cosmic-ray event in nuclear emulsions flown on sounding rockets from Fort Churchill (Fichtel & Guss 1961). They have subsequently been seen again in the November 1960 events, in the July 18, 1961 event, and in the September 2, 1966 event. For a detailed summary of the data (except for the most recent event) see Biswas & Fichtel (1965).

The evidence indicates that the energy per nucleon spectra of the medium nuclei and of the helium nuclei are the same. Further, since all the medium nuclei which occur in detectable amounts have the same charge-to-mass ratio as He^4, the rigidity spectra of these two nuclear species are also the same, assuming of course that the nuclei have been completely stripped of their electrons. In addition to having the same rigidity spectra, the relative abundances of helium and medium nuclei in the same rigidity intervals have been measured several times in three events and found to be generally the same within uncertainties. The relative abundances among the heavy nuclei for those nuclei which could be measured in the same energy per nucleon intervals have been found to be the same each time a measurement was made, namely, five times in two events, although the uncertainties in some cases are quite large. This result suggests that if these other nuclei had been present in sufficient numbers for measurement of a rigidity spectrum, their rigidity spectra would have been the same as those of the medium and helium nuclei.

The results summarized in the last paragraph indicate that it is possibly meaningful to speak of relative abundances of solar cosmic-ray multiply charged nuclei, and that the best estimates of these abundances would be obtained by taking the average composition in the same velocity intervals from all of the data available. Therefore, the average composition of the multiply charged nuclei in the same velocity intervals as measured in the experiments mentioned is presented in Table V, with a base of one having been chosen for oxygen. Among the heavy nuclei (those with nuclear charges >2), the medium nuclei (6≤nuclear charge≤9) are the most abundant, while Be and B are so rare that only upper limits can be set. A closer examination shows that the relative abundances of the energetic solar particles are

TABLE V

RELATIVE ABUNDANCES OF NUCLEAR SPECIES
BY NUMBER, BASED ON 1.0 FOR OXYGEN

Element	Solar cosmic rays[a]	Sun[b] photosphere	Sun[c] corona	Galactic cosmic rays[d]
$_2$He	107 ± 14	?	445	48
$_3$Li	—	$<10^{-5}$	—	0.3
$_4$Be-$_5$B	<0.02	$<10^{-5}$	—	0.8
$_6$C	0.59 ± 0.07	0.6	1.3	1.8
$_7$N	0.19 ± 0.04	0.1	.1	$\lesssim 0.8$
$_8$O	1.0	1.0	1.0	1.0
$_9$F	<0.03	0.001	—	$\lesssim 0.1$
$_{10}$Ne	0.13 ± 0.02	?	.11	0.30
$_{11}$Na	—	0.002	.01	0.19
$_{12}$Mg	0.043 ± 0.011	0.027	.20	0.32
$_{13}$Al	—	0.002	.01	0.06
$_{14}$Si	0.033 ± 0.011	0.035	.22	0.12
$_{15}$P-$_{21}$Sc	0.057 ± 0.017	0.032[e]	—	0.13
$_{22}$Ti-$_{28}$Ni	$\lesssim 0.02$	0.006	~.1	0.28

[a] Biswas et al (1962), Biswas et al. (1963), Biswas & Fichtel (1963), and Biswas et al. (1966)

[b] The uncertainty of the values in this column is probably of the order of $\frac{1}{2}$ of the value. Aller (1953) or Goldberg, Muller & Aller (1960).

[c] The uncertainty of the values in this column is hard to estimate, but is probably about a factor of 2. Pottasch (1964).

[d] The uncertainty of the values in this column varies from 10 to about 30 per cent.

[e] A 5/2 ratio for the abundance of $_{16}$S relative to $_{18}$A was assumed, the relative abundance of $_{18}$A being unknown.

the same within uncertainties as the solar photospheric abundances determined by spectroscopic means. Since the solar and universal abundances are similar, although not the same, the solar cosmic-ray composition is also similar to the universal abundances. The solar cosmic rays are, however, markedly different in composition from the galactic cosmic rays, which are well known to be rich in the heavy elements.

Table V also shows the best estimate of the composition of the lower corona or upper chromosphere obtained from ultraviolet observations. Although these measurements are less certain than those related to the photosphere, they are probably good to within a factor of two or perhaps slightly poorer for the higher charges. The coronal abundances, if correct, are markedly different from both those of the photosphere and those of solar cosmic rays. This feature suggests that the solar particles are more likely to have their origin in the photosphere (or the lower chromosphere), whose abundances they reflect, rather than in the corona. There is the possibility of unfavorable acceleration of the heavy nuclei, which would compensate for the overabundance of the heavier nuclei. However, this is not a likely explanation for at least three reasons: it is unlikely that the unequal acceleration would compensate in just the proper amount to reproduce the photospheric abundances; this effect would enhance the inverted carbon-to-oxygen ratio, if in fact, this inversion is accepted; and, finally, the difference between the helium-to-medium nuclei ($6 \leq Z \leq 9$) ratio in solar cosmic rays and the coronal abundances would be enhanced even further. The significance of the composition of solar cosmic rays seeming to indicate that the particles come from the photosphere or lower chromosphere will be discussed further in the section on acceleration.

It has been indicated that the energetic solar nuclei coming from the Sun with charges ranging from that of helium through at least 20 seem to reflect the composition of the solar surface. If the composition of these nuclei is accepted as representative of the Sun, the relative abundances given in Table V may be used to estimate the helium and neon abundances in the Sun, whereas it is not possible to obtain a good estimate of the abundance of these two elements spectroscopically in the photosphere. The average helium-to-oxygen ratio is 107 ± 14, and the average neon-to-oxygen ratio is 0.13 ± 0.02. The neon-to-oxygen ratio is similar to the universal abundances estimated by Suess & Urey (1956) and Cameron (1959), although a bit low. The helium-to-medium ratio is also typical, but the more interesting ratio is that of protons to helium. Because of the different energy spectra for particles with different charge-to-mass ratios, there is no simple reliable way to determine this ratio from solar cosmic rays alone. However, if the helium-to-medium ratio of 60 ± 7 is accepted as representative of the Sun, and the proton-to-medium value from spectroscopic data, namely 650 (Aller 1953; Goldberg, Muller & Aller 1960), is used, a proton-to-helium ratio of 11_{-5}^{+7} is obtained. The uncertainty in this number depends on the correctness of the assumption above and the uncertainty in the proton-to-medium ratio, hence the estimated error placed on this ratio is large. However, it is worth noting that this number agrees with structure calculations and is in the range of the hydrogen-to-helium ratio determined for stars and gaseous nebulae within our Galaxy ($9 \pm 3:1$). These ratios are all well below the $100:1$ value expected from current stellar activity, which suggests that the universe was quite different in the past (Hoyle 1965).

SOLAR-PARTICLE ACCELERATION

Although the acceleration of particles in a major solar flare is one of the most outstanding features of the solar event, the question of the means of acceleration to the observed solar cosmic-ray energies is far from fully solved. Although there are some general ideas concerning the way in which energy is transferred to particles, the specific details of the mechanism are yet to be determined. Therefore, before discussing specific accelerating processes, we shall examine a few general aspects of the problem. This discussion will be pointed specifically toward the major solar-flare event.

Time Scale

Within 10 to 20 minutes of the observation of a major solar flare and type IV radio noise, the highest-energy solar cosmic rays are already beginning to arrive at the Earth. Further, the time history and energy dependence of the energetic solar particles as they arrive at the Earth can be explained quite well by anisotropic diffusion theory with the release of the fully accelerated particles over a period of less than 10 minutes. Also, solar-particle events are well correlated with major solar flares which last from 20 to 90 minutes and have a rise time of from 5 to 20 minutes. Hence, one possible conclusion is that the time of the major fraction of the energy gain is less than 20 minutes and that the acceleration results from a relatively sudden dynamic process. There is the alternate possibility that the particles gain some, or all, of their energy over a longer period, perhaps days, are held trapped, and then are suddenly released and possibly further accelerated by another process (Schatzman 1966).

Source of Energy

The source of the solar-particle energy is normally thought to be in the region just above the visibly active area on the Sun. There are several reasons for this belief:

(a) the failure of most flares to be visible in white light, indicating that the radiation is not an enhancement of the normal solar thermal spectrum;
(b) the failure to observe violent disturbances in the photosphere beneath the flare;
(c) the failure to observe any evidence of an external source for the energy.

It is interesting that, although the location of type IV radiation, now thought to be due to the synchrotron radiation of electrons and closely associated with solar-particle events, is in the corona, the composition of solar particles seems to differ from that in the corona. It may be that the particles originate in the upper photosphere or lower chromosphere and move upward along field lines. Assuming that the particle energy comes from above the surface of the Sun, the only apparent choice for an energy source is magnetic energy. The magnetic fields are, of course, linked to the solar interior, so the ultimate source of energy is the solar interior, but perhaps not the immediate one.

Additional Properties of the Particle Flare

A particle flare usually occurs between spots of opposite polarity, which suggests some relationship between the magnetic-field lines joining the spots and particle acceleration. The brightening tends to occur along dark filaments between spots or spot groups (Howard 1963), and seems to be along the magnetic neutral line (Warwick et al. 1966). There is also brightening of plages at some distance from the flare on some occasions. The big solar-particle flares, in general, have areas of from 0.5 to 10 times 10^{19} cm². The emission region seems to be about as high as it is wide (Warwick 1955), i.e. about a few times 10^9 cm. When the flare appears on the solar limb, it often appears as the sudden brightening of active, loop prominences over a sunspot group. These features again suggest that the particle energy was stored above the solar surface, probably in the magnetic-field lines connecting the sunspots.

There also seems to be some indication that flares associated with energetic solar particles tend to have a double-filament structure. The available observations on this subject are summarized in some detail by Warwick et al. (1966). If this feature should prove truly characteristic, it is of special interest since it would provide the necessary picture for converging trains of shock waves which would accelerate particles by the Fermi mechanism (Wentzel 1965), as discussed later.

VISIBLE-FLARE ENERGY

Although many emission lines are strongly enhanced, the continuum increase is not easy to observe accurately; therefore, estimates of the radiated energy that have been calculated by integrating over the frequency spectrum of the electromagnetic radiation are not very accurate. Parker (1957) and Ellison (1963) have estimated that the energy released during the life of a very large flare is about 10^{32} ergs. If this energy were in the form of magnetic energy, i.e. $B^2/8\pi$, ΔB^2 would be of the order of 10^5 gauss².

BLAST-WAVE ENERGY

On the basis of particle density, energy, and the blast-wave volume, Parker (1963a, b) has estimated that the blast-wave energy is of the order of 10^{32} ergs. Hence it is comparable to the visible-flare energy.

SOLAR COSMIC-RAY PARTICLE ENERGY

An estimate of the particle energy in a large solar-particle event can be made in two ways. The first is to take a typical flux and assume that it extends over a reasonable volume, such as 1 cubic a.u. The other is to fit the observations at the Earth to a reasonable diffusion model, and then calculate the number of particles and their energies. Both approaches lead to estimates of the order of 10^{30} ergs for the biggest events, e.g. the one on February 23, 1956 or the one on November 12, 1960. It was mentioned earlier that the type IV radiation might be due to synchrotron radiation of electrons. The question arises as to whether or not this is a comparable energy source. It is not; the total type IV energy radiated is much smaller.

It is possible that much more energy is initially given to the kinetic energy of particles, many of which may then never escape from the environment near the Sun. If this were true, interactions of the downward particles in the solar atmosphere might create detectable secondaries such as neutrons, deuterons, tritons, and electromagnetic radiation. However, the magnetic-field configuration is probably such that the field strength decreases rapidly with altitude near the flare; this inhibits the downward motion of particles.

Of all the energy available, it is unlikely that the solar cosmic-ray particles have much more than 1 per cent. Thus, there seems to be no problem of the availability of the energy, and the next question is how the existing energy is converted to particle energy.

ACCELERATING MECHANISMS

There seem to be only a few fundamental accelerating mechanisms occurring outside the laboratory, and they are probably, at least in some senses, similar. These include the Fermi mechanism, the betatron effect, and the Sweet mechanism. Since it is not known which is the most effective, it is worthwhile to outline each mechanism and suggest what features of the solar-particle radiation it might explain and what characteristics seem difficult to explain with it.

Fermi mechanism.—The first accelerating mechanism was originally suggested by Fermi (1949, 1954) in an attempt to explain the acceleration of galactic cosmic rays. Basically the acceleration occurs when a charged particle is reflected by a magnetic region initially moving toward the particle. Since we know that there are shock fronts, moving plasmas, and moving magnetic fields in space, this mechanism is naturally one which must be considered. To second order, the total energy after a collision W_f is related to the initial energy by the relationship,

$$W_f = W_i(1 - 2\beta_p\beta_s \cos \nu + \beta_s^2 + \cdots) \qquad 3.$$

where β_p is the velocity of the particle in terms of the velocity of light, β_s is the velocity of the magnetic region, and ν is the angle between them. The second term is important only in cases where the motion of the magnetic clouds is random with respect to the particle velocity so that the total effect of the first term is small. Since a statistical process of this type is inherently slow and since the solar-flare phenomenon is most probably fast, this term is probably not significant in solar-particle acceleration.

In the case of a head-on collision, i.e. $\cos \nu = -1$, the average increase in energy, ΔW, is given by

$$\Delta W = 2\beta_p\beta_s\gamma W = 2\beta_s\gamma c p \qquad 4.$$

where γ is the fraction of collisions that will lead to a reflection, and p is the particle momentum. If γ is a function of particle rigidity (as it probably is since the reflection would depend primarily on the particle's radius of curvature), the rate of acceleration will be the same for particles of the same charge-to-mass ratio, but different for particles with different charge-to-mass ratios.

Biswas et al. (1963) originally suggested that this feature probably explains the similar energy/nucleon spectra of solar helium and medium nuclei, but the different spectra of protons and helium nuclei.

When various different assumptions are made about the way in which particles are accelerated, it is possible to obtain different types of spectra. Wentzel (1965) has explored this problem in some detail. It is instructive to look at a few specific cases. If the probability of escape, $\Delta t/T$, is a constant, and γ is 1, then

$$J \sim p^{-d/2T\beta_s c} \qquad\qquad 5.$$

where d is the distance between shocks and J is the particle flux. In order to introduce a rigidity dependence into γ and T, let

$$\gamma T = T_0/\{1 + (R/R_0)^a\} \qquad\qquad 6.$$

This function has the advantage of being simple, yet it has the desired property of being essentially constant at low rigidities where all particles should be reflected, and decreasing at higher rigidities, determined by R_0, with a rate determined by a. With this assumption

$$J = J_0(p^{-d/2T_0\beta_s C}) \exp\left\{-(R/R_0)^a d/2T_0\beta_s ca\right\} \qquad\qquad 7.$$

Notice that if $a = 1$ and $d/2T_0\beta_s c \ll 1$, i.e. the characteristic time of escape, T_0, is large compared to $d/\beta_s c$ (a plausible assumption), then

$$J \approx J_0 \exp\left(-R/R_0'\right) \qquad\qquad 8.$$

where R_0' is a constant determined from the previous equation. This form seems to be characteristic of some events.

Thus, the Fermi mechanism can explain the spectral shape and the fact that the energy/nucleon spectra of protons and heavier nuclei are different, but is it reasonable from other points of view? In favor of the Fermi mechanism, there are several additional points:

(a) Electrons would also be accelerated, but to lesser total energies, and would lose most of their energy by synchrotron radiation in the magnetic fields near the flare. It has been shown by Stein & Ney (1963) that an analysis of the type IV radiation indicates that the electrons might have an initial spectrum similar to the protons.

(b) In the middle of the flare region about 12,000 km above the photosphere, the particle density is of the order of $10^9/cm^3$, and so the energy losses from that cause are completely negligible.

(c) Supersonic motions are possible and could in principle lead to sufficient shock-wave phenomena to heat at least some small portion of the medium to a temperature such that there would be enough particles in the thermal tail, which could then gain energy by the Fermi mechanism.

(d) There could be approaching magnetic fronts between which the first-order Fermi mechanism could operate. On the negative side, it is not yet at all clear that the proper conditions exist to allow particles to gain the observed energies in the existing time.

Betatron effect.—Another accelerating mechanism is the betatron effect. If the magnetic field increases in some region, it is possible for a particle to gain energy, and conversely, if the field decreases, the particles lose energy according to the relationship $p^2/H =$ constant. There can, in principle, be sustained acceleration by an increase in magnetic-field strength, followed by an averaging of the momentum vector among its three components by scattering, a decrease in the magnetic field with an accompanying decrease in energy—which is, however, smaller than the increase—and then more scattering which gives a random distribution to the momentum vector (Alfvén 1959). Although this process may occur in principle, it probably does not in the solar-flare region where the energy transfer is rapid, and there is probably not a series of fluctuations of this type. If, however, the acceleration takes place over several days, this mechanism represents an interesting possibility.

Sweet mechanism.—The third possible acceleration mechanism, the Sweet mechanism (Sweet 1958, 1964), is one whereby the energy for the flare and the particles comes from the annihilation of oppositely directed magnetic fields coming together. Detailed theoretical studies, however, have yet to find a way in which this process can occur in the environment of a solar flare within a length of time as short as that of a solar flare. Further, there has been a failure to observe any rapid variations in magnetic field, although there is some indication of significant decreases with the time scale of a day or two. This area of research remains in need of further clarifying observations. For the lack of any strong theoretical or experimental support, the Sweet mechanism will not be pursued further here.

Thus, although solar cosmic rays appear to have only a small portion of the total energy which is involved in a major solar-flare event, it is not yet clear how the transfer to solar cosmic rays occurs. The problem of solar-particle acceleration remains as perhaps the most important unsolved problem in the study of solar cosmic rays.

PROPAGATION CHARACTERISTICS OF SOLAR COSMIC RAYS

Following their acceleration, energetic solar particles must escape from the source region and penetrate the interplanetary medium in order to reach the orbit of the Earth. Even before there were magnetic-field measurements, solar-wind theory (Parker 1960) predicted that the magnetic field in this latter region would be in the form of generally spiral lines emanating from the Sun, with superimposed small-scale irregularities. The spiral nature of the field results from the field lines being drawn out radially by the solar wind while the base of the field line at the Sun rotates with the solar surface. The theory predicted a quiet-day field of the order of 3×10^{-5} gauss at the Earth's orbit. The direct observations of the interplanetary field (Ness et al. 1964) have confirmed the general Archimedes spiral nature of the field, the presence of magnetic irregularities along the field line, and a magnetic-field strength of from 3 to 6×10^{-5} gauss.

This magnetic-field pattern has two principal effects on the particle propagation. The guiding centers of the particles tend to follow a generally spiral trajectory from the Sun at least to the orbit of the Earth, and the particle motion is complicated by the magnetic irregularities which act as scattering centers changing its pitch angle along the field line, or causing the guiding center to move to an adjacent field line. For the case of a highly disordered field, scattering will dominate and the interplanetary field will act as a storage reservoir. As the field becomes more ordered, channeling effects will dominate. To examine these two cases, consider a particle of velocity w, moving in a magnetic field B with scattering length L. Let Z represent the number of scatterings per unit time. Then

$$Z = w_\parallel/L \qquad\qquad 9.$$

where $w_\parallel$ is the component of w parallel to B.

The diffusion coefficient along the field is

$$k_\parallel \simeq \nu L^2 \qquad\qquad 10.$$

The radius of gyration of the particle around the field line is:

$$S = w_\perp/\Omega \qquad\qquad 11.$$

where Ω is the cyclotron frequency of the particle in the field. The diffusion coefficient across the field is

$$k_\perp \simeq \nu S^2 \qquad\qquad 12.$$

If $w_\parallel/L = \nu \ll \Omega$ the particle is closely tied to the line of force. However, if $L \approx S$ the particle is scattered many times in one cyclotron period and $k_\perp \approx k_\parallel$ (Parker 1965).

The channeling effect of the interplanetary field is evident from:

(a) the anisotropy early in several events, observed by McCracken (1962), wherein solar particles were arriving from about 50°W of the Earth-Sun line;

(b) the long-term anisotropic behavior of solar cosmic rays observed by Bartley et al. (1966) and Fan, Lamport et al. (1966);

(c) the solar-longitude distribution of electron events observed by Anderson & Lin (1966);

(d) the solar-longitude distribution of solar proton events over solar minimum (Table II).

However, the neutron-monitor data for several events and the low-energy satellite data from the September 28, 1961 event show that there is a strong tendency in flare-associated events for the particle distributions to become isotropic as the event progresses. This development requires an efficient storage of particles in the interplanetary medium.

The first suggestion of a diffusion model for the propagation of energetic particles was by Parker (1956) and Meyer, Parker & Simpson (1956). In this model the interplanetary space is assumed to be an isotropic uniform three-dimensional space. A scattering mechanism with a mean free path λ is assumed. The density of solar particles $n(r,t)$ then obeys the equation

$$\frac{dn}{dt} = k\nabla^2 n \qquad 13.$$

where k is the diffusion coefficient which, given from kinetic theory, is

$$k = \tfrac{1}{3}\lambda w \qquad 14.$$

For the case of an isotropic medium, if the injection time is short compared to the transit time to the point of observation and if the dimensions of the injection region are small compared to the distance from the source to the observing point, then for times appreciably longer than the rectilinear travel time from the source to the point of observation, the following equation is a good approximation for the solution of Equation 13:

$$n = \left\{ \frac{\pi N}{2} \Big/ (\pi kt)^{3/2} \right\} \exp \left\{ -r^2/4kt \right\} \qquad 15.$$

where N is the number of particles per unit solid angle emitted at the source. Hence, in this case, $\ln(nt^{3/2})$ plotted as a function of t^{-1} will be a straight line with a slope—$r_2/4k$. The time from the beginning of the event to that of maximum intensity, t_m, is given by $r^2/6k$. In general it is possible to fit observed intensity data to and somewhat beyond the time of maximum using this model. The values of k calculated in this way are plausible. However, it is generally observed that a simple power law is not followed throughout the decay phase. Good agreement has been obtained by Bryant et al. (1962) and Hofman & Winckler (1963) by assuming a sharp outer boundary r_0. The solution of the equation for the initial phase is not affected appreciably, and, for times late in the event, the solution becomes

$$n(t) = \left\{ \frac{2\pi N}{r_0^2 r} \right\} \sin \left\{ \frac{\pi r}{r_0} \right\} \exp \left\{ -\frac{\pi^2 kt}{r_0^2} \right\} \qquad 16.$$

where r_0 is the distance from the source to the outer boundary.

The effect of this perfectly transmitting boundary then is to impose an exponential decay during the latter part of the event. Generally, values of r_0 have been on the order of 2 to 3 a.u. These do not appear to be realistic, since theoretical considerations suggest that the boundary should occur at 10–100 a.u. (Parker 1963a).

A similar behavior can be obtained by assuming that the diffusion coefficient has a spatial dependence of the form

$$k = Mr^\beta \qquad 17.$$

where β and M are parameters independent of r but may be dependent on particle energy or rigidity.

The diffusion equation for this case is (Parker 1963b, Krimigis 1965)

$$\frac{\partial n}{\partial t} = \frac{M}{r^\alpha} \frac{\partial}{\partial r} \left[r^{\alpha+\beta} \frac{\partial}{\partial r} (r^\beta n) \right] \qquad 18.$$

where α specifies the dimensionality of the space to be used.

The quantity n is related to the directional intensity I by

$$n = 4\pi I/w \qquad\qquad 19.$$

where it is assumed that I is isotropic.

The solution of Equation 18 is indicated in Figure 21 along with semilogarithmic graphs of the time variations of the intensity of monoenergetic particles for various heliocentric radial distances. For this solution the graph of $\ln It^{(\alpha+1)/(2-\beta)}$ vs. t^{-1} should be a family of straight lines corresponding to various values of the ratio

$$\frac{\alpha+1}{2-\beta} \qquad\qquad 20.$$

The slope m of the lines in Figure 21 is given by the relation

$$m = -\frac{r^{(2-\beta)}}{(2-\beta)^2 M} \qquad\qquad 21.$$

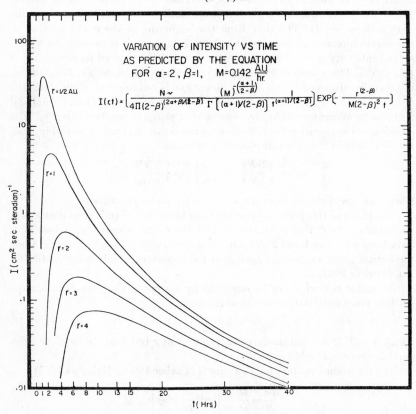

FIG. 21. Similogarithmic plots of the time variation of intensity of monoenergetic particles for various heliocentric radial distances (Krimigis 1965).

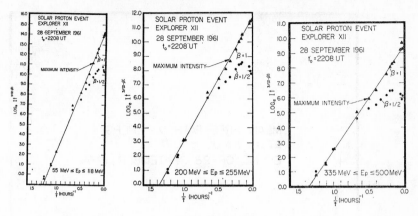

FIG. 22. Analysis of the Explorer XII data (Figure 5) for three different energy intervals. I is in $(cm^2\text{-}sec\text{-}ster)^{-1}$ (Krimigis 1965).

and the time of maximum intensity is given by

$$t_{max} = \frac{m(\beta - 2)}{\alpha + 1}$$ 22.

for radial diffusion in three-dimensional space $\alpha = 2$ and

$$t_{max} = \frac{m(c\beta - 2)}{3}$$

The application of this particular model to the solar-particle data has
been considered in detail by Krimigis. His approach was to assume $\alpha = 2$ and
vary β until a straight-line fit was obtained for the quantity $\ln (It^{(\alpha+1)/(2-\beta)})$
vs. $1/t$. Figure 22 shows his analysis for three different energy intervals during the September 28, 1961 event. An excellent fit is obtained at all three
energies for $\beta = 1$. At lower energies β decreases. If a mean free path is defined by the relation $\lambda = 3 k/w$, it is observed that λ is constant from 500–100
MeV with a value of .08 a.u. but increases to .15 a.u. at 20 MeV (Figure 23)
(Krimigis 1965).

One of the main objectives of any model of particle propagation in the
interplanetary field is to provide estimates of the intensity and energy spectra
of the solar particles at injection. With the model developed by Krimigis,
once β and M have been determined from the intensity vs. time profile
(Figure 21), it is possible to calculate the total number of particles of a given
energy emitted at $t = 0$ (Figure 24). A compilation of data at various energies
makes possible the determination of the energy spectra at $t = 0$. For the September 28, 1961 event Krimigis considers four spectral forms: a power law in
energy, an exponential in energy, a power law in magnetic rigidity, and an exponential in magnetic rigidity. In the region 100–500 MeV a reasonably

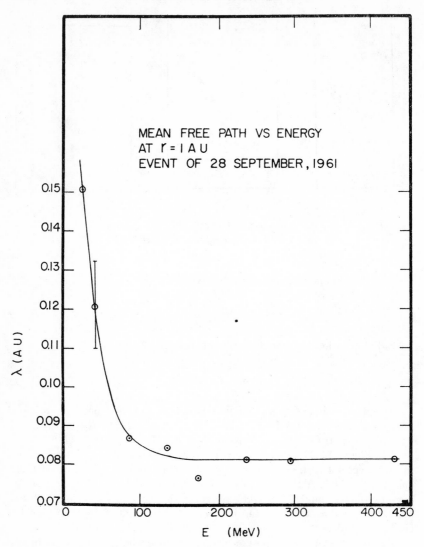

FIG. 23. The mean free path as a function of energy for the event of September 28, 1961 at $\gamma = 1$ a.u. (Krimigis 1965).

straight line first was obtained for all cases except the exponential energy. These had the form:

power law in kinetic energy $dN/dE = 2.23 \times 10^{32} E^{-(2.41 \pm 0.3)}$
power law in magnetic rigidity $dN/dR = 8.87 \times 10^{36} P^{-(3. \pm 0.3)}$ 23.
exponential in rigidity $dN/dR = 1.53 \times 10^{28} \exp(-P/173 \pm 30)$

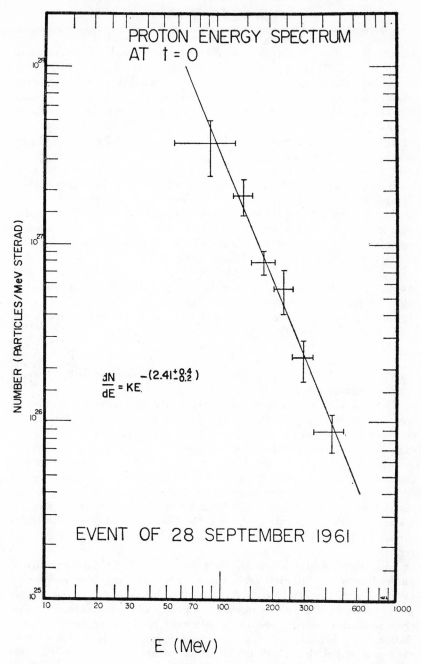

FIG. 24. The source spectra for the September 28, 1961 event derived by two independent methods (Krimigis 1965).

From these data Krimigis determines that the total flux of particles >23 MeV emitted at the Sun was $N \simeq 10^{31}$ with a total kinetic energy of $\sim 5 \times 10^{27}$ ergs. Obviously, this is a moderate-size event. However, the precision with which the energy spectrum was determined makes it suitable for detailed calculations. A more empirical method of analysis for the September 28 event was employed by Bryant et al. (1965a) who observed a very striking linear dependence on velocity. They determined the distance a particle had traveled between acceleration and observation by taking the product of particle velocity and time from the beginning of the event as defined by the solar X-ray burst. The intensity vs. time profile is then converted to an intensity vs. distance profile. The intensity vs. distance profile is effectively a distribution in distance traveled. This distribution is a property of the medium through which the particles have traveled. It is found that the distributions at all energies have the same form and with an arbitrary normalization factor along the intensity scale, all the intensity vs. distance profiles can be fitted to a single curve. Then the common curve is radiated for distances traveled over a range of 2–100 a.u. These authors point out that there is nothing to suggest that an extrapolation back to zero distance is valid. The relative intensity of two components at zero distances is by definition a measure of the energy spectrum. This method gives best agreement with a power law in kinetic energy of the form

$$dN/dE = \text{const} \cdot E^{-(2.2 \pm .2)} \qquad\qquad 24.$$

over the range 3.5–450 MeV. This is in excellent agreement with the more-detailed calculations of Krimigis.

The isotropic-diffusion model provides a good description for many events. However, it cannot explain the strong dependence of event characteristics on the solar longitude of the parent flare. The complex problem of anisotropic diffusion in the interplanetary field has been considered by several authors. Reid (1964) and Axford (1965) have postulated a model with homogeneous, isotropic diffusion in a thin layer between an inner reflecting boundary and an outer partially absorbing boundary in the solar chromosphere. Reid assumes that the particles leak out through the boundary outer layer into a magnetic bottle which extends beyond the Earth. Axford introduces diffusion along the interplanetary-field lines following their escape from the source region.

More-detailed consideration of the anisotropic diffusion has been carried out by Burlaga (1966), Fibich & Abraham (1965), Parker (1965), and Roelof (1966). Roelof has emphasized the importance of determining the spatial-power spectral density of the random component of the magnetic field. Discrete scattering centers as such do not exist in the interplanetary field. Rather, there is a wide spectrum of magnetic irregularities which seem to randomize the pitch-angle distribution. Roelof has shown that this one-dimensional diffusion along diverging field lines (where the magnitude falls off as r^{-2}) will produce roughly the same behavior as three-dimensional isotropic diffusion.

As additional measurements are made, it is becoming increasingly obvious that the process is even more complex. For lower-energy particles there is probably significant storage close to the Sun. As the new techniques are applied to the coming period of solar maximum, it is probable that even more complex solar-particle phenomena will be observed.

LITERATURE CITED

Alfvén, H. 1959, *Tellus*, **11**, 106

Aller, L. H. 1953, *Astrophysics, The Atmospheres of the Sun and Stars* (Ronald Press, New York)

Anderson, K. A. 1958, *Phys. Rev. Letters*, **1**, 336

Anderson, K. A., Lin, P. R. 1966, *Phys. Rev. Letters*, **16**, 1121

Axford, W. I. 1965, *Planetary Space Sci.*, **13**, 1301

Bailey, D. K. 1957, *J. Geophys.* **62**, 431

Bartley, W. C., Bukata, R. P., McCracken, K. G., Rao, U. R. 1966, *J. Geophys. Res.*, **71**, 3297

Biswas, S., Fichtel, C. E. 1963, *Ap. J.*, **139**, 941

Biswas, S., Fichtel, C. E. 1965, *Space Sci. Rev.*, **IV**, 709–36

Biswas, S., Fichtel, C. E., Guss, D. E. 1962, *Phys. Rev. Letters*, **128**, 2756

Biswas, S., Fichtel, C. E., Guss, D. E. 1966, *J. Geophys. Res.*, **71**, 4071

Biswas, S., Fichtel, C. E., Guss, D. E., Waddington, C. J. 1963, *J. Geophys. Res.*, **68**, 3109

Bryant, D. A., Cline, T. L., Desai, U. D., McDonald, F. B. 1962, *J. Geophys. Res.*, **67**, 4983

Bryant, D. A., Cline, T. L., Desai, U. D., McDonald, F. B. 1963, *Phys. Rev. Letters*, **11**, 144

Bryant, D. A., Cline, T. L., Desai, U. D., McDonald, F. B. 1965a, *Astrophys. J.*, **141**, 478

Bryant, D. A., Cline, T. L., Desai, U. D., McDonald, F. B. 1965b, *Phys. Rev. Letters*, **14**, 481

Burlaga, L. F. 1966, *Univ. Minnesota School Phys. Tech. Rept. CR88*

Cameron, A. G. W. 1959, *Ap. J.*, **129**, 676

Cline, T. L., McDonald, F. B. 1966 (Private communication)

Davis, L. R., Fichtel, C. E., Guss, D. E., Ogilvie, K. W. 1961, *Phys. Rev. Letters*, **6**, 492

Dorman, L. I. 1962, *Progr. Elem. Particles Cosmic Ray Phys.*, **VII** 1

Ellison, M. A. 1963, *Quart J. Roy. Astron. Soc.*, **4**, 62

Fan, C. Y., Gloeckler, G., Simpson, J. A. 1966, *Proc. Intern. Conf. Cosmic Rays, 9th, Inst. Phys. Soc. London*, **1**, 105

Fan, C. Y., Lamport, J. E., Simpson, J. A., Smith, D. R. 1966, *J. Geophys. Res.*, **71**, 3289

Fermi, E. 1949, *Phys. Rev.*, **75**, 1169

Fermi, E. 1954, *Ap. J.*, **119**, 1

Fibich, M., Abraham, P. B. 1965, *J. Geophys. Res.*, **70**, 2475

Fichtel, C. E., Guss, D. E. 1961, *Phys. Rev. Letters*, **6**, 495

Fichtel, C. E., Guss, D. E., Ogilvie, K. W. 1963, *Solar Proton Manual, NASA Tech. Rept. TR-R169*

Firor, J. 1954, *Phys. Rev.*, **94**, 1017

Freier, P. S., Ney, E. P., Winckler, J. R. 1959, *Proc. Moscow Cosmic Ray Conf.*

Freier, P. S., Webber, W. R. 1963, *J. Geophys. Res.*, **68**, 1605

Goldberg, L., Muller, E. A., Aller, L. H. 1960, *Ap. J. Suppl. 5*, 1

Gregory, J. B. 1962, *J. Geophys. Res.*, **67**, 3829

Hofmann, D. J., Winckler, J. R. 1963, *J. Geophys. Res.*, **68**, 2067

Howard, R. 1963, *Ap. J.*, **138**, 1312

Hoyle, F. 1965, *Nature*, **208**, 111

Hultqvist, B. 1959, *Tellus*, **11**, 332

Krimigis, S. M. 1965, *J. Geophys. Res.*, **70**, 2943

Krimigis, S. M., Van Allen, J. A. 1966, *Phys. Rev. Letters*, **16**, 419

Lange, I., Forbush, S. E. 1942, *Terrest. Magnetism Atmospheric Elec.*, **47**, 185

Leinbach, H., Reid, G. C. 1959, *Phys. Rev. Letters*, **2**, 60

McCracken, K. G. 1962, *J. Geophys. Res.*, **67**, 423

McCracken, K. G., Ness, N. F. 1966, *J. Geophys. Res.*, **71**, 3325

McDonald, F. B. 1966 (Private communication)

Malitson, H. H., Webber, W. R. 1963, *Solar Proton Manual, NASA Tech. Rept. TR R169*, 1

Meyer, P., Parker, E. N., Simpson, J. A. 1956, *Phys. Rev.*, **104**, 768

Meyer, P., Vogt, R. 1962, *Phys. Rev. Letters*, **8**, 387

Ness, N. F., Scearce, C. S., Seek, J. B. 1964, *J. Geophys. Res.*, **69**, 3531

O'Gallagher, J. J., Simpson, J. A. 1966, *Phys. Rev. Letters*, **26**, 1212

Parker, E. N. 1956, *Phys. Rev.*, **103**, 1518

Parker, E. N. 1957, *Phys. Rev.*, **107**, 830

Parker, E. N. 1958, *Ap. J.*, **128**, 664

Parker, E. N. 1960, *Ap. J.*, **132**, 821

Parker, E. N. 1963a, *Ap. J. Suppl. 77, Ser.* VIII, 177

Parker, E. N. 1963b, *Interplanetary Dynamical Processes* (Interscience, New York)

Parker, E. N. 1965, *Planetary Space Sci.*, **13**, 9

Pottasch, S. R. 1964, *Space Sci. Rev.*, **3**, 816

Reid, G. C. 1964, *J. Geophys. Res.*, **69**, 2659

Reid, G. C., Collins, C. 1959, *J. Atmospheric Terrest. Phys.*, **14**, 63

Roelof, E. C. (Ph.D. thesis, Univ. of California, Berkeley, Dec. 17, 1966)

Rothwell, P., McIlwain, C. E. 1959, *Nature*, **184**, 138

Schatzman, E. 1966 (Private communication)

Stein, W. A., Ney, E. P. 1963, *J. Geophys. Res.*, **68**, 65

Steljes, J. F., Carmichael, H., McCracken, K. G. 1961, *J. Geophys. Res.*, **66**, 1363

Suess, H. E., Urey, H. C. 1956, *Rev. Mod. Phys.*, **28**, 53

Sweet, P. A. 1958, *Nuovo Cimento Suppl. 8, Ser. X*, 188

Sweet, P. A. 1964, *AAS-NASA Symp. Phys. Solar Flares, US GPO*, 409

Van Allen, J. A., Frank, L. A., Venkatesan, D. 1964, *Trans. Am. Geophys. Union*, **45**, 80

Van Allen, J. A., Krimigis, S. M. 1965, *J. Geophys. Res.*, **70**, 5737

Warwick, C., Agy, V. L., Haurwitz, M. W. 1966, *Natl. Bur. Std. (U.S.) Rept.* 8862

Warwick, J. W. 1955, *Ap. J.*, **121**, 376

Webber, W. R. 1963, *AAS-NASA Symp. Phys. Solar Flares, NASA SP50*, 215

Wentzel, D. G. 1965, *J. Geophys. Res.*, **70**, 2716

Wilcox, J. M., Ness, N. F. 1965, *J. Geophys. Res.*, **70**, 5793

QUASI-STELLAR OBJECTS

By E. Margaret Burbidge

University of California, San Diego

INTRODUCTION

The first quasi-stellar radio source was discovered in 1960, and during the past six years many papers, both observational and theoretical, have been published. Since this is the first review article on the subject, I have attempted to give a complete bibliography up to December 1966 (somewhat earlier than this for overseas journals) and have included later references where possible. A fuller account of the subject will be found in the book by Burbidge & Burbidge (50).

The name "quasi-stellar object" or "QSO" has been adopted here, to cover quasi-stellar radio sources (or quasars) and radio-quiet quasi-stellar sources (also called blue stellar objects, quasi-stellar galaxies, and interlopers, in the literature).

DISCOVERY

In 1960 the Owens Valley radio interferometer began producing accurate declination measurements of sources (167). In previous surveys the errors in declination had been substantially larger than those in right ascension. There became available then some sources with positions good to about ±5" in both coordinates. Some of these were found to be of high radio surface brightness (small radio angular diameters), and with no obvious optical object (such as a galaxy) at the position. The first QSO to be discovered (by Sandage & Matthews) was 3C 48; within the small error rectangle around the radio position, the only optical object visible on a plate taken with the 200-inch Palomar telescope was a 16th mag "star" with a faint wisp of nebulosity attached. The first optical observations were reported at the 107th American Astronomical Society meeting in December 1960 (177). Sandage, followed by a number of astronomers, both at Palomar and elsewhere, obtained spectra of 3C 48 and saw broad emission lines at wavelengths that did not correspond with features normally seen in emission-line stars.

A radio position of 3C 273 accurate to better than 1" was measured by means of a lunar occultation of the object, observed with the Parkes 210-ft radio telescope by Hazard, Mackey & Shimmins (103). As in 3C 48, the brightness temperature was large, the source having a small angular diameter. The source had two components, A and B. Hazard et al. noted that the position of component B agreed with that of a 13th mag star, and Schmidt (192) obtained spectra of this and noted that, like 3C 48, it had some nebulosity associated with it, in the form of a faint jet extending from the stellar object in exactly the direction of the separation between the two radio components, with component A lying at the end of the optical jet. Schmidt found broad emission lines in 3C 273 and identified them with Balmer hydrogen lines and Mg II λ 2798, shifted to longer wavelengths by an amount $z = \Delta\lambda/\lambda_0 = 0.158$. Following this, Greenstein & Matthews (96) were able to identify the spectroscopic features in 3C 48 and obtained $z = 0.367$ for it.

General Properties

Schmidt (193) has described the optical properties of the QSO's as follows:

(a) they are starlike objects identified with radio sources;
(b) they are variable in light;
(c) they have a large ultraviolet flux of radiation;
(d) there are broad emission lines in the spectra;
(e) the spectrum lines have large redshifts.

Bearing in mind the discoveries made since these definitions were given, we replace (a) by (a'): they are starlike objects *often* identified with radio sources and (d) by (d'): there are broad emission lines in the spectra, with absorption lines sometimes present.

Regarding property (c), Sandage's photometry of 3C 48 in 1960 (177,142) gave $V = 16.06$, $B - V = 0.38$, $U - B = -0.61$, i.e. 3C 48 lies well above the locus of main-sequence stars in the two-color diagram, up in the region where some white dwarfs, old novae, and related highly evolved stars lie, and this has been found to be a very general property of QSO's.

IDENTIFICATION OF QSO'S
Quasi-Stellar Radio Sources

Of the properties of 3C 48 and 3C 273—small radio diameter, starlike optical object, ultraviolet excess in the optical radiation, and peculiar emission-line spectrum, it is the ultraviolet excess which has proved particularly valuable in making identifications of QSO's, once accurate radio positions are available. Objects identified in the same period were 3C 196 and 3C 286 (142).

As is often the case in a period of rapid discoveries, it is not possible to put the advances in chronological order by directly using the dates on references. Matthews & Sandage (142) gave the fundamental optical identifications and data on 3C 48, 196, and 286 in a paper completed in 1962. At that time it was generally believed that these objects were likely to be galactic stars. The discovery of 3C 273 came while their paper was in press and thus they added a note in proof concerning the extragalactic nature of these objects. Like 3C 48 and 273, 3C 196 and 286 have small radio diameters and in each case the only object within the error rectangle around the radio position was a 17th mag star. A wisp of nebulosity was seen associated with the stellar object identified with 3C 196 but none was found in the case of 3C 286. The U, B, V colors were measured and the objects lie in the same region as 3C 48 in the two-color diagram.

A list of positions of optical objects found close to 42 radio sources of fairly small angular diameter and accurate radio positions was published by Griffin (98). In addition to 3C 48, 3C 196, and 3C 286, this list included stellar objects at the positions of 3C 147 and 3C 298, which were later proved to be QSO's. No wisps of associated nebulosity were found around these objects, and in fact this feature is absent in the majority of QSO's now known and has not proved to be of use in making identifications after the first few cases. The U, B, V colors of 3C 147 and 3C 298 were not measured at that time; Griffin's identifications were made solely on the basis of a starlike image being the only object seen within the error rectangle around an accurate radio position.

Hazard, Mackey & Nicholson (102) obtained very accurate radio positions of more sources by the lunar occultation method, and suggested optical identifications for them; of these, 3C 245 and MSH 14–12*1* were later proved, on the basis of photometry and spectroscopy, to be QSO's as Hazard et al. had suggested.

Since the first QSO's were found to have ultraviolet excesses, Ryle & Sandage made a search for identifications at good radio positions by taking two successive direct exposures on the same plate, through a blue and an ultraviolet filter, and picking out stellar images that were stronger on the ultraviolet exposure. Longair (130) had searched the fields around 88 radio sources of small angular diameter with good radio positions (63) and found 4 possible QSO's. Ryle & Sandage (175) certainly identified 3C 9, 216, and 245 as QSO's; 3C 9 and 245 had been suggested by Longair. Next Schmidt & Matthews (196) identified 3C 47 with a stellar object appearing rather blue from comparison of the red and blue Palomar Sky Survey prints. It was confirmed as a QSO by its spectrum, as was also 3C 147, identified by Griffin (98).

Adgie (1) identified 3C 254, using a radio position determined with the Royal Radar Establishment interferometer at Malvern and Griffin's measures of positions of optical objects in the fields of radio sources. Sandage & Wyndham (188) identified 11 more QSO's, by means of positional agreement between a stellar object and a radio source together with either two-color photographic plates showing that the stellar object had an ultraviolet excess, or the observation that the stellar object appeared stronger on the blue Palomar Atlas plate than on the red.

The number of identified or possible QSO's then rapidly increased [Wyndham (232, 233); Sandage, Véron & Wyndham (186)]. At this time, new catalogues of radio sources from Parkes, Cambridge (the 4C), and NRAO were becoming available. Bolton et al. have identified many possible QSO's by comparing the Parkes radio positions with the positions of nearby blue stellar objects on the Palomar Sky Survey prints (26 to 31, 64, 79). Between $-20°$ and $+20°$ some 80 possible QSO's have been listed; optical confirmation in the form of U, B, V photometry and spectroscopic observations has been made for at least a third of these (27, 33, 34, 126a). QSO's in the first sections of the 4C catalogue to appear have been identified by Wyndham (234), Scheuer & Wills (191), and Wills (230).

Radio-Quiet QSO's

During the course of searches for identifications by blue and ultraviolet exposures, both Sandage and Lynds (with the Kitt Peak 84-inch telescope) sometimes found ultraviolet objects that did not lie close to the radio positions. Sandage found about 3 such objects per square degree to a limiting magnitude $B \approx 18^m.5$. These objects are of the same type as those found in previous surveys in high galactic latitudes undertaken by Humason, Zwicky, Luyten, Iriarte, Chavira, Haro, Feige, and others. The relevant references are listed by Sandage (179). These previous surveys had yielded a frequency of about 4 ultraviolet objects per square degree to $B \approx 19^m.0$.

Sandage found that in a two-color plot ($U-B$ against $B-V$), the objects divided into two groups, those brighter and those fainter than $V = 14^m.50$. The brighter objects lay mostly in the region where lie normal fairly high-temperature stars, with some in the region of metal-deficient old halo-population stars. Very few lay in the region where the QSO's lie. But for $V > 14^m.50$, while some were clearly normal high-temperature or metal-deficient stars, the majority lay well above the normal sequence, in the region where QSO's lie. In a plot of log $N(m)$ against m, where $N(m)$ is the number of objects brighter than magnitude m, Sandage found a change of slope at about 15th magnitude suggesting a change in the type of objects being sampled at this point, and concluded that the majority of the fainter blue stellar objects are extragalactic, with large redshifts, and in fact are QSO's that are radio-quiet down to the limits set by existing radio-source catalogues. Of 6 objects whose spectra were taken, 1 was a galactic star, 2 had continuous spectra with no emission or absorption features, and 3 were extragalactic. Of these, 1 had a nonstellar image and a small redshift, but 2 had completely stellar images, one with a spectrum indistinguishable from the radio QSO's and a large redshift; this was called BSO 1. The other had a similar kind of spectrum and a smaller redshift; it is Ton 256 (Ton = Tonanzintla).

Sandage concluded that virtually all high-latitude blue stellar objects fainter than 15th mag are QSO's that are radio-quiet down to limits set by the existing catalogues. Thus, with a frequency of 4 per square degree, these would be 500 times more numerous, spatially, than the QSO's identified with radio sources.

The validity of this conclusion hinges on Sandage's interpretation of the plot of log $N(m)$ against m, and alternative interpretations were given almost immediately by Kinman (125) and by Lynds & Villere (138). With more recent determinations of the density of horizontal-branch type stars in the halo of our Galaxy as a function of distance from the plane, and new estimates of the frequency of white dwarfs, Kinman found that the observed log $N(m)$ vs. m plot could be well represented by a combination of these. Lynds & Villere reached essentially the same conclusion. Thus the bulk of the objects in question are probably galactic stars—white dwarfs, evolved hot subwarfs, and halo stars of the horizontal-branch type. This conclusion was supported by a random sampling of spectra of 12 stars of about 16th apparent magnitude by Kinman (125); 7 were found to be white dwarfs, 4 were horizontal-branch stars, and 1 was a hot subdwarf star.

There are undoubtedly a number of compact blue galaxies among the blue stellar objects, as discovered by Humason & Zwicky (117), Zwicky (237), and Haro (100). Sandage & Luyten (183) have isolated a larger sample of possible QSO's, using colors and zero proper motion as criteria, and 10 of these objects have been confirmed as QSO's from spectra. Some of these objects have been seen to show light variations. Van den Bergh (223) has also prepared a list of faint blue objects with excess ultraviolet radiation, and some vary in light.

At present it is difficult to estimate the true frequency of QSO's among the fainter blue stellar objects in the high-latitude surveys. Kinman estimated that it was 20 per cent or less of Sandage's earlier estimate. Sandage & Luyten (183) estimated a frequency of 0.4 objects per square degree down to a limiting magnitude $B = 18$, and 1 to 3 per square degree down to $B = 19.7$, which would give 10^5 objects with $B < 19.7$ over the whole sky.

Table I lists known and probable QSO's with coordinates, photometry, and redshifts where measured. An object lettered PHL is from the Palomar-Haro-Luyten catalogue of blue stellar objects (101), and Ton denotes the various lists by Chavira & Iriarte (60, 61, 118). Not all suggested identifications are given in Table I, only those with some supporting evidence besides position coincidence.

LINE SPECTRA OF QUASI-STELLAR OBJECTS

The first spectrogram of 3C 48 was obtained in 1960 by Sandage. Early spectroscopic work was carried out by Greenstein, Münch, and others, on this object. Following his discovery of the redshift of 3C 273, Schmidt was largely responsible for all of the spectroscopic observations of the QSO's until 1964 when others entered the field, namely, Lynds and colleagues at Kitt Peak National Observatory with an image-tube spectrograph, Burbidge & Kinman at Lick with a conventional spectrograph and more recently with an image-tube spectrograph, Dibai & Yesipov with an image-tube spectrograph at the Crimean station of the Sternberg Institute, Andrillat & Andrillat at Haute Provence, Ford & Rubin with an image-tube spectrograph at Lowell and at Kitt Peak, and Hiltner and colleagues with an image-tube spectrograph at McDonald.

Line Identifications

The identification by Schmidt (192) of 4 broad emission lines in the brightest QSO, 3C 273, as the Balmer lines Hβ-Hϵ with a redshift $z = 0.158$, provided the breakthrough in understanding the line spectra of these objects by demonstrating that considerable redshifts are present. From the redshift given by the Balmer lines, Schmidt identified a broad emission feature as a blend of the Mg II doublet $\lambda\lambda$ 2796, 2803, which had hitherto only been seen

TABLE I

List of Quasi-Stellar Objects[a]

Object	α (1950)	δ (1950)	m_v	z	B − V	U − B
PHL 658	0ʰ 03ᵐ 25ˢ4 (radio)	+15° 53′ 10″	16.40	0.450	+0.11	−0.70
3C 2	0 03 48.70	−00 21 06.6	19.35	1.037	+0.79	−0.96
3C 9	0 17 49.83	+15 24 16.5	18.21	2.012	+0.23	−0.76
MSH 00 −29	0 22 01	−29 45.5	(20)			
PHL 6638	0 44 35.3 (radio)	−07 22.0	17.72		+0.18	−0.69
PHL 923	0 56 31.7	−00 09 16	17.33	0.717	+0.20	−0.70
PKS 0056 −17	0 56 36.8	−17 16 51	(17)	2.125		
PHL 938	0 58.2	+01 56	17.16	1.93	+0.32	−0.88
PKS 0106+01	1 06 04	+01 19.0	18.39	2.107	+0.15	−0.70
PKS 0114+07	1 14 49.7	+07 26.3	(18)			
AO 0118+03 (3C 39)	1 18 27.6	+03 28 19				
PKS 0119 −04	1 19 55.8	−04 37 08	16.88	1.955	+0.46	−0.72
PKS 0122 −00	1 22 55.5	−00 21 34	(16)	1.070		
3C 43	1 27 15.18	+23 22 52.0	(20.0)			
PHL 3375[a]	1 28.4	+07 28	18.02		+0.29	−0.51
PHL 1027[a]	1 30.5	+03 22	17.04		−0.03	−0.77
PHL 3424[a]	1 31.2	+05 32	18.25	1.847	+0.19	−0.90
3C 47	1 33 40.30	+20 42 16.0	18.1	0.425	+0.05	−0.65
PHL 1070[a]	1 34.8	+03 21	(17.6)			
3C 48	1 34 49.8	+32 54 20	16.2	0.367	+0.42	−0.58
PHL 1072[a]	1 35.2	+05 39	(18.3)			
PHL 1078	1 35 29.1 (radio)	−05 42.1	18.25	0.308	+0.04	−0.81
PHL 1093	1 37 22.9 (radio)	+01 16.3	17.07	0.260	+0.05	−1.02
PHL 1127[a]	1 41.5	+05 14	18.29	1.990	+0.14	−0.83
PHL 3740	1 44 14.9 (radio)	−05 54.2	18.61		+0.09	−0.65
PHL 1186[a]	1 47.6	+09 01	(18.6)			
PHL 1194[a]	1 48.7	+09 02	17.50	0.298	−0.07	−0.85
PHL 1222[a]	1 51.2	+04 48	17.63		+0.41	−0.78
PHL 1226[a]	1 51.8	+04 34	(18.2)			
3C 57	1 59 30.4	−11 47 00	16.40	0.68	+0.14	−0.73
PKS 0202 −17	2 02 34.4	−17 15 37	(18)			
PHL 1305	2 26 21.6 (radio)	−03 54.3	16.96	2.064	+0.07	−0.82
PKS 0229+13	2 29 02.3	+13 09 42	(18)	2.065		
PHL 1377 = 4C −4.6	2 32 36.4 (radio)	−04 16.9	16.46	1.436	+0.15	−0.89
PKS 0237 −23	2 37 53.4	−23 22 05	16.63	2.223	+0.15	−0.61
PKS 0336 −01	3 36 59.2	−01 56 19	18.41		+0.55	−0.82
3C 93	3 40 51.47	+04 48 21.6	18.09		+0.35	−0.50
PKS 0347+13	3 47 14.0	+13 10 01	(19)			
MSH 03 −19	3 49 09.5	−14 38 07	16.24	0.614	+0.11	−0.65
3C 94	3 50 04.1	−07 19 55	(17.5)	0.962		
PKS 0403 −13	4 03 14.0	−13 16 16	(18)	0.571		
MSH 04 −12	4 05 27.4	−12 19 34	(16)	0.574		
3C 119	4 29 07.84	+41 32 08.7	(>20.0)			
3C 138	5 18 16.5	+16 35 26	17.9	0.760	+0.23	−0.38 (reddened)
3C 147	5 38 43.5	+49 49 43	16.9	0.545	+0.35	−0.59
PKS 0541 −24	5 41 09.5	−24 22.7				
3C 172	6 59 04.5	+25 17 36	(17.2)			
3C 175	7 10 15.3	+11 51 30	(17.5)	0.768		
3C 175.1	7 11 14.3	+14 41 33	(18.0)			
3C 181	7 25 20.36	+14 43 47.2	18.92	1.382	+0.43	−1.02
PKS 0736+01	7 36 42.4	+01 43 57	(18)	0.191		

[a] Objects found in a selected field studied by Sandage & Luyten (183).

TABLE I (*Continued*)

Object	α (1950)			δ (1950)			m_v	z	$B-V$	$U-B$
3C 186	7	40	56.67	+38	00	31.9	17.60	1.063	+0.45	−0.71
3C 190	7	58	44.1	+14	23	0	17.46		−0.20	−0.90
3C 191	8	02	03.78	+10	23	58.1	18.4	1.946	+0.25	−0.84
3C 196	8	09	59.4	+48	22	08	17.6	0.871	+0.60	−0.43
PKS 0812+02	8	12	47.2	+02	04	11	(17)	0.402		
PKS 0825−20	8	25	03.4	−20	16	31	(18)			
4C 37.24	8	27	55.0	+37	52	20	(18.2)	0.914		
3C 204	8	33	18.23	+65	24	05.9	18.21	1.112	+0.55	−0.99
3C 205	8	35	10.6	+58	04	46	(17.8)			
3C 207	8	38	01.7	+13	23	05.4	18.15	0.684	+0.43	−0.42
3C 208	8	50	22.79	+14	03	58.3	17.42	1.110	+0.34	−1.00
PKS 0859−14	8	59	55	−14	03	37	(17.8)	1.327		
4C 22.22	9	01	56.5	+22	31	36	(19.0)			
3C 215	9	03	44.2	+16	58	16	18.27	0.411	+0.21	−0.66
3C 217	9	05	41.0	+38	00	27	18.50		+0.25	−0.86
3C 216	9	06	17.26	+43	05	59.0	18.48		+0.49	−0.60
PKS 0922+14	9	22	22.27	+14	57	26.2	17.96	0.895	+0.54	−0.52
4C 39.25	9	23	55.4	+39	15	24	(17.3)	0.699		
3C 230	9	49	25.5	+00	12	57	(17.5)			
3C 232 = Ton 469	9	54	31 (radio)	+32	37		15.78	0.534	+0.10	−0.68
AO 0952+17	9	52	11.92	+17	57	46.6	(17.7)	1.471		
PKS 0957+00	9	57	43.84	+00	19	50.0	17.57	0.906	+0.47	−0.71
3C 239	10	08	37.5	+46	43	15	(17.5)			
3C 245	10	40	06.11	+12	19	15.1	17.25	1.029	+0.45	−0.83
PKS 1049−09	10	48	59.5	−09	02	12	16.79	0.344	+0.06	−0.49
3C 249.1	11	00	30.56	+77	15	08.1	15.72	0.311	−0.02	−0.77
3C 254	11	11	53.35	+40	53	42.0	17.98	0.734	+0.15	−0.49
PKS 1116+12	11	16	20.79	+12	51	06.3	19.25	2.118	+0.14	−0.76
PKS 1127−14	11	27	35.6	−14	32	57	16.90	1.187	+0.27	−0.70
3C 261	11	32	16.31	+30	22	01.0	18.24	0.614	+0.24	−0.56
PKS 1136−13	11	36	38.6	−13	34	09	(17)	0.554		
3C 263	11	37	09.38	+66	04	25.9	16.32	0.652	+0.18	−0.56
PKS 1148−00	11	48	10.2	−00	07	15	17.60	1.982	+0.17	−0.97
4C 31.38	11	53	44.4	+31	44	47	(19.4)	1.557		
3C 268.4	12	06	41.7	+43	56	05	18.42	1.400	+0.58	−0.69
PKS 1217+02	12	17	38.35	+02	20	20.9	16.53	0.240	+0.02	−0.87
3C 270.1	12	18	04.00	+33	59	50.0	18.61	1.519	+0.19	−0.61
4C 21.35	12	22	23.5	+21	39	27	(18.0)	0.434		
Ton 1530	12	22	57	+22	53		(16.8)	2.051		
3C 273	12	26	33.35	+02	19	42.0	12.8	0.158	+0.21	−0.85
PKS 1229−02	12	29	25.9	−02	07	31	16.75	0.388	+0.48	−0.66
PKS 1233−24	12	32	59.4	−24	55	46	(17)			
PKS 1237−10	12	37	07.3	−10	07	04	(18.2)			
3C 275.1	12	41	27.68	+16	39	18.7	19.00	0.557	+0.23	−0.43
BSO 1	12	46	29	+37	46	25	16.98	1.241	+0.31	−0.78
3C 277.1	12	50	15.31	+56	50	37.0	17.93	0.320	−0.17	−0.78
PKS 1252+11	12	52	07.86	+11	57	20.8	16.64	0.871	+0.35	−0.75
3C 279	12	53	35.94	−05	31	08.0	17.8	0.538	+0.26	−0.56
3C 280.1	12	58	14.15	+40	25	15.4	19.44	1.659	−0.13	−0.70
3C 281	13	05	22.52	+06	58	16.4	17.02		+0.13	−0.59
4C 22.38	13	24	29.9	+22	58	22	(18.9)			
PKS 1326+06	13	26	43	+06	56.4		(16)			
PKS 1327−21	13	27	23.2	−21	26	34	16.74	0.528	+0.10	−0.54
3C 287	13	28	16.12	+25	24	37.1	17.67	1.055	+0.63	−0.65
3C 286	13	28	49.74	+30	45	59.30	17.30	0.849	+0.22	−0.84
MSH 13−011	13	35	31.34	−06	11	57.4	17.68	0.625	+0.14	−0.66
3C 288.1	13	40	30.4	+60	36	55	18.12	0.961	+0.39	−0.82

TABLE I (*Continued*)

Object	α (1950)			δ (1950)			m_v	z	$B-V$	$U-B$
PKS 1354+19	13	54	42.3	+19	33	41	16.02	0.720	+0.18	−0.55
3C 298	14	16	38.59	+06	42	21	16.79	1.439	+0.33	−0.70
4C 20.33	14	22	37.5	+20	13	49	(17.1)	0.871		
MSH 14−121	14	53	12.22	−10	56	39.9	17.37	0.940	+0.44	−0.76
PKS 1454−06	14	54	02.7	−06	05	45	18.0	1.249	+0.60	
3C 309.1	14	58	57.6	+71	52	19	16.78	0.904	+0.46	−0.77
PKS 1510−08	15	10	08.9	−08	54	48	16.52	0.361	+0.17	−0.74
PKS 1514+00	15	14	14.8	+00	26	01	(19)			
3C 323.1	15	45	31.2	+21	01	34	(15.8)	0.264		
MSH 16+03	16	03	39.5	+00	07	55	(18.0)			
Ton 256	16	12.0		+26	13		15.91	0.131	+0.57	−0.84
3C 334	16	18	07.40	+17	43	30.5	16.41	0.555	+0.12	−0.79
3C 336	16	22	32.45	+23	52	00.7	17.47	0.927	+0.44	−0.79
3C 345	16	41	17.70	+39	54	11.1	16−17.30	0.595	+0.29	−0.50
3C 351	17	04	03.58	+60	48	29.9	15.28	0.371	+0.13	−0.75
3C 380	18	28	13.38	+48	42	39.3	16.81	0.692	+0.24	−0.59
PKS 2115−30	21	15	11.1	−30	31	50	16.47		+0.49	−0.54
3C 432	21	20	25.64	+16	51	46.0	17.96	1.805	+0.22	−0.79
3C 435	21	26	37.6	+07	19	49	(19.5)			
PKS 2128−12	21	28	52.5	−12	20	19	15.99		+0.13	−0.67
PKS 2135−14	21	35	01.1	−14	46	27	15.53	0.200	+0.10	−0.83
PKS 2144−17	21	44	17.7	−17	54	05	(19.5)			
PKS 2145+06	21	45	35.9	+06	43	43	(17.5)			
PKS 2146−13	21	46	46.1	−13	18	24	(20)	1.800		
PKS 2154−18	21	54	12.5	−18	28.5		(16.5)			
PKS 2203−18	22	03	25.8	−18	50	16	(19)			
PKS 2216−03	22	16	16.3	−03	50	43	(17)	0.901		
3C 446	22	23	11.05	−05	12	17.0	18.39	1.403	+0.44	−0.90
PHL 5200	22	25	50.6	−05	30.6		(18.2)	1.981		
CTA 102	22	30	07.71	+11	28	22.8	17.32	1.037	+0.42	−0.79
3C 454	22	49	07.86	+18	32	46.6	18.40	1.757	+0.12	−0.95
3C 454.3	22	51	29.61	+15	52	53.6	16.10	0.859	+0.47	−0.66
PKS 2251+11	22	51	40.6	+11	20	39	15.82	0.323	+0.20	−0.84
4C 29.68 ≡CTD 141	23	25	41.3	+29	20	36	(17.3)	1.012		
PKS 2344+09	23	44	03.4	+09	14	04	15.97	0.677	+0.25	−0.60
PKS 2345−16	23	45	27.6	−16	47	50	(18)			
PKS 2354+14	23	54	44.7	+14	29	26	(18)	1.810		
PKS 2354−11	23	54	57.1	−11	42	23	(18)			

in solar spectra taken outside the Earth's atmosphere. Oke (155) detected Hα in 3C 273.

The λ 2798 feature was then identified by Greenstein & Matthews (96) in 3C 48, and this, together with several other emission lines, gave a larger redshift for this object. A detailed analysis of 3C 48 and 3C 273 was then carried out by Greenstein & Schmidt (97).

The identification of 17th and 18th magnitude stellar objects with radio sources, in conjunction with the redshifts found for 3C 273 and 3C 48, and, later, for 3C 47 and 3C 147 by Schmidt & Matthews (196), suggested that objects with really large redshifts might be found (38). In these spectra the ultraviolet wavelength region normally inaccessible from ground-based instruments, and therefore at that time unobserved except in the Sun, could be expected to be shifted into the visible region, posing a considerable problem in identifying the lines. The lines found in 3C 48 and 3C 273 (all emission lines, and all very broad) besides the Balmer lines and Mg II, are due to [O II], [O III], [Ne III], [Ne V]. These are the sort of features to be expected in hot gaseous nebulae with physical conditions similar to those of planetary nebulae, in radio galaxies like Cygnus A, and in nuclei of Seyfert galaxies. Osterbrock (158) had prepared a list of emission lines to be expected in the ultraviolet spectra of planetary nebulae, with computed relative intensities. Schmidt (194) compiled a search list of emission lines with which he identified emission features in 5 more QSO's, all of which proved to have very large redshifts, running up to $z = 2.012$ in 3C 9, in which, for the first time, Lyman α was seen as a strong emission feature. A search list similar to Schmidt's is given in (50).

The 4 strongest emission lines in the wavelength range extending shortward from about 5000 A are Ly-α, C IV λ 1549, C III] λ 1909, and Mg II λ 2798. Apart from λ 1909, these are resonance transitions involving the ground level of the ion in question. If a spectrum shows only 2 emission lines, the ratio of whose wavelengths agrees with any of the ratios of rest wavelengths from these 4 lines, then it is highly likely that they are the correct identifications. Although λ 1909 and λ 1549 give a ratio fairly near that of [Ne V] λ 3426 and Mg II λ 2798, the presence or absence of other lines at predicted wavelengths usually settles the question.

Table II lists in order of increasing z all the redshifts determined to date. The details of the emission features detected in the spectrum of each object, with approximate strengths and widths, were compiled by Burbidge et al. (51) and a later version is given in (50). The lines found belong to ions with a wide range of ionization potentials, in the light elements that are most abundant in the Sun and stars of the solar neighborhood. The elements H, He, C, N, O, Ne, Mg, Si, S, and Ar are all represented; of the elements heavier than Ar, permitted lines of Fe II have been identified by Wampler & Oke (226b) in 3C 273 with some weak emission bands first seen by Greenstein & Schmidt (97). They have tentatively identified Na I also.

ABSORPTION LINES

The first QSO's for which redshifts were determined all showed only emission lines in their spectra, and consequently the first models for explaining the line spectra did not include any provision for discussing absorption lines. It is now clear, however, that absorption lines are not uncommon in the spectra of QSO's.

The first recorded observation of absorption lines was in Schmidt's spectra of the radio-quiet QSO known as BSO 1 (179), in which the broad emission of C IV λ 1549 is bisected by a sharp absorption. Absorption components in the short-wavelength wing of both Ly-α and C IV λ 1549 were found in PHL 938, another radio-quiet QSO, by Kinman (126).

3C 191 was then found to show a large number of sharp absorption lines in its spectrum (45, 214), as well as some of the usual broad emission lines. These absorptions are almost all from the ground state of the relevant ions,

TABLE II
Quasi-Stellar Objects with Redshifts

Object	z	References	Object	z	References
Ton 256	0.131	179	3C 380	0.692	179, 40
3C 273	0.158	192	4C 39.25	0.699	135, 43
PKS 0736+01	0.191	134a	PHL 923	0.717	134a
PKS 0837−12	0.200	126b	PKS 1354+19	0.720	43
PKS 2135−14	0.200	126b	3C 254	0.734	194
PKS 1217+02	0.240	135, 87, 105	3C 138	0.760	135, 43
PHL 1093	0.260	134a, 126b	3C 175	0.768	134a
3C 323.1	0.264	[a]	3C 286	0.849	137
PHL 1194	0.298?	183	3C 454.3	0.859	134a, 77a
PHL 1078	0.308	134a	PKS 1252+11	0.871	137, 195,
3C 249.1	0.311	195, 87, 105			105
3C 277.1	0.320	195, 105	4C 20.33	0.871	135
PKS 2251+11	0.323	126b	3C 196	0.871	135
PKS 1049−09	0.344	126b	PKS 0922+14	0.895	135, 105
PKS 1510−08	0.361	43	PKS 2216−03	0.901?	134a
3C 48	0.367	96	3C 309.1	0.904	43, 134a
3C 351	0.371	137	PKS 0957+00	0.906	135, 105
PKS 1229−02	0.388	126b	4C 37.24	0.914	134a
PKS 0812+02	0.402	126b	3C 336	0.927	135
3C 215	0.411	135	MSH 14−121	0.940	39, 195
3C 47	0.425	196	3C 288.1	0.961	[a]
4C 21.35	0.434	43, 134a	3C 94	0.962	134a
PHL 658	0.450	134a	4C 29.68[b]	1.012	195, 134a,
PKS 1327−21	0.528	43			126b
3C 232	0.534	[a]	3C 245	1.029	194, 137
3C 279	0.538	137, 46	CTA 102	1.037	194
3C 147	0.545	196	3C 2	1.037	134a
PKS 1136−13	0.554	126b	3C 287	1.055	194
3C 334	0.555	137, 40	3C 186	1.063	135
3C 275.1	0.557	135	PKS 0122−00	1.070	134a
MSH 04−12	0.574	77a, 126b	3C 208	1.110	41, 195
PKS 0403−13	0.574	134a	3C 204	1.112	195
3C 345	0.595	137, 40	PKS 1127−14	1.187	43
MSH 03−19	0.614	195	BSO 1	1.241	179
3C 261	0.614	135, 43	PKS 1454−06	1.249	43
MSH 13−011	0.625	43	PKS 0859−14	1.327	43
3C 263	0.652	87	3C 181	1.382	195
PKS 2344+09	0.677	126b	3C 268.4	1.400	[a]
3C 57	0.68	77a	3C 446	1.403	40, 195
3C 207	0.684	134a	PHL 1377	1.436	105, 126b

[a] M. Schmidt, in preparation.
[b] 4C 29.68 = CTD 141.

TABLE II (*continued*)

Object	z	References	Object	z	References
3C 298	1.439	135	PHL 5200	1.98	134a
AO 0952+17	1.472	126b	PKS 1148−00	1.982	43
3C 270.1	1.519	195	PHL 1127	1.990	183
4C 21.38	1.557	43	3C 9	2.012	194
3C 280.1	1.659	135	Ton 1530	2.051	105
3C 454	1.757	195	PHL 1305	2.064	134a
PKS 2146−13	1.800	134a	PKS 0229+13	2.065	126b
3C 432	1.805	195	PKS 0106+01	2.107	41
PKS 2354+14	1.810	126b	PKS 1116+12	2.118	195, 136
PHL 3424	1.847	183	PKS 0056−17	2.125	126b
PHL 938	1.93	126	PKS 0237−23	2.223	12a, 41a,
3C 191	1.946	45, 214			97a
PKS 0119−04	1.955	126b			

and include Ly-α and C IV λ 1549. One of the absorption-line identifications by Burbidge et al. (45) and Stockton & Lynds (214) in 3C 191, namely N V λ 1240.1, was questioned by Bahcall (14), who believed it might be due instead to a blend of Mg II $\lambda\lambda$ 1239.9, 1240.4. Burbidge & Lynds (44), however, replied that Stockton's Kitt Peak spectrogram of 3C 191 resolves the line into fine-structure components at the right wavelengths for N V; moreover, the Mg II lines are the second member of the principal series in Mg II, and the transition probability can be only a few per cent of that of Mg II λ 2798.

A strong absorption component appears in the short-wavelength wing of Ly-α in PKS 1116+12 on a spectrum taken by Lynds & Stockton (136), about 26 Å from the center of the Ly-α emission. It was apparently not seen by Schmidt (195) on his spectra of PKS 1116+12, but Bahcall, Peterson & Schmidt (15) detected two other absorptions, probably Ly-α and C IV, at a redshift corresponding to some 17,000 km/s less than the respective emission-line centers. Lynds & Stockton saw the one corresponding to Ly-α, but not the other.

Ford & Rubin (87) found redshifted Ca II H and K absorption lines in 3C 263; their redshifts agreed with that obtained from the emission lines. On the other hand, Lynds did not see the Ca II absorption lines in his spectra of 3C 263. Possibly absorption lines can be variable on a short time scale in the spectra of QSO's.

Other QSO's for which there is published evidence for absorption lines are 3C 270.1 (195) and 3C 298 (135) (both in C IV λ 1549), and PKS 1510-08 (in the shortward wing of Mg II λ 2798) (43).

The absorption lines described above are mostly presumed to arise in gas associated with the QSO and not in the intergalactic medium. Bahcall & Salpeter (16, 17) considered the possibility that resonance lines due to atoms and ions of the most common elements might be produced by gas in clusters of galaxies lying between the QSO and observer; in cases where the ground level had fine structure, only absorptions from the zero energy state would be expected. In 3C 191, however, there appears to be normal occupancy of all the fine-structure states in the ground levels of Si II and N V. Bahcall, Peterson & Schmidt (15) suggested that the absorptions in PKS 1116+12 might be intergalactic.

Recently, attention has been focused anew on absorption lines. Lynds (134a) found broad absorption bands, adjacent on the violet side to broad emission bands, in PHL 5200, which was identified with 4C −5.93 by Scheuer & Wills (191). The emission lines have three separate (though closely spaced) redshifts; the two additional ones might be produced by the dissipation of shock waves passing through an expanding shell of gas which gives the absorption. PKS 0237-23 was found by Arp, Bolton & Kinman (12a) to have the largest emission-line redshift yet measured, $z = 2.223$, and an unusual absorption-line spectrum. They found $z_{abs} = 2.20$. Burbidge (41a) gave different identifications for the absorption lines with $z = 1.95$. Greenstein & Schmidt (97a) found that actually both absorption-line redshifts seem to be present. Burbidge (48a) realized that there was a collection of QSO's with values of z_{abs} close to 1.95, and raised again the possibility of gravitational redshifts.

PKS 0119-04, identified by Bolton & Ekers (31a) and confirmed with photometry by Kinman et al. (126a), had its redshift determined by Kinman & Burbidge (126b), and a set of absorption lines with $z = 1.966$ was found; this time the emission-line redshift is slightly smaller, $z = 1.955$. Burbidge & Burbidge (50a) collected the evidence on 7 QSO's with several absorption lines that give a redshift near 1.95, and 15 QSO's with only 1 or 2 absorption lines, in most of which z_{abs} is approximately the same as z_{em}. Of 14 QSO's with $z_{em} > 1.9$, 10 show absorption lines, while of 87 QSO's with $z_{em} < 1.9$, only a further 10 show absorptions. Two of the 22 showing absorptions have not had redshifts determined yet.

THE NATURE OF THE REDSHIFTS

There are two mechanisms which are known to cause spectral line shifts, line-of-sight velocities and strong gravitational fields. Doppler shifts can be either redward or blueward, but outside the nearby galaxies in the expanding universe we see only receding objects, and hence redshifts.

Gravitational redshifts.—Greenstein & Schmidt (97) discussed the possibility that the redshifts of 3C 48 and 3C 273 might be gravitational, arising in either (*a*) a collapsed star in our Galaxy or (*b*) a collapsed mass of galactic size outside our Galaxy. In either case, the gradient of gravitational potential across the region where the spectrum lines arise will produce a broadening of the lines. They took the observed linewidths as limiting the size of this gradient. For a collapsed star of solar mass, which must not be very nearby because of the absence of measured proper motions, this sets an upper limit on the linear thickness of the shell and hence a lower limit to the density of gas in the shell (since a determined number of emitting atoms are required to give the observed emission line fluxes). The limiting density of $N_e \gtrsim 6 \times 10^{18}$ cm^{-3} for 3C 273 was incompatible with the presence of the forbidden line [O III] λ 5007 in 3C 273.

If 3C 273 were a massive object, outside our Galaxy but not at a large distance, Greenstein & Schmidt obtained limits by the condition that the mass must not be great enough nor the distance small enough to produce detectable perturbations on stellar motions, i.e. this effect must be less than 10 per cent of the gravitational acceleration of the whole Galaxy. Taking $N_e \sim 10^7$ cm^{-3}, this gave limits on the distance d and mass M as follows:

$$d \geq 2,500 \text{ pc}, \quad M/M_\odot \geq 7 \times 10^8 \text{ (3C 273)}$$
$$d \geq 25,000 \text{ pc}, \quad M/M_\odot \geq 7 \times 10^{10} \text{ (3C 48)}$$

Thus collapsed massive extragalactic objects on this scale could account for the observed redshifts in these 2 objects, but the conditions are rather stringent. For objects with $z \simeq 2$, the above limits will be changed, but not by a large factor.

From the theoretical standpoint the situation is not clear. The Schwarzschild interior solution for a perfect fluid sphere gives $z_{max} = 2$, but this corresponds to a transmission velocity in the fluid exceeding c. If appropriate conditions on the equation of state and stability are applied, Bondi (35) showed that $z_{max} \leq 0.62$ [see also Buchdahl (36, 37)]. Hoyle & Fowler (112a) suggested a model that avoids the difficulties pointed up by Greenstein & Schmidt (97), and in which z may exceed 2. A large number of collapsed masses are packed in a volume at the center of which, in a deep potential well, gas collects which gives rise to the redshifted spectrum lines. The collapsed objects are small relative to the total volume, so the light from the center escapes without interception. No gradient in gravitational potential, with consequent line broadening, is experienced in this model. The redshift at the center, z_c, can be much larger than the redshift from the surface, z_s.

Doppler shifts.—Most workers have generally agreed that the redshifts are due to the motion of the QSO's relative to the observer, and most have gone further and assumed that the redshifts are cosmological in origin and are due to the expansion of the universe. Alternatively it has been suggested that the redshifts are Doppler shifts due to the translational motion of a local object. Terrell (221, 222) argued that the objects were ejected from our Galaxy at relativistic speeds. Hoyle & Burbidge (107) considered the possibility that the objects have been ejected from comparatively nearby radio galaxies in which violent explosions have occurred. These theories will be discussed below.

Other causes.—Arp (10, 11) has argued that the redshifts are "nonvelocity" shifts, but has not suggested any previously neglected physical mechanism.[1]

---

[1] Boccaletti et al. (24) have suggested that the observed wavelengths are produced in transitions in atoms and ions containing one or two quarks, but the calculated wavelengths do not agree well with the observed ones.

INTERPRETATION OF THE LINE SPECTRA

Despite the fact that it is not yet possible to study the whole spectral region of any single QSO, from putting together the observations of many QSO's with various redshifts, it is clear that there are differences in the relative intensities of emission lines and in the intensities of lines relative to the continuum from one object to another (50, 51, 54). The task of constructing a model of the line-producing region can therefore be applied either to one particular QSO, or to some generalized "average" QSO. Since the emission lines are of the sort seen in gaseous nebulae, the standard methods of analyzing these can be applied.

This type of analysis was applied by Greenstein & Schmidt (97) to 3C 273 and 3C 48, and by Shklovsky (206) and Dibai & Pronik (76) to a spectroscopic study of 3C 273 made with the telescope in Crimea. When spectroscopic observations became available for other objects, Osterbrock & Parker (159) used them to derive the physical conditions in an average or composite of 9 QSO's.

Greenstein and Schmidt *assumed* that the abundances of the elements in 3C 273 and 3C 48 are the same as the average Population I abundances of gaseous nebulae and young stars in the solar neighborhood in our Galaxy, and estimated $T_e = 16,800°$K for the electron temperature. Then they calculated the relative intensities of the lines that were measured, for various values of the electron density N_e, and found the value which gave the best fit to the observations. Somewhat ambiguous estimates of the electron densities resulted (in that the same calculations applied to planetary nebulae yielded relative intensities that needed correction by arbitrary factors to agree with the observations, and the same "correction factors" were applied to the QSO's). The results are as follows:

$$N_e = 3 \times 10^4 \text{ cm}^{-3} \text{ for 3C 48}$$
$$N_e = 3 \times 10^6 \text{ cm}^{-3} \text{ for 3C 273}$$

with uncertainty factors of about an order of magnitude either way.

Dibai & Pronik derived a higher value for the electron density in 3C 273, $N_e \approx 10^7$ cm^{-3}, with a temperature $T_e \approx 10,000°-15,000°$. Shklovsky pointed out that the smallness of the Balmer discontinuity in the continuous spectrum of 3C 273 suggests that T_e is at least as high as $20,000°$, while the strength of the Hβ emission line relative to the continuum shows that $T_e < 60,000°$. Thus Shklovsky's value of $T_e \approx 30,000°$ seems a good choice. The absence of [O II] 3727 and [Ne V] lines in 3C 273 shows that N_e must be higher in 3C 273 than in 3C 48; later observations have indicated that 3C 48 is more typical of most QSO's than 3C 273.

Cameron (56) estimated a still higher value for N_e in 3C 273, namely 2×10^8 cm^{-3}, but Greenstein & Schmidt thought an upper limit to the acceptable electron density in this object would be 3×10^7 cm^{-3}. $N_e = 10^7$ cm^{-3} is probably the best estimate. Osterbrock & Parker, for their composite average

QSO, found a good match with the observations for $T_e = 15,000°$ and $N_e = 3 \times 10^6$ cm^{-3}, for an assumed level of ionization (again assuming normal or Population I relative element abundances).

The fact that the studies just described yield relative intensities that match the measures reasonably well shows that the abundances must be similar to those found in young Population I stars and gaseous nebulae in the solar neighborhood. If the QSO's are indeed extremely distant objects dating back billions of years in time in an evolving universe, this is an extremely surprising result; Shklovsky (206) was the first to draw attention to this. Osterbrock & Parker (159) found that the only possible exception to this is helium; they predicted that the line of He II λ 1640 should be of comparable intensity with C IV λ 1549, while actually it is weaker in all cases and in some QSO's it has not been seen. They suggested that helium might be less abundant, with respect to hydrogen and the heavier elements, in QSO's than in the "normal" abundance distribution.

Burbidge et al. (51), however, pointed out that an ionization equilibrium for QSO's had not been calculated; a distribution among the various ionization stages which seemed reasonable had been assumed, with helium taken to be 50 per cent doubly ionized. This might not be the case; there might well be less doubly ionized helium than this.

Dimensions and mass of region giving emission lines.—If the redshifts are of cosmological origin, the luminosity distances of QSO's may be derived from them for any chosen cosmological model. Greenstein & Schmidt used a model with the deceleration parameter $q_0 = 0$ and from the equivalent widths of Hβ in 3C 48 and 3C 273 obtained Hβ luminosities of 6.4×10^{42} erg/s (3C 48) and 8.8×10^{43} erg/s (3C 273). Shklovsky, using the measures of Dibai & Pronik, found the Hβ flux in 3C 273 to be 8×10^{43} erg/s.

From these values, with T_e and N_e determined, using the hydrogen recombination formula, Greenstein & Schmidt obtained values for the radius R, and mass M_H, of the hydrogen-emitting region, as follows:

$$3C\ 48: \quad R = 11\ \text{pc}, \quad M_H/M_\odot = 5 \times 10^6$$
$$3C\ 273: \quad R = 1.2\ \text{pc}, \quad M_H/M_\odot = 6 \times 10^5$$

If the redshifts are not of cosmological origin, and if the objects are at about 10 Mpc distance, as has been suggested (107), then the values will be much reduced, as follows:

$$3C\ 48: \quad R = 0.48\ \text{pc}, \quad M_H/M_\odot = 410$$
$$3C\ 273: \quad R = 0.092\ \text{pc}, \quad M_H/M_\odot = 270$$

Widths of the emission lines.—Emission lines in 3C 48 and 3C 273 are some 20–30 Å wide. Still broader lines have been found in other QSO's; sometimes the observed widths are as great as 100 Å (see 51). To obtain widths as emitted in the rest frame at the source, these should be divided by $(1+z)$. In general the resonance lines Ly-α, C IV λ 1549, and Mg II λ 2798 are the broadest, and the forbidden lines are narrowest. In PKS 1217+02 Lynds (51) found that the forbidden lines of [O III] at λ 4959 and λ 5007 are much narrower than [O III] λ 4363, which has a higher excitation for the upper level of the transition. Such effects suggest stratification in the region producing the spectral lines. The absorption lines, when present, are usually very much narrower than the emission lines, again suggesting stratification, In 3C 191, Stockton's spectrum (214) shows that those lines which are not blended have about the same width as the weaker comparison lines, or about 8 Å between half-intensity points. This sets an upper limit of about 3 Å to their width at the source. The absorption features in PHL 5200, however, are 100–150 Å wide (134a).

The emission linewidths were first assumed to be due to mass motions of the gas, of the order of thousands of km/s (97). These greatly exceed the velocities of escape set by the masses of the hydrogen emission region which are only $\sim$100 km/s. If the gas were free to escape, the time scales would be only 10^3–10^4 years. Therefore the gas should be anchored by the gravitational attraction of a large mass at the center of the gas cloud, and to do this, a mass of about $10^9\ M_\odot$ is required (97).

In one of the physical models that have been proposed for QSO's (see later), Colgate & Cameron (69) suggested that supernova explosions might be occurring with great frequency in the centers of dense star clusters. Then the emission lines would come from shells of gas ejected from the supernovae with characteristic speeds of thousands of km/s; these shells would collide with each other and with quiescent gas in the cluster, and emission lines very much broadened by Doppler shifts would be produced.

Another possible broadening agent is electron scattering (51). That electron scattering might be very important in QSO's was realized by Shklovsky (206). In 3C 273, with the dimensions of the hydrogen-emission region determined for a cosmological distance, the optical depth in electron scattering, τ_e, must be about 10. This holds for a thick spherical shell of radius 1.2 pc. But there is presumed to be a small central object inside this, which is giving the continuum radiation, and this radiation varies. An arbitrarily short pulse of radiation produced in the center would take 10–20 years to escape through such an electron-scattering region, because the velocity of diffusion of the quanta would be c/τ_e. To avoid this difficulty, we should have $\tau_e \leq 1$, and to do this with the physical parameters deduced for 3C 273, the hydrogen-emission region must be spread out over a much larger dimension. Shklovsky suggested it might be in a comparatively thin closed spherical shell of much larger radius.

Schmidt (193) suggested that the hydrogen-emission region might be in the form of filaments with an electron density N_e embedded in a large region of much lower density. Again, the light from the central object would only cross a few such small filaments and the electron scattering could be kept low.

Shklovsky also considered a nonuniform distribution of the line-emitting gas, but rejected this possibility because a nonuniform distribution of gas would not lead to complete absorption of the ultraviolet quanta from the energy source. Even if τ_e is considerably less than 10, there can be considerable broadening of the emission lines by electron scattering; this was considered quantitatively by Burbidge et al. (51).

The widest lines that are seen are Ly-α, C IV λ 1549, and Mg II λ 2798. In the case of permitted resonance lines, quanta are emitted and reabsorbed in sequence until electron scattering out of the line center takes place. The opacity within about $\pm 3\text{Å}$ from the line center is large and not until electron scattering takes the quanta sufficiently off resonance can they escape. Using the computations of electron scattering profiles by Münch they found that, even with small optical depths along a direct linear path through the gas, a width of about 20 Å is obtained for an electron temperature of 10,000°K, $\sim$30 Å for $T_e \approx 30,000°\text{K}$, and $\sim$50 Å for $T_e \approx 100,000°\text{K}$. For really strong lines the widths can be nearly double this.

In the case of forbidden lines the ion which gives rise to the transition does not supply a high opacity near the central frequency. To broaden a forbidden line appreciably, τ_e must be $\sim$1. Again using the results of Münch, they found for $\tau_e = 0.8$ a width of about 40 Å at $T_e \approx 30,000°\text{K}$, and about 60 Å at $T_e = 100,000°\text{K}$. In the cases in which narrow forbidden lines are seen, τ_e must be small.

The early line identifications, including [O II] and [Ne V] in 3C 48, showed that a wide range of ionization potential is encompassed, and Shklovsky (206) pointed out that the line-producing region should be stratified, with [O II], for example, coming from a different layer from that giving rise to [Ne V]. According to Osterbrock & Parker (159), however, a wide range of ionization could result from photoionization by thermal radiation from a very hot supermassive star or by ultraviolet synchrotron radiation, since both have a great abundance of high-energy photons.

However, the great strength of the Mg II λ 2798 emission line and the narrow absorption lines in several QSO's provide more cogent reasons for considering a stratified model, and Burbidge et al. (51) proposed such a model with three distinct zones: I, optically thick to quanta with $h\nu > 13.60$ eV; II, optically thick to quanta with $h\nu > 54.40$ eV; and III, optically thick to quanta with $h\nu > 24.58$ eV.

Zone I is necessary to explain the presence of the Mg II emission. The ionization potential of Mg$^+$ is only 15.03 eV and this ion can exist in appreciable abundance only in an H I region. Zone II will be responsible for most of the emission lines, including C IV, He II, O III, O IV, Ne III, and Ar IV, while the absorption lines, and C II, C III, and O II can arise in Zone III. In Zone III, all Mg will be doubly ionized. Zone I was suggested to lie innermost (to account for the breadth of Mg II λ 2798 and lack of absorption in this line), and Zone III on the outside (to explain narrow absorption lines and a tendency for [O II] λ 3727 to be narrower than other emission lines).

A further reason for placing Zone I on the inside is the absence of [O I] λλ 6300, 6363 in 3C 273. Andrillat & Andrillat (9) did not see these features on their infrared spectra, yet they are commonly seen in the spectra of the nuclei of Seyfert galaxies. If the H I region is innermost, then the electron density here may be sufficient to cause considerable collisional deexcitation of the upper level of the [O I] lines, and these lines will be suppressed.

Woltjer (231b) has constructed a model in which small dense regions ($N_e \sim 10^8$ cm^{-3}) are embedded in a tenuous background region with $N_e \sim 10^6$ cm^{-3}. The ionization would be lower in the dense blobs and H and Mg II lines could be produced here, while C IV and other higher-ionization lines, as well as the forbidden lines, would come from the tenuous region in which $T_e = 5 \times 10^4$ °K. Two different temperatures of the dense regions are considered; $T_e = 2.5 \times 10^4$ °K, for which heavy element abundances lower than normal by a factor of not more than 20 would give the observed line strengths, and $T_e = 1 \times 10^4$ °K, for which normal abundances would give agreement. With this model, Woltjer believes the objects could be at cosmological distances without giving too great a pathlength in electron scattering; it would be sufficiently large sometimes to broaden lines and small enough not to affect the light variations much. Possibly a combination of a zoned model (51) and an inhomogeneous model (231b) may best account for the observations. Some such model is required, whatever the distances of the QSO's are.

An interesting question is whether higher Lyman lines—Ly-β and Ly-γ—will be visible in QSO's with z appreciably greater than 2. The usual analysis of ultraviolet radiation transfer in gaseous nebulae postulates that all the higher Lyman lines will be transformed by repeated absorptions and cascade re-emissions into lines of the Balmer, Paschen, etc., series and Ly-α. Bahcall (13) showed, however, that this will not be the case; there will actually be appreciable emission in Ly-β and Ly-γ due to leakage from the surface layers of the line-producing region.

CONTINUOUS ENERGY DISTRIBUTION
THE OBSERVATIONS

In the optical wavelength region, knowledge of the continuous energy distribution comes from measurements with spectrum scanners for a few objects and U, B, V measures of a considerable number of objects. 3C 273 has been measured by scanner by Oke (156, 157), as have 3C 9, 48, 245, 286, 446, and CTA 102. 3C 9 has been measured by scanner by Field, Solomon & Wampler (86). The U, B, V measures collected in Table I are from Matthews & Sandage (142), Ryle & Sandage (175), Sandage (179), Sandage & Wyndham (188), Sandage & Véron (185), Bolton et al. (34), Bolton & Kinman (33), Sandage (180), and Kinman et al. (126a). Divan (77b) made a spectrophotometric study of 3C 273 with the Chalonge spectrograph, and Wampler (226a) measured the continua of 3C 9, 345, 380, and MSH 14-12*1* with the scanner.

Oke (157a) made scanner measurements of 3C 279 and 446 during a period of rapid variation (see below).

In order to compare scanner measures of objects with different redshifts, the observed absolute-energy distribution has to be shifted back to the wavelength in a system at rest with respect to the observer. The formula for doing this is given by Oke (157); it depends on the cosmological model adopted, and its form for a deceleration parameter $q_0 = +1$ is the same as for local objects moving at relativistic speeds. Oke adopted the $q_0 = +1$ model.

For 3C 273 the observations have been extended into the infrared by Johnson (120) and by Johnson & Low (121) and both 3C 273 and 3C 279 have been measured in the millimeter region by Low (132) and by Epstein (81).

Radio fluxes are given in the various radio-source catalogues: MSH (145–147); 3C (78, 23); Parkes (32, 166, 72, 205) 4C (165, 95).

Form of the Continuous Energy Spectrum

Only for 3C 273 is a spectrum available covering almost the whole observable frequency range, from 85 Mc/s to 10^{15} c/s, with an upper limit to an X-ray flux at 10^{18} c/s; a plot of this has been given by Stein (213). Oke (157) compared his scanner measures with the computed hydrogen spectra at $T_e = 14,000°$ and $160,000°$ and showed that neither fitted the observations; the Balmer discontinuity at $T_e = 14,000°$ was too large (only a very small jump is observed), and the observed continuum rises into the infrared, suggesting that synchrotron radiation is becoming important here [however, see also (77b)]. A similar comparison of the observations with computed fluxes at these two temperatures was made by Hoyle, Burbidge & Sargent (110), with the far-infrared measurement added; the steep rise noted by Oke continues as the wavelength increases. Such a steep rise must represent flux emitted by a nonthermal process, either synchrotron emission or emission by the inverse Compton process. In either case a flux of high-energy electrons must be present; in the first case a magnetic field must be present, and in the second there must be an intense radiation field of low-energy quanta. In other QSO's Oke (157) thought it possible that the continuum could arise from a very hot gas, $T \sim 10^5$.

Pacholczyk & Wisniewski (161a) measured the infrared radiation of 3C 273 and NGC 1068 and found very similar steep increases toward lower frequency for the two objects. Burbidge & Burbidge (50b) and Shklovsky (206a) had earlier pointed out the similarity between QSO's and Seyfert nuclei in the general optical properties.

Matthews & Sandage (142) discussed the form of the energy spectrum in the optical region that would give the observed $U - B$, $B - V$ colors of QSO's, and gave formulae by means of which the colors that would be given by any theoretical spectrum can be computed. They showed that the correct colors would be obtained for energy distributions both with an exponential dependence on frequency and with a power-law dependence of the form $F(\nu) \propto \nu^{-n}$, with n lying in the range 0 to 2.

The U, B, V measures necessarily include the flux put out in any emission lines occurring in the bandpass of the filter together with the pure continuum radiation admitted through each filter. Yet, aside from this complication, the U, B, V measures clearly provide the ingredients of a coarse integral equation from which the original energy distribution $F(\nu)$, very much blurred by the breadth of the U, B, V bandpasses, can be recovered. This is so because the $(B - V)$ and $(U - B)$ colors are related to the first derivatives of the energy distribution function. This was realized by McCrea (144) and by Kardashev & Komberg (123), and they set out to derive $F(\nu)$ from the U, B, V measures. Sandage (180) made a more detailed study using more extensive data.

Correlations between colors and redshifts.—McCrea (144) pointed out that the colors of QSO's are correlated with their redshifts, and Kardashev & Komberg (123) and Barnes (19) independently discussed these correlations. McCrea realized that a relationship between colors and redshift must be due to the intrinsic form of the energy distribution emitted by the objects. When $F(\nu)$ is recovered from the U, B, V measures, the QSO's divide into two distinct groups, one in which $F(\nu)$ is concave to the ν axis or not noticeably curved, and one in which $F(\nu)$ is convex to the ν axis. Of the QSO's with redshifts available at the time of McCrea's work, all those in the "concave" group were found to have $z \leq 0.7$, with an average $z = 0.5$, and all in the "convex" group were found to have $z > 0.8$, with an average $z = 1.0$. The correlation becomes even more striking if one plots $Q = (U - B) - (B - V)$ against z; the scatter is less than in the plots of $(U - B)$ and $(B - V)$ alone.

Kardashev & Komberg (123) arrived independently at the conclusion that the colors give information on the energy distribution being received throughout the U, B, V filters. Sandage (180), with more extensive photometric data, assumed that there is an intrinsic $F(\nu)$ distribution, similar for all QSO's whatever their redshift, and derived its form from the measured $(U - B)$ and $(B - V)$. With an $F(\nu)$ that has some structure in it—changes of slope, maxima and minima—then one samples different parts of such a curve in looking at QSO's with different redshifts through the U, B, V filters with fixed bandpasses. The curve for $F(\nu)$ derived by Sandage is called Sandage's compromise composite (SCC); it is the average from 43 QSO's. It has a maximum near $\lambda_0 = 2800$ Å and a depression near $\lambda_0 = 2100$ Å, features which are not present in the observed spectrum scans by Oke (157).

The form of the SCC flux distribution was suggested by Strittmatter & Burbidge (217) and by Lynds (134) to be due to the inclusion of the emission lines in the bandpasses of the U, B, V filters, the difference from the spectral scans lying in the fact that in the latter, the emission lines are excluded as far as is possible.

One can take a simple form for the basic continuous energy distribution, i.e. either Equation 1 or Equation 2:

$$F(\nu) \propto \nu^{-n} \qquad\qquad\qquad 1.$$

$$F(\nu) \propto e^{-\nu/\nu_0} \qquad\qquad\qquad 2.$$

The first gives a continuum whose slope at a given frequency is independent of redshift z for any n. The second gives a continuum whose slope increases linearly with $(1+z)$. Strittmatter & Burbidge (217) took the form of Equation 1, with $n = 1$, and adopted a set of emission-line equivalent widths based on measures by Oke (157) and eye estimates in a number of QSO's (51). They computed the run of $U-B$, $B-V$, and Q with z and found that the computed curves represented the observed run of these quantities quite well, though the observed points had considerable scatter about the computed curves. The scatter in the plots of $(U-B)$ and $(B-V)$ was much larger than that in the plot of Q, and was deduced to be primarily due to the effect of variations in the slope of the continuum among the QSO's. Since Q is a difference, most of the scatter due to fluctuations in the continuum slope should disappear, and indeed the observed scatter in the plot of Q was found to be less (217). The remaining scatter may be attributed to differences in line strengths among the QSO's. A computation giving results similar to those in (217) has been made by Komberg (128a).

The K correction for QSO's.—For any class of distant objects which may be used for tests of various cosmological theories, such as the normal or radio galaxies, or QSO's, the measured apparent magnitudes have to be corrected for the so-called "K effect" [Humason, Mayall & Sandage (116)], which arises because the standard filters admit different regions of the intrinsic energy curves of objects with different redshifts. Sandage (180) used his derived curve SCC to compute K corrections as a function of z, and tabulated them normalized to $z = 1.0$ which falls in the middle range of the observational data. He pointed out that the magnitude corrections are very small except near $z = 0$, because, except for the effect of the emission lines, the SCC curve is fairly close to $F(\lambda) \propto \lambda^{-1.0}$, though it departs most at the longer wavelengths. If the form were exactly that given above, the correction would be zero at all redshifts.

Polarization of continuum radiation.—Kinman et al. (127) measured a degree of polarization amounting to about 10 per cent in 3C 446, after its rapid increase in brightness which will be described in the next section. The position angle of the polarization was found to have changed markedly between observations one month apart. Kinman and his colleagues (127a) have also recently detected optical polarization in 3C 279 and 3C 345, during periods when both of these were showing rapid variations.

MECHANISMS FOR PRODUCING CONTINUUM RADIATION

Several possible physical mechanisms which may give rise to the continuous radiation in different parts of the spectrum have been suggested: thermal emission from a hot gas, coherent plasma oscillations, synchrotron radiation, Cerenkov radiation, and the inverse Compton process. As just described, in 3C 273, and probably in other QSO's, most of the radiation appears to be emitted by a nonthermal process, though the possibility cannot be excluded that some part of the optical continuum is thermal bremsstrahlung.

The radio properties of the QSO's are very similar to those of the radio galaxies, and it is well established that most of this radiation is due to the synchrotron process. Many authors have considered that the synchrotron mechanism is responsible for the bulk of the radiation all the way from radio to optical frequencies (see e.g. 142). Strong evidence that the optical flux is of synchrotron origin is provided by recent observations of 3C 446 by Kinman et al. (127) which show that a high degree of linear polarization is present. As will be seen when the theories are discussed, for some classes of models difficulties arise if it is supposed that the radio and millimeter flux (in 3C 273) is emitted by the synchrotron process. Therefore Ginzburg & Ozernoy (93) have suggested that coherent oscillations may be responsible for this radiation. Stein (213) discussed Cerenkov radiation as well as the other possibilities, in considering the millimeter-wavelength radiation in 3C 273. Hoyle et al. (110) and Shklovsky (206a) discussed the inverse Compton process.

Given a large flux of low-energy photons, then, in inverse Compton collisions with high-energy electrons or positrons a fraction of these photons can be lifted in energy. Thus some part of the flux observed may originally have been emitted at much lower frequencies and may have been lifted by this process. More discussion of all these mechanisms will be given later in the section on theories.

VARIATIONS IN THE FLUX EMITTED BY QUASI-STELLAR OBJECTS

OPTICAL VARIATIONS

Following the identification of the first QSO, 3C 48, Smith & Hoffleit (210) looked back on old plates of the Harvard plate collection to attempt to determine whether it was variable in light. They concluded that there was no detectable variation within the rather large errors ($\sim 0^m\!.3$) which are present in the old plate material. However, the first observations using accurate photoelectric methods by Matthews & Sandage (142) showed that 3C 48 is varying in optical flux, by about $0^m\!.4$ over ~ 13 months. In addition to this they reported that a variation of the order of $0^m\!.04$ in a period of 15 min in October 1961 had been measured, and they concluded that night to night variations, as well as the variations over periods of months, were real.

Following the discovery of 3C 273, Smith & Hoffleit (211) and Sharov & Efremov (204) attempted to measure the light-curve of this object using the old plate collections going back ~ 70 years at the Harvard and Pulkova Observatories. They found that variations by a factor of ~ 2 over periods of years are seen while shorter-period "flashes" with time scales of months or weeks may be present. Smith & Hoffleit suggested that a characteristic period of ~ 13 years could be detected in the observations, but this result is still in doubt. While many spectra of this object have been taken since 1963, over the past 3 years there has been no evidence for any spectroscopic variations.

With the identification of considerable numbers of QSO's there is the opportunity to look closely into the question of variability. Sandage has shown (178, 186, 182) that variations are present as follows:

3C 2: $B \geq 21^m.1$ September 1954, $\sim 20^m.5$ September 1960, $\sim 19^m.5$ August 1963, and $\sim 19^m.5$ November 1964.

3C 43: $B \sim 19^m.0$ in 1954 and $B = 20^m.5$ in 1965 (based on 2 plates taken 11 years apart).

3C 47: $\Delta B = 0^m.20$ based on observations on 2 nights in 9 months.

3C 48: $\Delta B = 0^m.30$ based on observations on 16 nights over 4 years (178). More recently (182) the object has remained fairly constant.

3C 196: $\Delta B = 0^m.27$ based on observations on 9 nights over 45 months. The object has gradually brightened and there was a sudden increase in December 1963.

3C 216: $\Delta B = 0^m.37$ observed on 3 nights in 14 months.

3C 273: The most recent observations in the last 2 years show that the object has not varied by more than $\Delta B \sim 0^m.4$ though larger variations were detected in the past.

3C 454.3: $B = 15^m.7$ in August 1954, $16^m.75$ in September 1965, and $16^m.55$ in October 1965.

In the case of 3C 2, significant color changes probably occurred. 3C 279 has also varied between the time the Palomar Sky Survey plates were taken, when it was about $16^m.8$, and the time when the first spectra were taken in 1965, when it was about $17^m.8$ (46).

Thus, although the QSO's have only been observed intermittently, variations have frequently been found and it seems probable that optical variability is a common property.

More detailed investigations have been made of 3C 345 and 3C 446. Goldsmith & Kinman (94) observed 3C 345 fairly continuously over a period of ~ 100 days in the interval June–September 1965. They found that the object increased in brightness by $\sim 0^m.4$ in a period of ~ 20 days and then decreased more slowly, but showed smaller variations in time scales less than a week, until early October 1965. A single observation of this object was made later in October by Sandage (180) who found that it had brightened again by ~ 1 mag. Thus in this object it has been established that large variations occur on a time scale of weeks or less. During the period June–September 1965 it was observed spectroscopically (42) and the structure of Mg II λ 2798 was seen to change though there was no detectable change in the total strength of this line relative to the continuum level. Dibai & Yesipov (77) also observed a change in the structure of Mg II λ 2798 in 3C 345 between May 31 and June 1, 1965. Wampler (226a) found from scanner observations [with lower resolution than the observations in (42) and (77)] that changes appear to occur in the red wing of the line alone, that there is no simple correlation between the strength of the line and the brightness of 3C 345 itself, and that the total emission in the line may be variable over about 10 days, but the observations are not conclusive on this.

3C 446 had an apparent magnitude of $\sim 18^m_.4$ in 1964 and 1965, when spectra of it were being obtained (40, 195). However, it brightened by $3^m_.2$ sometime between October 1965 and June 24, 1966, when Sandage (181) observed it to be about $15^m_.2$. In the period July–September 1966 it has been studied in detail by Sandage, Westphal & Strittmatter (187) and by Kinman, Lamla & Wirtanen (127). In about 10 days in late July it dropped in brightness by ~ 2 mag but by early August it was rising steeply again. From the middle of August until late September it was bright, near 16^m, but varied on several occasions by $0^m_.5$ to $0^m_.8$ in time scales of the order of a day. These are the largest short-period variations yet observed in a quasi-stellar object. During July 1966 when the object was very bright Sandage obtained spectra and it was found that the lines appeared to be extremely weak, as compared with the strengths in the spectra taken when the object was much fainter. Sandage et al. (187) showed that this could be explained by supposing that the intrinsic line strengths have remained constant and that this apparent weakness is due entirely to the increased continuum emission corresponding to the increase of brightness of about 3 magnitudes. Cannon & Penston (56a) found a continuation of large and rapid variations in the optical flux in 3C 446.

Kinman et al. (127a) found that 3C 279 has also varied during 1966–67 by large amounts on a short time scale. Oke (157a) made scanner measurements of 3C 279 and 3C 446 during May–July 1966, during the time of rapid variation. The absolute strengths of Mg II λ 2798 in 3C 279 and C IV λ 1549 in 3C 446 remained constant during this time independent of changes in their continua.

Radio Variations

Early in 1965 Sholomitsky (208) announced that CTA 102 was showing rapid cyclic variations at radio frequencies near 1000 Mc/s with a period ~ 100 days. Since this result was announced several groups have attempted to check on the reality of this observation, e.g., Maltby & Moffet (140), who checked their records of observation of this object during or adjacent to the period observed by Sholomitsky and close to his observing frequency. No confirmation of the variations has been obtained, and thus it is not generally accepted at the time of writing. However, also in 1965, Dent (73) announced that he had found a secular variation in 3C 273B at a frequency of 8000 Mc/s such that it was increasing about 17 per cent per year. He also had weaker evidence for variation in 3C 279 and 3C 345. Maltby & Moffet (141) then investigated this object at a number of frequencies and showed that there appeared to be a secular increase in flux down to frequencies of about 970 Mc/s, below which there is no significant variation observed.

Pauliny-Toth & Kellermann (162a) have looked at variations of a number of QSO's at frequencies of 750 and 1360 Mc/s, using observations made in 1962 and 1963, and in 1965 and 1966. They have combined them with other observations made at higher frequencies 2700, 5000, and 15000 Mc/s at the

National Radio Astronomy Observatory and at other observatories to find variations as follows. 3C 279, in which Dent (74) first reported variation, shows very large variations and its flux has been increasing steeply. The variations in 3C 273, found by Dent, have also been confirmed, and 3C 345, 3C 418, and 3C 454.3 have been found to be variable. In addition to this the object 3C 84, which is a source in the nucleus of the Seyfert galaxy NGC 1275, is also shown to vary. This latter variation was first shown to exist by Dent (74) at 8000 Mc/s. Other radio sources which are variable are NRAO 140, 190, and 530, with no optical identification, and 3C 120 which is a radio galaxy. The radio spectra of 3C 273 and 279 have varied during the period of observation.

In 3C 273 Epstein and his collaborators (81, 82) have shown that significant changes in the flux at 3.4 mm are taking place on a time scale of months or less. In general, however, as far as the observations have been carried out, radio variations have not been found to take place on such short time scales as those found for the optical variations. Aller & Haddock (6a), however, found short-period variations in the radio polarization (both polarized flux density and position angle) in 3C 273, 3C 279, and 3C 345.

INTERPRETATION OF VARIATIONS

The condition that a large change in flux takes place in a time τ sets a limit to the size of the region R through the inequality $R \leq c\tau$. This condition can only be relaxed if the matter which is giving rise to the radiation is itself moving at relativistic speed.

This question has been considered by Terrell (221, 222), Williams (227), and Noerdlinger (152) for cases in which the emitting surface is moving non-relativistically. The results by Terrell will be summarized briefly. He considered a fluctuating source consisting of a spherical surface of radius R oscillating in brightness with a period $\tau_0 = 2\pi/\omega_0$ measured in its own reference frame, with all parts fluctuating in phase. He showed that

$$R \leq \frac{2c\tau_0}{\pi} \left(\frac{\overline{L}}{\Delta L} \right) \qquad 3.$$

where $\overline{L}$ is the mean luminosity and ΔL the fluctuation. If we wish to consider fluctuations as rates of change of luminosity, then since for sinusoidal fluctuations $|dL/dt| < \pi \Delta L/\tau_0$,

$$R \leq 2c\overline{L} \left(\left| \frac{dL}{dt} \right| \right)^{-1} \qquad 4.$$

and this expression applies to fluctuations which are not sinusoidal. For an observer not in the reference frame of the source, the observed period is $\tau = \tau_0(1+z)$ so that Equation 4 becomes

$$R \lesssim \frac{2c\overline{L}}{(1+z)} \left(\left| \frac{dL}{dt} \right| \right)^{-1} \qquad 5.$$

For variations in flux to be well established observationally, it is usually the case that $\Delta L \sim \overline{L}$. Under these conditions we see that it is appropriate to use the simple condition $R < c\tau$. Thus the importance of the flux variations is that they set limits to the sizes of the objects which are now in the case of 3C 446 about one light day. The fact that the line-producing region did not change during the period of variations of 3C 446 shows that this region is much larger than the source giving rise to the continuum. In the case of 3C 345, however, changes in structure and possibly line strength in Mg II λ 2798 occurred during a period when the continuum was changing in strength. Thus in this case it may be that the line-producing region and the continuum-producing region are comparable in size, though the uncertainties are greater because the total change in light was smaller in 3C 345 than it was in 3C 446.

In the section discussing theories, it will be seen that there are severe problems posed by the limit $R < c\tau$, if the QSO's are at cosmological distances. For this reason, Rees (168) has proposed that the emitting surface is moving relativistically. In this case the surface will appear to be moving faster than the speed of light and the limitation can be relaxed to the form $R < 2c\tau\gamma$, where $\gamma = [1 - (V^2/c^2)]^{-1/2}$

Model to explain variability of Mg II λ 2789.—Burbidge & Burbidge (42) and Dibai & Yesipov (77) observed variation in the structure and intensity of the line Mg II λ 2798 in 3C 345. Shklovsky (207) has proposed a model to account for this. He suggests a dimension of $R \sim 10^{16}$ cm for a region in which Mg is mostly in the singly ionized state. As long as this is the case, such a region will be optically thick to Mg II λ 2798 radiation. If a powerful flux of relativistic particles impinges on such a condensation, then, since the cross section for excitation of the upper level of Mg II λ 2798 is some orders of magnitude greater than the cross section for second ionization, a large flux of Mg II λ 2798 quanta will first be generated inside the condensation. This flux cannot immediately escape, as long as most of the Mg exists as Mg^+, but an increasing degree of double ionization will follow. The plasma inside the condensation will be intensely heated in less than 10^5 sec, and, once the Mg is in the form Mg^{++}, it will become transparent for the resonance quanta of Mg II λ 2798. A burst of line radiation can thus be emitted by such small regions, followed by fading, the time scale of the consequent variation being $\sim 10^5$ sec.

RADIO PROPERTIES OF QSO'S

The majority of the quasi-stellar objects were first identified through their radio-emitting properties, and at the frequencies at which the radio surveys have been conducted they appear over a wide range of flux levels, as do the radio galaxies. If they are at the distances indicated by their redshifts, then, because the redshifts are large compared with those of the radio galaxies, many of them are emitting at the power levels of the strongest radio emitters, in the range 10^{43}–10^{45} erg/s.

The distribution of radio brightness for most of the objects has not been determined. A number of QSO's show structures similar to those found for radio galaxies (150), i.e., they are double with large separations between the two components. Examples are 3C 47 (196), MSH 14–12*1* (102, 225, 39) and 3C 9 (194). The separations between the two components in these three cases, assuming the objects are at cosmological distances with $q_0 = 1$, are: 3C 47, 207 kpc; MSH 14–12*1*, 133 kpc; 3C 9, 130 kpc. If the objects are local ($d = 10$ Mpc), the corresponding separations are: 3C 47, 3.4 kpc; MSH 14–12*1*, 1.8 kpc; 3C 9, 1.9 kpc.

In many cases also, the QSO's have at least one radio component with a diameter corresponding, at a cosmological distance, to a size comparable to those found for radio galaxies in general (150), i.e., ∼50 kpc. Such sizes, together with the high power levels, mean that the energy content of the sources in relativistic particles and magnetic flux has minimum values ∼10^{60} ergs (48, 49). Many of the QSO's, however, unlike most radio galaxies, have at least one exceedingly small radio component which may, or may not, be the only component (the radio galaxy NGC 1275 has a small central component).

The suggestion that some of the radio sources would have very small angular diameters came first from attempts to interpret the radio spectra. It is well known that the radio sources have spectra of the form $P(\nu) \propto \nu^{\alpha}$, where α is an index with a median value near -0.7 (71). In a fraction of the sources examined, the spectra show a pronounced curvature, getting flatter as one goes to longer wavelengths. These sources have large brightness temperatures. Frequently in these cases a maximum is reached in the spectrum and the flux appears to decrease at longer wavelengths. To explain this property LeRoux (129), Slish (209), and Williams (228) proposed that the curvature of the low-frequency part of the spectrum is probably due to synchrotron self absorption.

From formulae given by Dent & Haddock (75) and Williams (228), if the radio spectrum is known, so that the frequency at which synchrotron self-absorption sets in can be estimated, the angular size can be calculated as a function of the magnetic-field strength. For example, in the cases of the quasi-stellar sources 3C 48, 119, 147, 298, CTA 21, and CTA 102, with assumed values of $B = 10^{-4}$ G, Williams obtained maximum angular diameters of 0.4″, 0.3″, 0.2″, 0.6″, 0.01″, respectively.

Methods which are being used to investigate the structure of the sources of small angular size are:

(*a*) Long base-line interferometry at Jodrell Bank and at Malvern (6, 2, 8, 18). These groups are now working with effective baselines of up to 604,000 wavelengths at a frequency of 1422 Mc/s. The National Radio Astronomy Observatory astronomers at Greenbank, West Virginia, have also carried out similar work (62).

(*b*) The method of lunar occultations (101a–103, 106, 103a).

(*c*) The method of using the irregularities in the interplanetary plasma which cause scintillations of small-diameter sources (104, 65).

Detailed studies have been carried out, particularly for 3C 273 (103, 106), and a considerable number of sources have been studied using long base-line interferometry. We discuss first the results achieved by this method.

The following QSO's have been shown to have sizes $<0.1''$ by the observations at 1422 Mc/s of Barber et al. (18): 3C 279, 3C 345, and 3C 380. These objects all show evidence for large fluxes at high frequencies (75). A number of other QSO's have been studied by the interferometer technique at 408 Mc/s and 3C 286, 147, 48, 273, 287, CTA 21, CTA 102, 3C 119, 3C 138, 3C 279, 3C 345, and 3C 380 all have significant fluxes coming from sizes ≤ 0.5.

The method of lunar occultations is being used to determine positions of sources and structural characteristics by a number of groups, notably by Hazard, and also by von Hoerner; they have worked extensively on 3C 273. Hazard, Mackey & Shimmins (103) first showed that this object was double with a separation of 19.6 and that the two components had very different spectra. Component A had a value of α of about -0.9, while 3C 273B which is associated with the optical object and agrees in position with it to about 0.1 had a spectral index with $\alpha \sim 0$. Both components are elongated along the line joining them. It appears now from the work of von Hoerner that the structure of both components depends strongly on frequency. B is a single strong component about $2''$ wide at 2695 Mc/s; at 735 Mc/s it shows a halo about $6''$ long and a core $\leq 1''$ which emits 30 per cent of the flux. The core is less pronounced at 405 Mc/s, but at the lower frequencies B shows a deep minimum at the center. A seems to have a central dip at 2695 Mc; at 735 Mc it has a halo about $10''$ long and a core $<1''$ (with 30 per cent of flux) at the lower frequencies. A has a length at half-power points of about $4''$, but shows a faint, very flat extension about $23''$ long. According to this more recent work the spectrum of A is a straight line with $\alpha = -0.68 \pm 0.08$. The spectrum of B is a straight line only above 400 Mc/s, with $\alpha = +0.25 \pm 0.08$; it cuts off toward lower frequencies. Each component is then divided into an outer part and a central part 4.8 long. The central spectrum of A is still straight with $\alpha = -0.82 \pm 0.22$ while the low-frequency cutoff for the center of B is very steep. Both outer halos are very similar and give $\alpha = -0.45 \pm 0.20$. Such a complex structure has so far been found only in 3C 273, but it may be expected to be generally present.

Method (c) has been used to show that the angular sizes of the QSO's 3C 48, 119, 138, and 147 have components that lie in the size range 0.3 to 0.8 (104). Scintillation measurements have now been carried out at Arecibo (66, 67) for a considerable number of sources. Also, according to Bolton (25) a considerable amount of scintillation data has been obtained by Ekers with the 210-ft Parkes radio telescope. Cohen & Gundermann (65a) have found a component in 3C 279 with a size $<0.005''$.

Kellermann (124) showed that the radio spectra of QSO's can be explained if relativistic electrons which have an initial energy distribution $N(E) \propto E^{-1.5}$ are injected into a magnetic field in a series of recurring bursts. At low frequencies energy losses are not important and the spectral index α has its initial value of -0.25. At intermediate frequencies energy losses are balanced by the injection of new particles, and the spectral index reaches an equilibrium value of -0.75. At high frequencies radiation losses steepen the spectrum, and the limiting index becomes -1.33, in good agreement with the cutoff near -1.3 in the observed spectral index distribution.

Heeschen (103b) plotted intrinsic radio luminosity against surface brightness at 1400 Mc/s, for radio galaxies and QSO's on the assumption that the latter are at cosmological distances. The QSO's form a continuation of the main radio-galaxy sequence, while the peculiar objects NGC 1068 and M82 and the cores of core-halo objects lie off this sequence. If the QSO's are not at cosmological distances, they move down to the level of NGC 1068 and the core-sources, but spread out in the direction of high surface brightness. The continuity of the QSO's with radio galaxies has been used as an argument in favor of their being at cosmological distances.

PROPER MOTIONS

If the QSO's were really local objects, within (say) a few hundred parsecs of the Sun, as had to be considered in the original discussion by Greenstein & Schmidt (97) of whether they could be collapsed stellar objects with large gravitational redshifts, then proper motions might have been expected to be detectable.

Jeffreys (119) made a study of the proper motion of 3C 273, using 14 plates covering the period 1887–1963. The plate material was not homogeneous, but an attempt was made to reduce the systematic errors in such a determination by using a large number of reference stars. The absolute proper motion was found to be:

$$\mu_\alpha = +\ 0''.0009 \pm 0''.0025/\text{year}$$
$$\mu_\delta = -\ 0''.0012 \pm 0''.0025/\text{year}$$

Because of the position of 3C 273, in a direction approximately at right angles to the direction of the Sun's peculiar motion, Jeffreys concluded that the object was likely to be more distant than 2000 pc (unless it were traveling parallel to the Sun at the same speed). Its high galactic latitude would then mean that, if it belonged to our Galaxy, it might well have a different galactocentric velocity from that of the Sun, which should in turn lead to a detectable proper motion unless it were at an even greater distance, effectively outside the Galaxy.

Luyten & Smith (133) made a study of 12 QSO's, and essentially the same result was found for all 12 objects: the relative proper motions were of the same order of magnitude as the estimated mean errors. The 12 objects were averaged in 3 groups, and the mean proper motions were found to be the inverses of the expected parallactic motions of the comparison stars, as would be expected if the QSO's have zero absolute proper motion. Luyten & Smith concluded that the QSO's, as effectively stationary objects with stellar images, are ideal objects for determining the corrections from relative to absolute proper motions.

SPATIAL DISTRIBUTION

With only about 150 QSO's positively identified it is too early to gain much information by looking at the distribution of these objects in the sky. The work of the Cambridge radio astronomers has shown that there is a fairly high degree of isotropy present as far as the radio sources as a whole are concerned. However, in the large-scale surveys complete identifications into the two nongalactic categories of quasi-stellar objects and radio galaxies have not been made, and it is premature to claim (131) that the results obtained so far suggest that the QSO's are at cosmological distances. It is of some interest to see whether the QSO's are distributed in such a way as to suggest that they may have some physical relationship with other extragalactic systems. The evidence bearing on this comes from different directions and is very preliminary.

The relation between quasi-stellar objects and clusters of galaxies.—If the objects are at cosmological distances they are $\sim 3^m$ brighter than the brightest galaxies in clusters. In addition, there will be an appreciable K correction for the galaxies which will not apply for the QSO's and this, for a redshift of $z = 0.5$, amounts to nearly 2^m (184). Consequently the QSO's will appear approximately 5 mag brighter than the brightest galaxies in clusters in which they might lie for redshifts near 0.5. The majority of identifications of QSO's have been made with the 48-inch Palomar Sky Survey plates and the normal limit of these plates is $20^m_{\cdot}0$ (visual). A preliminary conclusion is, therefore, that clusters would only be identified on these plates if the QSO's were brighter than 15^m. Only 3C 273 is brighter than this and certainly no cluster is associated with it. However, it is known that there is only a weak correlation of redshift with apparent magnitude, and there are 23 QSO's now known besides 3C 273 with redshifts $z < 0.5$, and of these 7 have magnitudes brighter than 16^m and 13 have magnitudes brighter than 17^m. (Also, 3C 232 has a redshift $z > 0.5$, but is brighter than 16^m.) In these cases the K correction for the galaxies, if they were present, would be less than $1^m_{\cdot}9$, and we might therefore expect that the difference between the apparent brightness of the QSO and cluster galaxies would often be less than 4 mag, so that galaxies would be detectable. However, none has been seen.

In a number of cases plates reaching 1–2 mag fainter than Sky Survey plates have been taken for fields incorporating the QSO's, and inspection of those plates shows no galaxies. In the case of 3C 48 Sandage & Miller (184) have used a special emulsion with which they were able to reach apparent mag $24^m_{\cdot}5$ (blue), and no galaxies were found.

Are quasi-stellar objects associated with individual bright galaxies?—If it could be established that QSO's are associated with a particular kind of galaxy, then it would be strongly presumed that they have some genetic relationship with these galaxies. Evidence bearing on this possibility has recently been presented by Arp (10, 11). He concluded that galaxies in his Atlas of

Peculiar Galaxies (12) lie closer to the radio sources than would be expected if the radio sources were distributed at random with respect to the peculiar galaxies. He noticed that radio sources with similar flux densities tend to form pairs separated by from 2° to 6° on the sky and that there was a tendency for a certain class of peculiar galaxy to fall approximately on the line joining the pair. These peculiar galaxies are often elliptical galaxies which he believes show evidence for structures that may have been ejected from them. In some cases there is evidence that more than 2 radio sources are associated with a given peculiar galaxy.

The double nature of radio sources is well established. In fact Arp's hypothesis is in a sense only an extension of the previously well-known observation that the bulk of the radio sources are double, but Arp has attempted to establish the existence of double sources with much wider angular separations than accepted before. However, in Arp's case some of the objects have previously been identified with QSO's and apparently more distant radio galaxies.

Unless all of the evidence presented by Arp is due to chance coincidence, it must be concluded that the QSO's do not lie at cosmological distances. At present most workers are somewhat skeptical of the data, and intensive work on this aspect of the QSO's is under way (e.g. 106a).

Similarities in pairs of radio sources.—There is some evidence concerning the similarity of pairs of radio sources. Moffet (149) has shown that the radio sources 3C 343 and 3C 343.1 have very similar power levels and very similar radio spectra, though they are separated by 29′. They are both exceedingly small radio sources with angular sizes $\leq 10''$ so that the ratio of separation to size exceeds 200. No optical identifications have been made for these sources, but the curved radio spectra strongly suggest that they are quasi-stellar objects. Moffet has estimated that there is a probability $\sim 10^{-6}$ that the objects are not physically connected. If they are at cosmological distances their separation in space would be so large that physical association is quite unreasonable. If they are physically connected, one might argue that the objects therefore have a much smaller separation (at a distance of 10 Mpc this would be ~ 90 kpc) and both sources might have been ejected from a local QSO, or might be a pair of identical QSO's which have been ejected from a galaxy. However, the case of the pair of radio sources forming 3C 33 [Moffet (148)] should also be mentioned. Here a similar situation prevails. The two sources are small and very similar and the ratio of separation to size is ~ 16. In this case an optical galaxy is identified as the object giving rise to the pair of radio sources, and if this identification is correct the pair lies at a modest distance.

There is only one QSO known so far lying very close in the plane of the sky to a bright galaxy with some optical peculiarities. This is 3C 275.1; it lies very close to NGC 4651, so close that the optical identification was originally thought to be with NGC 4651. The galaxy has a faint tail or jet extending approximately in the direction of the QSO.

CORRELATIONS AND STATISTICS

REDSHIFT–APPARENT MAGNITUDE RELATION

If the quasi-stellar objects are at cosmological distances, the m-log z relation may be used to investigate cosmological models. Sandage (179) made a plot of the relation between apparent magnitude and redshift for the 10 QSO's whose redshifts were known at that time, alongside a similar plot for the radio galaxies, and the QSO's already showed considerable scatter. No correction for K effect was applied to the magnitudes of the QSO's; as already discussed, this correction is very small for QSO's unless their continua have slopes differing considerably from that given by $F(\nu) \propto \nu^{-1}$ (180).

An alternative way of correlating optical luminosity with redshift is that adopted by Schmidt (194). For a chosen cosmological model (evolutionary, with $q_0 = +1$) he calculated the monochromatic flux from 9 QSO's at an emitted frequency of 10^{15} c/s, i.e. at a rest wavelength of 3000 Å, and examined whether these fluxes were approximately the same for all the objects. Since spectral scans covering the whole wavelength range, from near 3000 Å to 9000 Å, did not exist, he used the UBV photometry and found that the fluxes of the 9 objects were consistent with an evolutionary cosmology with $q_0 = +1$, with a scatter whose extreme range is a factor 20.

As new redshifts have become available, the scatter in the apparent magnitude–redshift relation has become larger instead of smaller (108). A plot of the present data will be found in (50). What little correlation there is shows a scatter of a similar order of magnitude to the total span of the relation. Indeed, with the luminosity variations shown by individual QSO's (more than 3 mag or a factor of nearly 20 in the case of 3C 446), such scatter is not surprising. The plot represents, in fact, the quality of the intrinsic luminosity function of the QSO's, rather than a meaningful distance–redshift relation.

REDSHIFT–RADIO–FLUX CORRELATION

Hoyle & Burbidge (108) plotted the logarithm of the radio flux (or radio apparent magnitude) at 178 Mc/s against $(1 + \log z)$ for the QSO's in the 3CR catalogue and this showed even more scatter than the similar plot of redshifts against optical apparent magnitudes. There was no sign of any correlation, and this implies that the range in apparent radio fluxes is not determined by a range in distances but entirely by the intrinsic spread in the radio properties of the QSO's.

Bolton (25) suggested that such a plot made with radio measures at the comparatively low frequency of 178 Mc/s, contains objects with a variety of radio spectra, and he plotted $(1 + \log z)$ against radio magnitude at 1410 Mc, for QSO's in the Parkes catalogue. While such a plot shows an equally large overall scatter, one can make a distinct separation into two groups if one considers objects with flat or relatively flat radio spectra as one group and objects with steep radio spectra as another group. The objects with steep radio spectra show no positive correlation, and Bolton suggested that this is simply a reflection of a very large dispersion in the relevant intrinsic radio luminosities. The QSO's with flat radio spectra may all undergo synchrotron self-absorption and the radio emission may be emanating from a very small volume; Bolton suggested that such objects might have less intrinsic spread in their high-frequency radio radiation, and, while there is no very distinct relationship, the scatter for the flat-spectrum sources is considerably lower than for the others.

LOG N–LOG S CURVE

For a uniform distribution of sources of radiation in Euclidean space, with a luminosity distribution that is independent of distance, a plot of log N against log S (N = number of sources brighter than flux S) should have a slope of $-3/2$. For large redshifts this will be modified by (a) the K correction (but this is small), (b) the fact that the redshift reduces the brightness over and above the inverse square law, and (c) an increased density of sources in the past in an evolutionary universe.

In the steady-state universe this last effect is not present, but in the evolutionary models the second effect overwhelms the third. Thus cosmological effects will tend to reduce the slope of the log N–log S plot.

In the early work on the source counts of all radio sources by Ryle & Scheuer (176), the slope of the log N–log S curve was ~ -3, much steeper than any predicted values. Next a survey by Mills, Slee & Hill (145) gave a slope of ~ -1.8. The next survey was made in 1959 (78) and gave a slope of ~ -2. A further survey was then carried out with greatly increased precision by the Cambridge group (202) and gave a slope of -1.8. Most recently a survey has been carried out by Bolton, Gardner & Mackey (32) and this gives a slope of -1.85. Thus there is now good evidence from several independent groups that the slope of the log N–log S curve for all sources is near -1.8.

Since the cosmological effects all tend to flatten the curve to values below -1.5, it is obvious that other effects must be present to explain the observed slope. Either an excess of faint sources, or a deficit of bright sources is required. If an excess of faint sources is present, this could be caused by the intrinsic brightness of a source being a function of its time of formation in an evolving universe, and the effect has been used as a strong argument against the steady-state cosmological theory. If the slope were due to a deficit of bright sources, this could be explained by postulating a local irregularity in distribution.

These investigations were carried out at a time when the majority of the radio sources remained unidentified, though they were largely thought to be associated with galaxies. Since a considerable fraction of the 3CR sources are now identified, Véron has plotted the log N–log S curve separately for radio galaxies and QSO's (224), and has found that for the radio galaxies the slope is -1.5, while for QSO's it is about -2.2. The slope for all 296 sources in the revised 3C taken together was made earlier by Ryle & Neville (174) and gives a slope of -1.85. The value of -1.5 obtained for the galaxies is presumably due to the fact that the size of the region over which the 3C survey sources have been found is small enough so that the Euclidean space approximation holds. The greatest redshift known so far for a radio galaxy is that for 3C 295, $z = 0.46$, and the majority of the radio galaxies so far studied spectroscopically have $z \leq 0.2$. Thus, in this survey at least, the departure from the $-3/2$ law arises from the QSO's. There are two possible explanations for this. If the QSO's are cosmological, then we must attribute this steeper slope to evolutionary effects in an evolving universe, as has been done by Longair (131). Alternatively, if they are local, this slope must be attributed to the local conditions under which these objects were ejected from galaxies.

A discussion of effects to be expected in the log N–log S plot for much fainter sources on the cosmological and the local hypotheses for QSO's was given in (50).

Another type of argument to explain the observations and preserve the steady-state theory has been proposed (197, 198, 201). This involves the idea that another class of intrinsically very faint objects—QSO's with zero redshift, lying at characteristic distances of ~ 100 pc—is present. There is no direct evidence, however, for the existence of such objects.

The assumption underlying the studies of the log N–log S curve for cosmological purposes is that the slope is to be accounted for by a distance-volume effect. Hoyle & Burbidge (108) tested this for a sample of about 30 QSO's from the 3C catalogue for which redshifts were known, and showed that the log N–log S plot for these objects has a slope near to -1.5, not quite as steep as the value of -1.8 obtained from the counting of all radio sources and not as steep as the value of about -2.2 obtained by Véron for all QSO's in the 3C catalogue. The particular value of the slope, depending at the bright end on very few points, is not significant. What does appear to be significant, however, is that even this small sample of QSO's gives a log N–log S slope near -1.5. The distance-volume interpretation requires that the objects with smaller S are at greater distances; and if redshifts are cosmological in origin, small S must be correlated with large z. From the redshift–apparent radio magnitude plot for these objects it is obvious that no such correlation exists. Thus if it is assumed that the distance-volume interpretation of the log N–log S relation holds, it must be concluded that the redshifts have nothing to do with distances.

This investigation was criticized by Longair (131), Sciama & Rees (199), and Roeder & Mitchell (172), who thought Hoyle & Burbidge had concluded unconditionally that redshift had nothing to do with distance. However, the conclusion stated above can be inverted to the form: If the redshifts are related to distance in the usual cosmological sense, then the distance-volume interpretation of $NS^{3/2} \approx$ constant must be abandoned for the sources in this particular sample. Consider the sources in a shell between distances r and $r+$

dr. Provided these sources have an intrinsic scatter in their radio emission, they will exhibit a log N–log S curve. If all such shells have the same log N–log S curve, then summation of all shells will give a curve related to intrinsic scatter, not to distance. This is the point made by Bolton (25). In order that all shells give the same log N–log S curve, however, it is necessary for the average emission to vary in a special way from one shell to another. This indeed is the suggestion of Longair and of Roeder & Mitchell. It requires the average emission to be a function of r and hence of the epoch. Such an interpretation is evidently in disagreement with the strict steady-state theory (199).

Finally, the status of the radio sources which remain unidentified should be mentioned, since they affect the statistics. In the 3C, many sources with good radio positions, in unobscured fields, lie in apparently empty fields, to the limits of the Palomar Sky Survey. Véron (224) believed that these are probably QSO's, while Bolton (25) believed that they are probably radio galaxies. At present, it is the reviewer's opinion that Bolton's view is the more likely to be correct. Sandage is carrying out a search of the empty fields with the 200-inch Palomar telescope, which goes to a limit some 3 magnitudes fainter than the Sky Survey.

THE NATURE OF THE QUASI-STELLAR OBJECTS

When large redshifts were first found in QSO spectra, and shown (97) to be probably Doppler shifts, it was assumed by nearly everyone that they were due to the expansion of the universe and the QSO's were consequently at very large distances. Terrell (221), however, suggested that QSO's are entities that have been ejected from the nucleus of our Galaxy in an explosive event like those giving rise to radio galaxies. Hoyle & Burbidge (107) extended this to the possibility that they might have been ejected from the strong nearby radio galaxy NGC 5128. Arp (10, 11) further extended it to suggest that pairs or groups of radio sources, including QSO's, are ejected from a class of peculiar galaxies. There are difficulties in interpreting the observations in terms of all of these possibilities, and these problems will be briefly discussed.

Cosmological Hypothesis

The difficulties here stem mainly from the observed optical and radio variations and the consequent small dimensions set by the short time scales and the condition $R \leq c\tau$, coupled with the fact that, at cosmological distances, the objects are emitting very large amounts of radiation (40 times the optical radiation of the brightest galaxies, on average, and, including the microwave and infrared radiation in 3C 273, some 10^{47} erg/s).

The first indications that there were difficulties with simple homogeneous models came from the discovery of the time variations of 3C 273B at 8000 Mc/s by Dent (73) and interpretation of the data. The frequency at which synchrotron self-absorption sets in is related to the size of the object, the flux radiated at that frequency, and the magnetic-field strength. For the parameters thought to be appropriate for 3C 273B at that time, the time scale for variation came out to be $\tau \geq 23$ years, for a frequency $\nu = 400$ Mc/s and magnetic field $B \geq 10^{-5}$G. In his original calculation of this result Dent made a numerical error which was corrected by Field (85). If the radio emission at $\nu = 400$ Mc/s to which the spectrum extends with no cutoff arises from the same region as that giving the secular variations at 8000 Mc/s, then the value $\tau \geq 23$ years was just compatible with Dent's result that the flux increased by 40 per cent in 2.5 years. Otherwise it must be concluded either that the synchrotron mechanism is not operating, or else that the object is not at a cosmological distance. If $\nu = 8000$ Mc/s is used in the self-absorption equation, the relevant dimension is 0.5 lyears for 3C 273B at a cosmological distance. Rees & Sciama (169) assumed that the source 3C 273 contains an exceedingly small component with a dimension $\sim 10^{-3}$ sec of arc buried in a much larger source with a dimension $\sim 0.''5$. Then for a large enough magnetic field in the central source, $B \geq 1$ G, they were able to account for the variable flux at 8000 Mc/s. [See also (170) for an earlier model proposed for CTA 102.]

Hoyle & Burbidge (107) considered a more general class of models in which the magnetic field in the object is not assumed constant and the radio spectrum is controlled by variations in the magnetic field. It can then be shown that the flat form of the spectrum observed in 3C 273B can be obtained, and that it continues down to 200 Mc without being self absorbed. This type of model leads to a situation in which different shells are contributing to the radio flux in different energy ranges. Hoyle & Burbidge concluded that such a model for 3C 273B could satisfy the requirement that the process operating be the synchrotron process and that it could be at its cosmological distance.

Ginzburg & Ozernoy (93) concluded that under certain conditions, provided that there is equipartition between particle and magnetic energies, in an inhomogeneous model more energy is required to give the observed flux. They suggested coherent plasma oscillations. For the QSO's in general it appears that the very small source sizes require that if the synchrotron mechanism is operating, the conditions are actually very far from the equipartition condition (229).

Stein (213) considered the varying microwave radiation in 3C 273, and showed that magnetic fields near 10^5 G must be invoked for plasma oscillations to be responsible. A consistent model producing this radiation by the synchrotron mechanism, at a cosmological distance, does not seem possible.

A further problem caused by the limit $R < c\tau$ arises through the very large radiation density within the small volume and the consequent competition between the synchrotron mechanism and the inverse Compton process as the dominating source of energy loss of the high-energy particles. Hoyle et al. (110) showed that, in 3C 273, for the synchrotron mechanism to dominate, $B \geq 15$ G, while in 3C 446, with a much shorter variation period and consequently smaller dimension, $B \geq 400$ G. The electron lifetimes would then be very short so that a large number of co-phased sources of injection or acceleration of charged particles throughout the object would be necessary.

Woltjer (231a) considered the effect of a nonisotropic radiation field. The Compton process becomes less important for close alignment of the light beam with the magnetic field. For an assumed set of conditions, Woltjer showed that the synchrotron process can dominate over the inverse Compton effect for values of the magnetic field about an order of magnitude smaller than those given above. The magnetic field has to be radial, and the electrons must have very small pitch angles.

Another kind of model in which the flux is emitted from coherent blobs of matter moving at relativistic speed was suggested by Hoyle & Burbidge (109).

As already mentioned, Rees (168) pointed out that a relativistically expanding cloud can increase its apparent angular size at γ_V times the rate given by the usual nonrelativistic formula, where

$$\gamma_V = \left(1 - \frac{V^2}{c^2}\right)^{-1/2}$$

V being the expansion velocity, which relaxes the condition $R < c\tau$. This model was suggested to explain a steep rise in radio emission; probably a model could be constructed involving the acceleration of emitting material that gave a similar result for optical radiation. However, the fluctuation data involve decreases of light as well as increases, and the situation for a fall in apparent magnitude is not so clear in a simple radially expanding model.

It has been suggested (231a) that since at least one galaxy (NGC 1275) is known to show variations in high-frequency radio flux, similar to the variations seen in a number of QSO's, this is evidence supporting the view that the QSO's are at cosmological distances. However, the difficulties encountered with the cosmological models of the QSO's stem, as just shown, from the very high densities of radiation which, in turn, arise because of the very great distances of the objects. No difficulties are encountered if the objects are closer by, and none are encountered in the case of NGC 1275 simply because it lies at a distance of only about 50 Mpc.

Hypothesis That QSO's Were Ejected from Our Galaxy or a Nearby Galaxy

Terrell (221, 222) originated this idea, suggesting our Galaxy as the seat of origin. Hoyle & Burbidge developed it, suggesting NGC 5128. Unless a nearby galaxy is chosen, and unless the explosion occurred long enough ago for the objects to have passed the observer, some objects would be seen approaching the observer, and consequently they would have blueshifted spectra. No blueshifts have been observed. Also, there would not be an isotropic distribution of objects. Preliminary evidence (218) does suggest that there may be an anisotropic distribution of redshifts on the celestial sphere, but more work is needed in this field.

The main objections to this form of the local hypothesis are twofold: (a) no mechanism leading to the ejection of coherent blobs of matter moving at relativistic speeds has been suggested; and (b) the total energy release needed in an event leading to such ejection would be very large.

Regarding (b), the kinetic energies of the relativistically moving objects depend on the masses. Setti & Woltjer (203) estimated the total masses by supposing that the emission linewidths were due to large random motions of gas and that there must be a central mass large enough to stabilize the object gravitationally. From this argument they obtained masses of the order of 10^7-10^8 $M_\odot$ for 3C 273 and 3C 48 on the local hypothesis. However, their argument is vitiated if the line broadening is not due to mass motions, and, as discussed in the section on line spectra, electron scattering may well be the dominant line-broadening mechanism.

Another method of estimating the mass of a local QSO is to suppose that it does not conserve its mass and that gas is continuously escaping from it. Bahcall, Peterson & Schmidt (15) considered the absorption lines in some QSO's, particularly PKS 1116+12, as evidence for such mass loss. Estimating lifetimes of 10^8 years for QSO's as local objects, and supposing the average absorption-line phase lasts 1/10 of this, with continuous ejection at 17,000 km/s in PKS 1116+12, the total mass of gas which has escaped was found to be $\sim10^9 M_\odot$. Counter arguments can clearly be raised in which the various assumptions underlying this calculation are questioned; until a definitive model accounting for the absorption lines is produced, the question is controversial.

HYPOTHESIS THAT QSO's WERE EJECTED FROM MANY GALAXIES

Arp (10, 11) proposed that QSO's (and other radio sources) are ejected from a class of peculiar galaxies. The main problem here is that no QSO's with blueshifted spectra have been seen. It has been shown that, for local objects moving at relativistic speeds, the observer should see a number of blueshifted objects that is $(1+z)^4$ times the number of redshifted ones, where z is the numerical value of the maximum observed shift. For a detailed discussion of the distribution among randomly moving objects one may turn to (50), which gives a summary of work by a number of authors (216, 83, 153, 235).

Selection effects which might modify this result by discriminating against the detection of QSO's with blueshifts have been considered (55), e.g. the relative weakness of emission lines in the infrared which could be blueshifted into the visible region, and the rising intensity of the continuous spectrum in the infrared, but the factor $(1+z)^4$ is a large one and it seems likely that some blueshifts with $z < 0.5$ should have been seen if they occurred in nature.

Partly to avoid this difficulty, Arp suggested that the redshifts are not of Doppler origin, but no new physical possibility has been proposed.

Concluding Remarks

It is clear that there are difficulties with all three hypotheses, and the problem reduces to that of producing a complete model which will satisfactorily account for all the observations. For the cosmological hypothesis, many limitations on possible models are set by the observed variability; for the local hypothesis, a mechanism for ejecting coherent blobs and providing a large energy source is needed; for the more extended local hypothesis, a new physical cause of redshifts is needed.

QUASI-STELLAR OBJECTS AS PROBES OF THE INTERGALACTIC MEDIUM

Mg II absorption.—Shklovsky (206) pointed out that intergalactic Mg^+ should produce absorption, corresponding to the resonance doublet of Mg II at λ 2798, to the blue of the position of this emission line in QSO's. Since no absorption is detectable, he concluded that the density of $Mg^+ < 8 \times 10^{-13}$ cm^{-3}, and an upper limit $n = 2 \times 10^{-7}$ cm^{-3} is set to the intergalactic hydrogen density.

Lyman-α absorption.—Immediately following the discovery that 3C 9 has Ly-α shifted to a wavelength of 3666 Å, Scheuer (190) pointed out that if the spectrum showed a continuum below this, either the mean density of neutral atomic hydrogen is exceedingly low, or else the ionization is nearly complete. He also did not exclude the third possibility, that the object is comparatively nearby, in which case no appreciable absorption is to be expected. A more detailed investigation using Schmidt's observational material on 3C 9 was made by Gunn & Peterson (99). They estimated a depression of 40 per cent in the continuum, giving a number density of neutral atomic hydrogen of $n = 6 \times 10^{-11}$ cm^{-3}, or a density $\rho = 1 \times 10^{-34}$ g/cm^3. However, since the first spectra of 3C 9 were obtained, a considerable number of QSO's with Ly-α in the photographic region have been observed and scanner observations of 3C 9 have been made by Oke & Wampler. The observers are now agreed that there is little evidence for any significant depression to the blue of Ly-α. Thus the upper limit to the density is somewhat lower than the value above. As already described, absorption lines seen in PKS 1116+12 (15) may be intergalactic, produced by gas in a cloud or cluster of galaxies in the pathlength traversed by the light.

Bahcall & Salpeter (16, 17), Rees & Sciama (171), and Bahcall et al. (15) considered more generally the possibility that resonance lines due to intergalactic absorption by various ions of the more abundant elements might be detectable in the spectra of distant QSO's. As pointed out in the section on absorption lines in the spectra of QSO's, occupancy of the upper fine-structure levels in the ground states of various ions seen proves in most cases that the absorption lines are intrinsic to the QSO's (see 17).

Sciama & Rees (200) attempted to interpret some features in the spectral region of 3C 9 below 3300 Å as being due to Ly-α absorption and an absorption line of N V with a redshift $z = 1.62$, caused by a hot intergalactic cloud. These identifications were made, however, from a reproduction of the spectral intensity in this region (86) without taking into account noise in the observations near the atmospheric cutoff.

21-cm absorption.—Koehler & Robinson (128) reported the detection of 21-cm absorption in 3C 273 at a wavelength corresponding to $z = 0.0037$. This is almost exactly the mean redshift of the galaxies in the Virgo cluster and thus they concluded that this absorption is caused by a neutral hydrogen cloud associated with the Virgo cluster, 10 Mpc distant. However, the observation is marginal, and so far it has not been confirmed by other radio astronomy groups.

Molecular H.—Bahcall & Salpeter (16) pointed out that limits on intergalactic H_2 may be set by considering the absorption produced in the Lyman band shortward of 1108 Å. Again objects with $z \geq 2$ are required. This test has been attempted using 3C 9 by Field et al. (86), who detected no absorption. This led to an intergalactic density of $H_2 < \sim 10^{-32}$ g/cm^3.

Thomson scattering.—Bahcall & Salpeter (16) discussed scattering by electrons in an ionized intergalactic medium, and derived formulae that depend on the cosmological model. For $z \sim 2$ the effect is not yet important, but for large z it will be significant.

Thus all the observations made so far show no good firm evidence for any intergalactic absorption or scattering. Most investigators have used this result to set very low limits to the density of neutral gas, and have then concluded that the bulk of the mass-energy in the universe (assumed by most to be near 10^{-29} g cm^{-3}) is present in the form of ionized gas, mainly hydrogen. An alternative interpretation would be that the QSO's are local rather than cosmological. Or thirdly, most of the intergalactic matter might be in the form of stars, uniformly distributed, or condensed into small clusters or low-luminosity galaxies, or in the form of solid matter, or even in the form of neutrinos.

THEORIES FOR THE ENERGY REQUIREMENTS OF THE QUASI-STELLAR OBJECTS

From the radio spectra of the QSO's, the synchrotron mechanism is most likely responsible for the emission, as is the case for the radio galaxies. If the QSO's are at cosmological distances, then the minimum total energies which are required to give rise to the radio emission are of the same order of magnitude as those for the radio galaxies, i.e. $\sim 10^6$ ergs (48, 49). The total energy released may be above this minimum value, in the range 10^{62}-10^{63} ergs, largely in the form of relativistic particles. In the case of the radio sources of very small size associated with QSO's, the minimum total energy required can be much smaller since, for a given synchrotron flux, this total energy is proportional to $R^{-6/7}$ where R is the dimension of the system, so that it might be as low as 10^{58} ergs. However, there are other arguments in this case to suggest that this minimum total energy condition cannot be fulfilled, and that the total energy, largely in the form of particles, must be $\sim10^{60}$-10^{61} ergs confined in a dimension of a few parsecs or less. All of these arguments are based on the assumption that the magnetic field is fairly homogeneous. If one considers a nonhomogeneous model (107), somewhat lower total energies may be feasible.

If the QSO's are comparatively nearby, then the minimum total energy in the relativistic particles and magnetic fields in a given object is much reduced—to perhaps 10^{55}-10^{56} ergs in an individual source. However, a very large amount of kinetic energy must then be contained in the relativistically moving QSO. For a mass $\sim10^4$ $M_\odot$ and $v = 0.6$ c, for example, the total energy $\approx 10^{58}$ ergs, and a number of QSO's must be ejected in a galactic explosion. Thus in any case very large energy releases are required.

As for other wavelength regions, 3C 273 at the cosmological distance emits $\sim2\times10^{47}$ erg/s, mostly in the range 10^{11}-10^{13} c/s. At a "local" distance of ~10 Mpc, a flux less by a factor of 10^4-10^6 is needed. If the QSO's are emitting optical, infrared, and millimeter radiation by the synchrotron process, then in the most extreme situation in which the magnetic fields are as large as 100 G (206), the total energy present in the electrons must be $\sim2\times 10^{52}$ ergs, and this must be renewed every 10^5 sec. Thus, if the objects last for $\sim10^6$ years, the total energy release is $\sim10^{61}$ ergs. If the inverse Compton process is operating (and it is not clear whether such a model can be devised), two components are required: an intense source of low-frequency photons and a supply of high-energy electrons. Since in such a model the photons must be raised many octaves in frequency, most of the energy must reside in the electrons. If, for example, a "machine" generating photons with frequencies as low as 10^4 c/s were operating, it would require electrons with energies of 10 GeV to lift the photons by the inverse Compton process to $\nu \sim 10^{12}$ c/s. It is really not possible to estimate how the energy is divided, but the total energy release must be at least $\sim10^{47}$ t where t, the lifetime of the source, may be $\sim10^6$ years, so again the total will be $\sim10^{61}$ ergs.

The various theories proposed to account for the energy emitted in the radio sources have been summarized earlier (52, 49). Since then, a number of new investigations and suggestions have been put forward, which will be summarized here. The majority of the theories attempt to account for the QSO's as objects at cosmological distances, but some are compatible with the idea that they are objects thrown out of galaxies. Some attempt is made to explain the properties of the radio galaxies as well. In very few, if any, of the theories is the physical mechanism which leads to the release of energy largely in the form of relativistic matter described satisfactorily. If the QSO's are coherent objects ejected from radio galaxies, or indeed objects ejected in galactic explosions in general, the underlying mechanism is obscure.

A fundamental assumption made in nearly all of the hypotheses so far put forward is that matter is present in a highly condensed form. We consider first the theories based on the idea that dense galactic nuclei are present.

Supernova Theories

For several years the group at Livermore, under the direction of Colgate, have been considering the hydrodynamics of the final stages of evolution of a supernova, with their starting point the final stages of nuclear evolution discussed by others (cf. 90). They start with a hot evolved star of mass 10 $M_\odot$ which is dense enough to be unstable against gravitational collapse, and follow it as it falls inward (70). A shock is formed, heating results, and nuclear reactions take place. The core falls in rapidly and the outer part more slowly. The energy released in the collapse is emitted largely in the form of neutrinos. These escape if, and until, the outer parts of the star have collapsed to a high enough density so that a significant opacity to the neutrino flux is produced. At this point the neutrinos exert sufficient force to halt the collapse of the outer part of the star, which is ejected. At the same time the inner parts of the star continue to collapse and can form a stable neutron configuration, if the infalling mass is less than the critical mass for a neutron star. With the approximations chosen and using Newtonian gravitational theory, Colgate & White (70) concluded that about 10^{-3} of the total mass energy could be ejected, i.e., about 10^{52} ergs from a star of 10 $M_\odot$, while $\sim 2 \times 10^{51}$ ergs was ejected in the form of relativistic particles. This is more than 100 times the energy for which we have direct evidence of release in a supernova. If correct, this much larger energy release makes the supernova hypothesis for strong radio sources and QSO's more attractive again. It should be remembered that the energy is gravitational in origin.

Colgate & Cameron (69) (see also 56) first applied the argument to attempt to account for the light variations in QSO's. They suggested that the very large luminosities were produced by the ejected gas heating surrounding interstellar gas. A more ambitious attempt to explain the flux radiated by a cosmologically distant QSO (3C 273 being taken as the prototype) was recently made by Colgate (68), starting with an assumed star density of $\sim 10^{10}$ /pc^3. Massive stars (50 $M_\odot$) will be formed by inelastic collisions between the original stars. These evolve to the supernova stage in times $\sim 10^6$ years and supernova explosions occur at a rate ~ 3 per year. The kinetic energy ejected from the supernovae then heats the gas remaining from previous explosions, and it is this excited gas which gives rise to the high luminosity with a variable component. The radio emission is supposed to arise from the material at very high kinetic energy (0.1 c^2/g). This passes through the bulk of the gas cloud with little energy loss, but at the boundary of the dense cloud a two-stream plasma instability occurs in which ions and electrons share kinetic energy. The radio emission is then supposed to arise from electrostatic bremsstrahlung and this is scattered from coherent plasma oscillations giving the spectral characteristics of 3C 273B. It is claimed that this model avoids the difficulties associated with synchrotron emission models. A still higher-energy component of gas ejected from supernovae at relativistic speeds is then invoked to give rise to the radio source 3C 273A, and such components will also be required for any QSO's which have extended radio sources. Although many of the details of this model are not easily understood, the whole concept is highly ingenious. The underlying model involving many supernovae may also be considered for the strong radio galaxies.

Aizu et al. (3) considered a possible mechanism of explosion in a galactic nucleus which contains a high density of stars together with gas. They supposed that the gas may speed up the evolution of the stars, and induce collective explosions of stars. They call this a "pile theory" but have not worked out the consequences in detail.

STELLAR COLLISIONS

Given a very high star density in a galactic center, the stars interact more and more rapidly, through inelastic collisions. The velocity dispersion of the stars gets larger as the cluster shrinks; the violence of the star collisions increases; and at high enough energy the stars will completely disrupt. Throughout this process some stars will be ejected from the cluster at higher and higher velocities. These stages have been considered in some detail recently by Spitzer & Saslaw (212), following earlier work summarized elsewhere. Very high star densities, $\sim 10^{11}$ stars/pc^3, may be required in such a model. If the collision velocities are $\sim 10^4$ km/s, the kinetic energy available is some 10^{51} erg/$M_\odot$. In this class of models, however, the violent phase in which most of the energy is released (the QSO phase) appears to be very short, $\sim 10^9$-10^{10} sec.

In such violent stellar collisions the major part of the kinetic energy will be dissipated by radiation processes and some mass will be ejected from the cluster. However, most of the matter will fall back together and give rise to a massive cloud, with a small net angular momentum. This is not perhaps the only way in which a very condensed object can be produced. However, it is the final evolutionary phase of a dense galactic nucleus of stars. How long it takes to evolve to this state depends on the initial density assumed. If this is low, comparable to the densities seen in nearby galaxies, the total time involved may be much longer than the Hubble time.

MASSIVE SUPERSTARS

The previous discussion leads naturally to the investigations of Hoyle & Fowler (111, 112; see also 113), who first considered the problem of the release of gravitational energy in the collapse of a superstar. The basic idea of Hoyle & Fowler is well known; it is that in the gravitational collapse of a massive object a small but significant fraction of the rest-mass energy may be released. Close to the Schwarzschild radius it is very difficult for energy to be emitted in a spherically symmetrical collapse, unless it is supposed that the theory of general relativity is modified in this extreme condition. The role of rotation in the relativistic regime is unclear, though it has been suggested (226) that expulsion of matter can occur in extreme configurations.

Modification of the theory of relativity has been attempted by Hoyle & Narlikar (114) who introduced a field of negative energy which they had considered in their cosmological investigations. This halts gravitational collapse, and the object might carry out radial oscillations which would take it for some part of the time outside the Schwarzschild radius; in this phase, energy could be emitted.

In gravitational collapse of a massive object it is not clear in what form the energy will be emitted. The most efficient process would be if it were in the form of high-energy particles, since many arguments suggest that processes converting energy from any other form to high-energy particles are inefficient.

If the QSO's are local objects ejected from galaxies, then the energy must be emitted in coherent lumps with high-density cores. These would be most likely to be produced in a process of fragmentation, perhaps due to rotation in the final collapse phases. Obviously a proper theoretical treatment of such ideas will be required, if the local hypothesis for QSO's is to be pursued.

While the gravitational collapse theory for the large energy of radio sources remains popular despite the difficulties, Fowler (88, 89), following his earlier work, has also attempted to account for the optical properties of the QSO's by considering the early collapse phases of a massive object, in which normal hydrogen burning gives a luminosity of 2×10^{46} erg/s for a mass of $10^8 M_\odot$. Fowler had estimated that the total thermonuclear energy available in a massive star would enable it to radiate for $\sim 10^6$ years, but the general relativistic instability of a nonrotating star means that it cannot be stable for this long but must undergo free-fall collapse (57–59, 89). However, it has now been shown (89, 173) that a small amount of rotation is able to stabilize the massive star against gravitational collapse for a limited period, and that turbulent or magnetic forces will also be able to stabilize the star as long as it contains nuclear energy sources. Thus a massive star may be able to exist for $\sim 10^6$ years in its thermonuclear burning phase, provided that its mass does not exceed 10^8-$10^9 M_\odot$.

It should be remembered that for any QSO's in which very extended radio sources are seen, at least two massive objects are required. One must have formed, evolved to the gravitational collapse stage, and emitted enough high-energy particles to give rise to the extended source, while a second must currently be passing through its thermonuclear phase to produce the quasi-stellar component. To explain the strong radio galaxies by this mechanism the massive object must have evolved and collapsed to give rise to the extended radio source.

THE ROLE OF MAGNETIC FIELDS IN MASSIVE OBJECTS

Little attention has been paid until recently to the problem of the magnetic fields which are an integral part of the phenomenon to be explained. It has commonly been assumed that the conditions which give minimum total energy requirements for synchrotron emission hold, i.e. homogeneous uniform fields of 10^{-4}-10^{-5} G. However [see discussion by Woltjer (231)], it is difficult for magnetic fields as strong as this to have been either ejected in the relativistic plasma cloud in the central explosion, or amplified from a much weaker intergalactic field by the explosion. If the fields are smaller, then the total energy requirement of the source is greater than the minimum total energy, and most of the energy is in the relativistic particles.

Layzer (128b) suggested that any gas cloud, provided that it is sufficiently massive, will convert rotational energy into magnetoturbulence while it contracts, and, before it becomes unstable, it will generate a magnetoturbulent field whose energy is comparable with its own rest energy. It must then be presumed that a considerable part of this energy is converted with high efficiency into relativistic particles, which then radiate in local strong magnetic fields.

In the highly condensed QSO's there are strong arguments for believing that a nonhomogeneous magnetic field is present, and there is no reason to believe that the equipartition condition is fulfilled. In order to avoid many difficulties, particularly if the QSO's are at cosmological distances, it may be necessary to invoke very strong magnetic fields ≥ 100 G in some regions.

If the relativistic particles gain energy by a conventional acceleration mechanism, the magnetic field will play an important role in energy transfer, and its configuration is one of the main factors determining the synchrotron-radiating properties of the source. A number of authors (91, 92, 122, 160, 161, 163, 164, 219, 220) have explored some possibilities following the hypothesis that gravitational energy released in the condensation and collapse of a massive object is converted to magnetic energy and through it to the relativistic particles.

Both Piddington & Sturrock discussed, qualitatively, the condensation of a mass of galactic size out of the intergalactic medium. Piddington (163) considered the condensation of galaxies from gas clouds with frozen-in magnetic fields and argued that when the rotational (ω) and magnetic-field (B) vectors are orthogonal, radio galaxies result. Stars form while the uncondensed gas continues to shrink, giving rise to a condensed mass, while if star formation is inhibited by some process the whole galactic mass will shrink to "nuclear" dimensions and a QSO will result.

Sturrock's argument is based on a model of his (220) for solar flares which must be scaled up by an enormous factor. A galaxoid or protogalaxy, a compact elliptical galaxy, a galactic nucleus, or a QSO, is condensed from the intergalactic gas so that it maintains a connection with the intergalactic magnetic field and accretes matter from it. The intergalactic material is supposed to be at least partially ionized and the magnetic field has an hourglass configuration through the object; accreted matter will be "funneled" into the condensed object. The angular momentum problem is discussed in (220a).

The situation is reached in which the galaxoid or QSO is highly condensed and contains magnetic energy that supports it against gravitational collapse, which condition implies that $\Phi \approx 10^{-3}\ M$ where Φ is the flux in G cm^2 and M is the mass in grams of the condensed object. If now the values of Φ for the extended radio sources calculated by Maltby et al. (139) are inserted, masses of the order 10^{10} to 10^{12} $M_\odot$ are obtained, by using the virial theorem and equating the gravitational binding energy with the total magnetic energy. These masses are much greater than those given by Hoyle & Fowler on the basis of energy arguments.

Sturrock then argues that energy release from the QSO will be the tearing mode instability in the region of the sheet pinch which, in his assumed magnetic-field configuration, is perpendicular to the axis of the hourglass. This instability will give rise to ejection of a pocket of magnetic field and high-energy particles which comprise the jets seen in QSO's and radio galaxies. Such a jet will eventually divide into two clouds which will be the double radio source. The main interest in this model lies in the plasma instability mechanism. Given the condensed object together with the necessary magnetic-field configuration, the model is attractive.

Piddington (163, 164) paid more attention to the problems of rotation in a contracting cloud. When ω and B are orthogonal, rotational energy is converted to magnetic energy so that a continuing and steady collapse can occur. The object eventually reaches a spherical state followed by explosions along the $\pm\omega$ axis. The particles are accelerated in a neutral sheet at the expense of magnetic energy in a process bearing some resemblance to that of Sturrock. Piddington has argued that the hourglass magnetic-field model is one that gives rise to a normal spiral galaxy (ω parallel to B) and thus he reaches quite a different conclusion from Sturrock's in this respect.

While flare models of this type are attractive and promising, they require initial conditions that cannot be deduced from observation nor unambiguously derived by theoretical argument.

Ginzburg & Ozernoy (92) started from the idea that QSO's and strong radio galaxies get their energy from the gravitational collapse of a massive superstar. A very large spherically symmetric mass condenses into a small volume; they consider the effect of further collapse on the magnetic dipole moment which is trapped within the object. The magnetic energy is assumed to be small compared with the gravitational energy of the initial star. As the star collapses, the magnetic field grows, reaching enormous values in the star's interior. In the lower-density atmosphere surrounding the star, conditions may develop so that a current-carrying shell may become detached. Thus the collapsing magnetic star may develop very powerful "radiation belts" within a magnetosphere, and it is argued that it is these regions which give rise to the flux of optical and radio emission in quasi-stellar objects. Ozernoy (160, 161) developed a model of a "magnetoid"—a quasi-stationary configuration in stable rotation along the lines of force of a toroidal magnetic field; such a model can explain some of the variable features of QSO's, particularly if they are quasi-periodic. While Ginzburg & Ozernoy stress that the development and structure of this model involve many unsolved complex problems, the model has some attractive features. It may be able to account for the very strong magnetic fields that are required to avoid some of the difficulties discussed earlier. It is not clear, however, how most of the gravitational energy is to be transformed to the high-energy particles which radiate in this field.

The classes of theory outlined above involve either the evolution and shrinkage to high densities of a galactic nucleus of stars, or the formation of a dense gas cloud by condensation in the intergalactic medium, an idea first proposed by Field (84) who suggested that QSO's were galaxies in the process of formation. But the formation of a dense object is not well understood. McCrea (143) suggested that the problem of the formation of condensations may be man-made and that there is no evidence that nature poses such problems. The dense phase might be a remnant of the very early evolution of the universe in which all the mass-energy was contained in a very small volume, or else it might be supposed that matter is continuously created in conditions of high density. These ideas tie the QSO's and the radio galaxies directly to cosmological theories. Various models are described next.

Theories Using the Concept That Massive Objects Have a Cosmological Origin

Proposals along this line were made by McCrea (143), Hoyle & Narlikar (115), Stothers (215), Novikov (154), and Ne'eman (151).

McCrea considered a modification of the steady-state theory in which matter is created only in the presence of already existing matter. Thus all matter is contained in galaxies, and continuous creation simply increases the galactic mass. Occasionally a galaxy may eject a fragment, which is then the embryo out of which a new galaxy is grown. Such embryos are closely related to the quasi-stellar objects, and the phenomena of outburst and ejection from galaxies to produce radio sources are to be associated with the creation and ejection process. Such a model could explain the QSO's if they have a local origin.

A rather similar proposal has been made by Hoyle & Narlikar. They show that the growth in "pockets" of creation is likely to be an unstable process, since the growth of a mass to infinity takes only twice the time for the mass to double itself. Such a runaway process will be prevented by a fragmentation of the growing mass, and each fragment will serve as an individual pocket of creation—the embryo in McCrea's formulation.

Stothers, on the other hand, argued that matter is created between clusters of galaxies just because matter is lacking here, and the QSO's (which may not occur in clusters) are the manifestation of this creation. However, for a variety of reasons the McCrea-Hoyle-Narlikar hypothesis is more attractive.

If the universe is evolutionary, then the superdense objects could be inhomogeneities remaining in the general expansion of the universe, and again the problem of forming condensations out of a diffuse intergalactic background is bypassed. Ne'eman proposed that the QSO's, which he supposed are cosmological objects, are in an expanding state which have lain near the singular state for some 10^{10} years. Thus in its own coordinate system the object is behaving as a miniature expanding universe. Extremely large energies must have accumulated from strong interactions in the superdense state with no outlet for the decay of various mesons. Having attained a lower density in the slow expansion, these mesons and hence high-energy particles may be released.

The argument of Novikov is rather similar to this, though he suggested that energy is released when shells of matter moving outward in the expansion collide with shells which have been ejected earlier, or with matter falling onto the object from outside (154).

In this whole class of theory, conventional acceleration mechanisms for high-energy particles are not invoked, and the high-energy particles are directly injected. The ideas contained in these theories were foreshadowed in the early papers of Ambartsumian (7).

A problem not covered in these theories is associated with the apparently normal composition of the gas cloud giving the line radiation in the QSO's. It must be supposed either that extensive nucleosynthesis has gone on in the evolution of these objects leading to normal composition—an unlikely situation—or that the object has managed to accrete material with normal composition.

Classes of theory involving fundamental particles will be outlined next.

QUARKS AS ENERGY SOURCES IN MASSIVE OBJECTS WITH A COSMOLOGICAL ORIGIN

If quarks exist, they might form the major constituent of the universe in a stage in which matter has not evolved. Thus bare quarks might have predominated early in an evolving universe, while in a steady-state universe, matter might be continuously created in the form of quarks. The situation in an evolving universe has been discussed by Zeldovich et al. (236) and by Saslaw (189). The former discussed a situation in which the quarks burned out in the first few microseconds of the expansion, and Saslaw considered that cosmological information might be obtained by considering the role of the quarks in this very short epoch.

If quarks are of importance in these cosmologically dense configurations and if massive objects containing bare quarks are left behind in the initial expansion, or are present in newly created material, then as the objects evolve they can give rise to the major constituents of the radio sources and QSO's. The first mention of this possibility was made by Pacini (162); it is discussed further in (50, 54).

MATTER—ANTIMATTER ANNIHILATION

This mechanism was suggested ten years ago for radio galaxies (47, 53), but dropped because of the difficulty in understanding how matter and antimatter could be created or evolve and remain separated, and then later be brought into interaction in isolated events.

Alfvén & Klein (5) proposed a model universe in which matter and anti-matter enter in a symmetric way. Alfvén (4) developed the concept of an "ambiplasma" which contains both kinds of matter and he showed that a mechanism of separation involving the magnetic field was possible. If one accepts this type of cosmology, then the difficulties just mentioned can be removed. However, this type of cosmological model based on the old theoretical model of Charlier is probably not compatible with present views. Setting aside this objection for the moment, we consider the consequences of the annihilation proposal further. Ekspong et al. (80) attempted to account for the radio spectra of some QSO's and radio galaxies using this idea. They calculated the electron-positron spectrum through the decays of the π and μ mesons produced in $p\bar{p}$ annihilation, calculated the resulting synchrotron spectra, and compared these with observations of 3C 48, 3C 147, 3C 286, CTA 102, and the radio galaxies Cygnus A and 3C 295. The calculated spectra are curved and show a tendency to flatten at low frequencies. As discussed earlier, the generally accepted reason for this flattening is that it is an approach to synchrotron self-absorption in sources containing very small components. An objection to this theory is that to give radiation in the observed frequency range, large magnetic fields are required, $\sim 10^{-2}$ G. For the extended sources which include 3C 147, Cygnus A, and 3C 295, the minimum total energy condition gives magnetic fields in the range 10^{-4}-10^{-5} G. If the magnetic field is much larger than this, as it has to be in this theory using very low-energy electrons, then there is a strong departure from the minimum total energy condition. This in itself does not mean that it is not a plausible hypothesis, but it does mean that the total energy in particles and magnetic field is many orders of magnitude greater than the minimum, and most of it is in the magnetic field. Thus one is left with the difficulty of explaining the origin of an enormous amount of energy in the magnetic field which cannot arise from annihilation. This model is therefore unsatisfactory.

Gravitational Focusing

Barnothy (20–22) suggested that the very high luminosities of the QSO's, assumed to be at cosmological distances, are due to amplification by gravitational focusing by a large concentration of mass lying very close to the light path between the QSO and the observer. The focusing objects must be dark; otherwise their own radiation would swamp the flux which is to be focused. Moreover, they must occur quite frequently in space for the chance of this effect occurring to be nonnegligible. Both of these conditions are inherently implausible. In addition it is necessary still to postulate the existence of objects with the spectral properties of the QSO's, since this effect only increases the apparent luminosities and does not change the spectra. Therefore Barnothy argued that the QSO's are the nuclei of very distant Seyfert galaxies. However, while the latter bear some spectral resemblance to the QSO's, the two classes of objects are clearly distinguishable. These are strong arguments against the proposal.

CONCLUSION

The task of writing a review on a subject changing as rapidly as this one is somewhat difficult; new observations are perhaps being made at this moment which will make the review already out of date, and likewise new theoretical ideas are perhaps being formulated at the same time.

In the observational sections, I have devoted more space to the optical than to the radio data, and this is merely a reflection of where my own work lies.

In conclusion, I would like to express my gratitude to the many colleagues who have sent preprints ahead of publication, and comments and criticisms, and particularly to Geoffrey Burbidge, who did much of the assembling of data for this review.

I would also like to thank Mrs. Jean Fox for typing the manuscript and Mrs. Del Crowne for ordering the references and checking the manuscript.

This work has been supported in part by the National Science Foundation and in part by NASA through grant NsG-357.

LITERATURE CITED

1. Adgie, R. L., *Nature*, **204**, 1028 (1964)
2. Adgie, R., Gent, H., Slee, O. B., Frost, A., Palmer, H. P., Ronson, B., *Nature*, **208**, 275 (1965)
3. Aizu, K., Fujimoto, Y., Hasegawa, H., Kawabata, K., Taketani, M., *Progr. Theoret. Phys. (Kyoto) Suppl. No. 31*, 35 (1964)
4. Alfvén, H., *Rev. Mod. Phys.*, **37**, 652 (1965)
5. Alfvén, H., Klein, O., *Arkiv Fysik*, **23**, 187 (1962)
6. Allen, L. R., Anderson, B., Conway, R., Palmer, H. P., Reddish, V. C., Ronson, B., *Monthly Notices Roy. Astron. Soc.*, **124**, 477 (1962)
6a. Aller, H. D., Haddock, F. T., *Ap. J.*, **147**, 833 (1967)
7. Ambartsumian, V. A., *Astron. J.*, **66**, 536 (1961)
8. Anderson, B., Donaldson, W., Palmer, H. P., Ronson, B., *Nature*, **205**, 375 (1965)
9. Andrillat, Y., Andrillat, H., *Compt. Rend.*, **258**, 3199 (1964)
10. Arp, H. C., *Science*, **151**, 1214 (1966)
11. Arp, H. C. (Private communication, 1966), *Ap. J.*, **148** (1967)
12. Arp, H. C., Atlas of Peculiar Galaxies, *Ap. J. Suppl.* (In press) (1966)
12a. Arp, H. C., Bolton, J. G., Kinman, T. D., *Ap. J.*, **147**, 840 (1967)
13. Bahcall, J. N., *Ap. J.*, **145**, 684 (1966)
14. Bahcall, J. N., *ibid.*, **146**, 615
15. Bahcall, J. N., Peterson, B. A.,

Schmidt, M., *Ap. J.*, **145**, 369 (1966)
16. Bahcall, J. N., Salpeter, E. E., *Ap. J.*, **142**, 1677 (1965)
17. Bahcall, J. N., Salpeter, E. E., *ibid.*, **144**, 847 (1966)
18. Barber, D. R., Donaldson, W., Miley, G. K., Smith, H., *Nature*, **209**, 753 (1966)
19. Barnes, R. C., *Ap. J.*, **146**, 285 (1966)
20. Barnothy, J. M., *Astron. J.*, **70**, 666 (1965)
21. Barnothy, J. M., *ibid.*, **71**, 154 (1966)
22. Barnothy, M. F., Barnothy, J. M., *Astron. J.*, **71**, 155 (1966)
23. Bennett, A. S., *Mem. Roy. Astron. Soc.*, **68**, 163 (1962)
24. Boccaletti, D., de Sabbata, V., Gualdi, C., *Nuovo Cimento*, **45A**, 513 (1966)
25. Bolton, J. G., *Nature*, **211**, 917 (1966)
26. Bolton, J. G., Clarke, M. E., Ekers, R. D., *Australian J. Phys.*, **18**, 627 (1965)
27. Bolton, J. G., Clarke, M. E., Sandage, A. R., Véron, P., *Ap. J.*, **142**, 1289 (1965)
28. Bolton, J. G., Clarke, M. E., Sandage, A. R., Véron, P., *ibid.*, **144**, 860 (1966)
29. Bolton, J. G., Ekers, J., *Australian J. Phys.*, **19**, 275 (1966)
30. Bolton, J. G., Ekers, J., *ibid.*, 471
31. Bolton, J. G., Ekers, J., *ibid.*, 559
31a. Bolton, J. G., Ekers, J., *ibid.*, 713
32. Bolton, J. G., Gardner, F. F., Mackey,

M. B., *Australian J. Phys.*, **17,** 340 (1964)

33. Bolton, J. G., Kinman, T. D., *Ap. J.*, **145,** 951 (1966)

34. Bolton, J. G., Shimmins, A. J., Ekers, R. D., Kinman, T. D., Lamla, E., Wirtanen, C. A., *Ap. J.*, **144,** 1229 (1966)

35. Bondi, H., *Lectures on General Relativity* (*Brandeis Summer Institute in Theoretical Physics*), 456 (Prentice-Hall, Englewood Cliffs, N. J., 1964)

36. Buchdahl, H. A., *Phys. Rev.*, **116,** 1027 (1959)

37. Buchdahl, H. A., *Ap. J.*, **146,** 275 (1966)

38. Burbidge, E. M., *Proc. Liège Symp.*, *Ann. Ap.*, **28,** 164 (1964)

39. Burbidge, E. M., *Ap. J.*, **142,** 1291 (1965)

40. Burbidge, E. M., *ibid.*, 1674

41. Burbidge, E. M., *ibid.*, **143,** 612 (1966)

41a. Burbidge, E. M., *ibid.*, **147,** 845 (1967)

42. Burbidge, E. M., Burbidge, G. R., *Ap. J.*, **143,** 271 (1966)

43. Burbidge, E. M., Kinman, T. D., *Ap. J.*, **145,** 654 (1966)

44. Burbidge, E. M., Lynds, C. R., *Ap. J.*, **147,** 388 (1967)

45. Burbidge, E. M., Lynds, C. R., Burbidge, G. R., *Ap. J.*, **144,** 447 (1966)

46. Burbidge, E. M., Rosenberg, F. D., *Ap. J.*, **142,** 1673 (1965)

47. Burbidge, G. R., *Ap. J.*, **124,** 416 (1956)

48. Burbidge, G. R., *Paris Symp. Radio Astron. Paris, 1958*, 541 (1959)

48a. Burbidge, G. R., *Ap. J.*, **147,** 851 (1967)

49. Burbidge, G. R., Burbidge, E. M., *The Structure and Evolution of Galaxies*, *13th Solvay Congress*, 137 (Wiley, New York, 1965)

50. Burbidge, G. R., Burbidge, E. M., *Quasi-Stellar Objects* (Freeman, New York, in press) (1967)

50a. Burbidge, G. R., Burbidge, E. M., *Ap. J.*, **148** (In press) (1967)

50b. Burbidge, G. R., Burbidge, E. M., *Quasi-Stellar Sources and Gravitational Collapse* (Robinson, I., Schild, A., Schücking, E. L., Eds., Univ. of Chicago Press, 1963)

51. Burbidge, G. R., Burbidge, E. M., Hoyle, F., Lynds, C. R., *Nature*, **210,** 774 (1966)

52. Burbidge, G. R., Burbidge, E. M., Sandage, A. R., *Rev. Mod. Phys.*, **35,** 947 (1963)

53. Burbidge, G. R., Hoyle, F., *Nuovo Cimento*, **4,** 558 (1956)

54. Burbidge, G. R., Hoyle, F., *Sci. Am.* (December 1966)

55. Burbidge, G. R., Strittmatter, P. A., Sargent, A. (Private communication)

56. Cameron, A. G. W., *Nature*, **207,** 1140 (1965)

56a. Cannon, R. D., Penston, M. V., *Nature*, **214,** 256 (1967)

57. Chandrasekhar, S., *Phys. Rev. Letters*, **12,** 114, 437E (1964)

58. Chandrasekhar, S., *Ap. J.*, **140,** 417 (1964)

59. Chandrasekhar, S., *Phys. Rev. Letters*, **14,** 241; *Ap. J.*, **142,** 1488, 1519 (1965)

60. Chavira, E., *Bol. Tonantzintla Tacubaya*, No. 17, 15 (1958)

61. Chavira, E., *ibid.*, No. 18, 3 (1959)

62. Clark, B. G., Hogg, D. E., *Ap. J.*, **145,** 21 (1966)

63. Clarke, M. E., *Monthly Notices Roy. Astron. Soc.*, **127,** 405 (1964)

64. Clarke, M. E., Bolton, J. G., Shimmins, A. J., *Australian J. Phys.*, **19,** 375 (1966)

65. Cohen, M. H., *Nature*, **205,** 277 (1965)

65a. Cohen, M. H., Gundermann, E. J., *Ap. J.* (In press) (1967)

66. Cohen, M. H., Gundermann, E. J., Hardebeck, H. E., Harris, D. E., Salpeter, E. E., Sharp, L. E., *Science*, **153,** 744 (1966)

67. Cohen, M. H., Gundermann, E. J., Hardebeck, H. E., Sharp, L. E., *Ap. J.*, **147,** 449 (1967)

68. Colgate, S. A. (Preprint, 1966)

69. Colgate, S. A., Cameron, A. G. W., *Nature*, **200,** 870 (1963)

70. Colgate, S. A., White, R. H., *Ap. J.*, **143,** 626 (1966)

71. Conway, R. G., Kellermann, K. I., Long, R. J., *Monthly Notices Roy. Astron. Soc.*, **125,** 261 (1966)

72. Day, G. A., Shimmins, A. J., Ekers, R. D., Cole, D. J., *Australian J. Phys.*, **19,** 35 (1966)

73. Dent, W. A., *Science*, **148,** 1458 (1965)

74. Dent, W. A., *Ap. J.*, **144,** 843 (1966)

75. Dent, W. A., Haddock, F. T., *Ap. J.*, **144,** 568 (1966)

76. Dibai, E. A., Pronik, V. I., *Astron. Tsirk. No. 286* (1964)

77. Dibai, E. A., Yesipov, V. F., *Astron. Zh.*, **44,** 55 (1967)

77a. Dibai, E. A., Yesipov, V. F., *Astron. Tsirk. Akad. Nauk SSSR No. 403,* 7 (1967)

77b. Divan, L., *Ann. Ap.*, **28,** 70 (1965)

78. Edge, D. O., Shakeshaft, J. R., Mc-Adam, W. B., Baldwin, J. E., Archer, S., *Mem. Roy. Astron. Soc.*, 67, 37 (1959)

79. Ekers, R. D., Bolton, J. G., *Australian J. Phys.*, 18, 669 (1965)

80. Ekspong, A. G., Yandagni, N. K., Bonnevier, B., *Phys. Rev. Letters*, 16, E564(c) (1966)

81. Epstein, E., *Ap. J.*, 142, 1282; 1285 (1965)

82. Epstein, E., Oliver, J. P., Schorn, R. A., *Ap. J.*, 145, 367 (1966)

83. Faulkner, J., Gunn, J. E., Peterson, B., *Nature*, 211, 502 (1966)

84. Field, G. B., *Ap. J.*, 140, 1434 (1964)

85. Field, G. B., *Science*, 150, 78 (1965)

86. Field, G. B., Solomon, P. M., Wampler, E. J., *Ap. J.*, 145, 351 (1966)

87. Ford, W. K., Rubin, V. C., *Ap. J.*, 145, 357 (1966)

88. Fowler, W. A., *Proc. Ann. Sci. Conf. Belfer Grad. School, Yeshiva Univ.* (Academic Press, New York, 1965)

89. Fowler, W. A., *Ap. J.*, 144, 180 (1966)

90. Fowler, W. A., Hoyle, F., *Ap. J. Suppl.*, 9, 201 (1965)

91. Ginzburg, V. L., *Soviet Phys. Dokl.*, 9, 329 (1964)

92. Ginzburg, V. L., Ozernoy, L. M., *Zh. Theoret. Phys.*, 47, 1030; *JETP*, 20, 489 (1964)

93. Ginzburg, V. L., Ozernoy, L. M., *Ap. J.*, 144, 599 (1966)

94. Goldsmith, D. W., Kinman, T. D., *Ap. J.*, 142, 1693 (1965)

95. Gower, J. F. R., Scott, P. F., Wills, D., *Mem. Roy. Astron. Soc.* (In press) (1967)

96. Greenstein, J. L., Matthews, T. A., *Nature*, 197, 1041 (1963)

97. Greenstein, J. L., Schmidt, M., *Ap. J.*, 140, 1 (1964)

97a. Greenstein, J. L., Schmidt, M., *Ap. J.*, 148, L13 (1967)

98. Griffin, R. F., *Astron. J.*, 68, 421 (1963)

99. Gunn, J. E., Peterson, B. A., *Ap. J.*, 142, 1633 (1965)

100. Haro, G., *Bol. Obs. Tonantzintla Tacubaya*, 2, No. 14, 8 (1956)

101. Haro, G., Luyten, W. J., *Bol. Obs. Tonantzintla Tacubaya*, 3, 37 (1962)

101a. Hazard, C., *Quasi-Stellar Sources and Gravitational Collapse* (See Ref. 50b), 135

102. Hazard, C., Mackey, M. B., Nicholson, W., *Nature*, 202, 227 (1964)

103. Hazard, C., Mackey, M. B., Shimmins, A. J., *Nature*, 197, 1037 (1963)

103a. Hazard, C., Mackey, M. B., Sutton, J., *Ap. J.* (In press) (1967)

103b. Heeschen, D. S., *Ap. J.* 146, 517 (1966)

104. Hewish, A., Scott, P. F., Wills, D., *Nature*, 203, 1214 (1964)

105. Hiltner, W. A., Cowley, A. P., Schild, R. E., *Publ. Astron. Soc. Pacific*, 78, 464 (1966)

106. von Hoerner, S., *Ap. J.*, 144, 483 (1966)

106a. Holden, D. J., *Observatory*, 86, 229 (1966)

107. Hoyle, F., Burbidge, G. R., *Ap. J.*, 144, 534 (1966)

108. Hoyle, F., Burbidge, G. R., *Nature*, 210, 1346 (1966)

109. Hoyle, F., Burbidge, G. R., *ibid.*, 212, 1334 (1966)

110. Hoyle, F., Burbidge, G. R., Sargent, W. L. W., *Nature*, 209, 751 (1966)

111. Hoyle, F., Fowler, W. A., *Monthly Notices Roy. Astron. Soc.*, 125, 169 (1963)

112. Hoyle, F., Fowler, W. A., *Nature*, 197, 533 (1963)

112a. Hoyle, F., Fowler, W. A., *Nature*, 213, 373 (1967)

113. Hoyle, F., Fowler, W. A., Burbidge, G. R., Burbidge, E. M., *Ap. J.*, 139, 909 (1964)

114. Hoyle, F., Narlikar, J. V., *Proc. Roy. Soc. A*, 277, 1; 278, 465 (1964)

115. Hoyle, F., Narlikar, J. V., *ibid.*, 290, 143, 162 (1966)

116. Humason, M. L., Mayall, N. U., Sandage, A. R., *Astron. J.*, 61, 97, Appendix B (1956)

117. Humason, M. L., Zwicky, F., *Ap. J.*, 105, 85 (1947)

118. Iriarte, B., Chavira, E., *Bol. Tonantzintla Tacubaya*, No. 16, 3 (1957)

119. Jeffreys, W. H., *Astron. J.*, 69, 255 (1964)

120. Johnson, H. L., *Ap. J.*, 139, 1022 (1964)

121. Johnson, H. L., Low, F. E., *Ap. J.*, 141, 336 (1965)

122. Kardashev, N. S., *Astron. Zh.*, 41, 807; *Soviet Astron.*, 8, 643 (1964)

123. Kardashev, N. S., Komberg, B. V., *Astron. Tsirk. No. 357* (1966)

124. Kellermann, K. I., *Ap. J.*, 146, 621 (1966)

125. Kinman, T. D., *Ap. J.*, 142, 1241 (1965)

126. Kinman, T. D., *ibid.*, 144, 1232 (1966)

126a. Kinman, T. D., Bolton, J. G., Clarke, R. W., Sandage, A., *Ap. J.*, 147, 848 (1967)

126b. Kinman, T. D., Burbidge, E. M., *Ap. J.*, 148 (In press) (1967)

127. Kinman, T. D., Lamla, E., Wirtanen, C. A., *Ap. J.*, **146**, 964 (1966)
127a. Kinman, T. D., Lamla, E., Wirtanen, C. A. (Private communication 1967)
128. Koehler, J., Robinson, B. J., *Ap. J.*, **146**, 488 (1966)
128a. Komberg, B. V., *Astron. Tsirk. Akad. Nauk SSSR No. 379* (1966)
128b. Layzer, D., *Ap. J.*, **141**, 837 (1965)
129. LeRoux, E., *Ann. Ap.*, **24**, 71 (1961)
130. Longair, M. S., *Monthly Notices Roy. Astron. Soc.*, **129**, 419 (1965)
131. Longair, M. S., *Nature*, **211**, 949 (1966)
132. Low, F. E., *Ap. J.*, **142**, 1287 (1965)
133. Luyten, W. J., Smith, J. A., *Ap. J.*, **145**, 366 (1966)
134. Lynds, C. R., Discussion at Byurakan Conference (1966)
134a. Lynds, C. R., *Ap. J.*, **147**, 396 (1967)
134b. Lynds, C. R., *Ap. J.*, **147**, 837 (1967)
135. Lynds, C. R., Hill, S. J., Heere, K., Stockton, A. N., *Ap. J.*, **144**, 1244 (1966)
136. Lynds, C. R., Stockton, A. N., *Ap. J.*, **144**, 446 (1966)
137. Lynds, C. R., Stockton, A. N., Livingstone, W. L., *Ap. J.*, **142**, 1667 (1965)
138. Lynds, C. R., Villere, G., *Ap. J.*, **142**, 1296 (1965)
139. Maltby, P., Matthews, T. A., Moffet, A. T., *Ap. J.*, **137**, 153 (1963)
140. Maltby, P., Moffet, A. T., *Ap. J.*, **142**, 409 (1965)
141. Maltby, P., Moffet, A. T., *Science*, 63 (October 1, 1965)
142. Matthews, T. A., Sandage, A. R., *Ap. J.*, **138**, 30 (1963)
143. McCrea, W. H., *Monthly Notices Roy. Astron. Soc.*, **128**, 336 (1964)
144. McCrea, W. H., *Publ. Astron. Soc. Pacific*, **78**, 49; *Ap. J.*, **144**, 516 (1966)
145. Mills, B. Y., Slee, O. B., Hill, E., *Australian J. Phys.*, **11**, 360 (1958)
146. Mills, B. Y., Slee, O. B., Hill, E., *ibid.*, **13**, 676 (1960)
147. Mills, B. Y., Slee, O. B., Hill, E., *ibid.*, **14**, 497 (1961)
148. Moffet, A. T., *Science*, **146**, 764 (1964)
149. Moffet, A. T., *Ap. J.*, **141**, 1580 (1965)
150. Moffet, A. T., *Ann. Rev. Astron. Ap.*, **4**, 145 (1966)
151. Ne'eman, Y., *Ap. J.*, **141**, 1303 (1965)
152. Noerdlinger, P., *Ap. J.*, **143**, 1004 (1966)
153. Noerdlinger, P., Jokipii, J., Woltjer, L., *Ap. J.*, **146**, 523 (1966)

154. Novikov, I. D., reported by V. A. Ambartsumian in *Structure and Evolution of Galaxies, 13th Solvay Congress*, 172; *JETP* (March 1966); also *Soviet Astron.*, **41**, No. 6 (1964)
155. Oke, J. B., *Nature*, **197**, 1040 (1963)
156. Oke, J. B., *Ap. J.*, **141**, 6 (1965)
157. Oke, J. B., *ibid.*, **145**, 668 (1966)
157a. Oke, J. B., *ibid.*, **147**, 901 (1967)
158. Osterbrock, D. E., *J. Planetary Space Sci.*, **11**, 621 (1963)
159. Osterbrock, D. E., Parker, R. A. R., *Ap. J.*, **143**, 268 (1966)
160. Ozernoy, L. M., *Astron. Zh.*, **43**, 300 (1966)
161. Ozernoy, L. M., *Soviet Phys. Dokl.*, **10**, 581 (1966)
161a. Pacholczyk, A. G., Wisniewski, W. Z., *Ap. J.*, **147**, 394 (1967)
162. Pacini, F., *Nature*, **209**, 389 (1966)
162a. Pauliny-Toth, I. K., Kellermann, K. I., *Ap. J.*, **146**, 634 (1966)
163. Piddington, J. H., *Monthly Notices Roy. Astron. Soc.*, **128**, 345 (1964)
164. Piddington, J. H., *ibid.*, **133**, 163 (1966)
165. Pilkington, J. D. H., Scott, P. F., *Mem. Roy. Astron. Soc.*, **69**, 183 (1965)
166. Price, R. M., Milne, D. K., *Australian J. Phys.*, **18**, 329 (1965)
167. Read, R. B., *Ap. J.*, **138**, 1 (1963)
168. Rees, M. J., *Nature*, **211**, 468 (1966)
169. Rees, M., Sciama, D. W., *Nature*, **208**, 371 (1965)
170. Rees, M. J., Sciama, D. W., *ibid.*, **207**, 738 (1965)
171. Rees, M. J., Sciama, D. W., *Ap. J.*, **145**, 6 (1966)
172. Roeder, R. C., Mitchell, G. F., *Nature*, **212**, 165 (1966)
173. Roxburgh, I., *Nature*, **207**, 363 (1965)
174. Ryle, M., Neville, A., *Monthly Notices Roy. Astron. Soc.*, **125**, 9 (1963)
175. Ryle, M., Sandage, A. R., *Ap. J.*, **139**, 419 (1964)
176. Ryle, M., Scheuer, P. A. G., *Proc. Roy. Soc. A*, **230**, 448 (1955)
177. Sandage, A. R., *Sky Telescope*, **21**, 148 (1961)
178. Sandage, A. R., *Ap. J.*, **139**, 416 (1964)
179. Sandage, A. R., *ibid.*, **141**, 1560 (1965)
180. Sandage, A. R., *ibid.*, **146**, 13 (1966)
181. Sandage, A. R., *I.A.U. Circ. No. 1961* (1966)
182. Sandage, A. R., *Ap. J.*, **144**, 1234 (1966)
183. Sandage, A. R., Luyten, W. (In preparation, 1967)
184. Sandage, A. R., Miller, W. C., *Ap. J.*, **144**, 1240 (1966)

185. Sandage, A. R., Véron, P., *Ap. J.*, **142**, 412 (1965)
186. Sandage, A. R., Véron, P., Wyndham, J. D., *Ap. J.*, **142**, 1306 (1965)
187. Sandage, A. R., Westphal, J. A., Strittmatter, P. A., *Ap. J.*, **146**, 322 (1966)
188. Sandage, A. R., Wyndham, J. D., *Ap. J.*, **141**, 328 (1965)
189. Saslaw, W. C., *Nature*, **211**, 729 (1966)
190. Scheuer, P. A. G., *Nature*, **207**, 963 (1965)
191. Scheuer, P. A. G., Wills, D., *Ap. J.*, **143**, 274 (1966)
192. Schmidt, M., *Nature*, **197**, 1040 (1963)
193. Schmidt, M. (Paper read at Second Texas Conf. Relativistic Astrophysics, Austin, Texas, December 1964)
194. Schmidt, M., *Ap. J.*, **141**, 1295 (1965)
195. Schmidt, M., *ibid.*, **144**, 443 (1966)
196. Schmidt, M., Matthews, T., *Ap. J.*, **139**, 781 (1964)
197. Sciama, D. W., *Monthly Notices Roy. Astron. Soc.*, **126**, 195 (1963)
198. Sciama, D. W., *Science Progr.*, **53**, 1 (1965)
199. Sciama, D. W., Rees, M. J., *Nature*, **211**, 1283 (1966)
200. Sciama, D. W., Rees, M. J., *Nature*, **212**, 1001 (1966)
201. Sciama, D. W., Saslaw, W. C., *Nature*, **210**, 348 (1966)
202. Scott, P. F., Ryle, M., *Monthly Notices Roy. Astron. Soc.*, **122**, 381 (1961)
203. Setti, G., Woltjer, L., *Ap. J.*, **144**, 838 (1966)
204. Sharov, A. S., Efremov, Yu. I., *Intern. Bull. Variable Stars*, No. 23 (Com. 27, I.A.U., 1963)
205. Shimmins, A. J., Day, G. A., Ekers, R. D., Cole, D. J., *Australian J. Phys.*, **19**, 837 (1966)
206. Shklovksy, I. S., *Astron. Zh.*, **41**, 801 (1964); *Soviet Astron. AJ*, **8**, 638 (1965); *Astron.Tsirk. No. 303* (1964)
206a. Shklovsky, I. S., *Astron. Zh.*, **42**, 893 (1965)
207. Shklovsky, I. S., *Proc. Byurakan Symp.* (1966)
208. Sholomitsky, G. B., *Intern. Bull. Variable Stars*, No. 83 (Com. 27, I.A.U., 1965)
209. Slish, V. I., *Nature*, **199**, 682 (1963)
210. Smith, H. J., Hoffleit, D., *Publ. Astron. Soc. Pacific*, **73**, 292 (1961)
211. Smith, H. J., Hoffleit, D., *Nature*, **198**, 650; Smith, H. J., *Dallas Symp.*, 227 (1963)
212. Spitzer, L., Saslaw, W. C., *Ap. J.*, **143**, 400 (1966)
213. Stein, W., *Ap. J.* (In press) (1967)
214. Stockton, A. N., Lynds, C. R., *Ap. J.*, **144**, 451 (1966)
215. Stothers, R., *Monthly Notices Roy. Astron. Soc.*, **132**, 217 (1966)
216. Strittmatter, P. A. (Private communication, 1966)
217. Strittmatter, P. A., Burbidge, G. R., *Ap. J.*, **147**, 13 (1967)
218. Strittmatter, P. A., Faulkner, J., Walmsley, M., *Nature*, **212**, 1441 (1966)
219. Sturrock, P. A., *Nature*, **208**, 861 (1965)
220. Sturrock, P. A., *ibid.*, **211**, 697 (1966)
220a. Sturrock, P. A., *Stanford Univ. Inst. Plasma Res. Preprint No. 99* (1966)
221. Terrell, J., *Science*, **145**, 918 (1964)
222. Terrell, J., *Ap. J.*, **147**, 827 (1967)
223. van den Bergh, S., *Ap. J.*, **144**, 866 (1966)
224. Véron, P., *Nature*, **211**, 724; *Ann. Ap.*, **29**, 231 (1966)
225. Véron, P., *Ap. J.*, **141**, 1284 (1965)
226. Wagoner, R., *Phys. Rev. Letters*, **16**, V503(c) (1965)
226a. Wampler, E. J., *Ap. J.*, **147**, 1 (1967)
226b. Wampler, E. J., Oke, J. B., *Ap. J.* (In press) (1967)
227. Williams, I. P., *Phys. Letters*, **14**, 19 (1965)
228. Williams, P. J. S., *Nature*, **200**, 56 (1963)
229. Williams, P. J. S., *ibid.*, **210**, 285 (1966)
230. Wills, D., *Observatory*, **86**, 245 (1966)
231. Woltjer, L., *Structure and Evolution of Galaxies, 13th Solvay Congress*, 30 (Wiley, New York, 1965)
231a. Woltjer, L., *Ap. J.*, **146**, 597 (1966)
231b. Woltjer, L. (Paper read at 3rd Texas Conf. Relativistic Ap., New York, January 1967)
232. Wyndham, J. D., *Astron. J.*, **70**, 384 (1965)
233. Wyndham, J. D., *Ap. J.*, **144**, 459 (1966)
234. Wyndham, J. D. (Preprint, 1966)
235. Zapolsky, H. S., *Science*, **153**, 635 (1966)
236. Zeldovich, Ya. B., Okun, L., Pikelner, S. B., *Phys. Letters*, **17**, 164 (1965)
237. Zwicky, F., *Ap. J.*, **142**, 1293 (1965)

THE DYNAMICS OF DISK-SHAPED GALAXIES[1]

By C. C. Lin

Massachusetts Institute of Technology, Cambridge, Massachusetts

INTRODUCTORY REMARKS

The present article is a report on certain theoretical aspects of the recent developments in the dynamics of highly flattened galaxies. In view of the recent publication (1965) of Volume V of *Stars and Stellar Systems: Galactic Structure*, which contains several articles related to the subject under discussion (e.g., those by Woolley, Sharpless, Kerr & Westerhout, Oort, Schmidt, and Woltjer), this article will be restricted to those aspects of the problem which are not fully covered in that book or in the recent articles of Perek (1962), King (1963), and Wentzel (1963). The reader is referred to the article by Lebovitz (1967) in this volume for other aspects of the dynamics of rotating masses, to articles by Contopoulos (1966) for investigations related to the third integral, and to the 1967 report of I.A.U. Commission 33 prepared by Bok and Contopoulos for further references and general information.

It turns out that a major part of the efforts in this area during the past few years centered around the problem of the spiral structure, but there are other noteworthy developments. We shall give a brief overall survey of all these problems in Part I and devote the balance of this article (Part II) to a more detailed discussion of the issues and the work concerned with the dynamical theory of the spiral structure.

PART I: SOME PROBLEMS OF DYNAMICAL INTEREST

The thin disk model.—About 70 per cent of the known galaxies are normal spirals consisting primarily of a prominent spiral pattern over a disk-shaped structure, but having also a nuclear region, and an associated halo (spherical system). To examine the dynamical behavior of such a galaxy, we usually adopt, as a first step, a simple disk model, with small or even no thickness. Such an idealization is convenient in order to arrive at a relatively simple description of the essential behavior of a basically complicated system; but we must also realize the limitations imposed.

(*a*) The nuclear region is relatively dense compared to the rest of the galaxy. The possibility even remains that (see e.g., Spitzer & Saslaw 1966) "at least some of the very luminous radio sources may be galactic nuclei" which are going through a certain stage of evolution. To counteract gravitational collapse, the peculiar velocity of the stars in this heavy nuclear region must be quite large. Therefore, in a simple disk model, it may be quite

---

[1] The survey of literature for this review was concluded in January 1967.

appropriate to acknowledge a singular behavior at the center. At the same time, it is to be expected that not all the observed features in the very central part of the galaxy can be adequately described in a pure disk model. A separate effort must be made to deal with the nuclear region. Indeed, the observed gas motion in this region appears to be very complicated. It is not unlikely that the rotation of the nucleus is so fast that a barlike structure would result in a manner similar to the formation of a Jacobi ellipsoid. This may have a bearing on the two-armed structure of the spiral pattern; but no extensive systematic investigation seems to be available.

(b) In view of the observed gaseous clouds at high galactic latitudes, Oort (e.g. 1962, 1966), Blaauw (1966), and their collaborators have suggested the possible existence of substantial amount of gas moving from the halo region into the disk. It has not yet been firmly established, however, whether such gaseous clouds are located nearby (say, within 1 kpc), at the edge of the galactic system, or are entirely extragalactic. There is also some uncertainty in the determination of the amount of gas involved (Shklovsky 1966).[2] If there does exist such a flow of gas at a substantial rate between the halo and the disk, it would contribute significantly to the mass and momentum balance in the disk. Indeed, in a certain picture of gas circulation in the spiral arms, it would play an essential role. On the other hand, if we adopt the latest density-wave theory for the spiral structure (Lin & Shu 1966a,b), this problem can be relegated to a separate investigation, since it is then not necessary to depend on such a gaseous flow in order to arrive at a consistent dynamical picture. The possibility of course still remains open that this influx of gas is actually present at a substantial rate. It is in principle not difficult to incorporate such a flux into a dynamical theory for the disk, and it is not expected that this will upset the basic density-wave mechanism.

Equilibrium models.—Consider therefore only a very thin disk of stars and gas in differential rotation, without a halo region. The first dynamical problem is indeed to construct equilibrium models for such a system, based primarily on the balance between centrifugal acceleration and gravitational attraction, including both the stars and the gas. One can easily convince oneself that the hydromagnetic forces in the gas are of secondary importance in such a balance.

In the simplest model, all matter is supposed to be in exactly circular motion, and there is consequently no difference between the gas and the stars. A comprehensive review of the methods for constructing such self-gravitating rotating disks was given by Perek (1962). Since then, Toomre (1963) has described a new method for deducing the mass density in an

---

[2] Shklovsky suggests that the maser effect in the interstellar hydrogen clouds of the galactic halo might lead to a substantial overestimate of the amount up to a factor of $\sim$200. In a private communication (Jan. 11, 1967), C. H. Townes has shown that the danger of this overestimate is exaggerated.

explicit manner once the rotating curve is known. Disks of infinite radius are considered. In particular, the method yields simple analytic expressions for a family of galactic models.

Hunter (1963) developed a method involving the use of oblate spheroidal coordinates. General expressions for the law of rotation and the density distribution are found as related series of Legendre functions. Simple particular disk models involving only a finite number of terms in these series can be found, the simplest being the well-known uniformly rotating disk.

Vandervoort (1967) developed a systematic method for constructing a model of finite thickness in dynamic equilibrium. Starting with any model of the elementary kind described above, he gives an asymptotic procedure for inclusion of the effect of the dispersion velocity of the stars. In the first approximation, the method shows that the dispersion velocity should follow the ellipsoidal distribution with the major axis pointing in the direction of rotation. In the direction perpendicular to the disk, the distribution in phase space is found to be as expected, that for a nonrotating stellar sheet in gravitational equilibrium.

Ng (1967) considered a family of self-consistent models of disk galaxies, assuming that the distribution function depends on the energy and angular momentum integrals in a certain particular manner, involving three parameters.

For a model of our own Galaxy, Schmidt (1965) has constructed an improved model for circular velocity and three-dimensional mass distribution, based on more and newer observational data. The results differ from his 1956 model in some respects. The Sun is now located at 10 kpc from the center (instead of 8.2 kpc), and the projected surface density in the solar vicinity is 114 (instead of 52) $M_\odot/pc^2$. As in the 1956 model, no attempt was made to describe the detailed dynamical balance perpendicular to the disk; but it is implied that gravitational attraction, dispersive motion of the stars, etc. are in equilibrium in a manner similar to that described by Vandervoort. For further details, the reader is referred to Schmidt's article.

Collective modes.—In a self-gravitating system, the motion is determined by the collective gravitational field of all the matter involved, i.e., by interactions involving long-range forces. This is true of the equilibrium configuration as well as for collective modes on smaller scales. For these modes, it is no longer obvious that hydromagnetic forces can be neglected, for their relative importance becomes enhanced with decreasing scale. In a proper dynamic theory, one must put into proper perspective the relative importance of gravitational forces and hydromagnetic forces. One should however note that hydromagnetic containment of interstellar gas is not the answer to the well-known winding dilemma of material arms by differential rotation, for it is basically a kinematical difficulty. Rather, hydromagnetic waves might play a role, if the magnetic field is sufficiently strong. Wentzel (1964, unpublished) investigated this problem, and came to the conclusion that the field is not sufficiently strong by itself to maintain a wave pattern. It now appears that the structure of the spiral pattern over the whole disk is essentially controlled by gravitational forces, whereas hydromagnetic forces may be important in the structure of an individual part of a spiral arm. We shall have occasion to come back to the magnetic field later.

For the present, let us consider collective modes associated with gravitational forces alone. These are well known in connection with the study of gravitational instability (of rotating gaseous masses, for example). In the case of rotating disks, such studies were recently undertaken by Hunter (1963) for a uniformly rotating disk, and by Toomre (1964), Hunter (1965), and Rehm (1965) for a disk in differential rotation. They found that the disk tends to collapse into rings if the material is in strictly circular motion. Nonaxisymmetrical modes are also considered by Hunter and by Rehm, with similar indications of instability.

Toomre (1964) considered also a disk of stars with noncircular velocity and showed that the instability may then be removed if the dispersion velocity is sufficiently large. He then advanced the idea that the stars in the galactic disk (and indeed also the gas before their formation) would tend to develop noncircular motion from the gravitational energy via this tendency toward gravitational collapse. Further attempts to develop a theory for the enhancement of random velocities have been carried out by Julian (1967).

Of special interest are collective modes of the spiral form, for they may be used as a basis for explaining the occurrence of spiral patterns over the galactic disk. Such a theory was developed in detail by Lin & Shu (1964, 1966a,b). The basic ideas of this theory will be outlined in Part II. It is related to that of Bertil Lindblad (see his paper of 1963 and earlier references) and P. O. Lindblad (1960, 1962): the similarities and differences will be discussed.

It is indeed not difficult to get spiral modes, for they can be obtained by the method of separation of variables if we do not restrict ourselves to axisymmetrical modes. As mentioned above, Hunter (1963, 1965) and Rehm (1965) considered nonsymmetrical small disturbances over a rotating disk and found them to be generally very unstable when the basic motion is strictly circular. Lin & Shu (1966a,b) considered the neutral density waves of the spiral form when the stellar dispersion velocity and gas turbulent velocity are sufficiently large to prevent such instabilities.

The possibility for the pattern to take on the spiral form is obviously an issue of great interest. It would be too complicated to go into a detailed discussion here. We should however refer briefly to the "antispiral theorem" of Lynden-Bell & Ostriker (1966):

This theorem applies to the normal modes of steady flows of inviscid gas in which the velocity is in circles about [a common] axis and magnetic forces are absent. Further it is assumed that small relative changes in pressure and density are proportional. Our theorem is that under these conditions we may choose a complete set of normal modes such that no stable normal mode has a spiral structure. Further, it is only possible to have stable normal modes with spiral structure if stable normal modes of the same symmetry are degenerate. Such degenerate spiral modes occur in conjugate pairs, one leading wherever the other is trailing.

While this theorem is very interesting, it cannot be directly compared with the conclusions of Lin & Shu (1966a,b); for it is restricted to a linear gas dynamical theory with uniform velocity of sound. For one thing, the work of Lin & Shu is based on stellar dynamics. There are indeed substantial differences when Lindblad resonance (see below) is involved. Secondly, in order to get neutral density waves of small but finite amplitudes, Lin & Shu did consider introducing nonlinear effects.

It should also be pointed out that a barlike singularity at the center is not to be excluded. After all, the fast-rotating heavy nucleus has a tendency to assume such a form. Rehm (1965) considered the case of a regular rotating disk, allowing for a "dipole" (barlike) singularity in the disturbances. The role of the heavy nucleus is thus suitably represented, though not explicitly included. He found that both leading and trailing spirals were possible solutions, depending on the angular velocity of the pattern. In the central part of the pattern, where the asymptotic calculations of Lin & Shu (1964) are expected to yield reliable results, good agreement was found.

Toomre (1964) considered the possibility of having density concentrations similar to the shape of a spiral arm which travel at approximately the local circular velocity of the stars and which are subjected to the wrapping process due to differential rotation. This type of density concentration is substantially different in nature from those considered by Lin & Shu as density waves. The two types of modes of density concentration, however, are not mutually exclusive, and presumably coexist in an actual galaxy, which often shows "branches" in the spiral structure. Indeed, it is not out of the question that the Sun should turn out to be located on such a "branch." Excellent detailed mathematical investigations of phenomena of this kind have been made by Goldreich & Lynden-Bell (1965) for the case of a gaseous disk, and by Julian and Toomre (1967) for the case of the stellar disk.

Lynden-Bell (1965) studied the free precession of the Galaxy. He showed that a one-degree deviation between the direction of the Galaxy's symmetry axis and that of its angular momentum may account for the observed bending of the galactic "plane." Hunter and Toomre (not yet published) studied more general bending modes. The tidal distortion of the Galaxy by the Magellanic clouds has also been considered by Habing & Viser (1966), by Toomre and Hunter, and by others. The problem has been studied by Elwert & Hablick (1965) without considering collective interaction. It has also been considered from the point of view of interaction with interglactic medium (Kahn & Woltjer 1959). It would appear that a theory based on gravitational factors alone would promise to account for all the principal features observed.

PART II: SPIRAL STRUCTURE

Evolution of ideas.—The dilemma of winding (or unwinding) of spiral arms and the consequent change of spacing is well known. In an excellent survey, Oort (1962) discussed these difficulties and the various kinematical possibilities of gaseous circulation. For a number of years, the late Bertil Lindblad (see his paper of 1963 and the references to previous papers) had advocated the concept of *density waves* as the dynamical basis for the spiral structure in highly flattened galaxies. Such a theory is free from the dilemma of differential rotation. It is visualized that a density wave with a spiral structure may be maintained by the self-gravitational field with a spiral structure superposed on the symmetrical gravitational field of the basic symmetrical disk. By studying the orbits of the individual stars, B. Lindblad was able to show, in a general manner, that a spiral gravitational field will *tend* to produce a density redistribution, which in turn *tends* to support the imposed field. In particular, he noted the possibility of resonance between the density wave and the epicyclic motion of the stars. Indeed, if the spiral structure has *two* arms, this resonance can even extend over a substantial portion of the galactic disk, for a *particular* distribution of the circular velocity as a function of the distance from the galactic center. B. Lindblad noted that this distribution is at least approximately satisfied for a substantial inner part of our own Galaxy.

Clearly, both the symmetrical field and the spiral field are due to the *collective* gravitational effect of the stars and of the gas. The method adopted by B. Lindblad is unfortunately not suited to the treatment of such collective behavior. He was therefore unable to give much-needed quantitative results; for example, a relationship between the wavelength and the frequency of the suggested waves. He was also led to place an undue emphasis on the occurrence of *exact* gravitational resonance over a finite portion of the galactic disk.

In trying to describe this collective behavior, P. O. Lindblad (1960, 1962) studied the motion of a system of 192 stars (representing Population I) in a given central field of force (due to Population II), with the aid of a high-speed computing machine. The calculated results show an evolving spiral structure with a typical time scale of a few hundred million years. On the other hand, B. Lindblad (1963) specifically suggested a *quasi-stationary spiral structure.*

In 1961, a conference (initiated and organized by B. Strömgren) was held at Princeton, N.J. to discuss problems of the distribution and motion of interstellar matter in galaxies. After attending this conference, the present writer also arrived at the concept of a quasi-stationary spiral structure maintained by density waves. It was visualized that a direct study of the collective behavior might allow us to describe density waves of a spiral form which are initiated by a certain form of dynamical instability, but which evolve only slowly (with a time scale of perhaps a few billion years) after it has reached a small but finite amplitude. Discussions with B. Strömgren and

L. Woltjer helped to consolidate and to develop these ideas. An elementary form of the theory was worked out in collaboration with Frank H. Shu (Lin & Shu 1964, Shu 1963). Instead of dealing with the behavior of individual stars, we adopted a method that deals directly with the density distribution of matter in the galaxy, and with the associated gravitational field. A more complete form of the theory, based on the laws of stellar dynamics and using the distribution function in phase space, was published in 1966 (Lin 1966a,b; Lin & Shu 1966a,b). In both forms of the theory, the hypothesis of quasi-stationary spiral structure (often abbreviated as the *QSSS hypothesis*) was shown to be strongly suggested by a simple mathematical procedure, and was adopted as the central working hypothesis.

So far, Lin & Shu have considered in detail only waves in the form of tightly wound spirals. Kalnajs (1965) emphasized the role of loosely wound spirals. Quantitative results are more difficult to obtain in this latter case. We shall therefore not go into a discussion of his work in this review; the reader is referred to his thesis for further details.

Mechanism for self-sustained density waves.—It is easy to be led to the hypothesis of the presence of a quasi-stationary spiral structure in disk-shaped galaxies from a casual examination of their photographs (Sandage 1961) and from a consideration of the winding dilemma in the face of differential rotation, as discussed above. It is thus natural to take the QSSS hypothesis as a focal point in the development of a theory. On the one hand, we may proceed to examine the dynamical basis for a mechanism to maintain such a structure. On the other hand, we may explore the consequences of such a hypothesis and compare them with observations. Both have been done by Lin (1965, 1966a,b) and by Lin & Shu (1964, 1966a,b). Work is still continuing in both directions. We shall outline their studies of the dynamical basis in this section and present a general comparison with observations in the next.

Lin & Shu considered a disk of stars with velocity dispersion and a co-existent gaseous disk with turbulent motion, simulated by pressure. The effect of hydromagnetic forces has not yet been included. They first demonstrated the plausibility of a purely gravitational theory for density waves by a continuum treatment applied to gas and stars of low velocity dispersion, neglecting the effect of pressure and velocity dispersion. These were later included in their calculations. In the present form, their work essentially amounts to establishing the possible existence of certain density waves of the spiral form,[3] sustained by self gravitation. These are collective modes in the combined gaseous and stellar system. Even in the gas, the long-range gravitational forces play the primary role. The pressure force or turbulent motion provides a stabilizing influence against gravitational collapse.

Even in the absence of velocity dispersion, Lin & Shu (1964) found certain density waves of a spiral form, but Toomre (1964) pointed out that

---

[3] The spiral form is simply a mathematical consequence of solution by the method of separation of variables.

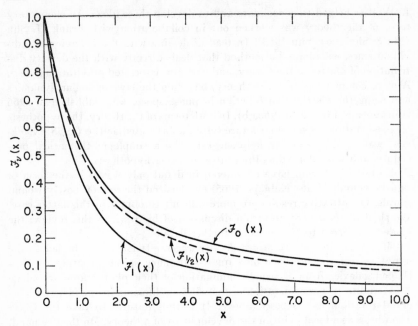

FIG. 1. The reduction factor. The abscissa is a measure of the mean square value of the radial component of the peculiar velocity of the stars.

there would be violent instability unless the velocity dispersion of the stars exceeds a certain minimum amount. By incorporating this velocity dispersion, Lin (1966), Lin & Shu (1966a) found that the waves still exist, but the effectiveness of the material to participate in the motion has to be decreased by a reduction factor, which can be explicitly calculated. The result is shown in Figure 1. Briefly, the physical interpretation is the following. Stars which move in epicycles with radius at least twice as large as the spacing between spiral arms can hardly be influenced by the spiral gravitational field. Thus, they contribute only to the average axisymmetrical gravitational field, while only stars with comparatively low velocity dispersion can participate in the formation of the density wave.

With the inclusion of this reduction factor, the dispersion relation (i.e. frequency wavenumber relationship) becomes exactly the same as that in the elementary theory (Lin & Shu 1964), which gives us more confidence in the whole analysis.

The theory was first developed for an infinitesimally thin disk. The effect of thickness has been recently included by Shu (not yet published).

In order that these waves may be observable, they must be maintained at a small but finite amplitude. Lin & Shu visualized the mechanism as a balance of a mild instability by nonlinear effects. From their analysis, they concluded that this instability is primarily caused by the gradient of stellar dispersion velocity which decreases from the center of the galaxy outwards. This appears to bear an analogy with certain instabilities found in plasmas, but the detailed physical processes have not yet been fully described. It is interesting that this instability does favor trailing waves, as nonlinear effects would also be expected to do.

The basic processes involved in the theory described above can be conveniently divided into the following three parts:

1) the response of the gaseous disk to a spiral gravitational field;
2) the response of the stellar disk to a spiral gravitational field;
3) the density distribution associated with a spiral gravitational field according to Poisson's equation.

Calculations are now available for a satisfactory demonstration of the theory, but further work involving more careful analysis is very much desired. The linear theory for 1) is too simple to deserve a publication of details. Fujimoto (1966, not yet published) made numerical calculations of the nonlinear response, including isothermal shocks. Helfer (1966) attempted an analytic treatment. The linear theory for 2) was carried out by Lin & Shu by using the asymptotic theory. An exact treatment of the linear theory has also been carried out in principle; the implications of this analysis are still being worked out. A study of 3) in the spheroidal coordinate system, including some fairly extensive numerical investigation, was recently carried out by Barbanis & Prendergast (1966) (see also Hunter 1963, 1965). The earlier work of Lin & Shu (1964) was again based on an integral representation as well as an asymptotic analysis.

An appraisal of the theory.—In order to make an appraisal of the theory, we shall first briefly summarize some of the important characteristics found for the spiral waves. For further details, see Lin (1966a,b), Lin & Shu (1966a, b), and other pending publications.

(*a*) Trailing waves are preferred.

(*b*) Waves with pattern speed Ω_p around the center can be self sustained only for the range of values of the radial distance ϖ for which the inequality

$$\Omega - \frac{\kappa}{m} < \Omega_p < \Omega + \frac{\kappa}{m}$$

holds, where $\Omega(\varpi)$ and $\kappa(\varpi)$ are the angular velocity and the epicyclic frequency, and m is the number of arms. From an examination of the rotation curves, Lin & Shu have shown that only *two*-armed spirals can be expected to be prominent in galaxies similar to our own. At the end points of this range, the type of resonance discussed by B. Lindblad holds.

(*c*) The waves satisfy a dispersion relationship such as that exhibited in Figure 2. In this figure, the ordinate is the radial wavelength λ measured in terms of the length scale

$$\lambda_* = 4\pi^2 G \sigma_* / \kappa^2$$

and the abscissa is

$$\nu = m |\Omega - \Omega_p| / \kappa$$

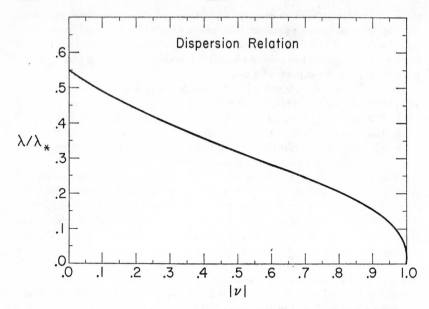

Fig. 2. Relationship between radial wavelength and the (angular) frequency at which the stars encounter the pattern (after Shu).

i.e., the (angular) frequency at which the stars meet with the pattern, measured in terms of the epicyclic frequency κ. In the above formula, σ^* (ϖ) is the projected surface density of the stars, and G is the gravitational constant.

To see the significance of these results, it should be emphasized that the primary purpose of the theory is to explain the occurrence of a spiral pattern over the *whole* galactic disk by providing a mechanism for its maintenance. For this primary purpose, the theory has produced the following conclusions which appear satisfactory from a general point of view.

(*a*) The theory is free from the kinematical difficulty of differential rotation, and it provides predictions on the systematic motion of the gas which can be checked against observations. (See below.)

(*b*) The theory enables us to provide a mechanism to explain the existence of a *spiral pattern over the whole disk* while allowing the individual spiral *arms* to be broken and fragmentary.

(*c*) The theory indicates that *two*-armed trailing patterns are generally to be expected, in agreement with observations.

(*d*) The scales of the patterns obtained from the theory are in general conformity with those observed.

There are at least two important implications of the theory of density waves, which we shall now describe.

The first is the systematic motion of the gas implied by the density-wave theory. This has a direct bearing on the reduction of data obtained from radio observations. Even without this systematic motion, the radio observations already suffer from a two-valued ambiguity in the assignment of distance (and hence also of gas content). For example, suppose there is a circular spiral arm of gas passing through our neighborhood; the observable motion in the line-of-sight is expected to be very small in any case, and care must be exercised to distinguish it from local features. Such difficulties obviously are greatly increased by the presence of systematic motions.

It would be very interesting to see if a better picture of the structure of the Galaxy can be arrived at from radio observations when we make use of the theory discussed here. In particular, the reconciliation of data obtained from the southern and the northern observations might be attempted in this light.

Another important implication of the theory is the following. If the distribution and motion of the gas is indeed stationary in the framework rotating with the angular velocity Ω_p, the magnetic lines of force would eventually also settle into a nearly stationary pattern, with a nearly definite magnitude. The lines of force would then be essentially running along the direction of the spiral arm. This conclusion appears to be in general agreement with observations. It would be interesting to explore a theory of star formation under these conditions. One would be inclined to think that gravitational condensation would tend to occur first *along* the magnetic lines of force. When sufficient matter density is built up, contraction *across* a spiral arm would be facilitated. All these considerations belong to a study of the phenomena on or below the scale of a spiral arm, and a separate investigation is needed.

LITERATURE CITED

Barbanis, B., Prendergast, K. H. 1966 (Preprint, Columbia Univ.)

Blaauw, A. 1966, *IAU-URSI Symp. No. 31, Noordwijk*

Contopoulos, G. 1966, Problems of Stellar Dynamics, *Proc. Space Math. Seminar* (AMS), **1**, 169

Elwert, G., Hablick, D. 1965, *Z. Ap.*, **61**, 273–84

Fujimoto, M. 1966 (Preprint, Columbia Univ.)

Goldreich, P., Lynden-Bell, D. 1965, *Monthly Notices Roy. Astron. Soc.*, **130**, 97–124, 125–58

Habing, H. J., Viser, H. C. D. 1966, *IAU-URSI Symp. No. 31, Noordwijk*

Helfer, H. L. 1966, *IAU-URSI Symp. No. 31, Noordwijk*

Hunter, C. 1963, *Monthly Notices Roy. Astron. Soc.*, **126**, 299–315

Hunter, C. 1965, *ibid.*, **129**, 321–43

Julian, W. N. 1967, *Ap. J.*, **148**, 175–84

Julian, W. H., Toomre, A. 1966, *Ap. J.*, **146**, 810

Kahn, F. D., Woltjer, L. 1959, *Ap. J.*, **130**, 705–17

Kalnajs, A. J., 1965 (Ph.D. thesis, Harvard Univ.)

Kerr, F. J., Westerhout, G. 1965, in *Galactic Structure*, 167–200 (Univ. of Chicago Press)

King, I. R. 1963, *Ann. Rev. Astron. Ap.*, **1**, 179–202

Lebovitz, N. 1967, *Ann. Rev. Astron. Ap.*, **5**, 465–80

Lin, C. C. 1965, *Am. Math. Soc. Seminar Relativity and Ap.* (Cornell Univ.)

Lin, C. C. 1966a, *J. Siam Appl. Math.*, **14**, 876–927

Lin. C. C. 1966b, *Vetlesen Symp.* (Columbia Univ., New York)

Lin, C. C., Shu, F. H. 1964, *Ap. J.*, **140**, 646–55

Lin, C. C., Shu, F. H. 1966a, *Proc. Natl. Acad. Sci. U. S.*, **55**, 229–34

Lin, C. C., Shu, F. H. 1966b, *IAU-URSI Symp. No. 31, Noordwijk*

Lindblad, B. 1963, *Stockholm Obs. Ann.*, **22**, 3–20

Lindblad, P. O. 1960, *Stockholm Obs. Ann.*, **21**, 3–73

Lindblad, P. O. 1962, in *Interstellar Matter in Galaxies*, 222–33 (W. A. Benjamin, New York)

Lynden-Bell, D. 1965, *Monthly Notices Roy. Astron. Soc.*, **129**, 299–307

Lynden-Bell, D., Ostriker, J. P. 1966, *14th Liège Symp.*

Ng, E. W. K. 1967 (Ph.D. thesis, Columbia Univ.)

Oort, J. H. 1962, in *Interstellar Matter in Galaxies*, 3–22, 234–44 (W. A. Benjamin, New York)

Oort, J. H. 1965, in *Galactic Structure*, 455–512 (Univ. of Chicago Press)

Oort, J. H. 1966, *IAU-URSI Symp. No. 31, Noordwijk*

Perek, L. 1962, *Advan. Astron. Ap.*, **1**, 165

Rehm, R. G. 1965 (Ph.D. thesis, Mass. Inst. of Technol., Cambridge, Mass.)

Sandage, A. 1961, *The Hubble Atlas of Galaxies* (Carnegie Inst. of Washington, Washington, D.C.)

Schmidt, M. 1965, in *Galactic Structure*, 513–30 (Univ. of Chicago Press)

Sharpless, S. 1965, in *Galactic Structure*, 131–56 (Univ. of Chicago Press)

Shklovsky, I. S. 1966, *IAU-URSI Symp. No. 31, Noordwijk*

Shu, F. H. 1963 (B.Sc. thesis, Mass. Inst. of Technol., Cambridge, Mass.)

Spitzer, L., Saslaw, W. C. 1966, *Ap. J.*, **143**, 400

Toomre, A. 1963, *Ap. J.*, **138**, 385–92

Toomre, A. 1964, *ibid.*, **139**, 1217–38.

Townes, C.H. 1967 (Private communication)

Vandervoort, P. O. 1967, *Ap. J.*, **147**, 91–111, 384

Wentzel, D. G. 1963, *Ann. Rev. Astron. Ap.*, **1**, 195

Wentzel, D. G. 1964 (Unpublished)

Woltjer, L. 1965, in *Galactic Structure*, 531–88 (Univ. of Chicago Press)

Woolley, R. 1965, in *Galactic Structure*, 85–110 (Univ. of Chicago Press)

ROTATING FLUID MASSES[1]

By N. R. Lebovitz

University of Chicago, Chicago, Illinois

As a practical matter, discussions of the effect of rotation on self-gravitating fluid masses divide into two categories: the structure of steady-state configurations, and the oscillations and the stability of these configurations. This division will be reflected in the present article. Consideration will be restricted to configurations described by the hydrodynamical equations in which appears a nonvanishing, isotropic pressure. Although magnetic fields may have important effects when combined with rotation (cf. Chandrasekhar 1961, p. 389; Ferraro 1937; Roxburgh 1966a), they will not be discussed here.

STEADY-STATE CONFIGURATIONS

The problem of finding steady-state configurations is as follows: in cylindrical coordinates ϖ, φ, and z, to find the pressure p and the density ρ as functions of position, when the velocity components have the forms

$$v_{\varpi} = 0, \qquad v_{\phi} = \varpi\Omega(\varpi, z), \qquad v_z = 0 \qquad\qquad 1.$$

The pressure and density must satisfy the hydrodynamical equations

$$\mathrm{grad}\ p = \rho[\mathrm{grad}\ \mathfrak{B} + \tfrac{1}{2}\Omega^2\ \mathrm{grad}\ \varpi^2] \qquad\qquad 2.$$

where $\mathfrak{B}$, the gravitational potential, is given by the formula

$$\mathfrak{B}(x) = G \int_{\mathfrak{R}} \frac{\rho(x')}{|x - x'|}\ dx' \qquad\qquad 3.$$

In addition, equations of state and of energy conservation must be specified, as well as conditions at the boundary of the domain $\mathfrak{R}$.

There is implicit in the formulation above a mathematical difficulty of a kind not encountered when rotation is absent: the integration in Equation 3 extends over a domain $\mathfrak{R}$ that is not given a priori. Moreover, conditions (e.g., the vanishing of the pressure) must be imposed on $\partial\mathfrak{R}$, a surface whose discovery constitutes part of the problem. The difficulty of such free-boundary problems prompted Volterra (1903) to complain that the rigorous results in the theory were mostly negative. Since the time of Volterra's complaint, positive progress has been made on both the purely mathematical and the practical fronts. The classical analyses of Liapounov and Lichtenstein (for an account of their work, together with bibliographies on these and other matters, see Jardetzky 1958) have established existence theorems for classes of problems included in the formulation above. In recent years, many approximate models have been found by numerical and by analytic means. It

---

[1] The survey of literature for this review was concluded in December 1966.

465

nevertheless remains true that among the principal general results, several are of a negative character.

General Results

We here give a résumé of the small but important group of results that apply to wide classes of steady-state configurations.

Existence of a centrifugal potential.—A necessary and sufficient condition for Ω (cf. Equation 1 above) to be independent of z is that the surfaces of constant pressure coincide with the surfaces of constant density, i.e., that p be a function of ρ only. Further, in this case the total potential U, where

$$U = \mathfrak{B} + \int^{\tilde{\omega}} \Omega(\tilde{\omega})^2 \tilde{\omega} d\tilde{\omega} \qquad 4.$$

is also a function of ρ only (see, for example, Wavre 1932). The integral in Equation 4 is the centrifugal potential. When it exists, the equations of state and of energy conservation may be thought of as determining the form of the p-ρ relationship. Hence, by prescribing a p-ρ relationship, one avoids the complications of those further equations. This effects a major simplification of the formal problem of constructing rotating configurations. This procedure will, of course, be inadequate for certain objectives, and much effort has been made (cf. Roxburgh, Griffith & Sweet 1965; Roxburgh 1964) in order to include the equations of state and of energy conservation explicitly.

Poincaré's estimate.—The size of a rotating configuration is limited by the condition that the centrifugal acceleration of rotation not exceed the centripetal acceleration of gravity. This leads, in the case of uniform rotation, to the following estimate by Poincaré (1903) relating the angular velocity to the mean density $\bar{\rho}$:

$$\Omega^2 \leq 2\pi G\bar{\rho} \qquad 5.$$

It is evident from the derivation of this inequality that equality cannot be attained, or even very closely approximated, for figures topologically like those one envisions for stars (equality can be attained for an infinite cylinder; cf. Tassoul & Cretin 1965).

Lichtenstein's theorem.—Another condition that a rotating mass must satisfy is that expressed by Lichtenstein's theorem: the configuration has a plane of symmetry perpendicular to the axis of rotation. This theorem, first proved for rigid-body rotation and uniform density (cf. Lichtenstein 1933), has been extended to angular velocities independent of z and a wide class of density stratifications, by Wavre (1932). A corollary is that a nonrotating configuration is necessarily spherical.

von Zeipel's theorem.—In 1924 von Zeipel derived a striking result by explicitly taking into account the equations of state and of energy conservation. Under the assumptions of uniform rotation and of purely radiative equilibrium in which the nuclear energy generation ϵ balances the radiative losses, he showed that ϵ would have to follow a very special law, namely (cf. Mestel 1965)

$$\epsilon = \text{const} \left(1 - \frac{\Omega^2}{2\pi G\rho} \right) \qquad\qquad 6.$$

Since the energy generation cannot conform to this formula, at least one of the assumptions leading to it must be relaxed.

It was observed (see, for example, Eddington 1926) that if, at a given instant, the energy flux should conform to the assumptions leading to von Zeipel's theorem, the subsequent development of the star would result in differences of temperature over a given level surface, which would cause fluid motions to be superimposed on the uniform rotation. We shall mention two possible consequences of this. First, one can suppose the fluid motions still obey Equation 1 above. For an axisymmetric star this continues to imply that all the energy transport is by radiation. One should expect (Roxburgh (1966b)

$$\frac{\partial \Omega}{\partial z} \neq 0 \qquad\qquad 7.$$

so that surfaces of constant pressure and of constant density no longer coincide. The partial derivative in the inequality above is positive or negative according as the equatorial temperature is higher than or lower than the polar temperature (Randers 1942).

A second possible consequence of the temperature differences is the setting up of a steady-state, meridional circulation, which is then partly responsible for the transport of energy. The meridional motions are extremely slow compared to the rotational motion (Sweet 1950), but could have important evolutionary effects (for a review, see Mestel 1965).

Ellipsoidal stratifications.—Further "negative results" have been emphasized by Wavre and Dive (cf. Dive 1952, where a bibliography and résumé are given), who have shown that if Ω is independent of z, ellipsoidal stratifications of density and pressure are impossible unless the density is constant.

SOLUTIONS IN CLOSED FORM

We shall discuss ellipsoidal configurations of constant density (central condensation of unity) and the Roche model (infinite central condensation). These two closed-form solutions have influenced the thinking of astronomers not only because of the inherent tractability of closed-form solutions, but also because, together, they bracket the range of central condensation.

The Riemann ellipsoids.—The most general formulation of the motion of an ellipsoidal mass of uniform density is due to Riemann (1860; see also Lebovitz 1965, Chandrasekhar 1965, 1966). The underlying assumption, due to Dirichlet (1860), is that in Cartesian coordinates, the components of the fluid velocity are linear functions of position, with (possibly) time-dependent coefficients. The application of the physical and kinematical requirements leads to a system of nine ordinary, nonlinear differential equations in the time, which serve to determine the coefficients in the velocity-position relationship. These in turn determine the semiaxes $a_1(t)$, $a_2(t)$, and $a_3(t)$; the angular velocity Ω (t) of the frame of reference in which the free surface appears to be at rest; and the vorticity ζ (t) of the fluid motion. The fluid motion does not in general conform to that given by Equation 1 above.

Certain results of this formulation may be singled out for mention here, since they go without mention in the usual astronomical references (e.g., Lyttleton 1953, Jardetzky 1958). One of these is Dedekind's (1860) theorem: if, to a given ellipsoidal figure, there corresponds a permissible fluid motion specified by Ω (t) and ζ (t), then, to the same figure, there corresponds a second fluid motion specified (say) by $\tilde{\Omega}$ (t) and $\tilde{\zeta}$ (t). The second, or "adjoint" (cf. Chandrasekhar 1965), motion can be obtained from the first by a simple transformation.

When all the relevant variables are assumed independent of time, Riemann found that Ω and ζ must lie in a principal plane of the ellipsoid. This means, first, that at least one of the three numbers Ω_1, Ω_2, and Ω_3 is zero; second, that if (say) Ω_1 is zero, then ζ_1 is also zero. This is a step toward solving the problem of determining the possible steady-state Riemann ellipsoids. The solution of this larger problem is further facilitated by the observation that the equations determining the geometry are homogeneous in a_1, a_2, and a_3. One can therefore put a_1 (say) equal to one, and map out those portions of the a_2a_3 plane that correspond to permitted ellipsoids. This was done partly by Riemann (1860), and completed by Chandrasekhar (1965, 1966). Four distinct areas in the a_2a_3 plane, designated S, I, II, and III, are found in this way. The ellipsoids of type S are those for which only Ω_3 and ζ_3 are different from zero; in the remaining three cases, Ω_2 and ζ_2 are different from zero as well.

The ellipsoids of type S include the classical sequences of Maclaurin and Jacobi as special cases, obtained by putting ζ_3 equal to zero. The more general framework into which Riemann's formulation fits these sequences clarifies certain aspects of the classical theory. In particular, the well-known point of bifurcation where the Jacobi sequence intersects the Maclaurin sequence appears, not as an isolated phenomenon, but as a special instance of a general phenomenon. Indeed, every Maclaurin spheroid is the intersection of the Maclaurin sequence with a sequence of Riemann ellipsoids of type S, provided the eccentricity e of its meridian section does not exceed 0.9529, where "ordinary" instability sets in (see p. 476). Since, however, a sequence of Riemann ellipsoids (other than the Jacobi sequence) has a nonvanishing vorticity relative to the rotating axes, the Maclaurin spheroid that it intersects must also be ascribed a nonvanishing vorticity. For an axisymmetric figure like a Maclaurin spheroid, this is always possible. One need only describe it in a reference frame rotating with an angular velocity Ω different from Ω_3, the angular velocity of the reference frame in which it is in equilibrium; in this new frame, it has a vorticity $2(\Omega - \Omega_3)$.

The Maclaurin pattern.—It will be useful to recall the behavior of the sequence of Maclaurin spheroids as a function of the angular momentum J.

When J is zero, the Maclaurin figure is, of course, a sphere. As J increases, the figure flattens progressively, approaching an infinitely thin circular disk as J approaches infinity. Although there is no limit to the angular momentum, there is—as Poincaré's estimate requires—a limit to the angular velocity. It reaches a maximum Ω_M given by (cf. Lyttleton 1953)

$$\Omega_M{}^2 = 0.4494\pi G\rho \qquad 8.$$

This maximum is attained when the eccentricity of the meridian section is 0.9299. The value of $\Omega_M{}^2$ is not only substantially smaller than Poincaré's estimate, but also (what is more important) it is not attained through the circumstance contemplated in deriving Poincaré's estimate; i.e., the effective gravity grad U (see Equation 4 above) does not vanish at any point of the surface. In fact, since for ellipsoidal configurations of equilibrium,

$$U = \text{const} \left(1 - \sum_{i=1}^{3} \frac{x_i{}^2}{a_i{}^2}\right) \qquad 9.$$

one can easily see that grad U never vanishes on the surface.

To this basic "Maclaurin pattern" one sometimes adds the notion of the "secular instability" that affects the Maclaurin spheroids with eccentricities exceeding 0.8127, the value at the point of bifurcation where the Jacobi and Maclaurin sequences intersect. The initial part of the evolutionary picture due to Poincaré, Darwin, and others, which suggested a way of forming double stars, depended heavily on imagining an evolutionary sequence (with fixed angular momentum but increasing density) that proceeded along the Maclaurin sequence to the point of bifurcation, and then along the stable Jacobi sequence. There seems now to be reason to doubt even this initial part of the picture (Lynden-Bell 1964), whereas the later part has long since been shown impossible, at least as originally envisaged (cf. Lyttleton 1953).

The Roche pattern.—A Roche model consists of a core containing all the mass M, producing the gravitational potential GM/r in an overlying atmosphere of negligible mass. The level surfaces are those for which U (cf. Equation 4 above) is constant. Although a variety of assumptions may be made concerning the variation of Ω with ϖ, the shape of the core, . . . (cf. Jeans 1929), we shall suppose Ω uniform and the core spherical.

We consider a sequence of such Roche models, labeled by the angular momentum J. Starting with the spherically stratified configuration when J is zero, the figures flatten progressively until there occurs a maximum angular momentum J_M, beyond which no further equilibrium configurations exist. This is accompanied by a maximum angular velocity Ω_M, satisfying (Jeans 1929)

$$\Omega_M{}^2 = 0.7215\pi G\bar\rho \qquad 10.$$

For the Roche model having this angular velocity, centrifugal and gravitational acceleration balance at the equator, and the ratio of the polar and equatorial radii is 2/3. There is no point of bifurcation along this Roche sequence.

If one imagines a slowly contracting sequence of such figures (and there-fore a sequence of increasing $\bar{\rho}$ rather than of increasing J), one is led to sup-pose that once the critical figure corresponding to Equation 11 has been reached, further contraction must result in the separation of matter at the equator from the remainder of the mass: it must be left behind in a ring, since it cannot participate in the contraction. The occurrence of the maximum angular velocity and the attendant separation of matter at the equator are sometimes referred to as equatorial instability. However, the behavior de-scribed here has nothing to do with stability; it is simply a semiquantitative description of a sequence of equilibrium configurations, which, after a cer-tain point, consist of two detached pieces. Although neither the evolutionary scheme suggested nor the question of the stability of the individual members of the sequence seems to have been adequately investigated, a plausible "Roche pattern" emerges, very different from the Maclaurin pattern. In order to see which pattern (if either) a "real" star follows, one has to con-sider models with central condensation equal neither to one nor to infinity.

APPROXIMATE AND NUMERICAL SOLUTIONS

We shall discuss here, in addition to some numerical solutions of the steady-state equations, some techniques that have been used to give approxi-mate solutions. We shall give special, though not sole, attention to whether the Maclaurin or the Roche pattern is followed.

Rapid, uniform rotation.—Jeans conceived the idea that the central con-densation of a sequence of uniformly rotating configurations should deter-mine into which pattern it fits. With this in mind, he constructed (cf. Jeans 1919) a sequence of uniformly rotating polytropes, the polytropic index being a measure of central condensation. In the expansion procedure that he used in solving the equations of equilibrium, the first approximation to the level surfaces was a set of spheroids. This means that the first approximation may be made exact for the case when n, the polytropic index, vanishes, for this corresponds to uniform density. It further means that the Maclaurin pattern persists in the first approximation. However, in the second approximation, the Roche pattern already prevails for all n exceeding a critical value n_c, where (roughly)

$$n_c = 0.83 \qquad\qquad 11.$$

The Roche pattern might in fact prevail for smaller values of n as well, but this is not discussed, for the following reason: for values of n smaller than n_c, a point of bifurcation, where a nonaxisymmetric sequence branches off, occurs before the sequence terminates by the balancing of gravitational and centrifugal acceleration at the equator. Jeans therefore concludes that if values of n smaller than n_c could occur, evolution would proceed along an analogue of the Maclaurin-Jacobi sequence, culminating in the formation of a double star. However, the values of n required for this evolutionary pattern are much too low (giving a central condensation of about three), so the con-clusion is that the Roche pattern prevails in all cases.

Jeans' problem has recently been solved numerically by James (1964), for a variety of polytropic indices less than three. The Roche pattern was found in all cases tried. Once again a critical value n_c was obtained (and by a more straightforward method than that adopted by Jeans), such that a point of bifurcation occurs for smaller values of n. James found

$$n_c = 0.808 \qquad\qquad 12.$$

It is remarkable how close Jeans came with his, apparently, less accurate methods. James also worked out the white-dwarf case, where no point of bifurcation was found.

Another approach to the problem of the rapidly rotating polytropes is that of Roberts (1963a, b) and Hurley & Roberts (1964, 1965). They have used a variational principle to find the density stratification under the assumption that the level surfaces are spheroids; in one case the eccentricities of the spheroids were assumed all to be the same; in a second case they were allowed to vary and were determined by the variational principle along with the density stratification. With the aid of the models of constant eccentricity, Roberts (1963b) inferred that n_c would be "less than, but quite near to, unity," in adequate agreement with the results of Jeans and James. All sequences conform to the Roche pattern, except for a slight deviation by some of the less centrally condensed models of constant eccentricity: for these Ω^2 has a maximum prior to the termination of the sequence. A conclusion of a slightly different kind is that, for n between 1.5 and 5.0, the maximum angular velocity agrees with that given by Equation 10 above, within about 1 per cent.

Rapid, nonuniform rotation.—The effect of nonuniform rotation was investigated by Stoeckly (1965). He considered the polytrope with index n equal to 1.5, and with angular velocity given by the formula

$$\Omega(\tilde{\omega}) = \Omega_0 \exp\left(-c\tilde{\omega}^2/\tilde{\omega}_e^2\right) \qquad\qquad 13.$$

Here Ω_0 is the central angular velocity, c a parameter of nonuniformity, and $\tilde{\omega}_e$ the equatorial radius of the configuration. When c is 1.02, this formula gives a good fit to the distribution of angular velocity one might expect after an initially homogeneous and uniformly rotating mass has contracted to a polytrope of index 1.5.

Many choices of c were made, from zero (uniform rotation) to 1.15 (strongly nonuniform), and for each value of c a sequence was constructed by varying Ω_0. It was found that for c less than 0.54, the Roche pattern was followed, whereas for larger values of c, the Maclaurin pattern appeared to be fully restored: the configurations could become arbitrarily flattened for large enough angular momentum, which could increase without limit, whereas the parameter Ω_0 had a maximum along the sequence. No attempt was made to find points of bifurcation.

Ostriker, Bodenheimer & Lynden-Bell (1966) have calculated the figure of a white dwarf that is rotating nonuniformly. The principal conclusion to which they draw attention is that white dwarfs of mass considerably (perhaps arbitrarily) greater than Chandrasekhar's limiting mass may exist in the presence of a sufficiently large and sufficiently nonuniform angular velocity. That rotation can increase the limiting mass is clear from previous work (cf. Roxburgh 1965, Anand 1965) on the effect of uniform rotation. However, in the latter case, the effective gravity vanishes at the equator before the increase of mass exceeds a few per cent.

The work of Stoeckly on the one hand and that of Ostriker et al. on the other are complementary in the following sense. If one assumes that a necessary condition for a star to reach the white-dwarf state is that its mass lie below the Chandrasekhar limit, one is inclined to invoke the mechanism of equatorial mass loss to explain how initially more massive stars may become white dwarfs. The work of Stoeckly shows that this mechanism may be inoperative for nonuniformly rotating stars; that of Ostriker et al., that it may be unnecessary.

The perturbation method.—The idea of assuming the distribution of mass to differ only to first order in Ω^2 from a known, spherical distribution goes back at least as far as Clairaut. Chandrasekhar (1933; see also Chandrasekhar & Lebovitz 1962c) has worked out the theory of the distorted polytropes on this basis. Comparison with the work of James and of Hurley & Roberts (1964) indicates that the perturbation theory continues to give a good representation of the configurations for larger values of Ω^2 than one might expect.

The circumstance that the inner level surfaces are only slightly distorted, even for large rotation, has led Monaghan & Roxburgh (1964) to use a combination of the perturbation method and the Roche method to construct approximations to rapidly rotating polytropes. They use the perturbation method in the interior, which contains "all" the mass, and then adjoin an atmosphere of "negligible" mass. The question at what point to join the two solutions found in this way is answered somewhat arbitrarily; it could be made more definite by further combining this method with the variational principle. The results are compared with the work of James (1964), and found to be in satisfactory agreement. The procedure is applicable to a wide class of stellar models (see, for example, Roxburgh, Griffith & Sweet 1965).

OSCILLATIONS AND STABILITY

We shall divide the subject of the oscillations and the stability of rotating masses into three parts: the effect of a small, uniform rotation on the modes of oscillation that are present even in the absence of rotation; the instabilities that are expressly due to the presence of rotation; and the methods and principles for inferring stability or instability.

THE EFFECT OF A SMALL, UNIFORM ROTATION ON OSCILLATION MODES IN A SPHERICAL STAR

Suppose a spherical star to be undergoing small, adiabatic oscillations. These can be conveniently classified into four groups (Cowling 1941), which should retain their essential characters if we consider, instead of a spherical star, one that is slightly distorted by rotation. We shall discuss the effect of a small, uniform rotation group by group.

Radial pulsations.—If the modes of adiabatic oscillation of a spherical star are analyzed according to the spherical harmonics $Y_l^m(\vartheta, \varphi)$, those for which l is zero must be distinguished from all others. These radial pulsations, usually held responsible for the variations of variable stars (cf. Rosseland 1949, Ledoux & Walraven 1958), are described by a second-order differential equation whose eigenvalues (which are the squares of the characteristic frequencies of pulsation) are all positive (i.e., of stable type), provided the ratio of specific heats γ everywhere exceeds $4/3$. If, however, γ should fall below $4/3$ over a sufficiently large part of the star, the star would be "dynamically unstable." Since there are circumstances under which γ does fall substantially below $4/3$ (cf. Ledoux 1965), and other circumstances under which the dynamical instability occurs for values of γ slightly in excess of $4/3$ (Chandrasekhar 1964b), it is clearly of interest to find the effect of rotation on the lowest mode of radial pulsation (Fowler 1966).

In the case of the homogeneous, compressible model, where the equilibrium configuration is a Maclaurin spheroid, this problem has been solved exactly, i.e., without the assumption that Ω^2 is small (Chandrasekhar & Lebovitz 1962b). The result is that a sufficiently large amount of angular momentum can suppress the dynamical instability for any value of γ, no matter how small. For a centrally condensed configuration that is only slightly distorted by rotation, Ledoux (1945) has given an approximate formula indicating that if $\gamma - 4/3$ is negative, but of order Ω^2 in absolute value, rotation is capable of suppressing the instability. However, $\gamma - 4/3$ may have to be very small indeed to be "of order Ω^2," as the following result (derived from Chandrasekhar & Lebovitz 1962c) indicates: for a polytrope of index 3.5 and a value of γ of 1.328, the amount of rotation necessary for stability, as derived from the approximate formula, exceeds the maximum possible rotation consistent with equilibrium. There is, further, numerical evidence (Cowling & Newing 1949) that the approximate formula overestimates the stabilizing effect of rotation. There is a clear need for better estimates, preferably not based on the assumption that Ω^2 is a small parameter.

Kelvin modes (f *modes*).—These modes, which can occur in a nontrivial way when l is greater than or equal to two, are all of stable type in the absence of rotation. The oscillation frequencies, which are $(2l+1)$-fold degenerate for a given l, are given, in the case of uniform density, by the formula

$$\sigma_l^2 = \frac{8\pi G\rho}{3} \frac{l(l-1)}{2l+1} \qquad\qquad 14.$$

This formula gets changed somewhat for a compressible, centrally condensed configuration (cf. Chandrasekhar 1964a), but the modes in question are qualitatively the same. Since the squares of the oscillation frequencies are all safely positive, the stability of these modes is not altered by a small amount of rotation (that it can, on the other hand, be altered by a large amount is evident from the observation that these are the modes by which the Maclaurin spheroids become unstable).

A small rotation does, however, have the effect of lifting the degeneracy with respect to the spherical-harmonic index m. If $\sigma_{l,m}$ and σ_l denote the frequencies for the rotating and nonrotating masses, respectively, one finds (see, for example, Cowling & Newing 1949)

$$\sigma_{l,m} = \sigma_l + mk\Omega \qquad\qquad 15.$$

where m has the range $-l$ to $+l$, and k is a numerical factor (unity if the density is uniform) depending on the distribution of mass. The splitting expressed by Equation 15 and, more generally, the notion of lifting a degeneracy by means of rotation have played roles in two different discussions of the close periods and beats occurring in certain variable stars (Ledoux 1951, Chandrasekhar & Lebovitz 1962d), as well as in interpreting the spectrum of frequencies of oscillation of the Earth (cf. Pekeris, Alterman & Jarosch 1961; Backus & Gilbert 1961).

Sound waves (p *modes*).—Among the infinite sequence of modes possible for a compressible mass, one finds the p modes (Cowling 1941), which are like sound waves, somewhat modified by gravity. Like the f modes, they are safely stable, the effect of rotation being a splitting like that expressed by Equation 16 above.

Gravity waves (g *modes*).—These modes make up the remainder of the possible small, adiabatic oscillations of a spherical, gaseous mass. For a given l, the frequency spectrum is real, extending from zero to a finite maximum, provided the temperature gradient is everywhere subadiabatic. If it is somewhere superadiabatic, the spectrum is, at least in part, imaginary; it is through the associated unstable modes that convection begins in a configuration unstable by Schwarzschild's criterion. Since the frequency spectrum always has a part lying near the origin of the complex plane, it is of interest to see what effect a small rotation can have. We may dismiss the subadiabatic case, for here all indications are that rotation cannot have a destabilizing effect (cf. Eckart 1960). This leaves the question whether rotation can stabilize a configuration with a superadiabatic temperature gradient.

A stabilizing effect is expected on grounds of conservation of angular momentum (see, for example, Clement 1965b); but rotation cannot, by itself, stabilize a configuration unstable by Schwarzschild's criterion, as Cowling (1951) has shown. The reason is that the stabilizing effect of angular momentum conservation becomes less as the disturbance becomes less axisymmetric, so that for sufficiently small azimuthal wavelengths, any superadiabatic gradient is unstable, no matter how large the angular velocity. The bare conclusion is that rotation leaves the Schwarzschild criterion unchanged. The effect on the energy transport of suppressing the long-wavelength instabilities, and the combined effect of dissipation (which may suppress the short-wavelength instabilities) and rotation, do not seem to have been adequately investigated.

INSTABILITIES CAUSED BY ROTATION

Since a small, uniform rotation has a generally stabilizing effect, we may look in two directions for instabilities that are expressly due to the presence of rotation: nonuniform rotation, and large, uniform rotation.

Nonuniform rotation.—For an incompressible liquid contained between concentric cylinders and rotating with angular velocity Ω (ϖ), Rayleigh's criterion—that $\varpi^4\Omega^2$ increase outwards—is necessary and sufficient for stability to axisymmetric disturbances (see, for example, Chandrasekhar 1961). In a star rotating with angular velocity $\Omega(\varpi)$ if we continue to restrict attention to axisymmetric disturbances, this criterion must be modified by buoyancy effects; that is, some combination of the Rayleigh and Schwarzschild criteria should obtain. Such a combination has been found, for example, by Høiland (1941; see also Ledoux 1965), and indicates, as one would expect, that a stable stratification of angular velocity exerts a stabilizing influence on an unstable distribution of temperature, and vice versa. The combined criterion has not been placed on the solid analytical foundation of its two component criteria, however.

The theoretical situation is worse where nonaxisymmetric disturbances are concerned, for the possibility of a shear instability then arises. Most of the precise work on this subject has been done for parallel flows (but see Cowling 1951 for a discussion in the context of stellar interiors). The parallel-flow work suggests that a necessary (but not sufficient) condition for instability is that $\varpi\Omega(\varpi)$ have a point of inflection, i.e.,

$$\frac{d^2}{d\varpi^2}(\varpi\Omega(\varpi)) = 0 \qquad\qquad 16.$$

somewhere in the star. This criterion (also due to Rayleigh 1880) holds only in the absence of force fields. A gravitational field is known to have a stabilizing effect in an incompressible, but stably stratified, liquid, provided the Richardson number, which gives a quantitative measure of the stabilizing effect, is large enough (Chandrasekhar 1961; see also Drazin & Howard 1966, where an up-to-date review is given). This effect should be further modified for compressibility; Coriolis acceleration should have, at least, a weakly stabilizing influence. These further effects have not yet been incorporated in a general and precise criterion for instability.

Large uniform rotation.—We may, for examples of highly rotating configurations, refer to the ellipsoidal figures and, in particular, to the Maclaurin spheroids. The disturbances of the Maclaurin spheroids that belong to the second harmonics lead to instability for smaller values of the angular momentum than do those that belong to any higher harmonic (cf. Poincaré 1885, Chandrasekhar & Lebovitz 1963), and have therefore been the most intensively studied. The five oscillation frequencies associated with these disturbances are generalizations of the fivefold degenerate Kelvin frequency given by (cf. Equation 14 above)

$$\sigma^2 = \frac{16}{15}\pi G\rho \qquad\qquad 17.$$

Of the five frequencies, one in particular is important because, following it as a function of the eccentricity of meridian section e, one is able to distinguish two critical points (Lebovitz 1961). The first, when e is 0.8127, is the point of bifurcation where the Maclaurin and Jacobi sequences intersect. Although σ^2 vanishes at this critical point, it is positive on either side, so that instability does not set in here, at least in the absence of further effects. The second, when e is 0.9529, represents the onset of instability, and also the most flattened Maclaurin spheroid that can lie on a Riemann sequence of type S (Chandrasekhar 1965). These two critical points have been the subjects of recent investigations.

It has long been assumed (cf. Thomson & Tait 1912) that the slightest viscosity would render the Maclaurin spheroid unstable as soon as e surpassed 0.8127. This has now been verified, by means of a boundary-layer calculation, by Roberts & Stewartson (1963). More recently, Rosenkilde (1967) has succeeded in enormously simplifying this calculation with the aid of the virial-tensor method (see below).

Concerning the second critical point, one may ask: to what kind of configuration does the instability lead? Rossner (1967) has answered this question with the aid of Riemann's formulation of the nonlinear, ordinary differential equations describing the motion of a liquid ellipsoid. Integrating the equations numerically, Rossner found that the configuration neither gets disrupted nor finds its way to a new steady state, but performs a complicated unsteady motion. The importance of Rossner's work, aside from determining the fate of an unstable Maclaurin spheroid, is that this is a rare example of a noncontrived problem solved both exactly and by linearization, allowing one to follow the consequence, and check the accuracy, of the linear theory.

We may single out for special mention a further classical problem, that of the Roche ellipsoids. A liquid primary and a spherical secondary rotate with angular velocity Ω about their common center of mass, in a circular orbit. There is a maximum rotation rate, and an associated minimum distance (the Roche limit) between their centers of mass, at which, it is sometimes said (cf. Jeans 1929), instability sets in. This stability problem has recently been solved by Chandrasekhar (1963), who found that instability sets in, not at the Roche limit, but when the primary is somewhat more elongated than it is at the Roche limit. A further result is that a Jacobi ellipsoid, to which an elongated Roche ellipsoid reduces when the secondary becomes small, is unstable under the influence of a small tidal force directed along its major axis.

Finally, we may remark on Chandrasekhar's determination of the stability of the Riemann ellipsoids to second-harmonic disturbances. Among those that are stable to such disturbances there are some (of Type I) for which the ratio of a_3 to a_1 is arbitrarily small, whereas the ratio of a_2 to a_1 is between 1.35 and 2.00 (this conclusion of stability disagrees with that in Riemann's original paper, where an incorrect stability criterion was used; cf. Lebovitz 1966). These enormously flattened objects may be of some interest as self-consistent galactic models.

METHODS AND PRINCIPLES FOR INFERRING INSTABILITY

We take as the criterion for instability the condition that if, in the usual normal mode analysis (cf. Ledoux 1958), a time dependence of the form $\exp(\lambda t)$ is assumed, then the real part of λ is positive for at least one normal mode. The problems one encounters in this way are frequently so difficult that one is led to indirect methods of solution. We discuss three of them below.

Exchange of stabilities.—The principle of exchange of stabilities appears in Poincaré's famous memoir (Poincaré 1885). His purpose was to ascertain where, along a sequence of equilibrium configurations, instability sets in, without actually calculating the frequencies of small oscillation. Imagine a parameter μ (say) that labels the equilibrium figures, and suppose there is a critical value μ_c such that if μ is less than μ_c there is a single configuration for each μ, whereas for μ greater than μ_c there are two or more. At the "point of bifurcation" μ_c, stability may be transferred to the "new" sequence that branches off there. This presupposes that we can distinguish an "old" sequence for all μ, less than, equal to, and greater than μ_c, so that there is a well-defined new sequence branching off. In the familiar case of the Maclaurin and Jacobi figures, we can certainly do this. The idea of exchange of stabilities has been elaborated in two directions.

One of these directions starts with the observation that at the point of bifurcation, the linearized equations allow a neutral mode: the characteristic exponent vanishes, and the characteristic vector deforms a figure on the old sequence into a figure on the new sequence (cf. Chandrasekhar & Lebovitz 1964). Hence if one wants to find a point of bifurcation where an exchange of stabilities occurs, one can do it by putting λ equal to zero in the linearized equations, a substantial simplification of the latter. However, if one doesn't know a priori that there is an exchange of stabilities, this method is ambiguous, for the occurrence of a normal mode with vanishing characteristic exponent need not imply an exchange of stabilities. Moreover, even when the onset of instability is known to occur through the vanishing of a characteristic exponent, it may be necessary to assume an algebraic growth of the small perturbations, rather than equating λ to zero, in order to find the onset of instability. In short, the vanishing of a characteristic exponent is not equivalent to an exchange of stabilities (for examples in the context of rotating fluid masses, see Lebovitz 1963, 1967).

The other direction of elaboration is concerned with what inferences one should make if something other than a point of bifurcation occurs as the parameter μ is increased indefinitely. In particular, if to each value of μ less than a maximum value μ_M there correspond two configurations, whereas for values of μ larger than μ_M there are none, it is sometimes said (cf. Lyttleton 1953) that stability is lost at μ_M. There would appear to be many exceptions to this rule, however. We may mention the Maclaurin spheroids which do not become unstable when Ω reaches its maximum, the Roche ellipsoids which do not become unstable at the Roche limit, and the Darwin ellipsoids (Chandrasekhar 1964c).

The virial method.—The virial equations are obtained by multiplying the ith equation of motion (in Cartesian coordinates) by x_j, and integrating over the volume occupied by the fluid. The usual virial theorem, obtained by putting j equal to i and summing, has long been in use in astronomy, but the idea of retaining all nine equations is relatively recent (Chandrasekhar 1960, 1961; Lebovitz 1961). It has been found particularly appropriate in the study of the oscillations of liquid ellipsoids, because it reduces the equations of motion to a system of linear, ordinary differential equations with constant coefficients. Their solutions can be shown to be those belonging to the harmonics of orders zero and two (Lebovitz 1961). In a similar way, one can show that the resulting virial equations obtained for the small motions, by multiplying the ith equation by $x_j x_k$ and integrating, are those belonging to the harmonics of orders one and three; and so on. For other configurations than liquid ellipsoids, one must introduce a "trial function" into the virial equations (Chandrasekhar & Lebovitz 1962a).

One useful characteristic of the virial method, which it shares with other integral methods, is that the boundary conditions are incorporated in the formulation of the fundamental equations. This is especially important in the work of Rosenkilde (1967), where the trial function violates the boundary conditions. Clement (1965) has shown that for a particular choice of trial function, the virial method gives the same result as the variational principle.

The variational principle.—Chandrasekhar (1964a) has shown that the squares of the oscillation frequencies of spherical, gaseous masses may be obtained from a variational principle. Clement (1965a) has extended the variational principle to uniformly rotating masses. But in the rotating case the stationary property enjoyed by the characteristic exponents holds only in the stable case (see also Cowling & Newing 1949), a restriction not encountered in the absence of rotation. However, if one considers only axisymmetric disturbances of axisymmetric equilibrium configurations, this restriction is lifted. Since the investigation of the effect of moderate to large rotation on the "dynamical instability" requires only consideration of such modes, the variational principle promises improved estimates in this direction. Finally, Lynden-Bell & Ostriker (1967) have now given a very general formulation to the variational principle, making it applicable to a wide class of steady-state configurations.

ACKNOWLEDGMENT

This article was written with the support of the Air Force Office of Scientific Research, through grant AF-AFOSR-712-67.

LITERATURE CITED

Anand, S. P. S. 1965, *Proc. Natl. Acad. Sci. U. S.*, **54**, 23

Backus, G., Gilbert, F. 1961, *Proc. Natl. Acad. Sci U. S.*, **47**, 362

Chandrasekhar, S. 1933, *Monthly Notices Roy. Astron. Soc.*, **93**, 390

Chandrasekhar, S. 1960, *J. Math. Anal. Appl.*, **1**, 240

Chandrasekhar, S. 1961, *Hydrodynamic and Hydromagnetic Stability* (Clarendon Press, Oxford)

Chandrasekhar, S. 1963, *Ap. J.*, **138**, 1182

Chandrasekhar, S. 1964a, *Ap. J.*, **139**, 664

Chandrasekhar, S. 1964b, *Ap. J.*, **140**, 417

Chandrasekhar, S. 1964c, *Ap. J.*, **140**, 599

Chandrasekhar, S. 1965, *Ap. J.*, **142**, 890

Chandrasekhar, S. 1966, *Ap. J.*, **145**, 842

Chandrasekhar, S., Lebovitz, N. R. 1962a, *Ap. J.*, **135**, 248

Chandrasekhar, S., Lebovitz, N. R. 1962b, *Ap. J.*, **136**, 1069

Chandrasekhar, S., Lebovitz, N. R. 1962c, *Ap. J.*, **136**, 1082

Chandrasekhar, S., Lebovitz, N. R. 1962d, *Ap. J.*, **136**, 1105

Chandrasekhar, S., Lebovitz, N. R. 1963, *Ap. J.*, **137**, 1162

Chandrasekhar, S., Lebovitz, N. R. 1964, *Ap. Norveg.*, **9**, 323

Clement, M. 1965a, *Ap. J.*, **140**, 1045

Clement, M. 1965b, *Ap. J.*, **142**, 243

Cowling, T. G. 1941, *Monthly Notices Roy. Astron. Soc.*, **101**, 367

Cowling, T. G. 1951, *Ap. J.*, **114**, 272

Cowling, T. G., Newing, R. A. 1949, *Ap. J.*, **109**, 149

Dedekind, R. 1860, *J. Reine Angew. Math.*, **58**, 217

Dirichlet, P. G. Lejeune. 1860, *J. Reine Angew. Math.*, **58**, 181

Dive, P. 1952, *Bull. Sci. Math.*, **76**, 38

Drazin, P. G., Howard, L. N. 1966. *Advan. Appl. Mech.*, **9**, 1

Eckart, C. 1960, *The Hydrodynamics of Oceans and Atmospheres* (Pergamon, New York)

Eddington, A. S. 1926, *The Internal Constitution of the Stars* (Cambridge Univ. Press)

Ferraro, V. C. A. 1937, *Monthly Notices Roy. Astron. Soc.*, **97**, 458

Fowler, W. A. 1966, *Ap. J.*, **144**, 180

Høiland, E. 1941, *Avhandl. Norske Vidensk.-Akad. Oslo Mat.-Naturv. Kl.*, *No. 11*

Hurley, M., Roberts, P. H. 1964, *Ap. J.*, **140**, 583

Hurley, M., Roberts, P. H. 1965, *Ap. J. Suppl. 11*, 95

James, R. A. 1964, *Ap. J.*, **140**, 552

Jardetzky, W. S. 1958, *Theories of Figures of Celestial Bodies* (Interscience, New York)

Jeans, J. H. 1919, *Phil. Trans. Roy. Soc.*, **218**, 157

Jeans, J. H. 1929, *Astronomy and Cosmogony* (2nd ed., Cambridge Univ. Press)

Lebovitz, N. R. 1961, *Ap. J.*, **134**, 500

Lebovitz, N. R. 1963, *Ap. J.*, **138**, 1214

Lebovitz, N. R. 1965, *The Riemann Ellipsoids* (Lecture notes, Inst. Ap., Cointe-Sclessin, Belgium)

Lebovitz, N. R. 1966, *Ap. J.*, **145**, 878

Lebovitz, N. R. 1967, *Ap. J.* (To appear)

Ledoux, P. 1945, *Ap. J.*, **102**, 143

Ledoux, P. 1951, *Ap. J.*, **117**, 373

Ledoux, P. 1958, *Handbuch der Physik*, **51**, 605 (Flügge, S., Ed., Springer-Verlag, Berlin)

Ledoux, P. 1965, *Stellar Structure*, 499 (Aller, L. H., McLaughlin, D. B., Eds., Univ. of Chicago Press)

Ledoux, P., Walraven, Th. 1958, *Handbuch der Physik*, **51**, 353 (Flügge, S., Ed., Springer-Verlag, Berlin)

Lichtenstein, L. 1933, *Gleichgewichtsfiguren Rotierinder Flüssigkeiten* (Verlag von Julius Springer, Berlin)

Lynden-Bell, D. 1964, *Ap. J.*, **139**, 1195

Lynden-Bell, D., Ostriker, J. 1967, *Monthly Notices Roy. Astron. Soc.* (To appear)

Lyttleton, R. A. 1953, *The Stability of Rotating Liquid Masses* (Cambridge Univ. Press)

Mestel, L. 1965, *Stellar Structure*, 465 (Aller, L. H., McLaughlin, D. B., Eds., Univ. of Chicago Press, Chicago, Ill.)

Monaghan, J. J., Roxburgh, I. W. 1964, *Monthly Notices Roy. Astron. Soc.*, **131**, 13

Ostriker, J., Bodenheimer, P., Lynden-Bell, D. 1966, *Phys. Rev. Letters*, **17**, 816

Pekeris, C., Alterman. Z., Jarosch, H. 1961, *Proc. Natl. Acad. Sci. U.S.*, **47**, 91

Poincaré, H. 1885, *Acta Math.*, **7**, 259

Poincaré, H. 1903, *Figures d'Equilibre d'une Masse Fluide* (C. Naud, Paris)

Randers, G. 1942, *Ap. J.*, **95**, 454

Rayleigh, Lord. 1880, *Proc. London Math. Soc.*, **9**, 57

Riemann, B. 1860, *Abhandl. Konigl. Ges. Wis. Gottingen*, **9**, 3; 1892, *Gesammelte Mathematische Werke* (Verlag von B. G. Teubner, Leipzig; reprinted 1953, Dover Publ., New York)

Roberts, P. H. 1963a, *Ap. J.*, **137**, 1129

Roberts, P. H. 1963b, *Ap. J.*, **138**, 809

Roberts, P. H., Stewartson, K. 1963, *Ap. J.*, **137**, 777

Rosenkilde, K. 1967, *Ap. J.* (To appear)

Rosseland, S. 1949, *The Pulsation Theory of Variable Stars* (Clarendon Press, Oxford)

Rossner, L. 1967, *Ap. J.* (To appear)

Roxburgh, I. W. 1964, *Monthly Notices Roy. Astron. Soc.*, **128**, 157

Roxburgh, I. W. 1965, *Z. Ap.*, **62**, 134

Roxburgh, I. W. 1966a, *Rotation and Magnetism in Stellar Structure and Evolution.* Symposium lecture (Goddard Space Flight Center, New York)

Roxburgh, I. W. 1966b, *Monthly Notices Roy. Astron. Soc.*, **132**, 201

Roxburgh, I. W., Griffith, J. S., Sweet, P. A. 1965, *Z. Ap.*, **61**, 203

Stoeckly, R. 1965, *Ap. J.*, **142**, 208

Sweet, P. A. 1950, *Monthly Notices Roy. Astron. Soc.*, **110**, 548

Tassoul, J.-L., Cretin, M. 1965, *Ann. Ap.*, **28**, 982

Thomson, W., Tait, P. G. 1912, *Treatise on Natural Philosophy* (Cambridge Univ. Press)

Volterra, V. 1903, *Acta Math.*, **27**, 105

Wavre, R. 1932, *Figures Planetaires et Geodesie* (Gauthier-Villars, Paris)

GAMMA RADIATION FROM CELESTIAL OBJECTS[1]

By G. G. Fazio

*Smithsonian Astrophysical Observatory and Harvard College Observatory
Cambridge, Massachusetts*

1. INTRODUCTION

Information about the physical universe reaches us via matter (meteorites, cosmic rays, etc.) and via electromagnetic radiation. The advantage of using electromagnetic radiation to study celestial objects is that it travels in a straight line from the source to the Earth. This is not true of charged corpuscular radiation, which undergoes deflections in magnetic fields. Particles such as neutrons and neutral mesons are not deflected, but their lifetimes are so short that they cannot traverse appreciable distances. The neutrino is a neutral particle that does not decay, but is very difficult to detect because of its very small interaction cross section.

Of particular interest in the electromagnetic spectrum is the extreme high-frequency range (gamma radiation), $E > 100$ keV ($\nu > 2.42 \times 10^{19}$ sec^{-1}, $\lambda > 0.124$ Å).

Gamma radiation is a particularly important probe in cosmology because of its high penetrability through Galactic and intergalactic matter, its direct and simple relationship to nuclear reactions that act as fundamental energy sources, and its direct relationship to high-energy electrons and protons. Measurement of the flux, energy spectra, and arrival direction of gamma rays can help to solve some of the most fundamental problems of cosmology, such as:

(*a*) the origin of cosmic rays;

(*b*) the density of cosmic radiation in the Galaxy and in intergalactic space;

(*c*) the density and composition of Galactic and intergalactic matter;

(*d*) the presence of antimatter in the universe, and in particular the hypothesis of the continuous creation of matter;

(*e*) the strength of Galactic and intergalactic magnetic fields.

The detection of cosmic gamma rays is not in general feasible from the ground, since gamma radiation is absorbed by the atmosphere and the expected primary fluxes are very low compared to cosmic radiation. Experiments must be performed in high-altitude balloons or in satellites with long observation times. They are therefore difficult, but the importance of the results justifies a concerted effort.

Up until now, no photons of energy greater than 100 keV originating from

[1] The survey of literature for this review was concluded in January 1967.

TABLE I

CORPUSCULAR AND ELECTROMAGNETIC EMISSION DUE TO PARTICLE TRANSITIONS
UNDER VARIOUS FORCES

Particle	Type of interaction in particle transition		
	Weak	Electromagnetic	Strong
Atom	—	Photons	—
Nuclei	Leptons (β decay)	**Photons**	—
Baryon	Leptons (β decay)	**Photons**	**Mesons**
Lepton	Leptons	**Photons**	—

beyond the solar system have definitely been detected. However, experiments on the lunar probes Ranger 3 and Ranger 5 (Arnold et al. 1962) have shown evidence for a diffuse source of gamma rays in the energy region 100 keV to 1 MeV, and a recent balloon experiment by Duthie, Cobb & Stewart (1966) has indicated a possible source of Cygnus in the energy region above 100 MeV.

2. GAMMA-RAY PRODUCTION MECHANISMS

These can be grouped into two classes:

1) Quantum-state transitions of nuclei, baryons (protons, neutrons, hyperons), and leptons (electrons, positrons, neutrinos), including free-free and free-bound transitions. The interactions are electromagnetic or strong. These reactions convert particle kinetic energy into gamma radiation.

2) Annihilation. These reactions include matter-antimatter annihilation of nucleons, baryons, and leptons.

QUANTUM-STATE TRANSITIONS

In atoms, transitions between quantum states occur through coupling with the electromagnetic field, resulting in the emission and absorption of X rays, and ultraviolet and visible light. Energy differences between atomic states are too small to allow the emission of lepton pairs (electron-neutrino), for which a transition energy of at least 0.51 MeV is needed. In nuclei, the electromagnetic transitions are in the X-ray and gamma-ray regions; in addition, emission (β radioactivity) of lepton pairs can occur, but not emission of mesons, the smallest of which would require a transition energy of at least 140 MeV. The baryon (proton, neutron) can undergo weak and electromagnetic transitions as well as strong interactions. The latter transitions involve the emission and absorption of π and K mesons. A summary of the transitions of baryon states has recently been given by Weisskopf (1965). Leptons interact with other particles only via the electric field and the weak interaction.

Table I exhibits the type of reaction and the particles or photons emitted in transitions of atoms, nuclei, baryons, and leptons when various forces are involved. A fourth possible force, gravitational, is not considered. The entries in boldface type indicate those transitions that are energetic enough to emit gamma radiation. Gamma-ray emission has been neglected in weak interactions, as well as in other processes where no significant contribution can result.

Gamma rays are primarily produced by the following processes:

Electromagnetic transitions of leptons.—(a) Collisional bremsstrahlung: This radiation is produced by free-free transitions of high-energy electrons and positrons and induced by Coulomb scattering on nucleons, protons, or electrons. (b) Magnetic bremsstrahlung (synchrotron radiation): Relativistic electrons or positrons deflected in a magnetic field generate this radiation. (c) Compton scattering (inverse Compton effect): This effect consists of the transfer of energy by scattering from relativistic electrons and positrons to photons in the local radiation field.

Electromagnetic transitions of baryons.—(a) Collisional bremsstrahlung. (b) Magnetic bremsstrahlung. (c) Compton scattering. These reactions are identical to lepton transitions except that baryons are involved.

Electromagnetic transitions of nucleons.—(a) Nuclear de-excitation: It is possible for a nucleus to decay from an excited quantum state by gamma-ray emission. The excited state can result from an interaction with a high-energy nucleon, electron, or photon.

Strong-interaction transitions of baryons.—(a) π^0-Meson emission: A proton in an excited quantum state emits π and K mesons in transition to a lower state. The predominant source of gamma rays in such decays is π^0 mesons, which decay by electromagnetic interaction into two gamma-ray photons in 2×10^{-16} sec. K mesons also decay by the weak interaction into π^0 mesons. Proton excited states can result from collisions with high-energy protons and electrons and from collisions with photons. (b) Hyperon decay: Hyperons, which are also produced in high-energy collisions, may be viewed as excited proton states. Hyperons decay by the weak interaction into π^0 mesons, except for the Σ^0, which decays by the electromagnetic interaction $\Sigma^0 \to \Lambda^0 + \gamma$.

MATTER-ANTIMATTER ANNIHILATION

Electromagnetic interactions.—Two gamma-ray photons of energy 0.51 MeV result from electron-positron annihilation at rest: $e^+ + e^- \to 2\gamma$. Annihilation in flight results in a continuous spectrum.

Strong interactions.—Proton-antiproton annihilation results in the emission of π mesons. On the average, ~ 5 pions are produced per annihilation, resulting in a mean of 3.5 gamma-ray photons.

3. GENERAL TECHNIQUES FOR THE CALCULATION OF GAMMA-RAY INTENSITIES

Definitions

The gamma-ray source strength $S(E_\gamma, r)$ is defined as the rate of production of photons with energy between E_γ and $E_\gamma + dE_\gamma$ at the source per unit volume, per unit time, per steradian $(cm^3 \ sec \ sr \ GeV)^{-1}$, where r is the vector from the detector to the source-volume element $d\tau$.

For a dectector of solid angle Ω and unit area, the incident flux is given by

$$J_\gamma(E_\gamma) = \int \frac{S(E_\gamma, r)}{r^2} r^2 dr d\Omega \qquad (cm^2 \ sec \ GeV)^{-1} \qquad \qquad 1.$$

where the integral is over the source volume located in Ω.

If S is constant over the detector solid angle, the flux per unit solid angle is

$$I_\gamma(E_\gamma) = \int S(E_\gamma, r) dr \qquad (cm^2 \ sec \ sr \ GeV)^{-1} \qquad \qquad 2.$$

If S is not a function of r, Equation 2 becomes

$$I_\gamma(E_\gamma) = S(E_\gamma) L \qquad (cm^2 \ sec \ sr \ GeV)^{-1} \qquad \qquad 3.$$

where L is the length of the source along the line of sight.

In particular, for discrete sources whose angular dimensions are small compared to the solid angle of the detector, the flux is

$$J_\gamma(E_\gamma) = \frac{1}{R^2} \int_V S(E_\gamma, r) dV \qquad (cm^2 \ sec \ GeV)^{-1} \qquad \qquad 4.$$

where R is the distance to the source of volume V.

The above relationships give the differential photon intensity, i.e., the intensity per unit energy interval. Some detectors measure the photon flux above a given photon energy E_γ, thus it is sometimes more convenient to express the integral photon flux as

$$J_\gamma(>E_\gamma) = \int_{E_\gamma}^\infty J_\gamma(E_\gamma) dE_\gamma \qquad (cm^2 \ sec)^{-1} \qquad \qquad 5.$$

The differential energy flux at the detector is given by

$$\Phi_\gamma(E_\gamma) = E_\gamma J_\gamma(E_\gamma) \qquad GeV(cm^2 \ sec \ GeV)^{-1} \qquad \qquad 6.$$

and the integral energy intensity by

$$\Phi_\gamma(>E_\gamma) = \int_{E_\gamma}^\infty E_\gamma J_\gamma(E_\gamma) dE_\gamma \qquad GeV(cm^2 \ sec)^{-1} \qquad \qquad 7.$$

Calculation of the Source Strength S

The nature of the production mechanism used and the amount of information known about the source determine how the function S is to be calculated. Two different techniques will be described, either of which can be used for most processes.

Most of the gamma-radiation production mechanisms involve the reaction between an incident particle (or photon) of total energy E_i and a target particle (or photon). In this case, the function S can be expressed as

$$S(E_\gamma, r) = q(E_\gamma, r)n(r) \qquad (\text{cm}^3 \text{ sec sr GeV})^{-1} \qquad \qquad 8.$$

where $n(r)$ is the density of target particles and $q(E_\gamma, r)$ is the rate of production of gamma rays per target particle:

$$q(E_\gamma, r)dE_\gamma = dE_\gamma \int_{E_i} I_i(E_i, r)\sigma(E_i)f(E_\gamma, E_i)dE_i \qquad (\text{sec sr})^{-1} \qquad 9.$$

Here $I_i(E_i, r)dE_i$ is the differential intensity of incident particles in the energy interval between E_i and $E_i + dE_i$ in units of cm^2 per sec per sr per GeV, $\sigma(E_i)$ the total cross section for the event, and $f(E_\gamma, E_i)$ the distribution function for the gamma-ray emission spectrum in the observer's frame of reference, integrated over emission angles; σ is in units of cm^2 and $f(E_\gamma, E_i)dE_\gamma$ is dimensionless.

Another method (Gould & Burbidge 1965) for calculating an approximate value of S for a given production mechanism is based on knowledge of the incident particle-energy spectrum and the photon-emission spectrum as functions of E_i. The energy loss by a particle of energy E_i due to the emission of dN_γ photons in the energy interval dE_γ at E_γ and in the time interval dt is

$$\frac{dE_i}{dE_\gamma dt} = E_\gamma \frac{dN_\gamma}{dE_\gamma dt} = \phi(E_\gamma, E_i) \qquad \qquad 10.$$

where $\phi(E_\gamma, E_i)$ is the photon-emission spectrum. The particle spectrum can be expressed as a density function $n_i(E_i, r)$, where for relativistic particles,

$$n_i(E_i, r) = \frac{4\pi}{c} I_i(E_i, r) \qquad (\text{cm}^3 \text{ GeV})^{-1} \qquad \qquad 11.$$

Thus $S(E_\gamma, r)$ is given by

$$S(E_\gamma, r) = \frac{1}{4\pi} \int dE_i n_i(E_i, r) \left(\frac{dN_\gamma}{dE_\gamma dt} \right) \qquad \qquad 12.$$

For a given particle energy E_i the photon-emission spectrum, whether due to bremsstrahlung, synchrotron, or Compton processes, can be approximated by a δ function at a characteristic photon energy E_γ^*:

$$\frac{dN_\gamma}{dE_\gamma dt} = \frac{1}{E_\gamma} \phi(E_i, E_\gamma) \approx -\frac{1}{E_\gamma} \left(\frac{dE_i}{dt} \right) \delta(E_\gamma - E_\gamma^*) \qquad 13.$$

For a given process, (dE_i/dt) and E_γ^* are known functions of E_i; thus the right side of Equation 12 can be readily computed.

For production of gamma rays by π^0 mesons resulting from high-energy collisions, the quantity $(dN_\gamma/dE_\gamma dt)$ can be approximated by a power law based on the known pion-production spectrum.

Intensity Calculations for Discrete Radio Sources

Calculating the gamma-ray intensity from discrete sources by the above formulae is almost impossible since little is known of the internal structure of the source. However, in a source emitting synchrotron radio radiation, high-energy electrons are present that also emit gamma radiation. Thus, indirectly, the gamma-ray flux can be predicted from knowledge of the radio-frequency spectrum and intensity. Garmire & Kraushaar (1965) have calculated the gamma-ray flux from several discrete strong radio sources in this way. They assumed that the ratio of the detected radio power to the detected gamma-ray power was the same as the ratio of the corresponding emitted powers at the source. A given gamma-ray production mechanism is either assumed or can be deduced from the shape of the energy spectrum. An advantage of this method is that it eliminates uncertainties in source distance and in the cosmological model used.

To calculate the gamma-ray flux due to π^0-meson production, we must know the high-energy proton flux. A method of determining this flux is based on the assumption that all the high-energy electrons are secondary; that is, they are continually produced by charged-meson production in proton collisions. Present calculations of gamma-ray flux due to π^0 mesons in discrete radio sources are based on this assumption; however, experimental evidence already indicates that the assumption is false in the Crab Nebula. Another method is based on the assumption that the total energy of cosmic rays (protons and nuclei) is 100 times larger than the total energy of high-energy electrons.

Hayakawa et al. (1964) have estimated the gamma-ray flux from several radio sources by using the relation:

$$J_\gamma = g \left(\frac{w}{\overline{E}_e \tau_\gamma} \right) \bigg/ 4\pi R^2 \qquad (\text{cm}^2 \text{ sec})^{-1} \qquad 14.$$

where w is the energy of the relativistic electrons in the source, τ_γ is the mean time for gamma-ray emission, $\overline{E}_e$ is the average energy of the electrons, R is the distance to the source, and g is a constant; $w/\overline{E}_e$ is the total number of electrons and is approximately equal to the number of photons radiated, provided both are the result of π-meson decay. If a considerable part of the electrons are of primary origin, then $w/\overline{E}_e$ gives an upper limit to the gamma-ray flux. The actual number of photons is then given by multiplying by the factor g. Both g and τ_γ are very model dependent.

4. DENSITY, COMPOSITION, AND ENERGY SPECTRUM OF THE REACTANT PARTICLES

To apply the above considerations to the calculation of gamma-ray fluxes, we must have information about the reactant particles. Very high-energy protons, nuclei, or electrons interact with matter or electromagnetic radiation at thermal energies. In this section we estimate the density, composition, and energy spectrum of the reactants producing cosmic gamma radiation.

HIGH-ENERGY PARTICLE FLUXES AND SPECTRA IN THE GALAXY

Knowledge of high-energy particle fluxes in the Galaxy and metagalaxy is rather limited and is based on extrapolation of cosmic-ray data at the Earth and on radio-astronomy data.

Protons, nucleons, and antiprotons.—The measured integral spectrum of primary cosmic-ray protons is represented by the form:

$$I_p(E > E_0) = [K_p/(\Gamma_p - 1)]E_0^{-\Gamma_p+1} \qquad 15.$$

where $I_p(E > E_0)$ is the intensity of protons with total energy $E > E_0$, and K_p and Γ_p are constants within a defined energy range. The differential energy spectrum $I_p(E)$ can be obtained from Equation 15, assuming K_p and Γ_p are constants and $(I_p - 1) > 1$:

$$I_p(E) = K_p E^{-\Gamma_p} \qquad 16.$$

Table II (cf. references in Pollack & Fazio 1963) gives the differential energy spectrum as a function of proton kinetic energy T in GeV, where $E = m_p c^2 + T \approx 1 + T$. In the energy region 10^{15} to 10^{17} eV the exponent Γ_p increases to ~ 3.1 and then decreases to ~ 2.6 again above 10^{18} eV (Fujimoto, Hasegawa & Taketani 1964). The following integral fluxes result from the spectra in Table II:

$$I_p(T > 1 \text{ GeV}) = 0.15 \text{ cm}^{-2} \text{ sec}^{-1} \text{ sr}^{-1}$$

$$I_p(T > 10 \text{ GeV}) = 0.016 \text{ cm}^{-2} \text{ sec}^{-1} \text{ sr}^{-1}$$

TABLE II

DIFFERENTIAL ENERGY SPECTRUM OF COSMIC-RAY PROTONS

Intensity (cm² sec sr GeV)⁻¹	Proton kinetic energy (GeV)
$0.380/(1+T)^{-2}$	0.3 – 1.0
$0.437/(1+T)^{-2.2}$	1 – 10
$0.894/(1+T)^{-2.5}$	10 – 100
$1.4/(1+T)^{-2.6}$	10^2 – 10^5

The primary nucleon $(Z \geq 2)$ spectra, expressed as a function of the total energy per nucleon, is identical to the proton spectrum in the energy region of interest, but reduced in intensity. The relative intensities of nucleons in cosmic radiation are given in Table III (Hayakawa 1963). For $Z \geq 3$, the nucleons are grouped into light (L), medium (M), heavy (H), and very heavy (VH) divisions, according to the value of the atomic number.

Experimental upper limits to the cosmic-ray antiproton intensity have recently been summarized by Milford & Rosen (1965). Theoretical estimates of the flux of low-energy antiprotons indicate an antiproton-to-proton ratio $\ll 3 \times 10^{-3}$.

The proton and nucleon spectra and intensities in the Galactic disk and

TABLE III

RELATIVE COMPOSITION OF COSMIC RADIATION

Component	Atomic number Z	Relative intensity for given energy per nucleon (Waddington 1962)	Relative cosmic abundance (Cameron 1959)
p	1	100	100
α	2	5.00 ± 0.2	15
L	3–5	0.090 ± 0.03	5.7×10^{-7}
M	6–9	0.33 ± 0.02	0.15
H	≥ 10	0.11 ± 0.02	0.014
VH	≥ 20	0.03 ± 0.01	0.00071

halo are assumed to be equal to the values measured at the Earth. The reasons for this assumption are the observed isotropy and the continuous spectrum and chemical composition of cosmic radiation.

Electrons and positrons.—Recent measurements of the primary electron flux give a differential spectrum of the form

$$I_e(E) = K_e E^{-\Gamma_e} \qquad 17.$$

In general, the intensity is a few per cent of the primary proton flux, and the spectrum is rather flat. In the high-energy region, data exist from 10 MeV to 50 GeV, but are rather limited by statistical uncertainties (Waddington 1965). In the region from 0.5 to 5 GeV the flux can be expressed by (Meyer 1965)

$$I_e(E) = 11E^{-1.6 \pm 0.5} \qquad (\text{m}^2 \text{ sec sr GeV})^{-1} \qquad 18.$$

Other values of Γ_e are consistent with these results for lower energies, with some evidence of $\Gamma_e \approx 2$ or 3 at higher energies. Little is known about the effects of solar origin and modulation on the measured electron flux.

The electron spectrum in the Galaxy has also been derived, on the assumption that all the radio noise in the Galactic halo is the result of synchrotron radiation by relativistic electrons. The spectral index δ of the observed flux of radio noise $S(\nu)$ is related to the electron spectral index Γ_e by

$$\Gamma_e = 2\delta + 1 \qquad 19.$$

where $S(\nu) \propto \nu^{-\delta}$. The observed value of $\delta = 0.6 \pm 0.2$ gives a different, although not incompatible, value of (2.2 ± 0.4). The intensity of the radio emission calculated from the measured electron spectrum is not in disagreement with the measured intensity (Meyer 1965).

The ratio of positrons to electrons in cosmic radiation in the energy interval 0.3 to 3.0 GeV has been found to be $N_+/(N_+ + N_-) = 0.20 \pm 0.15$ (Meyer 1965). From this value the percentage of primary electron flux produced by proton collisions is between 10 and 50 per cent.

No direct information about high-energy particles in intergalactic space exists. The assumption usually made is that the flux for the ith-type particle is some fraction ξ_i of the Galactic flux, where $\xi_i \leq 1$.

As stated in Section 3, energetic electron fluxes and spectra in discrete radio sources can be inferred from data in the optical and radio-frequency region, by assuming the source to be synchrotron radiation. A given value of the magnetic field in the source must also be assumed. The proton intensity and spectrum are usually derived from the electron data, assuming the electrons were produced by proton collisions. If the electrons were directly accelerated, the proton spectrum can only be surmised and the proton-to-electron ratio is taken to be 100:1.

DISTRIBUTION OF MATTER AND RADIATION

To complete the calculation of gamma-ray intensities, we need knowledge of the distribution of matter in the Galaxy and metagalaxy. The results of this section are summarized in Table IV, which includes data from Ginzburg & Syrovatskii (1964a,b), Gould & Burbidge (1965), and Felten (1965).

Galactic gas density and composition.—The most important matter from the viewpoint of gamma-ray production in the Galaxy is the interstellar gas. This gas consists mostly of hydrogen ($\sim$90 per cent) and helium ($\sim$10 per cent), very unevenly distributed but predominantly in a disk about 14 kpc in radius and 300 pc thick. The average density of atomic hydrogen (H) is 0.7 atom cm^{-2}. Between 11 and 12 kpc from the Galactic center the average density begins to drop sharply. A contour map of the distribution of H is given by Garmire & Kraushaar (1965). Gould & Salpeter (1963) and Gould, Gold & Salpeter (1963) have considered the abundance and distribution of molecular hydrogen (H$_2$) and conclude that the ratio H$_2$/H may vary from 0.1 to 10 and have the same distribution as H. It is assumed that H$^+$/H is $\sim$10 per cent (Hill 1960) and that H$^+$ has the same distribution as H.

Unfortunately, our knowledge of the Galactic halo surrounding the disk is rather meager. If it exists, the halo has a radius of $\sim$15 kpc. The density of H is assumed to be uniform and to have a value $\leq 10^{-2}$ cm^{-3}.

The density of helium in intergalactic space is taken everywhere to be $\sim$10 per cent of the hydrogen density, and the effects of the other components of the gas are usually neglected.

Metagalactic gas density.—The amount of H$_2$ and H in the metagalaxy is negligible (Field, Solomon & Wampler 1966; Gunn & Peterson 1965), and the density of H$^+$ is estimated to be $\sim$10^{-5} cm^{-3}, corresponding to a gas density of 2×10^{-29} g cm^{-3}.

Photon densities.—The value of the thermal-photon density ρ due to starlight, averaged over the entire Galaxy, is taken to be 0.1 eV cm^{-3}. Since

$$\rho = 4.75 \times 10^{-3} T^4 \text{ eV cm}^{-3} \qquad 20.$$

this value of ρ corresponds to a mean source temperature of 4000°K. The average photon energy $\bar{\epsilon}$ is 1 eV, where the value of $\bar{\epsilon}$ is given by

$$\bar{\epsilon} = 2.7kT \qquad\qquad 21.$$

and where $k = 8.6 \times 10^{-5}$ eV deg^{-1}.

The value of the metagalactic starlight density has been considered by several authors (McVittie 1962; Ginzburg & Syrovatskii 1964a,b; Garmire 1965; Felten 1966). The mean value of ρ is assumed to be 10^{-2} eV cm^{-3}.

Recent measurements indicate the presence of a universal black-body radiation of 3.5°K, which corresponds to a microwave ($\bar{\epsilon} = 8.1 \times 10^{-4}$ eV) photon density of 0.7 eV cm^{-3}. If this radiation exists, its density is considerably greater than the average starlight density in the Galaxy and metagalaxy.

Magnetic-field intensities.—For purposes of calculation, the mean value of $H_\perp$, the component of the magnetic field perpendicular to the particle velocity, is taken to be 3×10^{-6} Oe in the Galaxy and 10^{-7} Oe in the metagalaxy.

TABLE IV

DENSITY OF GAS AND RADIATION IN THE GALAXY AND METAGALAXY

	L (cm)	$N(L)$ (nuclei cm^{-2})	$M(L)$ (g cm^{-2})	ρ_{ph} (eV cm^{-3})	$H_\perp$ (Oe)
Galaxy, in the direction of:					
Center	7×10^{22}	3×10^{22}	6×10^{-2}		
Anticenter	1.5×10^{22}	6×10^{21}	1.2×10^{-2}		
Pole, including plane		3×10^{20}	6×10^{-4}		
Galactic halo, in direction of pole	3.5×10^{22}	1.5×10^{20}	3×10^{-4}		
Average over Galactic plane and halo		8×10^{20}	1.6×10^{-3}	10^{-1}	3×10^{-6}
Local Galactic cluster	1.2×10^{23}	1.2×10^{18}	2.4×10^{-6}	10^{-2}	
Local supercluster	2×10^{25}	2×10^{20}	4×10^{-4}	10^{-2}	
Metagalaxy	5×10^{27}	5×10^{22}	10^{-1}	10^{-2} ($T = 4000°$K) 0.7 ($T = 3.5°$K)	10^{-7}

5. GAMMA-RAY ABSORPTION MECHANISMS

Although gamma rays may be absorbed or scattered in many ways by electrons, nucleons, electric fields, and meson fields (Segrè 1953, Evans 1955), there are three predominant types of interaction: the photoelectric effect, the Compton effect, and electron-positron pair production. The total absorption cross section (including elastic scattering) for a photon of frequency ν is given by

$$\sigma_\nu = \sigma_{\text{ph}} + \sigma_c + \sigma_{\text{pr}} \; \text{cm}^2 \qquad\qquad 22.$$

where σ_{ph} is the cross section for the photoelectric effect, σ_c for the Compton effect, and σ_{pr} for pair production. The absorption coefficient k_ν is defined by

$$k_\nu = \sigma_\nu n_a \; \text{cm}^{-1} \qquad\qquad 23.$$

where n_a is the number of agents per cm^3 with which an interaction can occur. The gamma-ray depth t_ν is defined by

$$t_\nu = \int_L k_\nu dx \qquad\qquad 24.$$

where L is the distance traversed by the photon beam. Photons of initial intensity I_0, after traversing a distance L of absorber, will have an intensity I given by

$$I = I_0 e^{-t_\nu} = I_0 e^{-N(L)\sigma_\nu} \qquad\qquad 25.$$

The mass absorption coefficient, defined as k_ν divided by the density, is given in Figure 1 for various substances as a function of photon energy. In general for $Z \lesssim 20$ and for $h\nu < 0.05$ MeV the photoelectric effect dominates. In the energy range 0.05 to 10 MeV the Compton effect is dominant, and pair production is important only above 50 MeV.

Photoelectric Effect

In the energy region above 100 keV an analytic approximation to the photoelectric cross section is

$$\sigma_{\text{ph}} \approx \text{const} \, \frac{Z^\eta}{(h\nu)^3} \qquad\qquad 26.$$

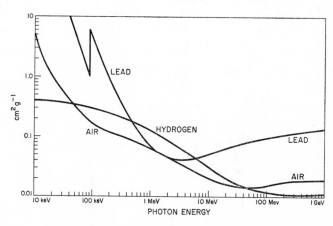

FIG. 1. Gamma-ray mass absorption coefficients.

where hv is the energy of the incident photon and Z is the effective nuclear charge of the absorbing material.

The index η is found to vary from about 4.0 to 4.6 as hv varies from 0.1 to 3 MeV. However, the energy exponent also varies continuously as hv increases. The above formula applies only to the ejection of electrons from the K shell of the atom (~ 80 per cent of the photoelectric effect). Empirical values of the photoelectric cross section are given by Davisson & Evans (1952) and White (1952).

COMPTON EFFECT

The total Compton cross section (elastic scattering and absorption) is given by the Klein-Nishina formula:

$$\sigma_c = \frac{3}{4}\sigma_T \left\{ \frac{1+\alpha}{\alpha^2} \left[\frac{2(1+\alpha)}{1+2\alpha} - \frac{1}{\alpha}\ln(1+2\alpha) \right] + \frac{1}{2\alpha}\ln(1+2\alpha) \right.$$
$$\left. - \frac{1+3\alpha}{(1+2\alpha)^2} \right\} \qquad \text{cm}^2 \text{ electron}^{-1} \qquad \qquad 27.$$

where $\alpha = hv/m_e c^2$,

$$\sigma_T = \frac{8\pi}{3}\left(\frac{e^2}{m_e c^2}\right)^2 = 6.651 \times 10^{-25} \text{ cm}^2$$

and $m_e c^2$ is the rest mass energy of the electron.

For $\alpha \ll 1$

$$\sigma_c = \sigma_T(1 - 2\alpha + 5.2\alpha^2 - 13.3\alpha^3 + \cdots) \qquad \text{cm}^2 \text{ electron}^{-1} \qquad 28.$$

and for $\alpha \gg 1$

$$\sigma_c = \left(\frac{3\sigma_T}{8\alpha}\right)\left[\frac{1}{2} + \ln 2\alpha\right] \qquad \text{cm}^2 \text{ electron}^{-1} \qquad \qquad 29.$$

Electrons are the dominant particles for Compton interactions. The cross section for particles of mass m_i can be calculated by replacing m_e by m_i.

Tables and graphs of the Compton cross section are given by Davisson & Evans (1952) and by Nelms (1953).

PAIR PRODUCTION IN THE ELECTROSTATIC FIELD OF NUCLEONS AND ELECTRONS

The threshold energy for electron-positron pair production is 1.02 MeV and the total cross section per nucleus is

$$\sigma_{\text{pr}} = \sigma_0 Z^2 \int_0^{hv-2m_e c^2} \frac{P dT_+}{hv - 2m_e c^2} = \sigma_0 Z^2 \overline{P} \qquad \text{cm}^2 \text{ nucleus}^{-1} \qquad 30.$$

where

$$\sigma_0 = \frac{1}{137}\left(\frac{e^2}{m_e c^2}\right)^2 = 5.80 \times 10^{-28} \qquad \text{cm}^2 \text{ nucleus}^{-1}$$

Here T_+ is the kinetic energy of the positron produced: Z is the atomic number of the absorber; P is a dimensionless quantity that is a complicated function of $h\nu$ and Z, varying between 0 (for $h\nu \leq 2\ m_e c^2$) and ~ 20 (for $h\nu = \infty$) for all values for Z; and $\bar{P}$ is the average value of P. Evans (1955) gives a graph of P as a function of $T_+(h\nu - 2m_e c^2)$. Analytical integration of the equation for σ_{pr} is possible only for relativistic cases and gives

$$\sigma_{\text{pr}} = \sigma_0 Z^2 \left(\frac{28}{9} \ln \frac{2h\nu}{m_e c^2} - \frac{218}{27} \right) \quad m_e c^2 \ll h\nu \ll 137\ m_e c^2 Z^{-1/3} \qquad 31.$$

when electron screening of the nuclear field can be neglected. Screening can also be neglected in the case of an ionized gas in the interstellar medium (e.g. a hydrogen plasma) independent of the photon energy.

In the case of complete screening, the cross section is

$$\sigma_{\text{pr}} = \sigma_0 Z^2 \left[\frac{28}{9} \ln (183\ Z^{-1/3}) - \frac{2}{27} \right] \quad h\nu \gg 137\ m_e c^2 Z^{-1/3} \qquad 32.$$

Thus at very high energies the pair-production cross section is constant.

The threshold energy for pair production in the field of atomic electrons is 2.04 MeV. Above this energy the pair-production cross section should be modified to account for the effect of atomic electrons by replacing Z^2 by $Z^2(1 + 1/cZ)$ where c varies from 2.6 at 6.5 MeV to 1.2 at 100 MeV and approaches unity as $h\nu$ approaches infinity.

PAIR PRODUCTION BY PHOTON-PHOTON INTERACTIONS

Cosmic gamma rays interact with thermal photons predominantly by elastic scattering and by electron-positron pair production. The threshold energy E_γ for pair production is

$$E_\gamma = \frac{(m_e c^2)^2}{E_{\text{photon}}} \qquad 33.$$

where E_{photon} is the energy of the thermal photon. The total cross section for pair production is given by

$$\sigma_{\text{pr}}{}^{\gamma\gamma} = \frac{1}{2} \left(\frac{e^2}{m_e c^2} \right)^2 \pi (1 - \beta^2) \left[(3 - \beta^4) \ln \frac{1+\beta}{1-\beta} - 2\beta(2 - \beta^2) \right] \qquad 34.$$

where

$$\beta = \frac{1}{\bar{E}} \sqrt{\bar{E}^2 - (m_e c^2)^2} \qquad 35.$$

and

$$\bar{E} = \sqrt{E_\gamma E_{\text{photon}}} \qquad 36.$$

In Equation 34 βc is the electron (or positron) velocity in the center-of-mass system. The cross section reaches its maximum value near the threshold energy and is approximately equal to 10^{-25} cm^2. Elastic scattering in most cases can be neglected as an absorption mechanism.

ABSORPTION OF GAMMA RAYS IN THE GALAXY AND METAGALAXY

The attenuation of gamma radiation over a distance L is given by:

$$I = I_0 e^{-t_\nu} \qquad\qquad 37.$$

where $t_\nu = k_\nu L$. For gamma-ray energies >100 keV, the Compton cross section decreases from its maximum value of 5×10^{-25} cm^2 at 100 keV. From Table IV, k_ν has the value of 2×10^{-25} cm^{-1} for the Galactic center and 5×10^{-30} cm^{-1} for the metagalaxy. In either case, $t_\nu \approx 2 \times 10^{-2}$, thus Compton absorption can be neglected. For pair production in atomic hydrogen, the maximum cross section is $\sim 10^{-26}$ cm^2; thus again $t_\nu \ll 1$ and absorption can be neglected at all energies.

Nikishov (1962) has computed the absorption coefficients k_ν for gamma rays with energy between 10^{11} eV and 5×10^{12} eV where pair-production interactions with starlight photons are important. For $\rho = 0.1$ eV cm^{-3}, the maximum value of k_ν is 7×10^{-27} cm^{-1} at 10^{12} eV. However, the value of ρ is overestimated, the more probable value being 10^{-2} to 10^{-3} eV cm^{-3}. In this latter case, in the metagalaxy ($L = 5 \times 10^{27}$ cm), $t_\nu = 3.5$ to 0.35.

For pair production on radio photons, Goldreich & Morrison (1964) have computed the values of k_ν for energies from 10^{18} to 10^{20} eV. In this case, the maximum value of k_ν was 4.7×10^{-27} cm^{-1}; thus $t_\nu = 24$ for $L = 5 \times 10^{27}$ cm.

The presence of a universal black-body photon gas at a temperature of 3.5°K would produce sufficient attenuation ($k_\nu \approx 10^{-23}$ cm^{-1}) that metagalactic gamma radiation greater than 10^{14} eV would be absorbed (Gould & Schréder 1966, Jelley 1966a).

In summary, metagalactic gamma radiation is absorbed in the region 10^{12} to 10^{13} eV by optical photons, above 10^{14} eV by microwave photons, and above 10^{20} eV by radio photons. In the energy region below 10^{12} eV and in the interval 10^{13} to 10^{14} eV, gamma rays are transmitted through the metagalactic medium. A partial transmission also occurs in the region 10^{18}–10^{20} eV (Gould & Schréder 1966).

Jelley (1966b) has also shown that the pair-production mechanism may cause considerable absorption of gamma radiation above 10^{11} eV in quasars and above 10^8 eV in "point" X-ray sources.

6. INTENSITY AND SPECTRUM FOR GAMMA-RAY SOURCE MECHANISMS

In this section the equations for the calculation of the cross section, rate of energy loss, gamma-ray intensity, and spectrum for the source mechanisms in Section 2 are given.

COLLISIONAL BREMSSTRAHLUNG

In an atomic or molecular gas, the differential cross section $\sigma_B(E, E_\gamma)$ for an electron of energy E to emit a photon with energy in dE_γ at E_γ by collisional bremsstrahlung in the field of a nucleus is given by

$$\sigma_B(E, E_\gamma)dE_\gamma = 4\sigma_0 Z^2 \frac{dE_\gamma}{E_\gamma} f(E_\gamma, E) \qquad 38.$$

where $f(E_\gamma, E)$ is a slowly varying function and σ_0 is defined in Equation 30.

In the extreme relativistic case, which includes complete screening of the nucleus by the atomic electrons, the cross section can be approximated by

$$\sigma_B(E, E_\gamma)dE_\gamma \approx \left(\frac{m}{X_0}\right) \frac{dE_\gamma}{E_\gamma} \text{ cm}^2 \qquad E \gg 137 \, m_e c^2 Z^{-1/3} \qquad 39.$$

where m is the mass of the target atom in grams, and Z the atomic number; X_0 is the radiation length (g cm^{-2}) in the target gas and is given by

$$X_0 = [4\sigma_0 N_0 Z^2 A^{-1}(\ln 183 \, Z^{-1/3})]^{-1} \qquad \text{g cm}^{-2} \qquad 40.$$

where N_0 is Avogadro's number and A is the mass number. For the Galactic and metagalactic gas the radiation length is $\sim$55 g cm^{-2}.

In the highly relativistic energy region, with no screening corrections the cross section is approximated by

$$\sigma_B(E, E_\gamma)dE_\gamma \approx 4\sigma_0 Z^2 \ln \left(\frac{2E}{m_e c^2}\right) \frac{dE_\gamma}{E_\gamma} \qquad m_e c^2 \ll E \ll 137 \, m_e c^2 Z^{-1/3} \qquad 41.$$

In a region that consists primarily of ionized hydrogen gas, Equation 41 with no screening corrections is applicable at all relativistic energies.

The contribution to bremsstrahlung radiation in electron-atomic electron collisions can be taken into account by changing Z^2 in the above equations to $Z(Z+1)$.

The rate of energy loss by electron bremsstrahlung is given by

$$\left(\frac{dE}{dt}\right)_B = -nc \int_0^E E_\gamma \sigma_B(E, E_\gamma)dE_\gamma \qquad 42.$$

and for $E \gg 137 \, m_e c^2 Z^{-1/3}$,

$$\left(\frac{dE}{dt}\right)_B = -\frac{nmc}{X_0} E \qquad 43.$$

The characteristic photon energy E_γ^* is therefore given by

$$E_\gamma^* \approx E \qquad 44.$$

The total rate of energy loss per unit volume is

$$P_B = \int_0^\infty \left(\frac{dE}{dt}\right) \frac{4\pi}{c} I_e(E)dE \qquad 45.$$

and for $E \gg 137 \, m_e c^2 Z^{1/3}$

$$P_B = -\frac{4\pi nm}{X_0} \int_0^\infty E I_e(E)dE \qquad 46.$$

Inserting Equation 39 into Equation 2, we find the differential gamma-ray flux to be

$$I_\gamma(E_\gamma)dE_\gamma = \frac{m}{X_0}N(L)\left(\frac{dE_\gamma}{E_\gamma}\right)\int_{E_\gamma}^{\infty}I_e(E)dE \qquad 47.$$

and if $I_e(E)dE = K_e E^{-\Gamma_e}dE$, then for $\Gamma_e > 1$

$$I_\gamma(E_\gamma) = \frac{M(L)}{X_0}\left(\frac{K_e}{\Gamma_e - 1}\right)E^{-\Gamma_e} \qquad 48.$$

and

$$I_\gamma(>E_\gamma) = \frac{M(L)}{X_0}\left[\frac{K_e}{(\Gamma_e - 1)^2}\right]E_\gamma^{-\Gamma_e + 1} \qquad 49.$$

Magnetic Bremsstrahlung (Synchrotron Radiation)

The rate of energy loss of an electron of energy E by synchrotron radiation in a magnetic field H is given by

$$-\left(\frac{dE}{dt}\right)_s = 2\sigma_T c\rho_H\left(\frac{E}{m_e c^2}\right)^2 \approx 10^{-3}H_\perp^2\left(\frac{E}{m_e c^2}\right)^2 \text{ eV sec}^{-1} \qquad 50.$$

where $\sigma_T = (8\pi/3)(e^2/m_e c^2)^2$, $\rho_H = H^2/8\pi$ is the magnetic-field energy density, and $H_\perp$ is the component of the magnetic field perpendicular to the particle velocity.

The photon-energy spectrum is continuous and has a maximum value at a frequency

$$\nu_{max} = 4.6 \times 10^{-6}H_\perp^-(E_{eV})^2 \text{ sec}^{-1} \qquad 51.$$

and

$$E_{\gamma_{max}} = h\nu_{max} = 4.14 \times 10^{-15}\nu_{max} \text{ eV} \qquad 52.$$

The characteristic photon energy is equal to $E_{\gamma_{max}}$.

The total rate of energy loss per unit volume is

$$P_S = -\frac{8\pi\sigma_T c\rho_H}{(m_e c^2)^2}\int_0^\infty E^2 I_e(E)dE \qquad 53.$$

For a random magnetic field and a uniform, isotropic distribution of electrons with a spectrum given by Equation 17, the differential flux is given by (Ginzburg & Syrovatskii 1965)

$$I_\gamma(E_\gamma) = 0.79a(\Gamma_e)L\left[\frac{4\pi}{c}K_e(6.3 \times 10^2)^{\Gamma_e - 1}\right]$$

$$\times H^{(\Gamma_e + 1)/2}\left(\frac{2.59 \times 10^4}{E_{\gamma_{eV}}}\right)^{(\Gamma_e + 1)/2} \text{ cm}^{-2} \text{ sec}^{-1} \text{ sr}^{-1} \text{ eV}^{-1} \qquad 54.$$

where $a(\Gamma_e)$ is a numerical coefficient with values $a(2.5) = 0.085$, $a(3) = 0.074$, and $a(4) = 0.072$, and K_e is given by Equation 17 in units of $(\text{GeV})^{\Gamma_e - 1} \text{ cm}^{-3}$.

Compton Scattering (Inverse Compton Effect)

This process has been considered in detail by Feenberg & Primakoff (1948), Donahue (1951), and Felten & Morrison (1963), and has recently been thoroughly discussed with respect to X-ray and gamma-ray astronomy by Felten (1965).

An electron of total energy E colliding with thermal photons of energy ϵ will produce a photon of average energy E_γ, where

$$E_\gamma \approx \epsilon \gamma_e \qquad \gamma_e \epsilon \ll m_e c^2 \qquad\qquad 55.$$

and

$$E_\gamma \approx \gamma_e (m_e c^2) \qquad \gamma_e \epsilon \gg m_e c^2 \qquad\qquad 56.$$

where $\gamma_e = E/m_e c^2$. The characteristic photon energy E_γ^* is taken equal to these average energies.

The theoretical cross section for this reaction is given by the Klein-Nishina formula, which has the asymptotic forms

$$\sigma_c = \sigma_T \left(1 - \frac{2\gamma_e \epsilon}{m_e c^2}\right) \qquad \gamma_e \epsilon \ll m_e c^2 \qquad\qquad 57.$$

and

$$\sigma_c = \frac{3}{8}\,\sigma_T \left(\frac{m_e c^2}{\gamma_e \epsilon}\right) \left[\ln\left(\frac{2\gamma_e \epsilon}{m_e c^2}\right) + \frac{1}{2}\right]_{\gamma_e \epsilon \gg m_e c^2} \qquad\qquad 58.$$

The rate of energy loss by an electron is given by

$$\left(\frac{dE}{dt}\right)_c = -\,\sigma_T c \rho \gamma_e^2 \qquad \gamma_e \epsilon \ll m_e c^2 \qquad\qquad 59.$$

and

$$\left(\frac{dE}{dt}\right)_c = -\frac{3}{8}\,\sigma_T (m_e c^2)^2 c \left(\frac{\rho}{\epsilon^2}\right) \ln\left(\frac{2\gamma_e \epsilon}{m_e c^2}\right)_{\gamma_e \epsilon \gg m_e c^2} \qquad\qquad 60.$$

At low energies the rate of energy loss is proportional to the photon density ρ, but at higher energies the loss is proportional to ρ/ϵ^2.

The total rate of energy loss per unit volume is

$$P_e = -\frac{4\pi\sigma_T \rho}{(m_e c^2)^2} \int_0^\infty E^2 I_e(E)dE \qquad \gamma_e \epsilon \ll m_e c^2 \qquad\qquad 61.$$

$$P_c = -\frac{3\pi}{2}\,\sigma_T(m_e c^2)^2 \left(\frac{\rho}{\epsilon^2}\right) \int_0^\infty \ln\left[\frac{2\gamma_e^2}{m_e c^2}\right] I_e dE \qquad \gamma_e \epsilon \gg m_e c^2 \qquad\qquad 62.$$

If a homogeneous, isotropic black-body photon gas at temperature T is assumed, then the photon density $n_{\rm ph}(\epsilon)$ is

$$n_{\rm ph}(\epsilon) = \frac{N_{\rm ph}}{2.404(kT)^3} \left(\frac{\epsilon^2}{e^{\epsilon/kT} - 1}\right) \qquad\qquad 63.$$

where $N_{\rm ph} = \int n_{\rm ph}(\epsilon)\,d\epsilon$ is the total number of photons per unit volume. The mean photon energy is

$$\bar{\epsilon} = \frac{1}{n_{\rm ph}} \int \epsilon n_{\rm ph}(\epsilon) d\epsilon = 2.7 kT \qquad\qquad 64.$$

For a homogeneous, isotropic electron-intensity distribution of the form

$$I_e(E) = K_e E^{-\Gamma_e}$$　　　　　　65.

the differential photon intensity is, from Equation 1,

$$I(E_\gamma) = \tfrac{1}{2} L \rho_{\mathrm{ph}} \sigma_T (m_e c^2)^{1-\Gamma_e} (\bar{\epsilon})^{(\Gamma_e - 3)/2} \times K_e E_\gamma^{-(1/2)(\Gamma_e + 1)} f(\Gamma_e)$$　　　　66.

where $f(\Gamma_e) \simeq 1$ for Γ_e in the range 1 to 2.

ELECTROMAGNETIC TRANSITIONS OF BARYONS

The cross section for bremsstrahlung by high-energy protons is similar to the above formulas for electrons, except that the cross section is reduced by the factor $(m_e/m_p)^2$. The details of the proton bremsstrahlung cross sections are given by Rossi (1952).

The equations for the differential gamma-ray intensity for both the inverse Compton effect and synchrotron radiation contain the factor γ^2/m^2, thus for protons the flux is reduced by a factor $(m_e/m_p)^4$ as compared to electrons.

Gamma-ray production by electromagnetic transitions of protons can usually be neglected when compared to similar interactions of electrons.

GAMMA-RAY EMISSION BY NUCLEAR DE-EXCITATION

Thermonuclear reactions.—If the kinetic temperature of a region is high enough, thermonuclear reactions may occur, and the energy liberated by mass conversion in these reactions can be in the form of gamma radiation. The reactions most likely to produce gamma rays, considering relative abundances and Coulomb-barrier potentials, are

(1) $H^1 + n^1 \rightarrow H^2 + \gamma$　　　($E_\gamma = 2.23$ MeV)

(2) $H^1 + H^2 \rightarrow H_e^3 + \gamma$　　　($E_\gamma = 5.5$ MeV)

The reaction rate between two unlike particles is given by

$$\tilde{S}(E_\gamma) = n_1 n_2 \langle \sigma v \rangle \ \mathrm{cm}^{-3} \ \mathrm{sec}^{-1}$$　　　　67.

where n_1 and n_2 are the number of reacting particles per cm³, and $\langle \sigma v \rangle$ is the average of the product of the reaction cross section and the relative velocity of the reacting particles.

If the particles have a Maxwell-Boltzmann velocity distribution, then (Gamow & Critchfield 1949, Salpeter 1952, Cameron 1961)

$$\tilde{S}(E_\gamma) = \frac{7.25 n_1 n_2}{Z_1 Z_2 A_1 A_2} (10^{-22})(A_1 + A_2) s \tau^2 e^{-\tau} \ \mathrm{cm}^{-3} \ \mathrm{sec}^{-1}$$　　　　68.

where

$$\tau = 42.48 \left(\frac{Z_1^2 Z_2^2 A_1 A_2}{A_1 + A_2} \right) T_6^{1/3}$$

$$s = \sigma E \exp \left\{ 988 Z_1 Z_2 \left[\frac{A_1 A_2}{(A_1 + A_2)E} \right]^{1/2} \right\} \ \mathrm{eVb}$$

where $Z =$ atomic number, $A =$ atomic mass number, $T_6 =$ temperature in units of $10^{6°}$K, $E =$ kinetic energy corresponding to the relative velocity v,

$\sigma =$ cross section for the reaction in barns (10^{-24} cm^2) at the energy E, and s is a slowly varying function of the energy E and can be considered constant.

The cross section for reaction 1, the neutron capture by a proton to form deuterium, is

$$\sigma = 7.30 \times 10^{-20} v^{-1} \text{ cm}^2 \qquad 69.$$

and therefore

$$\tilde{S}(E_\gamma) = (7.30 \times 10^{-20}) n_p n_n \text{ cm}^{-3} \text{ sec}^{-1} \qquad 70.$$

Neutrons, with a half life in the free state of 13 min, are not normally present, but originate from other nuclear reactions, primarily $He^4(p,pn)He^3$, $H^2(d,n)He^3$, and $H^3(d,n)He^4$. If high-energy protons and alpha particles are present, neutrons can also result from (p,n), (p,pn), $(p,2n)$, (α,n), (α,np), and $(\alpha,2n)$ reactions with helium, carbon, and oxygen, with a mean energy of a few MeV.

In reaction 2 the value of s is 7.8×10^{-2} eVb (Cameron 1961); however, owing to the expected low deuterium abundance, the gamma-ray flux is negligible.

High-energy nuclear reactions.—The cross section for the reaction of high-energy protons ($E_p > 2$ MeV) with target nuclei can be written as the sum of the cross sections of individual inelastic reactions. In the energy range of interest,

$$\sigma_r = \sigma(p,p') + \sigma(p,n) + \sigma(p,pn) + \sigma(p,2p) + \sigma(p,2n) + \sigma(p,\alpha) + \cdots \qquad 71.$$

where higher-order terms have been neglected as well as the reactions (p,γ) and (p,d).

Using the optical-model assumptions of Fernbach, Serber & Taylor (1949) for a uniform-density nucleus of radius $\rho = \rho_0 A^{1/3}$, we can approximate the reaction cross section by (Wattenburg 1957)

$$\sigma_r = \pi \rho^2 \left[1 - \frac{1 - (1 + 2K\rho)e^{-2K\rho}}{2K^2\rho^2} \right] \qquad 72.$$

where $K = 5 \times 10^{12}$ cm^{-1} and $\rho_0 = (1.28 \pm 0.3) \times 10^{-13}$ cm over the energy region where σ_r is approximately constant.

Direct gamma-ray line emission in the energy region 0.5 to 10 MeV results from these reactions because of decay of the excited states of the product nuclei. The decay usually occurs in less than 10^{-10} sec.

From the proton threshold kinetic energy of a few MeV to about 30 MeV, the predominant gamma-ray emission results from (p,p') reactions (inelastic scattering) on carbon, nitrogen, and oxygen. The gamma-ray production rate is proportional to $m\sigma_r$, where m is the gamma-ray multiplicity per interaction. The approximate values of $m\sigma_r$ are given in Table V. The gamma-ray production rate decreases at higher energies, although σ_r remains constant.

The predominant line emission is from carbon at 4.43 MeV and from oxygen at 6.15 MeV.

DECAY OF HEAVY RADIOACTIVE NUCLEI

Burbidge et al. (1956) have developed a theory of the exponential decrease of the light-curves of Type I supernova based primarily on the nuclear energy released by spontaneous fission of Cf^{254}. The assumed mechanism for synthesis of transbismuth elements in a supernova explosion is rapid neutron capture (r process). The induced radioactivity also results in line emission of gamma rays. Clayton & Craddock (1965) have calculated the expected gamma-ray intensity and spectrum from the Crab Nebula, assuming the Cf^{254} hypothesis is correct, and estimate that 2.7×10^{34} erg sec^{-1} are radiated by the Crab Nebula as complex line emission in the energy region 0.03 to 2 MeV. The strongest line flux is 9.7×10^{-5} photons cm^{-2} sec^{-1} for the 0.390-MeV line of Cf^{249}. The X-ray source in Scorpius would result in even higher fluxes if it is the remnant of a Type I supernova.

TABLE V

CROSS SECTIONS FOR GAMMA-RAY PRODUCTION BY NUCLEAR DE-EXCITATION

Target nucleus	Proton kinetic energy (MeV)	$m\sigma_r$ (10^{-27} cm^2)
He^4	> 4	0
C^{12}	4–30	242
	>30	23
N^{14}	4–30	558
	>30	27
O^{16}	6–30	560
	>30	50

PRODUCTION OF GAMMA RAYS FROM π^0-MESON DECAY

Proton-proton reaction.—The principal method of production of excited proton states that emit π mesons is the interaction of a high-energy proton with neutral or ionized hydrogen. The important reactions are

$$p + p \rightarrow p + p + a(\pi^+ + \pi^-) + b\pi^0$$
$$p + p \rightarrow p + n + \pi^+ + c(\pi^+ + \pi^-) + d\pi^0$$
$$p + p \rightarrow n + n + 2\pi^+ + f(\pi^+ + \pi^-) + g\pi^0$$
$$p + p \rightarrow D + \pi^+ + l(\pi^+ + \pi^-) + t\pi^0$$

where p indicates a proton, n a neutron, and D a deuteron, and a,b,c,d,f,g,l,t are positive integers. The threshold kinetic energy of the incident proton for the above reactions is 290 MeV. In the proton energy region from 290 MeV to 1 GeV, only single π-meson production is important:

$$p + p \rightarrow p + n + \pi^+$$
$$p + p \rightarrow p + p + \pi^0$$
$$p + p \rightarrow D + \pi^+$$

The $(pn\pi^+)$ reaction dominates in this energy region. When the kinetic energy of the incident proton exceeds 1 GeV, multiple-pion production becomes important. The minimum kinetic energy T_{min} of the incident proton needed to produce q mesons is given by

$$T_{min} = \frac{q^2 m_\pi^2 c^4}{2 m_p c^2} + 2 q m_\pi c^2 \approx q(280 + 10q) \text{ MeV} \qquad 73.$$

where $m_\pi c^2$ and $m_p c^2$ are the rest energy of the π meson and proton.

The detailed experimental cross sections and multiplicities for π-meson production have been summarized by Pollack & Fazio (1963) and are given in Table VI. At high energies ($E_p > 2$ GeV) the important properties of the reaction can be briefly stated. The total cross section for π-meson production is constant and equal to 27 mb. The charged-pion multiplicity is given by

$$m_\pm \approx 2 E_p^{0.25} \qquad 74.$$

where E_p is in units of GeV, and the ratio of π^0 mesons to charged mesons is

$$\frac{m_0}{m_\pm} \approx 0.5 \qquad 75.$$

The mean laboratory energy of the pions is

$$\langle E_\pi \rangle \approx m_\pi c^2 \gamma_p^{3/4} \approx 0.14 E_p^{3/4} \qquad 76.$$

where $\gamma_p = E_p / m_p c^2$, and E_p is in units of GeV.

In the rest frame of the π^0 meson each gamma ray produced has an energy

TABLE VI

PION-PRODUCTION CROSS SECTIONS AND MULTIPLICITIES IN
PROTON-PROTON COLLISIONS

Proton kinetic energy (GeV)	σ_{pp} (mb)	$m_0 \sigma_0$ (mb)	$m_+ \sigma_+$ (mb)	$m_- \sigma_-$ (mb)
0.34	0.5	0.01	0.5	0
0.50	7.4	0.6	6.8	0
0.80	27	4.7	23	0
0.93	27	6	27	0
2.0	27	9.1	27	3.1
2.9	27	13	27	4.6
6.2	27	26	39	12
9.0	27	30	43	16
24	27	46	58	31
100	27	62	76	49

of $m_\pi c^2/2 \approx 70$ MeV. In the laboratory frame the gamma-ray energy range is given by

$$\frac{m_\pi c^2}{2}\left(\frac{1-\beta_\pi}{\sqrt{1-\beta_\pi{}^2}}\right) \leq E_\gamma(\text{MeV}) \leq \frac{m_\pi c^2}{2}\left(\frac{1+\beta_\pi}{\sqrt{1-\beta_\pi{}^2}}\right) \qquad 77.$$

where $\beta_\pi c$ is the velocity of the π^0 meson, and the mean energy is

$$\langle E_\gamma \rangle \approx \frac{m_\pi c^2}{2}\gamma_\pi \qquad 78.$$

With the use of the above relationships the gamma-ray energy spectrum above 1 GeV can be easily approximated (Gould & Burbidge 1965, Ginzburg & Syrovatskii 1965). A more detailed calculation is given by Hayakawa et al. (1964). From Equation 9 the rate of production of gamma rays per target proton is given by

$$q_\gamma(E_\gamma)dE_\gamma = dE_\gamma \int_{E_p} dE_p I_p(E_p) \int_{E_{\pi\min}}^{\infty} dE_\pi \sigma(E_\pi, E_p)f_\gamma(E_\gamma, E_\pi) \qquad 79.$$

The π^0-meson distribution function can be approximated by a delta function, where

$$\sigma(E_\pi, E_p) = \sigma_0 m_0 f_\pi(E_\pi, E_p) \approx \sigma_0 E_p^{1/4}\delta[E_\pi - (0.14)E_p^{3/4}] \qquad 80.$$

The magnitude of $\sigma(E_\pi, E_p)$ includes the π^0-meson multiplicity function. The gamma-ray distribution function is given by (Stecker 1966):

$$f_\gamma(E_\gamma, E_\pi) = (E_\pi{}^2 - m_\pi{}^2 c^4)^{-1/2} \qquad 81.$$

for

$$\frac{E_\pi}{2}(1-\beta_\pi) \leq E_\gamma \leq \frac{E_\pi}{2}(1+\beta_\pi) \qquad 82.$$

and

$$f_\gamma(E_\gamma, E_\pi) = 0 \qquad 83.$$

otherwise. For a given gamma-ray energy the value of $E_{\pi\min}$ is given by

$$E_{\pi\min} = E_\gamma + \frac{m_\pi{}^2 c^4}{4E_\gamma} \qquad 84.$$

Since $I_p(E_p) = K_p E_p^{-\Gamma_p}$, Equation 79 becomes, in the high-energy limit, $E_\pi \gg m_\pi c^2$:

$$q_\gamma(E_\gamma) = 2\sigma_0 \int_{E_p} dE_p K_p E_p^{-(\Gamma_p - 1/4)} \int_{E_\gamma}^{\infty} dE_\pi \frac{\delta[E_\pi - (0.14)E_p^{3/4}]}{E_\pi} \qquad 85.$$

$$q_\gamma(E_\gamma) = \frac{2K_p\sigma_0}{0.14} \int_{(E_\gamma/0.14)^{4/3}}^{\infty} dE_p E_p^{-(\Gamma_p - 1/4)-3/4} \qquad 86.$$

$$q_\gamma(E_\gamma) = \frac{2K_p\sigma_0}{(0.14)(\Gamma_p - 1/2)}\left[\frac{E_\gamma}{0.14}\right]^{-4/3(\Gamma_p - 1/2)} \qquad 87.$$

$$a_\gamma(E_\gamma) = K_\gamma E_\gamma^{-4/3(\Gamma_p - 1/2)} \qquad 88.$$

Below 1 GeV the spectrum reaches a maximum at $\sim$70 MeV and falls rapidly below this energy.

Proton-alpha particle reactions.—The threshold for pion production in p-α reactions is $\sim$172 MeV. The cross sections for this reaction below proton kinetic energy 1 GeV have been summarized by Pollack & Fazio (1963). Above 1 GeV the p-α cross section is approximately twice the p-p cross section.

Considering the relative abundances of cosmic radiation and Galactic gas, the p-α reaction contributes $\sim$30 per cent of the meson production relative to the p-p reaction, and the α-p reaction $\sim$10 per cent, resulting in $\sim$40 per cent increase in the gamma-ray flux.

PRODUCTION OF GAMMA RADIATION FROM K-MESON AND HYPERON DECAY

In addition to π-meson decay, excited proton states may also decay by the emission of K mesons ($K^{\pm}, K^0$). The hyperons ($\Lambda^0, \Sigma^{\pm}, \Sigma^0, \Xi^-, \Xi^0$) may be viewed as an excited proton state or resonance. Both K mesons and hyperons can be produced in high-energy proton-proton (nuclei) reactions: $p+p \rightarrow \Lambda^0 + K^+ + p$; $p+p \rightarrow \Sigma^0 + K^+ + p$; and $p+p \rightarrow K^0 + \overline{K}^0 + p + p$. Table VII gives the threshold kinetic energies for the incident proton for various reaction products.

Direct gamma rays and neutral π mesons, which decay into gamma rays, are then produced by the decay ($\tau \sim 10^{-8}$ to 10^{-10} sec) of these particles. The most important decay modes are

$$K^{\pm} \rightarrow \pi^{\pm} + \pi^0 \qquad \text{(21.5 per cent)}$$
$$\rightarrow \pi^{\pm} + \pi^0 + \pi^0 \qquad \text{(1.7 per cent)}$$
$$\rightarrow \pi^0 + \mu^{\pm} + \nu \qquad \text{(3.4 per cent)}$$
$$\rightarrow \pi^0 + e^{\pm} + \nu \qquad \text{(4.8 per cent)}$$
$$K^0 \rightarrow K_1 + K_2$$
$$K_1 \rightarrow \pi^0 + \pi^0 \qquad \text{(31.1 per cent)}$$
$$K_2 \rightarrow 3\pi^0 \qquad \text{(26.5 per cent)}$$
$$\rightarrow \pi^+ + \pi^- + \pi^0 \qquad \text{(11.4 per cent)}$$
$$\Lambda^0 \rightarrow n + \pi^0 \qquad \text{(33.1 per cent)}$$
$$\Sigma^+ \rightarrow p + \pi^0 \qquad \text{(51.0 per cent)}$$
$$\Sigma^0 \rightarrow \Lambda^0 + \gamma \qquad \text{(100 per cent)}$$
$$\Xi^0 \rightarrow \Lambda^0 + \pi^0 \qquad \text{(100 per cent)}$$

where the quantities in parentheses are the branching ratios of the particle for the given decay mode. Results of high-energy accelerator experiments with protons of 25-GeV energy indicate that the K^+/π^+ ratio is 0.25 and could be greater at higher energies. Since $\sim$0.5 π^0 mesons are produced per K-meson decay, and $\sim$0.5 π^0 mesons are produced per hyperon decay, the above reactions relative to direct π^0-meson production contribute at least 25 per cent additional gamma-ray flux above a proton energy of 25 GeV. Above gamma-

TABLE VII

THRESHOLD KINETIC ENERGIES FOR K-MESON AND HYPERON
PRODUCTION IN PROTON-PROTON REACTIONS

Reaction products	Laboratory kinetic energy of incident proton (GeV)
ΛKN	1.58
ΣKN	1.78
$\Lambda KN\pi$	1.98
$\Sigma KN\pi$	2.17
$K\bar{K}NN$	2.49
$\Lambda\Lambda KK$	3.63

ray energies of 100 GeV, hyperon decay may predominate over π^0-meson decay as the source of gamma radiation.

PROTON-THERMAL PHOTON REACTIONS

Neutral π mesons may also be produced by photonuclear reactions of very high-energy protons with thermal photons, $p+\gamma \rightarrow p+\pi^0$. In the proton rest system the thermal photon of energy ϵ has an energy

$$\epsilon^* = \epsilon(1 + \cos\omega)\left(\frac{E_p}{m_p c^2}\right), \qquad \epsilon \ll m_p c^2 \ll E_p \qquad 89.$$

where ω is the angle the photon makes with respect to the direction of the head-on collision. The threshold energy for the above reaction is $\epsilon^* = 150$ MeV. The proton energy threshold in the laboratory system is given by

$$E_p^{\text{th}} = m_\pi c^2(m_\pi c^2 + 2m_p c^2)/2\epsilon(1 + \cos\omega) \qquad 90.$$

and for $\epsilon = 1$ eV and $\omega = 0$, $E_p^{\text{th}} = 7 \times 10^{16}$ eV.

For a proton energy of $\sim 10^{17}$ eV and $\epsilon = 1$ eV, the cross section for the reaction is $\sim 10^{-28}$ cm^2.

Hayakawa et al. (1964) have calculated the metagalactic gamma-ray flux due to the process assuming the black-body photon energy distribution and a proton differential energy spectrum $\sim E_p^{-2.5}$. The resulting flux for $E_\gamma > 10^{15}$ eV can be expressed as a gamma-ray source strength:

$$S(E_\gamma > 10^{15} \text{ eV}) = 2 \times 10^{-43}\rho \qquad (\text{cm}^3 \text{ sec sr})^{-1} \qquad 91.$$

where ρ is the thermal-photon density in eV per cm^3.

An integral energy spectrum was calculated numerically for $\rho = 5 \times 10^{-2}$ eV cm^{-3} and is given in the above reference.

GAMMA-RAY EMISSION IN MATTER-ANTIMATTER ANNIHILATION

Electromagnetic interactions.—The cross section for positron-electron annihilation into two gamma-ray photons has its maximum at a positron total energy of $E \approx m_e c^2$ and is given by

$$\sigma_a = \pi r_0^2 = 0.25 \times 10^{-24} \text{ cm}^2 \qquad\qquad 92.$$

where r_0 is the electron radius. For $E >> m_e c^2$ the cross section is small and is given by

$$\sigma_a = \pi r_0^2 \left(\frac{m_e c^2}{E}\right) \left(\frac{\ln 2E}{m_e c^2 - 1}\right) \qquad\qquad 93.$$

The mean annihilation time for a positron is

$$t_a = (n_e \ \sigma_a c)^{-1} \qquad\qquad 94.$$

where n_{e^-} is the electron density.

The gamma-ray source strength is

$$\tilde{S} = 2 n_{e^+} n_{e^-} \sigma_a c \qquad (\text{cm}^3 \text{ sec})^{-1} \qquad\qquad 95.$$

where n_{e^+} is the positron density and the gamma-ray multiplicity is 2. If $E/m_e c^2 \approx 1$ for the positron, then each of the two emitted photons has an energy of $m_e c^2 = 0.51$ MeV and they are emitted in opposite directions. At high positron energies in the laboratory system, the photon emitted in the forward direction has almost all the positron energy, while the other photon has an energy $\sim m_e c^2$.

Three-photon annihilation of the positron-electron system is also possible, but the reaction rate is negligible when compared to the two-photon annihilation.

Proton-antiproton annihilation directly into electromagnetic radiation competes very unfavorably with annihilation into mesons. The nonrelativistic cross section for electromagnetic annihilation in flight is given by

$$\sigma_a \approx \frac{3 \times 10^{-30}}{\beta} \text{ cm}^2 \qquad\qquad 96.$$

Strong interactions.—The proton-antiproton interaction can result in several processes that generate π^0 mesons:

(*a*) annihilation: $\bar{p} + p \rightarrow$ pions;

(*b*) inelastic collisions: $\bar{p} + p \rightarrow \pi^0 + p + \bar{p}$; $\bar{p} + p \rightarrow$ two nucleons + pions; $\bar{p} + p \rightarrow K$ mesons and hyperons.

The predominant mode of π^0 generation is the annihilation process that results in an average of 5.3 pions per interaction. The mean number of gamma rays thus produced is ~ 3.5. The annihilation cross section is given approximately by the relation

$$\sigma_a \approx \pi (r_0 + \lambda)^2 \qquad\qquad 97.$$

where $r_0 = 0.90 \pm 0.05 \times 10^{-13} \text{ cm}^2$, and $2\pi\lambda$ is the antiproton wavelength in the center of mass. The formula holds for the antiproton laboratory kinetic energy from 100 MeV, where the cross section is 84 mb, to 1 GeV, where the cross section is 51 mb.

The total inelastic cross section is ~ 10 mb, and the single π^0-meson production cross section is ~ 2 mb over the above energy range.

The average π-meson kinetic energy in annihilation is ~ 200 MeV, thus the gamma-ray energy resulting from the decay will be sharply peaked at 70 MeV. The gamma-ray spectrum from p-$\bar{p}$ annihilation at rest is given by Frye & Smith (1966).

The mean annihilation time for an antiproton in a proton gas of density n_p is

$$t_a \approx (n_p \sigma_a c)^{-1} \qquad 98.$$

and the gamma-ray source strength integrated over all photon energies is

$$\tilde{S} \approx n_{\bar{p}} n_p \sigma_a m_\gamma c \quad (\text{cm}^3 \text{ sec})^{-1} \qquad 99.$$

where $n_{\bar{p}}$ is the density of antiprotons, and m_γ is the gamma-ray multiplicity.

7. THEORETICAL PREDICTIONS OF CELESTIAL GAMMA-RAY FLUXES AND SPECTRA

On the basis of the formulas and data given in the previous sections, gamma-ray fluxes and spectra can be calculated for explicit sources. The results of several such computations by various authors are presented in Tables VIII through X. General reviews of the subject have been given by Morrison (1958), Greisen (1960), Milford & Shen (1962), Schatzman (1963), Ginzburg & Syrovatskii (1964a,b), Garmire & Kraushaar (1965), Gould & Burbidge (1965), Oda (1965), and Weekes (1966).

INTERSTELLAR MEDIUM

The theoretical diffuse gamma-ray fluxes and spectra for the most important source mechanisms in the Galaxy are summarized in Table VIII. Above 50 MeV and to energies $\sim 10^{15}$ eV, the primary source of photons is the decay of π^0 mesons produced by cosmic-ray interactions with the interstellar gas. The integral energy spectrum is given by an inverse power law of the form $E_\gamma^{-1.8}$. The calculations for this source are at present the most accurate predictions of a flux and spectrum of cosmic gamma rays. An important result of these calculations is the anisotropy of the flux. The ratio of the flux from the Galactic center to the flux from the Galactic pole is of the order of 30 to 100. Since the flux is proportional to the product of the cosmic-ray nucleon intensity and the interstellar gas density, the absence of such a flux would indicate the cosmic-ray intensity is not uniform in the Galaxy. A measurement of this flux is very important to an understanding of the origin and diffusion of cosmic radiation in the Galaxy. Underlying all these considerations is the real possibility that a large metagalactic isotropic flux might exist that would completely mask any Galactic source.

Table VIII also includes the flux values and spectra expected from cosmic-ray electrons interacting with the interstellar medium. The primary reactions are Compton scattering and bremsstrahlung, both of which contribute to the flux at all gamma-ray energies to 10^{11} eV. Below 10 MeV, Compton scattering is the principal source. The uncertainties in the quoted flux values result from the limited experimental data on the electron spectrum in the primary cosmic radiation and uncertainties in the magnitude of the electron

TABLE VIII

THEORETICAL DIFFUSE GAMMA-RAY FLUXES FROM THE GALAXY

Mechanism	Energy	Flux (cm² sec sr)⁻¹				Integral energy spectrum	References
		Center	Anticenter	Pole	Average		
π⁰-Meson decay	>50 MeV	3×10^{-4}	6×10^{-5}	3×10^{-6}	8×10^{-5}		Ginzburg & Syrovatskii (1964a, b)
		2.4×10^{-4}	1.2×10^{-5}	7.1×10^{-6a}			Pollack & Fazio (1963)
					3×10^{-5}	b	Hayakawa et al. (1964)
					2×10^{-5}	$E^{-1.8}$	Gould & Burbidge (1965)
					1.5×10^{-5}	b	Garmire & Kraushaar (1965)
	>1 GeV	2×10^{-5}	4×10^{-6}	2×10^{-7}		$E^{-1.8}$	Ginzburg & Syrovatskii (1964a, b)
	>10 GeV				5×10^{-7}	$E^{-2.2b}$	Lieber, Milford & Spergel (1965)
Bremsstrahlung	>50 MeV	10^{-4}	2×10^{-5}	10^{-6}	3×10^{-6}	b	Ginzburg & Syrovatskii (1964a, b)
					10^{-7}		Gould & Burbidge (1965)
					10^{-6}	$E^{-\Gamma_e+1b}$	Garmire & Kraushaar (1965)
	>1–10 GeV	5×10^{-6}	10^{-6}	5×10^{-8}	1.3×10^{-7}	E^{-1}	Ginzburg & Syrovatskii (1964a, b)
Compton scattering	>0.5 MeV	1.3×10^{-4}	2.7×10^{-5}	1.3×10^{-4}	1.3×10^{-4}	$E^{-0.5}$	Ginzburg & Syrovatskii (1964a, b)
					$10^{-3}–10^{-4}$	b	Felten & Morrison (1966)
					10^{-5}	b	Hayakawa et al. (1964)
					2×10^{-5}	b	Gould & Burbidge (1965)
					7×10^{-3}	$E^{-0.3}$	Fazio, Stecker & Wright (1965)
	>50 MeV	1.3×10^{-5}	3×10^{-6}	1.3×10^{-5}	$1.3\times10^{-}$		Ginzburg & Syrovatskii (1964a, b)
					10^{-6}	b	Hayakawa et al. (1964)
					4×10^{-9}	b	Gould & Burbidge (1965)
	>1 GeV	3×10^{-6}	6×10^{-7}	3×10^{-6}	3×10^{-6}	$E^{-(\Gamma_e-1)/2b}$	Ginzburg & Syrovatskii (1964a, b)
Nuclear gamma rays	>0.5 MeV	2×10^{-6}	4×10^{-7}	2×10^{-8}	5×10^{-8}	Line emission	Ginzburg & Syrovatskii (1964a, b)
					10^{-5}	Line emission	Hayakawa et al. (1964)
Positron annihilation	0.5 MeV	$(1.2-300)\times10^{-6}$	$(2.4-600)\times10^{-7}$	$(1.2-300)\times10^{-8}$	$(3.2-800)\times10^{-8}$	Line emission	Ginzburg & Syrovatskii (1964a, b)
		5×10^{-4}				Line emission	Pollack & Fazio (1963)
			2×10^{-6}	1.3×10^{-5a}	6×10^{-5}	Line emission	Hayakawa et al. (1964)

ᵃ Determined for a direction perpendicular to the Galactic plane.
ᵇ Data given in graphical form.

flux. The bremsstrahlung photon flux is proportional to the product of the electron flux and the interstellar gas density, and exhibits an anisotropy similar to photons from π^0-meson decay. The Compton flux, however, is proportional to the product of the electron flux and the interstellar thermal-photon density, and is therefore more isotropic, resulting in the largest average gamma-ray flux. From Equations 49 and 66 the two processes have integral photon spectra of the form $E_\gamma^{-\Gamma_e+1}$ and $E_\gamma^{-(\Gamma_e-1)/2}$, respectively, where Γ_e is the index of the differential electron spectrum.

The predominant form of gamma-ray line emission occurs at 0.5 MeV and is due to annihilation at rest of positrons generated from π^+-meson decay. The uncertainties in the calculation arise owing to the unknown leakage rate of positrons from the Galaxy. The ratio of the intensity of the π^0-meson gamma-ray flux to the 0.5-MeV flux is a measure of the cosmic-ray intensity 10^9 years ago, if the primary source of gamma rays is assumed to have been cosmic-ray interactions (Pollack & Fazio 1963).

Intergalactic Medium

Theoretical calculations of the flux and spectra of diffuse gamma rays from the metagalaxy indicate that the primary mechanisms are π^0-meson decay, bremsstrahlung, and Compton scattering. The results are summarized in Table IX. Below 10 MeV, Compton scattering dominates, and above 10 MeV all three mechanisms contribute. Compton scattering and bremsstrahlung are important to photon energies of 10^{11} eV, whereas π^0-meson and hyperon decay contribute to the highest energies. It is important to note that above 10^{15} eV, π^0 mesons are predominantly the result of the reaction of cosmic-ray nucleons and thermal photons (photoproduction).

The metagalactic densities of high-energy particles, gas, magnetic fields, and thermal photons are relatively unknown, thus the relative contributions of the various mechanisms are difficult to determine. With respect to the high-energy particle flux, the intensity has been assumed to be either equal to or some unknown factor ξ of the Galactic cosmic-ray flux, and the spectra have been taken equal to the Galactic spectra. Should the value of ξ for nucleons exceed 10^{-2} and for electrons 10^{-3}, the metagalactic gamma-ray flux would exceed the average Galactic flux. In contrast to the Galactic flux, the metagalactic flux will be isotropic. Through measurements of the isotropy it may be possible to separate the two contributions and thus determine the high-energy particle densities in intergalactic space.

Discrete Radio Sources

For discrete radio sources, calculations of gamma-ray fluxes are determined for three source mechanisms: π^0-meson decay, Compton scattering, and bremsstrahlung. As discussed in Section 3, the electron intensity in a given source is determined from the radio-flux density, assuming the radio radiation is due to synchrotron radiation. The proton-to-electron ratio at high energies is then taken to be 100 or the electrons are assumed to be secondary. The

TABLE IX
Theoretical Diffuse Gamma-Ray Fluxes from the Metagalaxy

Mechanism	Energy	Flux $(cm^2 sec sr)^{-1}$	Integral energy spectrum	References
π^0-Meson decay	>50 MeV	$5 \times 10^{-4}\xi_p$		Ginzburg & Syrovatskii (1964a, b)
(p-p reactions)		5×10^{-4}	a	Gould & Burbidge (1965)
		6×10^{-4}	a	Garmire & Kraushaar (1965)
	> 1 GeV	$3 \times 10^{-5}\xi_p$	$E^{-1.8}$	Ginzburg & Syrovatskii (1964a, b)
	>10 GeV		$E^{-2.2}$	Lieber, Milford & Spergel (1965)
	$>10^{14}$ eV	10^{-15}	a	Hayakawa et al. (1964)
π^0-Meson decay	$>10^{16}$ eV	10^{-16}	a	Hayakawa et al. (1964)
(photopion)				
Bremsstrahlung	>50 MeV	$1.6 \times 10^{-4}\xi_e$		Ginzburg & Syrovatskii (1964a, b)
		10^{-7}	a	Gould & Burbidge (1965)
	> 1 GeV	$8 \times 10^{-6}\xi_e$	E^{-1}	Ginzburg & Syrovatskii (1964a, b)
		5×10^{-6}	$E^{-\Gamma_e+1}$	Garmire & Kraushaar (1965)
Compton	> 0.5 MeV	$10^{-1}\xi_e$	$E^{-0.5}$	Ginzburg & Syrovatskii (1964a, b)
(electrons)		4×10^{-3}	a	Gould & Burbidge (1965)
		3×10^{-2}	a	Felten & Morrison (1966)
	>50 MeV	$9 \times 10^{-3}\xi_e$		Ginzburg & Syrovatskii (1964a, b)
		10^{-2}	$E_\gamma^{-(\Gamma_e-1)/2}$ a	Garmire & Kraushaar (1965)
		4×10^{-7}	a	Gould & Burbidge (1965)
	> 1 GeV	$2 \times 10^{-3}\xi_e$		Ginzburg & Syrovatskii (1964a, b)
Compton	$>10^{14}$ eV	10^{-18}	a	Hayakawa et al. (1964)
(proton)				
Nuclear gamma rays	> 0.5 MeV	$3 \times 10^{-6}\xi_p$	Line emission	Ginzburg & Syrovatskii (1964a, b)
Positron	0.5 MeV	$\ll 5 \times 10^{-4}\xi_p$	Line emission	Ginzburg & Syrovatskii (1964a, b)
annihilation		$10^{-3}\xi_p$	Line emission	Hayakawa et al. (1964)
Sum of all galaxies	>50 MeV	10^{-5}	a	Hayakawa et al. (1964)

a Data given in graphical form.

high-energy particles, gas, and radiation are usually assumed to fill the entire volume of the source. Because of the uncertain quantities used in the calculation of a gamma-ray flux, the theoretical flux values vary over a large range for any given source, as seen in Table X. The most detailed calculations have been performed for the Crab Nebula. Of particular interest in this object are the predictions of measurable fluxes of line emission in the energy region 0.03 to 2 MeV and continuum radiation due to Compton scattering above 10^{11} eV. Quasistellar radio sources (QSO) might possibly be a relatively strong source of gamma rays that are produced by Compton scattering in the region of high photon density. However, so little is known of the structure of these objects that any flux values are highly speculative.

As has happened in X-ray astronomy, a discrete source of gamma rays may possibly exist that is not a radio source.

Crude as the calculations of the expected gamma-ray fluxes from discrete radio sources may appear, they are important for the design of detectors. In particular, balloon-borne gamma-ray detectors are limited to searching primarily for discrete sources since the atmospheric background of gamma rays at high altitudes is nearly isotropic.

TABLE X

THEORETICAL GAMMA-RAY FLUXES FROM DISCRETE SOURCES

Object	Energy region (MeV)	Flux (cm² sec)⁻¹				References
		π^0 decay	Bremsstrahlung	Compton	Nuclear decay	
Taurus A (Crab Nebula)	0.03–2				10^{-2}	Morrison (1958)
					10^{-2}	Savedoff (1959)
					10^{-4}	Clayton & Craddock (1965)
	0.1–1		6×10^{-8}			Savedoff (1959)
	>1		2×10^{-7}	2×10^{-12}		Savedoff (1959)
	>10		3×10^{-6}	7×10^{-8}		Savedoff (1959)
	>100	1.5×10^{-5}				Garmire & Kraushaar (1965)
		5.7×10^{-6}				Gould & Burbidge (1965)
		10^{-4}				Hayakawa et al. (1964)
				10^{-6}		Gould (1965)
		2×10^{-9}	5×10^{-10}	2×10^{-9}		Ginzburg & Syrovatskii (1964a, b)
		7×10^{-5}				Savedoff (1959)
	>5×10⁵			4×10^{-10}		Gould (1965)
	>5×10⁶			10^{-12}		Gould (1965)
		8×10^{-9}				Gould & Burbidge (1965)
Cygnus A	>100	4×10^{-8}	8×10^{-9}	6×10^{-13}		Garmire & Kraushaar (1965)
		10^{-4}				Hayakawa et al. (1964)
		2×10^{-11}	5×10^{-12}	10^{-8}		Ginzburg & Syrovatskii (1964a, b)
		4×10^{-7}				Savedoff (1959)
Virgo A (M87)	>100	6.2×10^{-9}	1.2×10^{-9}	2×10^{-10}		Garmire & Kraushaar (1965)
		10^{-10}	3×10^{-11}	10^{-8}		Ginzburg & Syrovatskii (1964a, b)
Centaurus A	>100	3.4×10^{-8}	7×10^{-9}	2×10^{-9}		Garmire & Kraushaar (1965)
Sagittarius A	>100	10^{-4}				Hayakawa et al. (1964)
Andromeda Nebula (M31)	>100	8×10^{-10}	2×10^{-10}	10^{-7}		Ginzburg & Syrovatskii (1964a, b)
3C 273B	2.7–100			5×10^{-6}		Ginzburg & Syrovatskii (1964a, b)

SOLAR SYSTEM

Gamma rays emitted by the Sun can yield valuable information on the nature of solar flares (Dolan & Fazio 1965). Almost all gamma-ray activity will be associated with flare regions.The primary source of continuum radiation to 10 MeV appears to be electron bremsstrahlung. Line emission occurs at 0.51 MeV owing to positron annihilation, at 2.23 MeV owing to deuteron formation, and from 0.5 to 10 MeV owing to nuclear de-excitation, particularly at C^{14} and O^{16}. All the above line emission is indicative of nuclear reactions occurring in a flare. The intensity of the 2.23-MeV line is a measure of the neutron density in a flare. The flux above 50 MeV is primarily caused by π^0-meson decay, which causes a continuous spectrum peaked at about 70 MeV. The intensity of this radiation can yield much information about the accelerated particle spectrum in the flare.

The only gamma radiation that should emanate from a completely quiet Sun is the albedo radiation from the interaction of Galactic cosmic rays with the photosphere. The occurrence of many class 1⁻ flares and subflares, however, may result in small irregular fluxes of gamma rays.

Albedo gamma radiation is also expected from planets, owing to interaction of Galactic cosmic rays with atmospheric gases and surface material. Another source of gamma rays is the high-energy trapped radiation, which may generate bremsstrahlung in the planetary atmosphere.

8. EXPERIMENTAL TECHNIQUES

INTRODUCTION

Gamma radiation can be detected only by the charged particles it produces when it interacts with matter. The primary gamma-ray absorption mechanisms—the photoelectric effect, Compton scattering, and electron-positron pair production—were discussed in Section 5. Depending on the type of absorption interaction, detectors can be divided into three general energy ranges: (a) 0.1 to 10 MeV, (b) 10 MeV to 10 GeV, and (c) >10 GeV. In region (a) the detectors exploit the photoelectric effect and Compton scattering. In regions (b) and (c) the interaction detected is pair production accompanied by an electromagnetic cascade shower.

Several properties of this region of the electromagnetic spectrum make the problems of detection unique. Because gamma-ray photons are relatively penetrating and in order to absorb the radiation for energy measurements completely, the detectors are usually massive. The photon flux is also relatively low, thus requiring large-area detectors. Combining these factors with high energy resolution is very difficult. Gamma-ray detectors must also operate in a flux of charged particles (cosmic rays) that is at least 1000 times as intense and that produces a secondary flux of gamma rays. When good angular resolution is not possible, the secondary gamma-ray flux produced by matter near the detector becomes a serious background problem.

The Earth's atmosphere completely absorbs incident gamma radiation;

therefore, in general, detectors must be placed in high-altitude balloons and satellites. Rockets do not provide sufficient exposure time. Only in the region above 10^{11} eV can one use ground-based techniques based on the detection of electromagnetic cascade showers in the atmosphere. Balloon experiments are limited to searching for point sources rather than diffuse sources, owing to the almost isotropic background radiation produced in the Earth's atmosphere.

Performing an experiment that requires the measurement of low primary fluxes in the presence of large background radiation, with large and heavy detectors in a balloon or satellite, is not an easy task.

A recent survey of experimental gamma-ray astronomy has been given by Greisen (1966).

Detectors for the Energy Region 100 keV to 10 MeV

The primary detector for this energy range is the scintillation counter, which consists of a medium, in which light is produced by the moving charged particles resulting from the gamma-ray interaction, and a photomultiplier tube, in which this light is converted into an electrical signal. A common medium in this energy range is thallium-activated sodium or cesium iodide, Na I (Tl) or Cs I (Tl), which combines high photoelectric absorption coefficients with high light output. Detection of charged particles is eliminated by surrounding the absorbing material by another scintillation counter. This latter counter may also serve as a collimator for gamma rays if it is made thick enough and opened in one direction. The use of lead shielding to reject background and provide mechanical collimation has not been highly successful, owing to secondary gamma-ray production in the lead.

A detector designed by Frost & Rothe (1962) for balloon and satellite flights is shown in Figure 2. The angular response of the detector is about 30° half-angle. At 662 keV the efficiency is 18 per cent and the energy resolution is 16 per cent.

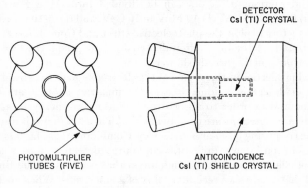

DETECTOR
CsI (TI) CRYSTAL

PHOTOMULTIPLIER
TUBES (FIVE)

ANTICOINCIDENCE
CsI (TI) SHIELD CRYSTAL

SCALE: ⊢——⊣
ONE INCH

Fig. 2. A low-energy gamma-ray detector designed by Frost & Rothe (1962).

Another detector, known as a Compton telescope, has been used by Peterson (1966). This detector consists of two Na I (Tl) detectors placed one behind the other. By selecting the amount of energy loss in the front detector, only forward Compton-scattered photons are selected and then absorbed in the rear detector. The Na I (Tl) crystal in the rear detector was surrounded by an anticoincidence plastic scintillator to eliminate direct charged-particle effects. The acceptance cone in this telescope was a 12° half-angle, and the efficiency was less than 1 per cent for 1-MeV gamma-ray-photons.

DETECTORS FOR THE ENERGY REGION 10 MeV TO 10 BeV

In this energy region a gamma-ray photon is detected by the electron-positron pair it produces when interacting with matter. Three types of instrument are primarily used to detect the pair of charged particles: nuclear emulsions, scintillation-Cerenkov counter systems, and spark chambers.

A nuclear emulsion consists of a high concentration of silver bromide crystals uniformly immersed in a gelatine base. Charged particles traversing the emulsion activate the silver bromide crystals, producing a series of dark silver grains after emulsion development. The particle tracks are viewed with a microscope. The nuclear emulsion provides many advantages as a gamma-ray detector, such as: (a) high absorption coefficients for gamma rays; (b) high angular resolution; (c) high energy resolution, obtained by multiple-scattering measurements on the charged-particle pair; (d) low energy threshold ($\sim$10 MeV) for gamma rays; (e) relative simplicity and low cost. However, the emulsion technique also has some disadvantages: (a) continuous sensitivity, which results in recording unwanted background radiation, particularly during ascent in the atmosphere; (b) long and tedious scanning of the emulsion with a microscope. In spite of these disadvantages, nuclear emulsions have already served as a useful research tool in gamma-ray astronomy.

Another detector technique is the use of a scintillation counter and a Cerenkov counter in series to detect the electron-positron pair. Cerenkov light is emitted by a charged particle when it travels through a dispersive medium at a velocity greater than that of light in that medium. A great advantage of the Cerenkov counter is that it is insensitive to low-energy particles. A disadvantage is the small number of photons produced in the medium. In a typical gamma-ray detector a thin sheet of lead is used to convert the incoming gamma-ray photon into an electron-positron pair. A Cerenkov counter somewhat removed from the lead is used to detect and to define the direction of the electron-positron pair and to define the geometry of the detector. A scintillation counter is placed before the lead in anticoincidence to exclude charged particles, and a second scintillation counter is placed between the lead and the Cerenkov counter to ensure detection only when a gamma-ray conversion occurs in the lead. A similar detector of this type was used on the satellite Explorer 11 (Kraushaar et al. 1965), with the modifica-

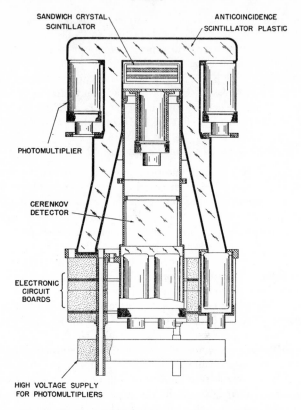

SANDWICH CRYSTAL
SCINTILLATOR

ANTICOINCIDENCE
SCINTILLATOR PLASTIC

PHOTOMULTIPLIER

CERENKOV
DETECTOR

ELECTRONIC
CIRCUIT
BOARDS

HIGH VOLTAGE SUPPLY
FOR PHOTOMULTIPLIERS

FIG. 3. Explorer 11 high-energy gamma-ray detector.

tion that the lead sheet and the adjacent scintillation counter for detecting the pair of charged particles were replaced by alternate layers of dense Na I (Tl)-Cs I (Tl) scintillators that served as both a converter and a detector (Figure 3). The Cerenkov detector in some cases is made of dense lead glass to absorb the entire energy of the electron-positron pair and thus to serve as an energy spectrometer.

The advantages of such a counter system are its high efficiency of detection and its large sensitive area ($\sim$100 cm^2). The disadvantages are poor angular resolution (15° to 30° half-angle) and susceptibility to counting background radiation as primary gamma-ray photons.

Within the last two years the spark chamber has become the principal detector in this energy region. The chamber is a device for making visible the paths of the electron-positron pair. This is achieved by a spark discharge, between parallel plates in a gas, along the ionized paths of the particles. A scintillation-Cerenkov counter system is used to detect the presence of the electron-positron pair and to initiate the spark discharge. The spark patterns

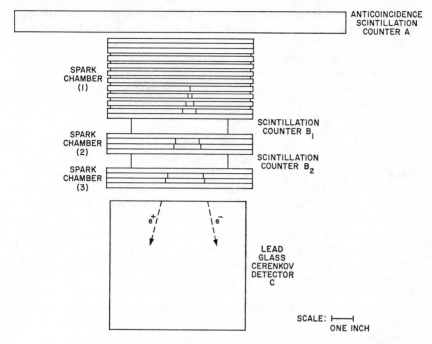

FIG. 4. Geometry of a Vidicon spark chamber for high-energy gamma-ray
astronomy (Helmken & Fazio 1966).

are recorded stereoscopically, usually on photographic film. Recently, film-
less techniques of recording the spark positions, such as vidicon cameras,
sonic detectors, and wire chambers, have been used. The spark chamber has
the following advantages: large sensitive area (10^2 to 10^4 cm^2); large sensitive
solid angle with the ability to determine photon-arrival direction to within a
few degrees; chamber sensitive only to a preselected type of interaction;
visual system, permitting verification of interaction; high signal-to-noise
ratio; simplicity; versatility; and reliability. The geometry of the vidicon
spark chamber flown in a balloon by Helmken & Fazio (1966) is shown in
Figure 4. The detector has a sensitive area of 150 cm^2 and an acceptance solid-
angle factor of 0.84. The determination of gamma-ray photon directions is
dependent on photon energy, but is typically of the order of $\pm 3°$. The lead-
glass Cerenkov counter in the detector permits energy measurements to be
performed.

Other detectors in this energy region being used or developed are nuclear
emulsion-spark chamber combinations and large-area gas Cerenkov counters.

DETECTORS FOR PHOTON ENERGIES GREATER THAN 10^{10} EV

Gamma-ray photons with energy $> 10^{11}$ eV can be detected by the cascade
shower of particles they produce upon interaction with the atmosphere. The

initiating reaction is primarily an electron-positron pair of charged particles that produce additional photons and electrons (positrons) by bremsstrahlung, Compton scattering, and pair production. The cascade shower is detected by two different techniques, both ground based: the detection of the Cerenkov radiation in the atmosphere produced by the shower of particles; the use of a large array of scintillation counters to detect the arrival of the burst of charged particles in the shower.

Atmospheric Cerenkov light.—Electrons with velocity βc in a cascade shower can generate Cerenkov light in the visible spectrum when $\beta > 1/n$, where n is the index of refraction of light. The light is generated in a narrow cone, and the radius of the pool of light at ground level is ~ 100 m. The burst of Cerenkov light is of the order of 10^{-8} sec, and the number of photons generated per unit pathlength per unit wavelength interval is proportional to $1/\lambda^2$. Gamma-ray detection by this method has many advantages; the primary one is that the detector is ground based. Also, the flux sensitivity is high, owing to the large area over which a gamma ray may be detected, yet the angular resolution of the light burst is about 2°. Since protons and nucleons in the primary cosmic radiation can also produce similar cascade showers, a disadvantage of the technique is the isotropic background bursts produced by these charged particles. Gamma-ray events can be distinguished only by an increase in burst rate, most probably in the direction of a suspected source. Thus, for gamma-ray astronomy this detection technique can be used only for discrete sources.

Two experimental groups have searched for gamma rays with energy $>5 \times 10^{12}$ eV by this technique. Chudakov et al. (1962) used twelve 1.5-m-diameter parabolic mirrors, each with a phototube at the focus. The mirrors were mounted in a 3- by 20-m array consisting of four telescopes of three mirrors each. Scans of possible sources were achieved by transits due to the Earth's rotation. The instrument used by the second group, Fruin et al. (1964) at University College, Dublin, consisted of two 3-ft mirrors mounted on a rotatable gun mount (Figure 5).

The Smithsonian Astrophysical Observatory is attempting measurements with the 28-ft square mirror of the solar furnace at the U.S. Army Natick Laboratories, Natick, Massachusetts, and is currently constructing a 34-ft mirror mounted on an alt-azimuth mount at Mt. Hopkins, Arizona. At both installations the mirrors consist of mosaics of 2-ft-diameter mirrors.

A general discussion of detection of showers by this technique is given by Jelley (1958).

Scintillation-counter techniques.—The thin disk of charged particles produced in the atmosphere in a cascade shower can be detected by coincidence pulses from a large scintillation-counter array. The arrival direction of the shower is determined by noting the firing sequence of the counters. The radius of the particle shower at ground level is considerably smaller than that of the Cerenkov light pool. Gamma-ray-initiated showers can be distinguished from charged-particle showers by the detection of an anisotropy in the

FIG. 5. A telescope for detecting atmospheric Cerenkov light due to high-energy cosmic gamma rays (Fruin et al. 1964).

direction of discrete sources, and in principle by assuming that gamma-ray showers are predominantly electromagnetic interactions, lacking mesons and nucleons in their core. The presence of a penetrating μ-meson component in the shower at ground level has been used to eliminate cascade showers initiated by charged particles.

Two groups are using this technique to perform gamma-ray astronomy at energies $>10^{14}$ eV. The Bolivian Air Shower Joint Experiment (BASJE) is located on Mt. Chacaltaya, Bolivia, at an altitude of 5200 m (Toyoda et al. 1965). The scintillation-counter array has a diameter of 300 m and the μ-meson detector consists of 60 m² of scintillator under lead and concrete shielding. The array is also capable of determining the arrival direction of showers with an uncertainty of $< \pm 5°$. A Polish-French group in Lodz, Poland (Gawin et al. 1965) uses a smaller array of detectors, in which the μ-meson detector is 21 m² in area.

9. EXPERIMENTAL RESULTS

DIFFUSE FLUX OF COSMIC GAMMA RAYS

A diffuse gamma-ray flux has been measured to energies of only 1 MeV. In the region from 0.1 to 1 MeV, measurements of an isotropic gamma-ray flux have been reported by Arnold et al. (1962), from experiments on the lunar probes Ranger 3 and Ranger 5, using a $2\frac{3}{4}$-inch Cs I (Tl) crystal surrounded by an anticoincidence shield of plastic scintillator. The detector had no directionality. The results of this experiment are shown in Figure 6, with

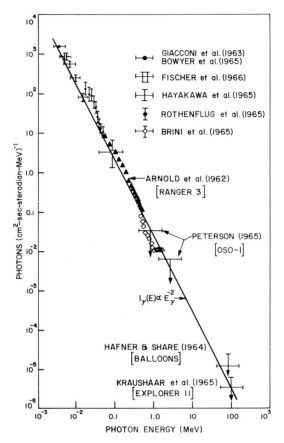

FIG. 6. Measured diffuse cosmic gamma-ray spectrum (Gursky et al. 1963, Bowyer et al. 1965, Fisher et al. 1966, Hayakawa et al. 1965, Rothenflug et al. 1965, Brini et al. 1965).

diffuse flux measurements from the X-ray region. These measurements, when combined with the cosmic X-ray measurements, form a differential photon-intensity curve given by $I_\gamma dE_\gamma \propto E_\gamma^{-2} dE_\gamma$. Peterson (1966), on the Orbiting Solar Observatory-1, used a Na I (Tl) crystal with a lead collimating shield as a detector in the energy region 50 to 150 keV and a Na I (Tl) crystal surrounded by an anticoincidence shield of plastic scintillator in the energy region 0.5 to 4.5 MeV. The latter detector had an isotropic response. The experiment gave upper limits that are not inconsistent with those of the Ranger 3 experiment.

At higher energies, only upper limits to the flux exist. The University of Rochester group (Duthie et al. 1963, Hafner & Share 1964), using a scintillator-Cerenkov counter telescope in high-altitude balloons, have determined an upper limit of 6×10^{-4} photons (cm² sec sr)$^{-1}$ for energies >50 MeV. This upper limit was determined by the extrapolation of counting rates at balloon altitudes to the top of the atmosphere. Likewise, Kraushaar et al. (1965), using a scintillator-Cerenkov counter telescope in the satellite Explorer 11, have placed an upper limit on the diffuse flux above 100 MeV of 3×10^{-4} photons (cm² sec sr)$^{-1}$. This upper limit to the flux is approximately an order of magnitude above the theoretical gamma-ray diffuse flux due to π^0 decay in the direction of the Galactic center.

Large air-shower experiments, based on the detection of showers with a low number of μ mesons, have given upper limits to the frequency of gamma rays in the energy region 10^{14} to 10^{16} eV. Toyoda et al. (1965), in the energy region 10^{15} to 10^{16} eV, determined an upper limit of 10^{-13} photons (cm² sec sr)$^{-1}$, based on a frequency of low μ-meson showers of 2×10^{-4} of the charged-particle flux at the same energy. Gawin et al. (1965) determined a frequency of 7×10^{-3} in the same energy interval, giving an upper limit to the gamma-ray flux above 10^{15} eV of about 10^{-12} (cm² sec sr)$^{-1}$.

DISCRETE SOURCES OF COSMIC GAMMA RAYS

Above photon energies of 1 MeV the only discrete source of cosmic gamma rays reported has been by Duthie et al. (1966) from the direction of the constellation Cygnus. The source was identified to be in a circle of 6° radius, centered at 35° N declination and 20^h15^m right ascension. The detector was a spark chamber flown in high-altitude balloons. The source exhibited an anomalously high counting rate associated with an energy spectrum that appears to differ significantly from the background spectrum. The interval flux above 100 MeV was measured to be $(1.5 \leq 0.8) \times 10^{-4}$ photons (cm² sec)$^{-1}$. Frye & Wang (1967) recently reported the results of three balloon flights using a spark chamber that observed the same region in Cygnus. No evidence of a gamma-ray source was found.

The spectra of two discrete sources of X rays have been measured to energies approaching the gamma-ray region. The measured spectrum of the Crab Nebula is shown in Figure 7. The highest energies measured are $\sim$100 keV. The spectrum of the strongest X-ray source, Scorpius XR-1, has been measured to 50 keV by Peterson & Jacobson (1967); it exhibits an exponential differential energy spectrum.

Numerous other attempts have been made to measure the gamma-ray flux over a wide energy range from known X-ray and radio sources. These experiments are summarized in Table XI.

The first observation of gamma rays from the Sun was made in 1958 by Peterson & Winckler (1959). The measured burst occurred in less than 18 sec and was associated with a solar flare (class 2) and a solar radio burst. The intensity of the radiation was 2×10^{-5} erg (cm² sec)$^{-1}$, peaked in the region 200 to 500 keV. This is the highest photon energy detected from the Sun. A summary of experimental data and theoretical predictions of gamma-ray fluxes from the Sun is given by Dolan & Fazio (1965). A summary of upper limits to the gamma-ray flux from the quiet Sun is given in Table XI.

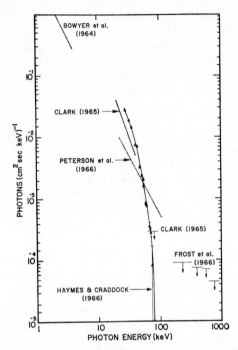

FIG. 7. High-energy X rays from the Crab Nebula (graph given by Haymes & Craddock 1966).

10. CONCLUSIONS

Although as yet only upper limits have been obtained for the intensity of diffuse celestial gamma rays, several important conclusions can be drawn from these upper limits:

(a) We infer that if matter is being created, as required by steady-state cosmology, matter and antimatter cannot be created in the metagalaxy in comparable amounts in the same region. The results of the Explorer 11 experiment indicate that the density ratio of antiprotons to protons is $<10^{-6}$. Likewise, the results of the low-energy gamma-ray experiments on Rangers 3 and 5 and OSO-1 indicate that the density ratio of positrons to electrons is $<10^{-6}$. Thus, if appreciable quantities of antimatter exist in the metagalaxy, they must be located in regions that are separated from matter.

(b) The results of the Explorer 11 experiment indicate that the high-energy intergalactic electron intensity is less, at least by a factor of 10, than the Galactic density. This conclusion is based on the expected gamma-ray flux due to Compton scattering by high-energy electrons of visible light photons in intergalactic space.

(c) Upper limits to the gamma-ray flux $>5 \times 10^{12}$ eV (Chudakov et al.

TABLE XI

EXPERIMENTAL UPPER LIMITS TO GAMMA-RAY FLUXES FROM DISCRETE SOURCES

Source	Energy threshold	Theoretical flux (cm^{-2} sec^{-1})	Flux upper limit (cm^{-2} sec^{-1})	Experiment
Taurus A	10 MeV		4.2×10^{-3}	Fichtel & Kniffen (1965)
	30 MeV		1.5×10^{-4}	Frye & Smith (1966)
	100 MeV	10^{-4} to 10^{-9}	6.6×10^{-4}	Kraushaar et al. (1965)
	100 MeV		7×10^{-5}	Cobb, Duthie & Stewart (1965)
	3×10^{11} eV		2.3×10^{-6}	Sekido et al. (1963)
	5×10^{12} eV	10^{-8} to 10^{-12}	5×10^{-11}	Chudakov et al. (1962)
	5×10^{12} eV		1×10^{-11}	Fruin et al. (1964)
	10^{15} eV		10^{-14}	Toyoda et al. (1965)
Cygnus A	10 MeV		2.6×10^{-3}	Fichtel & Kniffen (1965)
	20 MeV		1.2×10^{-2}	Frye, Reines & Armstrong (1966)
	50 MeV		5×10^{-3}	Braccesi, Ceccarelli & Salandin (1960)
	60 MeV		1.2×10^{-3}	Cline (1961)
	60 MeV		2×10^{-3}	Duthie et al. (1963)
	100 MeV	10^{-4} to 10^{-8}	5×10^{-4}	Kraushaar et al. (1965)
	1 GeV		6.0×10^{-4}	Ögelman, Delvaille & Greisen (1966)
	1×10^{11} eV		7.5×10^{-6}	Sekido et al. (1963)
	5×10^{12} eV		5×10^{-11}	Chudakov et al. (1962)
Cassiopeia A	20 MeV		1.6×10^{-2}	Frye et al. (1966)
	60 MeV		6×10^{-3}	Cline (1961)
	100 MeV		2.3×10^{-3}	Kraushaar et al. (1965)
	1 GeV		2.9×10^{-4}	Ögelman et al. (1966)
	5×10^{12} eV		5×10^{-11}	Chudakov et al. (1962)
Virgo A (M87)	100 MeV	10^{-8}	2.7×10^{-4}	Kraushaar et al. (1965)
	3×10^{11} eV		1.2×10^{-6}	Sekido et al. (1963)
	5×10^{12} eV		5×10^{-11}	Chudakov et al. (1962)
Andromeda (M31)	20 MeV		1.9×10^{-2}	Frye et al. (1966)
	100 MeV	10^{-7}	1.6×10^{-3}	Kraushaar et al. (1965)
	1 GeV		1.5×10^{-4}	Ögelman et al. (1966)
3C 147	30 MeV		1.5×10^{-4}	Frye & Smith (1966)
	5×10^{12} eV		1×10^{-10}	Fruin et al. (1964)
3C 196	30 MeV		1.5×10^{-4}	Frye & Smith (1966)
	5×10^{12} eV		5×10^{-11}	Fruin et al. (1964)
Sun (quiet)	1 MeV		2×10^{-3}	Peterson, Jacobson & Pelling (1966)
	20 MeV		2.8×10^{-2}	Frye et al. (1966)
	50 MeV		5×10^{-3}	Fazio & Hafner (1967)
			10^{-3}	Fichtel & Kniffen (1965)
	100 MeV		5.3×10^{-3}	Cobb et al. (1965)
	200 MeV		8×10^{-3}	Danielson (1960)

1962, Fruin et al. 1964) from the Crab Nebula have shown that the high-energy electrons in the Crab Nebula, which generate synchrotron radiation, are not the product of high-energy proton interactions and therefore must have been accelerated. If the electrons were the result of π-meson decay, the gamma-ray flux would have been approximately 100 times the measured upper limit.

(*d*) If we neglect the possibility that gamma rays with energy $>10^{16}$ eV could be absorbed, information on the intensity of extragalactic protons with energy $>10^{17}$ eV can be obtained from the BASJE experiment. The lack of evidence for gamma rays above 10^{16} eV, which are produced by photo-pion production, i.e., protons interacting with thermal photons, indicates that the extragalactic proton intensity above 10^{17} eV must be $<6.5\times10^{-2}$ of the cosmic-ray intensity at the Earth. This value was based on a metagalactic thermal ($5800°K$) photon density of 0.1 eV cm^{-3}. Or likewise, assuming the cosmic-ray intensity measured at the Earth exists in the metagalaxy, the upper limit of the photon density is 10^{-2} eV cm^{-3}.

The most accurate prediction of a diffuse flux of cosmic gamma rays is that due to high-energy proton interactions in the Galactic plane. The sensitivity of present experiments is within an order of magnitude of being able to detect this flux, at energies >50 MeV. Knowledge of either the presence or the absence of this flux is essential for an understanding of the origin and diffusion of cosmic rays.

One of the most interesting divisions of the gamma-ray spectrum yet to be explored is the region 1 to 50 MeV. Experiments in this region are difficult, yet in this region alone can the major source mechanisms be distinguished easily by their spectra.

Future experiments in gamma-ray astronomy are not going to be performed easily. However, the knowledge of the universe to be gained by the detection of a flux of gamma rays is so important that more sensitive experiments are essential.

The author is indebted to Dr. Henry F. Helmken for his assistance in reviewing this paper.

LITERATURE CITED

Arnold, J. R., Metzger, A. E., Anderson, E. C., Van Dilla, M. A. 1962, *J. Geophys. Res.*, **67**, 4878–80; Metzger, A. E., Anderson, E. C., Van Dilla, M. A., Arnold, J. R. 1964, *Nature*, **204**, 766–67

Bowyer, S., Byram, E. T., Chubb, T. A., Friedman, H. 1964, *Nature*, **201**, 1307–8

Bowyer, S., Byram, E. T., Chubb, T. A., Friedman, H. 1965, *Science*, **147**, 394–98

Braccesi, A., Ceccarelli, M., Salandin, G. 1960, *Nuovo Cimento*, **17**, 691–94

Brini, D., Ciriegi, V., Fuligni, F., Gandolfi, A., Moretti, E. 1965, *Nuovo Cimento*, **38**, 130–40

Burbidge, G. R., Hoyle, F., Burbidge, E. M., Christy, R. F., Fowler, W. A. 1956, *Phys. Rev.*, **103**, 1145–49

Cameron, A. G. W. 1959, *Ap. J.*, **129**, 676–99

Cameron, A. G. W. 1961, *AFCL 454* (Chalk River Project, Chalk River, Ontario)

Chudakov, A. E., Dadykin, V. L., Zatsepin, V. I., Nesterova, N. M. 1962, *J. Phys. Soc. Japan*, **17**, Suppl. A-III, 106 (also in *Cosmic Rays*, Proc. (Trudy) Lebedev Phys. Inst., Skobel'tsyn, D. V., Ed., Consultants Bureau, New York, **26**, 99–118, 1965)

Clark, G. W. 1965, *Phys. Rev. Letters*, **14**, 91–94

Clayton, D. D., Craddock, W. L. 1965, *Ap. J.*, **142**, 189–200

Cline, T. C. 1961, *Phys. Rev. Letters*, **7**, 109–13

Cobb, R., Duthie, J. G., Stewart, J. 1965, *Phys. Rev. Letters*, **15**, 507–11

Danielson, R. E. 1960, *J. Geophys. Res.*, **65**, 2055–59

Davisson, C. M., Evans, R. D. 1952, *Rev. Mod. Phys.*, **24**, 79–107

Dolan, J. F., Fazio, G. G. 1965, *Rev. Geophys.*, **3**, 319–43

Donahue, T. M. 1951, *Phys. Rev.*, **84**, 972–80

Duthie, J. G., Cobb, R., Stewart, J. 1966, *Phys. Rev. Letters*, **17**, 263–67

Duthie, J. G., Hafner, E. M., Kaplon, M. F., Fazio, G. G. 1963, *Phys. Rev. Letters*, **10**, 364–67

Evans, R. D. 1955, *The Atomic Nucleus* (McGraw-Hill, New York, 972 pp.)

Fazio, G. G., Hafner, E. M. (Preprint)

Fazio, G. G., Stecker, F. W., Wright, J. P. 1965, *Ap. J.*, **144**, 611

Feenberg, E., Primakoff, H. 1948, *Phys. Rev.*, **73**, 449–69

Felten, J. E. 1965, *Omnidirectional Inverse Compton and Synchrotron Radiation from Cosmic Distributions of Fast Electrons and Thermal Photons* (Ph.D. thesis, Cornell Univ.)

Felten, J. E. 1966, *Ap. J.*, **144**, 241–43

Felten, J. E., Morrison, P. 1963, *Phys Rev. Letters*, **10**, 453–57

Felten, J. E., Morrison, P. 1966, *Ap. J.*, **146**, 686–708

Fernbach, S., Serber, R., Taylor, T. B. 1949, *Phys. Rev.*, **75**, 1352–55

Fichtel, C. E., Kniffen, D. A. 1965, *J. Geophys. Res.*, **70**, 4227–34

Field, G. B., Solomon, P. M., Wampler, E. J. 1966, *Ap. J.*, **145**, 351–54

Fisher, P. C., Johnson, H. M., Jordan, W. C., Meyerott, A. J., Acton, L. W. 1966, *Ap. J.*, **143**, 203–17

Frost, K. J., Rothe, E. D. 1962, *IRE Trans. Nucl. Sci. NS-9*, No. 3, 381–85

Frost, K. J., Rothe, E. D., Peterson, L. E. 1966, *J. Geophys. Res.*, **71**, 4079–89

Fruin, J. H., Jelley, J. V., Long, C. O., Porter, N. A., Weekes, T. C. 1964, *Phys. Letters*, **10**, 176–77

Frye, G. M., Jr., Reines, F., Armstrong, A. H. 1966, *J. Geophys. Res.*, **71**, 3119–23

Frye, G. M., Jr., Smith, L. H. 1966, *Phys. Rev. Letters*, **17**, 733–36

Frye, G. M., Jr., Wang, C. P. 1967, *Phys. Rev. Letters*, **18**, 132–34.

Fujimoto, Y., Hasegawa, H., Taketani, M. 1964, *Suppl. Progr. Theoret. Phys.*, No. 30, 32–85

Gamow, G., Critchfield, C. L. 1949, *Theory of the Atomic Nucleus and Nuclear Energy Sources* (Clarendon Press, Oxford, 344 pp.)

Garmire, G. 1965, *Proc. Ninth Intern. Conf. Cosmic Rays (London)*, **1**, 315–16

Garmire, G., Kraushaar, W. 1965, *Space Sci. Rev.*, **4**, 123–46

Gawin, J., Hibner, J., Wdowczyk, J., Zawadzki, A., Maze, R. 1965, *Proc. Ninth Intern. Conf. Cosmic Rays (London)*, **2**, 639–41; see also Firkowski, R., Gawin, J., Zawadski, A., Maze, R. 1962, *J. Phys. Soc. Japan*, **17**, *Suppl. A-III*, 123–27

Ginzburg, V. L., Syrovatskii, S. I. 1964a. *Soviet Phys.—Usp.*, **84**, 201–42 [1965, *Soviet Phys.—Usp. (Engl. Transl.)*, **7**, 696–720]

Ginzburg, V. L., Syrovatskii, S. I. 1964b, *The Origin of Cosmic Rays* (ter Haar, D., Ed.; Massey, H. S. W., Transl., Macmillan, New York)

Ginzburg, V. L., Syrovatskii, S. I. 1965, *Ann. Rev. Astron. Ap.*, **3**, 297–350

Goldreich, P., Morrison, P. 1964, *Soviet Phys. JETP (Engl. Transl.)*, **18**, 239

Gould, R. J. 1965, *Phys. Rev. Letters*, **15**, 577

Gould, R. J., Burbidge, G. R. 1965, *Ann. Ap. Suppl. (France)*, **28**, 171–201

Gould, R. J., Gold, T., Salpeter, E. E. 1963, *Ap. J.*, **138**, 408–25

Gould, R. J., Salpeter, E. E. 1963, *Ap. J.*, **138**, 393–407

Gould, R. J., Schréder, G. 1966, *Phys. Rev. Letters*, **16**, 252–54; also *Phys. Rev.* (In press)

Greisen, K. 1960, *Ann. Rev. Nucl. Sci.*, **10**, 63–108

Greisen, K. 1966, in *Perspectives in Modern Physics*, 355–82 (Marshak, R. E., Ed., Interscience, New York)

Gunn, J. E., Peterson, B. A. 1965, *Ap. J.*, **142**, 1633–36

Gursky, H., Giacconi, R., Paolini, F. R., Rossi, B. B. 1963, *Phys. Rev. Letters*, **11**, 530–35

Hafner, E. M., Share, G. 1964, *High Energy Cosmic Gamma Radiation* (Univ. of Rochester preprint, 45 pp.)

Hayakawa, S. 1963, *Brandeis Summer Institute in Theoretical Physics*, **2**, 1–164 (Brandeis Univ. Press, Waltham, Mass.)

Hayakawa, S., Matsuoka, M., Yamashita, K. 1965, *Proc. Ninth Intern. Conf. Cosmic Rays (London)*, **1**, 119–22

Hayakawa, S., Okuda, H., Tanaka, Y., Yamamoto, Y. 1964, *Progr. Theoret.*

Phys. Suppl. (Japan), No. 30, 153–203

Haymes, R. C., Craddock. W. L., Jr. 1966, *J. Geophys. Res.*, **71**, 3261–64

Helmken, H. F., Fazio, G. G. 1966, *IEEE Trans. Nucl. Sci. NS-13*, No. 1, 486–92

Hill, E. R. 1960, *Bull. Astron. Inst. Neth.*, **15**, 1–44

Jelley, J. V. 1958, *Čerenkov Radiation, and Its Applications* (Pergamon, New York, 304 pp.)

Jelley, J. V. 1966a, *Phys. Rev. Letters*, **16**, 479–81

Jelley, J. V. 1966b, *Nature*, **211**, 472–75

Kraushaar, W., Clark, G. W., Garmire, G., Helmken, H., Higbie, P., Agogino, M. 1965, *Ap. J.*, **141**, 845–63

Lieber, M., Milford, S. N., Spergel, M. S. 1965, *NASA CR-149*, 113 pp.

McVittie, G. C. 1962, *Phys. Rev.*, **128**, 2871–78

Meyer, P. 1965, *Proc. Ninth Intern. Conf. Cosmic Rays (London)*, **1**, 61–67

Milford, S. N., Rosen, S. 1965, *Nature*, **205**, 582

Milford, S. N., Shen, S. P. 1962, *Nuovo Cimento*, **23**, 77–87

Morrison, P. 1958, *Nuovo Cimento*, **7**, 858–65

Nelms, A. T. 1953, *Natl. Bur. Standards (U.S.) Circ. 542*, 89 pp.

Nikishov, A. I. 1962, *Soviet Phys. JETP (Engl. Transl.)*, **14**, 393–94

Oda, M. 1965, *Proc. Ninth Intern. Conf. Cosmic Rays (London)*, **1**, 68–80

Ögelman, H. B., Delvaille, J. P., Greisen, K. I. 1966, *Phys. Rev. Letters*, **16**, 491–94

Peterson, L. E. 1966, in *Space Res., Proc. Intern. Space Sci. Symp.*, **6**, 53–66 (Macmillan, New York)

Peterson, L. E., Jacobson, A. S. 1967 (To be published)

Peterson, L. E., Jacobson, A. S., Pelling, R. M. 1966, *Phys. Rev. Letters*, **16**, 142–44

Peterson, L. E., Schwartz, D. A., Pelling, R. M., McKenzie, D. 1966, *J. Geophys. Res.*, **71**, 5778–81

Peterson, L. E., Winckler, J. R. 1959, *J.*

Geophys. Res., **64**, 697–707

Pollack, J. B., Fazio, G. G. 1963, *Phys. Rev.*, **131**, 2684–91

Rossi, B. 1952, *High-Energy Particles* (3rd ed., Prentice-Hall, Englewood, N. J., 488 pp.)

Rothenflug, R., Rocchia, R., Koch, L. 1965, *Proc. Ninth Intern. Conf. Cosmic Rays (London)*, **1**, 446–48

Salpeter, E. E. 1952, *Phys. Rev.*, **88**, 547–53

Savedoff, M. P. 1959, *Nuovo Cimento*, **13**, 12–18

Schatzman, E. 1963, *Space Sci. Rev.*, **1**, 774–80

Segrè, E. 1953, *Experimental Nuclear Physics*, **1** (Wiley, New York, 789 pp.)

Sekido, Y., Kondo, I., Murayama, T., Kamiya, Y., Ueno, H., Mori, S., Okuda, H., Makino, T., Sakakibara, S., Fujimoto, K. 1963, *Proc. Intern. Conf. Cosmic Rays (Jaipur, India)*, **4**, 194–98

Stecker, F. W. 1966, *Smithsonian Ap. Obs. Spec. Rept. No. 220*, 64 pp.

Toyoda, Y., Suga, K., Murakami, K., Hasegawa, H., Shibata, S., Domingo, V., Escobar, I., Kamata, K., Bradt, H., Clark, G., La Pointe, M. 1965, *Proc. Ninth Intern. Conf. Cosmic Rays (London)*, **2**, 708–11; see also Suga, K., Escobar, I., Clark, G. W., Hazen, W., Hendel, A., Murakami, K. 1962, *J. Phys. Soc. Japan*, **17**, *Suppl. A-III*, 128–30

Waddington, C. J. 1962, *J. Phys. Soc. Japan*, **17**, *Suppl. A-III*, 63–68 [*Proc. Intern. Conf. Cosmic Rays and the Earth Storm* (IC-CRES)]

Waddington, C. J. 1965, *Proc. Ninth Intern. Conf. Cosmic Rays (London)*, **1**, 462–68

Wattenburg, A. 1957, in *Handbuch der Physik*, **40**, 450–537 (Flügge, S., Ed., Springer-Verlag, Berlin)

Weekes, T. C. 1966, *Sci. Progr. (Oxford)*, **54**, 543–60

Weisskopf, V. F. 1965, *Science*, **149**, 1181–89

White, G. R. 1952, *Natl. Bur. Standards (U.S.) Rept. 1003*

THERMONUCLEAR REACTION RATES[1]

By William A. Fowler,[2] Georgeanne R. Caughlan,[2,3] and
Barbara A. Zimmerman[2]

*California Institute of Technology, Pasadena, California
and Montana State University, Bozeman, Montana*

The changing of Bodies into Light, and Light into Bodies, is very conformable to
the Course of Nature, which seems delighted with Transmutations.

Sir Isaac Newton, Knt. (1704)

INTRODUCTION

Modern computer technology has revolutionized theoretical calculations
on stellar structure and stellar evolution. Physical information of great complexity, previously ignored or crudely treated, can now be introduced in detail into stellar computations and can be treated with great precision and
completeness. In particular it is no longer necessary (Iben 1965, Wagoner,
Fowler & Hoyle 1967) to approximate hard-won laboratory results on nuclear
reaction rates by crude power-law dependences on temperature.

In this article we review the available experimental data on cross sections
for the nuclear interactions of neutrons, protons, and alpha particles with a
number of light and intermediate-mass nuclei and we present calculations on
the resulting reaction rates, nuclear lifetimes, and energy generation rates
under astrophysical conditions. We restrict our considerations to nondegenerate, nonrelativistic circumstances for the interacting nuclei. Table I lists
the general nuclear processes discussed, along with the units used in presenting numerical results and the notation used in representing nuclear reactions.
Previous reviews of nuclear reaction rates, somewhat more limited in scope,
have been given by Fowler (1954, 1960), Burbidge, Burbidge, Fowler &
Hoyle (1957) (hereafter referred to as B²FH), Caughlan & Fowler (1962), and
Parker, Bahcall & Fowler (1964). The last previous overall survey of nuclear
reaction rates which has come to our attention is that of Reeves (1965).

A primary motivation in the preparation of this article has been the desire
to extend where possible the calculated reaction rates to quite high temperatures of the order of 10 billion °K corresponding to interaction energies of the
order of 1 MeV.[4] This extension is clearly required if applications to the ad

---

[1] The survey of literature for this review was concluded in December 1966.

[2] Work supported in part by the Office of Naval Research [Nonr-220(47)] and
the National Science Foundation [GP-5391] at the California Institute of Technology.

[3] Work supported in part by the National Science Foundation [GP-4693 and
GP-6309] at Montana State University.

[4] It is for this reason that we have chosen 10⁹ °K as the unit of temperature and
1 MeV as the unit of energy in the numerical results given in Tables II to VI. Similar
tables employing 10⁶ °K and keV can be obtained from the authors on request.
We of course use the erg in expressing energy generation rates in erg g⁻¹ sec⁻¹.

TABLE I

Nuclear Processes Discussed

Hydrogen burning
  Proton-proton chain
  CNO bi-cycle
Helium burning
  $3He^4 \rightarrow C^{12}$, $C^{12}(\alpha,\gamma)O^{16}$, $O^{16}(\alpha,\gamma)Ne^{20}$, $Ne^{20}(\alpha,\gamma)Mg^{24}$
Silicon burning
  Photodisintegration and radiative-capture rates for S^{32}, Si^{28}, Mg^{24}, etc.
Interactions of neutrons, protons, and alpha particles with light nuclei

---

Units
  cm, g, sec; barn $= 10^{-24}$ cm^2
  10^9 °K, erg, MeV $= 1.6021 \times 10^{-6}$ erg

Nuclear notation for incident particles and the lighter products in reactions:
  n, p, d, t, τ, α, Li^6, Li^7, Be^9, etc., etc.
Atomic notation for target particles and the heavier products in reactions; also used in referring to mass fractions:
  H or H^1, D or D^2, T or T^3, He^3, He^4, Li^6, Li^7, Be^9, etc., etc. (Nuclear and atomic notations are identical for $A \geq 5$.)

Nondegenerate, nonrelativistic conditions for nuclei

---

vanced stages of stellar evolution, giants, supergiants, and supernovae are to be made. Furthermore there has recently been much interest once again in nuclear processes during the early stages of the expanding universe (Wagoner, Fowler & Hoyle 1967) and in the implosion-explosion of supermassive stars (Fowler 1966).

In order to make extensions to high temperature, we have reanalyzed much of the experimental data in the literature and have fitted the data with semi-empirical parameters frequently different from those given by the experimental investigators. At high temperatures the nuclear products of a given reaction react to reverse the original reaction. This is particularly true near the end of burning of nuclear fuels. Photodisintegration rates become comparable to those for radiative capture. Endoergic reactions replace exoergic ones and energy is absorbed in nuclear breakup rather than released in nuclear synthesis. Thus we make extensive use of the reciprocity theorem for nuclear reactions with detailed attention to statistical and identical-particle factors and we present rates in both directions for all reactions discussed.

There is one important aspect of nuclear reaction rates which we have not treated. This is the enhancement of the rates by the screening effects of electrons surrounding the interacting nuclei. The calculation of screening factors has been treated very comprehensively by Salpeter (1954) and Wolf (1965) and has recently been reviewed by Reeves (1965). *Nor have we considered the possible modifications of reaction rates due to inelastic collisions between the compound nucleus and other particles of the gas.* Clayton & Shaw (1967) have shown that these processes become relevant at high temperature and density. Finally we have abstained from any analysis requiring the choice of nuclear interaction radii. This has seriously limited the theoretical approaches available to us but, as a result, our numerical values reflect as closely as possible the experimental data obtained in the laboratory.

Nuclear Reaction Rates Under Astrophysical Conditions

Interaction rates and mean lifetimes.—The interaction rate between two nuclei, 0 and 1, with number densities n_0 cm^{-3} and n_1 cm^{-3} is

$$P_{01} = \frac{n_0 n_1}{1 + \delta_{01}} \langle 01 \rangle \text{ reactions cm}^{-3} \text{ sec}^{-1} \qquad 1.$$

where

$$\langle 01 \rangle \equiv \langle \sigma v \rangle_{01} \text{ in cm}^3 \text{ sec}^{-1} \qquad 2.$$

is the product of cross section and velocity averaged over the appropriately normalized velocity distribution (see Equation 32 below), and δ_{01} is the Kronecker delta. In terms of the *atomic* abundances by mass, X_0 and X_1, one has

$$n_0 = \rho N_A \frac{X_0}{A_0}, \qquad n_1 = \rho N_A \frac{X_1}{A_1} \qquad 3.$$

where ρ = matter density in g cm^{-3}, $N_A = M_U^{-1} = 6.02252 \times 10^{23}$ g^{-1} is Avogadro's number, $M_U = 1.66043 \times 10^{-24}$ g is the atomic (and nuclear) mass unit on the C^{12}(atom) = 12 scale, $A_0 = M_0/M_U$, $A_1 = M_1/M_U$, M_0 = *atomic* mass of 0, and M_1 = *atomic* mass of 1.[5] In general, interacting nuclei are labeled such that $A_0 > A_1$. In referring to laboratory coordinates, we refer to A_1 as the moving incident nucleus and A_0 as the stationary target nucleus.

The mean lifetimes of the nuclei to the interaction are given by

$$\frac{1}{\tau_1(0)} = \lambda_1(0) = -\frac{1}{n_0}\left(\frac{dn_0}{dt}\right)_1 = -\frac{1}{X_0}\left(\frac{dX_0}{dt}\right)_1 = n_1 \langle 01 \rangle = \rho N_A \frac{X_1}{A_1} \langle 01 \rangle = \frac{X_1}{A_1}[01] \text{ sec}^{-1}$$
$$4.$$

$$\frac{1}{\tau_0(1)} = \lambda_0(1) = -\frac{1}{n_1}\left(\frac{dn_1}{dt}\right)_0 = -\frac{1}{X_1}\left(\frac{dX_1}{dt}\right)_0 = n_0 \langle 01 \rangle = \rho N_A \frac{X_0}{A_0} \langle 01 \rangle = \frac{X_0}{A_0}[01] \text{ sec}^{-1}$$
$$5.$$

where

$\tau_1(0)$ = mean lifetime of 0 for interaction with 1,
$\tau_0(1)$ = mean lifetime of 1 for interaction with 0,

---

[5] The following atomic masses should be used: $A_n = 1.008665$, $A_H = 1.007825$, $A_D = 2.014102$, $A_T = 3.016050$, $A_{He^3} = 3.016030$, $A_{He^4} = 4.002603$ rounded off to the user's taste. All others are equal to the appropriate integral mass number within an error of no more than 3 parts in 1000.

and the λ's are the corresponding interaction rates. The quantity $(dn_0/dt)_1$ designates the partial depletion rate in n_0 due to the interaction of 0 with 1. Analogous definitions can be made for the other partial depletion rates. When more than one set of reaction products results from the interaction of 0 and 1, additional items of notation must in principle be included. The explicit notation we have employed is discussed in connection with Equations 37 and 38 in what follows.

In Equations 4 and 5 we have introduced a convenient quantity

$$[01] \equiv \rho N_A \langle 01 \rangle \ \text{sec}^{-1} \qquad\qquad 6.$$

in terms of which

$$P_{01} = \rho N_A \frac{X_0 X_1}{A_0 A_1} \frac{[01]}{(1 + \delta_{01})} \qquad\qquad 7.$$

Note that $\tau_1(0)$ and $\tau_0(1)$ do not contain $(1+\delta_{01})^{-1}$; when particle $0 \equiv$ particle 1, two are lost in each reaction, canceling the factor $(1+\delta_{01})^{-1} = \frac{1}{2}$.

Reverse reactions.—As discussed in the Introduction, at elevated temperatures the interaction of the product nuclei in reversing the original reaction is important. With sufficient generality we consider, at most, three reaction products 2, 3, and 4, usually in order of increasing mass, $A_2 < A_3 < A_4$, but with emitted photons always taken last. The energy release in $0+1 \rightarrow 2+3+4$ is given by

$$
\begin{aligned}
Q &= E_{234} - E_{01} \\
&= (M_0 + M_1 - M_2 - M_3 - M_4)c^2 \\
&= 931.478(A_0 + A_1 - A_2 - A_3 - A_4) \ \text{MeV} \qquad\qquad 8. \\
&= 1.49232 \times 10^{-3}(A_0 + A_1 - A_2 - A_3 - A_4) \ \text{erg}
\end{aligned}
$$

where E_{01} and E_{234} are the center-of-momentum energies for the incident and outgoing particles respectively. The ratio Q/k appears in numerous equations and is numerically equal to $11.605 \ Q_6$ in $10^9 \ ^\circ\text{K}$ where $Q_6 = Q$ in MeV. In the great majority of the reactions to be discussed only two nuclear reaction products emerge; in this case, set $A_4 = 0$ and $E_{234} = E_{23}$. This also holds if a photon is emitted along with two nuclei. In the case of two reaction products, one of which is a photon, set $A_3 = 0$.

It will be noted that *atomic* rather than *nuclear* masses have been used in calculating Q, in keeping with common practice in nuclear physics. Except in reactions involving positron emission, the difference involves only changes in atomic binding energies which are in general quite small. The energy release in astrophysical circumstances is complicated by the energy of the Coulomb interactions between nuclei and bound or free electrons which can be larger or smaller than ordinary atomic binding energies. In any case nuclear energy differences are no more accurate than atomic differences, so, to avoid confusion with tabulated Q values, the *atomic* values have been used. In the case of positron emission the atomic mass differences equal the nuclear differences to high approximation *plus* the energy, $2m_e c^2 = 1.0220 \ \text{MeV} = 1.6374 \times 10^{-6}$ erg, which results from the eventual annihilation of the positron with an electron. Since this is available thermonuclear energy, it should indeed be included in the effective Q value. In the case of electron emission, the atomic mass differences automatically incorporate the mass of the electron without explicit introduction into the expression for Q.

In the case of two interacting nuclei and two reaction product nuclei, the reciprocity theorem for nuclear reactions (Blatt & Weisskopf 1952), with due regard for spin statistical factors and possible particle identities, yields in the nonrelativistic approximation

$$\frac{\sigma(23)}{\sigma(01)} = \frac{(1 + \delta_{23})}{(1 + \delta_{01})} \frac{g_0 g_1}{g_2 g_3} \frac{A_0 A_1 E_{01}}{A_2 A_3 E_{23}} \qquad 9.$$

where $\sigma(01)$ is the cross section for $0+1 \rightarrow 2+3$ and $\sigma(23)$ is that for $2+3 \rightarrow 0+1$. Thus (Fowler & Vogl 1964)

$$\frac{[23]}{[01]} = \frac{\langle 23 \rangle}{\langle 01 \rangle} = \frac{(1 + \delta_{23})}{(1 + \delta_{01})} \frac{g_0 g_1}{g_2 g_3} \left(\frac{A_0 A_1}{A_2 A_3}\right)^{3/2} \exp(-Q/kT) \qquad 10.$$

where $g_I = 2J_I + 1$ and J_I is the spin of the appropriate nucleus. Where convenient we employ I to represent 0, 1, 2, 3, 4. The introduction of the $(1+\delta)$ factors reflects the fact that cross sections between identical particles are twice those between different particles, other factors being equal.

In nuclear processes at high temperatures the excited states of nuclei frequently take part. If equilibrium between the excited state and the ground state has not been attained, then the excited nucleus must be treated as a component of the overall composition in the same way as the ground-state nucleus. In most cases of interest there will be excited states in equilibrium with the ground state, even though general equilibrium has not been attained. A test can be made by comparing the lifetime $\tau(I \rightarrow I^*)$ for photoexcitation from the ground to the excited state with the time scale for the astrophysical circumstance under consideration, t. Equilibrium will have been attained if $\tau(I \rightarrow I^*) \lesssim t$. The lifetime $\tau(I \rightarrow I^*)$ can be computed from the *spontaneous* decay lifetime of the excited state, $\tau_{sp}(I^* \rightarrow I)$, which is frequently known, by using

$$\frac{\tau(I \rightarrow I^*)}{\tau_{sp}(I^* \rightarrow I)} = \frac{g_I}{g_{I^*}} \left[\exp(E_{I^*}/kT) - 1\right] \qquad 11.$$

where E_{I^*} is the excitation energy of the excited state and T is the ambient temperature. The effects of *induced* emission on $I^* \rightarrow I$ are incorporated by the inclusion of the -1 in Equation 11.

The reactions involving all excited states in equilibrium with the ground state can be included in Equation 9 if the g_I are replaced by the partition functions

$$G_I = \sum_{I^*} g_{I^*} \exp(-E_{I^*}/kT) \qquad 12.$$

where the sum over I^* includes the ground state and runs through all excited states for which $\tau(I \rightarrow I^*) \leq t$. We make the nonrelativistic approximation that $A_{I^*} \approx A_I$. The quantity $\langle 01 \rangle$ must include sums over all transitions to the relevant excited states in nuclei 2 and 3 and must be appropriately averaged over all possible products, $n_0 n_1$, $n_0^* n_1$, $n_1^* n_0$, $n_1^* n_0^*$ . . . . , of the number densities of the ground and excited states of nuclei 0 and 1 without regard to the possible identity of 0 and 1. The factor $(1 + \delta_{01})^{-1}$ in Equation 9 automatically corrects for the case of identical particles. Then n_0 and n_1 in Equation 1 become the sums over all states of nucleus 0 and 1 respectively. Similar statements, appropriately modified, apply to n_2, n_3, and $\langle 23 \rangle$, and in Equation 9 to the factor $1 + \delta_{23}$.

In Equation 9 the A_I now represent the *nuclear* masses on the atomic-nuclear mass scale. Problems do arise concerning which mass, *atomic* or *nuclear*, to use in equations involving *dynamical* considerations such as Equation 9 above and many of the equations to follow. The matter has been discussed in some detail by Christy (1961). Whereas we used *atomic* masses in Equation 3 for n_0 and n_1, in Equation 8 for Q, and in Equations 28–31 for $\mathcal{E}$ below, in all other cases we have elected to use *nuclear* masses which can be computed from the tabulated *atomic* masses by the use of the relation

$$A_I \text{ (nuclear)} = A_I \text{ (atomic)} - 0.5486 \times 10^{-3} Z_I + 1.68 \times 10^{-8} Z_I^{7/3} \qquad 13.$$

where Z_I is the atomic number of nucleus I. The second term on the right-hand side corrects for the atomic electron masses while the last term represents the Fermi-Thomas binding energy ($15.6\ Z_I^{7/3}$ eV.) The differences are small and we have chosen to use the same notation for *atomic* and *nuclear* masses and to treat them as equal when both appear in the same equation, as in Equations 18 and 19 below. Our calculations have been carried out with the difference taken into account.

The mean lifetimes $\tau_2(3)$ and $\tau_3(2)$ can be calculated from [23] in the same way that $\tau_1(0)$ and $\tau_0(1)$ were obtained from [01] above. The overall reaction rate for $0 + 1 \rightleftharpoons 2 + 3$ is given by

$$P_{01} - P_{23} = \frac{n_0 n_1}{1 + \delta_{01}} \langle 01 \rangle - \frac{n_2 n_3}{1 + \delta_{23}} \langle 23 \rangle = \frac{\rho N_A}{1 + \delta_{01}} \frac{X_0 X_1}{A_0 A_1} [01] - \frac{\rho N_A}{1 + \delta_{23}} \frac{X_2 X_3}{A_2 A_3} [23] \qquad 14.$$

so that at *equilibrium*, when $P_{01} = P_{23}$,

$$\frac{n_2 n_3}{n_0 n_1} = \frac{X_2 X_3}{A_2 A_3} \frac{A_0 A_1}{X_0 X_1} = \frac{g_2 g_3}{g_0 g_1} \left(\frac{A_2 A_3}{A_0 A_1} \right)^{3/2} \exp(+Q/kT) \qquad 15.$$

In the case of radiative capture, i.e., when particle 3, say, is a photon, it can be shown that the reverse photodisintegration rate for nucleus 2 is

$$\frac{1}{\tau_\gamma(2)} = \lambda_\gamma(2) = \frac{g_0 g_1}{(1 + \delta_{01}) g_2} \left(\frac{A_0 A_1}{A_2} \right)^{3/2} \left(\frac{M_u k T}{2 \pi \hbar^2} \right)^{3/2} \langle 01 \rangle \exp(-Q/kT) \text{ sec}^{-1}$$

$$= 0.98677 \times 10^{10} \frac{g_0 g_1}{(1 + \delta_{01}) g_2} \left(\frac{A_0 A_1}{A_2} \right)^{3/2} \rho^{-1} T_9^{3/2} [01] \exp(-11.605\ Q_6/T_9) \qquad 16.$$

where $T_9 = T$ in 10^9 °K, and $Q_6 = Q$ in MeV. The overall reaction rate is given by

$$P_{01} - P_{2\gamma} = \frac{n_0 n_1}{1 + \delta_{01}} \langle 01 \rangle - \frac{n_2}{\tau_\gamma(2)}$$

$$= \frac{\rho N_A}{1 + \delta_{01}} \frac{X_0 X_1}{A_0 A_1} [01] - \rho N_A \frac{X_2}{A_2} \lambda_\gamma(2)$$

17.

so that at equilibrium, when $P_{01} = P_{2\gamma}$,

$$\frac{n_2}{n_0 n_1} = \frac{X_2 A_0 A_1}{\rho N_A X_0 X_1 A_2} = \frac{g_2}{g_0 g_1} \left(\frac{A_2}{A_0 A_1} \right)^{3/2} \left(\frac{2\pi \hbar^2}{M_u k T} \right)^{3/2} \exp(+Q/kT) \text{ cm}^3$$

18.

and

$$\frac{X_2}{X_0 X_1} = 1.0134 \times 10^{-10} \, \rho T_9^{-3/2} \left(\frac{g_2}{g_0 g_1} \right) \left(\frac{A_2}{A_0 A_1} \right)^{5/2} \exp(11.605 \, Q_6/T_9)$$

19.

In the case of three nuclear reaction products, it can be shown that

$$\frac{\langle 234 \rangle}{\langle 01 \rangle} = \frac{g_0 g_1}{g_2 g_3 g_4} \left(\frac{A_0 A_1}{A_2 A_3 A_4} \right)^{3/2} \left(\frac{2\pi \hbar^2}{M_u k T} \right)^{3/2} \left(\frac{1 + \Delta_{234}}{1 + \delta_{01}} \right) \exp(-Q/kT) \text{ cm}^3$$

20.

and

$$\frac{[234]}{[01]} = \rho N_A \frac{g_0 g_1}{g_2 g_3 g_4} \left(\frac{A_0 A_1}{A_2 A_3 A_4} \right)^{3/2} \left(\frac{2\pi \hbar^2}{M_u k T} \right)^{3/2} \left(\frac{1 + \Delta_{234}}{1 + \delta_{01}} \right) \exp(-Q/kT)$$

$$= 1.0134 \times 10^{-10} \frac{g_0 g_1}{g_2 g_3 g_4} \left(\frac{A_0 A_1}{A_2 A_3 A_4} \right)^{3/2} \left(\frac{1 + \Delta_{234}}{1 + \delta_{01}} \right) \rho T_9^{-3/2}$$

$$\times \exp(-11.605 \, Q_6/T_9)$$

21.

where

$$\Delta_{234} = \delta_{23} + \delta_{34} + \delta_{42} + 2\delta_{234}$$

22.

In the above expressions $[234] = \rho^2 N_A^2 \langle 234 \rangle \text{ sec}^{-1}$ and these new quantities are defined in the equations

$$P_{234} = \frac{n_2 n_3 n_4}{1 + \Delta_{234}} \langle 234 \rangle \quad \text{reactions cm}^{-3} \text{ sec}^{-1}$$

$$= \rho N_A \frac{X_2 X_3 X_4}{A_2 A_3 A_4} \frac{[234]}{1 + \Delta_{234}}$$

23.

The overall reaction rate is given by

$$P_{01} - P_{234} = \frac{n_0 n_1}{1 + \delta_{01}} \langle 01 \rangle - \frac{n_2 n_3 n_4}{1 + \Delta_{234}} \langle 234 \rangle$$

$$= \frac{\rho N_A}{1 + \delta_{01}} \frac{X_0 X_1}{A_0 A_1} [01] - \frac{\rho N_A}{1 + \Delta_{234}} \frac{X_2 X_3 X_4}{A_2 A_3 A_4} [234]$$

24.

so that at equilibrium, when $P_{01} = P_{234}$,

$$\frac{n_2 n_3 n_4}{n_0 n_1} = \rho N_A \frac{X_2 X_3 X_4}{A_2 A_3 A_4} \frac{A_0 A_1}{X_0 X_1} = \frac{g_2 g_3 g_4}{g_0 g_1} \left(\frac{A_2 A_3 A_4}{A_0 A_1} \right)^{3/2} \left(\frac{M_u k T}{2\pi \hbar^2} \right)^{3/2} \exp(+Q/kT)$$

25.

and

$$\frac{X_2X_3X_4}{X_0X_1} = 0.98677 \times 10^{10} \, \rho^{-1}T_9^{3/2} \left(\frac{g_2g_3g_4}{g_0g_1}\right)\left(\frac{A_2A_3A_4}{A_0A_1}\right)^{5/2} \exp(11.605 \, Q_6/T_9) \quad 26.$$

Energy generation.—The energy generation in the forward reaction $0+1 \rightarrow 2+3+4+Q$, usually *exoergic* with positive Q, is given by

$$\mathcal{E}_{01} = \frac{P_{01}}{\rho} Q(\text{ergs})$$

$$= 1.6021 \times 10^{-6} \frac{P_{01}}{\rho} Q_6 \text{ erg g}^{-1} \text{ sec}^{-1} \quad 27.$$

so that

$$\mathcal{E}_{01} = 9.6487 \times 10^{17} \frac{X_0X_1}{A_0A_1} \frac{[01]}{1+\delta_{01}} Q_6 \text{ erg g}^{-1} \text{ sec}^{-1} \quad 28.$$

Similarly for the *endoergic* reactions:

$$\mathcal{E}_{23} = -9.6487 \times 10^{17} \frac{X_2X_3}{A_2A_3} \frac{[23]}{1+\delta_{23}} Q_6 \text{ erg g}^{-1} \text{ sec}^{-1} \quad 29.$$

$$\mathcal{E}_{2\gamma} = -9.6487 \times 10^{17} \frac{X_2}{A_2} \lambda_\gamma(2) Q_6 \text{ erg g}^{-1} \text{ sec}^{-1} \quad 30.$$

$$\mathcal{E}_{234} = -9.6487 \times 10^{17} \frac{X_2X_3X_4}{A_2A_3A_4} \frac{[234]}{1+\Delta_{234}} Q_6 \text{ erg g}^{-1} \text{ sec}^{-1} \quad 31.$$

At high temperatures the reverse reactions must be taken into account and the overall energy generation is $\mathcal{E}_{01}+\mathcal{E}_{23}$, $\mathcal{E}_{01}+\mathcal{E}_{2\gamma}$, or $\mathcal{E}_{01}+\mathcal{E}_{234}$ as the case may be. In Tables II to VI to be discussed later, the factors $(1+\delta_{01})^{-1}$, $(1+\delta_{23})^{-1}$, and $(1+\Delta_{234})^{-1}$ have been included in the numerical coefficients for $\mathcal{E}_{01}$, $\mathcal{E}_{23}$, and $\mathcal{E}_{234}$.

In reactions in which neutrinos are produced, the energy carried away by the neutrinos must be subtracted from Q to obtain the available thermonuclear energy generation under most circumstances. It must always be borne in mind that neutrino losses from a given reaction increase with temperature since from $\sim\frac{1}{2}$ to all of the thermal energy of the interacting particles is carried away by neutrinos. Thus, for example, the neutrino losses given in footnotes a, b, and c of Table III (to be discussed later) hold only in the limit of zero temperature.

The determination of $\langle\sigma v\rangle$.—The mean product of cross section and velocity appearing in Equations 1 and 2 enters into all the calculations discussed in the previous section. In the applications under discussion in this article it can be assumed that all nuclei and radiation are in thermodynamic equilibrium under nondegenerate, nonrelativistic conditions insofar as velocity and energy distribution are concerned. For a discussion of lepton and nuclear processes under degenerate or relativistic (or both) conditions, the reader is referred to Bahcall & Wolf (1964), Wolf (1966), Fowler & Hoyle (1964), Wagoner, Fowler & Hoyle (1967), and references therein. Using the Maxwell-Boltzmann distribution in particle velocities, transforming to the center-of-momentum energy E as the integration variable, and dropping superfluous subscripts, we found for two interacting particles that

$$\langle \sigma v \rangle = \frac{(8/\pi)^{1/2}}{M^{1/2}(kT)^{3/2}} \int \sigma E \exp(-E/kT)dE$$

$$= 6.1968 \times 10^{-14} A^{-1/2} T_9^{-3/2} \int \sigma_b E_6 \exp(-11.605 \ E_6/T_9)dE_6 \ \text{cm}^3 \ \text{sec}^{-1} \quad 32.$$

where σ_b is the cross section in barns (10^{-24} cm^2), E_6 is the energy in MeV, and $A = A_0 A_1/(A_0 + A_1)$, for example, is the reduced mass.

In only a few cases are sufficient experimental data available to carry out the calculation prescribed in Equation 32 over the wide range of temperatures encountered in astrophysical applications. It is customary to evaluate $\langle \sigma v \rangle$ at low temperatures using nonresonant cross sections extrapolated from measurements made in the laboratory at the lowest energies at which the reaction can be detected. Indirect measurements on the states of the compound nucleus involved must be made to ascertain whether or not resonances occur and are important at low temperatures. At higher temperatures *measured* nonresonant cross sections are relevant and in addition, resonances occur in profusion. Nonresonant cross sections constitute the slowly varying background on which resonances are superimposed. We will first present the standard treatment of resonant and nonresonant cross sections and will then make a comparison with the exact integrals of Equation 32 in two cases where this can be done.

Nonresonant cross sections, neutrons.—The energy dependence of neutron-interaction cross sections has been discussed by Fowler & Vogl (1964). At low energy the s wave ($l = 0$) interactions dominate, and the cross section is proportional to the square of the De Broglie wavelength λ and the partial width for neutron emission Γ_n. Since λ is proportional to v^{-1} while Γ_n is proportional to v, where v is the relative velocity of the neutron and interacting nucleus, it follows that σ is proportional to v^{-1} or σv is constant. Deviations from this occur at higher energies when other partial waves become important and σ no longer depends linearly on Γ_n as this width becomes comparable to other partial widths. When significant deviations occur, it is convenient to express the nonresonant, slowly varying behavior of $\langle \sigma v \rangle$ as the first three terms of a Maclaurin series in the velocity v of the neutron or, alternatively, a similar series in the square root of the neutron energy $E^{1/2}$, in the center-of-momentum system. Thus

$$\sigma v = S(E^{1/2}) = S(0) \left(1 + \frac{\dot{S}(0)}{S(0)} E^{1/2} + \frac{1}{2} \frac{\ddot{S}(0)}{S(0)} E \right) \ \text{cm}^3 \ \text{sec}^{-1} \quad 33.$$

where $S(0)$, $\dot{S}(0)$, and $\ddot{S}(0)$ are empirical constants and the dot indicates differentiation with respect to $E^{1/2} \propto v$. Substitution in Equation 32 then yields

$$\langle \sigma v \rangle = S(0) \left[1 + 2\pi^{-1/2} \frac{\dot{S}(0)}{S(0)} (kT)^{1/2} + \frac{3}{4} \frac{\ddot{S}(0)}{S(0)} kT \right]$$

$$= S(0) \left[1 + 0.3312 \frac{\dot{S}(0)}{S(0)} T_9^{1/2} + 0.06463 \frac{\ddot{S}(0)}{S(0)} T_9 \right]$$

$$34.$$

TABLE II

Neutron Nonresonant-Reaction Data

Reaction	$H^1(n,\gamma)D^2$	$D^2(n,\gamma)T^3$	$He^3(n,\gamma)He^4$	$He^3(n,p)T^3$	$Li^6(n,\gamma)Li^7$	$Li^6(n,t)He^4$
Parameters in typical equations indicated in parentheses						
Q (8)	2.225	6.257	20.578	0.764	7.253	4.785
$(Q/k)_9$ (39)(43)	2.582E 01	7.262E 01	2.388E 02	8.864	8.417E 01	5.553E 01
σ_{th} (35)	3.32 E−01	5.00 E−04	5.00 E−05	5.33 E 03	4.50 E−02	9.45 E 02
	±0.02 E−01	±1.00 E−04	±5.00 E−05	±0.01 E 03	±1.00 E−02	±0.10 E 02
$S(0)$ (36)	7.30 E−20	1.10 E−22	1.10 E−23	1.17 E−15	9.90 E−21	2.08 E−16
$\dot{S}(0)/S(0)$ (36)	−2.60	—	—	−1.80	—	—
$\tfrac{1}{2}\ddot{S}(0)/S(0)$ (36)	3.32	1.46 E 02	7.00 E 03	1.42	—	—
Coefficients in typical equations indicated in parentheses						
$T_9^{1/2}$ (38)	−8.60 E−01	—	—	−5.97 E−01	—	—
T_9 (38)	4.29 E−01	1.89 E 01	9.05 E 02	1.83 E−01	—	2.08 E−16
$\langle 01 \rangle$ (37)	7.30 E−20	1.10 E−22	1.10 E−23	1.17 E−15	9.90 E−21	2.08 E−16
$[01]$ (38)	4.40 E 04	6.62 E 01	6.62	7.06 E 08	5.96 E 03	1.25 E 08
$\langle 23 \rangle$ (44)	—	—	—	1.17 E−15	—	2.22 E−16
$[\lambda_\gamma]$ (39)(43)	2.07 E 14	1.08 E 12	1.73 E 11	7.07 E 08	7.10 E 13	1.34 E 08
$\tau_1(0)$ (40)	2.29 E−05	1.52 E−02	1.52 E−01	1.43 E−09	1.69 E−04	8.06 E−09
$\tau_0(1)$ (41)	2.29 E−05	3.04 E−02	4.55 E−01	4.27 E−09	1.01 E−03	4.80 E−08
$\mathcal{E}_{01}$ (42)	9.29 E 22	1.97 E 20	4.32 E 19	1.71 E 26	6.88 E 21	9.53 E 25
Uncertainty	±10%	±30%	FAC 2	±10%	±30%	±10%
T_9 limits	0 to 5	0 to 3	0 to 3	0 to 10	0 to 2	0 to 10
References	Hu 58, Be 56	St 64, Gr 63	Zu 63	St 64	St 64	Hu 58

Reaction	$Li^7(n,\gamma)Li^8$	$Be^7(n,p)Li^7$	$Be^9(n,\gamma)Be^{10}$	$B^{10}(n,\gamma)B^{11}$	$B^{10}(n,\alpha)Li^7$	$B^{11}(n,\gamma)B^{12}$
Parameters in typical equations indicated in parentheses						
Q (8)	2.033	1.644	6.815	11.456	2.792	3.369
$(Q/k)_9$ (39)(43)	2.359E 01	1.908E 01	7.909E 01	1.329E 02	3.240E 01	3.910E 01
σ_{th} (35)	3.70 E−02	5.10 E 04	9.50 E−03	5.00 E−01	3.84 E 03	5.00 E−03
	±0.40 E−02	±0.60 E 04	±1.00 E−03	±2.00 E−01	±0.01 E 03	±3.00 E−03
$S(0)$ (36)	8.14 E−21	1.12 E−14	2.09 E−21	1.10 E−19	8.44 E−16	1.10 E−21
Coefficients in typical equations indicated in parentheses						
$\langle 01 \rangle$ (37)	8.14 E−21	1.12 E−14	2.09 E−21	1.10 E−19	8.44 E−16	1.10 E−21
$[01]$ (38)	4.90 E 03	6.76 E 09	1.26 E 03	6.62 E 04	5.08 E 08	6.62 E 02
$\langle 23 \rangle$ (44)	—	1.12 E−14	—	—	6.37 E−16	—
$[\lambda_\gamma]$ (39)(43)	6.41 E 13	6.77 E 09	8.59 E 13	2.01 E 15	3.84 E 08	1.55 E 13
$\tau_1(0)$ (40)	2.06 E−04	1.49 E−10	8.01 E−04	1.52 E−05	1.98 E−09	1.52 E−03
$\tau_0(1)$ (41)	1.43 E−03	1.04 E−09	7.16 E−03	1.51 E−04	1.97 E−08	1.66 E−02
$\mathcal{E}_{01}$ (42)	1.36 E 21	1.51 E 27	9.11 E 20	7.25 E 22	1.36 E 26	1.94 E 20
Uncertainty	±30%	±30%	±30%	±50%	±10%	FAC 2
T_9 limits	0 to 5	0 to 3	0 to 5	0 to 2	0 to 10	0 to 0.03
References	La 66	La 66	St 64	Hu 58	St 64	Aj 67

Units: Energy in MeV; σ_{th} and ± standard deviation in barns; $S(0)$ in cm³ sec⁻¹, $\dot{S}(0)/S(0)$ in MeV$^{-1/2}$, $\tfrac{1}{2}\ddot{S}(0)/S(0)$ in MeV⁻¹; $\langle\ \rangle$ in cm³ sec⁻¹; $[\ \ ]$ in sec⁻¹; τ in sec; $\mathcal{E}_{01}$ in erg g⁻¹ sec⁻¹; $T_9 = T/10^9$.

TABLE II (*concluded*)

NEUTRON NONRESONANT-REACTION DATA

Reaction		$C^{12}(n,\gamma)C^{13}$	$C^{13}(n,\gamma)C^{14}$	$N^{14}(n,\gamma)N^{15}$	$N^{14}(n,p)C^{14}$	$N^{15}(n,\gamma)N^{16}$	$O^{16}(n,\gamma)O^{17}$
		Parameters in typical equations indicated in parentheses					
Q	(8)	4.947	8.176	10.835	0.626	2.487	4.143
$(Q/k)_9$	(39)(43)	5.741E 01	9.488E 01	1.257E 02	7.269	2.886E 01	4.807E 01
σ_{th}	(35)	3.40 E−03 ±0.30 E−03	9.00 E−04 ±2.00 E−04	7.50 E−02 ±0.80 E−02	1.81 ±0.05	2.40 E−05 ±0.80 E−05	1.78 E−04 ±0.25 E−04
$S(0)$	(36)	7.48 E−22	1.98 E−22	1.65 E−20	3.98 E−19	5.28 E−24	3.92 E−23
$\tfrac{1}{2}\ddot{S}(0)/S(0)$	(36)	—	7.70 E 02*	—	—	7.70 E 02*	7.70 E 02*
		Coefficients in typical equations indicated in parentheses					
T_9	(38)	—	1.00 E 02*	—	—	1.00 E 02*	1.00 E 02
(01)	(37)	7.48 E−22	1.98 E−22	1.65 E−20	3.98 E−19	5.28 E−24	3.92 E−23
$[01]$	(38)	4.50 E 02	1.19 E 02	9.94 E 03	2.40 E 05	3.18	2.36 E 01
$\langle 23 \rangle$	(44)	—	—	—	1.20 E−18	—	—
$[\lambda_\gamma]$	(39)(43)	3.99 E 12	4.27 E 12	2.69 E 14	7.21 E 05	2.31 E 10	7.17 E 10
$\tau_1(0)$	(40)	2.24 E−03	8.46 E−03	1.02 E−04	4.21 E−06	3.17 E−01	4.28 E−02
$\tau_0(1)$	(41)	2.66 E−02	1.09 E−01	1.41 E−03	5.84 E−05	4.72	6.78 E−01
$\mathcal{E}_{01}$	(42)	1.78 E 20	7.17 E 19	7.36 E 21	1.03 E 22	5.04 E 17	5.84 E 18
Uncertainty		±30%	FAC 2	±30%	±20%	FAC 2	FAC 2
T_9 limits		0 to 5	0 to 3	0 to 5	0 to 5	0 to 3	0 to 3
References		St 64	St 64, Fo 67	St 64	St 64	Hu 58, Fo 67	St 64, Fo 67

Reaction		$O^{17}(n,\alpha)C^{14}$	$O^{18}(n,\gamma)O^{19}$	$F^{19}(n,\gamma)F^{20}$	$Ne^{21}(n,\alpha)O^{18}$	$Ne^{22}(n,\gamma)Ne^{23}$	$Na^{23}(n,\gamma)Na^{24}$
		Parameters in typical equations indicated in parentheses					
Q	(8)	1.819	3.956	6.597	0.699	5.195	6.962
$(Q/k)_9$	(39)(43)	2.111E 01	4.591E 01	7.656E 01	8.114	6.029E 01	8.079E 01
σ_{th}	(35)	2.35 E−01 ±0.10 E−01	2.10 E−04 ±0.40 E−04	9.80 E−03 ±0.70 E−03	9.60 E 01 ±3.30 E 01	3.60 E−02 ±1.50 E−02	5.34 E−01 ±0.05 E−01
$S(0)$	(36)	5.17 E−20	4.62 E−23	2.16 E−21	2.11 E−17	7.92 E−21	1.17 E−19
$\dot{S}(0)/S(0)$	(36)	—	—	5.40 E 03	—	—	5.00 E 01
$\tfrac{1}{2}\ddot{S}(0)/S(0)$	(36)	7.70 E 02*	7.70 E 02*	−0.97 E 04	—	—	−9.00 E 01
		Coefficients in typical equations indicated in parentheses					
$T_9^{1/2}$	(38)	—	—	1.79 E 03	—	—	1.66 E 01
T_9	(38)	1.00 E 02*	1.00 E 02*	−1.25 E 03	—	—	−1.16 E 01
(01)	(37)	5.17 E−20	4.62 E−23	2.16 E−21	2.11 E−17	7.92 E−21	1.17 E−19
$[01]$	(38)	3.11 E 04	2.78 E 01	1.30 E 03	1.27 E 07	4.77 E 03	7.08 E 04
$\langle 23 \rangle$	(44)	1.05 E−19	—	—	2.69 E−17	—	—
$[\lambda_\gamma]$	(39)(43)	6.32 E 04	8.55 E 10	9.61 E 12	1.62 E 07	1.49 E 13	5.90 E 14
$\tau_1(0)$	(40)	3.24 E−05	3.63 E−02	7.77 E−04	7.93 E−08	2.11 E−04	1.43 E−05
$\tau_0(1)$	(41)	5.46 E−04	6.47 E−01	1.46 E−02	1.65 E−06	4.61 E−03	3.25 E−04
$\mathcal{E}_{01}$	(42)	3.19 E 21	5.85 E 18	4.31 E 20	4.05 E 23	1.08 E 21	2.05 E 22
Uncertainty		FAC 2	FAC 2	FAC 2	±50%	±50%	FAC 2
T_9 limits		0 to 3	0 to 3	0 to 2	0 to 3	0 to 2	0 to 2
References		St 64, Fo 67	Hu 58, Fo 67	St 64, Ma 65	St 64	Hu 58	St 64, Ma 65

Units: Energy in MeV; σ_{th} and ± standard deviation in barns; $S(0)$ in cm³ sec⁻¹, $\dot{S}(0)/S(0)$ in MeV⁻¹ᐟ², $\tfrac{1}{2}\ddot{S}(0)/S(0)$ in MeV⁻¹; ⟨ ⟩ in cm³ sec⁻¹; [ ] in sec⁻¹; τ in sec; $\mathcal{E}_{01}$ in erg g⁻¹ sec⁻¹; $T_9 = T/10^9$.

* Estimated.

where, for $E_6 = E$ in MeV, $\dot{S}(0)/S(0)$ is expressed in the unit MeV$^{-1/2}$, and $\ddot{S}(0)/S(0)$ in MeV^{-1}. In the great majority of cases, the available data do not warrant detailed analysis, but in a few cases, least-squares analysis of accurate, extensive data has been performed to determine $\dot{S}(0)$ and $\ddot{S}(0)$. In all cases $S(0)$ has been predetermined from the accurately measured thermal-neutron cross section σ_{th} at $v_{th} = 2.20 \times 10^5$ cm sec^{-1} or $E_{lab} = 2.53 \times 10^{-8}$ MeV. Thus

$$S(0) = (\sigma v)_{th}$$
$$= 2.20 \times 10^{-19} \, \sigma_{th} \, \text{(barn) cm}^3 \, \text{sec}^{-1} \qquad 35.$$

Whenever $S(0)$ is anomalously low, indicating the operation of some selection rule for the s-wave interaction, we have estimated a p-wave ($l=1$) contribution which arbitrarily increases $\langle \sigma v \rangle$ by a factor of 100 at $T_9 = 1$. We have marked the estimates made in this way in Table II with an asterisk.

The results for a number of neutron-induced reactions are tabulated in Table II. Empirical values for $\langle \sigma v \rangle$ as a function of temperature have been given for a large number of intermediate and heavy nuclei in a comprehensive paper by Macklin & Gibbons (1965) and are not repeated here, except for a very approximate treatment by our methods of their results for $F^{19}(n,\gamma)F^{20}$ and $Na^{23}(n,\gamma)Na^{24}$. The use of Table II will be illustrated by using the reaction $H^1(n, \gamma)D^2$ as an example. For this reaction $Q = 2.225$ MeV and $Q/k = 25.82 \times 10^{9}{}^\circ$K. The latter number occurs in the exponential in Equation 39 below. The thermal-capture cross section is 0.332 ± 0.002 barn (Hughes & Schwartz 1958) so that $S(0) = 7.30 \times 10^{-20}$ cm^3 sec^{-1}. We have used the theoretical calculations and data on the photodisintegration of deuterium, the reverse reaction, as discussed by Bethe & Morrison (1956) and Evans (1955), to determine $\langle \sigma v \rangle$ as a function of $E_6^{1/2}$ and have found from a least-squares analysis that $\dot{S}(0)/S(0) = -2.60$ MeV$^{-1/2}$ and $\frac{1}{2}\ddot{S}(0)/S(0) = 3.32$ MeV^{-1} as given in the Table. Thus Equation 33 for this case becomes

$$\sigma v = 7.30 \times 10^{-20} \, (1 - 2.60 \, E_6^{1/2} + 3.32 \, E_6) \text{ cm}^3 \text{ sec}^{-1} \qquad 36.$$

With the coefficients of $T_9^{1/2}$, T_9, and $\langle 01 \rangle$ listed in Table II, Equation 34 becomes

$$\langle pn \rangle_\gamma = 7.30 \times 10^{-20} \, (1 - 0.860 \, T_9^{1/2} + 0.429 \, T_9) \text{ cm}^3 \text{ sec}^{-1} \qquad 37.$$

while with the coefficient of [01] listed in Table II one finds

$$[pn]_\gamma = 4.40 \times 10^4 \, \rho \, (1 - 0.860 \, T_9^{1/2} + 0.429 \, T_9) \text{ sec}^{-1} \qquad 38.$$

Note that we append the lighter of the reaction products, in this case a photon, as a subscript in Equations 37 and 38 and in many equations to follow.

Equation 16 and the appropriate entries in the Table can now be employed to determine the photodisintegration rate for deuterium which becomes

$$\frac{1}{\tau_\gamma(D^2)} = \lambda_\gamma(D^2) = 2.07 \times 10^{14}\ T_9{}^{3/2}\ (1 - 0.860\ T_9{}^{1/2} + 0.429\ T_9)$$

$$\times \exp(-25.82/T_9)\ \text{sec}^{-1}$$

39.

Similar use of Equations 4 and 28 and appropriate entries in Table II yields

$$\tau_{n\gamma}(H^1) = \frac{2.29 \times 10^{-5}}{\rho X_n}\ (1 - 0.860\ T_9{}^{1/2} + 0.429\ T_9)^{-1}\ \text{sec}$$

$$= \frac{1.37 \times 10^{19}}{n_n}\ (1 - 0.860\ T_9{}^{1/2} + 0.429\ T_9)^{-1}\ \text{sec}$$

40.

In the second form of Equation 40 the number density of neutrons rather than the abundance by mass appears since for neutrons this is usually the convenient quantity. For the neutron lifetime one has

$$\tau_{p\gamma}(n) = \frac{2.29 \times 10^{-5}}{\rho X_H}\ (1 - 0.860\ T_9{}^{1/2} + 0.429\ T_9)^{-1}\ \text{sec}$$

41.

and for the energy generation rate

$$\mathcal{E}(pn)_\gamma = 9.29 \times 10^{22}\ \rho X_H X_n\ (1 - 0.860\ T_9{}^{1/2} + 0.429\ T_9)\ \text{erg g}^{-1}\ \text{sec}^{-1}$$ 42.

For the reverse of a reaction such as $He^3(n,p)T^3$, for which there are two nuclei as reaction products, Equation 39 is replaced by

$$[pT^3]_n = 7.07 \times 10^8\ \rho(1 - 0.597\ T_9{}^{1/2} + 0.183\ T_9)\ \exp(-8.864/T_9)\ \text{sec}^{-1}$$ 43.

and it is also possible to write

$$\langle pT^3 \rangle_n = 1.17 \times 10^{-15}\ (1 - 0.597\ T_9{}^{1/2} + 0.183\ T_9)\ \exp(-8.864/T_9)\ \text{cm}^3\ \text{sec}^{-1}$$ 44.

Note that the density ρ occurs in Equations 38 and 43 but not in Equations 37, 39, and 44. In equations for [234] and $\mathcal{E}_{234}$, ρ^2 appears. Equations 36 through 44 are the typical nonresonant neutron relations referred to in Table II. With this "do-it-yourself kit" the reader can construct appropriate equations for other reactions, proceeding in either direction, using the data given in Table II. It will be noted that in many cases either $\overset{.}{s}$ or $\overset{..}{s}$ or both are zero, in which case the terms in $T_9{}^{1/2}$ or T_9 or both will not occur in Equations 36 through 43.

In Table II and subsequent tables we use the symbol $[\lambda_\gamma]$ which is to be read as [23], [234], or λ_γ, whichever is appropriate to the reaction under consideration. All of these quantities have the dimensions sec^{-1}. In the case of two nuclear reaction products the tables contain coefficients for both $\langle 23 \rangle$ and [23]. In the case of reactions producing photons, the tables contain coefficients for only λ_γ. In the case of reactions with three final products the tables contain coefficients for only [234]. The quantity $\langle 234 \rangle$ in units $\text{cm}^6\ \text{sec}^{-1}$ can be calculated from $\langle 234 \rangle = [234]/\rho^2 N_A{}^2$.

We emphasize that the tables give only the coefficients and that the complete formulae must be constructed in the manner in which we have just obtained Equations 37 through 44. In reactions involving identical particles it is most important to include the factor $(1+\delta_{01})^{-1}$ in calculating the reaction rate P_{01}

from $\langle 01 \rangle$ using Equation 1 or from $[01]$ using Equation 7. Similar remarks apply in calculating P_{23} using Equation 14 and P_{234} using Equation 23.

The last entries in the tables require some comment. Our estimates for the uncertainty (standard deviation *not* limit of error) in the reaction rates and lifetimes have been chosen rather pessimistically and apply for the full range of temperature listed immediately below the uncertainty. The abbreviation FAC 2 is to be read as *factor of 2*. It will be clear from the experimental standard deviations given for the σ_{th} that in many cases the uncertainty is considerably smaller over a limited range which might typically cover $0 < T_9 \lesssim 0.1$. Detailed study of the available experimental data is required if it is desired to use uncertainties substantially smaller than the ones we have listed.

The upper limit cited in Table II for the temperature range over which the rates and lifetimes are valid is in most cases determined by lack of knowledge of the cross section above a maximum energy E_{max}, up to which observations have been made, or by failure of the expansion in Equation 33 to hold above this energy. The resonant energy of the first excited state in the compound nucleus above threshold, whose properties are unknown, sets a limiting E_{max}. In any case we take $T_{max} \sim E_{max}/k$; a more conservative choice would be a fraction, say one half, of this value. In ideal cases the background nonresonant cross section can be isolated from resonance effects up to large E_{max} and thus large T_{max}. In this case the data of Table II must be supplemented by those in Table IV which will be discussed below. In fact, it is in general necessary to employ both Table II and Table IV to obtain full information on the rate of a given reaction.

The references listed in the Table are designated by a notation in common use, e.g. Ja 57, and can be identified in detail by recourse to the list of literature cited at the end of the article. In many cases we have made a systematic least-squares analysis of available data so that our tabulated results may not agree with the results for $S(0)$, for example, given by the authors quoted. In other cases we have made the first analysis of experimental data in which case we list this present paper as Fo 67. In order to keep the number of references to a minimum, we have in general cited nuclear-data tabulations where critical references to the original literature are given. In a few cases we cite references on proton-induced mirror reactions. Thus, for example, data on $D^2(p,\gamma)He^3$ given by Griffiths, Lal & Scarfe (1963) have been analyzed by us in such a way that the Coulomb-barrier factor could be removed to yield the energy dependence of the rate of the $D^2(n,\gamma)T^3$ reaction.

Nonresonant cross sections, charged particles.—At low energy, charged-particle interactions are dominated by Coulomb-barrier penetration factors. This has been well known since the pioneer work of Gamow (1928) and of Gurney & Condon (1928, 1929) many years ago. The low-energy relation for neutrons, $\sigma = S(v)/v$, is replaced by

$$\sigma = \frac{S(E)}{E} \exp - (E_G/E)^{1/2}$$

45.

where

$$E_G = (2\pi\alpha Z_0 Z_1)^2 \ (Mc^2/2) \qquad\qquad 46.$$

is the "Gamow energy." Numerically one finds

$$E_G^{1/2} = 0.98948 \ Z_0 Z_1 A_1^{1/2} \ \text{MeV}^{1/2} \ \text{LAB}$$
$$= 0.98948 \ Z_0 Z_1 A^{1/2} \ \text{MeV}^{1/2} \ \text{CM} \qquad 47.$$

It is most important to note that charged-particle cross sections vary in the same way at low energy for all partial waves, in marked contrast to neutron cross sections where $\Gamma_n(l) \propto (kR)^{2l+1} \propto (E/E_R)^{l+1/2}$ where $k = Mv/\hbar$ is the wavenumber, R the radius of interaction, and E_R the centrifugal barrier energy $= \hbar^2/2MR^2$. In charged-particle widths the l dependence is separable as an approximately energy-independent factor which decreases fairly rapidly with increasing l. In addition the v dependence of the partial width is canceled by a term in v^{-1} in the Coulomb-barrier penetration factor, leaving the E^{-1} term from $\pi\lambda^2$ and the Gamow exponential in first approximation. In Equation 45 the *cross-section factor* $S(E)$ includes the energy dependence of the penetration factor not included in the Gamow exponential plus that of the intrinsic nuclear factors governing the cross section. Far from a nuclear resonance, $S(E)$ is a slowly varying function of E and can be conveniently expressed as the first three terms of a Maclaurin series in the center-of-momentum energy E. Thus

$$S(E) = S(0) \left(1 + \frac{S'(0)}{S(0)} E + \frac{1}{2} \frac{S''(0)}{S(0)} E^2\right) \qquad 48.$$

where the prime indicates differentiation with respect to E. If E is in MeV and σ is in barns in Equation 45, then $S(0)$ is in MeV-barn, $S'(0)$ is in barns, and $S''(0)$ in barn-MeV^{-1}. In this article all tabulated values for S refer to center-of-momentum coordinates.

Substitution of Equations 45 and 48 into Equation 32 yields (Caughlan & Fowler 1962, Bahcall 1966)

$$\langle\sigma v\rangle = \frac{(8/\pi)^{1/2}}{M^{1/2}(kT)^{3/2}} \int S(E) \exp(-E_G^{1/2}/E^{1/2} - E/kT)dE$$
$$= \left(\frac{2}{M}\right)^{1/2} \frac{\Delta E_o}{(kT)^{3/2}} S_{\text{eff}} \exp(-\tau) \qquad\qquad 49.$$

so that

$$\langle 01 \rangle = \left\{1.3006 \times 10^{-14} \ (Z_0 Z_1/A)^{1/3} S_{\text{eff}}\right\} \ T_9^{-2/3} \exp(-\tau) \ \text{cm}^3 \ \text{sec}^{-1} \qquad 50.$$

and

$$[01] = \left\{7.8327 \times 10^9 \ (Z_0 Z_1/A)^{1/3} S_{\text{eff}}\right\} \ \rho \ T_9^{-2/3} \exp(-\tau) \ \text{sec}^{-1} \qquad 51.$$

where

$$S_{\text{eff}} = S(0) \left[1 + \frac{5}{12\tau} + \frac{S'(0)}{S(0)} \left(E_0 + \frac{35}{36} kT \right) \right.$$

$$\left. + \frac{1}{2} \frac{S''(0)}{S(0)} \left(E_0^2 + \frac{89}{36} E_0 kT \right) \right] \text{MeV-barn} \tag{52.}$$

$$= S(0) \left[1 + 9.807 \times 10^{-2} W^{-1/3} T_9^{1/3} + 0.1220 \frac{S'(0)}{S(0)} W^{1/3} T_9^{2/3} \right.$$

$$+ 8.378 \times 10^{-2} \frac{S'(0)}{S(0)} T_9 + 7.447 \times 10^{-3} \frac{S''(0)}{S(0)} W^{2/3} T_9^{4/3}$$

$$\left. + 1.300 \times 10^{-2} \frac{S''(0)}{S(0)} W^{1/3} T_9^{5/3} \right] \tag{53.}$$

$$W = Z_0^2 Z_1^2 A \tag{54.}$$

$$\tau = 3E_0/kT = 3[\pi \alpha Z_0 Z_1 (Mc^2/2kT)^{1/2}]^{2/3}$$
$$= 4.2487 \, W^{1/3} T_9^{-1/3} \tag{55.}$$

$$E_0 = [\pi \alpha Z_0 Z_1 kT \, (Mc^2/2)^{1/2}]^{2/3}$$
$$= 0.12204 \, W^{1/3} T_9^{2/3} \text{ MeV} \tag{56.}$$

$$kT = 0.08617 \, T_9 = T_9/11.605 \text{ MeV} \tag{57.}$$

$$\Delta E_0 = 4(E_0 kT/3)^{1/2} = \frac{2^{11/6}}{3^{1/2}} (\pi \alpha Z_0 Z_1)^{1/3} (Mc^2)^{1/6} (kT)^{5/6}$$
$$= 0.23682 \, W^{1/6} T_9^{5/6} \text{ MeV} \tag{58.}$$

$$\alpha = e^2/\hbar c = (137.0388)^{-1} \tag{59.}$$

In the above equations E_0 is the effective interaction energy given by the energy at which the maximum occurs in the product of the Gamow and Maxwell-Boltzmann exponentials in the integrand in the first form of Equation 49. It is frequently convenient to know the temperature corresponding to a given E_0 in MeV. This temperature is given by

$$T_0 = 23.46 \, W^{-1/2} (E_6)_0^{3/2} \tag{60.}$$

The quantity ΔE_0 is the effective energy interval given by the full width between the points at $1/e$ times the maximum of the Gamow-Maxwell-Boltzmann product. Reference should be made to Fowler & Vogl (1964) for further details. Data for the nonresonant portion of certain charged-particle reactions are given in Table III. The numerical coefficients listed for $\langle 01 \rangle$ and [01] in Table III are evaluated by use of the factors in curly brackets in Equations 50 and 51 respectively with S_{eff} replaced by $S(0)$. The data in Table III must frequently be supplemented by the resonant and continuum data given respectively in Tables V and VI discussed below. Illustrative examples of the use of Tables III, V, and VI will be discussed after all necessary preliminaries have been given.

Resonant cross sections, neutrons, and charged particles.—A single resonance in the cross section of a nuclear reaction $(0+1 \rightarrow 2+3+Q)$ can be represented most simply as a function of energy in terms of the classical, Breit-Wigner formula (see p. 544):

TABLE III

Charged-Particle Nonresonant-Reaction Data

Reaction		$H^1(p,e^+\nu)D^2$	$H^1(pe^-,\nu)D^2$	$He^3(p,e^+\nu)He^4$	$D^2(p,\gamma)He^3$	$D^2(d,n)He^3$	$D^2(d,p)T^3$
Parameters in typical equations indicated in parentheses							
Q	(8)	1.442^a	1.442^b	19.796^c	5.494	3.269	4.033
$(Q/k)_9$	(80)(83)	$1.674E\ 01$	$1.674E\ 01$	$2.297E\ 02$	$6.375E\ 01$	$3.794E\ 01$	$4.680E\ 01$
$E_G^{1/2}$ LAB	(47)	$9.931E{-}01$	$9.931E{-}01$	1.986	$9.931E{-}01$	1.404	1.404
$E_G^{1/2}$ CM	(47)	$7.022E{-}01$	$7.022E{-}01$	1.720	$8.108E{-}01$	$9.928E{-}01$	$9.928E{-}01$
$E_0/T_9^{2/3}$	(56)	$9.709E{-}01$	$9.709E{-}01$	$1.764E{-}01$	$1.069E{-}01$	$1.223E{-}01$	$1.223E{-}01$
$T_0/E_0^{3/2}$	(60)	$3.305E\ 01$	$3.305E\ 01$	$1.350E\ 01$	$2.863E\ 01$	$2.338E\ 01$	$2.338E\ 01$
$\Delta E_0/T_9^{5/6}$	(58)	$2.112E{-}01$	$2.112E{-}01$	$2.847E{-}01$	$2.216E{-}01$	$2.371E{-}01$	$2.371E{-}01$
$\tau T_9^{1/3}$	(55)(78)	3.380	3.380	6.141	3.720	4.258	4.258
$S(0)$	(53)	$3.36\ E{-}25$	$1.15\ E{-}30^d$	$3.70\ E{-}23$	$2.50\ E{-}07$	$5.30\ E{-}02$	$5.30\ E{-}02$
$S'(0)/S(0)$	(53)	8.04	—	—	$3.16\ E\ 01$	4.95	4.95
Coefficients in typical equations indicated in parentheses							
$T_9^{1/3}$	(78)	$1.23\ E{-}01$	—	$6.78\ E{-}02$	$1.12\ E{-}01$	$9.79\ E{-}02$	$9.79\ E{-}02$
$T_9^{2/3}$	(78)	$7.80\ E{-}01$	—	—	3.38	$6.06\ E{-}01$	$6.06\ E{-}01$
T_9	(78)	$6.73\ E{-}01$	—	—	2.65	$4.15\ E{-}01$	$4.15\ E{-}01$
$\langle 01 \rangle$	(82)	$5.49\ E{-}39$	$1.88\ E{-}44^d$	$6.66\ E{-}37$	$3.71\ E{-}21$	$6.88\ E{-}16$	$6.88\ E{-}16$
$[01]$	(78)	$3.31\ E{-}15$	$1.13\ E{-}20^d$	$4.01\ E{-}13$	$2.24\ E\ 03$	$4.14\ E\ 08$	$4.14\ E\ 08$
$\langle 23 \rangle$	(83)	—	—	—	—	$1.19\ E{-}15$	$1.19\ E{-}15$
$[\lambda_\gamma]$	(80)(83)	—	—	—	$3.65\ E\ 13$	$7.17\ E\ 08$	$7.19\ E\ 08$
$\tau_1(0)$	(79)	$3.05\ E\ 14$	$8.91\ E\ 19^e$	$2.51\ E\ 12$	$4.51\ E{-}04$	$4.86\ E{-}09$	$4.86\ E{-}09$
$\tau_0(1)$	(79)	$3.05\ E\ 14$	$8.91\ E\ 19^e$	$7.52\ E\ 12$	$9.01\ E{-}04$	$4.86\ E{-}09$	$4.86\ E{-}09$
$\mathcal{E}_{01}$	(81)	$2.27^a E\ 03$	$7.75^b E{-}03^d$	$2.52^c E\ 06$	$5.84\ E\ 21$	$1.61\ E\ 26$	$1.99\ E\ 26$
Uncertainty		$\pm 10\%$	$\pm 50\%$	FAC 3	$\pm 10\%$	$\pm 10\%$	$\pm 10\%$
T_9 limits		0 to 5	0 to 5	0 to 5	0 to 5	0 to 10	0 to 10
References		Pa 64	Ba 64	We 67	Pa 64	Pa 64	Pa 64

Reaction		$D^2(d,\gamma)He^4$	$T^3(p,\gamma)He^4$	$T^3(d,n)He^4$	$He^3(d,p)He^4$	$He^4(t,\gamma)Li^7$	$He^4(\tau,\gamma)Be^7$
Parameters in typical equations indicated in parentheses							
Q	(8)	23.847	19.814	17.590	18.354	2.467	1.587
$(Q/k)_9$	(80)(83)	$2.767E\ 02$	$2.299E\ 02$	$2.041E\ 02$	$2.130E\ 02$	$2.863E\ 01$	$1.842E\ 01$
$E_G^{1/2}$ LAB	(47)	1.404	$9.931E{-}01$	1.404	2.808	3.437	6.872
$E_G^{1/2}$ CM	(47)	$9.928E{-}01$	$8.598E{-}01$	1.087	2.174	2.595	5.190
$E_0/T_9^{2/3}$	(56)	$1.223E{-}01$	$1.111E{-}01$	$1.299E{-}01$	$2.063E{-}01$	$2.321E{-}01$	$3.684E{-}01$
$T_0/E_0^{3/2}$	(60)	$2.338E\ 01$	$2.699E\ 01$	$2.135E\ 01$	$1.067E\ 01$	8.944	4.472
$\Delta E_0/T_9^{5/6}$	(58)	$2.371E{-}01$	$2.260E{-}01$	$2.444E{-}01$	$3.079E{-}01$	$3.266E{-}01$	$4.115E{-}01$
$\tau T_9^{1/3}$	(55)(78)	4.258	3.869	4.524	7.181	8.080	$1.283E\ 01$
$S(0)$	(53)	$2.22\ E{-}10$	$2.56\ E{-}06$	$1.10\ E\ 01$	7.00	$6.40\ E{-}05$	$4.70\ E{-}04$
$S'(0)/S(0)$	(53)	$1.19\ E\ 02$	$1.51\ E\ 01$	$1.38\ E\ 01$	-4.97	—	$-5.96\ E{-}01$
$\frac{1}{2}S''(0)/S(0)$	(53)	$1.37\ E\ 02$	$4.46\ E\ 01$	$6.23\ E\ 02$	$7.10\ E\ 01$	—	$1.97\ E{-}01$
Coefficients in typical equations indicated in parentheses							
$T_9^{1/3}$	(78)	$9.79\ E{-}02$	$1.08\ E{-}01$	$9.21\ E{-}02$	$5.80\ E{-}02$	$5.16\ E{-}02$	$3.25\ E{-}02$
$T_9^{2/3}$	(78)	$1.45\ E\ 01$	1.68	1.80	-1.03	—	$-2.19\ E{-}01$
T_9	(78)	9.92	1.26	1.16	$-4.16\ E{-}01$	—	$-4.99\ E{-}02$
$T_9^{4/3}$	(78)	2.04	$5.51\ E{-}01$	$1.05\ E\ 01$	3.02	—	$2.67\ E{-}02$
$T_9^{5/3}$	(78)	3.56	1.06	$1.72\ E\ 01$	3.12	—	$1.54\ E{-}02$
$\langle 01 \rangle$	(82)	$2.88\ E{-}24$	$3.65\ E{-}20$	$1.34\ E{-}13$	$1.08\ E{-}13$	$8.75\ E{-}19$	$8.10\ E{-}18$
$[01]$	(78)	1.73	$2.20\ E\ 04$	$8.09\ E\ 10$	$6.49\ E\ 10$	$5.27\ E\ 05$	$4.88\ E\ 06$
$\langle 23 \rangle$	(83)	—	—	$7.44\ E{-}13$	$5.97\ E{-}13$	—	—
$[\lambda_\gamma]$	(80)(83)	$7.86\ E\ 10$	$5.74\ E\ 14$	$4.48\ E\ 11$	$3.60\ E\ 11$	$5.87\ E\ 15$	$5.43\ E\ 16$
$\tau_1(0)$	(79)	1.16	$4.58\ E{-}05$	$2.49\ E{-}11$	$3.10\ E{-}11$	$5.72\ E{-}06$	$6.18\ E{-}07$
$\tau_0(1)$	(79)	1.16	$1.37\ E{-}04$	$3.73\ E{-}11$	$4.65\ E{-}11$	$7.59\ E{-}06$	$8.21\ E{-}07$
$\mathcal{E}_{01}$	(81)	$4.92\ E\ 18$	$1.38\ E\ 23$	$2.26\ E\ 29$	$1.89\ E\ 29$	$1.04\ E\ 23$	$6.19\ E\ 23$
Uncertainty		FAC 2	$\pm 20\%$	0 to 20%	-10 to $+70\%$	$\pm 20\%$	$\pm 20\%$
T_9 limits		0 to 10	0 to 10	0 to 0.15	0 to 0.65	0 to 10	0 to 10
References		Zu 63	Ja 57	Ja 57	Ja 57	Gr 61	Pa 64

Units: Energy in MeV; $S(0)$ in MeV-barn, $S'(0)/S(0)$ in MeV^{-1}, $\frac{1}{2}S''(0)/S(0)$ in MeV^{-2}; $\langle\ \rangle$ in $cm^3\ sec^{-1}$; $[\ ]$ in sec^{-1}; τ in sec; $\mathcal{E}_{01}$ in $erg\ g^{-1}\ sec^{-1}$; $T_9 = T/10^9$.

^{a} Neutrino loss ≈ 0.26 MeV. $\quad^b$ Neutrino loss ≈ 1.44 MeV. $\quad^c$ Neutrino loss ≈ 9.64 MeV.

(Nota bene: tabulated values for Q and $\mathcal{E}_{01}$ include neutrino energy.)

^{d} Include an additional factor $\rho(1+X_H)/2T_9^{1/2}$. $\quad^e$ Include an additional factor $2T_9^{1/2}/\rho(1+X_H)$.

TABLE III (continued)

CHARGED-PARTICLE NONRESONANT-REACTION DATA

Reaction		$T^3(t,2n)He^4$	$He^3(t,np)He^4$	$He^3(\tau,2p)He^4$	$He^3(t,d)He^4$	$Li^6(p,\tau)He^4$	$Li^7(p,\alpha)He^4$
		Parameters in typical equations indicated in parentheses					
Q	(8)	11.332	12.096	12.860	14.321	4.021	17.347
$(Q/k)_9$	(80)(83)	1.315E 02	1.404E 02	1.492E 02	1.662E 02	4.667E 01	2.013E 02
$EG^{1/2}$ LAB	(47)	1.718	3.437	6.872	3.437	2.979	2.979
$EG^{1/2}$ CM	(47)	1.215	2.430	4.860	2.430	2.757	2.786
$E_0/T_9^{2/3}$	(56)	1.399E−01	2.221E−01	3.526E−01	2.221E−01	2.417E−01	2.433E−01
$T_0/E_0^{3/2}$	(60)	1.910E 01	9.552	4.776	9.552	8.418	8.331
$\Delta E_0/T_9^{5/6}$	(58)	2.536E−01	3.195E−01	4.025E−01	3.195E−01	3.333E−01	3.344E−01
$\tau T_9^{1/3}$	(55)(78)	4.872	7.733	1.228E 01	7.733	8.413	8.471
$S(0)$	(53)	1.60 E−01	6.50 E−01	5.00	4.50 E−01	5.50	1.25 E−01
		Coefficients in typical equations indicated in parentheses					
$T_9^{1/3}$	(78)	8.55 E−02	5.39 E−02	3.39 E−02	5.39 E−02	4.95 E−02	4.92 E−02
$\langle 01 \rangle$	(82)	1.81 E−15	9.29 E−15	9.00 E−14	6.43 E−15	1.08 E−13	2.45 E−15
$[01]$	(78)	1.09 E 09	5.59 E 09	5.42 E 10	3.87 E 09	6.53 E 10	1.47 E 09
$\langle 23 \rangle$	(83)	—	—	—	1.03 E−14	1.16 E−13	1.15 E−14
$[\lambda_\gamma]$	(84)(83)	3.70 E−01	1.90	1.84 E 01	6.19 E 09	6.97 E 10	6.91 E 09
$\tau_1(0)$	(79)	2.76 E−09	5.39 E−10	5.56 E−11	7.79 E−10	1.54 E−11	6.84 E−10
$\tau_0(1)$	(79)	2.76 E−09	5.39 E−10	5.56 E−11	7.79 E−10	9.22 E−11	4.76 E−09
$\mathcal{E}_{01}$	(81)	6.57 E 26	7.18 E 27	3.70 E 28	5.88 E 27	4.18 E 28	3.49 E 27
Uncertainty		±30%	±30%	±40%	±30%	±10%	±10%
T_9 limits		0 to 10	0 to 10	0 to 10	0 to 10	0 to 5	0 to 5
References		Go 62	Yo 61	Wi 67	Yo 61	Ja 57	Ja 57

Reaction		$C^{12}(p,\gamma)N^{13}$	$C^{12}(p,\gamma)N^{14}$	$N^{13}(p,\gamma)O^{14}$	$N^{14}(p,\gamma)O^{15}$	$N^{15}(p,\gamma)O^{16}$	$N^{15}(p,\alpha)C^{12}$
		Parameters in typical equations indicated in parentheses					
Q	(8)	1.944	7.550	4.626	7.293	12.126	4.965
$(Q/k)_9$	(80)(83)	2.256E 01	8.762E 01	5.369E 01	8.463E 01	1.407E 02	5.761E 01
$EG^{1/2}$ LAB	(47)	5.958	5.958	6.952	6.952	6.952	6.952
$EG^{1/2}$ CM	(47)	5.723	5.740	6.697	6.714	6.729	6.729
$E_0/T_9^{2/3}$	(56)	3.932E−01	3.940E−01	4.367E−01	4.374E−01	4.381E−01	4.381E−01
$T_0/E_0^{3/2}$	(60)	4.056	4.043	3.466	3.457	3.449	3.449
$\Delta E_0/T_9^{5/6}$	(58)	4.251E−01	4.255E−01	4.480E−01	4.484E−01	4.487E−01	4.487E−01
$\tau T_9^{1/3}$	(55)(78)	1.369E 01	1.372E 01	1.520E 01	1.523E 01	1.525E 01	1.525E 01
$S(0)$	(53)	1.40 E−03	5.50 E−03	2.75 E−03[a]	2.75 E−03	2.74 E−02	5.34 E 01
$S'(0)/S(0)$	(53)	3.04	2.43	—	—	6.79	1.54 E 01
$\tfrac{1}{2}S''(0)/S(0)$	(53)	1.33 E 01	8.97	—	—	—	—
		Coefficients in typical equations indicated in parentheses					
$T_9^{1/3}$	(78)	3.04 E−02	3.04 E−02	2.74 E−02	2.74 E−02	2.73 E−02	2.73 E−02
$T_9^{2/3}$	(78)	1.19	9.58 E−01	—	—	2.97	6.74
T_9	(78)	2.54 E−01	2.04 E−01	—	—	5.69 E−01	1.29
$T_9^{4/3}$	(78)	2.06	1.39	—	—	—	—
$T_9^{5/3}$	(78)	1.12	7.53 E−01	—	—	—	—
$\langle 01 \rangle$	(82)	3.39 E−17	1.33 E−16	7.00 E−17	6.99 E−17	6.95 E−16	1.35 E−12
$[01]$	(78)	2.04 E 07	8.01 E 07	4.21 E 07	4.21 E 07	4.19 E 08	8.16 E 11
$\langle 23 \rangle$	(83)	—	—	—	—	—	9.56 E−13
$[\lambda_\gamma]$	(80)(83)	1.81 E 17	9.53 E 17	1.50 E 18	1.14 E 18	1.52 E 19	5.76 E 11
$\tau_1(0)$	(79)	4.93 E−08	1.26 E−08	2.39 E−07	2.40 E−08	2.41 E−09	1.24 E−12
$\tau_0(1)$	(79)	5.87 E−07	1.62 E−07	3.09 E−07	3.33 E−07	3.58 E−08	1.84 E−11
$\mathcal{E}_{01}$	(81)	3.17 E 24	4.45 E 25	1.44 E 25	2.10 E 25	3.24 E 26	2.58 E 29
Uncertainty		0 to −20%	0 to −20%	FAC 10	±10%	±20%	±20%
T_9 limits		0 to 0.55	0 to 0.65	0 to 1.5	0 to 2	0 to 1	0 to 1
References		Vo 63	Vo 63	Fo 67	He 63	He 60, Ca 62	He 60, Ca 62

Units: Energy in MeV; $S(0)$ in MeV-barn, $S'(0)/S(0)$ in MeV^{-1}, $\tfrac{1}{2}S''(0)/S(0)$ in MeV^{-2}; $\langle\ \rangle$ in cm^3 sec^{-1}; $[\ ]$ in sec^{-1}; τ in sec; $\mathcal{E}_{01}$ in erg g^{-1} sec^{-1}; $T_9 = T/10^9$.

[a] Assumed equal to $S(0)$ for $N^{14}(p,\gamma)O^{15}$.

TABLE III (concluded)

CHARGED-PARTICLE NONRESONANT-REACTION DATA

Reaction	$Be^7(p,\gamma)B^8$	$Be^9(p,d)Be^8$	$Be^9(p,\alpha)Li^6$	$B^{11}(p,2\alpha)He^4$	$N^{14}(\alpha,\gamma)F^{18}$	$O^{18}(\alpha,\gamma)Ne^{22}$
Parameters in typical equations indicated in parentheses						
Q (8)	0.135	0.559	2.126	8.682	4.416	9.667
$(Q/k)_9$ (80)(83)	1.564	6.492	2.468E 01	1.008E 02	5.125E 01	1.122E 02
$EG^{1/2}$ LAB (47)	3.972	3.972	3.972	4.965	2.771E 01	3.167E 01
$EG^{1/2}$ CM (47)	3.715	3.767	3.767	4.753	2.444E 01	2.864E 01
$E_0/T_9^{2/3}$ (56)	2.948E−01	2.976E−01	2.976E−01	3.474E−01	1.035	1.151
$T_0/E_0^{3/2}$ (60)	6.248	6.161	6.161	4.884	9.498E−01	8.103E−01
$\Delta E_0/T_9^{5/6}$ (58)	3.681E−01	3.698E−01	3.698E−01	3.996E−01	6.897E−01	7.272E−01
$\tau T_9^{1/3}$ (55)(78)	1.026E 01	1.036E 01	1.036E 01	1.209E 01	3.603E 01	4.006E 01
$S(0)$ (53)	4.00 E−05	1.50 E 01	1.50 E 01	1.00 E 02	8.73 E 06	7.77 E 07
Coefficients in typical equations indicated in parentheses						
$T_9^{1/3}$ (78)	4.06 E−02	4.02 E−02	4.02 E−02	3.45 E−02	1.16 E−02	1.04 E−02
$\langle 01 \rangle$ (82)	8.61 E−19	3.20 E−13	3.20 E−13	2.28 E−12	1.87 E−07	1.71 E−06
$[01]$ (78)	5.19 E 05	1.93 E 11	1.93 E 11	1.38 E 12	1.13 E 17	1.03 E 18
$\langle 23 \rangle$ (83)	—	3.61 E−13	1.98 E−13	—	—	—
$[\lambda_\gamma]$ (80)(83)	6.77 E 11	2.17 E 11	1.19 E 11	4.82 E 02	6.11 E 27	6.04 E 28
$\tau_1(0)$ (79)	1.94 E−06	5.23 E−12	5.23 E−12	7.33 E−13	3.55 E−17	3.88 E−18
$\tau_0(1)$ (79)	1.35 E−05	4.68 E−11	4.68 E−11	8.00 E−12	1.24 E−16	1.74 E−17
$\mathcal{E}_{01}$ (81)	9.54 E 21	1.15 E 28	4.35 E 28	1.04 E 30	8.58 E 33	1.34 E 35
Uncertainty	± 20%	± 30%	± 30%	± 10%	FAC 3	FAC 3
T_9 limits	0 to 10	0 to 2	0 to 2	0 to 3	0 to 2	0 to 2
References	Pa 66	Mi 54	Mi 54	Sc 51	Ca 64	Ca 64

Reaction	$B^{10}(p,\alpha)Be^7$	$C^{13}(\alpha,n)O^{16}$	$O^{16}(p,\gamma)F^{17}$	$O^{16}(\alpha,\gamma)Ne^{20}$	$O^{17}(p,\alpha)N^{14}$	$Ne^{20}(p,\gamma)Na^{21}$
Parameters in typical equations indicated in parentheses						
Q (8)	1.148	2.214	0.601	4.730	1.193	2.432
$(Q/k)_9$ (80)(83)	1.332E 01	2.570E 01	6.969	5.489E 01	1.384E 01	2.823E 01
$EG^{1/2}$ LAB (47)	4.965	2.375E 01	7.945	3.167E 01	7.945	9.931
$EG^{1/2}$ CM (47)	4.733	2.077E 01	7.706	2.832E 01	7.719	9.690
$E_0/T_9^{2/3}$ (56)	3.464E−01	9.286E−01	4.795E−01	1.142	4.800E−01	5.586E−01
$T_0/E_0^{3/2}$ (60)	4.904	1.118	3.012	8.195E−01	3.007	2.395
$\Delta E_0/T_9^{5/6}$ (58)	3.990E−01	6.533E−01	4.694E−01	7.244E−01	4.697E−01	5.067E−01
$\tau T_9^{1/3}$ (55)(78)	1.206E 01	3.233E 01	1.669E 01	3.976E 01	1.671E 01	1.945E 01
$S(0)$ (53)	9.12	5.48 E 05	1.03 E−02	1.00 E−01	1.20 E−01	5.50 E−02
$S'(0)/S(0)$ (53)	−1.44	2.20	−2.73	—	—	7.64 E−01
$\frac{1}{2}S''(0)/S(0)$ (53)	2.50	—	2.93	—	—	—
Coefficients in typical equations indicated in parentheses						
$T_9^{1/3}$ (78)	3.45 E−02	1.29 E−02	2.50 E−02	1.05 E−02	−3.68[a]	2.14 E−02
$T_9^{2/3}$ (78)	−4.98 E−01	2.04	−1.31	—	—	4.27 E−01
T_9 (78)	−1.21 E−01	1.84 E−01	−2.29 E−01	—	—	6.40 E−02
$T_9^{4/3}$ (78)	3.00 E−01	—	6.73 E−01	—	—	—
$T_9^{5/3}$ (78)	1.85 E−01	—	2.99 E−01	—	—	—
$\langle 01 \rangle$ (82)	2.09 E−13	1.12 E−08	2.73 E−16	2.22 E−15	3.17 E−15	1.56 E−15
$[01]$ (78)	1.26 E 11	6.77 E 15	1.64 E 08	1.34 E 09	1.91 E 09	9.41 E 08
$\langle 23 \rangle$ (83)	1.57 E−13	6.51 E−08	—	—	2.14 E−15	—
$[\lambda_\gamma]$ (80)(83)	9.48 E 10	3.92 E 16	4.98 E 17	7.57 E 19	1.29 E 09	4.36 E 18
$\tau_1(0)$ (79)	8.01 E−12	5.91 E−16	6.14 E−09	2.99 E−09	5.27 E−10	1.07 E−09
$\tau_0(1)$ (79)	7.96 E−11	1.92 E−15	9.74 E−08	1.19 E−08	8.89 E−09	2.12 E−08
$\mathcal{E}_{01}$ (81)	1.38 E 28	2.78 E 32	5.90 E 24	9.55 E 25	1.28 E 26	1.10 E 26
Uncertainty	± 20%	± 30%	± 30%	FAC 3	FAC 2	± 50%
T_9 limits	0 to 5	0 to 1	0 to 2	0 to 0.2	0.006 to 0.02	0 to 2
References	Ja 57	Da 66	Ch Du 61	Fo 64	Br 62b	Ta 59

Units: Energy in MeV; $S(0)$ in MeV-barn, $S'(0)/S(0)$ in MeV⁻¹, $\frac{1}{2}S''(0)/S(0)$ in MeV⁻²; $\langle\ \rangle$ in cm³ sec⁻¹; $[\ ]$ in sec⁻¹; τ in sec; $\mathcal{E}_{01}$ in erg g⁻¹ sec⁻¹; $T_9 = T/10^9$.

[a] Best empirical adjustment.

$$\sigma = \frac{\pi\hbar^2}{2ME} \frac{\omega_r \Gamma_1 \Gamma_2}{(E - E_r)^2 + \Gamma^2/4}$$

$$= \frac{0.6566}{AE} \frac{\omega_r \Gamma_1 \Gamma_2}{(E - E_r)^2 + \Gamma^2/4} \text{ barn } (E \text{ in MeV})$$

61.

where E_r is the resonance energy in the center-of-momentum system for particles $0+1$, Γ_1 is the partial width for the decay of the resonant state by re-emission of $0+1$, Γ_2 is the partial width for emission of $2+3$, $\Gamma = \Gamma_1 + \Gamma_2 + \ldots$ is the sum over all partial widths, and $\omega_r = (1+\delta_{01})g_r/g_0g_1$ with $g_r = 2J_r+1$, J_r being the spin of the resonant state.

In Tables IV and V a given resonance is most directly identified in the literature in terms of the resonance energy in the laboratory system (E_r LAB), but it must be remembered that the resonance energy in the center-of-momentum system (E_r CM) has been used in determining $(E_r/k)_9$ in the tables. A second means of identification is the excitation energy of the state in the compound nucleus of the reaction and this is listed as E_x in Tables IV and V.

It is customary to evaluate $\langle \sigma v \rangle$ for the cross section given by Equation 61 in the approximation that the resonance is sharp; that is, the full width at resonance, Γ_r, is considerably less than the effective spread in energy of the interacting particles. This effective spread in energy is kT when neutrons are involved, ΔE_0 for two charged particles. In this approximation it is a matter of elementary integration to show that

$$\langle \sigma v \rangle = \left(\frac{2\pi\hbar^2}{MkT}\right)^{3/2} \frac{(\omega\gamma)_r}{\hbar} \exp(-E_r/kT)$$

62.

so that

$$\langle 01 \rangle = \{2.557 \times 10^{-13} A^{-3/2}(\omega\gamma)_r\} T_9^{-3/2} \exp(-11.605 \, E_r/T_9) \text{ cm}^3 \text{ sec}^{-1}$$

63.

and

$$[01] = \{1.540 \times 10^{11} A^{-3/2}(\omega\gamma)_r\} \rho T_9^{-3/2} \exp(-11.605 \, E_r/T_9) \text{ sec}^{-1}$$

64.

where $(\omega\gamma)_r = \omega_r\gamma_r = (\omega\Gamma_1\Gamma_2/\Gamma)_r$.

The numerical coefficients listed for $\langle 01 \rangle$ and $[01]$ in Tables IV and V are evaluated by the use of the factors in curly brackets in Equations 63 and 64 respectively. A useful expression for $\lambda_\gamma(2)$ for resonant capture reactions, which can be derived for Equations 16 and 62, is

$$\lambda_\gamma(2) = \frac{g_r\gamma_r}{g_2\hbar} \exp\left(-\frac{Q+E_r}{kT}\right) = \frac{(\omega\gamma)_r}{\omega_2\hbar} \exp\left(-\frac{Q+E_r}{kT}\right) \text{ sec}^{-1}$$

65.

where $\omega_2 = (1+\delta_{01})g_2/g_0g_1$.

In the above equations the quantity $(\omega\gamma)_r$ is evaluated at resonance and can be determined even in a "poor" resolution experiment where the uncertainty in energy is greater than Γ_r, by experimentally determining the integral of the cross section over energy and then using

$$(\omega\gamma)_r = (\pi^2\hbar^2/ME_r)^{-1}\int_r \sigma dE \qquad 66.$$

This expression can be derived easily by using Equation 61 and the definition of $(\omega\gamma)_r$ above. In "good" resolution experiments $(\omega\gamma)_r$ can be determined in terms of the energy E_r, the cross section σ_r, and the full width at half-maximum Γ_r, all as measured at resonance by the use of

$$(\omega\gamma)_r = \frac{\sigma_r M \Gamma_r E_r}{2\pi\hbar^2} = 0.3807 \ \sigma_r A (\Gamma_r E_r)_{\text{CM}} \text{ MeV}$$

$$= 0.3807 \left(\frac{A_0}{A_0 + A_1}\right)^3 \sigma_r A_1 (\Gamma_r E_r)_{\text{LAB}} \text{ MeV} \qquad 67.$$

for σ_r in barns and Γ_r, E_r in MeV. In Tables IV and V, E_r, and $(\omega\gamma)_r$ are always listed and $J_r{}^\pi$, ω_r, σ_r, and Γ_r are listed when known. The superscript π designates the parity ($+$ for even, $-$ for odd) of the resonant state. Along

TABLE IV

Neutron Resonant-Reaction Data

Reaction		$\text{He}^3(n,p)\text{T}^2$	$\text{Li}^6(n,t)\text{He}^4$	$\text{Li}^7(n,\gamma)\text{Li}^8$	$\text{B}^{11}(n,\gamma)\text{B}^{12}$	$\text{N}^{14}(n,p)\text{C}^{14}$	$\text{N}^{14}(n,p)\text{C}^{14}$
Parameters in typical equations indicated in parentheses							
Q	(8)	0.764	4.785	2.033	3.369	0.626	0.626
E_x		22.354	7.471	2.258	3.387	11.297	11.431
E_r LAB	(71)	2.370	0.255	0.258	0.020	0.495	0.639
E_r CM	(64)(67)	1.776	0.218	0.226	0.018	0.462	0.596
$(E_r/k)_9$	(72)	2.061E 01	2.534	2.618	0.213	5.358	6.917
E_r' CM	(68)	2.540	5.004	2.258	3.387	1.088	1.222
$(E_r'/k)_9$	(73)	2.947E 01	5.807E 01	2.621E 01	3.931E 01	1.263E 01	1.419E 01
$J_r{}^\pi, l_r$		2^-, 1	$5/2^-$, 1	3^+, 1	(3^+), (1)	$1/2^-$, 1	$1/2^+$, 0
ω_r	(64)	1.250	1.000	0.875	0.875	0.333	0.333
σ_r	(67)(70)	4.40 $E-$01	2.60	5.004	—	3.94 $E-$01	2.58 $E-$01
Γ_r	(67)	2.45	1.02 $E-$01	3.06 $E-$02	—	7.00 $E-$03	4.01 $E-$02
$(\omega\gamma)_r$	(64)(65)	5.51 $E-$01	1.90 $E-$02	5.36 $E-$08	3.66 $E-$07	4.56 $E-$04	2.21 $E-$03
v_r	(71)	2.13 E 09	6.98 E 08	7.03 E 08	1.96 E 08	9.73 E 08	1.11 E 09
S_r	(70)	9.37 $E-$16	1.82 $E-$15	1.62 $E-$20	—	3.84 $E-$16	2.86 $E-$16
$S(0)$	(69)	0	0	0	0	0	3.23 $E-$19
Coefficients in typical equations indicated in parentheses							
$\langle 01\rangle$	(72)	2.14 $E-$13	6.07 $E-$15	1.65 $E-$20	1.05 $E-$19	1.28 $E-$16	6.20 $E-$16
$[01]$	(72)	1.29 E 11	3.65 E 09	9.96 E 03	6.35 E 04	7.69 E 07	3.73 E 08
$\langle 23\rangle$	(73)	2.15 $E-$13	6.49 $E-$15	—	—	3.84 $E-$16	1.86 $E-$15
$[\lambda_\gamma]$	(73)	1.29 E 11	3.91 E 09	1.30 E 14	1.49 E 15	2.31 E 08	1.12 E 09
$\tau_1(0)$	(74)	7.81 $E-$12	2.76 $E-$10	1.01 $E-$04	1.59 $E-$05	1.31 $E-$08	2.70 $E-$09
$\tau_0(1)$	(74)	2.34 $E-$11	1.65 $E-$09	7.05 $E-$04	1.73 $E-$04	1.82 $E-$07	3.75 $E-$08
$\mathcal{E}_{01}$	(75)	3.13 E 28	2.78 E 27	2.76 E 21	1.86 E 22	3.29 E 24	1.60 E 25
Uncertainty		$\pm 10\%$	$\pm 10\%$	$\pm 50\%$	$\pm 50\%$	$\pm 50\%$	$\pm 50\%$
T_9 limits		3 to 10	0.4 to 10	1 to 5	0.01 to 3	0.5 to 5	0.5 to 5
References		Hu 58	Hu 60	Hu 60, La 66	Aj 67	Aj 59	Aj 59

Units: Energy in MeV; σ_r in barns; v_r in cm sec^{-1}; (S) and $\langle\ \rangle$ in cm^3 sec^{-1}; $[\ ]$ in sec^{-1}; τ in sec; $\mathcal{E}_{01}$ in erg g^{-1} sec^{-1}. $T_9 = T/10^9$.

TABLE V

Charged-Particle Resonant-Reaction Data

Reaction		$T^3(d,n)He^4$	$He^3(d,p)He^4$	$Li^7(p,\gamma)Be^8$	$Li^7(\alpha,\gamma)B^{11}$	$Li^7(\alpha,\gamma)B^{11}$	$Li^7(\alpha,\gamma)B^{11}$
Parameters in typical equations indicated in parentheses							
Q	(8)	17.590	18.354	17.252	8.664	8.664	8.664
E_x		16.709	16.672	17.638	8.920	9.186	9.274
E_r LAB		0.128[a]	0.474[b]	0.441	0.401	0.819	0.958
E_r CM	(64)(77)	0.077	0.284	0.386	0.255	0.522	0.610
$(E_r/k)_9$	(78)	0.891	3.297	4.479	2.963	6.052	7.079
E_r' CM	(68)	17.666	18.638	17.638	8.920	9.186	9.274
$(E_r'/k)_9$	(80)(83)	2.050E 02	2.163E 02	2.047E 02	1.035E 02	1.066E 02	1.076E 02
J_r^π, l_r		$3/2^+$, 0	$3/2^+$, 0	1^+, 1	$5/2^-$, 2	$7/2^+$, 3	$5/2^+$, 1
ω_r	(64)	0.667	0.667	0.375	1.500	2.000	1.500
σ_r	(67)(77)	5.00	7.10 E−01	6.00 E−03	4.54 E−02	3.51 E−01	1.31 E−03
Γ_r	(67)	5.90 E−02	2.40 E−01	1.07 E−02	4.00 E−06	3.10 E−06	4.50 E−03
$(\omega\gamma)_r$	(64)(65)	1.04 E−02	2.23 E−02	8.29 E−06	4.50 E−08	5.50 E−07	3.50 E−06
S_r	(77)	1.94 E 01	1.19 E 01	2.05 E−01	1.62 E 06	9.15 E 04	1.49 E 02
$S(0)$	(76)	2.50	1.81	3.92 E−05	9.93 E−05	8.08 E−07	2.03 E−03
Coefficients in typical equations indicated in parentheses							
$\langle 01 \rangle$	(82)	2.01 E−15	4.29 E−15	2.56 E−18	2.83 E−21	3.46 E−20	2.20 E−19
$[01]$	(78)	1.21 E 09	2.59 E 09	1.54 E 06	1.70 E 03	2.08 E 04	1.33 E 05
$\langle 23 \rangle$	(83)	1.11 E−14	2.38 E−14	—	—	—	—
$[\lambda_\gamma]$	(80)(83)	6.69 E 09	1.43 E 10	1.01 E 17	6.85 E 13	8.37 E 14	5.32 E 15
$\tau_1(0)$	(79)	1.67 E−09	7.79 E−10	6.53 E−07	2.35 E−03	1.92 E−04	3.02 E−05
$\tau_0(1)$	(79)	2.50 E−09	1.17 E−09	4.55 E−06	4.12 E−03	3.37 E−04	5.30 E−05
$\mathcal{E}_{01}$	(81)	3.38 E 27	7.54 E 27	3.63 E 24	5.07 E 20	6.20 E 21	3.94 E 22
Uncertainty		+20% to −65%	+30% to −50%	±10%	±20%	±20%	±20%
T_9 limits		≥0.15	≥0.65	≥0.1	≥0.1	≥0.5	≥0.5
References		Ja 57	Ja 57	La 66	Aj 67	Aj 67	Aj 67

Reaction		$Be^7(p,\gamma)B^8$	$Be^7(\alpha,\gamma)C^{11}$	$Be^7(\alpha,\gamma)C^{11}$	$Be^9(p,d)Be^8$	$Be^9(p,d)Be^8$	$Be^9(p,d)Be^8$
Parameters in typical equations indicated in parentheses							
Q	(8)	0.135	7.545	7.545	0.559	0.559	0.559
E_x		0.778	8.105	8.427	6.884	7.433	8.071
E_r LAB		0.735	0.879	1.385	0.330	0.940	1.650
E_r CM	(64)(77)	0.643	0.560	0.882	0.297	0.845	1.484
$(E_r/k)_9$	(78)	7.462	6.499	1.024E 01	3.445	9.812	1.722E 01
E_r' CM	(68)	0.778	8.105	8.427	0.856	1.405	2.043
$(E_r'/k)_9$	(80)(83)	9.026	9.406E 01	9.780E 01	9.936	1.630E 01	2.371E 01
J_r^π, l_r		(1^+), (1)	$(3/2^-)$, (0)	$5/2^-$, 2	1^-, 0	2^-, 0	2^-, 0
ω_r	(64)	0.375	1.000	1.500	0.375	0.625	0.625
σ_r	(67)(77)	2.21 E−06	7.07 E−02	2.80 E−01	4.56 E−01	2.02 E−01	7.69 E−02
Γ_r	(67)	3.67 E−02	5.00 E−05	2.00 E−05	1.40 E−01	1.40 E−01	8.00 E−01
$(\omega\gamma)_r$	(64)(65)	1.75 E−08	1.92 E−06	4.80 E−06	6.53 E−03	8.23 E−03	3.15 E−02
S_r	(77)	1.46 E−04	8.52 E 05	1.72 E 05	1.36 E 02	1.03 E 01	2.51
$S(0)$	(76)	1.19 E−07	1.70 E−03	2.22 E−05	7.18	6.98 E−02	1.70 E−01
Coefficients in typical equations indicated in parentheses							
$\langle 01 \rangle$	(82)	5.41 E−21	1.21 E−19	3.02 E−19	1.94 E−15	2.44 E−15	9.33 E−15
$[01]$	(78)	3.26 E 03	7.27 E 04	1.82 E 05	1.17 E 09	1.47 E 09	5.62 E 09
$\langle 23 \rangle$	(83)	—	—	—	2.18 E−15	2.75 E−15	1.05 E−14
$[\lambda_\gamma]$	(80)(83)	4.25 E 13	2.92 E 15	7.30 E 15	1.31 E 09	1.66 E 09	6.33 E 09
$\tau_1(0)$	(79)	3.09 E−04	5.51 E−05	2.20 E−05	8.64 E−10	6.86 E−11	1.79 E−10
$\tau_0(1)$	(79)	2.15 E−03	9.65 E−05	3.86 E−05	7.73 E−09	6.13 E−09	1.60 E−09
$\mathcal{E}_{01}$	(81)	6.00 E 19	1.88 E 22	4.71 E 22	6.93 E 25	8.73 E 25	3.34 E 26
Uncertainty		±20%	FAC 3	FAC 3	±20%	±20%	±50%
T_9 limits		≥1	≥0.1	≥1	≥0.2	≥2	≥3
References		Pa 66, La 66	Aj 67	Aj 67	Mo 56	La 66	La 66

Units: Energy in MeV; σ_r in barns; S in MeV-barn; $\langle\ \rangle$ in cm³ sec⁻¹; [ ] in sec⁻¹; τ in sec; $\mathcal{E}_{01}$ in erg g⁻¹ sec⁻¹; $T_9 = T/10^9$.
[a] Tabulated in La 66 as E_r LAB = 0.107 MeV.   [b] Tabulated in La 66 as E_r LAB = 0.430 MeV.

TABLE V (continued)

CHARGED-PARTICLE RESONANT-REACTION DATA

Reaction		$Be^9(p,\alpha)Li^6$	$Be^9(p,\alpha)Li^6$	$Be^9(\alpha,n)C^{12}$	$Be^9(\alpha,n)C^{12}$	$Be^9(\alpha,n)C^{12}$	$Be^9(\alpha,n)C^{12}$
		Parameters in typical equations indicated in parentheses					
Q	(8)	2.126	2.126	5.704	5.704	5.704	5.704
E_x		6.884	7.433	11.011	11.066	11.980	12.209
E_r LAB		0.330	0.940	0.520	0.600	1.920	2.250
E_r CM	(64)(77)	0.297	0.845	0.360	0.415	1.330	1.558
$(E_r/k)_9$	(78)	3.445	9.812	4.179	4.822	$1.543E\ 01$	$1.808E\ 01$
E_r' CM	(68)	2.423	2.972	6.064	6.119	7.033	7.262
$(E_r'/k)_9$	(80)(83)	$2.812E\ 01$	$3.449E\ 01$	$7.037E\ 01$	$7.101E\ 01$	$8.162E\ 01$	$8.427E\ 01$
J_r^π, l_r		$1^-, 0$	$2^-, 0$	$(1/2^+), (1)$	—	$(7/2^-), (2)$	$(7/2^-), (2)$
ω_r	(64)	0.375	0.625	0.500	—	2.000	2.000
σ_r	(67)(77)	$3.52\ E-01$	$2.49\ E-01$	$1.80\ E-04$	—	$2.90\ E-01$	$1.75\ E-01$
Γ_r	(67)	$1.40\ E-01$	$1.40\ E-01$	$5.54\ E-02$	—	$1.38\ E-01$	$2.77\ E-01$
$(\omega\gamma)_r$	(64)(65)	$5.05\ E-03$	$1.02\ E-02$	$3.79\ E-06$	$8.60\ E-07$	$5.63\ E-02$	$7.97\ E-02$
S_r	(77)	$1.05\ E\ 02$	$1.27\ E\ 01$	$2.23\ E\ 05$	—	$3.54\ E\ 04$	$1.05\ E\ 04$
$S(0)$	(76)	5.55	$8.64\ E-02$	$1.31\ E\ 03$	—	$9.58\ E\ 01$	$8.21\ E\ 01$
		Coefficients in typical equations indicated in parentheses					
$\langle 01 \rangle$	(82)	$1.50\ E-15$	$3.02\ E-15$	$2.10\ E-19$	$4.77\ E-20$	$3.12\ E-15$	$4.42\ E-15$
$[01]$	(78)	$9.02\ E\ 08$	$1.82\ E\ 09$	$1.26\ E\ 05$	$2.87\ E\ 04$	$1.88\ E\ 09$	$2.66\ E\ 09$
$\langle 23 \rangle$	(83)	$9.25\ E-16$	$1.86\ E-15$	$2.16\ E-18$	$4.90\ E-19$	$3.21\ E-14$	$4.54\ E-14$
$[\lambda\gamma]$	(80)(83)	$5.57\ E\ 08$	$1.12\ E\ 09$	$1.30\ E\ 06$	$2.95\ E\ 05$	$1.93\ E\ 10$	$2.74\ E\ 10$
$\tau_1(0)$	(79)	$1.12\ E-09$	$5.54\ E-10$	$3.17\ E-05$	$1.39\ E-04$	$2.13\ E-09$	$1.50\ E-09$
$\tau_0(1)$	(79)	$9.99\ E-09$	$4.96\ E-09$	$7.13\ E-05$	$3.14\ E-04$	$4.79\ E-09$	$3.39\ E-09$
$\mathcal{E}_{01}$	(81)	$2.04\ E\ 26$	$4.11\ E\ 26$	$1.93\ E\ 22$	$4.38\ E\ 21$	$2.87\ E\ 26$	$4.06\ E\ 26$
Uncertainty		$\pm 20\%$	$\pm 20\%$	$\pm 10\%$	$\pm 10\%$	$\pm 30\%$	$\pm 30\%$
T_9 limits		≥ 0.2	≥ 2	≥ 0.5	≥ 1	≥ 1	≥ 2
References		Mo 56	La 66	Da 66	Da 66	Gi 65	Gi 65

Reaction		$B^{10}(p,\alpha)Be^7$	$B^{10}(p,\alpha)Be^7$	$B^{11}(p,\gamma)C^{12}$	$B^{11}(p,\gamma)C^{12}$	$B^{11}(p,\gamma)C^{12}$	$B^{11}(p,2\alpha)He^4$
		Parameters in typical equations indicated in parentheses					
Q	(8)	1.148	1.148	15.957	15.957	15.957	8.682
E_x		9.756	10.086	16.106	16.575	17.228	16.106
E_r LAB		1.170	1.533	0.163	0.675	1.388	0.163
E_r CM	(64)(77)	1.063	1.393	0.149	0.618	1.272	0.149
$(E_r/k)_9$	(78)	$1.234E\ 01$	$1.616E\ 01$	1.733	7.177	$1.476E\ 01$	1.733
E_r' CM	(68)	2.211	2.540	16.106	16.575	17.228	8.832
$(E_r'/k)_9$	(80)(83)	$2.565E\ 01$	$2.948E\ 01$	$1.869E\ 02$	$1.924E\ 02$	$1.999E\ 02$	$1.025E\ 02$
J_r^π, l_r		$(5/2^+), (0)$	$7/2^+, 0$	$2^+, 1$	$2^-, 0$	$1^-, 0$	$2^+, 1$
ω_r	(64)	0.429	0.571	0.625	0.625	0.375	0.625
σ_r	(67)(77)	$2.05\ E-01$	$3.00\ E-01$	$1.58\ E-04$	$5.00\ E-05$	$5.30\ E-05$	$1.02\ E-02$
Γ_r	(67)	$2.73\ E-01$	$2.27\ E-01$	$6.41\ E-03$	$2.95\ E-01$	1.16	$6.41\ E-03$
$(\omega\gamma)_r$	(64)(65)	$2.07\ E-02$	$3.31\ E-02$	$5.30\ E-08$	$3.21\ E-06$	$2.76\ E-05$	$3.43\ E-06$
S_r	(77)	$2.15\ E\ 01$	$2.31\ E\ 01$	5.16	$1.30\ E-02$	$4.56\ E-03$	$3.34\ E\ 02$
$S(0)$	(76)	$3.47\ E-01$	$1.52\ E-01$	$2.38\ E-03$	$7.01\ E-04$	$7.89\ E-04$	$1.54\ E-01$
		Coefficients in typical equations indicated in parentheses					
$\langle 01 \rangle$	(82)	$6.04\ E-15$	$9.66\ E-15$	$1.53\ E-20$	$9.24\ E-19$	$7.95\ E-18$	$9.90\ E-19$
$[01]$	(78)	$3.64\ E\ 09$	$5.82\ E\ 09$	$9.20\ E\ 03$	$5.57\ E\ 05$	$4.79\ E\ 06$	$5.96\ E\ 05$
$\langle 23 \rangle$	(83)	$4.55\ E-15$	$7.28\ E-15$	—	—	—	—
$[\lambda\gamma]$	(80)(83)	$2.74\ E\ 09$	$4.38\ E\ 09$	$6.46\ E\ 14$	$3.90\ E\ 16$	$3.36\ E\ 17$	$2.09\ E-04$
$\tau_1(0)$	(79)	$2.77\ E-10$	$1.73\ E-10$	$1.09\ E-04$	$1.81\ E-06$	$2.11\ E-07$	$1.69\ E-06$
$\tau_0(1)$	(79)	$2.75\ E-09$	$1.72\ E-09$	$1.20\ E-03$	$1.98\ E-05$	$2.30\ E-06$	$1.85\ E-05$
$\mathcal{E}_{01}$	(81)	$3.99\ E\ 26$	$6.38\ E\ 26$	$1.28\ E\ 22$	$7.72\ E\ 23$	$6.64\ E\ 24$	$4.50\ E\ 23$
Uncertainty		$\pm 20\%$	$\pm 20\%$	$\pm 10\%$	$\pm 20\%$	$\pm 20\%$	$\pm 10\%$
T_9 limits		≥ 1	≥ 2	≥ 0.03	≥ 1	≥ 2	≥ 0.2
References		Aj 59	Aj 59	Aj 59	Aj 59	Aj 59	Aj 59

Units: Energy in MeV; σ_r in barns; S in MeV-barn; $\langle\ \rangle$ in cm³ sec⁻¹; [ ] in sec⁻¹; τ in sec; $\mathcal{E}_{01}$ in erg g⁻¹ sec⁻¹; $T_9 = T/10^9$.

TABLE V (continued)

CHARGED-PARTICLE RESONANT-REACTION DATA

Reaction		$B^{11}(p,2\alpha)He^4$	$B^{11}(p,2\alpha)He^4$	$C^{12}(p,\gamma)N^{13}$	$C^{13}(p,\gamma)N^{14}$	$C^{13}(p,\gamma)N^{14}$	$Ne^{20}(p,\gamma)Na^{21}$
Parameters in typical equations indicated in parentheses							
Q	(8)	8.682	8.682	1.944	7.550	7.550	2.432
E_x		16.575	17.228	2.368	8.065	8.710	2.810
E_r LAB		0.675	1.388	0.460	0.555	1.250	0.397
E_r CM	(64)(77)	0.618	1.272	0.424	0.515	1.160	0.378
$(E_r/k)_9$	(78)	7.177	1.476E 01	4.925	5.978	1.346E 01	4.381
E_r' CM	(68)	9.301	9.954	2.368	8.065	8.710	2.810
$(E_r'/k)_9$	(80)(83)	1.079E 02	1.155E 02	2.748E 01	9.359E 01	1.011E 02	3.261E 01
$J_r{}^\pi, l_r$		2⁻, 0	1⁻, 0	1/2⁺, 0	1⁻, 0	0⁻, 0	(1/2⁺), (0)
ω_r	(64)	0.625	0.375	1.000	0.750	0.250	1.000
σ_r	(67)(77)	6.00 E−01	1.56 E−01	1.29 E−04	1.07 E−03	6.30 E−05	—
Γ_r	(67)	2.95 E−01	1.16	3.25 E−02	4.05 E−02	5.10 E−01	—
$(\omega\gamma)_r$	(64)(65)	3.85 E−02	8.11 E−02	6.29 E−07	7.94 E−06	1.33 E−05	1.00 E−06
S_r	(77)	1.56 E 02	1.34 E 01	3.58 E−01	1.64	1.51 E−02	—
$S(0)$	(76)	8.42	2.32	5.23 E−04	2.53 E−03	6.96 E−04	—
Coefficients in typical equations indicated in parentheses							
$\langle 01 \rangle$	(82)	1.11 E−14	2.34 E−14	1.80 E−19	2.25 E−18	3.76 E−18	2.72 E−19
$[01]$	(78)	6.68 E 09	1.41 E 10	1.08 E 05	1.35 E 06	2.26 E 06	1.64 E 05
(23)	(83)	—	—	—	—	—	—
$[\lambda\gamma]$	(80)(83)	2.34	4.93	9.56 E 14	1.61 E 16	2.69 E 16	7.60 E 14
$\tau_1(0)$	(79)	1.51 E−10	7.16 E−11	9.32 E−06	7.45 E−07	4.46 E−07	6.15 E−06
$\tau_0(1)$	(79)	1.65 E−09	7.82 E−10	1.11 E−04	9.62 E−06	5.75 E−06	1.22 E−04
$\mathcal{E}_{01}$	(81)	5.04 E 27	1.06 E 28	1.68 E 22	7.52 E 23	1.26 E 24	1.91 E 22
Uncertainty		± 20%	± 20%	−20% to +10%	−20% to +10%	+10% to −10%	FAC 3
T_9 limits		≥0.5	≥2	0.25 to 7	0.3 to 10	2.5 to 10	0.1 to 4
References		Aj 59	Aj 59	Vo 63	Vo 63	Se 52	En 62

Reaction		$N^{14}(p,\gamma)O^{15}$	$N^{14}(p,\gamma)O^{15}$	$N^{14}(p,\gamma)O^{15}$	$C^{13}(\alpha,n)O^{16}$	$C^{13}(\alpha,n)O^{16}$	$C^{13}(\alpha,n)O^{16}$
Parameters in typical equations indicated in parentheses							
Q	(8)	7.293	7.293	7.293	2.214	2.214	2.214
E_x		7.552	8.282	9.718	7.170	7.351	7.382
E_r LAB		0.278	1.060	2.600	1.063	1.300	1.340
E_r CM	(64)(77)	0.259	0.989	2.425	0.813	0.994	1.025
$(E_r/k)_9$	(78)	3.010	1.148E 01	2.815E 01	9.433	1.154E 01	1.189E 01
E_r' CM	(68)	7.552	8.282	9.718	3.027	3.208	3.239
$(E_r'/k)_9$	(80)(83)	8.764E 01	9.611E 01	1.128E 02	3.513E 01	3.723E 01	3.759E 01
$J_r{}^\pi, l_r$		1/2⁺, 0	3/2⁺, 0	—	5/2	—	5/2
ω_r	(64)	0.333	0.667	—	3.000	—	3.000
σ_r	(67)(77)	9.17 E−05	3.73 E−04	2.87 E−05	2.88 E−02	2.51 E−03	6.92 E−02
Γ_r	(67)	1.59 E−03	2.80 E−03	1.21	2.70 E−03	2.29 E−01	2.00 E−03
$(\omega\gamma)_r$	(64)(65)	1.35 E−08	3.69 E−07	3.01 E−05	7.37 E−05	6.68 E−04	1.65 E−04
S_r	(77)	1.27 E 01	3.15 E−01	5.20 E−03	2.37 E 08	2.78 E 06	5.77 E 07
$S(0)$	(76)	1.18 E−04	6.32 E−07	3.03 E−04	6.54 E 02	3.66 E 04	5.50 E 01
Coefficients in typical equations indicated in parentheses							
$\langle 01 \rangle$	(82)	3.79 E−21	1.04 E−19	8.46 E−18	3.52 E−18	3.19 E−17	7.89 E−18
$[01]$	(78)	2.28 E 03	6.24 E 04	5.09 E 06	2.12 E 06	1.92 E 07	4.75 E 06
(23)	(83)	—	—	—	2.04 E−17	1.85 E−16	4.57 E−17
$[\lambda\gamma]$	(80)(83)	6.16 E 13	1.68 E 15	1.37 E 17	1.23 E 07	1.11 E 08	2.75 E 07
$\tau_1(0)$	(79)	4.42 E−04	1.62 E−05	1.98 E−07	1.89 E−06	2.08 E−07	8.43 E−07
$\tau_0(1)$	(79)	6.14 E−03	2.24 E−04	2.75 E−06	6.14 E−06	6.77 E−07	2.74 E−06
$\mathcal{E}_{01}$	(81)	1.14 E 21	3.11 E 22	2.54 E 24	8.70 E 22	7.88 E 23	1.95 E 23
Uncertainty		± 20%	± 20%	± 30%	± 30%	± 30%	± 30%
T_9 limits		≥0.2	≥3	≥3	≥1	≥1	≥3
References		He 63	He 63	Ca 65	Wa 57	Wa 57	Wa 57

Units: Energy in MeV; σ_r in barns; S in MeV-barn; $\langle\ \rangle$ in cm³ sec⁻¹; $[\ ]$ in sec⁻¹; τ in sec; $\mathcal{E}_{01}$ in erg g⁻¹ sec⁻¹; $T_9 = T/10^9$.

TABLE V (concluded)

CHARGED-PARTICLE RESONANT-REACTION DATA

Reaction		$N^{15}(p,\gamma)O^{16}$	$N^{15}(p,\gamma)O^{16}$	$N^{15}(p,\alpha)C^{12}$	$N^{15}(p,\alpha)C^{12}$	$N^{15}(p,\alpha)C^{12}$	$O^{17}(p,\alpha)N^{14}$
Parameters in typical equations indicated in parentheses							
Q	(8)	12.126	12.126	4.965	4.965	4.965	1.193
E_x		12.443	13.072	12.443	13.072	13.260	5.674
E_r LAB		0.338	1.010	0.338	1.010	1.210	0.069
E_r CM	(64)(77)	0.317	0.946	0.317	0.946	1.134	0.065
$(E_r/k)_9$	(78)	3.676	1.098E 01	3.676	1.098E 01	1.316E 01	0.754
E_r' CM	(68)	12.443	13.072	5.281	5.911	6.098	1.258
$(E_r'/k)_9$	(80)(83)	1.444E 02	1.517E 02	6.129E 01	6.860E 01	7.077E 01	1.460E 01
J_r^π, l_r		$1^-,0$	$1^-,0$	$1^-,0$	$1^-,0$	$3^-,2$	$1^-,1$
ω_r	(64)	0.750	0.750	0.750	0.750	1.750	0.250
σ_r	(67)(77)	6.91 $E-06$	9.83 $E-04$	7.66 $E-02$	4.58 $E-01$	6.87 $E-01$	7.51 $E-09$
Γ_r	(67)	8.82 $E-02$	1.31 $E-01$	8.82 $E-02$	1.31 $E-01$	2.11 $E-02$	2.68 $E-04$
$(\omega\gamma)_r$	(64)(65)	6.93 $E-08$	4.39 $E-05$	7.68 $E-04$	2.04 $E-02$	5.90 $E-03$	4.74 $E-14$
S_r	(77)	3.41 $E-01$	9.39 $E-01$	3.78 E 03	4.37 E 02	4.33 E 02	6.88 E 03
$S(0)$	(76)	6.48 $E-03$	4.49 $E-03$	7.19 E 01	2.09	3.74 $E-02$	2.92 $E-02$
Coefficients in typical equations indicated in parentheses							
$\langle 01 \rangle$	(82)	1.93 $E-20$	1.22 $E-17$	2.14 $E-16$	5.70 $E-15$	1.65 $E-15$	1.31 $E-26$
$[01]$	(78)	1.16 E 04	7.36 E 06	1.29 E 08	3.43 E 09	9.91 E 08	7.86 $E-03$
$\langle 23 \rangle$	(83)	—	—	1.51 $E-16$	4.02 $E-15$	1.16 $E-15$	8.82 $E-27$
$[\lambda_\gamma]$	(80)(83)	4.22 E 14	2.67 E 17	9.11 E 07	2.42 E 09	7.00 E 08	5.31 $E-03$
$\tau_1(0)$	(79)	8.66 $E-05$	1.37 $E-07$	7.81 $E-09$	2.94 $E-10$	1.02 $E-09$	1.28 E 02
$\tau_0(1)$	(79)	1.29 $E-03$	2.04 $E-06$	1.16 $E-07$	4.37 $E-09$	1.51 $E-08$	2.16 E 03
$\mathcal{E}_{01}$	(81)	9.01 E 21	5.70 E 24	4.09 E 25	1.09 E 27	3.14 E 26	5.28 E 14
Uncertainty		$\pm 20\%$	$\pm 20\%$	$\pm 20\%$	$\pm 20\%$	$\pm 20\%$	$\pm 30\%$
T_9 limits		≥ 0.2	≥ 1	≥ 0.2	≥ 1	≥ 3	0.01 to 0.2
References		He 60	He 60	He 60	He 60	He 60	Br 62b

Reaction		$C^{12}(\alpha,\gamma)O^{16}$	$C^{12}(\alpha,\gamma)O^{16}$	$C^{12}(\alpha,\gamma)O^{16}$	$O^{16}(\alpha,\gamma)Ne^{20}$	$O^{16}(\alpha,\gamma)Ne^{20}$	$O^{16}(\alpha,\gamma)Ne^{20}$
Parameters in typical equations indicated in parentheses							
Q	(8)	7.161	7.161	7.161	4.730	4.730	4.730
E_x		9.583	9.850	10.356	5.622	5.785	6.721
E_r LAB		3.229	3.585	4.260	1.116	1.319	2.490
E_r CM	(64)(77)	2.422	2.688	3.194	0.893	1.055	1.992
$(E_r/k)_9$	(78)	2.810E 01	3.120E 01	3.707E 01	1.036E 01	1.224E 01	2.311E 01
E_r' CM	(68)	9.583	9.850	10.356	5.622	5.785	6.721
$(E_r'/k)_9$	(80)(83)	1.112E 02	1.143E 02	1.202E 02	6.525E 01	6.713E 01	7.800E 01
J_r^π, l_r		$1^-,1$	$2^+,2$	$4^+,4$	$3^-,3$	$1^-,1$	$0^+,0$
ω_r	(64)	3.000	5.000	9.000	7.000	3.000	1.000
σ_r	(67)(77)	3.70 $E-08$	1.54 $E-05$	4.20 $E-06$	4.60 $E-01$	7.18 $E-04$	7.16 $E-07$
Γ_r	(67)	6.45 $E-01$	7.50 $E-04$	2.70 $E-02$	2.24 $E-09$	1.04 $E-05$	1.52 $E-02$
$(\omega\gamma)_r$	(64)(65)	6.60 $E-08$	3.55 $E-08$	4.14 $E-07$	1.12 $E-09$	9.60 $E-09$	2.64 $E-08$
S_r	(77)	4.92 $E-02$	1.16 E 01	1.34	4.29 E 12	7.16 E 08	7.41 E 02
$S(0)$	(76)	8.58 $E-04$	2.26 $E-07$	2.38 $E-05$	6.75 $E-06$	1.74 $E-02$	1.08 $E-02$
Coefficients in typical equations indicated in parentheses							
$\langle 01 \rangle$	(82)	3.25 $E-21$	1.75 $E-21$	2.04 $E-20$	5.00 $E-23$	4.29 $E-22$	1.18 $E-21$
$[01]$	(78)	1.96 E 03	1.05 E 03	1.23 E 04	3.01 E 01	2.58 E 02	7.10 E 02
$\langle 23 \rangle$	(83)	—	—	—			
$[\lambda_\gamma]$	(80)(83)	1.00 E 14	5.40 E 13	6.29 E 14	1.70 E 12	1.46 E 13	4.01 E 13
$\tau_1(0)$	(79)	2.05 $E-03$	3.81 $E-03$	3.26 $E-04$	1.33 $E-01$	1.55 $E-02$	5.64 $E-03$
$\tau_0(1)$	(79)	6.14 $E-03$	1.14 $E-02$	9.79 $E-04$	5.31 $E-01$	6.20 $E-02$	2.25 $E-02$
$\mathcal{E}_{01}$	(81)	2.81 E 20	1.51 E 20	1.76 E 21	2.15 E 18	1.84 E 19	5.06 E 19
Uncertainty		$\pm 30\%$	$\pm 20\%$	$\pm 20\%$	$\pm 30\%$	$\pm 30\%$	$\pm 30\%$
T_9 limits		2 to 6	2 to 6	2 to 6	≥ 0.2	≥ 0.3	≥ 3
References		La 64	La 64	La 64	Va 65	Va 65	Va 65

Units: Energy in MeV; σ_r in barns; S in MeV-barn; $\langle \ \rangle$ in cm³ sec⁻¹; $[\ ]$ in sec⁻¹; τ in sec; $\mathcal{E}_{01}$ in erg g⁻¹ sec⁻¹; $T_9 = T/10^9$.

with J_r^π we list l_r, the minimum orbital angular momentum in units $\hbar$ with which the interacting nuclei $0+1$ can form the resonant state in a manner consistent with the conservation of angular momentum and parity. When placed in parentheses the values tabulated are uncertain.

When used in Equations 62 to 64, the quantity $(\omega\gamma)_r$ must be evaluated in center-of-momentum coordinates. Thus, in the first form of Equation 67, Γ_r and E_r are the values in these coordinates as indicated. In actual measurements Γ_r and E_r are initially determined in laboratory coordinates in which case the second form of Equation 67 can be employed. Frequently the factor $[A_0/(A_0+A_1)]^3$ is taken as equal to unity. As measurements become more precise, we hope that this practice will be abandoned.

It will be clear from Equations 10, 16, 21, and 62 that the reaction rates for reverse reactions $\langle 23 \rangle$, [23], $\lambda_\gamma(2)$, and [234] will contain an exponential factor with exponent $-E_r'/kT$ where

$$E_r' = Q + E_r \qquad\qquad 68.$$

is the resonance energy in center-of-momentum coordinates in the reverse system $(2+3, 2+\gamma,$ or $2+3+4)$. The quantities E_r' CM and $(E_r'/k)_9$ are also listed in Tables IV and V.

It is frequently of interest to know the contribution made by the wing of a given resonance to the low-energy off-resonance region. In some cases the nonresonant reaction rate can be identified with the contribution from the wings of one or more specific resonances. An accurate calculation requires the use of the most sophisticated nuclear theory (Lane & Thomas 1958 or Humblet & Rosenfeld 1961), but a useful approximation for $E_r > E$ can be obtained through the use of Equation 61. It is assumed that $\Gamma_2 = \Gamma_2 (Q+E)$ does not vary rapidly with the energy E of the interacting nuclei $0+1$ on the basis that Q is positive for $0+1 \rightarrow 2+3$ and is in general larger than E in the nonresonant range or at resonance E_r. Thus Γ_2 can be evaluated at resonance. The same assumption is made for any other partial width except Γ_1 which is small at low energies and in most cases does not make a contribution to Γ comparable to E_r. In other words we are essentially assuming that the resonance is sharp. Thus Γ can be evaluated at resonance Γ_r. With these approximations it is easily shown for s-wave $(l_r = 0)$ neutron reactions that

$$\frac{\mathcal{S}(0)}{\mathcal{S}_r} = \frac{\Gamma_r^2/4}{E_r^2 + \Gamma_r^2/4} \qquad\qquad 69.$$

where

$$\mathcal{S}_r = \sigma_r v_r \qquad\qquad 70.$$

and

$$v_r = 1.383 \times 10^9 \, (E_r \text{ LAB})^{1/2} \text{ cm sec}^{-1} \qquad\qquad 71.$$

For $l_r > 0$ the neutron partial width includes the factor E^{l_r} and thus $\mathcal{S}(0) = 0$ and at low energies $\mathcal{S}$ is negligible. Table IV lists v_r, $\mathcal{S}_r$, and $\mathcal{S}(0)$. An interesting example is the case of the reaction $N^{14}(n,p)C^{14}$. In this case the $l_r = 0$ resonance with neutron laboratory energy equal to 0.639 MeV makes a contribution at low energy given approximately by $S(0) = 3.23 \times 10^{-19}$ cm^3 sec^{-1} while the accurately determined thermal-neutron cross section and velocity yield $S(0) = 3.98 \times 10^{-19}$ cm^3 sec^{-1}. To within the accuracy of our approximation the thermal-reaction rate can be attributed entirely to the low-energy wing of the 0.639-MeV resonance.

From Tables II and IV one has the following examples[6]

$$[He^3n]_p = 6.0225 \times 10^{23} \, \rho \langle He^3n \rangle_p$$
$$= 7.06 \times 10^8 \, \rho(1 - 0.597 \, T_9^{1/2} + 0.183 \, T_9) \quad 0 \leq T_9 \leq 10 \qquad 72.$$
$$+ 1.29 \times 10^{11} \, \rho T_9^{-3/2} \exp(-20.61/T_9) \text{ sec}^{-1} \quad 3 \leq T_9 \leq 10$$

$$[pT^3]_n = 6.0225 \times 10^{23} \, \rho \langle pT^3 \rangle_n$$
$$= 7.07 \times 10^8 \, \rho(1 - 0.597 \, T_9^{1/2} + 0.183 \, T_9) \exp(-8.864/T_9) \quad 0 \leq T_9 \leq 10 \quad 73.$$
$$+ 1.29 \times 10^{11} \, \rho T_9^{-3/2} \exp(-29.47/T_9) \text{ sec}^{-1} \quad 3 \leq T_9 \leq 10$$

$$\frac{1}{\tau_{np}(He^3)} = \frac{1}{1.43 \times 10^{-9}} \rho X_n \, (1 - 0.597 \, T_9^{1/2} + 0.183 \, T_9) \quad 0 \leq T_9 \leq 10$$
$$+ \frac{1}{7.81 \times 10^{-12}} \rho X_n T_9^{-3/2} \exp(-20.61/T_9) \text{ sec}^{-1} \quad 3 \leq T_9 \leq 10 \qquad 74.$$

$$\mathcal{E}(He^3n)_p = 1.71 \times 10^{26} \, \rho X_n X_{He^3}(1 - 0.597 \, T_9^{1/2} + 0.183 \, T_9) \quad 0 \leq T_9 \leq 10$$
$$+ 3.13 \times 10^{28} \, \rho X_n X_{He^3} T_9^{-3/2} \exp(-20.61/T_9) \text{ erg g}^{-1} \text{sec}^{-1} \quad 3 \leq T_9 \leq 10 \qquad 75.$$

For charged particles the equations corresponding to 69 and 70 are

$$\frac{S(0)}{S_r} = \frac{\Gamma_r^2/4}{E_r^2 + \Gamma_r^2/4} \qquad 76.$$

where

$$S_r = \sigma_r E_r \exp(E_G/E_r)^{1/2} \qquad 77.$$

These expressions hold for any partial wave in the case of charged particles. Table V lists S_r and $S(0)$. In addition to the approximation inherent in using Equation 61, we have neglected any variation in S other than the simple resonance behavior exhibited in Equation 76. In particular we have neglected variations in the penetration factor other than the term in $\exp-(E_G/E)^{1/2}$ used in Equation 45. For resonances at energies greater than 20 per cent of the Coulomb-barrier energy this results in a value for $S(0)$ given by Equation 76 which could be low by as much as an order of magnitude. In most cases independent measurements on S are available in the low-energy range and it is not necessary to calculate $S(0)$ from resonance data.

There is an additional point of considerable importance: the amplitude

[6] When the nonresonant term is available, it is given first in all equations. The additional terms for one or more resonances then follow. In using the references to numbered equations in the tables, use the nonresonant terms for Tables II and III and the appropriate resonant terms for Tables IV and V.

contributions of states of the same spin and parity must be summed coherently. In general the phase is zero or π between two fairly well-separated states so that the overall contribution to $S(0)$ is the square of the sum or difference of the square roots of the $S(0)$ for the two states. In the case of $N^{15}(p,\gamma)O^{16}$ the resonances at laboratory proton energies 0.338 MeV and 1.010 MeV both have $J^{\pi} = 1^-$ and are known to interfere constructively at low energies. With the data for $S(0)$ in Table V our method of calculation yields $S(0) = [(6.48 \times 10^{-3})^{1/2} + (4.49 \times 10^{-3})^{1/2}]^2 = 2.18 \times 10^{-2}$ MeV-barn. A more sophisticated treatment of the problem by Hebbard (1960) (see also Caughlan & Fowler 1962) yields $S(0) = 2.74 \times 10^{-2}$ MeV-barn as given in Table III. In the case of $N^{15}(p,\alpha)C^{12}$ the two states interfere destructively, yielding $S(0) = [(71.9)^{1/2} - (2.09)^{1/2}]^2 = 49.5$ MeV-barn compared to the more accurate value, 53.4 MeV-barn (Hebbard 1960), given in Table III. It is also possible to calculate resonance contributions to $S'(0)$ by similar methods and this was done by Caughlan & Fowler (1962) using the data of Hebbard (1960) to give the values for $N^{15}(p, \gamma)O^{16}$ and $N^{15}(p,\alpha)C^{12}$ given in Table III.

The $C^{12}(p,\gamma)N^{13}$ and $C^{13}(p,\gamma)N^{14}$ reactions.—These reactions have been chosen to illustrate in detail the use of the charged-particle nonresonant and resonant data given in Tables III and V. The experimentally determined cross section for $C^{12}(p,\gamma)N^{13}$ is shown in Figure 1 from the work of Hebbard & Vogl (1960) and Vogl (1963). Earlier sources are cited in these references. The cross section for $C^{13}(p,\gamma)N^{14}$ is shown in Figure 2 and has been taken from Seagrave (1952) as well as from the two references just cited. The measurements of Seagrave on $C^{13}(p,\gamma)N^{14}$ have been adjusted to those of Vogl in the region of energy overlap. The solid curves have been calculated by the present authors with the full apparatus of the R-matrix theory described by Lane & Thomas (1958) using only the empirical parameters which can be determined at or near the observed resonances. In the case of the 0.460-MeV resonance in $C^{12}(p,\gamma)$, the Thomas (1952) factor has been used. Similar success in fitting the data has been achieved by Mahaux (1965) using the theory of Humblet & Rosenfeld (1961).

The extrapolations of our calculated curves through the low-energy experimental data are shown in Figure 3. It will be clear that the $C^{12}(p,\gamma)N^{13}$ reaction can be accounted for entirely in terms of the properties of the 0.460-MeV state ($E_x = 2.368$ MeV in N^{14}) over the energy range of interest. Similarly, with the exception of the regions near the sharp resonances in Figure 2, the $C^{13}(p,\gamma)N^{14}$ reaction is accounted for in terms of the properties of the relatively wide 0.555-MeV and 1.250-MeV resonances over the entire energy range from 0 to 1.5 MeV. These two resonances have $J^{\pi} = 1^-$ and 0^- respectively, so their cross-section contributions simply add at low energies.

The calculated curves given in Figures 1 and 2 are not expressible in terms of simple analytic formulae. However, it is possible in the case of these reactions to make a fairly accurate determination of the parameters entering into the analytic expressions for nonresonant and resonant cross sections under

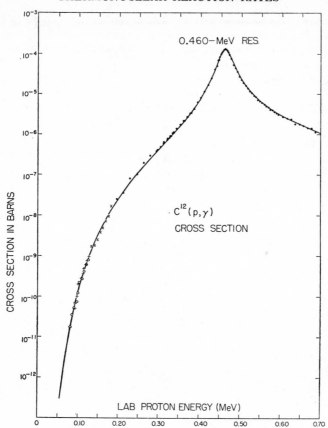

Fig. 1. The dependence on energy of the cross section for the reaction $C^{12}(p,\gamma)N^{13}$. Sources of the data are cited in the text. The solid curve is a theoretical curve based on the R-matrix theory described by Lane & Thomas (1958) using empirical parameters determined at or near the 0.460-MeV resonance.

discussion in this paper. We have carried out the necessary numerical integrations over the observed cross sections to determine $\langle\sigma v\rangle$ as a function of temperature. This was followed by a least-squares analysis using Equations 50 and 52 to determine $S(0)$, $S'(0)/S(0)$, and $\frac{1}{2}S''(0)/S(0)$ for both $C^{12}(p,\gamma)N^{13}$ and $C^{13}(p,\gamma)N^{14}$. Similarly a least-squares analysis was made to determine E_r, σ_r, and Γ_r for the 0.460-MeV, 0.555-MeV, and 1.250-MeV resonances. The percent deviation of the final analytical expressions given in Equations 78 and 82 below from the integrated values is shown as a function of temperature in Figure 4. By judicious choice of the temperature limits on the nonresonant and resonant expressions it has been found possible to keep the errors below ~ 20 per cent. The deviation-curve for $C^{12}(p,\gamma)N^{13}$ in Figure 4 illustrates the error which arises in using the approximate resonance expression given in Equation 62 for broad resonances. Note that the analytic expression falls below the actual integrated value on either side of the temperature range within which E_r is approximately equal to $E_0(T)$.

All of the numerical parameters have been recorded in Tables III and V and we now employ these to illustrate the use of the tables in forming desired analytic expressions:

$$
\begin{aligned}
[C^{12}p]_\gamma &= 6.0225 \times 10^{23}\, \rho \langle C^{12}p\rangle_\gamma \\
&= 2.04 \times 10^7\, (1 + 0.0304\, T_9^{1/3} + 1.19\, T_9^{2/3} + 0.254\, T_9 + 2.06\, T_9^{4/3} \\
&\quad + 1.12\, T_9^{5/3})\, \rho\, T_9^{-2/3} \exp(-13.69/T_9^{1/3}) \qquad\qquad 0 \le T_9 \le 0.55 \qquad 78. \\
&\quad + 1.08 \times 10^5\, \rho\, T_9^{-3/2} \exp(-4.925/T_9)\ \sec^{-1} \qquad\qquad 0.25 \le T_9 \le 7
\end{aligned}
$$

$$
\begin{aligned}
\frac{1}{\tau_{p\gamma}(C^{12})} &= \frac{1}{4.93 \times 10^{-8}}\, (1 + 0.0304\, T_9^{1/3} + 1.19\, T_9^{2/3} + 0.254\, T_9 + 2.06\, T_9^{4/3} \\
&\quad + 1.12\, T_9^{5/3})\, \rho X_H T_9^{-2/3} \exp(-13.69/T_9^{1/3}) \qquad\qquad 0 \le T_9 \le 0.55 \qquad 79.^7 \\
&\quad + \frac{1}{9.32 \times 10^{-6}}\, \rho X_H T_9^{-3/2} \exp(-4.925/T_9)\ \sec^{-1} \qquad\qquad 0.25 \le T_9 \le 7
\end{aligned}
$$

$$
\begin{aligned}
\lambda_{\gamma p}(N^{13}) &= 1.81 \times 10^{17}\, (1 + 0.0304\, T_9^{1/3} + 1.19\, T_9^{2/3} + 0.254\, T_9 + 2.06\, T_9^{4/3} \\
&\quad + 1.12\, T_9^{5/3})\, T_9^{5/6} \exp(-13.69/T_9^{1/3} - 22.56/T_9) \qquad 0 \le T_9 \le 0.55 \qquad 80. \\
&\quad + 9.56 \times 10^{14} \exp(-27.48/T_9)\ \sec^{-1} \qquad\qquad 0.25 \le T_9 \le 7
\end{aligned}
$$

$$
\begin{aligned}
\mathcal{E}(C^{12}p)_\gamma &= 3.17 \times 10^{24}\, (1 + 0.0304\, T_9^{1/3} + 1.19\, T_9^{2/3} + 0.254\, T_9 + 2.06\, T_9^{4/3} \\
&\quad + 1.12\, T_9^{5/3})\, \rho X_H X_{C^{12}}\, T_9^{-2/3} \exp(-13.69/T_9^{1/3}) \qquad 0 \le T_9 \le 0.55 \qquad 81. \\
&\quad + 1.68 \times 10^{22}\, \rho X_H X_{C^{12}}\, T_9^{-3/2} \exp(-4.925/T_9) \qquad\qquad 0.25 \le T_9 \le 7 \\
&\quad \mathrm{erg\ g^{-1}\ sec^{-1}}
\end{aligned}
$$

$$
\begin{aligned}
\langle C^{13}p\rangle_\gamma &= 1.6604 \times 10^{-24}\, \rho^{-1}[C^{13}p]_\gamma \\
&= 1.33 \times 10^{-16}\, (1 + 0.0304\, T_9^{1/3} + 0.958\, T_9^{2/3} + 0.204\, T_9 + 1.39\, T_9^{4/\nu} \\
&\quad + 0.753\, T_9^{5/3})\, T_9^{-2/3} \exp(-13.72/T_9^{1/3}) \qquad\qquad 0 \le T_9 \le 0.65 \quad 82. \\
&\quad + 2.25 \times 10^{-18}\, T_9^{-3/2} \exp(-5.978/T_9) \qquad\qquad 0.3 \le T_9 \le 10 \\
&\quad + 3.76 \times 10^{-18}\, T_9^{-3/2} \exp(-13.46/T_9)\ \mathrm{cm^3\ sec^{-1}} \qquad 2.5 \le T_9 \le 10
\end{aligned}
$$

etc. etc.

It is necessary to emphasize that our procedures in determining the resonance parameters E_r, Γ_r, and $(\omega\gamma)_r$ from fitting $\langle \sigma v\rangle$ as a function of temperature do not necessarily result in agreement with the usual values obtained by fitting σ as a function of energy. This is notably true in the cases just discussed in detail and in the cases $T(d,n)He^4$ and $He^3(d,p)He^4$. In general, resonances which are not sharp are not accurately treated by use of the approximation inherent in Equation 62. We find, however, that some adjust-

---

[7] Note that the coefficients listed in the tables occur here in the denominator for $1/\tau$ and are thus the correct coefficients in the numerator for τ when only one term is employed.

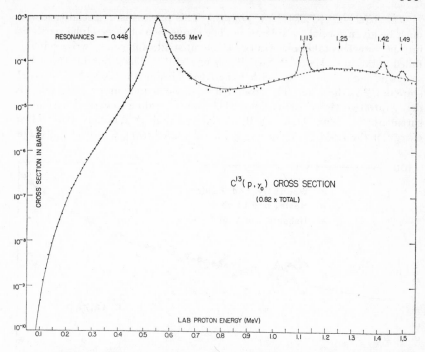

FIG. 2. The dependence on energy of the cross section for the reaction $C^{13}(p,\gamma_0)N^{14}$, where γ_0 designates the gamma-ray transition to the ground state of N^{14}. This transition accounts for 82 per cent of the total cross section which includes cascade transitions through excited states of N^{14}. Sources of the data are cited in the text. The solid curve, exclusive of the resonant features at 0.448, 1.113, 1.42, and 1.49 MeV, is a theoretical curve based on the R-matrix theory described by Lane & Thomas (1958) using empirical parameters determined at or near the 0.555-MeV and 1.250-MeV resonances.

ment in the empirical parameters as just described results in a tolerable uncertainty in the analytic expression for the reaction rate.

Examples of reverse reactions.—The reaction rates for $Be^7(\alpha,p)B^{10}$ which is the reverse of $B^{10}(p,\alpha)Be^7$ and for $He^4(2p,\tau)He^3$ which is the reverse of $He^3(\tau,2p)He^4$ can be found from

$$
\begin{aligned}
[\alpha Be^7]_p &= 6.0225 \times 10^{23} \, \rho \langle \alpha Be^7 \rangle_p \\
&= 9.48 \times 10^{10} \, (1 + 0.0345 \, T_9^{1/3} - 0.498 \, T_9^{2/3} - 0.121 \, T_9 + 0.300 \, T_9^{4/3} \\
&\quad + 0.185 \, T_9^{5/3}) \, \rho \, T_9^{-2/3} \exp(-12.06/T_9^{1/3} - 13.32/T_9) \quad 0 \leq T_9 \leq 5 \\
&\quad + 2.74 \times 10^9 \, \rho T_9^{-3/2} \exp(-25.65/T_9) \quad\quad\quad\quad\quad\quad 1 \leq T_9 \leq 10 \\
&\quad + 4.38 \times 10^9 \, \rho T_9^{-3/2} \exp(-29.48/T_9) \ \sec^{-1} \quad\quad\quad\quad 2 \leq T_9 \leq 10
\end{aligned}
$$
83.

$$
[2p He^4]_\tau = 18.4 \, \rho^2 T_9^{-13/6} \exp(-12.28/T_9^{1/3} - 149.2/T_9) \ \sec^{-1} \quad\quad 0 \leq T_9 \leq 10
$$
84.

The $3He^4 \to C^{12} + 7.274$-MeV process.—This process is a special case and must be given special consideration. The conversion of three alpha particles into a C^{12} nucleus takes place in two resonant stages and can be written in detailed notation as $He^4(\alpha)Be^8(\alpha)C^{12*}(\gamma\gamma$ or $e^+e^-)C^{12}$. The ground state of Be^8 serves as one resonance and either the 7.644-MeV or the 9.638-MeV excited state in C^{12} as the other. The ground state of Be^8 is unbound with respect to decay into two alpha particles by 0.092 MeV according to very precise measurements by Benn, Dally, Müller, Pixley, Staub & Winkler (1966). The energy of the 7.644-MeV state in C^{12} is calculated using C^{12}-Be^8-$He^4 = 0.278$

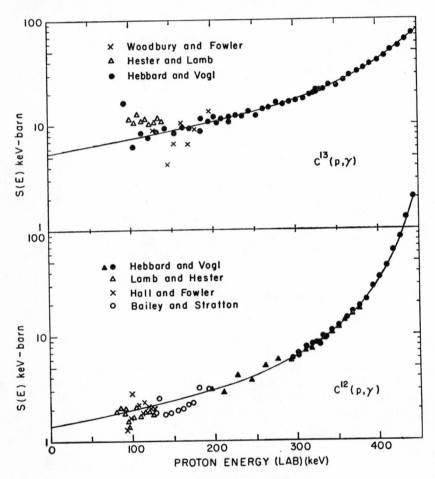

FIG. 3. The dependence on energy of the cross-section factors for the reactions $C^{12}(p,\gamma)N^{13}$ and $C^{13}(p,\gamma)N^{14}$.

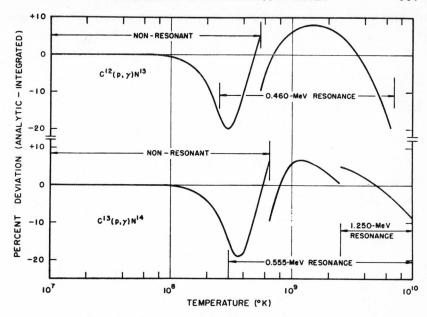

FIG. 4. *Upper curve:* The percent deviation of the analytical expressions for $\langle \sigma v \rangle$ for $C^{12}(p,\gamma)N^{13}$ given by Equation 78 from the values given by actual integration over the cross section illustrated in Figure 1.

Lower curve: The percent deviation of the analytical expression for $\langle \sigma v \rangle$ for $C^{13}(p,\gamma)N^{14}$ given by Equation 82 from the values given by actual integration over the cross section illustrated in Figure 2.

MeV/c^2 as determined by Cook, Fowler, Lauritsen & Lauritsen (1957). The 7.644-MeV state cascades by gamma-ray emission through the excited state in C^{12} at 4.433 MeV and also emits an electron-positron pair directly to the ground state. The 9.638-MeV state most probably cascades by an electric dipole transition through the 4.433-MeV state. This transition is inhibited by an isotopic-spin selection rule. Detailed measurements on the 7.644-MeV state are summarized in Seeger & Kavanagh (1963). The pair-emission width of the 7.644-MeV state enters directly into the determination of the total radiation width Γ_{rad} for this state. Seeger & Kavanagh used $(5.5 \pm 3) \times 10^{-11}$ MeV. Recent determinations are $(6.5 \pm 0.7) \times 10^{-11}$ MeV (Crannell & Griffy 1964), $(6.2 \pm 0.5) \times 10^{-11}$ MeV (Crannell 1965), and $(7.3 \pm 1.3) \times 10^{-11}$ MeV (Gudden & Strehl 1965). We take $(6.5 \pm 0.6) \times 10^{-11}$ MeV and find $\Gamma_{rad} = (2.8 \pm 0.5) \times 10^{-9}$ MeV. The statistical weight factor is $\omega = 2J + 1 = 1$ since the state has $J^{\pi} = 0^{+}$. For the 9.638-MeV state, $J^{\pi} = 3^{-}$ and $\omega = 7$. On the basis of the isotopic-spin selection rule, Hoyle & Fowler (1960) estimate $\Gamma_{\gamma} \sim 10^{-8}$ MeV (4 per cent of the full electric dipole width) so that $\omega \Gamma_{\gamma} \sim 7 \times 10^{-8}$ MeV. The excess energies of the two excited states in C^{12} over $3He^4$ are 0.370 MeV and 2.364 MeV respectively. With these data the rate of the $3\alpha \to C^{12}$ process is found to be

$$P_{3\alpha} = \tfrac{1}{6} n_{He^4}^3 \langle \alpha\alpha\alpha \rangle = 1.565 \times 10^{21} \, \rho X_{He^4}^3 [\alpha\alpha\alpha] \text{ reactions cm}^{-3} \text{ sec}^{-1} \qquad 85.$$

while the lifetime of He^4 to the process is given by

$$\frac{1}{\tau_{3\alpha}(He^4)} = -\frac{1}{n_{He^4}} dn_{He^4}/dt = -\frac{1}{X_{He^4}} dX_{He^4}/dt = \frac{1}{X_{He^4}} dX_{C^{12}}/dt$$
$$86.$$

$$= 3P_{3\alpha}/n_{He^4} = \tfrac{1}{2} n_{He^4}^2 \langle \alpha\alpha\alpha \rangle = \frac{1}{32.04} X_{He^4}^2 [\alpha\alpha\alpha]$$

where

$$\langle \alpha\alpha\alpha \rangle = 5.861 \times 10^{-56} \, T_9^{-3} \exp(-4.294/T_9) \qquad\qquad 0.03 \le T_9 \le 8$$
$$87.$$
$$+ 1.465 \times 10^{-54} \, T_9^{-3} \exp(-27.433/T_9) \text{ cm}^6 \text{ sec}^{-1} \qquad 4 \le T_9 \le 8$$

$$[\alpha\alpha\alpha] = \rho^2 N_A^2 \langle \alpha\alpha\alpha \rangle = 3.627 \times 10^{47} \rho^2 \langle \alpha\alpha\alpha \rangle$$
$$= 2.126 \times 10^{-8} \rho^2 T_9^{-3} \exp(-4.294/T_9) \qquad\quad 0.03 \le T_9 \le 8 \quad 87'.$$
$$+ 5.315 \times 10^{-7} \rho^2 T_9^{-3} \exp(-27.433/T_9) \text{ sec}^{-1} \qquad 4 \le T_9 \le 8$$

The photodisintegration rate for C^{12} becomes

$$\frac{1}{\tau_{\gamma\alpha}(C^{12})} = \lambda_{\gamma\alpha}(C^{12}) = 4.257 \times 10^{12} \exp(-88.71/T_9) \qquad 0.03 \le T_9 \le 8$$
$$88.$$
$$+ 1.064 \times 10^{14} \exp(-111.85/T_9) \text{ sec}^{-1} \qquad 4 \le T_9 \le 8$$

The energy generation rate is

$$\mathcal{E}(3He^4 \to C^{12}) = 3.88 \times 10^8 \, \rho^2 X_{He^4}^3 T_9^{-3} \exp(-4.294/T_9) \qquad 0.03 \le T_9 \le 8$$
$$89.$$
$$+ 9.70 \times 10^9 \, \rho^2 X_{He^4}^3 T_9^{-3} \exp(-27.433/T_9) \qquad 4 \le T_9 \le 8$$
$$\text{erg g}^{-1} \text{ sec}^{-1}$$

The 3 in the numerator of the first term on the right-hand side of Equation 86 occurs because three He^4 nuclei are destroyed in each reaction. The 6 in the denominator of Equation 85 arises from the identity of the three alpha particles. Equation 88 assumes that the 4.433-MeV state in C^{12} is in equilibrium with the ground state. The uncertainty in the above equations is approximately ± 60 per cent for the term due to the 7.644-MeV state and a factor of at least 3 for the term due to the 9.638-MeV state. Fortunately, in the applications currently thought to be of importance, the contribution of the 9.638-MeV state can be neglected.

The $C^{12}(\alpha,\gamma)O^{16}+7.161$-MeV reaction.—This reaction is also a special case and must be given special consideration. At low energy the cross section is dominated by the contribution of the 1^- excited state of O^{16} at 7.115 MeV which is bound by $|E_r| = 0.046$ MeV. The partial widths are small compared to $|E_r|$, and the term $\Gamma^2/4$ in the resonance denominator of Equation 61 can be neglected. In addition, in determining $\langle \sigma v \rangle$ from σ it is sufficiently accurate to substitute E_0 for $\dot{E}$ in the resonance denominator of this equation. This procedure was employed by B^{2}FH (1957), which should be consulted for additional details. The result which was obtained can be expressed as

$$S_{eff} = 2.67 \times 10^6 \frac{\theta_\alpha{}^2\Gamma_\gamma}{(E_o + |E_r|)^2} \text{ MeV-barn}$$

$$= 3.14 \times 10^6 \frac{\theta_\alpha{}^2\Gamma_\gamma}{T_9{}^{4/3}(1 + 0.050 \ T_9{}^{-2/3})^2}$$

$$= \frac{1.76 \times 10^{-2}}{T_9{}^{4/3}(1 + 0.050 \ T_9{}^{-2/3})^2}$$

90.

where $\theta_\alpha{}^2$ is the dimensionless, reduced alpha-particle width for the 7.12-MeV excited state and Γ_γ is the radiation width in MeV. The final numerical result has been obtained by using $E_o = 0.9226 \ T_9{}^{2/3}$ MeV, $\Gamma_\gamma = 6.6 \times 10^{-8}$ MeV from the measurements of Swann & Metzger (1957), and $\theta_\alpha{}^2(7.12) = (0.1 \pm 0.05) \times \theta_\alpha{}^2(9.58) = 0.085 \pm 0.040$ from a recent theoretical calculation by Stephenson (1966). The reduced alpha-particle width for the 9.58-MeV excited state of O^{16} has been measured to be equal to 0.85 (Ajzenberg-Selove & Lauritsen 1959). Stephenson's calculation employs a specific model for the O^{16} nucleus but his result has been confirmed by indirect measurements by Loebenstein, Mingay, Winkler & Zaidins (1967) who find the result $0.06 \le \theta_\alpha{}^2 \le 0.14$.

When Equation 90 is substituted into Equation 51 one obtains

$$[C^{12}\alpha]_\gamma = 6.0225 \times 10^{23} \rho \langle C^{12}\alpha \rangle_\gamma$$

91.

$$= 2.19 \times 10^8 \ \rho T_9{}^{-2}(1 + 0.050 \ T_9{}^{-2/3})^{-2} \exp(-32.12/T_9{}^{1/3}) \text{ sec}^{-1}$$

$$\text{nonres,} \quad 0 \le T_9 \le 6$$

so that

$$\frac{1}{\tau_{\alpha\gamma}(C^{12})} = \frac{1}{1.83 \times 10^{-8}} \rho X_{He^4} T_9{}^{-2}(1 + 0.050 \ T_9{}^{-2/3})^{-2} \exp(-32.12/T_9{}^{1/3}) \text{ sec}^{-1}$$

92.

$$\text{nonres,} \quad 0 \le T_9 \le 6$$

$$\lambda_{\gamma\alpha}(O^{16}) = 1.12 \times 10^{19} \ T_9{}^{-1/2}(1 + 0.050 \ T_9{}^{-2/3})^{-2}$$

$$\times \exp(-32.12/T_9{}^{1/3} - 83.11/T_9) \text{ sec}^{-1} \qquad \text{nonres,} \quad 0 \le T_9 \le 6$$

93.

and

$$\mathcal{E}(C^{12}\alpha)_\gamma = 3.15 \times 10^{25} \ \rho X_{He^4} X_{C^{12}} T_9{}^{-2}(1 + 0.050 \ T_9{}^{-2/3})^{-2}$$

$$\times \exp(-32.12/T_9{}^{1/3}) \text{ erg g}^{-1} \text{ sec}^{-1} \qquad \text{nonres,} \quad 0 \le T_9 \le 6$$

94.

The uncertainty in the numerical coefficients in the above expressions is ± 50 per cent. At high temperatures the resonances in $C^{12}(\alpha,\gamma)O^{16}$ listed in Table V make substantial contributions and *these contributions must be added to the nonresonant terms given in Equations 91 to 94.* Data on subsequent helium-burning reactions $O^{16}(\alpha,\gamma)Ne^{20}$ and $Ne^{20}(\alpha,\gamma)Mg^{24}$ are given in Tables III, V, and VI (to be discussed in what follows).

Continuum-reaction rates.—In this section the interactions of protons, alpha particles, and photons with nuclei in the mass range $19 \le A \le 40$ are discussed. These interactions occur under astrophysical circumstances at temperatures in excess of 10^9 °K and thus at effective energies in excess of the order of 1 MeV. Under these circumstances numerous resonances contribute to the reaction cross section, and in some cases these resonances are broad enough to overlap and produce a continuum cross section characterized by broad maxima and minima. Even in the case of sharp, well-isolated resonances, the great breadth of the effective range in interaction energy means that the smoothed-out cross section obtained by averaging over all resonances within the appropriate energy range is of primary interest. In any case it is a simple computational matter to calculate the total $\langle \sigma v \rangle$ by summing over the contributions of individual resonances as given by Equation 62. This is facilitated by the fact that most authors express their experimental results in terms of the quantities E_r and $(\omega \gamma)_r$ required to evaluate Equation 62.

We have carried out such a computational program to obtain $\langle \sigma v \rangle$ as a function of temperature over the range $1 \leq T_9 \leq 5$ for the reactions listed in Table VI. We have supplemented the data given in Ajzenberg-Selove & Lauritsen (1959) and Endt & Van der Leun (1962) by a fairly extensive search of the recent literature. A least-squares analysis has been made in each case of the fit of $\langle \sigma v \rangle$ versus T to equations of the form $T^n \exp(-E_{th}/kT)$, where E_{th} is an effective "threshold" energy and $n = -\frac{1}{2}$, 0, $\frac{1}{2}$, and $\frac{3}{2}$. Our analyses showed that in almost all cases the best fit was obtained by the use of $n = 0$, and because of the inherent simplicity of the resulting equation we have, in all cases, used

$$[01] = \rho N_A \langle 01 \rangle = C\rho \exp(-E_{th}/kT) \qquad 95.$$

where C as well as E_{th} are semi-empirical constants determined by a least-squares analysis.

The threshold energy is that energy above which barrier penetration effects in Γ_1 are no longer of importance in determining $(\omega \gamma)_r$ as defined by Equation 65. This always occurs as the energy approaches the top of the Coulomb barrier, but it also occurs in the (p, γ), (p, α), and (α, γ) reactions of interest here at considerably lower energy. In all these reactions Γ_1 rises rapidly until it is the major partial width in $\Gamma = \Gamma_1 + \Gamma_2 + \ldots$ so that $\Gamma \approx \Gamma_1$ and $(\omega \gamma)_r \approx (\omega \Gamma_2)_r$, where Γ_2 varies from resonance to resonance but shows no systematic rapid increase or decrease with the energy of the interacting particles. The corresponding threshold energy in the reverse reaction $2 + 3 \rightarrow 0 + 1$ is given by

$$E_{th}' = E_{th} + Q \qquad 96.$$

Our results are tabulated in Table VI which yields, as examples, the following results for $Ne^{20}(\alpha, \gamma)Mg^{24}$, $Mg^{24}(\gamma, \alpha)Ne^{20}$, and $Ne^{20}(\alpha, p)Na^{23}$:

$$[Ne^{20}\alpha]_\gamma = 6.0225 \times 10^{23} \, \rho \langle Ne^{20}\alpha \rangle_\gamma$$
$$= 8.68 \times 10^3 \, \rho \, \exp(-15.43/T_9) \, \text{sec}^{-1} \text{ continuum, } 1 \leq T_9 \leq 5 \qquad 97.$$

$$\tau_{\alpha\gamma}(Ne^{20}) = \frac{4.61 \times 10^{-4}}{\rho X_{He^4}} \, \exp(15.43/T_9) \, \text{sec continuum, } 1 \leq T_9 \leq 5 \qquad 98.$$

TABLE VI

CHARGED-PARTICLE CONTINUUM-REACTION DATA

Reaction		$F^{19}(p,\gamma)Ne^{20}$	$Ne^{20}(\alpha,\gamma)Mg^{24}$	$Na^{23}(p,\gamma)Mg^{24}$	$Na^{23}(p,\alpha)Ne^{20}$	$Mg^{24}(\alpha,\gamma)Si^{28}$	$Al^{27}(p,\gamma)Si^{28}$
Parameters in typical equations indicated in parentheses							
Q	(8)	12.844	9.317	11.694	2.377	9.981	11.583
$(Q/k)_9$	(100)	$1.491E\ 02$	$1.081E\ 02$	$1.357E\ 02$	$2.759E\ 01$	$1.158E\ 02$	$1.344E\ 02$
E_{th} CM	(95)	0.537	1.330	0.435	0.380	1.329	0.526
$(E_{th}/k)_9$	(97)	6.235	$1.543E\ 01$	5.042	4.408	$1.543E\ 01$	6.100
E_{th}' CM	(96)	13.382	10.647	12.129	2.757	11.311	12.108
$(E_{th}'/k)_9$	(100)	$1.553E\ 02$	$1.236E\ 02$	$1.407E\ 02$	$3.200E\ 01$	$1.313E\ 02$	$1.405E\ 02$
Coefficients in typical equations indicated in parentheses							
$\langle 01 \rangle$	(97)	$1.24\ E-19$	$1.44\ E-20$	$1.13\ E-19$	$1.82\ E-18$	$2.85\ E-21$	$1.92\ E-19$
$[01]$	(97)	$7.46\ E\ 04$	$8.68\ E\ 03$	$6.79\ E\ 04$	$1.10\ E\ 06$	$1.72\ E\ 03$	$1.16\ E\ 05$
$\langle 23 \rangle$	(101)	—	—	—	$2.27\ E-18$	—	—
$[\lambda_\gamma]$	(100)(101)	$2.76\ E\ 15$	$5.22\ E\ 14$	$5.08\ E\ 15$	$1.37\ E\ 06$	$1.08\ E\ 14$	$1.31\ E\ 16$
$\tau_1(0)$	(98)	$1.35\ E-05$	$4.61\ E-04$	$1.49\ E-05$	$9.18\ E-07$	$2.33\ E-03$	$8.71\ E-06$
$\tau_0(1)$	(98)	$2.55\ E-04$	$2.30\ E-03$	$3.39\ E-04$	$2.09\ E-05$	$1.40\ E-02$	$2.33\ E-04$
$\mathcal{E}_{01}$	(99)	$4.83\ E\ 22$	$9.75\ E\ 20$	$3.30\ E\ 22$	$1.09\ E\ 23$	$1.72\ E\ 20$	$4.76\ E\ 22$
Uncertainty		±50%	FAC 2	±50%	±50%	±50%	±50%
T_9 limits		1 to 5	1 to 5	1 to 5	1 to 5	1 to 5	1 to 5
References		Aj 59	Sm 65	Pr 62, En 62 Gl 62	Ku 63, En 62	Sm 62, En 62 We 64	An 63, En 62

Reaction		$Al^{27}(p,\alpha)Mg^{24}$	$Si^{28}(\alpha,\gamma)S^{32}$	$P^{31}(p,\gamma)S^{32}$	$P^{31}(p,\alpha)Si^{28}$	$Cl^{35}(p,\gamma)Ar^{36}$	$K^{39}(p,\gamma)Ca^{40}$
Parameters in typical equations indicated in parentheses							
Q	(8)	1.601	6.948	8.864	1.917	8.506	8.333
$(Q/k)_9$	(100)	$1.858E\ 01$	$8.063E\ 01$	$1.029E\ 02$	$2.224E\ 01$	$9.871E\ 01$	$9.671E\ 01$
E_{th} CM	(95)	0.947	1.900	0.521	1.031	0.551	0.910
$(E_t/k)_9$	(97)	$1.099E\ 01$	$2.205E\ 01$	6.044	$1.197E\ 01$	6.392	$1.056E\ 01$
E_t' CM	(96)	2.549	8.848	9.385	2.948	9.057	9.243
$(E_{th}'/k)_9$	(100)	$2.958E\ 01$	$1.027E\ 02$	$1.089E\ 02$	$3.421E\ 01$	$1.051E\ 02$	$1.073E\ 02$
Coefficients in typical equations indicated in parentheses							
$\langle 01 \rangle$	(97)	$7.05\ E-18$	$6.71\ E-22$	$3.25\ E-20$	$1.31\ E-17$	$4.43\ E-20$	$9.61\ E-20$
$[01]$	(97)	$4.24\ E\ 06$	$4.04\ E\ 02$	$1.96\ E\ 04$	$7.92\ E\ 06$	$2.67\ E\ 04$	$5.78\ E\ 04$
$\langle 23 \rangle$	(101)	$1.27\ E-17$	—	—	$7.74\ E-18$	—	—
$[\lambda_\gamma]$	(100)(101)	$7.68\ E\ 06$	$2.61\ E\ 13$	$7.44\ E\ 14$	$4.66\ E\ 06$	$2.04\ E\ 15$	$4.44\ E\ 15$
$\tau_1(0)$	(98)	$2.37\ E-07$	$9.91\ E-03$	$5.15\ E-05$	$1.27\ E-07$	$3.78\ E-05$	$1.74\ E-05$
$\tau_0(1)$	(98)	$6.36\ E-06$	$6.93\ E-02$	$1.58\ E-03$	$3.91\ E-06$	$1.31\ E-03$	$6.74\ E-04$
$\mathcal{E}_{01}$	(99)	$2.41\ E\ 23$	$2.42\ E\ 19$	$5.36\ E\ 21$	$4.69\ E\ 23$	$6.21\ E\ 21$	$1.18\ E\ 22$
Uncertainty		±50%	FAC 3	±50%	±50%	±50%	±50%
T_9 limits		1 to 5	1 to 5	1 to 5	1 to 5	1 to 5	1 to 5
References		Ab 63, En 62	Sm 64	Sm 64, Sp 65	Ku 63, En 62	Er 65	Le 66

Units: Energy in MeV; $\langle\ \rangle$ in $cm^3\ sec^{-1}$; $[\ ]$ in sec^{-1}); τ in sec; $\mathcal{E}_{01}$ in erg $g^{-1}\ sec^{-1}$; $T_9 = T/10^9$.

$$\mathcal{E}(\text{Ne}^{20}\alpha)_\gamma = 9.75 \times 10^{20}\, \rho X_{\text{He}^4}\, X_{\text{Ne}^{20}}\, \exp(-15.43/T_9)\ \text{erg g}^{-1}\ \text{sec}^{-1} \qquad\qquad 99.$$

$$\text{continuum},\ 1 \leq T_9 \leq 5$$

$$\frac{1}{\tau_{\gamma\alpha}(\text{Mg}^{24})} = \lambda_{\gamma\alpha}(\text{Mg}^{24}) = 5.22 \times 10^{14}\, T_9^{3/2}\, \exp(-15.43/T_9 - 108.1/T_9)$$

$$= 5.22 \times 10^{14}\, T_9^{3/2}\, \exp(-123.6/T_9)\ \text{sec}^{-1}\text{continuum},\, 1 \leq T_9 \leq 5 \quad 100.$$

$$\ulcorner\text{Ne}^{20}\alpha\urcorner_p = 6.0225 \times 10^{23}\, \rho\, \langle\text{Ne}^{20}\alpha\rangle_p$$

$$= 1.37 \times 10^6\, \rho\, \exp(-32.00/T_9)\ \text{sec}^{-1}\,\text{continuum},\, 1 \leq T_9 \leq 5 \quad 101.$$

GENERAL DISCUSSION

The section just concluded has presented in tabular form the currently known empirical data on the reaction rates for a number of nuclear reactions involving neutrons, protons, and alpha particles. In this section we discuss the relevance of these data to thermonuclear processes occurring under astrophysical circumstances.

The proton-proton chain.—Energy generation in the Sun and other stars with central temperature less than $\sim2\times10^7$ °K occurs through the pp chain, the rate for which is basically determined by the cross section for the $\text{H}^1(p,e^+\nu)\text{D}^2$ reaction which must be calculated theoretically. However, the calculations are based on well-determined empirical parameters for proton-proton scattering and for Gamow-Teller beta decay, and we have assigned an uncertainty of only ±10 per cent to the calculated rate.

Bahcall (1964) has estimated the rate of $\text{H}^1(pe^-,\nu)\text{D}^2$ to be $\sim3.4\times10^{-6}$ $\cdot\rho(1+X_\text{H})/2T_9^{1/2}$ times that for $\text{H}^1(p,e^+\nu)$ so that the pep process contributes significantly to hydrogen burning at $T_9\sim0.01$ only under the rather special circumstances that $\rho>10^4$ g cm^{-3}. Werntz & Brennan (1967) have shown that another "weak" interaction process which may occur in the pp chain, namely $\text{He}^3(p,e^+\nu)\text{He}^4$, has an upper limit calculated from the rate of the analogous reaction $\text{He}^3(n,\gamma)\text{He}^4$ which makes it unimportant under most circumstances. Bahcall & Wolf (1964) have shown that $\text{He}^3(e^-,\nu)\text{T}^3$ and subsequent tritium-burning reactions are important only in the late stages of hydrogen burning in stars of small mass.

An important and competitive weak interaction is the capture of electrons by Be^7. The rate for this capture under astrophysical circumstances has been calculated from the decay rate of atomic Be^7 by Bahcall (1962) who gives an approximate expression

$$\lambda_e(\text{Be}^7) = 6.70 \times 10^{-11}\, \rho(1 + X_\text{H})T_9^{-1/2}\ \text{sec}^{-1} \qquad T_9 \leq 1 \qquad 102.$$

which is valid under circumstances in which electrons are nondegenerate and nonrelativistic.

The rates of all "strong" reactions (Fowler & Vogl 1964) in the pp chain have now been measured with considerable precision. Recent measurements by Parker (1966) show that the cross section for the first reaction in the overall process, $\text{Be}^7(p,\gamma)\text{B}^8(e^+\nu)\text{Be}^{8*}(\alpha)\text{He}^4$, is somewhat greater than heretofore thought. This result is important in determining the competition of this process with $Be^7(e^-,\nu)Li^7(p,\alpha)He^4$ in the burning of Be^7. Even more recent measurements of Winkler & Dwarakanath (1967) yield a considerably larger and more precise cross-section factor, $S(0) = 5 \pm 1$ MeV-barn, for the $He^3(\tau,2p)He^4$ reaction than the value $S(0) \sim 1$ MeV given by earlier measurements of Good, Kunz & Moak (1954). This changes the competition of this reaction with $He^4(\tau,\gamma)Be^7$.

It will be clear that important modifications have been introduced in the production rates of Be^7 and B^8. The neutrinos emitted in the decay of these two nuclei in the Sun are those which will be detected at the new solar-neutrino observatory under construction by Davis (1964). A detailed re-evaluation of the neutrino flux from the Sun for comparison with the imminent observations by Davis (1964) is now under way by Shaviv, Bahcall & Fowler (1967). Calculations based on previous results have been made by Sears (1964) and Bahcall (1966) and have been discussed by Bahcall & Davis (1966).

The CNO bi-cycle.—The new results reported here do not significantly change the equilibrium abundances calculated for the CNO bi-cycle by Caughlan & Fowler (1962) although, for example, $(C^{12}/C^{13}) = 3.4$ rather than 4.0 at equilibrium at 2×10^7 °K. More importantly, the conclusion is still inescapable that the CNO bi-cycle under ordinary circumstances processes C^{12} and O^{16} primarily into N^{14} with the production of only relatively small amounts of C^{13}, N^{15}, and O^{17}. Measurements by Honsaker (1960), Clayton (1962), and Brown (1962a) exorcised a number of ghostly resonances in the bi-cycle. Recent measurements by Hensley (1967) have shown that the sixth excited state in O^{15}, found to be perilously close to the $N^{14}(p,\gamma)O^{15}$ threshold by Warburton, Olness & Alburger (1965), is actually *bound* by 21.6 ± 1.1 keV and requires d-wave ($l = 2$) protons for formation of $N^{14}+p$ in any case. Consequently the wing of the state makes a negligible contribution to the rate of $N^{14}(p,\gamma)O^{15}$, and the conclusions stand that the overall rate of the bi-cycle is basically determined by the rate of this reaction and that N^{14} is the main constituent (>92 per cent) of the CNO isotopes at equilibrium under ordinary circumstances. Caughlan (1965) has shown that the production of excess N^{14} can be avoided only by limiting C^{12} or O^{16} to an exposure of less than one proton per nucleus involved.

Helium burning and subsequent processes.—It is important to emphasize that the rate of the Salpeter-Hoyle process, $3He^4 \rightarrow C^{12}$, is basically determined by beautiful but difficult measurements on the monopole transition in the inelastic scattering of electrons by C^{12} involving the key excited state in this nucleus at 7.644 MeV (Crannell & Griffy 1964). From the monopole matrix element for $C^{12} \rightarrow C^{12*}$ the partial width for electron-positron pair emission from $C^{12*} \rightarrow C^{12}$ can be calculated by reciprocity arguments. The widths for alpha-particle and gamma-ray emission by the excited state are then empirically determined relative to the pair-emission width. The electron-positron pair emission is relatively unimportant in the decay of $C^{12}*(7.644$ MeV) but the "inverse" monopole transition is the only one for which an absolute determination has been made, and all other transition widths are determined relative to the monopole matrix element. This is a superb illustration of the importance of detailed nuclear spectroscopy in astrophysics. The information given on page 558 reflects the latest in empirical results but does not greatly change the rate of the $3He^4 \rightarrow C^{12}$ process from the previously accepted value.

The rate of the next step in helium burning, $C^{12}(\alpha,\gamma)O^{16}$, has long been very uncertain because of the impossibility of directly measuring the reduced alpha-particle-emission width θ_α^2 for the bound excited state in O^{16} at 7.115 MeV. B^{2}FH (1957) made an order-of-magnitude estimate, $\theta_\alpha^2 \sim 0.1$, primarily on the basis that the reduced width is not expected to exceed unity and should not be less than 1 per cent on any reasonable model of the O^{16} nucleus. Fowler & Hoyle (1964) showed that attempts to evaluate θ_α^2 for the 7.115-MeV state by averaging over measured widths for other states were fallacious in principle. As discussed above, a valid theoretical calculation and an indirect measurement in $Li^6(C^{12},d)O^{16}$ are consistent with $\theta_\alpha^2 = 0.085 \pm 0.04$.

The uncertainty in the value for θ_α^2 is still relatively large, but reasonable conclusions can now be drawn concerning the end result of helium burning. This problem has been investigated most recently by Deinzer & Salpeter (1964). With the new input data their results indicate approximately equal production of C^{12} and O^{16} at the end of helium burning in stars with mass up to 10 $M_\odot$. The C^{12} production is less for greater masses, dropping to 10 per cent by mass at 100 $M_\odot$. Small amounts of Ne^{20} and Mg^{24} are also produced in stars with $M > 10$ $M_\odot$ but O^{16} is the major product, 60 per cent by mass in stars with $M = 100$ $M_\odot$.

It will be clear that helium burning will be succeeded by carbon burning $(C^{12}+C^{12})$ and eventually by oxygen burning $(O^{16}+O^{16})$ in giant stars for all stellar masses of practical interest. We have abstained from tabulating the rates of these processes on the grounds that it is not presently feasible to extrapolate the experimental results obtained at relatively high bombardment energies by Bromley, Kuehner & Almqvist (1960) and by Almqvist, Bromley & Kuehner (1960) into the relevant stellar-energy range. Efforts to extend the experimental determinations to lower energies are now being made by Patterson, Winkler & Zaidins (1967). We believe that the question whether carbon burning can occur in red supergiants at a temperature low enough to avoid large neutrino-energy loss through $e^+ + e^- \rightarrow \nu + \bar{\nu}$ is still open. The problem has been discussed in detail by Hayashi, Hōshi & Sugimoto (1962).

Silicon burning.—Carbon and oxygen burning lead eventually to the production mainly of Si^{28} but with smaller amounts of other intermediate-mass nuclei. Silicon burning proceeds through a complicated chain of events involving the photoproduction and radiative capture of alpha particles, protons, and neutrons. The process was originally dubbed the alpha process by B^2FH but a better terminology is just silicon burning. Tables II through VI contain data on the key photodisintegration rates for S^{32}, Si^{28}, Mg^{24}, Ne^{20}, O^{16}, and C^{12}. The buildup from Si^{28} to the iron-group nuclei through a quasi-equilibrium process is under investigation by Bodansky, Clayton & Fowler (1967).

In concluding this paper we wish to acknowledge help from Fay Ajzenberg-Selove, Neta Bahcall, Donald D. Clayton, Charles C. Lauritsen, Thomas Lauritsen, Peter B. Lyons, Georges J. Michaud, Edmund A. Milne, James W. Toevs, and Joseph L. Vogl. We also wish to express our appreciation to Jan Rasmussen for special care in typing the mansucript and the numerical tables.

We have taken every possible care in this research to avoid errors in the accumulation and presentation of the numerical data tabulated in Tables II through VI. In this connection, however, we wish to draw the reader's attention to the following quotation:

Research, of course, is no substitute for wisdom.

William Manchester (1966)

SUPPLEMENTARY TABLES

At the request of many colleagues we append supplements to Tables II through VI. These supplements give the numerical coefficients in the reaction rate equations for P_{01}, P_{23}, $P_{2\gamma}$, or P_{234} in reactions cm^{-3} sec^{-1}.

In form the equations are very similar, except for an additional factor ρ, to those throughout the main text for $\mathcal{E}$, the energy generation. For example, compare the following expression with Equation 81:

$$P(C^{12}p)_\gamma = 1.02 \times 10^{30} \, (1 + 0.0304 \, T_9^{1/3} + 1.19 \, T_9^{2/3} + 0.254 \, T_9$$
$$+ 2.06 \, T_9^{4/3} + 1.12 \, T_9^{5/3})$$
$$\times \rho^2 X_H X_{C^{12}} T_9^{-2/3} \exp(-13.69/T_9^{1/3}) \qquad 0 \leq T_9 \leq 0.55 \quad 103.$$
$$+ 5.38 \times 10^{27} \, \rho^2 X_H X_{C^{12}} \, T_9^{-3/2} \exp(-4.925/T_9) \qquad 0.25 \leq T_9 \leq 7$$

$$\text{reactions cm}^{-3} \text{ sec}^{-1}$$

TABLE II SUPPLEMENT

Neutron Nonresonant-Reaction Data
Coefficients in Reaction Rates per cm³ sec

Reaction	$H^1(n,\gamma)D^2$	$D^2(n,\gamma)T^3$	$He^3(n,\gamma)He^4$	$He^3(n,p)T^3$	$Li^6(n,\gamma)Li^7$	$Li^6(n,t)He^4$
P_{01}	2.61 E 28	1.96 E 25	1.31 E 24	1.40 E 32	5.92 E 26	1.24 E 31
$P_{23}\,P_{2\gamma}\,P_{234}$	6.20 E 37	2.16 E 35	2.61 E 34	1.40 E 32	6.09 E 36	6.68 E 30

Reaction	$Li^7(n,\gamma)Li^8$	$Be^7(n,p)Li^7$	$Be^9(n,\gamma)Be^{10}$	$B^{10}(n,\gamma)B^{11}$	$B^{10}(n,\alpha)Li^7$	$B^{11}(n,\gamma)B^{12}$
P_{01}	4.17 E 26	5.75 E 32	8.34 E 25	3.95 E 27	3.03 E 31	3.59 E 25
$P_{23}\,P_{2\gamma}\,P_{234}$	4.81 E 36	5.77 E 32	5.17 E 36	1.10 E 38	8.23 E 30	7.76 E 35

Reaction	$C^{12}(n,\gamma)C^{13}$	$C^{13}(n,\gamma)C^{14}$	$N^{14}(n,\gamma)N^{15}$	$N^{14}(n,p)C^{14}$	$N^{15}(n,\gamma)N^{16}$	$O^{16}(n,\gamma)O^{17}$
P_{01}	2.24 E 25	5.48 E 24	4.24 E 26	1.02 E 28	1.27 E 23	8.80 E 23
$P_{23}\,P_{2\gamma}\,P_{234}$	1.85 E 35	1.83 E 35	1.08 E 37	3.08 E 28	8.68 E 32	2.54 E 33

Reaction	$O^{17}(n,\alpha)C^{14}$	$O^{18}(n,\gamma)O^{19}$	$F^{19}(n,\gamma)F^{20}$	$Ne^{22}(n,\alpha)O^{18}$	$Ne^{22}(n,\gamma)Ne^{23}$	$Na^{23}(n,\gamma)Na^{24}$
P_{01}	1.09 E 27	9.23 E 23	4.08 E 25	3.62 E 29	1.30 E 26	1.84 E 27
$P_{23}\,P_{2\gamma}\,P_{234}$	6.80 E 26	2.71 E 33	2.89 E 35	1.36 E 29	3.89 E 35	1.48 E 37

TABLE III SUPPLEMENT

Charged-Particle Nonresonant-Reaction Data
Coefficients in Reaction Rates per cm³ sec

Reaction	$H^1(p,e^+\nu)D^2$	$H^1(pe^-,\nu)D^2$	$He^3(p,e^+\nu)He^4$	$D^2(p,\gamma)He^3$	$D^2(d,n)He^3$	$D^2(d,p)T^3$
P_{01}	9.81 E 08	3.35 E 03	7.94 E 10	6.63 E 26	3.07 E 31	3.07 E 31
$P_{23}\,P_{2\gamma}\,P_{234}$	—	—	—	7.29 E 36	1.42 E 32	1.42 E 32

Reaction	$D^2(d,\gamma)He^4$	$T^3(p,\gamma)He^4$	$T^3(d,n)He^4$	$He^3(d,p)He^4$	$He^4(t,\gamma)Li^7$	$He^4(\tau,\gamma)Be^7$
P_{01}	1.29 E 23	4.36 E 27	8.02 E 33	6.43 E 33	2.63 E 28	2.43 E 29
$P_{23}\,P_{2\gamma}\,P_{234}$	1.18 E 34	8.64 E 37	6.68 E 34	5.37 E 34	5.04 E 38	4.66 E 39

Reaction	$T^3(t,2n)He^4$	$He^3(t,np)He^4$	$He^3(\tau,2p)He^4$	$He^3(t,d)He^4$	$Li^6(p,\tau)He^4$	$Li^7(p,\alpha)He^4$
P_{01}	3.62 E 31	3.70 E 32	1.79 E 33	2.56 E 32	6.48 E 33	1.25 E 32
$P_{23}\,P_{2\gamma}\,P_{234}$	2.73 E 22	2.81 E 23	1.36 E 24	4.62 E 32	3.48 E 33	1.30 E 32

Reaction	$C^{12}(p,\gamma)N^{13}$	$C^{13}(p,\gamma)N^{14}$	$N^{13}(p,\gamma)O^{14}$	$N^{14}(p,\gamma)O^{15}$	$N^{15}(p,\gamma)O^{16}$	$N^{15}(p,\alpha)C^{12}$
P_{01}	1.02 E 30	3.68 E 30	1.94 E 30	1.80 E 30	1.67 E 31	3.25 E 34
$P_{23}\,P_{2\gamma}\,P_{234}$	8.37 E 39	4.10 E 40	6.47 E 40	4.56 E 40	5.71 E 41	7.22 E 33

Reaction	$Be^7(p,\gamma)B^8$	$Be^9(p,d)Be^8$	$Be^9(p,\alpha)Li^6$	$B^{11}(p,2\alpha)He^4$	$N^{14}(\alpha,\gamma)F^{18}$	$O^{18}(\alpha,\gamma)Ne^{22}$
P_{01}	4.42 E 28	1.28 E 34	1.28 E 34	7.47 E 34	1.21 E 39	8.63 E 39
$P_{23}\,P_{2\gamma}\,P_{234}$	5.08 E 38	8.11 E 33	2.98 E 33	7.54 E 23	2.05 E 50	1.65 E 51

Reaction	$B^{10}(p,\alpha)Be^7$	$C^{13}(\alpha,n)O^{16}$	$O^{16}(p,\gamma)F^{17}$	$O^{16}(\alpha,\gamma)Ne^{20}$	$O^{17}(p,\alpha)N^{14}$	$Ne^{20}(p,\gamma)Na^{21}$
P_{01}	7.51 E 33	7.83 E 37	6.14 E 30	1.26 E 31	6.72 E 31	2.81 E 31
$P_{23}\,P_{2\gamma}\,P_{234}$	2.03 E 33	1.46 E 39	1.77 E 40	2.28 E 42	1.39 E 31	1.25 E 41

TABLE IV SUPPLEMENT

Neutron Resonant-Reaction Data
Coefficients in Reaction Rates per CM³ SEC

Reaction	$He^3(n,p)T^3$	$Li^6(n,t)He^4$	$Li^7(n,\gamma)Li^8$	$B^{11}(n,\gamma)B^{12}$	$N^{14}(n,p)C^{14}$	$N^{14}(n,p)C^{14}$
P_{01}	$2.56\ E\ 34$	$3.63\ E\ 32$	$8.47\ E\ 26$	$3.44\ E\ 27$	$3.28\ E\ 30$	$1.59\ E\ 31$
$P_{23}\ P_{2\gamma}\ P_{234}$	$2.56\ E\ 34$	$1.95\ E\ 32$	$9.78\ E\ 36$	$7.45\ E\ 37$	$9.87\ E\ 30$	$4.79\ E\ 31$

TABLE V SUPPLEMENT

Charged-Particle Resonant-Reaction Data
Coefficients in Reaction Rates per CM³ SEC

Reaction	$T^3(d,n)He^4$	$He^3(d,p)He^4$	$Li^7(p,\gamma)Be^8$	$Li^7(\alpha,\gamma)B^{11}$	$Li^7(\alpha,\gamma)B^{11}$	$Li^7(\alpha,\gamma)B^{11}$
P_{01}	$1.20\ E\ 32$	$2.56\ E\ 32$	$1.31\ E\ 29$	$3.65\ E\ 25$	$4.47\ E\ 26$	$2.84\ E\ 27$
$P_{23}\ P_{2\gamma}\ P_{234}$	$9.98\ E\ 32$	$2.14\ E\ 33$	$7.60\ E\ 39$	$3.74\ E\ 36$	$4.58\ E\ 37$	$2.91\ E\ 38$

Reaction	$Be^7(p,\gamma)B^8$	$Be^7(\alpha,\gamma)C^{11}$	$Be^7(\alpha,\gamma)C^{11}$	$Be^9(p,d)Be^8$	$Be^9(p,d)Be^8$	$Be^9(p,d)Be^8$
P_{01}	$2.78\ E\ 26$	$1.56\ E\ 27$	$3.90\ E\ 27$	$7.73\ E\ 31$	$9.74\ E\ 31$	$3.73\ E\ 32$
$P_{23}\ P_{2\gamma}\ P_{234}$	$3.19\ E\ 36$	$1.60\ E\ 38$	$3.99\ E\ 38$	$4.91\ E\ 31$	$6.19\ E\ 31$	$2.37\ E\ 32$

Reaction	$Be^9(p,\alpha)Li^6$	$Be^9(p,\alpha)Li^6$	$Be^9(\alpha,n)C^{12}$	$Be^9(\alpha,n)C^{12}$	$Be^9(\alpha,n)C^{12}$	$Be^9(\alpha,n)C^{12}$
P_{01}	$5.98\ E\ 31$	$1.21\ E\ 32$	$2.11\ E\ 27$	$4.79\ E\ 26$	$3.14\ E\ 31$	$4.44\ E\ 31$
$P_{23}\ P_{2\gamma}\ P_{234}$	$1.39\ E\ 31$	$2.81\ E\ 31$	$6.47\ E\ 28$	$1.47\ E\ 28$	$9.62\ E\ 32$	$1.36\ E\ 33$

Reaction	$B^{10}(p,\alpha)Be^7$	$B^{10}(p,\alpha)Be^7$	$B^{11}(p,\gamma)C^{12}$	$B^{11}(p,\gamma)C^{12}$	$B^{11}(p,\gamma)C^{12}$	$B^{11}(p,2\alpha)He^4$
P_{01}	$2.17\ E\ 32$	$3.47\ E\ 32$	$5.00\ E\ 26$	$3.02\ E\ 28$	$2.60\ E\ 29$	$3.24\ E\ 28$
$P_{23}\ P_{2\gamma}\ P_{234}$	$5.88\ E\ 31$	$9.40\ E\ 31$	$3.24\ E\ 37$	$1.96\ E\ 39$	$1.68\ E\ 40$	$3.27\ E\ 17$

Reaction	$B^{11}(p,2\alpha)He^4$	$B^{11}(p,2\alpha)He^4$	$C^{12}(p,\gamma)N^{13}$	$C^{13}(p,\gamma)N^{14}$	$C^{13}(p,\gamma)N^{14}$	$Ne^{20}(p,\gamma)Na^{21}$
P_{01}	$3.63\ E\ 32$	$7.65\ E\ 32$	$5.38\ E\ 27$	$6.21\ E\ 28$	$1.04\ E\ 29$	$4.90\ E\ 27$
$P_{23}\ P_{2\gamma}\ P_{234}$	$3.66\ E\ 21$	$7.72\ E\ 21$	$4.43\ E\ 37$	$6.92\ E\ 38$	$1.16\ E\ 39$	$2.18\ E\ 37$

Reaction	$N^{14}(p,\gamma)O^{15}$	$N^{14}(p,\gamma)O^{15}$	$N^{14}(p,\gamma)O^{15}$	$C^{13}(\alpha,n)O^{16}$	$C^{13}(\alpha,n)O^{16}$	$C^{13}(\alpha,n)O^{16}$
P_{01}	$9.74\ E\ 25$	$2.66\ E\ 27$	$2.17\ E\ 29$	$2.45\ E\ 28$	$2.22\ E\ 29$	$5.50\ E\ 28$
$P_{23}\ P_{2\gamma}\ P_{234}$	$2.47\ E\ 36$	$6.76\ E\ 37$	$5.52\ E\ 39$	$4.58\ E\ 29$	$4.15\ E\ 30$	$1.03\ E\ 30$

Reaction	$N^{15}(p,\gamma)O^{16}$	$N^{15}(p,\gamma)O^{16}$	$N^{15}(p,\alpha)C^{12}$	$N^{15}(p,\alpha)C^{12}$	$N^{15}(p,\alpha)C^{12}$	$O^{17}(p,\alpha)N^{14}$
P_{01}	$4.64\ E\ 26$	$2.93\ E\ 29$	$5.14\ E\ 30$	$1.37\ E\ 32$	$3.95\ E\ 31$	$2.76\ E\ 20$
$P_{23}\ P_{2\gamma}\ P_{234}$	$1.59\ E\ 37$	$1.00\ E\ 40$	$1.14\ E\ 30$	$3.04\ E\ 31$	$8.77\ E\ 30$	$5.71\ E\ 19$

Reaction	$C^{12}(\alpha,\gamma)O^{16}$	$C^{12}(\alpha,\gamma)O^{16}$	$C^{12}(\alpha,\gamma)O^{16}$	$O^{16}(\alpha,\gamma)Ne^{20}$	$O^{16}(\alpha,\gamma)Ne^{20}$	$O^{16}(\alpha,\gamma)Ne^{20}$
P_{01}	$2.45\ E\ 25$	$1.32\ E\ 25$	$1.54\ E\ 26$	$2.83\ E\ 23$	$2.43\ E\ 24$	$6.68\ E\ 24$
$P_{23}\ P_{2\gamma}\ P_{234}$	$3.78\ E\ 36$	$2.03\ E\ 36$	$2.37\ E\ 37$	$5.13\ E\ 34$	$4.39\ E\ 35$	$1.21\ E\ 36$

TABLE VI SUPPLEMENT

Continuum-Reaction Data
Coefficients in Reaction Rates per cm^3 sec

Reaction	$F^{19}(p,\gamma)Ne^{20}$	$Ne^{20}(\alpha,\gamma)Mg^{24}$	$Na^{23}(p,\gamma)Mg^{24}$	$Na^{23}(p,\alpha)Ne^{20}$	$Mg^{24}(\alpha,\gamma)Si^{28}$	$Al^{27}(p,\gamma)Si^{28}$
P_{01}	2.35 E 27	6.54 E 25	1.76 E 27	2.85 E 28	1.08 E 25	2.56 E 27
P_{23} $P_{2\gamma}$ P_{234}	8.31 E 37	1.31 E 37	1.28 E 38	1.03 E 28	2.32 E 36	2.82 E 38

Reaction	$Al^{27}(p,\alpha)Mg^{24}$	$Si^{28}(\alpha,\gamma)S^{32}$	$P^{31}(p,\gamma)S^{32}$	$P^{31}(p,\alpha)Si^{28}$	$Cl^{35}(p,\gamma)Ar^{36}$	$K^{39}(p,\gamma)Ca^{40}$
P_{01}	9.40 E 28	2.17 E 24	3.77 E 26	1.53 E 29	4.56 E 26	8.87 E 26
P_{23} $P_{2\gamma}$ P_{234}	4.81 E 28	4.92 E 35	1.40 E 37	2.51 E 28	3.42 E 37	6.70 E 37

LITERATURE CITED

Abuzeid, M. A., Aly, F. M., Antoufiev, Y. P., Baranik, A. T., Nower, T. M., Sorokin, P. V. 1963, *Nucl. Phys.*, **45**, 123

Ajzenberg-Selove, F., Lauritsen, T. 1959, *Nucl. Phys.*, **11**, 1

Ajzenberg-Selove, F., Lauritsen, T. 1967 (To be published)

Almqvist, E., Bromley, D. A., Kuehner, J. A. 1960, *Phys. Rev. Letters*, **4**, 515

Antoufiev, Y. P., El-Nadi, L. M., Darwish, D. A. E., Badawy, O. E., Sorokin, P. V. 1963, *Nucl. Phys.*, **46**, 1

Bahcall, J. N. 1962, *Phys. Rev.*, **128**, 1297

Bahcall, J. N. 1964, *ibid.*, **135**, B137

Bahcall, J. N. 1966, *Phys. Rev. Letters*, **17**, 398

Bahcall, J. N., Davis, R. Jr. 1966, *Stellar Evolution*, 241–43 (Stein, R. F., Cameron, A. G. W., Eds., Plenum Press, New York)

Bahcall, J. N., Wolf, R. A. 1964, *Ap. J.*, **139**, 622

Benn, J., Dally, E. B., Müller, H. H., Pixley, R. E., Staub, H. H., Winkler, H. 1966, *Phys. Letters*, **20**, 43

Bethe, H. A., Morrison, P. 1956, *Elementary Nuclear Theory*, 75 (2nd ed., Wiley, New York)

Blatt, J. M., Weisskopf, V. F. 1952, *Theoretical Nuclear Physics* (Wiley, New York)

Bodansky, D., Clayton, D. D., Fowler, W. A. 1967 (Private communication)

Bromley, D. A., Kuehner, J. A., Almqvist, E. 1960, *Phys. Rev. Letters*, **4**, 365

Brown, R. E. 1962a, *Ap J.*, **137**, 338

Brown, R. E., 1962b, *Phys Rev.*, **125**, 347

Burbidge, E. M., Burbidge, G. R., Fowler, W. A., Hoyle, F. 1957, *Rev. Mod. Phys.*, **29**, 547

Caughlan, G. R. 1965, *Ap. J.*, **141**, 688

Caughlan, G. R., Fowler, W. A. 1962, *Ap. J.* **136**, 453

Caughlan, G. R., Fowler, W. A. 1964, *ibid.*, **139**, 1180

Christy, R. F. 1961, *Nucl. Phys.*, **22**, 301

Christy, R. F., Duck, I. 1961, *Nucl. Phys.*, **24**, 89

Clayton, D. D. 1962, *Phys. Rev.*, **128**, 2254

Clayton, D. D., Shaw, P. B. 1967, *Ap. J.*, **148**, 301

Cook, C. W., Fowler, W. A., Lauritsen, C. C., Lauritsen, T. 1957, *Phys. Rev.*, **107**, 508

Crannell, H. L. 1965 (Private communication)

Crannell, H. L., Griffy, T. A. 1964, *Phys. Rev.*, **136**, B1580

Davids, C. 1966 (Private communication)

Davis, R., Jr. 1964, *Phys. Rev. Letters*, **12**, 302

Deinzer, W., Salpeter, E. E. 1964, *Ap. J.*, **140**, 499

Endt, P. M., Van der Leun, C. 1962, *Nucl. Phys.*, **34**, 1, 325

Erné, F. C., Endt, P. M. 1965, *Nucl. Phys.*, **71**, 593

Evans, R. D. 1955, *The Atomic Nucleus*, 336 (McGraw-Hill, New York)

Fowler, W. A. 1954, *Mém. Soc. Roy. Sci. Liège*, Ser. 4, **13**, 88

Fowler, W. A. 1960, *ibid.*, Ser. 5, **3**, 207

Fowler, W. A. 1966, *Ap. J.*, **144**, 180

Fowler, W. A., Caughlan, G. R., Zimmerman, B. A. 1967, *Ann. Rev. Astron. Ap.*, **5**, 525. References to this present paper in the tables indicate original calculations made by the authors.

Fowler, W. A., Hoyle, F. 1964, *Ap. J. Suppl. No. 91*, **9**, 201

Fowler, W. A., Vogl, J. L. 1964, *Lectures in Theoretical Physics*, **VI**, 379 (Univ. Colorado Press, Boulder, Colo.)

Gamow, G. 1928, *Z. Physik*, **51**, 204

Gibbons, J. H., Macklin, R. L. 1965, *Phys. Rev.*, **137**, B1508

Glaudemans, P. W. M., Endt, P. M. 1962, *Nucl. Phys.*, **30**, 30

Good, W. M., Kunz, W. E., Moak, C. D. 1954, *Phys. Rev.*, **94**, 87

Govorov, A. M. Youn, Li Ha, Osetinskii, G. M., Salatskii, V. I., Sizov, I. V. 1962, *Soviet Phys. JETP*, **14**, 508

Griffiths, G. M., Morrow, R. A., Riley, P. J., Warren, J. B. 1961, *Can. J. Phys.*, **39**, 1397

Griffiths, G. M., Lal, M., Scarfe, C. D. 1963, *Can. J. Phys.*, **41**, 724

Gudden, F., Strehl, P. 1965, *Z. Physik*, **185**, 111

Gurney, R. W., Condon, E. U. 1928, *Nature*, **122**, 439

Gurney, R. W., Condon, E. U. 1929, *Phys. Rev.*, **33**, 127

Hayashi, C., Hōshi, R., Sugimoto, D. 1962, *Progr. Theoret. Phys. Suppl.* **22**, 1

Hebbard, D. F. 1960, *Nucl. Phys.*, **15**, 289

Hebbard, D. F., Bailey, G. M. 1963, *Nucl. Phys.*, **49**, 666

Hebbard, D. F., Vogl, J. L. 1960, *Nucl. Phys.*, **21**, 652

Hensley, D. C. 1967, *Ap. J.*, **147**, 818

Honsaker, J. L. 1960, *Ap. J.*, **132**, 516

Hoyle, F., Fowler, W. A. 1960, *Ap. J.*, **132**, 565

Hughes, D. J., Magurno, B. A., Brussel, M. K. 1960, *Brookhaven Natl. Lab. Rept. BNL Suppl. 1* (2nd ed.)

Hughes, D. J., Schwartz, R. B. 1958, *Brookhaven Natl. Lab. Rept. BNL 325* (2nd ed.)

Humblet, J., Rosenfeld, L. 1961, *Nucl. Phys.*, **26**, 529

Iben, I., Jr. 1965, *Ap. J.*, **142**, 1447

Jarmie, N., Seagrave, J. D. 1957, *Los Alamos Rept. LA 2014*

Kuperus, J., Glaudemans, P. W. M., Endt, P. M. 1963, *Physica*, **29**, 1281

Lane, A. M., Thomas, R. G. 1958, *Rev. Mod. Phys.*, **30**, 257

Larson, J. D., Spear, R. H. 1964, *Nucl. Phys.*, **56**, 497

Lauritsen, T., Ajzenberg-Selove, F. 1966, *Nucl. Phys.*, **78**, 1

Leenhouts, H. P., Endt, P. M. 1966, *Physica*, **32**, 322

Loebenstein, H. M., Mingay, D. W., Winkler, H. C., Zaidins, C. S. 1967, *Nucl.*

Phys., **A91**, 481

Macklin, R. L., Gibbons, J. H. 1965, *Rev. Mod. Phys.*, **37**, 166

Mahaux, C. M. 1965, *Nucl. Phys.*, **71**, 241

Manchester, W. 1966, *Look*, **30**, No. 24, 89

Milne, E. A. 1954 (Ph.D. thesis, California Inst. of Technol., Pasadena)

Mozer, F. S. 1956 (Ph.D. thesis, California Inst. of Technol., Pasadena)

Newton, I., *Opticks: or, A Treatise of the Reflexions, Refractions, Inflexions and Colours of Light* (Smith & Walford, Printers to the Royal Society, London, 1704. The fourth edition, first published in 1730, has been reproduced by Dover Publ., New York, 1952. The introductory quotation is to be found on page 374 of the Dover edition.)

Parker, P. D. 1966, *Ap. J.*, **145**, 960

Parker, P. D., Bahcall, J. N., Fowler, W. A. 1964, *Ap. J.*, **139**, 602

Patterson, J. P., Winkler, H., Zaidins, C. S. 1967 (Private communication)

Prosser, F. W., Jr., Unruh, W. P., Wildenthal, B. H., Krone, R. W. 1962, *Phys. Rev.*, **125**, 594

Reeves, H. 1965, *Stellar Structure*, Chap. 2 (Aller, L. H., McLaughlin, D. B., Eds., Univ. of Chicago Press, Chicago)

Salpeter, E. E. 1954, *Australian J. Phys.*, **7**, 373

Schardt, A. W. 1951 (Ph.D. thesis, California Inst. of Technol., Pasadena)

Seagrave, J. D. 1952, *Phys. Rev.*, **85**, 197

Sears, R. L. 1964, *Ap. J.*, **140**, 477

Seeger, P. A., Kavanagh, R. W. 1963, *Nucl. Phys.*, **46**, 577

Shaviv, G., Bahcall, J. N., Fowler, W. A. 1967 (Private communication)

Smulders, P. J. M. 1964, *Physica*, **30**, 1197

Smulders, P. J. M. 1965, *ibid.*, **31**, 973

Smulders, P. J. M., Endt, P. M. 1962, *Physica*, **28**, 1093

Spring, E., Holmberg, P., Jungner, H. 1965, *Comm. Phys.-Math.*, **31**, No. 1

Stehn, J. R., Goldberg, M. D., Magurno, B. A., Weiner-Chasman, R. 1964, *Brookhaven Natl. Lab. Rept. BNL 325, Suppl. 2* (2nd ed.)

Stephenson, G. J., Jr. 1966, *Ap. J.*, **146**, 950

Swann, C. P., Metzger, F. R. 1957, *Phys. Rev.*, **108**, 982

Tanner, N. 1959, *Phys. Rev.*, **114**, 1060

Thomas, R. G. 1952, *Phys. Rev.*, **88**, 1109

Van der Leun, C., Sheppard, D. M., Smulders, P. J. M. 1965, *Phys. Letters*, **18**, 134

Vogl, J. L. 1963 (Ph.D. thesis, California Inst. of Technol., Pasadena)

Wagoner, R. V., Fowler, W. A., Hoyle, F. 1967, *Ap. J.*, **148**, 3

Walton, R. B., Clement, J. D., Boreli, F. 1957, *Phys. Rev.*, **107**, 1065

Warburton, E. K., Olness, J. W., Alburger, D. E. 1965, *Phys. Rev.*, **140**, B1202

Weinman, J. A., Meyer-Schützmeister, L., Lee, L. L., Jr. 1964. *Phys. Rev.*, **133**, B590

Werntz, C. W., Brennan, J. G. 1967, *Phys. Rev.*, (To be published)

Winkler, H. C., Dwarakanath, M. R. 1967, *Bull. Am. Phys. Soc.*, **12**, 16

Wolf, R. A. 1965, *Phys. Rev.*, **137**, B1634

Wolf, R. A. 1966, *Ap. J.*, **145**, 834

Youn, Li Ha, Osetinskii, G. M., Sodnom, N., Govorov, A. M., Sizov, I. V., Salatskii, V. I. 1961, *Soviet Phys. JETP*, **12**, 163

Zurmühle, R. W., Stephens, W. E., Staub, H. H. 1963, *Phys. Rev.*, **132**, 751

STELLAR EVOLUTION WITHIN AND OFF THE MAIN SEQUENCE[1,2]

BY ICKO IBEN, JR.

Massachusetts Institute of Technology
Cambridge, Massachusetts

INTRODUCTION

During the past five years, considerable progress has been made in our understanding of stellar evolution during and beyond the hydrogen-burning phases. Through ingenious approximations which reduce the labor of integration and yet retain the basic features of the physical description, Japanese workers have followed stellar evolution from the early stages of pre-main-sequence contraction through and somewhat beyond the shell helium-burning stage. In Germany and the United States several groups have exploited the speed and capacity of available digital computers to construct, by relaxation techniques, nearly uninterrupted sequences of models covering the same range but showing considerably more detail, particularly in the transitional phases between major periods of nuclear burning.

Among Population I stars outside the main-sequence band, the observed regularities for which one can now account in terms of the properties of helium-burning models have become sufficiently numerous that it is almost fair to say that we understand evolution through helium-burning phases for Population I stars nearly as well as we understand evolution through pure hydrogen-burning phases. This does not mean that the description of either the hydrogen- or helium-burning phases is complete. There are still serious gaps and uncertainties in the description of such stars, as well as quantitative inconsistencies between theory and observation.

The effects of mass loss, rotation, and neutrino processes on evolutionary characteristics during the hydrogen- and helium-burning phases have not yet been fully studied. The evolution of close binaries has barely been touched upon. As there is evidence that many "cataclysmic" variables owe their behavior to the presence of a close companion, it is clear that this latter area of research is important.

In this review, attention will be restricted to theoretical studies of stars in the hydrogen- and helium-burning phases and to comparisons between these studies and observation. Emphasis will be placed on work that has not been reviewed by Schwarzschild (1), by the Burbidges (2), by Hayashi, Hōshi & Sugimoto (3), and by Sears & Brownlee (4). An extensive bibliography of recent (≥1958) studies of stellar interiors has been prepared by Langer, Herz

[1] The survey of literature for this review was concluded December 31, 1966.

[2] Supported in part by the National Science Foundation (GP-6387) and in part by the National Aeronautics and Space Administration (NsG-496).

& Cox (5). The large number, 834, of entries in this compilation demonstrates the practical impossibility of doing justice, in a short review, to all work in the field of stellar evolution. The selection of papers for explicit citation has been necessarily somewhat arbitrary and has been centered about the reviewer's own special interests.

THE EVOLUTION OF A $5M_\odot$ POPULATION I STAR

A description of the evolution of a metal-rich star of $5M_\odot$ illustrates recent advances in the study of Population I stars that begin their nuclear-burning life on the upper main sequence. The evolution of models of $5M_\odot$ has been carried further than for models of any other mass (6–8). Moreover, evolution through the core helium-burning phase has been examined by several investigators with varying input physics (8–10); this has offered an opportunity for intercomparison and the assessment of theoretical uncertainties.

The evolutionary track of a $5M_\odot$ model of Population I characteristics is sketched in Figure 1. Numbered circles along the path denote boundaries of easily distinguished phases. Brief descriptions and approximate time scales for each phase are also supplied.

The track in Figure 1 is a rough, order-of-magnitude map which exhibits the basic features of all $5M_\odot$ models constructed to date (8–10) prior to the shell helium-burning phase. The track and time scales between points 1 and 12 are taken from Iben (10). The initial composition is $(X, Y, Z) = (0.71, 0.27, 0.02)$ and $(l/H_d) = $ (mixing length/density scale height) $= 0.5$. The track beyond point 12 is topologically equivalent to that given by Kippenhahn, Thomas & Weigert (8) but is at variance with tracks obtained by Hofmeister (9).

CORE HYDROGEN BURNING

During the main-sequence or core hydrogen-burning phase (points 1–2), hydrogen is converted into helium via the CN-cycle reactions near the center of a large convective core. Mixing causes the hydrogen concentration to drop steadily in a region large compared to the region of significant nuclear-energy production. Throughout the main-sequence phase, the mass fraction in the convective core decreases. For the particular choice of composition that gives the track in Figure 1, the convective-core mass fraction drops steadily from about 0.2 to 0.08 before diminishing rapidly to zero following the phase of overall contraction (near point 3).

If, as has been customary, one assumes that the abundance of N^{14} remains fixed in energy-generating regions, one finds that the central portion of the star contracts and heats throughout the main-sequence phase (8, 9). According to the usual interpretation of this phenomenon, as the average number of particles per gram decreases in the convective core, the balance between pressure forces outward and gravitational forces inward can be maintained only by an increase in density and hence, by virtue of the virial theorem, by

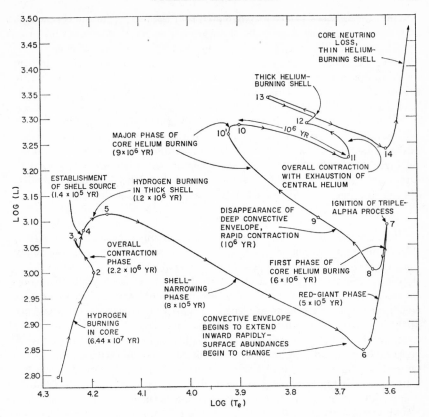

FIG. 1. The path of a metal-rich $5M_\odot$ star in the Hertzsprung-Russell diagram. Luminosity is in solar units, $L_\odot = 3.86 \times 10^{33}$ erg/sec, and surface temperature T_e is in deg K. Traversal times between labeled points are given in years.

heating. Since nuclear sources are highly temperature sensitive, rising core temperatures lead to an increase in stellar luminosity, in agreement with numerical results.

In a real Population I star the abundance of N^{14} does not remain fixed in energy-generating regions. As a result of the reactions $O^{16}(p,\gamma)F^{17}(\beta^+\nu)O^{17}(p,\alpha)N^{14}$, the abundance of N^{14} increases steadily during the main-sequence phase. Central temperatures *drop* during the early portions, and central densities *drop* for a significantly larger fraction, of the main-sequence phase (10). One can argue that in order to maintain a constant energy flux, the increase in the product of the N^{14} and hydrogen abundances must be compensated by a decrease in temperature or density (or both) in energy-generating regions. The increased luminosity during the early period of central cooling and expansion results from the increased concentration of CN-cycle elements.

Once O^{16} has been sufficiently depleted that the product of the H^1 and N^{14} abundances decreases with time, the central regions do contract and heat simultaneously. The rates of heating and condensing accelerate as the core-hydrogen abundance drops until, when the core abundance of hydrogen reaches approximately 0.05 (abundance by mass), the whole star begins to contract. During the ensuing phase of overall contraction (points 2–3), the rate of nuclear-energy production in the convective core remains nearly constant. The increase in luminosity is due primarily to the conversion, in the contracting envelope, of gravitational potential energy into heat and thence into escaping radiation. Toward the end of the phase of overall contraction, the mass fraction in the convective core begins to decrease at an accelerated rate. As the star reaches point 3, the hydrogen abundance in the now rapidly diminishing convective core is reduced to a fraction of 1 per cent.

Shell Hydrogen Burning

Between points 3 and 4, the region of major nuclear-energy production shifts from near the center, where hydrogen becomes exhausted, to a thick shell away from the center, where hydrogen is abundant. The details of this shift are rather complex. Within the central regions, where hydrogen is being exhausted, a finite temperature gradient carries outward a flow of energy which exceeds the rate of nuclear-energy production. The central regions call upon the only other resources immediately available—gravitational potential energy and thermal energy. The hydrogen-impoverished core contracts much more rapidly than during the preceding phase of overall contraction. At the same time, the core also cools; this helps to satisfy the demand for energy. Through cooling more rapidly at the center, the region of hydrogen exhaustion rapidly becomes nearly isothermal. As isothermality is approached, the demand for energy from the core decreases and the rates of core contraction and cooling diminish considerably. At point 4 in Figure 1, the nearly isothermal, hydrogen-exhausted core comprises about 7 per cent of the mass of the star.

While the core is still in the process of contracting and cooling rapidly, hydrogen-rich matter just outside of the region of hydrogen exhaustion is drawn inward to higher densities and temperatures and thereupon ignited. The ignition is mildly explosive, in the sense that matter is pushed away in both directions from the new hydrogen-burning region. Matter throughout the envelope expands outward and the stellar radius increases. Since energy is required to expand matter in the envelope, not all of the nuclear energy generated in the shell reaches the surface. This is the reason for the drop in luminosity immediately following point 3 in Figure 1. It is to be emphasized that between points 3 and 4, the total rate of nuclear-energy production increases with time. Had absorption in the envelope been omitted, as in earlier computations, no drop in luminosity following point 3 would have been detected.

At point 4 in Figure 1, reasonably stable conditions have again been established. The release of gravitational energy in the hydrogen-exhausted core and the absorption of energy in the stellar envelope are minimal. Between points 4 and 5, nuclear-energy production occurs in a fairly thick shell containing approximately 5 per cent of the star's mass. The shell moves outward (in mass fraction) through the star, adding more mass to the hydrogen-exhausted core, which continues to contract.

The distribution of several composition variables within the $5\,M_\odot$ star immediately following the phase of overall contraction, between points 3 and 4, is shown in Figure 2 (10). Only the features that are independent of the choice of initial Population I abundances are of interest here.

Wherever hydrogen has been converted almost completely into helium, oxygen has been converted almost completely into N^{14}. Within approximately the inner one half of the star's mass, C^{12} has been converted almost completely into N^{14}. Any initial lithium (not shown) has been destroyed over the inner 98–99 per cent of the star's mass. Similar statements could be made for boron, beryllium, and other light elements. It is worthwhile to point out that considerable quantities of He^3 are made during the main-sequence phase

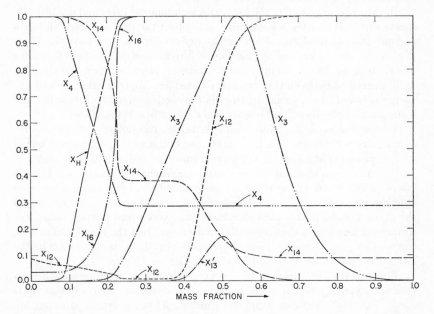

FIG. 2. The variation of composition with mass fraction in the $5\,M_\odot$ star immediately following the phase of overall contraction (point 3 in Figure 1). The X_i are abundances by mass of $H^1(X_H)$, $He^3(X_3)$, $He^4(X_4)$, $C^{12}(X_{12})$, $C^{13}(X'_{13})$, $N^{14}(X_{14})$, and $O^{16}(X_{16})$. Scale limits correspond to $0.0 \le X_H < 0.708$, $0.0 \le X_3 \le 1.30 \times 10^{-4}$, $0.0 \le X_4 \le 0.976$, $0.0 \le X_{12}$, $X'_{13} \le 3.61 \times 10^{-3}$, $0.0 \le X_{14} \le 1.45 \times 10^{-2}$, and $0.0 \le X_{16} \le 1.08 \times 10^{-2}$.

in approximately the middle third of the star's mass, far from regions of significant nuclear-energy production.

As the star approaches point 5 in Figure 1, the mass of the nearly isothermal, hydrogen-exhausted core approaches approximately 10 per cent of the star's mass. A satisfactory physical explanation of why the pressure balance cannot be maintained in an isothermal core when the mass fraction in the core reaches the Schönberg-Chandrasekhar limit (11)—approximately 0.1—has not yet been given. Nor is it understood why core contraction should not proceed rapidly enough to produce a much steeper temperature gradient in the core when the mass fraction is still less than 0.1.

At any rate, as the star evolves beyond point 5, the core begins to contract more rapidly and a sizable temperature gradient is built up as heating occurs preferentially toward the center. Matter in the nuclear-energy-producing shell is drawn more rapidly than before to higher temperatures and densities. Energy production in the shell begins again to be explosive and the stellar envelope expands rapidly. At the same time, the mass of this shell decreases rapidly: e.g. from 3.5 per cent of the star's mass, at point 5, to only 0.5 per cent of the star's mass, at point 6.

Between points 5 and 6, the rates at which temperatures and densities increase within the shell are not sufficient to offset the rate at which the shell narrows and the total rate of nuclear-energy production decreases with time. The decline in the rate of nuclear-energy production is not the only factor contributing to the drop in luminosity. Energy is required to expand the matter in the region between the shell and the surface. Approximately half of this energy is taken from thermal motions in the cooling envelope. The other half is supplied by the shell. In this way the envelope blankets the shell.

Within the expanding and cooling envelope the radiative opacity is increasing. Shortly before point 6, convection becomes the dominant mode of energy transport in a growing region extending inward from the surface.

When the base of the convective envelope reaches the level where lithium has been destroyed during the main-sequence phase, the surface abundance of lithium begins to drop. As the base of the convective envelope sweeps past the corresponding levels for other elements, convective mixing causes the surface abundances of these elements to decrease. Just the reverse, of course, happens to surface abundances of elements that have been formed in the interior during earlier phases, for example He^3 (see Figure 2).

Between the outer edge of the growing convective envelope and the surface, where energy still flows by radiation, the dominant source of opacity is the H^- ion, electrons being supplied by metals of low ionization potential. Instead of increasing with decreasing temperature as in subphotospheric regions, the opacity in the photosphere decreases with decreasing temperature.

Smaller opacities favor a larger energy flow. As the star's surface temperature decreases beyond point 6, its luminosity begins to increase. Within the convective portion of the envelope, the energy required for expansion is supplied entirely by thermal energy. Thus, as the mass of the convective envelope increases, the nuclear source is blanketed less strongly and delivers a larger fraction of its output to the surface. At the same time, the rate of nuclear-energy production in the shell increases.

As the star's luminosity increases along the red-giant branch, the mass within which nuclear energy is produced continues to decrease. The increased temperatures in the shell more than offset the decrease in shell size.

As the star evolves between points 6 and 7, convection covers more and more of the outer part of the star, reaching almost to the hydrogen-burning shell at point 7. As convection extends into the region where most of the original C^{12} has been converted into N^{14}, mixing begins to carry N^{14} outward and C^{12} inward. As the convective envelope grows, the abundance of N^{14} in the envelope gradually increases while the abundance of C^{12} drops. The ratio of N^{14} to C^{12} in the surface of the star, where it might be observed by the spectroscopist, therefore increases between points 6 and 7. Beyond point 7, the mass of the convective envelope begins to decrease and the surface ratio of N^{14} to C^{12} ceases to change.

During the ascent along the giant branch, central temperatures become sufficiently high that the reactions $N^{14}(\alpha,\gamma)F^{18}(\beta^+\nu)O^{18}$ are ignited. Within a small convective core, N^{14} is rapidly converted completely into O^{18}. Thus, near the stellar center, almost all of the original C^{12}, N^{14}, and O^{16} have been converted into O^{18} by the time the red-giant tip is reached. The only "observable" effect of the $N^{14} \rightarrow O^{18}$ reactions in the stellar core is to prolong the life of the star along the giant branch.

CORE HELIUM BURNING

When the star reaches point 7, central temperatures and densities have become high enough that nuclear burning near the center via the triple-alpha process $(3\alpha \rightarrow C^{12})$ alters the course of evolution.

The ignition of the triple-alpha process is explosive in the sense that temperatures and densities near the center at first rise to such values that the rate of nuclear-energy production exceeds the rate at which energy can be transported outward. The energy trapped near the center forces the central regions to expand, causing the temperatures and densities in the hydrogen-burning shell, which still provides most of the star's energy output, to increase less rapidly. Hence, immediately after the triple-alpha ignition at point 7, the luminosity of the star drops.

Between points 7 and 10', the stellar envelope contracts and the surface temperature increases. This behavior may be ascribed to the fact that the major source of nuclear-energy production is still the hydrogen-burning shell. When averaged over the entire track between points 7 and 10, the shell contributes about 85 per cent of the star's energy output. In order to maintain a balance between energy production and energy outflow, matter in the shell must be kept at sufficiently high densities and temperatures. Since the stellar core continues to expand because of the triple-alpha reactions, this is accomplished through envelope contraction. The contracting envelope tends to heat and compress the hydrogen-rich matter at the leading edge of the shell.

Envelope contraction is particularly rapid between points 8 and 9. This phase of contraction is not closely connected with nuclear-burning processes in the interior. It is associated with a recession of envelope convection which occurs as a result of decreasing opacity in deep portions of the heating and condensing envelope. As the dominant mode of energy flow switches from convection to radiation, a rapid readjustment of the matter in the envelope ensues. Core helium burning is thus broken into two major phases separated in the H-R diagram by a phase of rapid transit.

The energy flux produced by helium burning leads to the formation of a convective central region. In contrast with its behavior during the main-sequence phase, the convective core *grows* with time during most of the core helium-burning phase (points 7 to 10). At its maximum, the convective core occupies approximately 6 per cent of the star's mass.

As C^{12} increases in the convective core, it reacts with helium to form O^{16}. Whether or not C^{12} reaches an equilibrium abundance depends sensitively on the cross section for the $C^{12}(\alpha,\gamma)O^{16}$ reaction. During the core helium-burning phase, helium also reacts with O^{18} to form Ne^{22}.

Although central densities continue to drop over most of the core helium-burning phase, temperatures in the convective core continue to rise. As the star evolves between points 7 and 8 and between points 9 and 10, the rate of energy production in the core increases relative to the rate of energy production in the hydrogen-burning shell. Except for a brief drop during the phase of rapid envelope contraction (points 8 to 9), the ratio of helium-burning to hydrogen-burning energy-production rates increases steadily from about 0.05 to about 0.36 as the star evolves between points 7 and 10.

As the number of reacting particles in the core decreases, central densities begin to drop less rapidly until, near point 10′, the core begins to contract and the envelope begins to expand. The ensuing phase of core contraction and envelope expansion is reminiscent of the main-sequence phase, when the evolution is controlled by a strong source of energy near the center. The drop in luminosity following point 10 is a result of absorption in the expanding envelope and of a decrease in the strength of the hydrogen-burning shell.

Shell Helium Burning

As the star approaches point 11 and the central helium abundance decreases, the convective core begins to decrease in mass fraction and a larger and larger proportion of the energy produced by helium burning is released outside the convective core. Between points 11 and 12, all parts of the star move toward the center. This phase, associated with the exhaustion of central helium and the formation of a helium-burning shell, is analogous to the phase of overall contraction (points 2 to 3) which precedes the exhaustion of central hydrogen.

Between points 12 and 13, helium burning occurs in a thick shell outside a steadily condensing and heating helium-exhausted core. The strength of the helium-burning shell continues to increase relative to that of the narrow hydrogen-burning shell. As point 13 is approached, and the mass fraction of the helium-exhausted core reaches approximately 0.07, the helium-burning shell rapidly narrows. The matter between the helium-burning shell and the surface expands outward and cools. The cooling is so great at the hydrogen-helium interface that hydrogen burning temporarily ceases. This occurs shortly beyond point 13 and the hydrogen-helium discontinuity thereafter remains stationary at a mass fraction of about 0.23.

Evolution beyond point 13 is analogous to the earlier phase between points 5 and 7. The core contracts and heats rapidly, while the expanding envelope at first absorbs energy from the narrowing helium-burning shell, whose strength decreases. Over an increasing fraction of the cooling envelope, convective transport dominates over radiative transport. As photospheric opacities drop, the luminosity begins to rise and the star then proceeds again along a "Hayashi" track which is a more luminous extension of the track leading to point 7.

As the star progresses upward along this second giant branch, the base of the growing convective envelope extends deeper into the star, eventually reaching the hydrogen-helium discontinuity marking the position of the inactive hydrogen-burning shell. Mixing then enriches the helium content of the entire convective envelope (6, 8). Convection continues to grow inward until it reaches quite close to the helium-burning shell.

During the ascent along the giant branch following point 14, densities in the helium-exhausted core continue to rise until electrons in the core eventually become relativistically degenerate. When postulated neutrino-loss processes are included, Weigert (7) finds that central regions cool as they contract, most of the energy carried away by neutrinos being supplied from the region of most rapid contraction below the helium-burning shell. The maximum temperature occurs not at the center of the star but in the region of most rapid contraction, and the net rate at which energy escapes in the form of neutrinos is small compared to the rate of nuclear-energy production in the helium-burning shell. The major effect of the neutrino processes is thus to prolong the period of pure shell burning along the giant branch. The refrigerating action of neutrino losses from central regions prevents carbon in the core from burning.

When plasma- and photoneutrino-loss processes are not included, Kippenhahn et al. (8) find that central regions heat as the star ascends the giant branch beyond point 14. Eventually carbon burning is ignited explosively in the relativistically degenerate core. In the ensuing "carbon-flash," energy produced by the conversion of carbon into heavier elements remains in the central regions of the star, heating and expanding these regions until the degeneracy is lifted. Matter near the carbon-helium interface is carried out to such low densities and temperatures that the helium-burning shell becomes inactive. At the same time, the hydrogen-helium interface is heated and a hydrogen-burning shell is reactivated. In a period of 10,000 years the entire process is completed. Thereafter carbon burning continues near the center of a large convective core, in which electrons are no longer degenerate, and a hydrogen-burning shell moves outward, in mass fraction, forcing the base of the convective envelope to recede ahead of it. The helium-burning shell remains almost inactive.

Schwarzschild & Härm (12) have shown that helium burning in a sufficiently thin shell is thermally unstable. Energy produced in the shell is trapped there, raising the temperature and increasing the rate of energy production until the stored thermal energy is sufficient to expand the overlying material whose mass is large compared to the mass of the shell. The shell then cools and its source strength decreases. Eventually the matter above the shell falls back, compressing the shell until the process begins again.

Weigert (7) has followed the early growth of thermal instability in a $5M_\odot$ star in which the carbon core cools as the star evolves upward along the giant branch. Thermal instability occurs when the mass fraction in the helium-burning shell is roughly 2×10^{-4}. The time interval between successive pulses is roughly 4000 years and the amplitude of each new pulse is larger than that of earlier ones. Whether the amplitude asymptotically approaches a modest final value or whether pulses become sufficiently violent to result in the ejection of mass from the star has yet to be determined.

UNCERTAINTIES

How much confidence should one have in the theoretical results just described? There are three questions to be answered.

For a given initial composition, to what extent do uncertainties in the opacity, the convective-flow parameters, and the energy-generation rates affect the general features of the evolutionary track?

How sensitive are evolutionary features to composition changes within the limits of Population I?

Do "perturbations" such as rotation and mass loss alter evolutionary characteristics in an important way?

Answers to all of these questions must at present be partial and somewhat qualitative. A discussion of the third question will be deferred.

On the basis of models thus far published, one may infer that during pure hydrogen-burning phases (points 1–7 in Figure 1), the qualitative relationship between characteristic segments of the evolutionary track is both unaffected by uncertainties in the constitutive relations and invariant to changes in initial composition (within Population I limits). Over this range of evolution, the relationship between changes in observable characteristics and changes in interior processes is common to all nonrotating, constant-mass models of $5M_\odot$.

In contrast, the topology of the evolutionary track during the helium-burning phases, in particular during the phase of helium burning in a thick shell, appears to be highly sensitive to composition changes and is considerably affected by known uncertainties in the constitutive relations. Changes in the initial composition appear to have drastic consequences for the nature of the evolutionary track and hence for the relationship between this track and the state of the deep interior.

Thus, as evolution progresses to lower surface temperatures and through more advanced nuclear-burning stages, not only does the magnitude of the uncertainties increase, but the *effect* of all the uncertainties is magnified.

Several reasons for this state of affairs are clear. The description of energy flow is particularly incomplete at the low temperatures and densities found in stellar envelopes. The lower the surface temperature and the larger the stellar radius, the larger is the volume of the star affected by incompletely known opacities and unknown parameters in an *ad hoc* treatment of convection. Further, helium-burning reactions have not yet received the attention accorded to hydrogen-burning reactions and are less susceptible to direct experimental study. Finally, the more complex the internal structure of the star, the more precarious is the balance between the various factors and hence the greater is the sensitivity to changes in composition and to uncertainties in the constitutive relationships.

A sample of the variations in main-sequence properties obtained in recent investigations is presented in Table I. Luminosities and surface temperatures are given at the equivalents of points 1 and 2 in Figure 1. The main-sequence lifetime is given by t_{12} and $(M_{cc})_{max}$ represents the maximum mass fraction in the convective core during the main-sequence phase.

Presumably, the only major differences in input physics between cases b and c, and also between cases a and d, lie in the choice of opacities, which presumably differ significantly only in the outer layers, where bound-bound contributions to opacity are important. Line contributions to opacity are included in cases c and d, but not in cases a and b.

As may be seen from Table I, the actual choices of, or approximations to, interior opacities in cases b and c are sufficiently different that the maximum convective-core mass is 15 per cent smaller and the initial luminosity is 9 per cent larger in case c than in case b. These differences can occur only if interior opacities in case c are *smaller* than in case b, a rather surprising result if one supposes that including additional sources of opacity would lead to higher

TABLE I

VARIATIONS IN MAIN-SEQUENCE PROPERTIES

Investigator	Case	Composition		$\log (L/L_\odot)$		$\log (T_e)$		$(M_{cc})_{max}$	$t_{12}(10^7 \text{ yr})$
		X	Z	(1)	(2)	(1)	(2)		
10	a	0.708	0.02	2.79	3.00	4.269	4.195	0.225	6.44
8	b	0.602	0.044	2.78	3.00	4.245	4.194	0.212	5.37
9	c	0.602	0.044	2.82	2.99	4.236	4.173	0.185	4.83
9	d	0.739	0.021	2.66	2.87	4.223	4.150	—	8.71

opacities. One might also have expected a more pronounced difference in surface temperature between cases b and c. It has been shown (13) that, for the initial main-sequence model, the inclusion of line effects reduces log (T_e) by about 0.03. The differences between cases a and d are in the direction expected, but of sufficient magnitude to suggest that not only uncertainties in opacity but also uncertainties introduced by the *approximations* to opacities used by different investigators can have a surprisingly large effect on main-sequence properties.

Unfortunately, the effect of altering the initial hydrogen abundance or the initial heavy-element abundance separately cannot be deduced from Table I. However, one may note, on comparing cases c and d, that a variation of composition parameters within Population I limits can significantly alter the main-sequence lifetime.

Variations in observable properties following the main-sequence phase are compared quantitatively in Table II. Logarithms of luminosity and surface temperature are given at points corresponding to the numbering in Figure 1. Times (in units of 10^7 yr) refer to the interval between successive points in Table II. Only selective comments will be made.

The mean value of log T_e along the giant branch, $\langle \log T_e \rangle_{\mathrm{RG}}$, differs by $\sim$0.04 among the exhibited cases. Hofmeister (9) has shown that an increase of only 40 per cent in the ratio of mixing length to pressure scale height can increase $\langle \log T_e \rangle_{\mathrm{RG}}$ by 0.05. It may be concluded that for a given initial composition, $\langle \log T_e \rangle_{\mathrm{RG}}$ is uncertain by at least 0.1.

The giant-branch phase (points 6 to 7) lasts longer in case a than in all other cases because only in case a have the reactions $N^{14}(\alpha,\gamma)F^{18}(\beta^+\gamma)O^{18}$ been included. At the red-giant tip, in case a, convection occurs over the outer 70 per cent of the star's mass. In case b, envelope convection covers a maximum of only 50 per cent of the star's mass. The variation in surface abundances due to mixing is thus uncertain.

The lifetime for the core helium-burning phase is 1.67×10^7 yr for case c compared to only 1.1×10^7 yr, or a factor of 1.52 smaller, in case b. The reason for the difference presumably lies with the choice of interior opacities. In case b the convective core *decreases* in mass fraction from $\sim$0.04 to 0.03, giving an average mass fraction of $\sim$0.035. In case c, the mass fraction in the core *increases* from $\sim$0.04 to $\sim$0.05. The amount of fuel consumed during the core helium-burning phase is thus 0.05/0.035 $\sim$1.43 times greater in case c than in case b. In case a, the convective core increases to a maximum mass fraction of 0.06. Since no data concerning the convective core are available for case d, no comparison can be made.

Another feature of considerable importance is the maximum blueward extension of the evolutionary track during the core helium-burning phase. From entries in Table II corresponding to point 10' in Figure 1, it appears that this extension is tremendously sensitive not only to variations in both convective-flow parameters and opacities, but also to variations in the choice of Population I composition. In fact, in case d the entire evolutionary track

TABLE II

VARIATIONS IN OBSERVABLE CHARACTERISTICS

Point	(a)			(b)			(c)			(d)			Description ($i \rightarrow j$)
	$\log L$	$\log T_e$	t_{ij}	$\log L$	$\log T_e$	t_{ij}	$\log L$	$\log T_e$	t_{ij}	$\log L$	$\log T_e$	t_{ij}	
1	2.80	4.27		2.78	4.24		2.82	4.24		2.66	4.22		
2	3.00	4.20	6.44	3.00	4.19	5.37	2.99	4.17	4.83	2.87	4.15	8.71	Main sequence
3	3.06	4.23	0.22	3.08	4.24	0.25	3.08	4.22	0.23	2.93	4.19	0.25	Overall contraction
5	3.11	4.17	0.14	3.16	4.14	0.20	3.12	4.15	0.08	2.97	4.12	0.19	Hydrogen burning in thick shell
6	2.84	3.65	0.08	2.85	3.68	0.08	2.59	3.68	0.09	2.47	3.67	0.13	Shell narrowing
7	3.09	3.60	0.049	3.00	3.64	0.025	2.87	3.64	0.023	2.90	3.63	0.033	Red-giant branch
10	3.27	3.92	1.44	3.13	3.79	0.81	3.01	3.75	1.29	2.77	3.65	1.78	Core helium burning
11	3.22	3.68	0.19	3.10	3.66	0.29	2.99	3.64	0.38	2.96	3.63	0.42	
13	estimate		0.2–0.3	3.29	3.77	0.79	2.95	3.66	0.07	2.85	3.64	0.07	Helium burning in thick shell
>14	—	—	—	Giant branch		0.03	Giant branch		0.41	Giant branch		0.80	

during helium burning is confined to the giant branch. The major cause for the difference between cases c and d must be the difference in envelope hydrogen content, since decreasing Z for fixed initial X has the opposite effect from that obtained.

The maximum extention to the blue is also influenced by the relative rates at which C^{12} is created and destroyed. If the cross-section factor for the reaction $C^{12}(\alpha,\gamma)O^{16}$ is reduced from the theoretical maximum to one-tenth of the maximum, log T_e at point 10′ in Figure 1 is decreased by ~ 0.06 (10). Note that the center-of-mass cross-section factor for the reaction $C^{12}(\alpha,\gamma)O^{16}$ in case c is chosen to be about one tenth of the theoretical maximum, whereas $C^{12} \rightarrow O^{16}$ conversion is neglected entirely in case b. Yet log T_e at point 10′ is *smaller* in case c than in case b by ~ 0.04. This suggests that the inclusion of line absorption in the opacity can decrease the maximum extension to the blue during core helium burning by at least 0.1 in log T_e, unless of course the evolutionary track is already compressed to the giant branch.

The uncertainty in the magnitude of the reaction $C^{12}(\alpha,\gamma)O^{16}$ also introduces an uncertainty as to the end products of helium burning. If the maximum cross-section factor is chosen, the end product is entirely O^{16}. If the cross section is chosen ten times smaller, roughly equal amounts of C^{12} and O^{16} result as end products (9, 10).

Whether the evolutionary track extends to the blue during the phase of helium burning in a thick shell (following point 12 in Figure 1) or whether it proceeds directly to the giant branch appears also to be considerably affected by variations in opacity and in composition choices. The time spent in this phase clearly depends (among other things) on whether the convective core increases or decreases during the preceding core helium-burning phase. In case b the maximum blueward extension during the thick-shell phase exceeds the maximum blueward extension during the core helium-burning phase. In case c (same initial composition as in case b) the thick-shell phase is confined to the giant branch. In case d both major phases of helium burning are confined to the giant branch.

VARIATIONS WITH MASS

Evolutionary tracks for several metal-rich stars of different mass are shown in Figure 3. The initial composition and the input physics for all stars represented are the same as for the $5M_\odot$ star described in Figure 1. Solid portions of the tracks in Figure 3 are direct results of computation (10, 14–17); dashed portions are estimates. The times spent by each star in traversing intervals between successive labeled points are given in Tables III and IV.

Model tracks for stellar masses $M \geq 3M_\odot$ are terminated just before the phase of helium burning in a thick shell. Tracks for less massive stars are terminated just before the ignition of helium in the core. For stars of mass $M \leq 1.5 M_\odot$, these termination points are estimates.

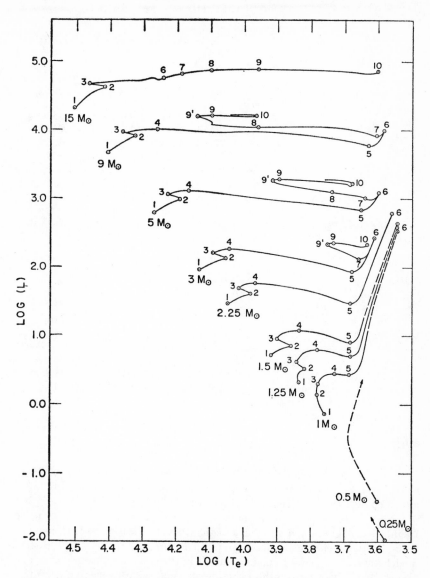

FIG. 3. Paths in the H-R diagram for metal-rich stars of mass $(M/M_\odot) = 15$, 9, 5, 3, 2.25, 1.5, 1.25, 1, 0.5, 0.25. Units of luminosity and surface temperature are the same as in Figure 1. Traversal times between labeled points are given in Tables III and IV. Dashed portions of evolutionary paths are estimates.

TABLE III

STELLAR LIFETIMES (yr)[a]

Mass ($M_\odot$) \ Interval ($i-j$)	(1–2)	(2–3)	(3–4)	(4–5)	(5–6)
15	1.010 (7)	2.270 (5)		7.55 (4)	
9	2.144 (7)	6.053 (5)	9.113 (4)	1.477 (5)	6.552 (4)
5	6.547 (7)	2.173 (6)	1.372 (6)	7.532 (5)	4.857 (5)
3	2.212 (8)	1.042 (7)	1.033 (7)	4.505 (6)	4.238 (6)
2.25	4.802 (8)	1.647 (7)	3.696 (7)	1.310 (7)	3.829 (7)
1.5	1.553 (9)	8.10 (7)	3.490 (8)	1.049 (8)	≥2 (8)
1.25	2.803 (9)	1.824 (8)	1.045 (9)	1.463 (8)	≥4 (8)
1.0	7 (9)	2 (9)	1.20 (9)	1.57 (8)	≥1 (9)

[a] Numbers in parentheses beside each entry give the power of ten to which that entry is to be raised.

TABLE IV

STELLAR LIFETIMES (yr)[a]

Mass ($M_\odot$) \ Interval ($i-j$)	(6–7)	(7–8)	(8–9)	(9–10)
15	7.17 (5)	6.20 (5)	1.9 (5)	3.5 (4)
9	4.90 (5)	9.50 (4)	3.28 (6)	1.55 (5)
5	6.05 (6)	1.02 (6)	9.00 (6)	9.30 (5)
3	2.51 (7)	4.08 (7)		6.00 (6)

[a] Numbers in parentheses beside each entry give the power of ten to which that entry is to be raised.

THE ZERO-AGE MAIN SEQUENCE

In a first approximation, the variation with mass of stellar properties averaged over the main-sequence phase is roughly the same as the variation of properties of homogeneous models along the zero-age main sequence (points 1 in Figure 3). A summary of the calculated variation with mass of several interior characteristics is given in Figure 4. The lower curve joins model stars on the zero-age main sequence in the mass-luminosity plane. The first three numbers in parentheses beside each mass give, respectively, central density (g/cm^3), central temperature ($10^{6\circ}$K), and stellar radius ($R_\odot$) corresponding to the initial main-sequence position. The fourth number gives the maximum mass fraction in the convective core during the main-sequence phase and the fifth number measures the lifetime of the model star in the main-sequence band.

In none of the initial low-mass models described in Figures 3 and 4 is electron degeneracy of overwhelming importance. At the center of the $1M_\odot$, $0.5M_\odot$, and $0.25M_\odot$ models, electron degeneracy is responsible for 1.7, 2.4, and 6.5 per cent of the total pressure. In the more massive homogeneous stars described in Figure 4, radiation pressure is not negligible. At the center of the $5M_\odot$, $9M_\odot$, and $15M_\odot$ stars, radiation pressure contributes 2.1, 5.1, and 11 per cent of the total pressure.

For stellar masses greater than $\sim 7M_\odot$, the calculated mass-luminosity relationship takes the form $L \propto M^3$. For very light stars, which are completely convective ($M \lesssim 0.4M_\odot$), $L \propto M^{1.68}$. In the neighborhood of $1M_\odot$, $L \propto M^{4.75}$.

The qualitative nature of trends along the zero-age main sequence can be established without reference to detailed computations. The assumption of hydrostatic equilibrium is sufficient to establish that, for the range in mass shown in Figure 4, $\rho_c/T_c^3 \propto \mu^{-3}M^{-2}$, where ρ_c and T_c are central density and temperature and μ is the mean molecular weight. If one further assumes that energy is transferred over most of the interior by radiative diffusion, it follows that $L \propto (\mu^4 M^3/\kappa)$, where κ is a measure of the mean interior opacity. Finally, with a rough knowledge of the magnitude and of the density and temperature dependence of opacity sources and of energy-generation rates, one can establish, still without recourse to detailed computations, that $L \propto M^3$ for stars sufficiently massive that electron scattering is the main source of opacity, whereas $L \propto M^5$ for stars sufficiently small that the p-p chains dominate over the CN-cycle reactions and free-free and bound-free absorption contribute predominantly to opacity.

The variation with mass of the initial central density is interesting. For masses above about $1.25M_\odot$ and below about $0.5M_\odot$, the central density decreases with increasing mass. In the intermediate region, the density increases with increasing mass. The calculated dependence of the central density on mass is consistent with a simple order-of-magnitude argument. Assuming that $\rho_c \propto T_c^3/M^2$ and that $L \propto \rho_c T_c^n M \propto M^{m+1}$, one obtains $(1+3/n)\log \rho_c = \text{const} - (2-3m/n) \log M$. In the high-mass range, where $n \sim 18$ and $m \sim 2$, ρ_c increases as M decreases. In the intermediate-mass range, where $n \sim 4$ and $m \sim 4$, ρ_c decreases as M decreases. In the low-mass range, where $n \sim 5$ and $m \sim 0.7$, density again increases as mass decreases. The values of m required for these statements are obtained, for the high- and intermediate-mass ranges, without detailed calculations. However, for the low-mass stars, which are completely convective except for the region near the photosphere, the mass-luminosity relationship can be obtained only by calculation.

Models along the initial main sequence may be classified, according to the relative importance of the CN cycle and the p-p chains, as belonging to the upper or the lower main sequence. For the composition and physical assumptions here considered, the two sources of energy contribute equally at about $1.9M_\odot$ ($\log L \sim 1.18$).

All upper main-sequence stars have a convective core. With decreasing stellar mass, the mass fraction in the convective core decreases as energy sources become less strongly concentrated toward the center and as radiation pressure becomes less important. Energy production occurs in a region small compared to the size of the core.

The size of the convective core is not a reliable indicator of the relative importance of the CN cycle. A sizable convective core may occur even along the lower main sequence, where the p-p chains provide more energy than the CN-cycle reactions. For example, in the initial $1.5 M_\odot$ model, convection extends over the inner 6 per cent of the star. Three quarters of the total energy output is produced in the core, but less than a quarter of this is contributed by CN-cycle reactions.

The mass at which central convection vanishes is sensitive to the temperature and density dependence of both the opacity and the energy-generation rates. Since interior opacity sources are quite complex for models in the vicinity of $1 M_\odot$, one should not rely on the results of a single calculation. For the particular physical assumptions and composition leading to the initial models in Figures 3 and 4, a convective core does not occur for masses and luminosities in the ranges $0.4 < M/M_\odot \lesssim 1.1$, $-1.6 \lesssim \log(L/L_\odot) \leq 0.1$.

Models along the lower main sequence with surface temperatures below 8000°K ($\log T_e \lesssim 3.9$) have relatively extended convective envelopes. The depth of and the mass fraction in the convective envelope increase as stellar mass is lowered until the entire star becomes convective ($M \gtrsim 0.4 M_\odot$, $\log L \gtrsim -1.6$). The uncertainties in surface temperature associated with an inadequate treatment of envelope convection reach a maximum near the center of the ranges $3.6 < \log T_e < 3.9$, $-1, < \log L < +1$. In the neighborhood of $1 M_\odot$, a decrease by a factor of 2 in the ratio of mixing length to density scale height lowers the surface temperature by $\Delta(\log T_e) \sim -0.03$ (13). However, the position of stars in the mass-luminosity diagram is not sensitive to changes in the treatment of envelope convection.

Uncertainties in the surface temperature of upper main-sequence models are similar in magnitude to those already described for the $5 M_\odot$ model. They are expected to result primarily from uncertainties in envelope opacity. The

---

≫→

FIG. 4. Positions in the mass-luminosity diagram, during the main-sequence phase, for the metal-rich stars described in Figure 3. Luminosity and mass are in solar units. The lower curve defines the locus of homogeneous, initial main-sequence models. The first three numbers in parentheses beside each point give central density (g/cm³), temperature (10^6 °K), and stellar radius ($R_\odot$) for the appropriate model. The upper solid curve represents the locus of models which have just terminated the main-sequence phase, as defined in the text. The dashed curve represents an estimated locus for low-mass models which have evolved for 10^{10} yr. The last two entries in parentheses beside each point give the maximum mass fraction in the convective core and the main-sequence lifetime in years.

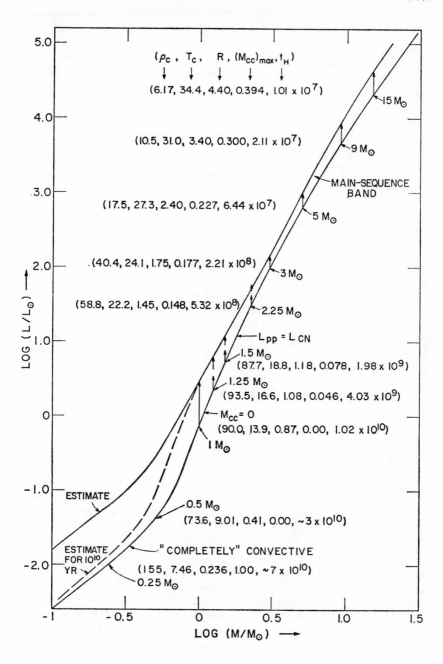

slope of the zero-age main sequence line is expected to be independent of these uncertainties. In particular, the inclusion of line contributions to the opacity lowers the surface temperature uniformly by $\Delta(\log T_e) \sim -0.03$ along the upper zero-age main sequence (13).

A brief summary of the effects of composition changes on the position of the zero-age main sequence is in order. A change in X influences luminosity and surface temperature primarily through the molecular weight. As one reduces the number of particles per gram throughout the star, the necessity of maintaining the pressure balance demands that much of the star contract. The virial theorem implies an increase in interior temperatures which in turn, owing to the temperature sensitivity of nuclear sources, implies an increase in luminosity. A change in Z affects L and T_e primarily through bound-free contributions to opacity.

For a fixed mass, increasing X or Z (or both) lowers model luminosity and surface temperature. For example (13), in the neighborhood of $\log (M) \sim 0.1$, $\Delta(\log L) \sim -(2.5\Delta X + 13\Delta Z)$, whereas near $\log (M) \sim 0.6$, $\Delta(\log L) \sim -(1.9\Delta X + 7\Delta Z)$.

In describing the effect of composition changes on the position of the zero-age main sequence in the H-R diagram, it is best not to focus on individual values of the mass. Over restricted portions of the H-R diagram, the slope of the zero-age line is unaffected by composition changes. One may therefore think of composition changes as shifting the zero-age line bodily either in the vertical direction (T_e = const) or in the horizontal direction (L = const). Along the lower main sequence, in the neighborhood of $\log L \sim 0.0$, the magnitude of the shift obeys the rough relationships (13): $\Delta(\log L) \sim \Delta X + 5\Delta Z$ when T_e = const; and $\Delta(\log T_e) \sim -(0.15\Delta X + 0.8\Delta Z)$ when L = const. In the vicinity of $\log L \sim 2.0$, one has: $\Delta(\log L) \sim 0.66\Delta X + 10.5\Delta Z$ when T_e = const; and $\Delta(\log T_e) \sim -(0.12\Delta X + 1.9\Delta Z)$ when L = const. A measure of the effect of uncertainties in opacities may be obtained by simply altering Z in these expressions by, say, 0.015.

The mass and luminosity level at which the convective core disappears is sensitive to composition (18). From more detailed investigations (19), it is found that the luminosity at which convection disappears varies with composition roughly like $\Delta(\log L) \sim 1.4\Delta X - 4.5\Delta Z$ and that the mass at which central convection disappears varies according to $\Delta(\log M) \sim .9\Delta X + 1.9\Delta Z$.

THE MAIN-SEQUENCE PHASE

The fact that relatively massive stars grow redder as they brighten while light stars initially grow bluer is related to the fact that significant helium production occurs over a much larger fraction of the lighter stars. The tendency toward contraction is most pronounced in regions where the molecular weight is increasing most rapidly. Hence, the more evenly molecular-weight changes are spread over the star, the greater is the fraction of the star that contracts and the smaller is the rate of increase of radius relative to the rate of luminosity increase.

For all the stars in Figure 3, the radius increases throughout the phase of core hydrogen burning. In the limit of extremely light stars which are completely convective, the molecular weight increases at the same rate throughout the star. The evolutionary track for such stars may be obtained simply by joining points of constant mass along zero-age main sequences characterized by successively higher values of molecular weight.

The slope of an evolutionary path, $|d \log L/d \log T_e|$, increases with mass along that portion of the upper main sequence shown in Figure 3. This tendency for the slope to increase with mass is reduced as radiation pressure becomes appreciable. In the limit of high masses, where radiation pressure dominates gas pressure, $L \propto M/(1+X_e)$, where X_e is the abundance of hydrogen in the envelope. Hence the luminosity tends to remain constant whatever the changes in internal composition. The tendency for the evolutionary track to become increasingly horizontal as the mass increases is demonstrated by a series of evolutionary models computed by Stothers (20) for $M = 45 M_\odot$ $-1000 M_\odot$.

The construction of models in which radiation pressure dominates gas pressure may be something of an academic exercise since Schwarzschild & Härm (21) have shown that such models are secularly unstable against radial pulsations in the linear approximation. On the other hand, there is as yet no guarantee that the pulsation amplitude will increase without limit when nonlinear terms are included, or that a star initially more massive than the stability limit might not survive for a significant fraction of its nominal main-sequence lifetime.

For all the models shown in Figure 3 with mass above $\sim 2.25 M_\odot$, the mass fraction in the convective core decreases with time as a consequence of decreasing interior opacity. In all of these stars, the major energy source is the CN cycle. For less massive stars, which possess convective cores at the beginning of the main-sequence phase, the mass fraction in the convective core increases as core temperatures and densities increase. This is because the CN-cycle contribution to energy production increases relative to the p-p chain contribution, and energy production becomes more concentrated toward the stellar center. Eventually the core mass reaches a maximum and thereafter decreases as the effect of decreasing interior opacities overwhelms the effect of the increasing central concentration of energy sources. Although no computations have yet been published to show this explicitly, it is clear that stars which possess no convective core at the start of the main-sequence phase (mass slightly less than $1.1 M_\odot$ in the present instance) may develop one as evolution progresses.

In stars more massive than about $8 M_\odot$, complications arise in the treatment of energy flow outside of the receding convective core. In several regions, matter becomes formally unstable against convection, but the extent to which mixing occurs in these regions is not clear. One approach (20, 22–25) has been to distribute composition parameters within such regions in such a way that the radiative temperature gradient equals the adiabatic temperature gradient. A satisfactory physical justification for this procedure has not been formulated.

The occurrence of an overall contraction phase (points 2 to 3 in Figure 3, where applicable) is correlated in a one-to-one fashion with the occurrence of a convective core during the period of core hydrogen burning. The occurrence of a strong shell-development stage—marked by an extremely rapid core contraction and, for not too massive stars, by a concurrent rapid cooling—is likewise tied to the presence of a relatively large convective core during the core hydrogen-burning phase. The relative magnitude of the decrease in central temperatures during the shell-development stage becomes smaller as stellar mass is increased, vanishing near $10 M_\odot$. Beyond $10 M_\odot$, central temperatures continue to rise following the exhaustion of central hydrogen. That central regions in sufficiently massive stars continue to heat following the exhaustion of central hydrogen is correlated with the fact that, in such stars, the mean convective-core mass exceeds the Schönberg-Chandrasekhar (S-C) limit during the phase of overall contraction.

For light stars, which do not possess a convective core while the central hydrogen abundance is finite, the phase of overall contraction is absent and the core helium-burning phase merges smoothly and uneventfully into the phase of hydrogen burning in a thick shell. The central temperature in these light stars does not drop during the shell-development stage.

The lifetime of a star in the thick-shell phase depends in an obvious way on the mean mass fraction in the convective core during the overall-contraction phase. If this mass fraction is larger than the effective S-C limit, then the thick-shell phase is quite rapid. In fact, the shell forms and narrows at the same time, while the core continues to heat as it condenses. For less massive stars, with mean convective cores smaller than the S-C limit, the lifetime in the thick-shell phase is proportional to the amount of fuel remaining, at the start of the phase, between the shell-base and the S-C limit. Since the mean mass fraction in the convective core decreases with stellar mass, the lifetime of the thick-shell phase τ_s, relative to the lifetime of the core hydrogen-burning phase τ_{HC}, increases with decreasing stellar mass. Two other factors contribute to the increase in τ_s/τ_{HC} with decreasing mass: the increasing importance of electron degeneracy, and the increasing smoothness in the variation of molecular weight with mass fraction. The additional contribution of electron degeneracy to core pressure permits the hydrogen-exhausted core to attain a larger mass prior to the onset of shell narrowing. In addition, the S-C limit is proportional to the square of the molecular weight at the hydrogen-rich outer edge of the shell (11). In the $1 M_\odot$ model, the mass fraction in the hydrogen-exhausted core reaches ~ 0.13 during the thick-shell phase, compared to a mass fraction of ~ 0.10 in upper main-sequence stars with mass below $\sim 10 M_\odot$.

When τ_s/τ_{HC} is large it makes sense to include the period of thick-shell burning as a part of the main-sequence phase. The theoretical main sequence is then defined as the region where pure hydrogen burning occurs on a long time scale rather than as just the region where hydrogen burning occurs in the core. In terms of the model paths exhibited in Figure 3, limits to the main-sequence band may be set as follows. The high-temperature limit corresponds, of course, to the locus of points 1. For all $\log L \geq 2.0$, the low-temperature limit is obtained by joining points labeled 2. For all $\log L \leq 1.0$, the low-temperature limit is obtained by joining points 4. As a rough average, the main-sequence band is characterized by a temperature spread $\Delta(\log T_e) \sim 1.2$ at any given luminosity and a luminosity spread $\Delta(\log L) \sim 0.75$ or $\Delta M_{bol} \sim 2$ at any given surface temperature. This compares with an average spread of $\Delta(\log T_e) \sim 0.045$ at constant L, and $\Delta(\log L) \sim .3$ or $\Delta M_{bol} \sim 0.8$ at constant T_e, associated with ranges in composition variables of about $(\Delta X, \Delta Z) \sim (0.2, 0.02)$.

The appearance of the main-sequence band in the mass-luminosity plane is also of interest. The arrows within the band in Figure 4 are evolutionary tracks for models during major hydrogen-burning phases. Where two arrows occur at a given mass, the lower arrow corresponds to evolution prior to the phase of overall contraction and the upper arrow corresponds to evolution in the phase of thick-shell burning. Below $\log L \sim 0.4$ the upper limit to the main-sequence band is an estimate. The dashed curve below $\log L \sim 0.4$ in Figure 4 is an estimate of the location of low-mass stars which have been allowed to evolve for only 10^{10} yr.

To the Red-Giant Tip

In all but the lightest stars, where the mass fraction in the convective envelope is already quite large when the shell-narrowing phase begins, the strength of the shell decreases and appreciable absorption occurs in the radiative portions of the envelope as the shell narrows. It is of interest that the base of the giant branch (points 5 in Figure 3, where applicable) occurs at approximately the same surface temperature for all stars of the same composition, regardless of mass.

The most important characteristic of the shell-narrowing phase is the rapid rise in central temperatures and densities. The more massive the star and, hence, the higher the core temperatures are to start with, the earlier do the core temperatures become high enough to ignite the $N^{14} \rightarrow O^{18}$ reactions. Ignition and burning occur along the giant branch (points 5–6) in the $5 M_\odot$ model. In the $9 M_\odot$ model, ignition occurs in the "Hertzsprung gap" (points 4–5), leading to a rise in luminosity not found along tracks of less massive stars. In the $15 M_\odot$ star, $N^{14} \rightarrow N^{18}$ reactions lead also to a brief luminosity rise (just prior to point 6). In the $3 M_\odot$ star, $N^{14} \rightarrow O^{18}$ reactions force the star to *descend* along the giant branch until N^{14} is exhausted in the core, whereupon reascent commences.

A major dichotomy in the character of giant-branch evolution occurs at about $2.25 M_\odot$. In stars more massive than this, core temperatures continue to rise monotonically as the star ascends the giant branch. Helium-burning reactions commence and terminate the ascent before electron degeneracy becomes appreciable in the core.

As stars less massive than $2.25 M_\odot$ ascend the giant branch, the degree of electron degeneracy grows to significant proportions before helium-burning reactions are ignited. In all cases, central temperatures drop temporarily when the electron Fermi-energy ϵ_F becomes comparable to kT. Two effects are operating in producing this drop: The *relative* importance of electron conduction increases with increasing degeneracy. The effective opacity drops and the available thermal energy flows out more readily, tending to make the core isothermal. As the electron Fermi-energy rises, energy is required to increase the nonthermal kinetic energy of the electrons. The most readily available energy is thermal.

After near-isothermality is approached in the core, central temperatures begin to rise again, but much less rapidly than in the absence of degeneracy. The efficiency of electron conduction ensures that central temperatures rise not much more rapidly than shell temperatures. As a result, low-mass stars rise quite far above the base of the giant branch before helium-burning reactions are ignited. Before helium burning commences in such stars, the mass fraction in the nearly pure helium core far exceeds the S-C limit. This is in strong contrast with stars more massive than $2.25 M_\odot$, within which the mass fraction in the hydrogen-exhausted core, when helium burning commences, is not much larger than it was at the end of the main-sequence phase.

Mestel (26) was the first to suggest that the ignition of nuclear fuel in a region of high electron degeneracy would be explosive. Schwarzschild & Härm (27) have demonstrated this actually to be the case when helium burning commences in the degenerate core of a light star. At the peak of the "flash," energy is released by nuclear reactions at the rate of $\sim 10^{11} L_\odot$. None of this energy escapes beyond a convective core which reaches almost to the hydrogen-burning shell (27, 28). Core temperatures rise until degeneracy is lifted. After degeneracy is lifted, the core expands and cools. Thereafter helium burning continues in a manner similar to that in more massive stars. It is clear from the trends exhibited by the tracks in Figure 3 that after degeneracy has been lifted, evolutionary tracks for light Population I stars will be compressed tightly against the giant branch, at least for stellar masses $M \gtrsim M_\odot$, for which the mass in the hydrogen-exhausted core is approximately $0.4 M_\odot$ at the start of the flash phase.

The mass in the hydrogen-exhausted core at the start of the flash phase appears to be relatively independent of the abundance of elements heavier than helium and also relatively independent of the total mass outside of the degenerate core. For values of initial $X \sim 0.6$–0.7, the helium-core mass is about $0.4 M_\odot$ (3, 17). For values of $X \sim 0.9$, the helium core mass is $\sim 0.6 M_\odot$ (3, 27).

These values for critical core mass apply if it is the triple-alpha process which initiates the flash. In the core of Population I stars, most of the initial oxygen and carbon has been converted into N^{14} which, on reacting with helium, produces energy rapidly at temperatures somewhat lower than required by the triple-alpha process. The N^{14}-flash in low-mass population models will therefore occur when the core mass is less than the critical mass for triple-alpha flashing. The difference in the brightness of the red-giant tip—depending on whether it is determined by the onset of N^{14} flashing or by the onset of triple-alpha flashing—amounts to several magnitudes (17).

The effect on giant-branch evolution of energy losses associated with the plasma-neutrino process (29, 30) has yet to be studied numerically. It is clear that plasma-neutrino losses are important (31) and may significantly alter the manner in which helium burning commences in the core of a light star. The inclusion of plasma-neutrino losses is expected to have the following effect. Energy loss will be most significant near the stellar center where densities are highest and the degree of degeneracy is greatest. On the other hand, the rate at which gravitational energy is released, as a result of the continued addition of helium to the partially degenerate core, will be larger in regions where the degree of degeneracy is smaller. Hence, in order to supply the neutrino loss from central regions, energy will tend to flow by conduction from regions away from the center (where the compression rate is large) toward the center (where the compression rate is small). The maximum in core temperatures will therefore not occur at the center but instead somewhere between the center and the shell. Helium burning will commence not at the center but in some region away from the center where the degree of degeneracy is smaller. The mass of the helium core will be larger when helium burning commences and the helium flash will be less violent than when neutrino losses are neglected—if indeed helium ignition is at all violently explosive.

Helium-Burning Phases

As may be seen from Figure 3, evolutionary tracks during core helium burning are topologically equivalent for stars in the mass range $3 \lesssim M/M_\odot \lesssim 9$. For stars in this range, several mass-dependent trends are of importance.

In the H-R diagram, the separation between the two major phases of core helium burning becomes more pronounced with increasing stellar mass. In both the $5M_\odot$ and $9M_\odot$ cases, the phase of rapid envelope contraction that is correlated with the decrease in the mass fraction of the convective envelope is bounded roughly by points 7 and 8. In the $9M_\odot$ case, the rapidity of the envelope contraction is clearly reflected by the increase in luminosity along the evolutionary track just beyond point 7. This brightening is the result of the release of gravitational energy within the contracting envelope. For stars less massive than $3M_\odot$, the phase of rapid envelope contraction is absent.

For all stars less massive than $\sim 9M_\odot$, then, the first major phase of core helium burning occurs near the giant branch (points 6 to 7), whereas the second major phase occurs (between points 8 and 9) in a region which becomes successively further removed from the giant branch as the mass increases. As the mass is increased, the time spent in the second phase of core helium burning increases relative to the time spent in the first phase.

In a star of $15\,M_\odot$, the core temperatures reach high enough values immediately following the main-sequence phase for the triple-alpha reactions to be ignited, and the long-term phase of core helium burning begins before the star reaches the red-giant phase. The star spends most of the core helium-burning phase between points 6 and 8.

A core helium-burning band or main sequence for massive stars may now be defined. Its high-temperature edge joins points 9' along the $3\,M_\odot$, $5\,M_\odot$, and $9\,M_\odot$ tracks with point 6 along the $15\,M_\odot$ track. Its low-temperature edge joins points 8 along the $15\,M_\odot$, $9\,M_\odot$, and $5\,M_\odot$ tracks with point 6 along the $3\,M_\odot$ track. Defined in this way, the core helium-burning band is roughly parallel to the hydrogen-burning main sequence and has roughly the same width. A subsidiary core helium-burning branch may be defined by joining first points 7 and then points 6 along the $(3–9)\,M_\odot$ tracks. For the models represented in Figure 3, the combined lifetime in the phases of core helium burning relative to the lifetime in the main-sequence phase decreases from about 0.3 near $3\,M_\odot$ to about 0.15 near $15\,M_\odot$.

One may consider the position in the H-R diagram of the major core helium-burning band as established only qualitatively. In contrast with the situation along the hydrogen-burning main sequence, known uncertainties in input physics make the position of this band, for any given initial composition, indeterminate by *at least* the width of the band. Further, a variation in the composition within Population I limits can alter the position of the band even more.

In particular, the luminosity at which the band strikes the line defined by the locus of red-giant tips (points 6 in Figure 3, where applicable) is highly uncertain and sensitive to composition changes. For example, calculations by Hofmeister (9) show that for the initial composition $X = 0.74$, $Z = 0.021$, *all* phases of helium burning are confined to the giant branch for stars at least as massive as $9\,M_\odot$. The core helium-burning band thus breaks away (if at all) from the locus of giant branches at luminosities in excess of $\log L \sim 4.0$. With an initial composition of $X = 0.60$, $Z = 0.044$ and the same input physics, Hofmeister finds that the core helium-burning band breaks away at $\log L \sim 2.7$.

The position in the H-R diagram of evolutionary paths during the period of helium burning in a thick shell is even more uncertain than the position of evolutionary paths during the preceding core helium-burning phase. A discussion of mass-dependent trends should therefore be deferred until the basic physical assumptions are better established and more model calculations are available. The same may be said of still more advanced phases.

COMPARISON WITH OBSERVATIONS

A quantitative comparison between model results and observational data requires, unfortunately, uncertain theoretical transformations of incomplete observational data. The statement of observational results in terms of such quantities as L, T_e, and M is as uncertain as the structure of model stars. At present, one can only hope to establish rough consistency, or discover major discrepancies, between theory and observation.

THE MAIN-SEQUENCE BAND

In Figure 5, the curves labeled $ZAMS_0$ and $ZAMS_1$ represent the theoretical zero-age main sequence defined by models of initial composition $(X, Y, Z) = (0.71, 0.27, 0.02)$ and $(l_{mix}/H_{dens}) = 0.5$, when the effect of line absorption on opacity is and is not neglected (13).

The curve labeled $TAMS$ represents the theoretical termination of the main-sequence phase as defined in the previous section, again with line absorption neglected. The dashed curve labeled overall contraction is the locus of models which are beginning the phase of overall contraction while hydrogen is burning in a convective core.

The crosses in Figure 5 mark the position of the standard zero-age main sequence, as defined by stars in the Hyades, the Pleiades, and NGC 2362. It is assumed that composition differences between the clusters may be neglected and that distances to Hyades cluster stars, as determined by the convergent-point method, are correct. Observational data and the conversion from $B - V$ and M_v to $\log T_e$ and $\log L$ are taken for most stars from Sandage (32). The ranges over which the cluster diagrams are forced to coincide are indicated along the left-hand portion of Figure 5.

Considering all of the uncertainties, the fit between the theoretical and the standard $ZAMS$ would appear to be remarkably good. However, comparisons in the mass-luminosity plane suggest that this agreement may be misleading.

In Figure 6, the theoretical main-sequence band for models of composition $(X, Y, Z) = (0.71, 0.27, 0.02)$ is bounded by the curves labeled $ZAMS$ and $TAMS$. The curve labeled $GAMS$ is an estimated locus of model stars which have evolved for 10^{10} years. For a galactic age of $\sim 10^{10}$ yr, the upper limit to the main-sequence band for the chosen composition should thus be the curve $GAMS$. The crosses denote the position of nearby main-sequence band stars selected by Schwarzschild (1).

Below $\log L \sim 2$, the *shape* of the theoretical main-sequence band agrees very nicely with the shape of the band defined by nearby stars. Deviations at very low luminosity could be due to inaccurate bolometric corrections. Since the models of very low luminosity are completely convective below the photosphere, their luminosity is very sensitive to changes in photospheric opacity. Hence, an equally likely explanation of the deviation is that the surface opacities are overestimated.

Above $\log L \sim 3$, the "empirical" main-sequence band falls consistently below the theoretical band. This deviation could also be attributed to an underestimate of bolometric corrections. It cannot easily be accounted for by a variation of composition with stellar mass and it is unlikely that faulty opacities can be blamed.

The squares in Figure 6 represent the position of components of several binaries in the Hyades cluster, again assuming that the convergent-point method of obtaining distance is applicable. Data are from Eggen (33) as treated by Iben (13). In order to simplify the discussion, a point characterized by $\log L = 0.0$, $\log M = -0.15$ shall be taken as representative of the Hyades cluster stars for which masses have been estimated.

It has been shown (13) that for a choice of Z in the range 0.01 to 0.04, the initial hydrogen abundance required to reproduce the representative point in the mass-luminosity plane obeys the relationship $X \simeq .5 - 4.65(Z - .01)$. Adherence to this relationship has the distressing consequence that for reasonable values of (Z/X), it is impossible to achieve consistency in the H-R diagram between models and the standard $ZAMS$ (13, 34). It has been suggested (13) that consistency may be achieved by giving up the assumption that Pleiades stars have the same composition as Hyades stars.

Hodge & Wallerstein (35) and Faulkner & Stefensen (36) have suggested that the convergent-point method of determining the distance to individual members of the Hyades cluster may be inapplicable, as it would be if the cluster were expanding or rotating. Hodge & Wallerstein conclude that the currently adopted distance modulus for the Hyades may be an underestimate. They point out that by increasing the distance modulus, much of the difficulty in achieving consistency between models and the observations in both the H-R and M-L planes is removed.

It is worth while to examine in some detail how this comes about. The total mass derived for a binary system is proportional to the cube of the adopted distance to the binary, whereas the luminosity is proportional to the square of the adopted distance. Hence a change in the adopted distance alters the position of a component star in the M-L diagram by $\Delta(\log M) = 3\Delta(\log d) = \frac{3}{2}\Delta(\log L)$, where d is the adopted distance. An increase of $\Delta(\log d) = 0.067$ brings the standard Hyades star, initially at $\log M = -0.15$, $\log L = 0.0$,

⋙→

FIG. 5. Theoretical predictions in the Hertzsprung-Russell diagram. Units are the same as in Figure 3. The curves labeled $ZAMS_0$ and $TAMS_0$ represent the theoretical boundaries of the main-sequence band, as defined in the text, for the composition $(X,Y,Z) = (0.71, 0.27, 0.02)$, when line effects are omitted. The curve $ZAMS_1$ is the estimated position of the zero-age main sequence when line effects are included. Crosses define the standard $ZAMS$ as determined by stars in the Hyades, Pleiades, and NGC 2362. The lines and vectors in the lower left-hand portion of the Figure describe possible effects of rotation on homogeneous model stars. Theoretical estimates for the core helium-burning bands and the position of the Cepheid instability strip are also shown.

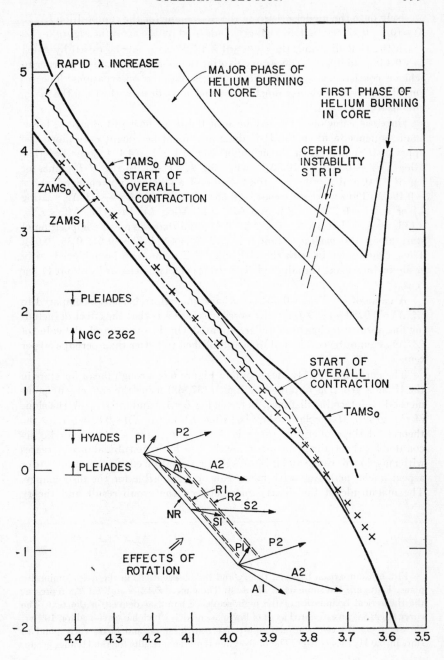

precisely onto the model $ZAMS$ in the M-L plane for the composition $(X, Y, Z) = (0.71, 0.27, 0.02)$. This change is indicated by the arrow in Figure 6.

In the H-R diagram, the standard $ZAMS$ must now be raised by $\Delta(\log L) = 0.133$, and it will be necessary to alter the model composition in order to achieve consistency again between models and the observations. The required composition change must not be such as to destroy the fit in the mass-luminosity plane.

Since the value of (l/H) may be adjusted to achieve a fit along the lower main sequence, a fit in the H-R diagram should be sought only along the upper main sequence. In the neighborhood of $\log T_e = 4.1$–4.2, one has that $\Delta(\log L) = -(0.66\Delta X + 10.5\Delta Z)$, when $\log T_e = $ const. In the M-L plane, for $\log M = 0.0$, and $X = 0.6$–0.7, the position of a star will not change if $\Delta Z \simeq -0.19\Delta X$. Thus we require that $-(0.66 - 10.5 \times 0.19)\Delta X = -1.34\Delta X = \Delta(\log L)$ or $\Delta X = -0.75\Delta(\log L)$, and $\Delta Z = 0.142\Delta(\log L)$. Inserting $\Delta(\log L) = 0.133$, we find that $\Delta X \sim -0.1$, $\Delta Z \sim 0.019$ and that the mean composition of stars along the *modified* standard $ZAMS$ is $(X, Y, Z) \sim (0.61, 0.35, 0.04)$. Thus, consistency in both the H-R and M-L planes can be achieved fairly easily by increasing the adopted distance to each Hyades star by about 17 per cent.

A value of $(Z/X)_{Hy} \sim 0.064$ is perhaps uncomfortably large compared to $(Z/X) \sim 0.028 \, [= (Z/X)_\odot?]$. One possible solution is that the effect of including line absorption has been underestimated. On the other hand, the value of $(Z/X)_\odot$ cannot be considered to be established to better than perhaps 50 per cent.

The curve labeled Hyades group in Figure 6 is a rough mean for stars in the Hyades group as obtained by Eggen (37, 38) using the convergent-point method to determine distance. Between $\log L = 1.0$ and $\log L = 1.4$, the slope of the group M-L relationship is $[\Delta(\log L)/\Delta(\log M)] \sim 0.8$, whereas the theoretical slope over this range is ~ 4.0. The group slope cannot be accounted for theoretically by assuming that the hydrogen abundance increases with mass in this range and it is unlikely that rotation effects (rotation rate or aspect angle increasing with mass) can be responsible for the discrepancy. The magnitude of the discrepancy between the group result and theory

$\ggg\!\!\rightarrow$

FIG. 6. Comparison between theory and the observations in the mass-luminosity plane. Units are the same as in Figure 4. The curves $ZAMS$ and $TAMS$ represent the theoretical boundaries of the main-sequence band, as defined in the text. The curve $GAMS$ is an estimated locus of low-mass models which have evolved for 10^{10} yr. Crosses represent observational results for nearby stars, and boxes correspond to stars in the Hyades cluster. The dashed line represents results for the Hyades group.

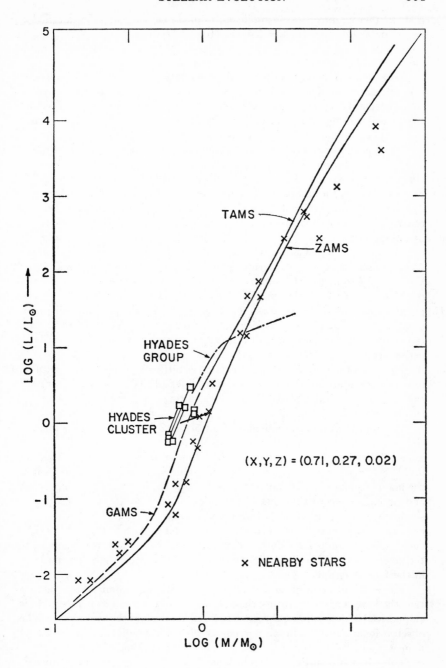

casts further doubt upon the applicability of the convergent-point method in obtaining distances to members of the Hyades group.

Effects of Rotation and "Be" Stars

Rotational effects have stimulated considerable recent work, both theoretical (39–43) and observational (44–47). Since the effects of rotation on stellar structure and on observable properties are sensitive to as yet uncertain theoretical assumptions and simplifications, it is useful to have a strong interplay between theoretical suggestions and the observations.

Of primary interest here is a recent analysis by Strittmatter (47). From a study of the relationship between M_v and $|V \sin i|$ for stars in the Praesepe cluster, Strittmatter finds that at a given color, rotating stars are displaced in magnitude above an arbitrary zero-age main sequence for nonrotating stars by an amount proportional to $|V \sin i|^2$. This is to be expected from the first-order theoretical treatments given by Sweet & Roy (48) and by Roxburgh & Strittmatter (49).

The dependence of ΔM_{bol} and $\Delta(\log T_e)$ on the parameter $\lambda = V_s^2/(GM/R)$ = (equatorial velocity)^2/(surface gravitational potential) is given in Table V for two different sets of assumptions. In Figure 5, the vectors $P1$ and $P2$ represent the displacement as predicted by the two theories for a star viewed pole-on when $\lambda = 1$. The vectors $A1$ and $A2$ represent the predicted displacement for $\lambda = 1$ when the star is viewed equator-on. The vectors $S1$ and $S2$ are estimated displacements for $\lambda = 1$ after an average over aspect angle has been performed.

The line in Figure 5 labeled NR is an arbitrarily placed zero-age main sequence for nonrotating stars. The dashed lines labeled $R1$ and $R2$ represent the displaced positions of the $ZAMS$ for rotating stars predicted by the two theories when $\lambda = \frac{1}{3}$ (for all stars).

It is significant that the displacement of the $ZAMS$ is relatively insensitive to aspect angle. According to the two theories, $\Delta(\log L)_1 \sim 0.33\lambda$ or $\Delta(\log L)_2 \sim 1.28\lambda$ when $T_e = $ const; and $\Delta(\log T_e)_1 \sim -0.066\lambda$ or $\Delta(\log T_e)_2 \sim -0.224\lambda$ when $L = $ const. Rotation thus introduces an additional spread into the main-sequence band comparable to that due to differences in composition and perhaps even to that due to evolution. Note that the luminosity of a single "average" star (averaged over aspect angle) is lowered by an insignificant amount for a reasonable rotation parameter (see the intersection of the curves NR, $R1$, and $R2$ with the vectors $S1$ and $S2$). Thus, rotation affects the position of a group of stars in the M-L plane much less than their position in the H-R diagram.

The effect of rotation on the standard $ZAMS$ is minimized by the fact that the lower envelope (rather than the mean) of the main-sequence band is emphasized in determining the $ZAMS$. Further, the fact that the shape of the observational $ZAMS$ is in fair agreement with that predicted on the basis of nonrotating models argues against a significant rotational effect.

The effect of rotation on evolutionary tracks has not yet been investi-

TABLE V

CHANGES IN BOLOMETRIC MAGNITUDE AND EFFECTIVE TEMPERATURE
DUE TO ROTATION

Model	Pole-on		Axis-on	
	$\Delta M_{bol}/\lambda$	$\Delta \log_{10} T_e/\lambda$	$\Delta M_{bol}/\lambda$	$\Delta \log_{10} T_e/\lambda$
Uniformly rotating star with circulation Sweet & Roy (1953)	-0.552	-0.0286	$+0.928$	-0.1208
Rotating magnetic star, zero field in core Roxburgh & Strittmatter (1966)	-0.662	-0.168	$+0.849$	-0.262

gated quantitatively. It is nevertheless interesting to discuss the question qualitatively. For simplicity, let us suppose that the position in the H-R diagram of a rotating star is displaced from that of a nonrotating star of the same age by a vector whose direction is constant but whose magnitude is proportional to the quantity $\lambda = V_s^2 R/GM$. Let us further suppose, again for simplicity, that the star rotates as a rigid body and that angular momentum is conserved. For each nonrotating model in an evolutionary sequence, the quantity R^3/I^2, where R is the radius and I is the solid-body moment of inertia, can be calculated. If the corresponding rotating model has an angular momentum J, then we can estimate $\lambda = (J^2/GM) (R^3/I^2)$.

In Figure 7, the time dependence of λ for the $1.25 M_\odot$ model depicted in Figure 3 is compared with several other $1.25 M_\odot$ characteristics. The value of J^2 is chosen in such a way that $\lambda_{max} \simeq 0.96$. Note that λ increases steadily throughout the phase of core hydrogen burning and also throughout the phase of hydrogen burning in a thick shell. The parameter λ increases most rapidly near the end of the phase of overall contraction, at the same time that the central temperature drops and core contraction is particularly rapid; it does not begin to decrease until near the end of the phase of hydrogen burning in a thick shell. As the star evolves toward the giant branch, λ drops rapidly and attains negligible size along the giant branch. Thus, if $J =$ const, rotational effects are expected to increase monotonically as the star evolves off the main sequence through most of the thick-shell stage and then to become much less important during the giant stage.

During phases of pure hydrogen burning, the relationship between changes in λ and variations in interior characteristics is much the same for all model stars with mass $M \geq 1.25 M_\odot$. In all cases, λ increases throughout the main-sequence phase and during most of the thick-shell phase. The most rapid rate of increase in λ occurs toward the end of the phase of overall con-

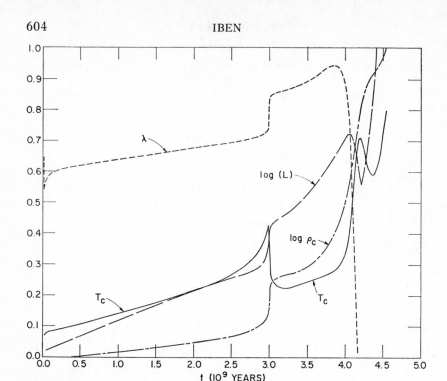

FIG. 7. The time dependences of the rotation parameter $\lambda = \text{const} \times (R^3/I^2)$, luminosity, central temperature, and central density for the $1.25 M_\odot$ model described in Figure 3. Here R and I represent the stellar radius and the moment of inertia. Luminosity is in solar units and temperature and density are in deg K and g/cm³, respectively. Scale limits correspond to $0 \leq \lambda \leq 1$, $0.35 \leq \log L \leq 0.975$, $15 \leq T_c \leq 35$, and $2.0 \leq \log \rho_c < 5.125$.

traction, and λ becomes quite small during the giant-branch phase. However, during the descent from the red-giant tip and during most of the phase of helium burning in the core, λ again increases.

A possible effect of rotation on the evolutionary track of a $3 M_\odot$ star is sketched in Figure 8. The slope of the adopted displacement vector is $d(\log L)/d(\log T_e) \simeq +0.60$ and its magnitude is proportional to λ. The slope of the displacement vector and its magnitude for $\lambda = 1$ are somewhat arbitrary choices, but are consistent with the vectors in Figure 5, labeled $S1$ and $S2$, which represent an average over aspect angle.

The speculation pictured in Figure 8 is not to be taken too seriously. A real star probably does not rotate even approximately like a rigid body (39) and the assertion that the effect of rotation on observable properties of an evolving star is proportional to the parameter λ, as determined from R^3/I^2 for nonrotating models, can at best be only a rough approximation. Further,

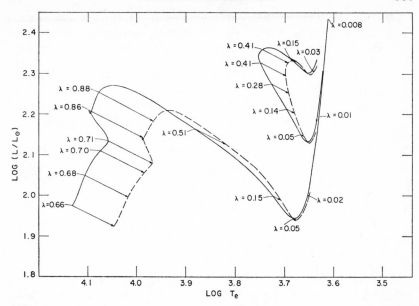

FIG. 8. A qualitative estimate of the effect of rotation on the evolutionary path of a $3M_\odot$ model if mass loss and angular momentum loss are neglected. The rotating model is assumed to be displaced by a vector which is proportional to the rotation parameter λ calculated for the nonrotating model. The direction of the displacement vector is constant and appropriate for some intermediate-aspect angle.

the assumption that total angular momentum (and also total mass) remains constant, even to first order, is undoubtedly wrong. In particular, if the star passes through the "Be" phase to be discussed shortly, angular momentum is probably transmitted to escaping matter through the intermediary of the magnetic field. Further, during phases characterized by a low surface temperature, when convective turbulence near the surface is particularly violent, mass loss and, concomitantly, a rapid loss of angular momentum from the star might be expected (50–52).

Crampin & Hoyle (53) were the first to attempt an explanation of Be stars in terms of the properties of calculated evolutionary models. They suggested that equatorial mass loss might occur when the parameter λ exceeds 1. Schmidt-Kaler (54) has found that extreme Be stars lie along a fairly narrow strip in the H-R diagram *parallel* to, but *above* the zero-age main sequence, and suggests that these stars are in the phase of rapidly increasing λ during or following (or both) the phase of overall contraction. Schild (55) has discussed the Be stars in h and χ Persei. He finds that extreme Be stars lie in a narrow strip lying above the $ZAMS$, but that ordinary Be stars occupy a very broad region ranging from the $ZAMS$ through the region of extreme Be stars, and considerably beyond.

On the basis of the reasoning of Crampin & Hoyle, the fact that Be stars lie above the $ZAMS$ is obvious from the theoretical prediction that all rotating stars lie above the $ZAMS$ for nonrotating stars. The fact that ordinary Be stars cover a broad range is consistent with the fact that the parameter λ increases during most of the pure hydrogen-burning phases prior to shell narrowing. The fact that extreme Be stars lie within a very narrow band is consistent with the fact that the most rapid increase in the parameter λ, and hence the most rapid rate of predicted equatorial mass loss, occurs near the end of the phase of overall contraction. The line labeled rapid λ increase in Figure 5 marks the end of the overall contraction phase for nonrotating stars and hence marks the predicted position of extreme Be stars if rotation did not displace the position of a star in the H-R diagram. Note that this line is essentially parallel to the $ZAMS$. Introducing the predicted displacement due to rotation, we see that the extreme Be stars still lie in a narrow strip, parallel to the $ZAMS$, but considerably above it, just as found by Schmidt-Kaler and Schild.

CEPHEIDS AND YOUNG CLUSTERS

Cepheid statistics.—Perhaps the most exciting recent development associated with studies of relatively massive stars has been the fairly secure identification of classical Cepheids as stars in helium-burning phases (56–58, 9, 10, 51).

An understanding of two statistical properties of galactic Cepheids—the existence of a rough period-luminosity relationship and the existence of a maximum in the distribution of Cepheid numbers versus luminosity—has been obtained on the basis of studies of envelope pulsation in conjunction with studies of stellar evolution through helium-burning phases. For a recent review of theoretical results concerning pulsation, see Christy (59).

The occurrence of large-amplitude pulsation is correlated with an extensive radiative region over which opacity increases with increasing temperature. In stellar envelopes of Population I composition, the region of partial helium ionization is the seat of the driving mechanism for large-amplitude pulsation.

For a given composition, and given interior mass, instability against radial pulsation is found to occur for values of mean luminosity and mean surface temperature which lie within a fairly narrow strip in the H-R diagram. A typical instability strip is sketched in Figure 5. Keeping stellar mass fixed, as luminosity is increased within the instability strip along a line parallel to the strip, one finds the pulsation period to increase.

It is clear that the existence of a very tight period-luminosity relationship would be predicted if most stars which evolve slowly through the instability strip were of similar mass and composition, or if a fairly close correlation existed between increasing mass and increasing luminosity. In the second case, the instability strip would show more curvature than in Figure 5, being constructed of segments appropriate for stars of different mass.

An inspection of the evolutionary tracks in Figure 3 reveals that during comparable phases, there is indeed a fairly tight correlation between mass and luminosity for stars of a given composition. In particular, this is true for stars in the major phase of helium burning in the core.

The region within which models spend most of their time during core helium burning is sketched in Figure 5. The intersection between the major helium-burning band and the instability strip represents the region where most Cepheids of the chosen composition are expected to be found. Not only does the mean mass of evolving models increase with luminosity along this intersection, but the variation of stellar mass over the intersection is relatively small ($\Delta M/M \sim 1/3$). Thus, on two counts, theory suggests a relatively tight correlation between period and luminosity, as observed not only for galactic Cepheids, but for Cepheids in M31 and in the Large and Small Magellanic Clouds.

In addition, it is evident that in the distribution of Cepheid numbers versus luminosity, there should be a maximum at some luminosity within the region of intersection between the instability strip and the major helium-burning band. The predicted low-luminosity cutoff to the distribution is due to the fact that for sufficiently light stars, the maximum extension to the blue during core helium burning does not reach the instability strip. The predicted decrease in Cepheid numbers at high luminosities is due in part to the fact that sufficiently massive stars pass rapidly through the instability strip during the phase of envelope contraction separating the two major phases of core helium burning and in part to the fact that the frequency of stars in selected mass intervals is a rapidly decreasing function of the mean mass characterizing that interval.

The maximum in the distribution is predicted (10) to occur at a luminosity appropriate to that model star which, at its furthest extension to the blue, just reaches the high-temperature edge of the instability strip. The *qualitative* accord between prediction and the distribution (in numbers versus luminosity) of Cepheids in selected fields has been demonstrated by several authors (9, 10, 51).

It does not seem reasonable to attempt a more quantitative comparison at the present time. The position of the major helium-burning band is extremely sensitive to composition changes and to variations in the input physics. The position of the theoretical instability strip is also affected by changes in composition and input physics. Further, additional crossings of the instability strip on a long time scale may be associated with the phase of helium burning in a thick shell. The exact nature of these later crossings is even more uncertain than is the nature of all previous crossings.

NGC 1866.—An extremely beautiful qualitative confirmation of the general features of evolution during phases of helium burning is afforded by stars in the cluster NGC 1866. The distribution of these stars in the color-magnitude diagram, Figure 9, is taken from Arp (58). The crosses mark the mean positions of Cepheids. The smooth curve in Figure 9 is an eye-fitted mean

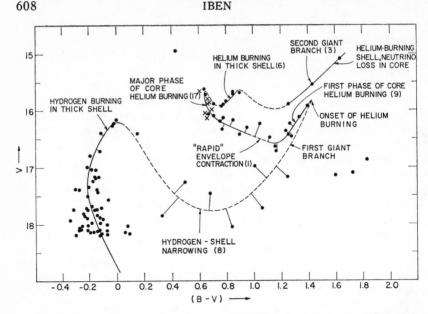

Fig. 9. The position, in the color-magnitude diagram, of stars in NGC 1866. The solid points are taken from Arp (58) and the smooth curves represent an eye-fitted mean cluster locus. Crosses denote the location of Cepheids.

cluster locus. Dashed portions along this curve represent suggested phases of rapid evolution (rapid relative to adjacent phases). Solid lines joining points to the mean curve indicate that portion of the mean curve with which each star has been identified.

The choice for the mean cluster locus in Figure 9 has been strongly influenced by the characteristics of evolutionary models in the neighborhood of $5M_\odot$, the mass thought by Arp to be most consistent with the luminosity level near the turnoff region of NGC 1866. The type of judgment employed is evidenced by the labels describing segments of the chosen mean curve.

Comparison with model results is essential to disentangle segments of the cluster locus near the red end of the color-magnitude diagram. For example, model results indicate that the rate of descent from the red-giant tip following the onset of helium burning in the core should be at least ten times slower than the rate of the preceding ascent. Hence, most of the stars near the first giant branch should be descending. The three stars along the segment labeled second giant branch have been associated with a phase more advanced than core helium burning. Otherwise, these three stars would suggest that helium burning begins in the core at a luminosity ~1 mag brighter than predicted by current evolutionary models of mass near $5M_\odot$.

According to the identifications made in Figure 9, there are twice as many stars in the major phase of core helium burning as there are in the first phase.

This ratio compares quite favorably with a ratio of about 5 to 3 suggested by the $5M_\odot$ model evolution described earlier and thereby strengthens the adopted assignment of stars to the first phase of core helium burning.

Six stars are assigned to the phase of helium burning in a thick shell. That these stars are not in the core helium-burning phase is suggested by the model result that the evolution to the red which occurs near the end of the phase of core helium burning is much more rapid than the preceding evolution to the blue. The assignment of the six stars to the phase of thick-shell burning is consequently quite probable and we thus have the very useful empirical result that for a $5M_\odot$ star, the phase of helium burning in a thick shell lasts about one-fifth as long as does the preceding phase of core helium burning.

The three stars assigned to the second giant branch are probably burning helium in a thin shell, following a rapid phase of helium-shell narrowing. The *relatively* high frequency of stars along this second giant branch can be considered as supporting evidence for Weigert's result (7) that neutrino losses in the core act as a refrigerant and may prolong the phase of pure shell burning.

One disturbing feature of the star distribution in Figure 9 diminishes somewhat the confidence one can place in arguments which depend on the relative density of stars along the mean cluster locus. The fact that there are so many (eight) stars along the dashed curve in Figure 9 between the segment labeled hydrogen burning in a thick shell and the point labeled onset of helium burning is in clear conflict with evolutionary theory. The $5M_\odot$ model spends five times as long in the first phase of core helium burning (points 7 to 8 in Figure 1) as it does in the shell-narrowing and ascending-giant phases combined (points 5 to 7 in Figure 1). There are thus at least four times as many stars along the lowest dashed curve in Figure 9 as theory predicts. No theoretical explanation (other than the occurrence of additional nuclear burning processes) for this rather serious discrepancy can be offered.

h and χ Persei.—Another highly interesting distribution in the color-magnitude diagram is presented by the stars in h and χ Persei. The distribution in Figure 10 is taken from Wildey (60). It is apparent (3, 61) that the majority of stars in h and χ Persei are quite young and that most of the stars to the red of the main-sequence band below $M_v \sim -3$ are contracting onto the main sequence. Because of the dense population all along the main sequence, it is further clear that star formation has occurred continuously over a period of at least 10^7 yr. This lower limit to the age is based on the fact that the main-sequence band is populated down to at least $M_v \sim +3.5$ (61). The existence of the most luminous blue stars near the main sequence is additional support for the view that star formation has been continuous. These luminous stars may be gauged to have masses of 45–$60M_\odot$, and the main-sequence lifetime for such stars is $(3$–$4) \times 10^6$ yr (20).

Hayashi & Cameron (62) have suggested that the red supergiants circled in Figure 10 are in a phase of evolution more advanced than that of helium burning in the core. They argue that these stars each have a mass of 15–$16M_\odot$. By comparing the number of red supergiants with the number of

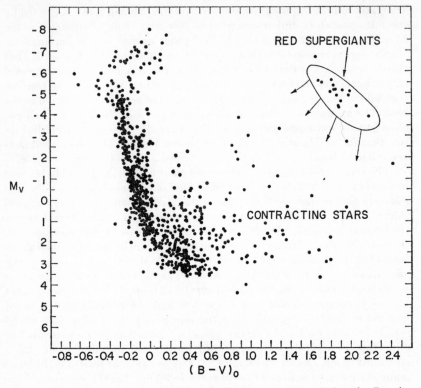

FIG. 10. The position, in the color-magnitude diagram, of stars in h and χ Persei. This Figure is reproduced from Wildey (60), except for the comments.

selected stars near the main sequence which they identify as 15–$16M_\odot$ stars in the phase of core helium burning, they conclude that the lifetime of 15–$16M_\odot$ stars in stages more advanced than core helium burning is comparable to the lifetime in the core helium-burning phase. On the basis of estimates, they then contend that model calculations for more advanced phases will give a total lifetime for these phases in accord with their interpretation of the observations, but only when neutrino loss from the core by the photoneutrino and pair annihilation processes is omitted. Objections can be raised concerning the interpretation of the observations and concerning the theory on which these conclusions are based. For example, neutrino loss from the stellar core may prolong the lifetime of relatively massive stars along the giant branch during the phase of combined hydrogen- and helium-shell burning (15). The red supergiants in h and χ Persei might be stars in a double-shell burning stage. The identification of stars near the upper main sequence in h and χ Persei (the early supergiant branch) as stars in the core helium-burning stage is also debatable. Variations in composition can shift the position of the core helium-burning phase radically in the H-R diagram. For example, an increase in initial X and/or in Z within Population I limits may shift the evolutionary track of a $15M_\odot$ star during this phase from the region between points 6 and 8 in Figure 3 over to the giant branch, near point 10 in Figure 3. Thus, the red supergiants might easily be stars in the core helium-burning phase.

The identification of the red supergiants in h and χ Persei as highly evolved stars and the supposition that luminous blue stars are all evolving away from the main sequence can also be questioned. The weight of evidence suggests that stars in h and χ Persei are very young, that star formation has occurred continuously, perhaps in spurts, and that it is impossible to tell in which direction stars are evolving near the main sequence. An alternate interpretation of the red supergiants is that they are the most recently formed stars in h and χ Persei, that they have masses which cover a large range, and that they are evolving downward and to the left in the color-magnitude diagram during the pre-main-sequence phase of gravitational contraction. No conflict with theory then remains.

OLD POPULATION I CLUSTERS

Schematic cluster loci for stars in M67 and NGC 188 are shown in Figure 11. These loci are based on data from Sandage (63) and from Johnson & Sandage (64) as treated by Iben (16). The two cluster loci have been forced to fit near the lower portion of the diagram.

When the cluster loci in Figure 11 are compared with the evolutionary tracks in Figure 3, it is obvious that stars in NGC 188 are older than stars in M67. Four major differences between the two cluster loci confirm this age separation and provide evidence in support of several mass-dependent trends predicted by the evolutionary calculations.

The most obvious difference is that in the region between the main-sequence band and the red-giant branches (points D to E and F to G), stars in M67 are more luminous than stars in NGC 188. This means, of course, that stars in M67 which have just left the main-sequence band are more massive and hence younger than the corresponding stars in NGC 188.

The existence of a gap (between points B and C) in the M67 locus suggests that stars near point C have almost the same mass as stars near point B and that evolution between points B and C is quite rapid relative to the rates of evolution toward point B and away from point C. One may conclude that all stars in M67 near and beyond point B possessed convective cores during the main-sequence phase, that stars near point B are just beginning the phase of overall contraction, and that stars near point C have begun to burn hydrogen in a thick shell. The apparent absence of a gap in the NGC 188 locus suggests that highly evolved stars in NGC 188 are too light to have possessed a convective core during the main-sequence phase. Once again, we conclude that the most massive visible stars in NGC 188 must be lighter and hence older than the most massive stars in M67.

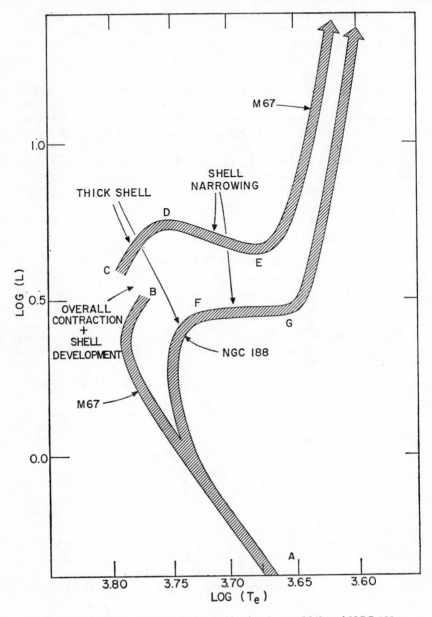

FIG. 11. Schematic cluster loci for stars in the clusters M67 and NGC 188.

The slope of the cluster locus for M67 between points D and E is steeper than that for NGC 188 between points F and G. From the tracks in Figure 3 it may be seen that as mass is decreased below $3M_\odot$, the luminosity drop during the shell-narrowing phase becomes less and less pronounced. This feature is reflected also in time-constant loci: the older the cluster, the lighter are the stars in the shell-narrowing phase and hence the shallower is the slope of that segment of the time-constant locus made up of stars in the shell-narrowing phase.

Finally, at a given luminosity, the giant branch of the M67 locus is distinctly bluer than is the giant branch of the NGC 188 locus. Inspection of Figure 3 shows that at a given luminosity, the giant branch of an individual heavy star is bluer than that of a lighter star. Again, we conclude that the most massive visible stars in M67 are heavier and younger than those in NGC 188.

The preceding arguments have tacitly assumed that composition differences between stars in NGC 188 and M67 are insignificant. Comparing cluster loci with models of composition $(X,Y,Z) = (0.71, 0.27, 0.02)$, one finds that the best fits (16) are obtained for assumed cluster ages of $(5.5 \pm 1) \times 10^9$ yr (for M67) and $(11 \pm 2) \times 10^9$ yr (for NGC 188). With these assignments, the most massive stars in NGC 188 are near $1M_\odot$ and no gap in the cluster locus is expected. In fact, the lack of a gap in Sandage's (63) data has been made use of in considerations of goodness of fit.

Sandage (65) has recently re-examined stars in NGC 188 and states that there is evidence for a gap near the turnoff point along the NGC 188 locus. If there is indeed a gap, and no change is made in model initial composition or in the input physics, then the cluster locus for NGC 188 must be raised by at least $\Delta(\log L) \sim 0.09$ above the luminosity level assigned by choosing an age of 11×10^9 yr. Upper limits to the ages of M67 and NGC 188 would then be 4×10^9 yr and 8×10^9 yr respectively. On the other hand, the existence of a small convective core is very sensitive to the temperature and density dependences of opacity and of nuclear-burning rates and no special virtue attaches to the composition $(X,Y,Z) = (0.71, 0.27, 0.02)$. The occurrence of a gap in the NGC 188 locus would therefore be far from an embarrassment. On the contrary, it would be an additional constraint which would help narrow the range of acceptable composition variables for stars in NGC 188 and would, it is to be hoped, shed more light on the adequacy of the basic physical assumptions.

THE EVOLUTION OF POPULATION II STARS

Studies of Population II evolution have been hampered by the facts that: (a) the helium-to-hydrogen ratio at the surface of most Population II stars is not obtainable spectroscopically; (b) the abundance of metals to hydrogen at the surfaces of individual stars may be uncertain by factors of 10; (c) there is no *direct* determination of the mass of a single extreme Population II star; (d)

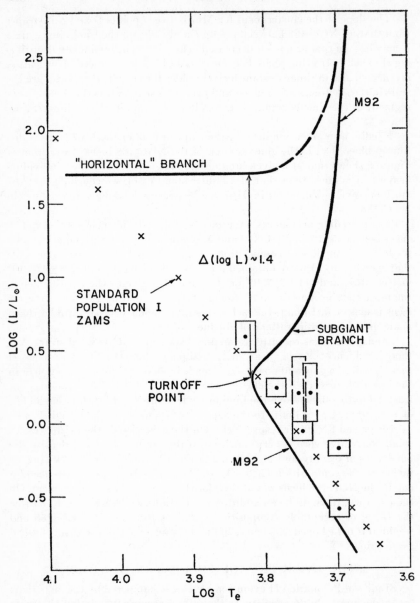

FIG. 12. Observational data for Population II stars. The crosses represent the standard zero-age main sequence for Population I stars. The boxes represent the positions of several subdwarfs as given by Strom, Cohen & Strom (67). The heavy curve is a schematic cluster locus for M92.

the absolute positions of globular cluster stars in the theoretical H-R diagram are highly uncertain; (e) the vast majority of Population II stars are in a region of the H-R diagram where the position of theoretical models varies most drastically with changes in the treatment of convective flow in the envelope; (f) the calculation of evolution along the giant branch and through the helium-flash phase is very time consuming—a fact which has restricted extensive theoretical exploration.

The type of observational information available for Population II stars is illustrated in Figure 12. Eggen & Sandage (66) have suggested that in the M_{bol}-log T_e plane, metal-deficient subdwarfs form a sequence which coincides, within observational and theoretical errors, with that portion of the zero-age main sequence determined by Hyades stars. The crosses in Figure 12, which define the standard Population I $ZAMS$, may thus represent the location of many Population II stars of low luminosity. Recently S. E. Strom, J. G. Cohen & K. M. Strom (67) have fitted model atmospheres to several subdwarfs for which spectrum scans and trigonometric parallaxes are available. Their results are given by the boxes in Figure 12.

The major features of the cluster locus defined by stars in a typical globular cluster M92 are also sketched in Figure 12. Observational data from Sandage & Walker (68) and from unpublished work by Sandage, as quoted by Sandage & Smith (69), have been used as a guide. The *relative* values of log T_e and log L along the cluster locus for log $L \lesssim 1.7$ are probably correct in order of magnitude. The only fairly reliable *relative* position along the horizontal branch is that of the RR Lyrae stars which lie approximately $\Delta(\log L)$ $\sim$1.4 above the cluster turnoff point, although the specification of turnoff luminosity is uncertain by perhaps $\pm$0.1 in log L. That portion of the horizontal branch to the blue of the RR Lyrae region has been drawn arbitrarily at constant luminosity. A fit between the lower portion of the globular-cluster locus and the field subdwarfs can be achieved by moving the sketched locus bodily either horizontally to the red or vertically to higher luminosity (or both). An attempt to specify the absolute position of the cluster locus has been purposely avoided. It is suggested that such an attempt should be made only after spectrum scans have been obtained for a number of cluster stars and model stellar atmospheres have been constructed to fit these scans.

The type of information which may be obtained from model evolution is shown in Figures 13 and 14. In Figure 13, the solid curves are evolutionary paths off the zero-age main sequence for model stars with an initial composition $(X, Y, Z) = (0.9, 0.1, 2 \times 10^{-4})$, when l/H_{dens} is chosen as 0.5 (70). The dashed curve is the evolutionary path of an $0.75 M_\odot$ model which is burning helium in a core of initial mass $0.6 M_\odot$ and of initial composition (X, Y, Z) $= (0, 1, 2 \times 10^{-4})$. The envelope composition of this model is identical with that of the initial main-sequence models. Matter is too cool at the hydrogen-helium interface for hydrogen burning to take place. The dotted curve is merely to indicate that the immediate progenitor of the initial $0.75 M_\odot$ model is presumed to be at the red-giant tip.

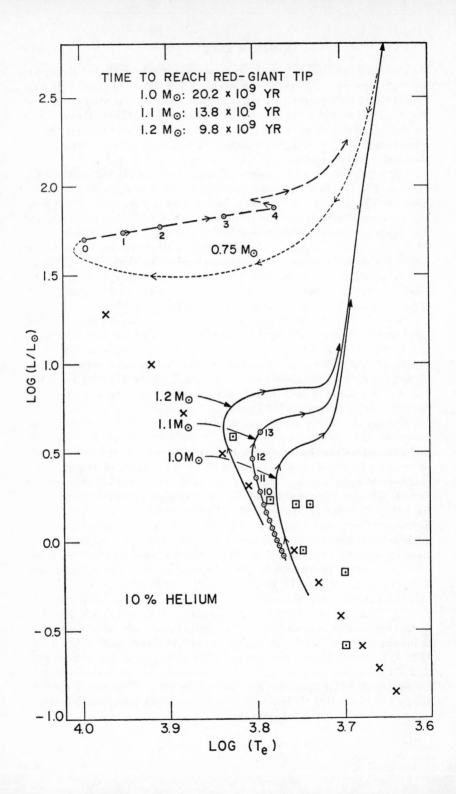

In Figure 14, the solid curves are evolutionary paths for models with an initial composition $(X, Y, Z) = (0.65, 0.35, 2 \times 10^{-4})$ and $(l/H_{\mathrm{dens}}) = 0.5$ (71). The dashed curve (70) is the path of an $0.65 M_\odot$ model with a hydrogen-exhausted core of mass $0.4 M_\odot$ and an envelope composition identical with that of the initial main-sequence models. Helium is burning in the core but hydrogen burning in the shell provides most of the star's energy output. Again, the dotted curve indicates only an assumed chronological relationship between calculated phases.

The total stellar mass and the mass in the hydrogen-exhausted core of the helium-burning models described in Figures 13 and 14 have been chosen primarily to achieve a mean-luminosity level near $\log L \sim 1.7$–1.8 and to achieve the maximum possible extension in $\log T_e$ during the entire phase of core helium burning. The core masses are roughly consistent with the masses in the hydrogen-exhausted core at the start of the helium flash as found by Schwarzschild & Härm (27) and by Hayashi, Hōshi & Sugimoto (3) for the relevant envelope ratios of hydrogen to helium. Further assumptions are that no mixing of hydrogen into the core occurs during the flash (28) and that a negligible amount of helium is burned in the core before degeneracy is lifted.

By a simple extrapolation of the model paths in Figures 13 and 14 to lower masses and luminosities, it is clear that a better fit with the mean of the seven subdwarfs shown can be achieved with the higher choice of initial hydrogen content and an even better fit can be achieved with a choice of 100 per cent initial hydrogen. For each composition, however, an age for several subdwarfs is found to exceed 15–20×10^9 yr.

Demarque (72) has constructed time-constant loci for several choices of initial helium content, the most relevant composition choices for the present discussion being $(X, Y, Z) = (0.75, 0.25, 10^{-3})$ and $(0.999, 0, 10^{-3})$. The mixing-length-to-pressure-scale-height ratio is chosen to be 1. Demarque's time-constant loci are compared with the standard $ZAMS$ in Figures 15 and 16. When $X = 0.75$, time-constant loci reach the $ZAMS$ after 17 billion years; whereas, when $X = 0.999$, time-constant loci reach the $ZAMS$ in only 11 billion years. Thus, if the standard observational $ZAMS$ actually coincides roughly with the lower portion of cluster loci, cluster ages more consistent with that estimated from the Hubble time are achieved with 100 per cent initial hydrogen. On the other hand, ages in excess of 20×10^9 yr are still

---

←⧏⧏⧏

FIG. 13. Evolutionary paths for metal-poor stars with initial composition $(X, Y, Z) = (0.9, 0.1, 2 \times 10^{-4})$. Solid tracks correspond to the phase of pure hydrogen burning. The dashed track describes the evolution of a core helium-burning model with $(M_s, M_{\mathrm{core}}, X_e) = (0.75, 0.6, 0.90)$. Intervals between consecutive circles along the $1.1 M_\odot$ model track correspond to evolutionary times of 10^9 yr. Intervals between circles along the $0.75 M_\odot$ track correspond to 10^7 yr. The crosses define the standard $ZAMS$ and the boxed points represent positions of field subdwarfs.

necessary to fit several of the extreme subdwarfs in Figure 12 with De-marque's models.

Faulkner & Iben (71) have shown that better agreement in shape with the turnoff region and the subgiant portions of M92 can be achieved with a choice of initial hydrogen content closer to $X = 0.65$ than to $X \geq 0.9$. That agreement in shape is better for the lower choice of initial X is evident from comparison between model tracks in Figures 13 and 14 and the cluster locus in Figure 12. This is demonstrated even more convincingly by the corre-sponding time-constant loci (70). Thus, for *fixed* values of $(l/H_{dens}) \sim 0.5$, better shape fits with some portions of cluster loci can be achieved with a lower value of initial X, whereas better fits with the estimated *absolute* posi-tion of extreme subdwarfs and the position of the lower portion of cluster loci, if coincident with the *ZAMS*, can be achieved with a higher value of initial X. This illustrates the danger of accepting conclusions based on at-tempted fits between model calculations and fragments of the available observational data.

The discussion has thus far proceeded on the assumption that model re-sults are reliable. Unfortunately, during the entire phase of hydrogen burn-ing, the position of every low-mass model of relevance here is highly sensitive to changes in the treatment of convective flow in envelope regions. There is no reason to suppose that a better representation of convective flow might not be achieved by allowing (l/H) to vary with surface temperature, with surface gravity, and with envelope composition. An appropriately chosen time variation of l/H and an improvement in opacities might well result, for any choice of initial $X \geq 0.65$, in adequate fits with extreme subdwarfs and globular-cluster loci and lead to considerably reduced derived ages.

In contrast with the surface temperature, the luminosity of a model star at any given age is unaffected by changes in the treatment of convection. The luminosity at the turnoff point of theoretical time-constant loci is not seriously affected by changes in (l/H). For any choice of composition and input physics, there is a nearly unique relationship between the age, 'the luminosity at the turnoff, and the mass (assuming no mass loss) of stars along the giant branch. From time-constant loci constructed for the composition and input physics which lead to the model paths in Figures 13 and 14, the entries in Table VI result (70). In this table, t_9 is the assigned age in 10^9 yr,

---

⋙→

FIG. 14. Evolutionary paths for metal-poor stars with initial composition $(X, Y, Z) = (0.65, 0.35, 2 \times 10^{-4})$. Solid tracks correspond to the phase of pure hydro-gen burning. The dashed track describes the evolution of a core helium-burning model with $(M_s, M_{core}, X_e) = (0.65, 0.4, 0.65)$. Intervals between consecutive circles along the $0.7 M_\odot$ model track correspond to evolutionary times of 10^9 yr. Intervals between circles along the $0.65 M_\odot$ track correspond to 10^7 yr. The crosses define the standard *ZAMS* and the boxed points represent positions of field subdwarfs.

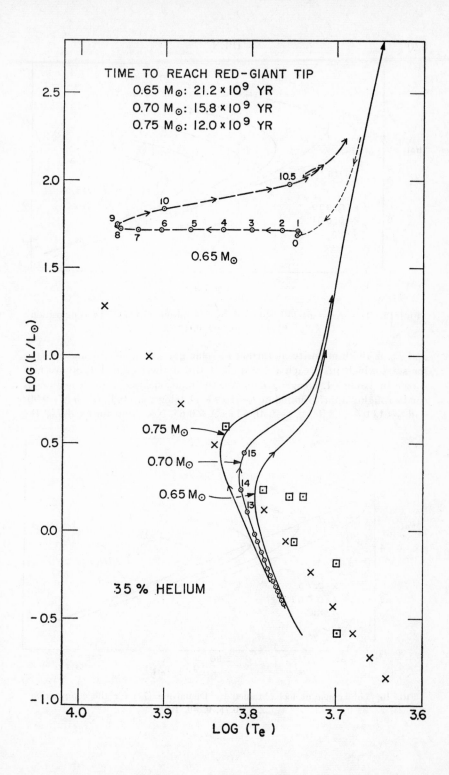

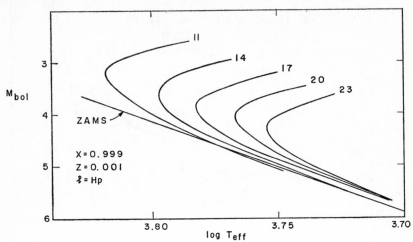

FIG. 15. Time-constant loci obtained by Demarque (72) for the composition
$(X,Y,Z) = (0.999, 0, 10^{-3})$.

and L_{to} is the luminosity at turnoff in solar units. The model masses M_{RG}
for stars which just reach the red-giant tip in the assigned time are also
shown in Table VI. Near log $L_{to} = 0.4$, the relationships between t_9 and L_{to}
can be roughly approximated by $t_9 \sim 16.9 - 39.2$ [log $L_{to} - 0.4$], when $X = 0.90$;
and $t_9 \sim 13.0 - 30.8$ [log $L_{to} - 0.4$], when $X = 0.65$. Note that for fixed L_{to}, the

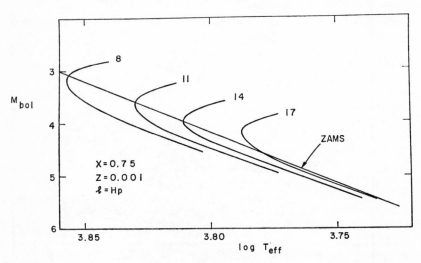

FIG. 16. Time-constant loci obtained by Demarque (72) for the composition
$(X,Y,Z) = (0.75, 0.249, 10^{-3})$.

TABLE VI

THE RELATIONSHIP BETWEEN CLUSTER AGE, TURNOFF LUMINOSITY,
AND STELLAR MASS ALONG THE GIANT BRANCH

t_9	$\log (L_{to})$		$M/M_\odot$	
	$X = 0.65$	$X = 0.90$	$X = 0.65$	$X = 0.90$
8	0.60	0.73	0.82	1.27
10	0.51	0.63	0.78	1.19
12	0.43	0.55	0.75	1.14
14	0.37	0.48	0.72	1.10
16	0.31	0.42	0.70	1.06
18	0.26	0.37	0.68	1.03
20	0.22	0.33	0.66	1.00

cluster age found with $X = 0.9$ is 3–4 billion years longer than that found with $X = 0.65$.

To illustrate the use of Table VI, suppose that relative surface temperatures along the cluster locus in Figure 12 are correct and that the locus can be moved freely in the vertical direction. If the cluster locus is raised by $\Delta(\log L) \sim 0.3$, the lower portion of the cluster locus lies roughly between the standard $ZAMS$ and the mean of the upper six subdwarfs, and the turnoff luminosity occurs at $\log L \sim 0.6$. For $X = 0.9$, the cluster age is then $t_9 \sim 11$ and the mass of a star along the giant branch is $M_{RG} \sim 1.17 M_\odot$. For $X = 0.65$, the corresponding numbers are $t_9 \sim 8$ and $M_{RG} \sim 0.82 M_\odot$. Remember, however, that near the turnoff region and along the subgiant branch, the shape of the cluster locus is not consistent with the theoretical shape for $X = 0.9$ and that models with $X = 0.65$ do not reach the lower portions of the cluster locus in the allotted time. Remember, too, that surface temperatures along the cluster locus in Figure 12 may well be in error by $\Delta(\log T_e) \sim \pm 0.02$. Shifting the cluster locus in Figure 12 over by $\Delta(\log T_e) \sim -0.02$ and then raising the locus by $\Delta(\log L) \sim 0.15$ would bring the lower portion of the locus into the approximate position achieved by keeping T_e fixed and raising the locus by $\Delta(\log L) \sim 0.3$. The turnoff luminosity would now, however, be at $\log L \sim 0.45$, and values for cluster age and red-giant mass would be $t_9 \sim 15$ and $M_{RG} \sim 1.08 M_\odot$, for $X = 0.90$; and $t_9 \sim 11.5$ and $M_{RG} \sim 0.76 M_\odot$, for $X = 0.65$.

As a result of theoretical studies of RR Lyrae pulsation, Christy (59) has suggested that RR Lyrae variables in very metal-weak clusters are at a luminosity given by $\langle \log L \rangle_{RR} \sim 1.66$. If we allow an error of ± 0.1 in the observational result (68, 69) that RR Lyrae variables in metal-poor clusters lie above the cluster turnoff by $\langle \log L \rangle_{RR} - \log L_{to} \sim 1.4$, then Christy's result places the cluster turnoff at $\log L_{to} \simeq 0.26 \pm 0.1$. Extrapolating from Table VI we conclude that, for $X = 0.90$, cluster age is $t_9 \sim 22.5 \pm 4$, whereas for $X = 0.65$, cluster age is $t_9 \sim 18 \pm 4$. Thus, ages long compared to that expected from the Hubble time are found for both choices of initial X, but smaller ages are found for smaller X. This is exactly the reverse of the conclusion reached by fitting time-constant loci to the $ZAMS$.

The most remarkable result of attempts to fit the horizontal branch with models characterized by an envelope helium content $Y \lesssim 0.1$ is that model mass must be chosen considerably smaller than that appropriate for approximate fits near and beyond the turnoff region. Hayashi et al. (3) were the first to discover this fact, subsequently confirmed by others (71, 73).

For the $0.75 M_\odot$ model shown in Figure 13, the mean luminosity during the phase of core helium burning is roughly $\langle \log L \rangle_{HB} \sim 1.8$. During the last quarter of the phase of core helium burning, where pulsation of the RR Lyrae type might be expected to occur, the mean luminosity is $\langle \log L \rangle_{RR} \sim 1.85$. Let us assume that $\log L_{to} = 1.85 - 1.4 = 0.45$. Then, if we agree that the $0.75 M_\odot$ model path in Figure 13 provides an adequate fit to the horizontal branch, cluster age is $t_9 \sim 15$, the mass of stars along the giant branch (without mass loss) is $M_{RG} \sim 1.08$, and it is necessary to assume that $\gtrsim 0.33 M_\odot$ has been lost by a star prior to reaching the horizontal branch.

For $X_e = 0.9$, an increase of $\sim 0.05 M_\odot$ in *either* M_s *or* M_{core} leads to increases of about 0.1 in both $\langle \log L \rangle_{HB}$ and $\langle \log L \rangle_{RR}$ (70, 71). If the turnoff luminosity is chosen as $\log L \sim 0.6$, then $\langle \log L \rangle_{RR} \sim 2.0$. A fit to the horizontal branch might then be achieved with a helium-burning model of total mass $M_s \sim 0.8 M_\odot$ and of core mass $M_{core} \sim 0.65 M_\odot$. Since the cluster characteristic M_{RG} is $\sim 1.17 M_\odot$, it must be assumed that a star loses $\sim 0.37 M_\odot$ at some time during the pure hydrogen-burning phases preceding the horizontal-branch phase.

When the envelope abundance of hydrogen is chosen to be $X_e = 0.65$, models burning helium in the core evolve to the blue. For the model characterized by $(M_s, M_{core}) = (0.65 M_\odot, 0.4 M_\odot)$, both $\langle \log L \rangle_{HB}$ and $\langle \log L \rangle_{RR}$ are about 1.72 (see Figure 14). If the age of globular-cluster stars is chosen to be $t_9 = 15$, then the turnoff luminosity occurs at $\log L_{to} \sim 0.32$ and the horizontal-branch model path in Figure 14 bears the correct relationship to the turnoff luminosity. For $t_9 = 15$, the mass of stars along the giant branch is $\sim 0.7 M_\odot$ and very little mass loss is demanded by fits to the two most distinctive characteristics of cluster loci—the steep subgiant branch and the horizontal branch. This is in strong contrast with the case $X_e = 0.9$ which requires that $0.3 - 0.4 M_\odot$ be lost by a star prior to reaching the horizontal branch.

If the age of globular-cluster stars is chosen to be $t_9 = 10$, and $X_e = 0.65$, then the mass of stars along the cluster giant branch is $M_{RG} \sim 0.78$, the luminosity at the cluster turnoff point is given by $\log L \sim 0.5$ and, therefore, $\langle \log L \rangle_{RR} \sim 1.9$. When $X_e = 0.65$, the value of $\langle \log L \rangle_{HB}$ is less sensitive to changes in M_s and M_{core} than is the case when $X_e = 0.9$. An increase in either M_s or M_{core} by $0.05 M_\odot$ increases $\langle \log L \rangle_{HB}$ and $\langle \log L \rangle_{RR}$ by about 0.06 (70). If, then, $\langle \log L \rangle_{RR} \sim 1.9$, the horizontal branch may be reached by models characterized by $(M_s, M_{core}, X_e) \sim (0.75 M_\odot, 0.45 M_\odot, 0.65)$ and essentially no mass loss is required to fit all portions of globular cluster loci.

Comparing the luminosity function for the cluster M13 with model results, Simoda & Kimura (74) have derived semi-empirical estimates for the lifetime of stars along the horizontal branch. For $X_e = 0.90$ and an assumed cluster age of 17×10^9 yr, they find $t_{HB} \sim 8 \times 10^7$ yr; for $X_e = 0.65$ and an assumed cluster age of 13.5×10^9 yr, $t_{HB} \sim 5 \times 10^7$ yr. Adopting a cluster age of $\sim 10^{10}$ yr, these results may be scaled down to $t_{HB} \sim 5 \times 10^7$ yr, for $X_e = 0.90$, and $t_{HB} \sim 4 \times 10^7$ yr, for $X_e = 0.65$.

For the choice $(M_s, M_{core}, X_e) \sim (0.75, 0.6, 0.9)$ the theoretical horizontal-branch lifetime is $t_{HB} \sim 4 \times 10^7$ yr and $\langle \log L \rangle_{RR} \sim 1.85$. An increase in $\langle \log L \rangle_{HB}$ by about 0.15 reduces this lifetime by about 10^7 yr to $t_{HB} \sim 3 \times 10^7$ yr (70). Thus, when $X_e \sim 0.9$, a choice of cluster age in the neighborhood of 10^{10} yr or larger leads to theoretical horizontal-branch lifetimes roughly two times smaller than the semi-empirical estimates of t_{HB}.

The theoretical horizontal-branch lifetime for the choice $(M_s, M_{core}, X_e) \sim (0.65, 0.4, 0.65)$ is $t_{HB} \sim 10^8$ yr. An increase in $\langle \log L \rangle_{HB}$ by 0.2 leads to a reduction in t_{HB} by a factor of 2 (70). Thus, for $X_e = 0.65$, consistency between the model determination and the semi-empirical determination of t_{HB} may be achieved when cluster age is chosen to be $\sim 10^{10}$ yr. Agreement becomes more difficult to achieve as the assumed cluster age is increased above 10^{10} yr.

It would be unwise to place too much weight on comparisons with semi-empirically determined values of t_{HB}. These might easily be wrong by factors of 2. In the case of $X_e = 0.9$, the semi-empirical determination of t_{HB} may be objected to on the ground that it is based on calculational results for hydrogen-burning models of constant mass, whereas it is necessary to assume significant mass loss at some point during the phase of pure hydrogen burning in order to account for the horizontal-branch stars. If the rate of mass loss is correlated with the violence of subphotospheric turbulence, one might expect mass loss to occur most prominently along the giant branch and lead to giant-branch lifetimes, and hence semi-empirically determined values of t_{HB}, considerably different from those calculated on the assumption of no mass loss.

Christy (59) finds that best fits to pulsational properties of cluster variables occur when model masses are chosen in the neighborhood of $0.5 M_\odot$, with a possible spread over the range $0.45 M_\odot$ to $0.6 M_\odot$. Such low masses are consistent with the values of M_s and M_{core} required to obtain $\langle \log L \rangle_{HB} \sim 1.66$ with evolutionary models characterized by an envelope hydrogen abundance $X_e = 0.65$. They are not consistent with evolutionary model predictions for $X_e \geq 0.9$, which suggest that the mass in the hydrogen-exhausted core is at least $M_{core} \sim 0.6 M_\odot$ (even larger when plasma neutrino losses are included).

Stars along the blue end of the horizontal branch are hot enough to permit a possibly meaningful determination of the surface ratio of helium to hydrogen. Helium lines in globular-cluster and halo B stars (75–77) are known to be weak, and a conventional analysis (75) suggests a surface ratio $(Y/X)_e \lesssim$ 0.03. Greenstein, Truran & Cameron (78) have argued that gravitationally induced diffusion inward may be responsible for the apparent deficiency in the surface helium abundance. Since elements heavier than helium are affected by diffusion even more strongly than is helium, it should be possible to verify or discard this suggestion by comparing the metal-to-hydrogen ratio in blue horizontal-branch stars with this same ratio in stars along the giant branch of the same cluster. Since convection insures thorough mixing over a large fraction of a giant-branch star's mass, the metal-to-hydrogen ratio in such stars should be much larger than it is in blue horizontal-branch stars—if diffusion is capable of effecting a mass segregation.

In this connection, it is interesting that a value of $(Y/X)_e \sim 0.03$ for matter in the envelope of helium-burning stars is consistent with model predictions when an initial hydrogen abundance of $X = 1$ is assumed. Faulkner & Iben (70) find that as a $1.3 M_\odot$ model with an initial value of $X \simeq 1.0$ evolves up the giant branch, envelope convection extends into the region where considerable helium has been formed during the main-sequence phase. Just before the base of the convective zone begins to recede outward as a consequence of the advancing hydrogen-burning shell, the helium abundance in the envelope reaches the value $Y \simeq 0.03$! If mass loss occurs along the giant branch, then the final surface value of Y may be even larger than this.

It should by now be apparent that a satisfactory interpretation of Population II evolution is not yet within our grasp. An attempt has been made to present as many divergent elements of the puzzle as possible, illustrating that in order to achieve consistency with some shreds of evidence it is necessary to forfeit consistency with other shreds. The strongest support for a very low initial helium content Y comes from the apparent position of extreme subdwarfs in the H-R diagram. The strongest support for an unconventionally high initial Y comes from theoretical studies which indicate that (a) pulsation of the RR Lyrae type requires a high envelope Y and a stellar mass small compared to the mass in the helium core of luminous, low-initial-Y red giants and (b) globular-cluster shapes are more easily matched and no significant mass loss is required for large values of initial Y. However, regardless of whether high or low Y is correct, a conflict exists between the theory of stellar evolution, the theory of stellar pulsation, and the Hubble time t_H, if, as conventionally supposed, $(2/3)t_H$ is indeed an upper limit to cluster age. Ages near $(2/3)t_H$ could be assigned by evolutionary theory to globular clusters if RR Lyrae stars had luminosities closer to $\langle \log L \rangle_{RR} \sim 1.9 - 2.0$ than to $\langle \log L \rangle_{RR} \sim 1.6 - 1.7$, as indicated by pulsation theory. But then the masses of RR Lyrae stars obtained from evolutionary calculations would be significantly larger than the masses suggested by pulsation theory.

LITERATURE CITED

1. Schwarzschild, M., *Stellar Structure and Evolution* (Princeton Univ. Press, Princeton, N. J., 1958)
2. Burbidge, E. M., Burbidge, G. R., *Handbuch der Physik*, **51**, 134 (1958)
3. Hayashi, C., Hōshi, R., Sugimoto, D., *Progr. Theoret. Phys. Suppl. No. 22*, 1 (1962)
4. Sears, R. L., Brownlee, R. R., in *Stellar Structure*, 575 (Aller, L. H., McLaughlin, D. B., Eds., Univ. of Chicago Press, Chicago, Ill., 1965)
5. Langer, E., Herz, M., Cox, J. P., *Joint Inst. Lab. Ap. Rept. No. 88* (Univ. of Colorado, Boulder, Colo., 1966)
6. Kippenhahn, R., *Mitteilungen der Astronomischen Gesellschaft*, 53 (1965)
7. Weigert, A., *Mitteilungen der Astronomischen Gesellschaft*, 61 (1965); *Z. Ap.*, **64**, 395 (1966)
8. Kippenhahn, R., Thomas, H. C., Weigert, A., *Z. Ap.*, **61**, 241 (1965); **64**, 373 (1966)
9. Hofmeister, E., *Z. Ap.*, **65**, 164, 194 (1967)
10. Iben, I., Jr., *Ap. J.*, **140**, 1631 (1964); **143**, 483 (1966)
11. Schönberg, M., Chandrasekhar, S., *Ap. J.*, **96**, 161 (1942)
12. Schwarzschild, M., Härm, R., *Ap. J.*, **142**, 855 (1965)
13. Iben, I., Jr., *Ap J.*, **138**, 452 (1963)
14. Iben, I., Jr., *ibid.*, **142**, 1447 (1965)
15. Iben, I., Jr., *ibid.*, **143**, 505, 516 (1966)
16. Iben, I., Jr., *ibid.*, **147**, 624 (1967)
17. Iben, I., Jr., *ibid.*, 650 (1967)
18. Iben, I., Jr., Ehrman, J. R., *Ap. J.*, **135**, 770 (1962)
19. Pearce, W. P., Bahng, J., *Ap. J.*, **142**, 164 (1965)
20. Stothers, R., *Ap. J.*, **144**, 959 (1966)
21. Schwarzschild, M., Härm, R., *Ap. J.*, **129**, 637 (1959)
22. Schwarzschild, M., Härm, R., *ibid.*, **128**, 348 (1958)
23. Sakashita, S., Ôno, Y., Hayashi, C., *Progr. Theoret. Phys. (Kyoto)*, **21**, 315 (1959)
24. Sakashita, S., Hayashi, C., *ibid.*, **22**, 830 (1959); **26**, 942 (1961)
25. Stothers, R., *Ap. J.*, **138**, 1074 (1963)
26. Mestel, L., *Monthly Notices Roy. Astron. Soc.*, **112**, 598 (1952)
27. Schwarzschild, M., Härm, R., *Ap. J.*, **136**, 158 (1962); **139**, 594 (1964); **145**, 496 (1966)
28. Saslaw, W. C., Schwarzschild, M., *Ap. J.*, **142**, 1468 (1965)
29. Adams, B., Ruderman, M., Woo, C., *Phys. Rev.*, **129**, 1383 (1963)
30. Inman, C. L., Ruderman, M., *Ap. J.*, **140**, 1025 (1964); **143**, 284 (1966)
31. Chiu, H. Y., *Ap. J.*, **137**, 343 (1963)
32. Sandage, A. R., *Ap. J.*, **125**, 435 (1967); **135**, 349 (1962)
33. Eggen, O. J., *Quart. J. Roy. Astron. Soc.*, **3**, 259 (1962)
34. Bodenheimer, P., *Ap. J.*, **142**, 451 (1965)
35. Hodge, P. W., Wallerstein, G., *Publ. Astron. Soc. Pacific*, **78**, 411 (1966)
36. Faulkner, J., Stefensen, G. (Private communication, 1966)
37. Eggen, O. J., *Ap. J. Suppl. No. 8*, 125 (1963)
38. Eggen, O. J., *Ann. Rev. Astron. Ap.*, **3**, 235 (1965)
39. Roxburgh, I. W., *Monthly Notices Roy. Astron. Soc.*, **128**, 157, 237 (1964)
40. Roxburgh, I. W., Griffith, J. S., Sweet, P. A., *Z. Ap.*, **61**, 203 (1965)
41. Roberts, P. H., *Ap. J.*, **137**, 1129 (1963); **138**, 809 (1963)
42. Collins, G. W., *Ap. J.*, **138**, 1134 (1963); (1965); **146**, 914 (1966)
43. Collins, G. W., Harrington, J. P., *Ap. J.*, **146**, 152 (1966)
44. Kraft, R. P., *Ap. J.*, **142**, 681 (1965)
45. Kraft, R. P., Wrubel, M. H., *Ap. J.*, **142**, 703 (1965)
46. Strittmatter, P. A., Sargent, W. L. W., *Ap. J.*, **145**, 130 (1966)
47. Strittmatter, P. A., *Ap. J.*, **144**, 430 (1966)
48. Sweet, P. A., Roy, A. E., *Monthly Notices Roy. Astron. Soc.*, **113**, 701 (1953)
49. Roxburgh, I. W., Strittmatter, P. A., *Monthly Notices Roy. Astron. Soc.*, **133**, 345 (1966)
50. Weymann, R., *Ann. Rev. Astron. Ap.*, **1**, 97 (1963)
51. Kraft, R. P., *Ap. J.*, **144**, 1008 (1966)
52. Kuhi, L. V., *Ap. J.*, **140**, 1409 (1964)
53. Crampin, J., Hoyle, F., *Monthly Notices Roy. Astron. Soc.*, **120**, 33 (1960)
54. Schmidt-Kaler, Th., *Bonn Veröffentl. No. 70*, 1 (1964)
55. Schild, R. E., *Ap. J.*, **146**, 142 (1966)
56. Hofmeister, E., Kippenhahn, R., Weigert, A., *Z. Ap.*, **60**, 57 (1964)
57. Baker, N., Kippenhahn, R., *Ap. J.*, **142**, 868 (1965)
58. Arp, H. C., *Ap. J.* (In press, 1967)
59. Christy, R. F., *Ann Rev. Astron. Ap.*, **4**, 353 (1966); *Ap. J.*, **144**, 108 (1966)

60. Wildey, R. L., *Ap. J. Suppl. No. 8*, 439 (1963)
61. Iben, I., Jr., Talbot, R., *Ap. J.*, **144**, 968 (1966)
62. Hayashi, C., Cameron, R. C., *Ap. J.*, **136**, 166 (1962); *Astron. J.*, **69**, 140 (1964)
63. Sandage, A. R., *Ap. J.*, **135**, 333 (1962)
64. Johnson, H. L., Sandage, A. R., *Ap. J.*, **121**, 616 (1955)
65. Sandage, A. R. (Private communication, 1966)
66. Eggen, O. J., Sandage, A. R., *Ap. J.*, **136**, 735 (1962)
67. Strom, S. E., Cohen, J. G., Strom, K. M., *Ap. J.*, **147**, 1038 (1967)
68. Sandage, A. R., Walker, M. F., *Ap. J.*, **143**, 313 (1966)
69. Sandage, A. R., Smith, L. L., *Ap. J.*, **144**, 886 (1966)
70. Iben, I,. Jr., Faulkner, J. (In preparation, 1967)
71. Faulkner, J., Iben, I., Jr., *Ap. J.*, **144**, 995 (1966)
72. Demarque, P., *Ap. J.* (In press, 1967)
73. Osaki, Y., *Publ. Astron. Soc. Japan*, **15**, 428 (1963)
74. Simoda, M., Kimura, H., *Ap. J.* (In press, 1967)
75. Greenstein, J. L., Munch, G., *Ap. J.*, **146**, 618 (1966)
76. Searle, L., Rodgers, A. W., *Ap. J.*, **143**, 809 (1966)
77. Sargent, W. L. W., Searle, L., *Ap. J.*, **145**, 652 (1966)
78. Greenstein, G. S., Truran, J. W., Cameron, A. G. W., *Nature*, **213**, 871 (1967)

COSMOLOGY[1]

By I. D. Novikov and Ya. B. Zeldovič

Institute of Applied Mathematics, USSR Academy of Sciences, Moscow

PART I. THE MODERN TRENDS—A GENERAL SURVEY

The first part of the paper gives a broad account of the recent development and the unsolved problems of modern "physical" cosmology. The second part, written in the form of short supplements, contains more technical detail and bibliographical references. Many questions have been discussed more fully in previous reviews (Zeldovič 1965, 1966; Zeldovič & Novikov 1966c). A complete account will be given in the authors' book *Relativistic Astrophysics* which will appear in Russian at the end of 1967.

1. THE HOT MODEL

Einstein formulated the theory of relativity and stressed its fundamental rôle for cosmology. The Soviet-Russian mathematician A. A. Friedmann (1922, 1924) gave the theoretical solution for the expanding universe that was confirmed by the American astronomer Hubble. In the recent development of cosmology the dominant interest is in the physical processes in the universe, involving elementary particles, light quanta, and nuclear reactions. In the past it was the geometry, kinematics, and dynamics of the universe which were in the foreground.

The hot model of the universe, proposed by Gamow (1946), was confirmed in the measurements of Penzias & Wilson 1965, the significance of which was understood by Dicke et al. (1965).

An isotropic radio noise on the wavelength of 7 cm was discovered. In thermodynamic equilibrium of radiation a single measurement defines the temperature equal to $3°K$ in the said experiment. It is possible, then, to predict the whole spectrum; these predictions are checked today on wavelengths 20, 3, and 0.26 cm. See Addendum I.

The radiation in this region cannot be ascribed to known types of sources (stars, radio galaxies, etc.). The hot-model theory supposes that in an early stage, when there were no stars, there existed thermodynamic equilibrium of ordinary matter (nuclei, electrons) with radiation. The discovered $3°K$-black-body radiation stems from the expansion and redshift of this equilibrium radiation, which has long ago lost thermic liaison (contact) with the ordinary matter, a part of which condensed in stars and another part of which perhaps still fills the intergalactic (and intercluster-of-galaxies) space. It is sometimes called *relict radiation*. The overall energy density of relict radiation is ~100 times the density of star radiation. The observed radiation contains 10^8 or 10^9 light quanta per single proton or electron (it is the ratio for the whole universe or the mean ratio for a great

[1] The survey of literature for this review was concluded in the fall of 1966.

volume containing several clusters); this ratio is invariant during the expansion.

2. Elementary Particles and Nuclear Reactions in the Early Phase

The temperature in the early phase dropped approximately like $T = 10^{10}/\sqrt{t}$, T °K, t sec, and there is no escape from assuming that tremendous temperatures up to $T \sim 10^{30}$ occurred.

In the early stage there was plenty of antimatter in equilibrium, but in the course of expansion it was annihilated and only the excess matter survived.[2]

If there exist *quarks* (*q*)—subelementary particles with fractional charge ($\pm\frac{2}{3}e$, $\pm\frac{1}{3}e$)—then the hot model predicts a not negligible content of quarks after expansion, of the order of 10^{-10} q per single nucleon. The search for such relict quarks is perhaps even more feasible than the search for quark production in cosmic rays and accelerators. The equilibrium between radiation and neutrinos is maintained up to a later stage $0 < t < 0.1$ sec. Later they are disconnected and expand independently. Still, the density of the relict neutrino must now be of the same order as the density of light quanta (and much greater than the density of the star neutrino); the predicted neutrino temperature is $\sim 2°$ as compared with 3° of the radiation. Unfortunately, the experimental detection of such neutrinos is a very difficult task; its significance for the whole hot-model theory cannot be overestimated.

At high temperatures the nucleons are also in thermodynamic equilibrium: there are no complex nuclei (which would be instantly photodissociated by γ quanta), and the n/p ratio corresponds to their rest-mass-energy difference. The adjustment of equilibrium to the decreasing temperature ends at $t \simeq 1$ sec, when there is ~ 16.5 per cent n, ~ 83.5 per cent p.

During further evolution two processes are competing: 1) the spontaneous decay of neutrons into protons and 2) a chain of nuclear reactions beginning with the neutron capture $n + p = d + \gamma$ and ending with formation of He^4.

The rather well-known data on nuclear reactions and the not so well-known present-day density of matter (which affects the competition of the two processes), together with the assumption of an isotropic and homogeneous expanding universe, lead to a very definite answer for the composition of primeval prestar matter: 28-30 per cent He^4, 72-70 per cent H (weight percentage), traces of D and He^3, no heavy elements.

Many observations of helium abundance confirm the theory, although

---

[2] It is a puzzling implication of the hot model that it predicts a short period, when the ratio of baryons to antibaryons $N/\overline{N} = 1 + a \cdot 10^{-8}$, $a \sim 1$ (at $t < 10^{-6}$ sec, $kT > Mc^2$), but the small departure from unity, $a \cdot 10^{-8}$ is of utmost importance. It is just these excess baryons which ensure the present charge asymmetry of the universe.

earlier and also in very recent times there are papers claiming a much smaller helium content.[3] Compare the remark in I.5.

3. The Intergalactic Matter

On the verge between cosmology and the other branches of astronomy is the theory of the intergalactic medium: what is its density? its temperature and degree of ionization? its chemical composition? Should it be considered as primeval prestar matter, never burned in stars?

The distant quasars with redshifts $Z > 2$ offer a unique possibility of obtaining some information on the absorption of light by the intergalactic gas. It follows that cosmologically important density of gas ($\sim 10^{-29}$ g/cm^3 for the present) can be reconciled with observations only if the hydrogen is fully ionized (H I $< 10^{-7}$H II). On the other hand, rocket measurements of the cosmic X-ray background limit the gas temperature $T < 5.10^6$ °K (Field & Henry 1964). Still there remains a range of temperature compatible with the mentioned gas density. In this case it would be 10 times more than the density of matter in galaxies. So it is possible that most of the matter is still not condensed in stars and galaxies.

A definite proof would be given by the observation of the free electrons in the intergalactic (exactly intercluster) space, perhaps by the scattering of light, but the effect is very small. The optical depth $\tau \sim 1$ corresponds to a source with $Z = 6$ (Kardashev & Sholomitsky 1965, Bahcall & Salpeter 1965). No such sources strong enough to be visible in the optical region are known. Definite conclusions on electron concentration are further made difficult by the absence of any spectral dependence of the free electron scattering. It is most important for the cosmology: the density of galaxies is of the order $(0.5+1) \cdot 10^{-30}$ g/cm^3; the density of electromagnetic radiation is still lower, $7 \cdot 10^{-34}$ g/cm^3; the density of neutrino is of the same order of magnitude as the density of radiation. So without intergalactic matter the universe is definitely open in the sense of Riemannian geometry of space in the theory of relativity. Only an adequate intergalactic density of ionized matter can make the universe closed (of course when assumptions are made about uniformity and isotropy, and $\Lambda = 0$). For a discussion of ultraviolet spectroscopic search of ionized matter see Kurt & Sjunjaev (1967a, b).

The classical, long-known method of determining ρ utilizes the investigation of apparent luminosity, angular size, number, and redshift of distant objects. Its strong point is that all forms of matter are included, because they all affect the curvature of space and the dynamics of expansion are included. But the investigation is based on definite assumptions on the properties of observed objects and their evolution.

---

[3] One of the authors (Zeldovič 1962) earlier advocated the "cold-universe model" leading to pure hydrogen. Since the discovery of relict radiation the cold model seems to be untenable. If the primeval helium content proves to have been low, this difficulty must be resolved by modification of the hot model (see below, anisotropic solutions).

The most fundamental results of Sandage (1961) give for the so-called *acceleration parameter* $\tilde{q} = 1 \pm 0.5$; if the intergalactic pressure is negligible, $\bar{p} \ll \bar{\rho}c^2$, then $\bar{p}/\rho_c = 2\tilde{q} = 2 \pm 1$. Because of many observational difficulties and possible evolutionary effects, the true margin of error is still uncertain.

The discovery of *quasars* extends the range of observation up to $Z = 2$, but till now no more definite values of $\tilde{q}$ or $\bar{p}/\rho_c$ are known. The observations prove the existence of a most powerful evolutionary effect (see for example Longair 1966). It is impossible, in practice as well as in principle, to untie the evolutionary effects and the influence of space curvature.

The observations can give the answer about the curvature only when some intrinsic properties have been established for quasars or radio galaxies: a unique relation between the spectrum or period and the absolute magnitude or linear dimensions (compare the spectrum-magnitude correspondence of main-sequence stars or the period-magnitude for Cepheids).

4. THE FORMATION OF GALAXIES

The next question is the theory of formation of stars, galaxies, and clusters of galaxies. There is a wide spectrum of theories ranging from Ambartsumian (1964, 1966) to Layzer (1964, 1965). We adhere to the point of view that the stars and galaxies came from the prestar matter, which is supposed to have been distributed more or less uniformly.

The hot model profoundly affects the growth of perturbations in the early stage, when the matter is ionized and interacts strongly with radiation.

The radiation pressure counteracts the condensation of matter, moving *together* with radiation. The condensation of matter, moving *relative* to the radiation (so as to leave the radiation pressure constant) is slowed by the friction of electrons against the radiation. There are no definite guesses about the initial degree of uniformity and initial spectrum of perturbations.

The plausible assumption is that at the end of the early stage the radiation temperature and pressure are uniform, the matter and the radiation are involved in the Hubble expansion, but there are no other motions; the density of matter is nonuniform but with $\delta\rho_m/\bar{\rho}_m < 1$. If follows that the nonuniformity of the matter density is constant during the whole early stage; in this stage the radiation density is far greater than the matter density and it follows that the nonuniformity of ρ_m did not affect the expansion at the early stage. The absence of motion (other than Hubble) and the uniformity of temperature follow from the theory (with some natural restrictions added); the nonuniformity of the matter density is taken *ad hoc* to obtain sensible results in the following stage.

This stage begins when the temperature drops under $\sim4000°$K, protons and electrons recombine, give neutral hydrogen, and are decoupled from the radiation.[4] See Addendum II.

---

[4] By a coincidence, near the same time the radiation density ρ_r is equal to the matter density ρ_m; $\rho_r \simeq \rho_m \simeq 10^{-21}$ g/cm³ (if $\rho_m = 5 \cdot 10^{-31}$ g/cm³ today).

The growth of perturbations with the characteristic mass of the order of $10^5 \, M_\odot$, corresponding to the critical Jeans wavelength of the neutral hydrogen at the given density and temperature, now is possible. In a very crude, tentative hypothesis *uhrstars* with such a mass are formed. During their formation at chance positions, where the initial perturbations $\delta\rho_m/\rho_m$ are maximal, the remainder of the matter is cooled further by Hubble expansion $T_m \lesssim T_r < 4000°$. But it is enough that a small part of matter, of the order of 0.01 per cent, condenses in uhrstars and burns in them, to give the amount of energy which is great enough to ionize the remaining 99.99 per cent of the matter and heat it up to $10^6 \div 10^7 °K$. After the heating of the gas the formation of uhrstars is no longer possible, there is a kind of autoregulation. It is from this time that the modern stage of evolution begins, for which it is characteristic that the temperature of matter (electrons, nuclei) is much higher than the temperature of radiation, which drops to the present-day 3° K.

The burning of nuclear fuel in uhrstars, and perhaps their outburst, heats the gas; but it also gives large-scale perturbations of the density of the remaining gas. Still possible is another hypothesis by which the formation of clusters is due to the growth of perturbations which are large scale ($M \gtrsim 10^{16} \div 10^{19} M_\odot$) enough not to be impeded by the radiation pressure. For details, see Doroshkevich, Zeldovič, Novikov (1967b). Now begins the phase of formation of galaxies and clusters of galaxies from hot gas clouds; this process has been investigated in many papers and is better understood.

5. Anisotropic and Magnetic Cosmological Solutions

We have given above a broad picture of the possible evolution of the universe in the framework of the isotropic homogeneous Friedmann cosmological solution. Now let us review the purely cosmological questions which are beyond this theory: (a) the wider class of anisotropic (but still homogeneous) cosmological solutions and (b) the singularity $\rho = \infty$, common to all cosmological solutions.

To begin with, the presently observed situation gives no direct indication of a definite anisotropy or inhomogeneity of the universe; in the scale L of the whole region which can be investigated, $L \sim c/H \sim 4000$ Mpc.

The observed isotropy (with a precision of 5 or 10 per cent) of the relict radiation at $\lambda = 7$ cm is the best proof of the isotropy of our universe. See Addendum III. Indirectly it also implies the homogeneity of the universe, because the isotropy is compatible with inhomogeneity only in the improbable case that the observer (i.e. our Sun or our Galaxy) is just in the center of a spherical-symmetrical distribution of matter and radiation.

The observed isotropy does not rule out an anisotropic beginning of the expansion; there are well-known anisotropic cosmological solutions which tend to isotropy as $\rho \rightarrow 0$, $t \rightarrow \infty$.

So the right question is *why*, having no real indication of any anisotropy, should one consider such solutions? Primarily, the anisotropic beginning of expansion would drastically change the conditions for nuclear reactions at the early phase.

Given the known temperature of relict radiation and the matter density today and assuming different values of the anisotropy parameter, one can obtain, as shown by Hawking & Tyler (1966) and more precisely by Thorne (1966), the composition of prestellar matter ranging from 100 per cent H, 0 per cent He^4 to near 100 per cent He^4 (but no heavy elements in either case).

So if it should be proved that early stars have a very low content of He^4 (and that He^4 did not disappear on the surface; see Sargent 1966) or, still better, if the intergalactic gas has no helium—then the hot model is compatible only with an anisotropic solution. One must remember an important caution: in the anisotropic model the evolution of relict neutrinos is also strongly affected; perhaps it will restrict the possible degree of anisotropy. In the anisotropic model the energy of neutrinos can be much higher than $2°K$; see Addendum IV. There is also a second line of thought leading to anisotropic solutions.

Piddington (1964, 1966) and Pikelner (1965) studied the formation of galaxies and quasars by the condensation of ionized gas with a frozen-in magnetic field which is assumed homogeneous on a large scale at the beginning of the process. The deformation, winding, and growth of this field during the condensation of the gas cloud is most important; it determines the spiral or elliptic structure of the galaxy, its radio emission, the radiation, and the jets of quasars. But what is the origin of the large-scale magnetic field? It is difficult to imagine its spontaneous birth by some dynamo mechanism. So one comes to the idea of a "magnetic universe" with a primeval, homogeneous throughout the universe, frozen-in magnetic field. The corresponding cosmological solution is easily written down, but obviously it must be anisotropic, because the magnetic field singles out a definite direction in space.

Such are the driving forces of the recent revival of interest in anisotropic solutions.

Still, up to now one feels some aversion to the anistropic solutions, with their wealth of adjustable parameters, as compared with the very rigid and definite predictions of the Friedmann cosmology. Seldom is the consideration of anisotropic models carried out completely with all its implications.

6. THE SINGULARITY

Most unclear—and most exciting—are questions concerning the singularity.

The adherents of steady-state theory have often claimed that this theory is not worse than the Friedmann solution: the steady state is bound to continuous creation of matter; in the Friedmann theory there is also creation of matter, but instantaneous, at $t=0$, with $\rho = \infty$.

As a matter of fact, the Friedmann theory is a description of the movement and evolution of matter, radiation, etc., together with evolution of the metric of space, going in accordance with all the known laws of physics and, among them, the law of conservation of baryons. In this respect the theory is definitely superior to the steady-state or C-field theories.

The Friedmann theory gives two solutions—one expanding and one contracting. From observations we know that we live in an expanding universe. But this does not imply the creation of the universe 10^{10} years ago (i.e. at $t=0$).

One can imagine that before this moment at $t<0$, there was a contracting universe and $t=0$ is the moment when the two solutions are matched; after the contraction begins the expansion (see for example Novikov 1966). It should be supposed that the conserved quantities of physics—baryon number, specific entropy—are also conserved in the passage from $t<0$ to $t>0$ through the singular point. But the jump itself at $t=0$ from one solution to the other is outside the limits of application of the Friedmann solution and the whole modern physics.

The breakdown of existing theory can begin not at $\rho=\infty$, but at $\rho\sim c^5/\hbar G^2 \approx 10^{93}$ g/cm^3 when mixed effects of quantum theory and general relativity are of the order of unity[5] (Wheeler 1960); it does not invalidate the analysis of quarks, neutrinos, and nuclear reactions given above.

The above treatment of the hot model takes from observations the initial values of entropy (given by the number of quanta per one baryon $\sim 10^8 \div 10^9$) and the initial perturbations. It may be that the clue for theoretical evaluation of these parameters[6] lies in the processes at $t<0$, before the expansion starts (see Zeldovič & Novikov 1966a). Finally, let us make a remark on oscillating models. As a result of the second law of thermodynamics they imply unlimited growth of specific entropy (per one baryon). So the oscillating model is not compatible with external existence of the universe from $t=-\infty$, even if we assume that the jump from contraction to expansion at $\rho=\infty$ (or $\rho\approx 10^{93}$ g/cm^3) is possible.

Of course if it should be demonstrated that the present-day density exceeds the critical value $\rho > \rho_c = 3H^2/8\pi G$, then we shall need some nontrivial new ideas.

7. THEORETICAL FOUNDATIONS OF COSMOLOGY AND THE MACH PRINCIPLE

Modern cosmology is based on the theory of relativity of Einstein. Today there are no observations or theoretical considerations which definitely demand a departure from this theory, or a new theory instead, or the introduction of new hypothetical fields [scalar field of Dicke (1962), C-field of Hoyle & Narlikar (1963)] which are not observed in physical laboratories. See Addendum V.

The only natural limitation of relativity is the region of ultrahigh densities with $\rho \geq 10^{93}$ g/cm^3, $l \leq 10^{-33}$ cm, $t \leq 10^{-43}$ sec. For all subsequent time

[5] For some other ideas see Sacharov (1966, 1967).
[6] And with it the solution of the antimatter puzzle referred to above.

there are no objections to the use of the theory. The cosmological constant Λ could be found in principle by careful comparison of actual density $\bar{\rho}$ and acceleration parameter $\bar{q}$.

Today both $\bar{\rho}$ and $\bar{q}$ are poorly known and the choice of $\Lambda = 0$ or $\Lambda \neq 0$ is made by personal aesthetic taste. The original suggestion $\Lambda > 0$ was made with the purpose of making possible a time-independent cosmological solution. But this motive is no longer valid since the discovery of the redshift. The observations give only $|\Lambda| < 10^{-56}$ cm^{-2}. Of course it is enough to ascertain that Λ does not affect the past story of evolution, when the density was great; in particular it does not affect the nuclear reactions and instability phenomena in the hot model.

In cosmology, one must use the original form of the theory of relativity, with curved space and nonlinear equations. The linearized theories, adapted for the discussion of weak gravitational interaction and its quantization, are inadequate for cosmology.

Finally—some comments on the *"Mach's principle,"* which in modern literature is pretty far from the original idea of E. Mach as published in 1883. Symptomatic is Dicke's expression "many-faced Mach."

As a rule, one believes that in nature are fulfilled the equations of the theory of relativity in which matter is composed of elementary particles; the rest masses of particles are definite constants for every type.

In this sense the "inertia" = mass of particles or macroscopic bodies does not depend on the existence of far (or near) galaxies and stars.

In accordance with the theory, other galaxies and their motion influence the choice of the locally inertial frame of reference.

A classical example is the theory of Thirring and Lense: as known, near a rotating heavy body the locally inertial frame of reference also rotates (slowly, when compared with the body) from the point of view of a distant observer. But here also the masses of particles, their momenta, etc., measured in this frame of reference are not affected.

The Thirring-Lense effect could be called *Machian;* but then Mach's principle is reduced to a sort of pedagogical comment on the formulae of relativity—then, it is true and harmless but also almost useless.

Without the theory of relativity, a naive application of Mach's principle could lead to the idea that in the vicinity of a definite point of space there is *one* privileged "rest"-frame of reference, in which the redshift of distant galaxies is isotropic, is the same in different directions. But relativity defines not a unique locally inertial frame of reference but a whole set of such frames of reference, Lorentz-transformed (moving relative to one another). Physical experiments confirm their equivalence.

So Mach's principle taken in accordance with the theory of relativity gives no new information, and Mach's principle taken without the theory can lead to erroneous conclusions.

It seems to us not convincing that Mach's principle can be used to make a choice between different solutions of the theory of relativity or to make a choice of a closed-world model instead of the open-world model.

PART II: SUPPLEMENTS

1. ELECTROMAGNETIC RELICT RADIATION

Gamow (1956) suggested a relict radiation temperature of 6°K with the conclusion that the energy density of relict radiation is much greater than it is for star radiation.

The overall spectrum of known sources (stars, radiogalaxies) with due account of the redshift and evolution was calculated by Doroshkevich & Novikov (1964). They also put the Planck spectrum for $T = 1°K$ on their graph for comparison and pointed out that the choice of a cold- or hot-universe model should be made by measurements at the wavelength between 10 and 0.03 cm.

The experiment of Penzias & Wilson (1965) and its interpretation by Dicke et al. (1965) have already been mentioned in Part I.

Measurements at 3 cm have been reported by Roll & Wilkinson (1966) and at 0.7 cm by Shakeshaft (1966). The optical absorption of interstellar CN (cyanogen) gives the population ratio $CN^*/CN° = \exp(\Delta E/kT)$ with ΔE corresponding to $\lambda = 0.26$ cm and $T = 3 \pm 0.5°K$ (McKellar 1941).

It is interpreted as a proof of the corresponding intergalactic density of radiation as mentioned by Field & Hitchcock (1966). The data are summarized on Figure 1. The left line is the averaged radiation of discrete radio sources; the maximum in the right part, at $\lambda \sim 10,000$ Å, is the averaged radiation of stars. Our Galaxy is excluded; the radiation of distant galaxies includes an extrapolation for unresolved objects with due account of the redshift. Different assumptions on the cosmological model and on the evolutionary effect do not greatly affect the integral curve (Doroshkevich & Novikov 1964).

The curve with high maximum at $\lambda \sim 0.1$ cm is the Planck black-body equilibrium for $T = 3°K$. The measurements at 20, 7.3, 3, and 0.25 cm are shown, in beautiful agreement with the Planck curve. The dashed line on Figure 1 shows the possible radiation of dust (in all galaxies except ours) which is heated by stars up to 10° K and re-emits the absorbed energy in another spectral interval (calculations of Doroshkevich & Novikov).

The overall density of relict radiation $E_r = \sigma T^4 = 6 \cdot 10^{-13}$ erg cm^{-3} is much more than the star light energy density $E_r \sim 10^{-14}$ erg cm^{-3}. Obviously the radiation of dust can only be a part of the radiation of stars, which feed the dust by energy.

G. Burbidge (1966) asked if the relict radiation should not be the radiation of some kind of uhrstars re-emitted by dust. The difficulty is that the early celestial bodies cannot give enough energy (the radiation energy degrades during expansion), and later the density of the dust is not enough to obtain an equilibrium spectrum.

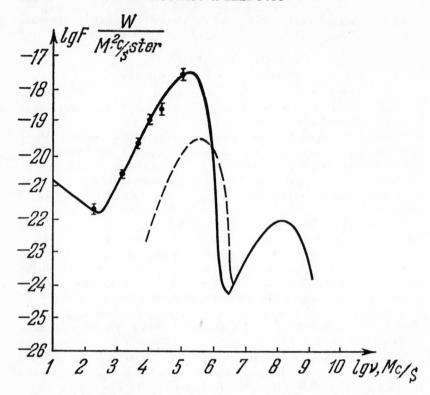

FIG. 1. Averaged spectrum of electromagnetic radiation in the universe. The maximum at lg $\nu \approx 8$ is the radiation of stars; the straight line at lg $\nu < 2.5$ is due to radio galaxies. The dotted line is the radiation of dust with $\bar{\rho} \approx 5 \cdot 10^{-34}$ g/cm^3 and $T = 10°$K. The full line with maximum at lg $\nu \approx 5.5$ is the black-body equilibrium radiation at 3°K—the relict radiation. The points give experimental results with their errors.

The interaction of radiation with ionized gas is considered by Weymann (1966) (for the general theory see Kompaneets 1956). But there are no energy sources adequate to feed the electrons so early that they could give the observed radiation.

One concludes that there is no trivial explanation of the observed radiation in the wavelength 21 cm–0.25 cm; the hot-model theory seems inescapable (Zeldovič & Novikov 1967).

The universal occurrence of relict radiation profoundly affects the cosmic rays of great energy. The interaction $\gamma + \gamma = e^+ + e^-$ limits the energy of cosmic-ray gammas by $E_\gamma < 10^{15} \div 10^{16}$ eV (Nikishov 1961, Goldreich & Morrison 1963, Ielley 1966). The protons lose energy mostly by $p + \gamma$

$= p$(or n) $+ \pi^0$ (or π^+); at the proton energy $\sim 10^{20}$ eV most thermal quanta have the needed energy in the proton rest frame (Greisen 1966, Zatsepin & Kuzmin 1966).

2. ELEMENTARY PARTICLES AT HIGH TEMPERATURES

At high temperatures the reaction rate is tremendously high and all elementary particles are in thermodynamic equilibrium, except perhaps gravitons. At every given T the role of particles with $Mc^2 > kT$ is small, of the order of exp $(-Mc^2/kT)$; for particles with $Mc^2 < kT$, neglecting interaction, one obtains the energy density $\epsilon = (g/2)\epsilon_\gamma$ for bosons and $\epsilon = (7/16)g\epsilon_\gamma$ for fermions. Here g is the statistical weight of the particle ($g_\gamma = 2$; $g_\pi = 3$; $g_e = 2$; $g_\mu = 2$; $g_\nu = 1$, etc.); $\epsilon_\gamma = \sigma T^4$ is the well-known formula for black-body radiation, $\sigma = 7.6 \cdot 10^{-15}$ erg/cm³ deg⁴. The entropies of different species are in the same ratio.

The equilibrium does not depend upon the previous history, so the equilibrium at, say, $T = 20$ MeV does not depend upon the existence and number of particles heavier than neutrons and protons, whose concentration at 20 MeV surely is negligible.

In the expansion governed by the Friedmann isotropic solution, the equilibrium between ν_μ, $\bar{\nu}_\mu$ and γ, e^+, e^- is not maintained at $T < 10\text{-}20$ MeV, and the equilibrium between ν_e, $\bar{\nu}_e$ and γ, e^+, e^- at T is less than 3 MeV. It follows that during subsequent cooling the entropy S (γ, e^+, e^-) is constant, as well as S (ν_e, $\bar{\nu}_e$) and S (ν_μ, $\bar{\nu}_\mu$). At $m_\mu c^2 = 100$ MeV $> T > M_e c^2 = 0.5$ MeV.

$$S(\gamma, e^+, e^-) = \left(1 + \frac{7}{4}\right) \cdot \frac{4}{3} \sigma T_\gamma^3 / \rho_m$$

$$S(\nu) = \frac{7}{16} \cdot \frac{4}{3} \sigma T_\nu^3 / \rho_m$$

Now, with $T_{\gamma(0)} = 3°$K, without e^+e^- pairs, $S(\gamma, e^+, e^-) = 4/3\sigma \, T_{\gamma(0)}^3 / \rho_{m(0)}$; $S(\nu) = 7/16 \cdot 4/3\sigma T_{\nu(0)}^3 / \rho_{m(0)}$. From $T_\gamma = T_\nu$ at high temperature it follows (Peebles 1966) that $T_{\nu(0)} = (4/11)^{1/3} T_{\gamma(0)} \approx 2°$. The joint relict energy density now is

$$\epsilon_\gamma + \epsilon_{\nu_e} + \epsilon_{\bar{\nu}_e} + \epsilon_{\nu\mu} + \epsilon_{\bar{\nu}\mu} = \left[1 + 4 \cdot \frac{7}{16} \cdot \left(\frac{4}{11}\right)^{4/3}\right] \sigma T_{\gamma(0)}^4$$

About gravitons see Zeldovič (1966). On the rest mass of ν_μ see Gershtein & Zeldovič (1966). The equilibrium density q of quarks drops like exp $(-Mc^2/kT)$. But the establishment of equilibrium needs ternary $3q \rightleftarrows N$+mesons or at least binary collisions $q + \bar{q} \rightleftarrows N + \bar{q}(\sigma_1)$, $q + q \rightleftarrows$ mesons (σ_2); cross sections are in parentheses. Here, $\bar{q} =$ antiquark. It follows that at low temperature when the vanishing of quarks is stronger than their production, the ratio $\bar{q}/q \rightarrow 1$. Their concentration obeys the equation

$$\frac{dq}{dt} = -\frac{3}{2}\frac{q}{t} - 2aq^2 - bq\bar{q} + a\bar{q}^2$$

$$\frac{d\bar{q}}{dt} = -\frac{3}{2}\frac{\bar{q}}{t} - 2a\bar{q}^2 - bq\bar{q} + aq^2$$

$$a = \sigma_1\bar{v}, b = \sigma_2\bar{v}$$

The first term is due to expansion. It follows that asymptotically $q = \bar{q} = rt^{-3/2}$; the ratio of surviving nonequilibrium q and $\bar{q}$ to electromagnetic quanta q/γ is constant.

The temperature at which the production of quarks is no longer important is proportional to M_qc^2, so that the resulting q/γ does not depend exponentially on M_q. Neglecting numerical factors one obtains

$$\frac{q}{\gamma} = \left(\frac{GM_q^2}{hc}\right)^{1/2}\left(\frac{h}{M_qc}\right)^2 c/\sigma\bar{v} \simeq \left(\frac{GM_q^2}{hc}\right)^{1/2} \simeq 10^{-20}$$

As stated, the ratio of γ to baryons is of the order of $\gamma/N = 10^9$. So one obtains the quoted $q/N \simeq 10^{-11}$ for the isotropic cosmological solution. For details and some comments on the further fate of quarks, see Okun, Pikelner & Zeldovič (1965).[7] In the anisotropic universe the q/N would be greater. If the quarks should be discovered in cosmic rays or on accelerators (and one kind of quark is stable), and no relict quarks should be found, it would be a a difficult puzzle for the cosmologist!

On the search of relict quarks see Chupka, Schiffer & Stevens (1965); Pikelner & Vainstein (1966); Becchi, Gallinaro & Morpurgo (1965) Braginski (1966); Braginski et al. (1967).

3. NUCLEAR REACTIONS

We consider the $p \rightleftarrows n$ transformation in a hot plasma through $p + e^-$ $\rightleftarrows n + \nu_e$; $p + \bar{\nu}_e \rightleftarrows n + e^+$; $p + e^- + \bar{\nu}_e \rightleftarrows n$ at 10 MeV $> T >$ 200 keV (1 MeV $= 1.17 \cdot 10^{10}$ °K).

The plasma is assumed to be fully neutral with equilibrium of γ, e^+, e^-, ν_e, $\bar{\nu}_e$, ν_μ, $\bar{\nu}_\mu$ plus a small admixture of the baryons p, n. The overall density is uniquely bound to temperature, approximated by $\rho = 4.5\sigma T^4/c^2 = 4 \cdot 10^{-35}T^4$ g/cm³. The baryon density is expressed through the present-day baryon density ρ_m corresponding to

$$\rho_b = \rho_m\frac{4}{11}\left(\frac{T}{3}\right)^3$$

$T = 3°$ is the present-day temperature.

Later, at $T < 200$ keV we consider the sequence

---

[7] The same method of calculation was applied to the remaining concentration of nucleons and antinucleons in a charge-symmetric hot universe (Zeldovič 1965, Chiu 1966). Evidently it gives the ratio $n/\gamma \approx 10^{-20}$, some 10^{-10} smaller than the observed; the homogeneous charge-symmetric universe is untenable.

$$p + n \rightleftarrows d + \gamma, \quad d + p \rightarrow \mathrm{He}^3 + \gamma, \quad d + n \rightarrow T + \gamma, \quad \mathrm{He}^3 + n \rightarrow T + p,$$

$$T + p \rightarrow \mathrm{He}^4 + \gamma, \quad d + d \quad \rightarrow \mathrm{He}^3 + n,$$

$$d + d \rightarrow T + p, \quad T + d \rightarrow \mathrm{He}^4 + n, \quad T + \mathrm{He}^3 \rightarrow \mathrm{He}^4 + p$$

and the spontaneous β decay $n \rightarrow p + e^- + \bar{\nu}_e$; later T also β-decays into He^3. The reactions shown with one arrow $\rightarrow$ (instead of $\rightleftarrows$) are practically irreversible, and the inverse reaction can be neglected under the conditions mentioned.

At this stage, the plasma consists of γ, ν_e, $\bar{\nu}_e$, ν_μ, $\bar{\nu}_\mu$ plus an admixture of all mentioned nuclei and a negligible amount of other particles. The temperature of the baryons and the reaction rate correspond to the temperature of photons T. The temperature of neutrinos is lower, $T_\nu = (4/11)^{1/3} T$, and the overall density is

$$\rho = \left[1 + \frac{7}{4} \left(\frac{4}{11} \right)^4 \right] \sigma T^4 / c^2$$

The baryon density, i.e. the sum of densities of all sorts of nuclei, is

$$\rho_b = \rho_m \left(\frac{T}{3} \right)^3$$

In the isotropic homogeneous Friedmann cosmological solution the overall density of hot plasma with $p = \epsilon/3$ depends on time as

$$\rho = \frac{3}{32\pi G t^2} = \frac{4.5 \cdot 10^5}{t^2}$$

Calculations of the first phase were made by Hayashi (1950). Some refinements (see Zeldovič 1965) give a final value $n/(n+p) = 0.165$; this result does not depend on ρ_m.

Calculations of the second phase made by Fermi and Turkevich are quoted by Alpher, Follin & Herman (1953). They were repeated by Smirnov (1964), Peebles (1966), Wagoner, Fowler & Hoyle (1966). On Figure 2 is shown the time dependence of a number of different nuclei[8] for $\rho_m = 2 \cdot 10^{-29}$ g/cm³ (Doroshkevich & Sjunjaev). Table I shows the final composition as depending on ρ_m. The plausible values of ρ_m are $5 \cdot 10^{-31} < \rho_m < 4 \cdot 10^{-29}$ corresponding to 30 per cent $> \mathrm{He}^4$ per cent > 28 per cent. The He^3 and d content is still sensitive to ρ_m but it is low and easily obscured by other mechanisms of production of He^3 and d.

In a wide class of anistropic homogeneous nonmagnetic cosmological solutions, the overall density varies in the early phase as $\rho_{an} = A t^{-4/3}$ instead of $\rho_{is} = 3/32\pi G t^2$ for the isotropic model, with the only limitation on A that $\rho_{an} > \rho_{is}$ at given t (see below). In the next phase $\rho_{an} \rightarrow \rho_{is}$. It follows that the time interval for a given drop $\rho \rightarrow \rho - \delta\rho$ is less in the anisotropic solution for large ρ; $\delta t = A^{4/3} \rho^{-7/3} \delta\rho$ instead of

[8] For other types of solutions see Taub (1951), Heckmann & Schücking (1962).

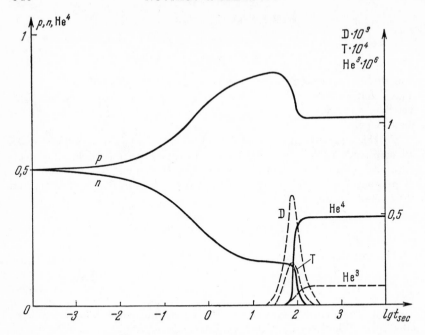

FIG. 2. The composition of matter during expansion in the hot isotropic universe.
Different species are given in different scales.

$$\delta t = \left(\frac{3}{128\pi G}\right)^{1/2} \rho^{3/2} \delta\rho$$

The final composition depends now on A, as pointed out semiqualitatively
by Hawking & Tyler (1966) and calculated by Thorne (1966). We take from
him Figure 3, giving the final composition (the H content, not shown, is
$H = 1 - He^4 - d - He^3$). The accelerated expansion first shifts $n/(n+p) \to 0.5$
(the equilibrium $p \rightleftarrows n$ is frozen at higher temperatures), and so the helium
formation, which was limited by the number of neutrons, is enhanced. With
further acceleration it is the rate of thermonuclear reaction which is not
great enough to build helium, and still further, the capture of neutrons with

TABLE I

COMPOSITION OF PRESTELLAR MATTER[a]

Present-day density ρ_m	H	He^4	D	He^3+T
10^{-33}	98	0.4	1.5	0.02
$3 \cdot 10^{-31}$	75	25	0.1	$4 \cdot 10^{-3}$
$3 \cdot 10^{-29}$	71	29	$< 10^{-10}$	$5 \cdot 10^{-4}$

[a] Data from Wagoner, Fowler & Hoyle (1966)—per cent by weight.

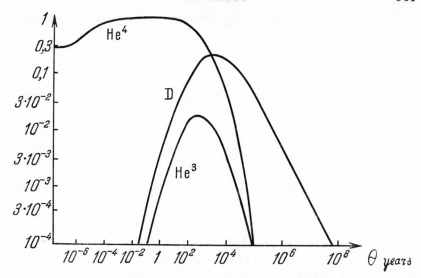

FIG. 3. The composition of prestellar matter in the hot anisotropic model, as depending on the parameter θ, characterizing the moment of effective isotropization.

deuteron formation is not fast enough. Note the high $(d/H)_{max} \sim 0.3$ and $He^3/H \sim 0.05$ in the intermediate region. The content of He^4, d, He^3 is given as a function of the time θ when the expansion ceases to be anisotropic ($\theta \simeq G^{-3/2}A^{3/2}$).

Many observational contributions to the helium content giving $He \sim 30$ per cent are summarized by Hoyle & Tyler (1964); the older papers on halo stars gave very small He (see Sargent & Searle 1966 and Greenstein 1966). Compare the remark in I.5.

Some suggestions on the detection of intergalactic He^4 and He^3 have appeared (see Sjunjaev 1966; Zeldovič, Novikov & Sjunjaev 1966). These references on observational evidence are far from complete.

4. Anisotropic and Magnetic Models, the Singularity

The first general-relativity anisotropic solution with Euclidean space

$$ds^2 = c^2dt^2 - a^2(t)dx^2 - b^2(t)dy^2 - e^2(t)dz^2 \qquad 1.$$

was considered by Schücking & Heckmann (1958) for dust matter ($p=0$). Asymptotically at $t \to \infty$ it tends to the isotropic Euclidean Friedmann solution; for this solution it is necessary for this period that the present-day density $\bar{\rho} = \rho_c \simeq 2 \cdot 10^{-29}$ g/cm^{-3}.

There is a class of solutions with two equivalent axes with Euclidean (Equation 2a) or non-Euclidean geometry (Equations 2b, 2c)

$$ds^2 = c^2dt^2 - dl^2$$
$$dl^2 = a^2(t)[dx^2 + dy^2] + e^2(t)dz^2 \equiv a^2(t)[dr^2 + r^2d\varphi^2] + e^2(t)dz^2 \qquad 2a.$$

$$dl^2 = a^2(t)[dr^2 + \sin^2 r d\varphi^2] + e^2(t)dz^2 \qquad \text{2b.}$$
$$dl^2 = a^2(t)[dr^2 + sh^2 r d\varphi^2] + e^2(t)dz^2 \qquad \text{2c.}$$

The first of them (2a) is a particular case of Equation 1; the other (2b, 2c) has $\rho > \rho_c$ (2b) or $\rho < \rho_c$ (2c) but does not tend to isotropy at $t \to \infty$.[8]

The qualitative properties of the solutions are the same for $p = \frac{1}{3}\rho c^2$ (Kompaneets & Chernov 1964, Doroshkevich 1965).

The expansion is characterized by a tensor Hubble constant

$$u_i = H_{ik} x_k, \quad H_{ik} = \begin{vmatrix} \dot{a}/a & 0 & 0 \\ 0 & \dot{b}/b & 0 \\ 0 & 0 & \dot{e}/e \end{vmatrix}$$

At small t the common property of the above-mentioned solutions is the power law

$$a \sim t^\alpha, \quad b \sim t^\beta, \quad e \sim t^\epsilon,$$
$$\alpha + \beta + \epsilon = \alpha^2 + \beta^2 + \epsilon^2 = 1$$

It results that one of exponents is negative, except in the degenerate case $\alpha = \beta = 1$, $\epsilon = 0$ (which is one of the two cases with equivalent axes, the other being $\alpha = \beta = \frac{2}{3}$, $\epsilon = -\frac{1}{3}$). The co-moving volume always depends on t like

$$V \sim abe \sim t^{\alpha+\beta+\epsilon} \sim t$$

so that the densities behave like

$$n \sim t^{-1}, \quad \rho \sim t^{-4/3}$$

where n is the conserved baryon density and ρ is the mass density of radiation; $\rho \sim n^{4/3}$.

Locally the matter feels strong $[(\ddot{a}/a) \sim (\ddot{b}/b) \sim (\ddot{c}/c) \sim t^{-2}$ and ∞ at $t = 0]$ external tidal forces, much greater than the self-gravitation.[9] There is always (except in the degenerate case) one axis along which the matter is compressed ($\epsilon < 0$), being distended along the two other axes. General methods of treating anisotropic and unhomogeneous motion in general relativity were developed by Zelmanov (1959).

The magnetic case was discussed by Hoyle (1958) and Zeldovič (1964) and analyzed by Doroshkevich (1965) and Thorne (1966), but always for two equivalent axes perpendicular to the magnetic field.

In the general case considered recently by Doroshkevich with three different axes (see Rosen 1964), there is always an initial compression along an axis perpendicular to the field.

The most stringent observational limits on anisotropy arise from the observed isotropy of relict radiation; see Partridge & Wilkinson (1967) [discussed by Thorne (1966)]; the same paper quotes other work on observations in the anisotropic case.

---

[9] The other condition is $\gamma < 1$ in $P = \gamma \rho c^2$. If $\gamma = 1$, $\rho \sim n^2 \sim t^{-2}$ and the situation is complicated.

On the subject of the singularity there is a controversy in the literature between Lifshitz & Khalatnikov (1963) and Penrose (1965), Hawking (1966), and Geroch (1966).

The first authors point out that the singular solution is not the most general one; but they consider the solution in the vicinity of a singular point.

The second group considers the expanding matter which is already contained inside its Schwarzschild sphere.[10] It follows from their arguments that there was a singularity, but the topological methods give no answer as to whether all matter did go through the singularity or occurred only in a space-time point. The relativistic singularity is of a very different origin as compared with a Newtonian singularity in the degenerate case of spherical symmetry with pure radial movement. The breakdown of the Newtonian singularity if random movement is allowed (Pachner 1966) does not invalidate the Penrose proof.

5. Gravitational Instability

The theory of gravitational instability in the expanding universe was made by Lifshitz (1946), and developed further and elucidated by Bonnor (1957) and, in the case of anisotropic universe, by Doroshkevich (1966).

The situation in the hot isotropic model is summarized on Figure 4. As a function of time t(ordinate) is given the instantaneous value of M_c—"Jeans mass" defined as the mass of baryons in the sphere with diameter equal to the critical wavelength, $M_c = \rho_m(\pi/6)l_c^3$. It is equal to the mass in the perturbed region when the mass of photons drops below the mass of baryons during expansion. The mass of the photons is taken into account, as well as their pressure,[11] in the calculation of l_c.

The line $M_c(t)$ on Figure 4 divides the region of stability (left)from the region where perturbations are growing (right). The equation of the line is $M_c = at^{3/2}$, it ends at

$$t = t^* = 10^{14} \text{ sec } (T = 3500°\text{K}, \quad \rho \approx 10^{-21} \text{ g·cm}^{-3})$$
$$M_c \approx 10^{19} M_\odot. \quad \text{At } t < t^*, \quad l_c = (c/\sqrt{3})t.$$

A given small perturbation with $M < 10^{19} M_\odot$ first grows on the piece of line segment AB, then stops growing and oscillates on the segment BC. If the initial amplitude was great enough, the perturbation could lead to gravitational collapse of the perturbed volume before the point of stability (B) is reached. The collapse is followed by accretion and leads to the formation of a geon full of photons; see Addendum VI. The absence of such geons in nature limits the initial amplitude of perturbations (Zeldovič & Novikov 1966b,c; see also a remark by Peebles 1965).

[10] It was stressed long ago by Novikov (1962) that the already investigated part of universe undoubtedly was in such a situation.

[11] On different types of instability of plasma in the hot model with prevailing radiation pressure, see Peebles (1965), and Ozernoy (1966); on the instability of hot stellar plasma, see Lebedinskii (1954), Gurevich (1954), Zeldovič & Novikov (1965).

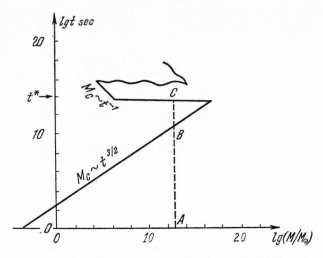

FIG. 4. The critical (Jeans) mass dividing the region of instability—on the right-hand side of the curve, and the region of stability—on the left-hand side of the curve. The time is given on the ordinate. On the abscissa is given the mass of ordinary matter (without radiation). The lower straight line gives the condition for stability against adiabatic perturbations of matter with radiation for the stage when the radiation density is greater than the matter density. At the time t^*, electrons and protons are tied in neutral hydrogen; the wavy line corresponds to new ionization. The exact position of the lines depends on the present value of density ρ_m. The figure is a schematic one! The line is plotted for the case when t^* coincides with the moment $\rho_r = \rho_m'$.

Returning to Figure 4 we see that later, at $t > t^*$, the matter is neutral, it does not interact with radiation.[12] The Jeans mass drops abruptly to $M_c = 10^5 M_\odot$ and diminishes further during the expansion and cooling after t^*. See Addendum VII.

The M_c is subsequently raised only after the formation of stars—or uhrstars?—gives the energy for the heating and ionization of the gas, whose stability is investigated. This new change of M_c is shown hypothetically by the wavy line.

6. CORRECTIONS FOR THE OBSERVABLES IN AN INHOMOGENEOUS UNIVERSE

The inhomogeneity of the matter distribution near the path of the light rays coming from the object to the observer affects the angular size and

[12] If the present-day density is equal to $\rho_m = 5 \cdot 10^{-31}$, then the moment of neutralization nearly coincides with the time when the radiation density is equal to the matter density. Of course such coincidence is not necessary; for $\rho_m = 2 \cdot 10^{-29}$ it is destroyed, the maximal Jeans mass is $10^{16} M_\odot$ instead of $10^{19} M_\odot$. Perhaps this type of perturbation is important for formation of clusters.

apparent luminosity of the object. This remark, made independently by Zeldovič (1964) and Feynman (unpublished, told to us by Münch), must be taken into account in the study of distant objects.

It is known that in a homogeneous Friedman universe the angular size of an object with definite absolute linear diameter l goes through minimum with increasing redshift. For example, if $\bar{\rho} = \rho_c$, $p = 0$

$$\theta = \frac{lH}{c} \cdot \frac{1+Z}{2[1 - \sqrt{1/(1+Z)}]}$$

where H is the Hubble constant and $Z = \Delta\lambda/\lambda$. This minimum disappears if the lightcone is empty. Then

$$\theta = \frac{lH}{c}\left\{\frac{2}{5}\left[1 - \frac{1}{(1+Z)^{5/2}}\right]\right\}^{-1}$$

It confirms the idea that the rise of θ at $Z > Z_{min}$ in the first case is due to gravitational focusing of light rays.

For detailed computations see Dashevski & Slysh (1965), Zeldovič (1965).

Addendum I: At the Texas symposium on relativistic astrophysics, in New York, January 1967, Penzias reported on further measurements at $\gamma = 21.1$ cm, giving $T = 3.2 \pm 1°$K, and at $\gamma = 1.5$ cm, giving $T = 2 + 0.8/0.7°$K.

Addendum II: Characteristic of modern ideas on galaxy formation is the recombination of hydrogen, followed by a subsequent heating and ionization of the gas. The question immediately arises: could not one devise a workable scheme without the neutral hydrogen phase? An analysis made by Sjunjaev, based on radio emission of plasma at low temperature and energy losses at high temperature, shows that at $1300 > z > 300$ the neutral hydrogen stage is inescapable.

Addendum III: Partridge & Wilkinson (1967) measured the isotropy of relict radiation at $\lambda = 3.2$ cm. They give the anisotropy smaller than 0.2 per cent; this result strengthens the statements made in the text. For all details and further literature see the original paper.

Addendum IV: The peculiarities of neutrino and other weak interacting particles are of interest in anisotropic models. At high densities, when the time of free path is small, the encounters of particles support thermodynamic equilibrium and, in particular, the isotropic impulse distribution even when the expansion is anisotropic. Later the neutrino moves freely. In the strong anisotropic deformation an expansion along two axes is accompanied by a compression along the third axis. The corresponding third component of momentum of freely moving particles grows ("blue shift"). After some time, all the neutrinos move along the third axis; their energy is much higher than the energy of other particles (γ, e^+, e^-) and their stress tensor is highly anisotropic, affecting the dynamics of further expansion. For details see Doroshkevich, Zeldovič & Novikov (1967a).

Addendum V: At the Texas symposium in New York, January 1967,

Dicke reported observations of the figure of the Sun. He claims that these observations confirm his scalar theory of gravitation. The question is still open.

Addendum VI: As pointed out by Lifshitz (1946), the perturbation of metric is not growing during the time when the perturbations of density are growing. In particular, to build the geons mentioned in the text, it is necessary that from the very beginning (the singularity), for some parts of the hot plasma, dimensionless quantities like the ratio of circumference to diameter differ from their mean values.

Addendum VII: A more detailed investigation of the recombination of protons and electrons led to the following conclusions: the recombination is slowed down first by the absorption of Ly-α photons and later by the small gas density. The concentration of electrons and ions tends toward a limit $e^-/\mathrm{H\,I} = p/\mathrm{H\,I} \approx 4 \cdot 10^{-5}$ at $t \to \infty$, $\rho \to 0$. The electrons achieve heat transfer between photons and neutral atoms; therefore up to $t \approx 10^{14}$ sec, $T \approx 600°\mathrm{K}$ (for $\rho = \rho_\mathrm{crit}$), the temperature of the gas is equal to the radiation temperature. During the interval $5 \cdot 10^{12}$ sec $> t > 10^{14}$ sec (for $\rho = \rho_\mathrm{crit}$) the Jeans mass of the neutral gas is constant.

COSMOLOGY

LITERATURE CITED

Alpher, R. A., Follin, J. W., Herman, R. C. 1953, *Phys. Rev.*, **92**, 1347

Ambartsumian, V. A. 1964, *Conseil Phys.*, *13th, Solvay, Brussels*

Ambartsumian, V. A. 1966, *I.A.U. Symp. No. 29, Byurakan*

Bahcall, J. N., Salpeter, E. E. 1965, *Ap. J.*, **142**, 1677

Becchi, C., Gallinaro, G., Morpurgo, G. 1965, *Nuovo Cimento*, **39**, 409

Bonnor, W. B. 1957, *Monthly Notices Roy. Astron. Soc.*, **117**, 104

Braginskii, V. B. 1966, *Pisma JETP (USSR)*, **3**, 69; Braginskii, V. B., Zeldovič, Ya. B., Martynov, V. K., Migulin, V. V. 1967, *JETP (USSR)*, **52**, 29

Burbidge, G. 1966 (Private communication, Byurakan)

Chiu, H. Y. 1966, *Phys. Rev. Letters*, **17**, 712

Chupka, W. A., Schiffer, J. P., Stevens, C. M., 1965, *Phys. Rev. Letters*, **17**, 60

Dashevski, V. M., Zeldovič, Ya. B. 1964, *Astron. J. (USSR)*, **41**, 1071

Dicke, R. H. 1962, *Phys. Rev.*, **126**, 1875

Dicke, R. H., Peebles, P. J. E., Roll, P. E., Wilkinson, D. T. 1965, *Ap. J.*, **142**, 414

Doroshkevich, A. G. 1965, *Astrophysika (USSR)*, **1**, 255

Doroshkevich, A. G., 1966, *ibid.*, **2**, 37

Doroshkevich, A. G., Novikov, I. D. 1964, *Dokl. Akad. Nauk SSSR*, **154**, 745

Doroshkevich, A. G., Zeldovič, Ya. B., Novikov, I. D. 1967a, *Pisma JETP (USSR)*, **5**, 119

Doroshkevich, A. G., Zeldovič, Ya. B., Novikov, I. D. 1967b, *Astron. J. (USSR)*, **44**, 295

Field, G. B., Henry, R. C. 1964, *Ap. J.*, **140**, 1002

Field, G. B., Hitchcock, J. L. 1966, *Ap. J.*, **146**, 1

Friedmann, A.A. 1922, *Z. Phys.*, **11**, 377

Friedmann, A. A. 1924, *ibid.*, **21**, 326

Gamow, G. 1946, *Phys. Rev.*, **70**, 572

Gamow, G. 1956, *Vistas Astron.*, **2**, 1726

Geroch, R. P. 1966, *Phys. Rev. Letters*, **17**, 445

Gershtein, S. S., Zeldovič, Ya. B. 1966, *Pisma JETP (USSR)*, **3**, 400

Ginzburg, V. L., Ozernoy, L. M. 1965, *Astron. J. (USSR)*, **42**, 943

Goldreich, P. G., Morrison, P. 1963, *JETP (USSR)*, **45**, 344; 1964, *Soviet Phys. JETP (Engl. Transl.)*, **18**, 239

Gould, R. J., Schreder, G. 1966, *Phys. Rev. Letters*, **16**, 252

Greenstein, J. L. 1966, *Ap. J.*, **144**, 496

Greisen, K. 1966, *Phys. Rev. Letters*, **16**, 748

Gurevich, L. E. 1954, *Vopr. Kosmogonii, Akad. Nauk SSSR*, **3**, 94

Hawking, S. W. 1966, *Phys. Rev. Letters*, **17**, 144

Hawking, S. W., Tyler, R. J. 1966, *Nature*, **209**, 1278

Hayashi, C. 1950, *Progr. Theoret. Phys. (Kyoto)*, **5**, 224

Heckmann, O., Schücking, E. 1962, *Gravitation, An Introduction to Current Research*, Chap. 11 (Academic Press, New York)

Hoyle, F. 1958, *Conseil Phys.*, *11th, Solvay, Brussels*, 66

Hoyle, F., Narlikar, S. 1963, *Proc. Roy. Soc. London*, **273**, 4

Hoyle, F., Tyler, R. J. 1964, *Nature*, **203**, 1108

Ielley, I. V. 1966, *Phys. Rev. Letters*, **16**, 479

Kardashev, N. S., Sholomitsky, G. B. 1965, *Astron. Circ. (USSR)*, No. 336

Kompaneets, A. S. 1956, *JETP (USSR)*, **31**, 877; 1957, *Soviet Phys. JETP (Engl. Transl.)*, **4**, 730

Kompaneets, A. S., Chernov, A. S. 1964, *JETP (USSR)*, **47**, 1939; 1965, *Soviet Phys. JETP (Engl. Transl.)*, **20**, 1303

Kurt, V. G., Sjunjaev, R. A. 1967a, *Pisma JETP (USSR)*, **5**, 299

Kurt, V. G., Sjunjaev, R. A. 1967b, *Kosmich. Issled. (USSR)*, **5**, 611

Layzer, D. 1964, *Ann. Rev. Astron. Ap.*, **2**, 341

Layzer, D. 1965 (Preprint)

Lebedinskii, A. I. 1954, *Vopr. Kosmogonii, Akad. Nauk SSSR*, **2**, 5

Lifshitz, E. M. 1946, *JETP (USSR)*, **16**, 587

Lifshitz, E. M., Khalatnikov, I. M. 1963, *Usp. Fiz. Nauk*, **80**, 391; *Soviet Phys.—Usp. (Engl. Transl.)*, **6**[4], 495

Longair, M. S., 1966, *Monthly Notices Roy. Astron. Soc.*, **133**, 421

McKellar, A. 1941, *Publ. Dominion Ap. Obs., Victoria, B.C.*, **7**, 251

Nikishov, A. I. 1961, *JETP (USSR)*, **41**, 549; 1962, *Soviet Phys. JETP (Engl. Transl.)*, **14**, 393

Novikov, I. D. 1962, *Vestn. Mosk. Univ., Ser. III. No. 6*, 66

Novikov, I. D., 1966, *Astron. J. (USSR)*, **43**, 911

Okun, L. B., Pikelner, S. B., Zeldovič, Ya.

B. 1965, *Phys. Letters*, **17**, 164; *Usp. Fiz. Nauk*, **87**, 113

Ozernoy, L. M. 1966 (Dissertation, Moscow Univ.)

Pachner, J. 1966 (Preprint)

Partridge, R. B., Wilkinson, D. T. 1967, *Phys. Rev. Letters*, **18**, 557

Peebles, P. J. E. 1966, *Phys. Rev. Letters*, **16**, 410

Peebles, P. J. E. 1965, *Ap. J.*, **142**, 1317

Penrose, R. 1965, *Phys. Rev. Letters*, **14**, 57

Penzias, A. A., Wilson, R. W. 1965, *Ap. J.*, **142**, 419

Piddington, J. H. 1964, *Monthly Notices Roy. Astron. Soc.*, **128**, 345

Piddington, J. H. 1966, *ibid.*, **132**, 163

Pikelner, S. B. 1965, *Astron. J. (USSR)*, **42**, 3

Pikelner, S. B., Vainstein, L. A. 1966, *Pisma JETP (USSR)*, **4**, 307

Roll, P. G., Wilkinson, D. T. 1966, *Phys. Rev. Letters*, **16**, 405

Rosen, G. 1964, *Phys. Rev.*, **136**, N1B, 297

Sacharov, A. D. 1966, *Pisma JETP (USSR)*, **3**, 439

Sacharov, A. D. 1967, *ibid.*, **5**, 32

Sandage, A. R. 1961, *Ap. J.*, **133**, 355

Sargent, W. L. W. 1966 (Preprint)

Sargent, L. W., Searle, L. 1966, *Ap. J.*, **145**, 652

Schücking, E., Heckmann, O. 1958, *Conseil Phys., 11th, Solvay, Brussels*

Shakeshaft, J. R. 1966, *I.A.U. Symp. No. 29, Byurakan*

Smirnov, Yu. N. 1964, *Astron. J. (USSR)*, **40**, 1084

Sjunjaev, R. A. 1966, *Astron. J. (USSR)*, **43**, 1242

Taub, A. H. 1951, *Ann. Math.*, **53**, 472

Thorne, K. S. 1966 (Preprint California Inst. Technol., Pasadena, Calif.)

Wagoner, R. V., Fowler, W. A., Hoyle, F. 1966 (Preprint)

Weymann, R. 1966, *Ap. J.*, **145**, 560

Wheeler, J. A. 1960, *Neutrinos, Gravitation and Geometry* (Bologna)

Zatsepin, G. T. Kuzmin, V. A. 1966, *Pisma JETP (USSR)*, **4**, 114

Zeldovič, Ya. B. 1962, *JETP (USSR)*, **43**, 1561; 1963, *Soviet Phys. JETP (Engl. Transl.)*, **16**, 1102

Zeldovič, Ya. B. 1964, *Astron. J. (USSR)*, **41**, 19

Zeldovič, Ya. B. 1965, *Advan. Astron. Ap.*, **3**, 241

Zeldovič, Ya. B. 1966, *Usp. Fiz. Nauk*, **89**, 647

Zeldovič, Ya. B., Novikov, I. D. 1965, *Usp. Fiz. Nauk*, **86**, 447

Zeldovič, Ya. B., Novikov, I. D. 1966a, *Pisma JETP (USSR)*, **4**, 117

Zeldovič, Ya. B., Novikov, I. D. 1966b, *Astron. J. (USSR)*, **43**, 758

Zeldovič, Ya. B., Novikov, I. D. 1966c, *I.A.U. Symp. No. 29, Byurakan*

Zeldovič, Ya. B., Novikov, I. D. 1967, *Astron. J. (USSR)*, **44**, 663

Zeldovič, Ya. B., Novikov, I. D., Sjunjaev, R. A. 1966, *Astron. Circ. (USSR)*, No. 371

Zelmanov, A. L. 1959, *Tr. G. Soveshch. Vopr. Kosmogenii, Moscow (USSR)*

SOME RELATED ARTICLES APPEARING
IN OTHER *ANNUAL REVIEWS*

Articles of direct interest to our readers

From the *Annual Review of Nuclear Science*, Volume **17** (1967):

Reynolds, John H.: Isotopic Abundance Anomalies in the Solar System
Friedman, Herbert: X Rays from the Stars

Articles of peripheral interest

From the *Annual Review of Physical Chemistry*, Volume **18** (1967):

Taylor, Howard S.: Quantum Theory of Atoms and Molecules
Dorfman, Leon M., and Firestone, Richard F.: Radiation Chemistry
Curtiss, Charles F.: Statistical Mechanics
Westrum, Edgar F., Jr.: Thermochemistry and Thermodynamics

AUTHOR INDEX

SUBJECT INDEX

SUBJECT INDEX

and cosmology, 632
Primary condensation
in the solar nebula, 270
Primary particle flux, 354,
378, 481, 486, 487, 488,
506, 508, 516
Primeval matter, 628, 629,
630
Primordial radiation, 200
Prisms
in astronomical optics,
57, 64-65
Production mechanics
for gamma rays, 482-83
Progressive modes, 73, 76
Projected orbit, 106
Projection-measuring devices
for binary stars, 88
Projection microscope, 33
Prominence
lines, 254
Prominences, 8
and the corona, 223, 225,
229, 231, 237, 253
and emission lines, 2, 308
Propagation effect, 362
Propagation modes, 72
Propagation of structures
on Sun, 9, 17
Propane
in interferometry, 162
Protogalaxy
QSO's, 443
Proton-alpha-particle reaction, 503
Proton-antiproton annihilation, 505
and production of gamma
rays, 483
Proton bremsstrahlung, 498
see also Collisional
bremsstrahlung;
Bremsstrahlung
Proton detection, 381
Proton-electron collisions
coronal, 250
Proton events, 352, 369
376, 390
Proton-to-helium ratio, 362,
363, 366, 384
Proton-induced reactions,
538
Proton-proton chain, 526,
562-63
Proton-proton reaction, 500-
3, 506
Protons
cosmic-ray, 170
nuclear reactions, 525
Proton temperature
of the corona, 251
Protoplanets, 268
Protostars
and binary systems, 37, 40
and OH observations, 203,
206

see also Solar system
Protosun, 272, 276
Pulsation
of fluid
radial, 473
stellar evolution, 606, 621,
623, 624
Pulsation frequency
of stellar atmospheres, 72
Pulse counting
electronic, 140
Pumping efficiency
UV pump, 209
Pumping mechanisms
for maser action, 200, 201,
202-6
P-wave interaction, 536
PW Her, 97
Pyrex as optical material,
62, 63
Pyroceram as optical material, 62, 63

Q

Quadrupole moment
solar
and cosmology, 646
Quantization
of theory of relativity, 634
Quantum defect method, 295
Quantum detectors, 140
Quantum efficiency
in interferometry, 157
of photographic plate, 53,
58
Quantum-state transitions
gamma-ray production,
482-83, 489
Quantum theory
and cosmology, 633
Quarks, 287
and cosmology, 628, 633,
637, 638
as QSO energy source, 446
Quartz
in astronomical optics, 53,
161
in prisms, 155
Quasars
and cosmology, 629, 630,
632
see also Quasi-stellar
objects
Quasi-periodic motion
of the interplanetary field,
375, 376
of the solar corona, 238
Quasi-stationary spiral
structure, 458
Quasi-stellar objects, 399-
452
continuum, 416-20
cosmological, 433-37
energy generation, 438-47
gamma rays, 494, 509
gravitational focusing, 447

intergalactic medium, 437-
38
line spectra, 403-16
matter, antimatter, 446-47
radio, 422-23
radio properties, 424-28
redshift-apparent magnitude relation, 430
redshifts, 410-16
spatial distribution, 428-30
stellar collisions, 441
supernova, 440-41
superstars, 441-42
variations, 420-24
Quasi-thermal X rays, 314,
316
Quiet Sun, 1
corona, 247
QSSS hypothesis, 459

R

Radar probing, 231, 243, 247
solar corona, 213
Radial pulsation
of rotating fluid, 473
stellar evolution, 591, 606
Radial velocity observations,
30, 31
of binary stars, 25, 88, 113,
118, 121, 124
OH, 145, 184, 207
optical, 57
Radiance
coronal, 216, 217
see also Coronal radiance
Radiance response, 143
Radiation-belt studies, 351
Radiation collector, 56
Radiation-monitoring satellites, 314
Radiation pressure
and cosmology, 630, 643
solar corona, 241
stellar evolution, 587, 591
Radiative capture rates
nuclear, 526, 530, 564
Radiative core
stellar
and binaries, 39
Radiative damping
in stellar atmospheres, 71,
73
Radiative diffusion, 587
Radiative equilibrium
in rotating fluids, 466
Radiative excitation mechanisms, 188
Radiative opacity
in stellar evolution, 576
Radiative recombination, 296,
297, 318
X-ray, 340
Radiative relaxation, 70, 74
Radiative smoothing, 69-70
Radiative transfer
in stellar atmospheres, 69

CUMULATIVE INDEXES

VOLUMES 1-5

INDEX OF CONTRIBUTING AUTHORS

692

INDEX OF CHAPTER TITLES

VOLUMES 1-5